Taphonomy

Process and Bias Through Time

second edition

Aims & Scope
Topics in Geobiology Book Series

Topics in Geobiology series treats geobiology - the broad discipline that covers the history of life on Earth. The series aims for high quality, scholarly volumes of original research as well as broad reviews. Recent volumes have showcased a variety of organisms including cephalopods, corals, and rodents. They discuss the biology of these organisms-their ecology, phylogeny, and mode of life and in addition, their fossil record their distribution in time and space.
Other volumes are more theme based such as predator-prey relationships, skeletal mineralization, paleobiogeography, and approaches to high resolution stratigraphy, that cover a broad range of organisms. One theme that is at the heart of the series is the interplay between the history of life and the changing environment. This is treated in skeletal mineralization and how such skeletons record environmental signals and animal-sediment relationships in the marine environment.
The series editors also welcome any comments or suggestions for future volumes.

For other titles published in this series, go to
www.springer.com/series/6623

Taphonomy

Process and Bias Through Time

second edition

Peter A. Allison · David J. Bottjer
Editors

Springer

Editors
Peter A. Allison
Department of Earth Science & Engineering
South Kensington Campus
Imperial College London
SW7 2AZ London
United Kingdom
p.a.allison@imperial.ac.uk

David J. Bottjer
Department of Earth Sciences
University of Southern California
90089-0740 Los Angeles
California
USA
dbottjer@usc.edu

ISBN 978-90-481-8642-6 e-ISBN 978-90-481-8643-3
DOI 10.1007/978-90-481-8643-3
Springer Dordrecht Heidelberg London New York

Cover illustration: Main Image Caption – Illustration of Lower Devonian *Hollardops* from Bou Tserfine, Morocco (see p. 131)
Small figure top left – *Eurypterus dekayi* from the Late Silurian Williamsville Formation in Ontario, Canada (see p. 202)
Small figure top middle – Small *Nummulites* from the late Eocene, Autochthonous Molasse of Upper Austria (see p. 345)
Small figure top right – Modern *Limulus* (see p. 202)

Printed on acid-free paper

Springer is part of Springer Science+Business Media (www.springer.com)

Preface

The study of taphonomy has evolved substantially in recent decades. A brief history of the subject is given in Chapter 1 and will not be repeated here, however it is fair to say that there is now a first-order understanding of taphonomic processes. It is particularly noteworthy that taphonomic research breaches the barriers of traditional research disciplines. The multi-disciplinarity of the subject is evidenced by the breadth of the publication base that supports the subject; consider for example, the quantity of vertebrate taphonomy research in the paleontological, archeological and forensic domains (e.g. see Chapter 8). The subject is also inter-disciplinary and this is particularly evidenced by work on inorganic and organic geochemistry (e.g. see Chapters 5, 6 and 11). It is also true that taphonomic research has always been quick to incorporate new approaches and techniques. This includes use of the latest data-bases (Chapters 2 and 16) and analytical methods (Chapters 13 and 14). Of course paleontological data is ultimately collected by field geologists and paleontologists and sedimentological and stratigraphic approaches continue to yield new insights (Chapters 3, 4 and 7).

The great challenges in paleontology are to deepen our understanding of the origins and evolution of life and elucidate the impact of global change on the biosphere. The first has obvious appeal because it is a basic fundamental question and the second is relevant to a modern world in the throes of climate change. Taphonomic research is pertinent to both of these grand challenges, not least because it is necessary to truly release the data locked in the fossil record. For example, the controversies surrounding the biogenicity of Archean fossils (see Chapters 13 and 14) are, in the broadest sense, taphonomic in nature. We also note that taphonomic research is now being used to evaluate our understanding of large-scale trends in biodiversity through time (see Chapters 2 and 3). It is also certainly feasible that global change and mass extinction could impact upon taphonomic processes. A reduction in the diversity of shell-destroying taxa, a change in the processing rate of bioturbating organisms, or a change in sedimentary/diagenetic environment could all influence fossil preservation. This emerging question is developed in Chapters 9 and 16.

The default assumption for paleontologists is that the fossil record is biased. The extent of the bias varies between extremes according to depositional circumstances (see Chapters 4, 7 and 8) and can be mitigated for by using appropriate research methodologies and statistical approaches, but it is still there. A deeper understanding of taphonomic process requires an evaluation of how taphonomic bias has changed over geological time. It is one thing to deal with a biased dataset and quite another to deal with a bias that has changed with time. This question lies at the heart of all of the chapters in this book. Chapters 5 and 6 deal with the impact of biomolecular innovations in the evolution of organic skeletons; Chapters 2, 3, 11 and 12, tackle the issue of secular changes in diagenesis; Chapters 3, 4, 9 and 10 explore the nature of temporal change in taphonomic processes in marine environments; Chapters 7 and 8 focus on terrestrial environments; and Chapters 14–16 evaluate the extent that taphonomic bias has changed during, or as a result of, major bio-events. This book as a whole does not define the extent to which taphonomic bias has changed through time. It does, however, go some way towards properly defining the questions that need to be asked before that can be done.

It is left to us as editors to thank: the contributors for their patience; the reviewers of the chapters for their valuable time and insight; our friends, family and colleagues who have supported us; and the forbearance and support of the staff at Springer who have published this work.

Peter A. Allison
David J. Bottjer

Contents

Contributors

Peter A. Allison
Department of Earth Science and Engineering, South Kensington Campus, Imperial College London, SW7 2AZ London, UK
p.a.allison@imperial.ac.uk

Jonathan B. Antcliffe
Department of Earth Sciences, University of Oxford, Parks Road, Oxford OX1 3PR, UK
jonathan.antcliffe@univ.ox.ac.uk

Davide Bassi
Dipartimento di Scienze della Terra, Università di Ferrara, Via Saragat 1, 44122 Ferrara, Italy
bsd@unife.it

David J. Bottjer
Department of Earth Sciences, University of Southern California, Los Angeles, CA 90089-0740, USA
dbottjer@usc.edu

Martin D. Brasier
Department of Earth Sciences, Oxford University, Parks Road, Oxford OX1 3PR, UK
martinbrasier@yahoo.co.uk

Carl E. Brett
Department of Geology, University of Cincinnati, Cincinnati, OH 45221, USA
C.brett@uc.edu

Derek E.G. Briggs
Department of Geology and Geophysics, Yale University, P. O. Box 208109, New Haven, CT 06520-8109, USA;
Peabody Museum of Natural History, Yale University, P.O. Box 208118, New Haven, CT 06520-8118, USA
Derek.briggs@yale.edu

Susan H. Butts
Division of Invertebrate Paleontology, Peabody Museum of Natural History, Yale University, P.O. Box 208118, New Haven, CT 06520-8118, USA
Susan.Butts@yale.edu

Richard H. T. Callow
Department of Earth Sciences, University of Oxford, Parks Road, Oxford, OX1 3PR, UK
richardc@earth.ox.ac.uk

Lesley Cherns
School of Earth and Ocean Sciences, Cardiff University, Park Place, Cardiff CF10 3YE, UK
cherns@cardiff.ac.uk

Matthew E. Clapham
Department of Earth and Planetary Sciences, University of California Santa Cruz, 1156 High Street, Santa Cruz, CA 95064, USA
mclapham@es.ucsc.edu

Margaret E. Collinson
Department of Earth Sciences, Royal Holloway University of London, Egham, Surrey, TW20 0EX, UK
m.collinson@es.rhul.ac.uk

Timothy M. Demko
Department of Geological Sciences, University of Minnesota Duluth, Duluth, MN 55812, USA;
ExxonMobil Exploration Company, Houston, TX 77210, USA
tdemko@d.umn.edu

Steve Q. Dornbos
Department of Geosciences, University of Wisconsin-Milwaukee, Milwaukee, WI 53201-0413, USA
sdornbos@uwm.edu

Margaret L. Fraiser
Department of Geosciences, University of Wisconsin-Milwaukee, Milwaukee, WI 53203, USA
mfraiser@uwm.edu

Robert A. Gastaldo
Department of Geology, Colby College, Waterville, ME 04901, USA
ragastal@colby.edu

Neal S. Gupta
Department of Geology and Geophysics, Yale University, P.O. Box 208109, New Haven, CT 06520–8109 USA;

Geophysical Laboratory, 5251 Broad Branch Road NW,
Washington, DC, 20015, USA
ngupta@ciw.edu

Austin J.W. Hendy
Center for Tropical Paleoecology and Archaeology Smithsonian Tropical Research Institute, Panamá, República de Panamá;
Department of Geology and Geophysics, Yale University, New Haven, CT 06510, USA
hendyaj@si.edu

Anatoliy B. Kudryavtsev
Institute of Geophysics and Planetary Physics (Center for the Study of Evolution and the Origin of Life) and NASA Astrobiology Institute, University of California, Los Angeles, CA 90095, USA
Kudryavtsev@ess.ucla.edu

Nicola McLoughlin
Department of Earth Sciences and centre of Excellence in Geobiology, University of Bergen, 5020 Bergen, Norway
Nicola.Mcloughlin@geo.uib.no

James H. Nebelsick
Institute for Geosciences, University of Tübingen, Sigwartstrasse 10, 72076 Tübingen, Germany
nebelsick@uni-tuebingen.de

Christopher R. Noto
Department of Biomedical Sciences, Grand Valley State University, Allendale, MI 49401, USA
crnoto@life.bio.sunysb.edu

Michael W. Rasser
Museum of Natural History Stuttgart, Rosenstein 1, 70191 Stuttgart, Germany
rasser.smns@naturkundemuseum-bw.de

J. William Schopf
Department of Earth and Space Sciences, Institute of Geophysics and Planetary Physics (Center for the Study of Evolution and the Origin of Life), Molecular Biology Institute, and NASA Astrobiology Institute, University of California, Los Angeles, CA 90095, USA
Schopf@ess.ucla.edu

Abhishek B. Tripathi
Advanced Projects Office, Constellation Program, NASA Johnson Spacecraft Center, 77058, Houston, TX, USA
Tripathi@ess.ucla.edu

David Wacey
Centre for Microscopy, Characterization and Analysis + School of Earth and Environment, The University of Western Australia, 35 Stirling Highway, Crawley, WA 6009, Australia
David.Wacey@uwa.edu.au

Rachel Wood
Grant Institute of Earth Sciences, School of Geosciences, University of Edinburgh, King's Buildings, West Mains Road, Edinburgh EH9 3JW, UK
rwood4@staffmail.ed.ac.uk

V. Paul Wright
BG-Group, 100 Thames Valley Park, Reading RG6 1PT, UK
v.vpw@btopenworld.com

J.R. Wheeley
School of Geography, Earth and Environmental Sciences, University of Birmingham, Edgbaston, Birmingham B15 2TT, UK

A.D. Czaja
Department of Earth and Space Sciences and Institute of Geophysics and Planetary Physics (Center for the Study of Evolution and the Origin of Life), University of California, 90095, Los Angeles, CA, USA

Chapter 1
Taphonomy: Bias and Process Through Time

Peter A. Allison and David J. Bottjer

Contents

Abstract It is now 18 years since the volume "Taphonomy: Releasing the Data Locked in the Fossil Record" was published by Plenum Press as part of the successful "Topics in Geobiology" series. The book was one of several published as the subject blossomed and diversified. The Plenum book was multi-disciplinary and focused on processes, including chapters on emerging concepts such as sequence stratigraphy, and rapidly developing fields such as organic and inorganic geochemistry. In a sense the book functioned as an entry point for those embarking upon interdisciplinary research and was quickly out-of-print. Taphonomic bias is now recognized as a pervasive feature of the fossil record. This is supported by a series of laboratory experiments and field studies during the last 20 years that have provided a sound first order understanding of the processes at work. A pressing concern, however, is how these processes have varied through time in different depositional environments. This second-order understanding is essential if we are to truly fully release the data locked in the fossil

P.A. Allison (✉)
Department of Earth Science and Engineering, South Kensington Campus, Imperial College London, SW7 2AZ, UK
e-mail: p.a.allison@imperial.ac.uk

D.J. Bottjer
Department of Earth Sciences, University of Southern California, Los Angeles, CA 90089-0740, USA
e-mail: dbottjer@usc.edu

P.A. Allison and D.J. Bottjer (eds.), *Taphonomy: Process and Bias Through Time*, Topics in Geobiology 32, DOI 10.1007/978-90-481-8643-3_1,

record. It is one thing to work with a biased data set and quite another to work with a bias that has changed with time. This new book for the "Topics in Geobiology" series focuses on the extent to which taphonomic bias has changed through time in different environments. The chapters include work from researchers who are using laboratory, field and data-base techniques. It does not provide the answers to these questions but does at least highlight some of the emerging questions.

1 Introduction

Taphonomic processes have exerted a profound and widespread bias to the fossil record and there are few, if any fossil biotas that are preserved bias-free. The most striking example of preservational bias is the rarity of fossilized soft parts and soft-bodied organisms. In "normal" marine near-shore communities such organisms can account for about two thirds of the species and individuals (Allison 1988a) and yet they are rarely preserved. There are of course, examples of biotas which preserve such tissues and organisms (Bottjer et al. 2002) but it would be fallacious to assume that the preservation of soft-tissues implied a minimal taphonomic bias. For example, the Iron-Age peat bogs of Europe preserve human carcasses that include exquisite preservation of soft-tissues (Brothwell 1986; Stead et al. 1986; Stankiewicz et al. 1997; Glob 2004). Preservation in this instance was enhanced by the action of organic acids in the peat. However, in some instances the acids which promoted soft-part decay also promoted mineral dissolution to the extent that some carcasses are now devoid of bone! The fact that soft-parts are preserved in preference to skeletal remains underscores the pervasive nature of taphonomic bias. That is not to say though, that taphonomic processes always result in signal degradation. Taphonomic bias is influenced by diverse biological, physical and geochemical processes which are, in turn dependent upon depositional environment. It is therefore possible to document the nature and extent of taphonomic bias and invert to infer something of depositional environment; "paleontology's loss is a sedimentologist's gain" (Thomas 1986)! Fundamentally, this aspect of taphonomic bias is incorporated into Walther's facies concept but was explicitly developed in the 1980s with the concepts of taphonomic feedback (Kidwell and Jablonski 1983) and taphofacies (Brett and Baird 1986). Taphonomic bias in marine environments is most active close to the sediment-water interface: the Taphonomically Active Zone (Davies et al. 1989), so that sedimentation rate exerts a strong control on the taphonomy of biogenic remains. Given that the net rate and episodicity of sedimentation in an aquatic system varies with distance from land and water depth it is easy to see how relative taphonomic trends can be used to define sea-level fluctuations (Kidwell 1991; Brett 1995, 1998; Brett and Baird 1993, 1997) and key trends and surfaces in sequence stratigraphy (e.g. Courville and Collin 2002; Brett et al. 2009).

Taphonomic research is clearly wide-ranging, and in the Earth sciences impacts upon all aspects of "soft-rock" research. To put the current work in context it is necessary to briefly review the history and diversity of research that forms the body of the subject.

1.1 Taphonomy: A Brief History

Although Efremov (1940) is credited with coining the word, the most obvious and influential early contributors to the current understanding of taphonomy are the various German researchers who published in the period between the first and second World Wars. That is not to say that these workers were the first to ponder or make deductions about fossil preservation (see Cadee 1991) but they were the first to make systematic actualistic observations. In 1927 Weigelt, for example, studied the fate of diverse modern vertebrate carcasses in and around Lake Smithers in Texas (Weigelt 1989). He noted the role of insects in carcass degradation and studied modern mass mortalities and these observations were used in his interpretations of fossil Lagerstätten. At this point the classic work of Zangerl and Richardson (1963) should also be highlighted. They conducted a meticulous field study of two Pennsylvanian Lagerstätten and augmented their interpretations with actualistic experiments. This was followed by the extensive observations of North Sea tidal flats made in the influential work of Schäfer (1972 and references therein). These broad tidal flats provided Schäfer with a low-tech approach for examining marine taphonomic processes on a daily basis. The abundant and sometimes dramatic observations that he made on taphonomic systems such as marine animal carcasses have spurred much additional research. In many ways his observations provided the modern foundation for actualistic studies of shallow marine systems.

Taphonomic studies assumed ever greater prominence in the 1970s, as demanded by the rapid growth of the field of paleoecology. Terrestrial studies moved from the purely observational to those conducted through a time series. One of the pioneers in this approach has been Behrensmeyer, who focused her earlier studies on the fate of modern bones in African terrestrial environments and what they can tell us about the paleoecology of fossil bone assemblages (e.g., Behrensmeyer 1978, 1986; Behremsmeyer and Hill 1980).

In the 1980s, as taphonomic understanding of different fossil systems matured, this knowledge was transferred to studies of how taphonomic processes affect aspects of sedimentary systems and the production of sedimentary deposits. This is exemplified in the concept of taphofacies coined by Brett and Baird (1986) whereby different taphonomic processes are considered to characterize particular sedimentary facies. Similarly, taphonomic and depositional processes affecting shell beds, and the paleoecological and paleobiological meaning of shell beds, have been extensively investigated through the pioneering work of Kidwell (1985, 1986, 1994, 2002; Kidwell and Jablonski 1983; Kidwell and Flessa 1996; Kidwell and Brenchley 1996; Kidwell et al. 1986). By the end of the 80s understanding of taphonomic processes had reached a level requiring broad syntheses of rapidly accumulating data. This need was met by overview volumes edited by Donovan (1991), and Allison and Briggs (1991) as well as texts by Lyman (1994) and Martin (1999), which still provide a useful entry point to the subject.

The concept that some rare fossil deposits have undergone exceptional preservation, including evidence for soft tissues, was first popularized by Seilacher (Seilacher et al. 1985). These *Fossil Lagerstätten*, many of which have exceptional paleobiological importance, also began to receive important systematic study in the 1980s (Allison 1986, 1988a, b; Allison and Briggs 1993). Such studies fostered extensive efforts to investigate already-known Lagerstätten and spurred searches for new Lagerstätten, and the desire to understand the taphonomic processes that lead to exceptional preservation (e.g., Poinar 1992; Bottjer et al. 2002).

The drive to understand how soft tissues are preserved opened up a new experimental field of taphonomy. This promoted a stronger focus on understanding process (e.g., Martin 1999). Progress developed from the early experiments of Plotnick (1986) and Allison (1986, 1988a) to more sophisticated levels driven by the work of Briggs (e.g., Briggs 2003; Briggs and Kear 1993, 1994; Sageman et al. 1999).

Innovative approaches have continually been developed, as taphonomic research has blossomed into a large discipline within paleontology and sedimentary geology. Numerous aspects of taphonomy encompassing paleoenvironmental reconstruction (e.g. Brett and Baird 1997; Martin et al. 1999; Rogers et al. 2007), paleoecology (e.g. Meldahl et al. 1997; Flessa and Kowalewski 2007), paleobiology (e.g. Kidwell and Behrensmeyer 1993) and stratigraphy (Kidwell and Holland 2002) are very active research areas. The latest development is the use of databases to quantify the impact of taphonomy upon past diversity (e.g., Behrensmeyer et al. 2005). In this context we embrace the most catholic definition of taphonomy and include the effects of sedimentation, lithification and rock preservation (e.g. Marshall 1997; Holland 2000; Crampton et al. 2003; Hendy 2009; Sessa et al. 2009; Wall et al. 2009).

2 Is Taphonomic Bias Uniform?

At its heart, paleontology addresses two key concerns that are relevant to mankind: the origins of life and biodiversity, and the history of past climate change. The first is relevant because it reveals the evolutionary history of life on the planet (e.g. see Alroy et al. 2008; Benton 2009; Wagner et al. 2006) and our origins, and the second is pertinent because the study of past climate change, biodiversity and extinction (Hallam and Wignall 1997) might warn us of future change. Taphonomy speaks to both of these endeavours. Given the pervasive nature of preservational bias, an understanding of that bias is essential to properly decipher the history of biodiversity (e.g. Powell and Kowalewski 2002) and the impact of climate change on past biological systems.

Process-based research in the field and in the lab in the last two decades has gone a long way towards understanding taphonomic bias in modern environments.

A crucial question that remains however, is the extent to which taphonomic bias has changed through time. It is one thing to work with a data-set where the bias varies with depositional environment. It is magnitudinally more challenging to work with data where the bias has also varied with time. There are many reasons to suspect that this is likely to have been the case, including:

Biomolecular innovation (evolution of the materials from which organisms are constructed): Some organic molecules and skeletons are more preservable than others and this has changed with time. The appearance of specific biomolecules such as lignin and sporopollenin has potentially imparted decay resistance to plants (but see the chapter by Collinson).

Secular trends in ocean chemistry and skeletal mineralogy: Ocean chemistry has changed through time and this has influenced the relative preservation of calcite and aragonite (Sandberg 1975, 1983; Montañez 2002; Cherns and Wright 2000).

Biological evolution: The evolution and diversification of organisms that burrow into and disturb sediment has clear potential to indirectly promote temporal shifts in taphonomic bias. Such organisms would disturb and potentially degrade carcasses that were buried. This bias can be expected to have increased as the depth of burrowing has increased with time (Thayer 1983; Bottjer and Ausich 1986). Equally as biodiversity has increased organisms have evolved whose ecology promotes the direct destruction of biogenic remains (e.g. insects, fungi and microbes that destroy plant material in the terrestrial realm, diverse borers that degrade shelly remains in aquatic habitats.

Conserving environments through time: Fossil Lagerstätten occur in preservational windows that are unevenly distributed in time and space (Allison and Briggs 1991) and clearly reflect temporal trends in fossilization. Similar but more frequently encountered biases result from variations in lithification! Much of the sedimentary rock record was deposited in vast shallow epicontinental seas which lack modern analogues. These seas may have been more prone to stratification and this could conceivably have enhanced fossil preservation.

Each of these effects can cause changes in taphonomic biases and are discussed each in turn.

2.1 Biomolecular Innovation

The vast majority of organisms that have lived are not preserved in the rock record. In a sense, this is fortunate as the complete preservation of biogenic molecules for a prolonged interval of time would lead to shifts in atmospheric and Earth surface chemistry. For example, the accumulation of organic carbon subsequent to, and during the Devonian-Carboniferous led to marked reductions in levels of atmospheric carbon dioxide (Berner 1991; Ehleringer et al. 2002). The evolutionary pressure for space in early terrestrial environments promoted the development of floral tiering which was facilitated by the complex aromatic molecule lignin (Kenrick and Edwards 1988). This molecule imparted great strength to early plants and allowed them to reach substantial heights (Esau 1977). The Carboniferous forests flourished in low-lying areas that were prone to flooding. Thus, as sea-level waxed and waned to the orbital beat, vast swathes of forest were periodically waterlogged or drowned. Lignin has traditionally been considered as particularly decay-resistant in oxygen deficient regimes (but see Collinson, herein). As well as allowing Carboniferous forests to become tall it is often considered to have facilitated the accumulation of vast peat deposits, which subsequently became coal. The carbon cycle was therefore, very different after the Carboniferous because it included an expanded terrestrial carbon reservoir and a new linking process connecting the atmospheric to the lithospheric reservoirs.

This is a striking example of how taphonomic processes have changed with time and shows the extent to which those changes can influence the chemical cycles on the Earth's surface.

The appearance of molecular novelties that impart some level of decay resistance has of course impacted upon the quality of the fossil record. Chitin is a polysaccharide that occurs in the exoskeleton of arthropods. The preservation potential of chitin has long been a source of debate. Prior to the 1950s it was thought that the biomolecule, chitin was significantly decay resistant (see Richards 1951 for discussion). Taphonomic research in the 1980s (Plotnick 1986; Allison 1988a) showed that arthropod cuticles were degraded over periods of months in laboratory experiments. In the 1990s however, detailed geochemical investigations (Baas et al. 1995; Briggs 1999) showed that Richards (1951) was at least partially correct: there is some evidence that chitin imparts decay resistance immediately after burial and that chitin derivatives are preserved in geologically ancient deposits (Flannery et al. 2001). However, in the majority of cases the chitin has been diagenetically altered to an aliphatic composition (Briggs 1999). The fossil record of non-mineralized arthropods may have been significantly enhanced as a result of this molecule. However, recent work is questioning these paradigms. Chapters by Gupta and Briggs, and Collinson highlight a growing body of evidence suggesting that selective preservation is not simply the result of biomolecular composition. These authors argue that plant and animal biomacromolecules provide a structural template that is subsequently diagenetically altered to a geomacromolecules in fossils. The authors of these chapters highlight the need for future research and suggest a tentative agenda of research goals.

2.2 Secular Trends in Ocean Chemistry and Skeletal Mineralogy

The notion that seawater chemistry has changed through time was first mooted by Sandberg (1975) based upon his work on the mineralogy of Mesozoic ooids. It was subsequently proposed that the Ca/Mg ratio of seawater influenced the mineralogy of the dominant abiotic carbonates during the Phanerozoic (Sandberg 1983). The oscillation between so-called "calcite and aragonite seas" coincides with

Fisher's (1981) icehouse and greenhouse cycles and this in turn has been linked to ridge spreading activity and atmospheric PCO_2 (Wilkinson and Given 1986; Wilkinson et al. 1985). Subsequent studies (Dickson 2002, 2004; Harper et al. 1997; Montañez 2002; Stanley and Hardie 1998; Taylor et al. 2009) have shown that calcareous skeletal mineralogies are also impacted by this secular trend although the relationship is by no means straightforward. For example clades whose skeletons evolved in the Ediacaran-Tommotian developed aragonitic skeletons whilst those that arose between the Tommotian and the Ordovician had a calcitic skeleton (Porter 2007; Zhuravlev and Wood 2008). Post-Ordovician patterns are more complex (Taylor 2008; Taylor et al. 2009). This secular variation in seawater chemistry and skeletal mineralogy clearly has the potential to impart a temporally variable taphonomic overprint on the fossil record (e.g. see Cherns and Wright 2000; Wright et al. 2003) although the magnitude and pattern of the bias remains a subject of debate (Bush and Bambach 2004). This theme is touched on in several of the following chapters but is most pertinent to the chapters by Wood, and Cherns et al.

Wood highlights the way that taphonomic processes affecting the preservation of reefs has changed. Many of these taphonomic processes involve biological destruction, and include an escalation of herbivorous grazers, carnivores, and bioerosion that began in the Mesozoic. Changing ocean water chemistry affecting cementation rates over time also strongly affects the preservation of primary reef structures. Modern climate change is predicted to strongly affect taphonomic processes in reef environments in the future.

The fidelity of the fossil record for paleoecological and paleobiological studies is affected by the response of skeletons of different original mineralogy to diagenesis. The chapter by Cherns et al. explores the well-known problem of differential preservation of calcitic and aragonitic molluscan fossil faunas. They demonstrate a number of depositional and diagenetic conditions that are capable of preserving aragonitic and calcitic shells.

2.3 Biological Evolution

The impact of predator–prey escalation through geological time (Stanley 1974, 1977, 2008; Vermeij 1977, 1987) clearly has the potential to impact upon fossil preservation. Innovations in predation could potentially lead to a bias against fossil preservation. Equally, this may have led to the evolution of defence mechanisms that included stronger more robust shells that were more likely to be preserved and more capable of withstanding extended time-averaging (Kidwell and Brenchley 1994, 1996). The unprecedented diversity of durophagous marine vertebrates that thrived in the Cretaceous is particularly noteworthy (Walker and Brett 2002). The crunching jaws of vertebrates are not the only agent of biological destruction of shells however. Shell borings by diverse invertebrates can significantly impact upon shell strength (Kelley 2008) and thereby reduce preservation potential. The evolutionary diversification of organisms with this mode of life is therefore likely to have impacted upon hard-part preservation through time.

Evolution has also impacted upon the depth and nature of burrowing organisms through time. Modern marine organisms burrow into soft-sediment seafloors to depths of a meter or more (Bottjer and Ausich 1986). The behavioral activities that lead to this burrowing range from open burrow systems in which organisms live, to movement on and through sediment in search of prey, to complex systems in which microbes are farmed (e.g., Seilacher 2007, 2008). The study of these preserved burrows, or trace fossils, and the overall fabric it imparts to sediment, or ichnofabric, has revealed a variety of trends through the Phanerozoic (e.g., Thayer 1983; Droser and Bottjer 1993). Prior to the Cambrian seafloors were commonly covered with microbial mats and only in the later part of the Ediacaran did bioturbation first appear, as trails found at the surface of the seafloor (e.g., Seilacher 1999, 2007). However, with the Cambrian explosion animals began to evolve the ability to burrow into the seafloor for a variety of activities (e.g., Droser et al. 1999; Bottjer et al. 2000). This trend of increasing depth and extent of bioturbation in subtidal environments continued from low levels in the Cambrian (Droser and Bottjer 1988, 1989) to where burrows reaching modern depths of one meter or more at the end of the Paleozoic (Bottjer and Ausich 1986).

The Cambrian is well-known for its exceptional preservation of soft-bodied faunas in Lagerstätte such as the Burgess Shale. Burgess Shale-type faunas are found preserved globally, and the Cambrian is a time that has an unusual number of Lagerstätte with preservation of soft tissues (e.g., Allison and Briggs 1993). The Cambrian was a time of relatively low depth and extent of bioturbation (Bottjer and Ausich 1986; Droser and Bottjer 1988, 1989), but with the Cambrian explosion it also saw a proliferation of soft-bodied organisms. Bioturbation can include scavenging and disruption of carcasses, and it is likely that the low levels of Cambrian bioturbation led to a greater chance for preservation of soft-bodied organisms, as compared to the post-Cambrian, when extent of bioturbation increased significantly (Allison and Briggs 1993; Orr et al. 2003). This intriguing example of taphonomic bias towards greater preservation under globally-reduced bioturbation levels is a fascinating example of how the evolution of biological processes, such as bioturbation, can affect taphonomic processes, and thus introduce bias through time.

The aftermaths of mass extinctions are also times when it might be expected that bioturbation is reduced, due to extinction of burrowing organisms, with a resultant effect upon taphonomic processes. This topic is considered as part of the analysis of the effects of mass extinctions on taphonomic processes in the chapter by Fraiser et al. Mass extinctions entail a dramatic change in the fossil record through a short time interval. The question is, how much is this a primary change, and how much could be due to changes in taphonomic conditions? In this chapter temporal patterns for Lazarus taxa and distribution of silicified benthic faunas are assessed for the Permian-Triassic. These analyses show that the fossil record of the end-Permian mass extinction and the Early Triassic aftermath reflects largely a primary signal, and is not significantly obscured by a taphonomic megabias due to skeletal mineralogy or fossil preservation. The impact of mass extinctions on taphonomic processes is also considered by Nebelsick et al. They document taphonomic attributes of carbonate grains through the Paleogene in a range of facies. They conclude that extinction events among larger foraminifera that dramatically influence the occurrence and distribution of facies at this time have little effect on the distribution of taphonomic features.

2.4 Temporal Trends in Conserving Environments

Fossil lagerstatten are unevenly distributed through time and most abundant in particular environments (Allison and Briggs 1991, 1993) and it has long been recognized that this could impact upon estimates of global diversity through time (Sepkoski 1981). There are for example, times in Earth history when diagenetic minerals were more likely to preserve fossils. This theme is developed in several chapters within the book.

Butts and Briggs review the conditions that lead to silicification of marine fossils. The process of silicification is a function of both taxonomic and environmental factors, which control the rates of carbonate dissolution and silica precipitation. Silicification is variable through the Phanerozoic, being common in the Paleozoic, but much less so in the Mesozoic and Cenozoic. This temporal distribution of silicification results in taphonomic biases in the record of biodiversity through time.

Chapters by Brasier et al. and Dornbos detail the nature of phosphatization in the Precambrian and Phanerozoic respectively. Phosphatization can preserve organisms ranging from vertebrates to bacteria at the cellular level. The Phanerozoic record of phosphatization is biased towards taxa with recalcitrant tissues, those with body parts enriched in phosphate, and those with small body size. Phosphatization is common in phosphogenic environments, but can also occur in local phosphatizing microenvironments created by a decaying organism. Phosphatization appears to have been particularly common from the Cambrian through Early Ordovician and Cretaceous through Eocene.

The issue of mineralization in the Precambrian is of course fundamental to our understanding of apostrophe Earth's earliest fossil biotas where the challenge can sometimes be to determine whether a particular structure is fossil or artifact. This issue is hotly argued and is addressed in chapters by Schopf et al. and Brasier et al. Preservation of fine-scale structure at the cellular level has not been adequately documented in the past because of the lack of appropriate technology to investigate its occurrence. Confocal laser scanning microscopy (CLSM) and two- and three-dimensional Raman imagery represent new technological approaches that have successfully been utilized to examine preservation at the cellular level in animals, plants, fungi, algal protists, and microbes, preserved variously in phosphorites, cherts, and carbonates. The wide applicability of this new technology promises to yield an understanding in the future of how such preservation at the cellular level has varied through time.

Brasier et al highlight a preservational paradox in the early rock record. They argue that cellular preservation and stromatolite complexity is reduced before the late Archaean and often considered controversial. They argue that this could be because scientists have largely been looking in the wrong places: they go on to identify some exciting and new taphonomic windows, including pillow lavas, hydrothermal vents and beach sandstones.

The impact of secular changes in bioturbation, geochemistry and climate on fossil preservation in small scale (10–100 kyr) sedimentary cycles (ubiquitous in offshore marine successions) is treated in the chapter by Brett et al. In particular, they characterize the taphonomy of such cycles from Phanerozoioc "greenhouse" times. The primary taphonomic moderator in these cycles is rate of sedimentation, which varies exponentially from sediment-starved concentrations to obrutionary deposits. The occurrence of a persistent motif over this time scale suggests that biological innovations, which might be expected to impact upon fossil preservation, have in fact been overprinted by the extremes of sedimentation preserved in these small-scale cycles. For example, having a skeleton, which is more resistant to abrasion, is of little import when sedimentation is dominated by the extremes: instant obrution or condensation.

Large scale databases, such as the Paleobiology Database (PBDB), can provide a unique perspective on the effects of taphonomy on the perceived fossil record. Hendy et al. present an analysis of Phanerozoic data from the PBDB and identify a variety of taphonomic biases. The availability of fossil assemblages from unlithified sediments, more typical of later Mesozoic and Cenozoic rocks, is likely related to increases in local as well as global diversity. The occurrence of phosphate and silica replacement, as well as Konservat-Lagerstätten, is time-restricted. Similarly, shell beds show increased frequency in middle Paleozoic and Cenozoic rocks, and fossil molds are most frequent in rocks of early Cambrian and early Mesozoic age. All of these taphonomic processes are likely to have strong effects on comparisons of diversity or ecologic complexity through the Phanerozoic.

The nature of terrestrial taphonomic windows is addressed in chapters by Gastaldo and Demko, and Noto on plants and vertebrates respectively. Gastaldo and Demko show that in terrestrial settings, plant material is preserved not only in areas where organic detritus accumulates, but also in burial sites where pore-water geochemistry retards or halts organic degradation. Thus, whereas previously, the lack of a plant fossil record was interpreted as a function of ecosystem reorganization, extirpation, or extinction; it is now apparent that this absence of plant fossils is due to variations in sediment supply and geochemistry interacting with landscape and climate. This new understanding of what controls the preservation of plant material will revolutionize our understanding of the meaning of trends in the plant fossil record through time.

Noto argues that taphonomic processes are influenced by multiple hierarchical factors. Every environment contains a specific set of taphonomic conditions and each biome thus contains a subset of taphonomic conditions termed a taphonomic regime. As biomes shift through time taphonomic regimes change. Such a perspective, applied here to the terrestrial vertebrate fossil record, provides a powerful tool for assessing genuine biotic change through space and time in Earth history.

3 Taphonomy: A Prospectus?

It is clear that our understanding of taphonomy has benefited from diverse approaches that vary in scale from laboratory and field based studies to the analyses of data-bases. The latter are growing in number and sophistication and will clearly continue to do so. That is not to say that there is no place for lab or field based studies. Field-based studies obviously supply the primary data for subsequent data-base analyses but have also highlighted potential biases (e.g. Cherns and Wright 2000; Wright et al. 2003; Bush and Bambach 2004). What though are the ongoing grand challenges for taphonomic research? We argue that they are the same as they are for paleontology in general and that is to advance our understanding of the diversification of life on Earth as it evolved and fluctuated in the face of environmental change.

Diversity can be considered to be composed of three components (Whittaker 1972); alpha (within communities), beta (diversity of different communities in a region), and gamma (diversity of regions). It is clearly important to know how temporal shifts in taphonomic bias have affected these three components of diversity. The goal is not simply to understand how taphonomic bias has affected the global headcount of Phanerozoic diversity but also to understand how it has influenced the preserved community structure and ecological evenness. The Paleobiology Database (PBDB) has of course been a fundamental facilitating endeavour that has supported the foundation efforts that have already been made in this direction (see Powell and Kowalewski 2002; Alroy et al. 2008).

An emerging issue relates to the nature of epicontinental seas. Most of the sedimentary rock that is available for paleontological study was deposited in vast shallow seas on flooded continents. These seaways lack suitably scaled modern counterparts and this has long been recognized as a potential problem for uniformitarian analysis (e.g. Hallam 1975; Irwin 1965; Shaw 1964). In essence these seaways were less likely to experience tidal mixing (Wells et al. 2005, 2007) and were more prone to stratification. This clearly has implications for paleoecology, and sediment accumulation (Allison and Wright 2005; Allison and Wells 2006) as well as taphonomic bias (Peters 2007; Smith and McGowan 2008). How this has biased estimates of diversity is an emerging question.

Predicting the future direction of research is challenging because the very best research sometimes produces unforeseen results. However, we note the impact of thorough data-base studies and we can at least predict that this valuable research tool will be used with greater frequency. We also highlight the need for detailed, thorough, statistically rigorous fieldwork, because fieldwork always inspires and is also the raw material for data-base research. But where are the biggest gaps in taphonomic knowledge? We highlight 3 areas:

1. Precambrian taphonomy: The deepest recesses of Precambrian time included environments and fossils that lack modern counterparts and are challenging to identify and interpret. A better understanding of the taphonomy of such systems will elucidate the early history of Earth and potentially inform the exploration of other planets.

2. Organic geochemistry: Collinson's chapter shows that there is still much to learn about the pathways between organic molecules and preservation of organic carbon.
3. Global biodiversity: The Earth has suffered several mass extinction events. To what extent do these events impact upon taphonomic processes? Further development of this work will shed further light on preservational biases and provide an enhanced understanding of the extinctions themselves.

References

Alroy, J., et al. (2008). Phanerozoic trends in the global diversity of marine invertebrates. *Science, 321*, 97–100.

Allison, P. A. (1986). Soft-bodied animals in the fossil record: The role of decay in fragmentation during transport. *Geology, 14*, 979–981.

Allison, P. A. (1988a). The role of anoxia in the decay and mineralization of proteinaceous macro-fossils. *Paleobiology, 14*, 139–154.

Allison, P. A. (1988b). Konservat-Lagerstätten: Cause and classification. *Paleobiology, 14*, 331–344.

Allison, P. A., & Briggs, D. E. G. (1991). *Taphonomy: Releasing the data locked in the fossil record*. New York: Plenum.

Allison, P. A., & Briggs, D. E. G. (1993). Exceptional fossil record: Distribution of soft-tissue preservation through the Phanerozoic. *Geology, 21*, 527–530.

Allison, P. A., & Wells, M. R. (2006). Circulation in large ancient epicontinental seas: What was different and why? *Palaois, 21*, 513–515.

Allison, P. A., & Wright, V. P. (2005). Switching off the carbonate factory: A-tidality, stratification and brackish wedges in epeiric seas. *Sedimentary Geology, 179*, 175–184.

Baas, M., Briggs, D. E. G., van Heemst, J. D. H., de Kear, A. J., & Leeuw, J. W. (1995). Selective preservation of chitin during the decay of shrimps. *Geochimica Cosmochimica Acta, 59*, 945–951.

Behrensmeyer, A. K. (1978). Taphonomic and ecologic information from bone weathering. *Paleobiology, 4*, 150–162.

Behrensmeyer, A. K. (1986). Tramping as a cause of bone surface damage and pseudo-cutmarks. *Nature, 319*, 768–771.

Behremsmeyer, A., & Hill, A. P. (Eds.). (1980). *Fossils in the making: Vertebrate taphonomy and paleoecology*. Chicago: University of Chicago Press.

Behrensmeyer, A. K., Fürsich, F. T., Gastaldo, R. A., Kidwell, S. M., Kosnik, M. A., Kowalewski, M., et al. (2005). Are the most durable shelly taxa also the most common in the marine fossil record? *Paleobiology, 31*, 607–623.

Benton, M. J. (2009). The Red Queen and the Court Jester: Species diversity and the role of biotic and abiotic factors through time. *Science, 323*, 728–732.

Berner, R. A. (1991). A model for atmospheric carbon dioxide over Phanerozoic time American. *Journal of Science, 291*, 339–76.

Bottjer, D. J., & Ausich, W. I. (1986). Phanerozoic development of tiering in soft substrata suspension-feeding communities. *Paleobiology, 12*, 400–420.

Bottjer, D. J., Etter, W., Hagadorn, J. W., & Tang, C. M. (Eds.). (2002). *Exceptional fossil preservation: A unique view on the evolution of marine life*. New York: Columbia University Press.

Bottjer, D. J., Hagadorn, J. W., & Dornbos, S. Q. (2000). The Cambrian substrate revolution. *GSA Today, 10*, 1–7.

Brett, C. E. (1995). Sequence stratigraphy, biostratigraphy, and taphonomy in shallow marine environments. *Palaois, 10*, 597–616.

Brett, C. E. (1998). Sequence stratigraphy, paleoecology, and evolution: Biotic clues and responses to sea-level fluctuations. *Palaois, 13*, 241–262.

Brett, C. E., & Baird, G. C. (1986). Comparative taphonomy: A key to paleoenvironmental interpretation based on fossil preservation. *Palaios, 1*, 207–227.

Brett, C. E., & Baird, G. C. (1993). Taphonomic approaches to temporal resolution in stratigraphy: Examples from Paleozoic marine mudrocks. In S. M. Kidwell & A. K. Behrensmeyer (Eds.), *Taphonomic approaches to temporal resolution in fossil assemblages: Paleontological Society Short Course 6* (pp. 250–274). Knox-ville, Boston: The Paleontological Society.

Brett, C. E., & Baird, G. C. (1997). *Paleontological events: Stratigraphic, ecological, and evolutionary implications*. New York: Columbia University Press.

Brett, C. E., Allison, P. A., DeSantis, M. K., Liddell, W. D., & Kramer, A. (2009). Sequence stratigraphy, cyclic facies, and lagerstatten in the Middle Cambrian Wheeler and Marjum Formations, Great Basin, Utah. *Palaeogeogr Paleoclimatol Palaeoecol, 277*, 9–33.

Briggs, D. E. G. (1999). Molecular taphonomy of animal and plant cuticles: Selective preservation and diagenesis. *Philosophical Transactions of the Royal Society of London, 354*, 7–17.

Briggs, D. E. G. (2003). The role of decay and minerallzation in the preservation of soft-bodied fossils. *Annual Review Earth Planet Science, 31*, 275–301.

Briggs, D. E. G., & Kear, A. J. (1993). Fossilization of soft tissue in the laboratory. *Science, 259*, 1439–1442.

Briggs, D. E. G., & Kear, A. J. (1994). Decay and mineralization of shrimps. *Palaios, 9*, 431–456.

Brothwell, D. (1986). *The bog man and the archaeology of people*. London: British Museum Publications.

Bush, A. M., & Bambach, R. K. (2004). Did alpha diversity increase during the Phanerozoic? Lifting the veils of taphonomic, latitudinal, and environmental biases. *The Journal of Geology, 112*, 625–642.

Cadee, G. C. (1991). The history of taphonomy. In S. K. Donovan (Ed.), *The processes of fossilization*. New York: Columbia.

Crampton, J. S., Beu, A. G., Cooper, R. A., Jones, C. M., Marshall, B., & Maxwell, P. A. (2003). Estimating the rock volume bias in paleobiodiversity studies. *Science, 301*, 358–360.

Cherns, L., & Wright, V. P. (2000). Missing molluscs as evidence of large-scale, early skeletal aragonite dissolution in a Silurian Sea. *Geology, 28*, 791–794.

Courville, P., & Collin, P. Y. (2002). Taphonomic sequences: A new tool for sequence stratigraphy. *Geology, 30*, 511–514.

Davies, D. J., Powell, E. N., & Stanton, R. J., Jr. (1989). Taphonomic signature as a function of environmental process: Shells and shell beds in a hurricane-influenced inlet on the Texas coast. *Palaeogeogr Paleoclimatol Palaeoecol, 72*, 317–356.

Dickson, J. A. D. (2002). Fossil echinoderms as monitor of the Mg/Ca ratio of Phanerozoic oceans. *Science, 298*, 1222–1224.

Dickson, J. A. D. (2004). Echinoderm skeletal preservation; calcite-aragonite seas and the Mg/Ca ratio of Phanerozoic oceans. *Journal of Sedimentary Research, 74*, 355–365.

Donovan, S. K. (Ed.). (1991). *The processes of fossilization*. New York: Columbia University Press.

Droser, M. L., & Bottjer, D. J. (1988). Trends in depth and extent of bioturbation in Cambrian carbonate marine environments, western United States. *Geology, 16*, 233–236.

Droser, M. L., & Bottjer, D. J. (1989). Ordovician increase in extent and depth of bioturbation: Implications for understanding early Paleozoic ecospace utilization. *Geology, 17*, 850–852.

Droser, M. L., & Bottjer, D. J. (1993). Trends and patterns of Phanerozoic ichnofabrics. *Annual Reviews of Earth and Planetary Sciences, 21*, 205–225.

Droser, M. L., Gehling, J. G., & Jensen, S. (1999). When the worm turned: Concordance of Early Cambrian ichnofabric and trace-fossil record in siliciclastic rocks of South Australia. *Geology, 27*, 625–628.

Efremov, E. A. (1940). Taphonomy: A new branch of paleontology. *Pan-American Geologist, 74*, 81–93.

Ehleringer, J. R., Cerling, T. E., & Dearing, M. D. (2002). *A history of atmospheric CO_2 and its effects on plants, animals, and ecosystems series: Ecological studies, 177* (p. 530). New York: Springer.

Esau, K. (1977). *Anatomy of seed plants*. New York: Academic.

Fisher, A. G. (1981). Climatic oscillations in the biosphere. In M. Nitecki (Ed.), *Biotic crises in ecological and evolutionary time*. New York: Academic.

Flannery, M. B., Stott, A. W., Briggs, D. E. G., & Evershed, R. P. (2001). Chitin in the fossil record: Identification and quantification of D-glucosamine. *Organic Geochemistry, 32*, 745–754.

Flessa, K. W., & Kowalewski, M. (2007). Shell survival and time-averaging in nearshore and shelf environments: Estimates from the radiocarbon literature. *Lethaia, 27*, 153–165.

Glob, P. V. (2004). *The bog people: Iron age man preserved*. New York: New York Review Book Classics.

Harper, E. M., Palmer, T. J., & Alphey, J. R. (1997). Evolutonary response by bivalves to changing Phanerozoic sea-water chemistry. *Geological Magazine, 134*, 403–407.

Hallam, A. (1975). *Jurassic environments*. Cambridge: Cambridge University Press.

Hallam, A., & Wignall, P. B. (1997). *Mass extinctions and their aftermath*. Oxford: Oxford University Press.

Hendy, A. J. W. (2009). The influence of lithification on Cenozoic marine biodiversity trends. *Paleobiology, 35*, 51–62.

Holland, S. M. (2000). The quality of the fossil record: A sequence stratigraphic perspective. In D. H. Erwin & S. L. Wing (Eds.), *Deep time: Paleobiology's perspective*. Lawrence, KA: The Paleontological Society.

Irwin, M. L. (1965). General theory of epeiric clear water sedimentation: American Association of Petroleum Geologists. *Bulletin, 49*, 445–459.

Kelley, P. H. (2008). Role of bioerosion in taphonomy: Effect of predatory drillholes on preservation of mollusc shells. In L. Tapanila & M. Wisshak (Eds.), *Current developments in bioerosion*. Heidleberg: Springer.

Kenrick, P., & Edwards, D. (1988). The anatomy of Lower Devonian Gosslingia breconensis Heard based on pyritized axes, with some comments on the permineralization process. *Botanical Journal of the Linnean Society, 97*, 95–123.

Kidwell, S. M. (1985). Palaeobiological and sedimentological implications of fossil concentrations. *Nature, 318*, 457–460.

Kidwell, S. M. (1986). Models for fossil concentrations: Paleobiologic implications. *Paleobiology, 12*, 6–24.

Kidwell, S. M. (1991). The stratigraphy of shell concentrations. In P. A. Allison & D. E. G. Briggs (Eds.), *Taphonomy, releasing the data locked in the fossil record*. New York: Plenum.

Kidwell, S. M. (1994). Patterns in bioclastic accumulation through the Phanerozoic: Changes in input or in destruction? *Geology, 22*, 1139–1141.

Kidwell, S. M. (2002). Time-averaged molluscan death assemblages: Palimpsests of richness, snapshots of abundance. *Geology, 30*, 803–806.

Kidwell, S. M., & Jablonski, D. (1983). Taphonomic feedback: Ecological consequences of shell accumulation. In M. J. S. Tevesz & P. L. McCall (Eds.), *Biotic interactions in recent and fossil benthic communities*. New York: Plenum.

Kidwell, S. K., & Holland, S. M. (2002). The quality of the fossil record: Implications for evolutionary analyses. *Annual Review of Ecology & Systematics, 33*, 561–588.

Kidwell, S. M., Fürsich, F. T., & Aigner, T. (1986). Conceptual framework for the analysis and classification of fossil concentrations. *Palaios, 1*, 228–238.

Kidwell, S. M., & Flessa, K. W. (1996). The quality of the fossil record: Populations, species, and communities. *Annual Review Earth Planet Science, 24*, 433–464.

Kidwell, S. M., Behrensmeyer, A. K., & eds. (1993). *Taphonomic approaches to time resolution in fossil assemblages. Short courses in paleontology no. 6*. Knoxville: Paleontological Society.

Kidwell, S. M., & Brenchley, P. J. (1994). Patterns of bioclastic accumulation through the Phanerozoic: Changes in input or in destruction? *Geology, 22*, 1139–1143.

Kidwell, S. M., & Brenchley, P. J. (1996). Evolution of the fossil record: Thickness trends in marine skeletal accumulations and their implications. In D. Jablonski, D. H. Erwin, & J. H. Lipps (Eds.), *Evolutionary paleobiology*. Chicago: University of Chicago Press.

Lyman, R. L. (1994). *Vertebrate taphonomy*. Cambridge: Cambridge University Press.

Marshall, C. R. (1997). Confidence intervals on strati- graphic ranges with nonrandom distribution of fossil horizons. *Paleobiology, 23*, 165–173.

Martin, R. E. (1999). *Taphonomy: A process approach*. Cambridge: Cambridge University Press.

Martin, R. E., Paterson, R. T., Goldstein, S. T., Kumar, A., & (Eds.). (1999). Taphonomy as a tool in paleoenvironmental reconstruction and environmental assessment. *Palaeography, Palaeoclimatology, Palaeoecol, 149*, 1–429.

Meldahl, K. H., Flessa, K. W., & Cutler, A. H. (1997). Time-averaging and postmortem skeletal survival in benthic fossil assemblages; quantitative comparisons among Holocene environments. *Paleobiology, 23*, 209–229.

Montañez, I. P. (2002). Biological skeletal carbonate records changes in major-ion chemistry of paleo-oceans: National Academy of Sciences, USA. *Proceedings, 99*, 15,852–15,854.

Poinar, G. O., Jr. (1992). *Life in amber*. Stanford, CA: Stanford University Press.

Orr, P. J., Benton, M. J., & Briggs, D. E. G. (2003). Post-Cambrian closure of the deep water slope-basin taphonomic window. *Geology, 31*, 769–772.

Peters, S. E. (2007). The problem with the Paleozoic. *Paleobiology, 33*, 165–181.

Plotnick, R. E. (1986). Taphonomy of a modern shrimp: Implications for the arthropod fossil record: PALAIOS, *1*, 286–293.

Porter, S. M. (2007). Seawater chemistry and early carbonate biomineralization. *Science, 316*, 1302.

Powell, M. G., & Kowalewski, M. (2002). Increase in evenness and sampled alpha diversity through the Phanerozoic: Comparison of early Paleozoic and Cenozoic marine fossil assemblages. *Geology, 30*, 331–334.

Richards, A. G. (1951). *The integument of arthropods*. Minneapolis: University of Minnesota Press.

Rogers, R. R., Eberth, D. A., & Fiorillo, A. R. (2007). *Bonebeds: Genesis, analysis, and paleobiological significance*. Chicago: University of Chicago Press.

Sageman, J., Bale, S. J., Briggs, D. E. G., & Parkes, R. J. (1999). Controls on the formation of authigenic minerals in association with decaying organic matter: An experimental approach. *Geochim et Cosmochim Acta, 63*, 1083–1095.

Sandberg, P. A. (1975). New interpretations of Great Salt Lake ooids and of ancient nonskeletal carbonate mineralogy. *Sedimentology, 22*, 497–538.

Sandberg, P. A. (1983). An oscillating trend in Phanerozoic non-skeletal carbonate mineralogy. *Nature, 305*, 19–22.

Schäfer, W. (1972). *Ecology and palaeoecology of marine environments*. Chicago: University of Chicago Press.

Seilacher, A., Reif, W. E., & Westphal, F. (1985). Sedimentological, ecological and temporal patterns of Fossil Lagerstätten. *Proceedings of the Royal Society of London, Series B, 311*, 5–23.

Seilacher, A. (1999). Biomat-related lifestyles in the Precambrian. *Palaios, 14*, 86–93.

Seilacher, A. (2007). *Trace fossil analysis*. Berlin: Springer.

Seilacher, A. (2008). Biomats, biofilms, and bioglue as preservational agents for arthropod trackways. *Palaeogeography, Palaeoclimatology, and Palaeoecology, 270*, 252–257.

Sepkoski, J. J., Jr. (1981). A factor analytic description of the Phanerozoic marine fossil record. *Paleobi ology, 7*, 36–53.

Sessa, J. A., Patzkowsky, M. E., & Bralower, T. J. (2009). The impact of lithification on the diversity, size distribution, and recovery dynamics of marine invertebrate assemblages. *Geology, 37*, 115–118.

Shaw, A. B. (1964). *Time in stratigraphy*. New York: McGraw-Hill.

Smith, A. B., & McGowan, A. J. (2008). Temporal patterns of barren intervals in the Phanerozoic. *Paleobiology, 34*, 155–161.

Stankiewicz, B. A., Hutchins, J. C., Thomson, R., Briggs, D. E. G. & Evershed, R. P. (1997). Assessment of bog-body tissue preservation by pyrolysis–gas chromatography/mass spectrometry. *Rapid Communications in Mass Spectrometry, 11*, 1884–1890.

Hutchins, S. B. A., Thomson, J. C., & Briggs DEG Evershed RP, R. (1997). Assessment of bog-body tissue preservation by Pyrolysis-Gas Chromatography/Mass Spectrometry. *Rapid Communications in Mass Spectrometry, 11*, 1884–1890.

Stanley, S. M. (1974). What has happened to the articulate brachiopods? *Geological Society of America Abstracts with Programs, 6*, 966–967.

Stanley, S. M. (1977). Trends, rates, and patterns of evolution in the Bivalvia. In A. Hallam (Ed.), *Patterns of evolution, as illustrated by the fossil record*. Amsterdam: Elsevier.

Stanley, S. M. (2008). Predation defeats competition on the seafloor. *Paleobiology, 34*, 1–21.

Stanley, S. M., & Hardie, L. A. (1998). Secular oscillations in the carbonate mineralogy of reef-building and sediment-producing organisms driven by tectonically forced shifts in seawater chemistry. *Palaeogeography, Palaeoclimatology, Palaeoecology, 144*, 3–19.

Stead, I. M., Bourke, J. B., & Brothwell, D. (1986). *Lindow man, the body in the bog*. London: British Museum Publications.

Taylor, P. D. (2008). Seawater chemistry, biomineralization and the fossil record of calcareous organisms. In H. Okada, S. F. Mawatari, N. Suzuki, & P. Gautam (Eds.), *Origin and evolution of natural diversity: Sapporo*. Japan: University of Hokkaido.

Taylor, P. D., James, N. P., Bone, Y., Kuklinski, P., & Kyser, T. K. (2009). Evolving mineralogy of cheilostome bryozoans. *Palaois, 24*, 440–452.

Thomas, R. D. K. (1986). Taphonomy: Ecology's loss is sedimentology's gain. *Palaois, 1*, 206.

Thayer, C. W. (1983). Sediment-mediated biological disturbance and the evolution of marine benthos. In M. J. S. Tevesz & P. L. McCall (Eds.), *Biotic interactions in recent and fossil benthic communities*. New York: Plenum.

Vermeij, G. J. (1977). The Mesozoic marine revolution: Evidence from molluscs, predation, and grazing. *Paleobiology, 3*, 245–258.

Vermeij, G. J. (1983). Shell breaking predation through time. In M. J. S. Tevesz & P. L. McCall (Eds.), *Biotic interactions in recent and fossil benthic communities*. New York: Plenum.

Vermeij, G. J. (1987). *Evolution and escalation*. Princeton, NJ: Princeton University Press.

Wagner, P. J., Kosnik, M. A., & Lidgard, S. (2006). Abundance distributions of post-Paleozoic marine ecosystems. *Science, 314*, 1289–1292.

Walker, S.E., & Brett, C.E. (2002). Post-Paleozoic patterns in marine predation: Was there a mesozoic and cenozoic marine predatory revolution? In M. Kowalewski & P. H. Kelley (Eds.), *The fossil record of predation. The Paleontological Society Papers* (Vol. 8, pp. 119–193).

Wall, P. D., Ivany, L. C., & Wilkinson, B. H. (2009). Revisiting Raup: Exploring the influence of outcrop area on diversity in light of modern sample-standardization techniques. *Paleobiology, 35*, 146–167.

Wells, M. R., Allison, P. A., Hampson, G. J., Piggott, M. D., & Pain, C. C. (2005). Modelling ancient tides: The upper carboniferous epi-continental seaway of Northwest Europe. *Sedimentology, 52*, 715–735.

Wells, M. R., Allison, P. A., Piggott, M. D., Gorman, G. J., Hampson, G. J., Pain, C. C., et al. (2007). Numerical modeling of tides in the late Pennsylvanian Midcontinent seaway of North America with implications for hydrography and sedimentation. *Journal of Sedimentary Research, 77*, 843–865.

Weigelt, J. (1989). *Recent vertebrate carcasses and their paleobiological implications*. Chicago: University of Chicago Press.

Whittaker, R. H. (1972). Evolution and measurement of species diversity. *Taxon, 21*, 213–251.

Wilkinson, B. H., & Given, K. R. (1986). Secular variation in abiotic marine carbonates: Constraints on Phanerozoic atmospheric carbon dioxide contents and oceanic Mg/Ca ratios. *Journal of Geology, 94*, 321–333.

Wilkinson, B. H., Owen, R. M., & Carroll, A. R. (1985). Submarine hydrothermal weathering, global eustacy, and carbonate polymorphism in Phanerozoic marine oolites. *Journal of Sedimentary Petrology, 55*, 171–183.

Wright, V. P., Cherns, L., & Hodges, P. (2003). Missing molluscs: Field testing taphonomic loss in the Mesozoic through early large-scale aragonite dissolution. *Geology, 31*, 211–214.

Zangerl, R., & Richardson, E. S., Jr. (1963). The paleoecological history of two Pennslvanian black shales. *Fieldiana Geology Memoir, 4*, 1–132.

Zhuravlev, A. Y., & Wood, R. A. (2008). Eve of biomineralization: Controls on skeletal mineralogy. *Geology, 36*, 923–926.

Chapter 2
Taphonomic Overprints on Phanerozoic Trends in Biodiversity: Lithification and Other Secular Megabiases

Austin J.W. Hendy

Contents

Abstract Taphonomic biases introduce heterogeneity into the quality of the fossil record and can skew paleontologists' perception of biodiversity. This paper reviews the temporal extent and consequences of major taphonomic biases, including lithification of sediments, skeletal replacement through silicification and phosphatization, concentration of skeletal hard-parts, and the exceptional preservation of soft-bodied faunas. The frequency of occurrence of particular biases, and their effects of fossil faunas is identified using occurrence-based datasets, such as the Paleobiology Database.

A.J.W. Hendy (✉)
Center for Tropical Paleoecology and Archaeology Smithsonian Tropical Research Institute, Panamá, República de Panamá
and
Department of Geology and Geophysics, Yale University, New Haven, CT 06510, USA
e-mail: hendyaj@si.edu

P.A. Allison and D.J. Bottjer (eds.), *Taphonomy: Process and Bias Through Time*, Topics in Geobiology 32, DOI 10.1007/978-90-481-8643-3_2,

Lithification of most Paleozoic and Mesozoic fossiliferous sediments has likely had a significant influence on perceptions of within-community diversity and paleoecological composition. The increased availability of unlithified sediments in rocks of late Mesozoic through Cenozoic age coincides with a two- to threefold increase in local diversity, a discrepancy that remains even after employing sampling-standardization techniques. Taxa that possess small body size and aragonitic skeletal mineralogy are preferentially lost or obscured following the cementation of host sediments. Additionally, morphological details are often obscured or not preserved in specimens obtained from lithified sediments, suggesting that taphonomic damage could hinder taxonomic practice and estimates of diversity at the global-scale. Silica replacement, which generally enhances diversity among groups composed of less stable skeletal composition, appears most frequently among Permian fossil assemblages. Phosphatic replacement, which plays a key role in the preservation of soft-bodied and small-shelly faunas, appears commonly in assemblages of Cambrian age. Konservat-lagerstätten, while providing a rich source of information on the rarely preserved soft-bodied biota, are infrequent in the fossil record, but perhaps are most notable in rocks of Cambrian age. Shell beds are well known as sources of tremendous diversity and although they are not easily defined these beds appear to increase in frequency in middle Paleozoic and Cenozoic age successions. Fossil molds, unlike previously mentioned biases, suggest lost diversity, and are most frequent in rocks of early Cambrian and early Mesozoic age.

The non-random nature of the above biases raises concerns regarding the comparison of diversity or ecological complexity over the course of the Phanerozoic or between contemporaneous faunal groups. Furthermore, a number of the biases have tremendous potential to affect community-scale patterns, either degrading (e.g., lithification, aragonite dissolution) or enhancing (e.g., silicification, phosphatization) the relative quality of fossil data. A number of approaches can be undertaken to minimize these biases, including the selective filtering of datasets to remove taphonomically vulnerable groups or the use of taphonomic control taxa that indicate the appropriate preservation state of fossil assemblages.

1 Introduction

Documenting trends in biodiversity through geological time is one of the basic goals of paleontology. Interpretation of these trends has broad implications for understanding of the evolution of Earth's environments, the history of life, and the responses of organisms to environmental change. The observed fossil record of all organisms is, however, a consequence of taphonomic processes, and diversity data is subject to a taphonomic overprint. Paleontologists have devoted considerable attention to the causes, recognition, and mitigation of deficiencies in the record (Donovan and Paul 1998; McKinney 1991; Kidwell and Flessa 1995; Behrensmeyer et al. 2000; Kidwell and Holland 2002). Attempts to minimize known taphonomic biases are therefore important to correctly establish underlying evolutionary and environmental signals.

As a complication, however, the taphonomic biases that affect marine invertebrate taxa have not only changed over time (Kidwell and Flessa 1995; Kidwell and Holland 2002), but vary significantly between major groups of taxa (Schopf 1978). Factors that generate temporal patterns in preservation may include gross changes in seawater geochemistry (e.g., calcite or aragonite saturation states), the gain or loss of depositional environments (e.g., microbial mats), changes in the production or concentration of skeletal sediments, depth and intensity of bioturbation, or the evolution of durophagous predatory organisms (e.g., piscivorous carnivores) (Thayer 1983; Wilkinson and Given 1986; Vermeij 1987). Factors that influence variation among taxonomic groups may include presence, absence, or robustness of skeletons, skeletal mineralogy (e.g., calcitic, aragonitic) (Fig. 1), and substantive variations in life-habit (e.g., epifaunal, infaunal modes of life) (Stanley 1968; Behrensmeyer et al. 2005).

Because of their relative ease of preservation, the fossil record is very heavily biased towards animals with a robust or chemically resistant skeleton (Forey et al. 2004). Nichol (1977) estimated that c. 8% of animal species had a skeleton and were therefore likely to be preserved. However, having a mineralized skeleton is no guarantee for preservation and/or fossilization (Smith and Nelson 2003), given the range of physical, chemical, and biological factors that combine to determine the ultimate fate of skeletal material. The fossilization potential among skeletonized organisms is variable but has been estimated to be around 89% of shelled molluscs, 76% of echinoids, and ~50% of crabs (Kidwell and Flessa 1995). While Valentine (1989) demonstrated that about 80% of shelled mollusc species found living in the Californian province are preserved in local Pleistocene fossil assemblages, it seems probable that this figure would be much lower for many non-molluscan groups, like crabs and echinoids. Alternatively, special deposits are scattered through the geological record in which soft-bodied and poorly skeletonized organisms are fossilized (Konservat-Lagestätten) (Allison and Briggs 1993). Paul (1998) therefore considered that it would be reasonable to assume that c. 10% of the biota might have entered the fossil record, while Forey et al. (2004) went further to conclude that maybe only 1–5% of species are preserved in the geological record that survives today. That record, however, is probably most representative for specific paleoenvironments, geographic regions, or tectonic settings through time.

Our knowledge of past biodiversity represents only a portion of Earth's former biota, although it is probably fairly representative for those taxa that possess a moderate to high preservation potential. A recent trend has been towards compiling comprehensive taxonomic databases for the estimation of diversity (e.g., Sepkoski 2002). The consensus among global analyses such as Sepkoski (1982, 1997) and Benton (1995) has led to growing confidence that they depict the true history of biodiversity. However, if there are systematic biases affecting the nature of the fossil record, then raw systematic counts will give a poor estimation of the underlying diversity (Alroy et al. 2001, 2008). It is therefore essential that potential biases affecting the rock record be properly accounted for if paleontologists are to obtain accurate estimates of past biodiversity. Electronic databases provide the only practical means for investigating large-scale paleontological patterns and provide direction for investigation of the processes responsible for such biases (Markwick and Lupia 2002).

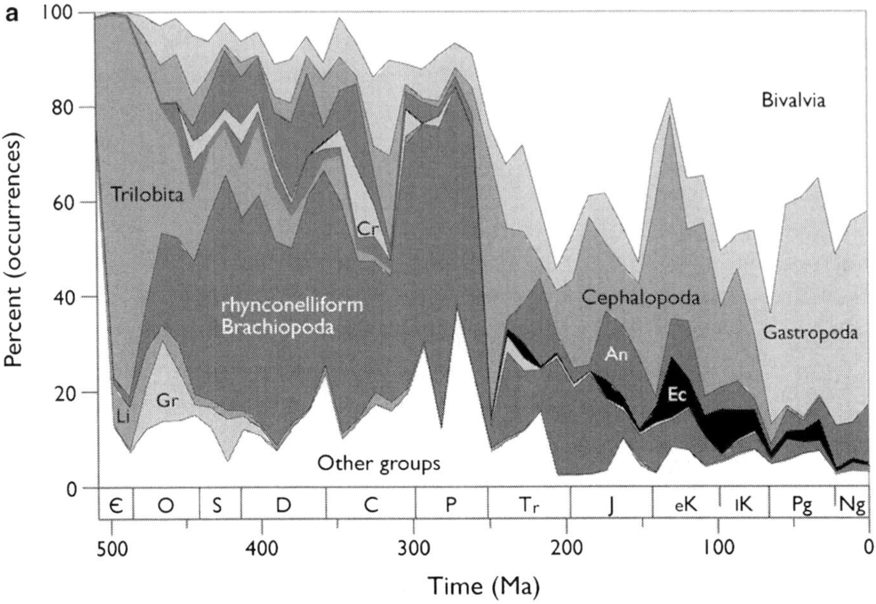

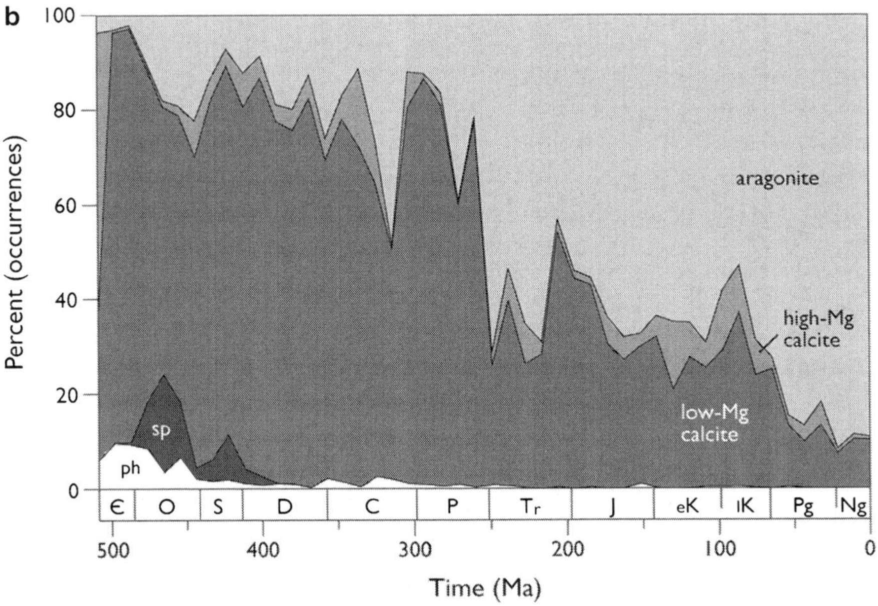

Fig. 1 Variations in the frequency of occurrences for major taxonomic groups of macrofauna (**a**) and their skeletal mineralogy (**b**) through the Phanerozoic. Data from the Paleobiology Database (downloaded 9/4/2007). Li, Lingulida; Gr, Graptolithina; Cr, Crinoidea; An, Anthozoa; Ec, Echinoidea; Mg, magnesium; sp, sclero-protein; ph, phosphatic

The development of sophisticated databases, such as the Paleobiology Database (www.paleodb.org) has not only encouraged the dissection of the fossil record among geographic regions and paleoenvironments (e.g., Miller 1997; Kiessling and Aberhan 2007; Bottjer et al. 2008; Wall et al. 2009), but also by the lithologic and taphonomic context of fossil occurrences (e.g., Foote 2006; Hendy 2009a).

This chapter outlines the nature of past and present assessments of biodiversity in the light of potential taphonomic biases and in particular the lithification of most pre-Cenozoic fossiliferous sediments. I not only use data from the Paleobiology Database (global-scale) and New Zealand Fossil Record Electronic Database (regional-scale), but also highlight a number of independent basin-scale case studies (Koch and Sohl 1983; Hendy 2009a; Sessa et al. 2009). These carefully designed field sampling and specimen-based studies accurately test how taphonomic and lithologic characteristics of the fossil record influence paleobiological patterns in the absence of potentially overprinting factors (i.e. paleoenvironmental or biogeographic heterogeneity of data). Other potentially secular variations in preservational biases are illustrated using data reposited in the Paleobiology Database. This exercise serves as both an initial quantitative assessment of these potential biases and provides an opportunity to critically assess the design and value of occurrence-based datasets for examining taphonomic trends. The final section of the chapter summarizes these results, critiques the fidelity of existing data and design of databases, and suggests future avenues of research.

2 Lithification and Diagenesis in the Fossil Record

A number of authors have noted the potential for secular changes in the nature of the fossil record associated with lithification and carbonate diagenesis (Raup 1976; Miller 2000; Bush and Bambach 2004; Cherns and Wright 2009). Broadly speaking, alteration of sedimentary rocks and fossils (i.e. dissolution and recrystallization of skeletal components) is least in younger assemblages. Raup (1976) commented that this may have the consequence of enhancing apparent species diversity in the younger rocks, but lamented that quantitative estimates of the effect are not available.

Recently, there has been considerable interest in investigations of the various spatial components of biodiversity, particularly alpha diversity, or the richness of individual benthic marine communities. Bambach's (1977) seminal work revealed a two- to threefold increase in median community richness from the Paleozoic to the Cenozoic, an increase also found in subsequent studies (e.g., Sepkoski 1997). A reinvestigation by Bush and Bambach (2004) not only confirmed this view, but also suggested that the increase could be even higher after mitigating for such biasing influences as secular variation in aragonite dissolution, environmental coverage, and latitudinal variation through the Phanerozoic. Nevertheless, previous assessments of alpha diversity, while noting the potential bias of secular taphonomic trends, did not attempt to mitigate for the considerable increase of unlithified fossiliferous sediments in strata of late Mesozoic and Cenozoic age. The following section uses several distinct data sources to evaluate the impact of lithification on biodiversity estimates. Specific focus is applied to three case studies at markedly different scales to identify lithification bias on (a) a global scale through the Mesozoic and Cenozoic, (b) regionally through the late Cenozoic, and (c) at basin scale with carefully collected data that limits the affect of confounding taphonomic, environmental, and geographic variables. The final section of this chapter focuses on the biases of lithification and carbonate diagenesis on taxonomic and morphologic data using a specimen-based dataset.

2.1 *Time-Series Analysis of Lithification and Alpha Diversity: A Global Perspective*

Using occurrence data from the Paleobiology Database the relationship between lithification and temporal trends in alpha diversity can be explored. The database has global, high density coverage of the Phanerozoic fossil record and provides an excellent resource for evaluating first-order patterns in paleontological data. Data were downloaded (9/4/2007) only for molluscan and brachiopod components of marine invertebrate collections for the Phanerozoic. Collections reposited in the database are typically assigned to one of four lithification categories (metamorphosed, lithified, poorly lithified, unlithified) during the entry of information on the geographic and stratigraphic provenance of faunal lists. The most appropriate characterization is selected if this information is stated or illustrated in the bibliographic source of the faunal list. Additional assignments were made for collections that lack this data, but possessed other lithological or taphonomic information that is indicative of one of these lithification states. Changes in relative availability of fossil assemblages assigned to these various lithification states are presented in Fig. 2.

For much of the Paleozoic, little or no unlithified or poorly lithified fossil material is available. Skeletal assemblages from metamorphosed sedimentary rock are also fairly scarce, for the intuitive reason that metamorphism generally destroys fossil evidence, though some examples are recorded from a range of facies of Cambrian-Devonian age. The earliest Phanerozoic poorly lithified assemblages of reasonable number are available from the Jurassic fossil record. Additionally, a number of unlithified assemblages are noted from the Late Jurassic of Greenland, Europe, and Middle East. Nevertheless, closer inspection of associated data on preservation quality of these collections, and isolated Paleozoic examples, reveals that many still show diagenetic alteration through calcite replacement. This suggests that the unlithified state of at least some older fossil assemblages may be the result of secondary dissolution of carbonate cement binding sedimentary rocks, perhaps associated with weathering processes. Notably (Fig. 2) there is an apparent paucity of Early Cretaceous unlithified assemblages, and, not many more in Late Cretaceous. The precise reasons for this decrease are not established at this stage, but perhaps relate to the swamping out of a global signal by large numbers of apparently lithified latest Cretaceous collections from the Gulf Coastal Plain (see Sohl 1960, 1964a, b; Sohl and Koch 1983, 1984,

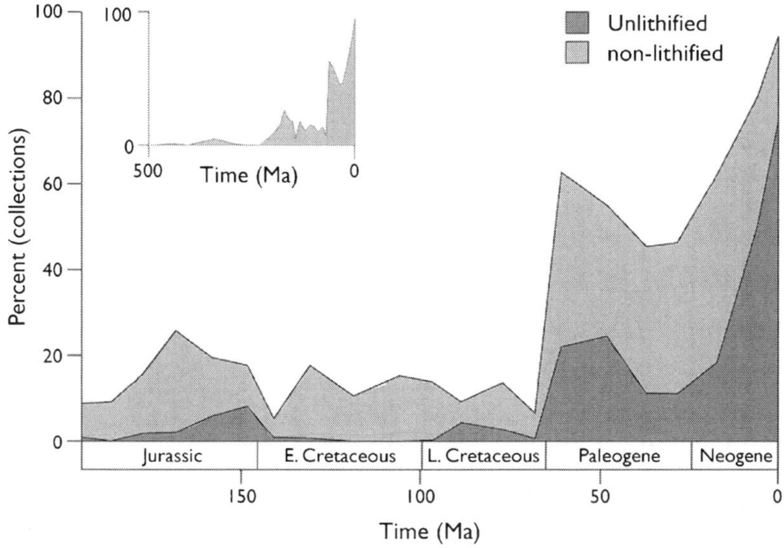

Fig. 2 Variation in proportion of collections derived from unlithified and non-lithified (combined unlithified and poorly lithified) sediments through the Jurassic-Cenozoic. *Inset* presents the variation in proportion of collections derived from non-lithified sediments through the Phanerozoic

1987). Easily observed, however, is the sharp increase in availability of unlithified and poorly lithified assemblages during Paleocene, and maintained through the Eocene. There is a drop in non-lithified assemblages during the Oligocene, but steady rise in availability of unlithified assemblages is observed for the remainder of the Cenozoic. The Oligocene is noted as an interval of increased carbonate production, as recorded by widespread limestone facies (typically well cemented or lithified) in many regions of the world (e.g., King et al. 1999; McGowran et al. 2004).

As much as 95% of Pleistocene assemblages are non-lithified (75% of which are unlithified). This percentage drops to 65% (25% unlithified) during the Middle Eocene and ranges between 5% and 15% for Late Cretaceous assemblages (typically less than 5% unlithified). An obvious question to ask is how this secular variation might affect perceived properties of these collections, such as sample size (large samples will on average yield more taxa than small samples), preservation of fossil material (well preserved samples generally appear richer in taxonomic diversity than poorly preserved samples), and composition of those collections (fragile skeletal elements or shells with reactive mineralogy are known to be more susceptible to destructive diagenetic processes).

The sample size of each of the collections used for this analysis (Fig. 2) is generally not known. In fact less than 3% of bibliographic references that contribute to biodiversity data used in these analyses give details of volume or area, although 16% provide abundance data from which the total number of specimens can be established. On average, unlithified assemblages within this subset contain more specimens than lithified assemblages (64% and 40% greater sample size for Neogene and Paleogene collections, respectively). While lithified sediments do not

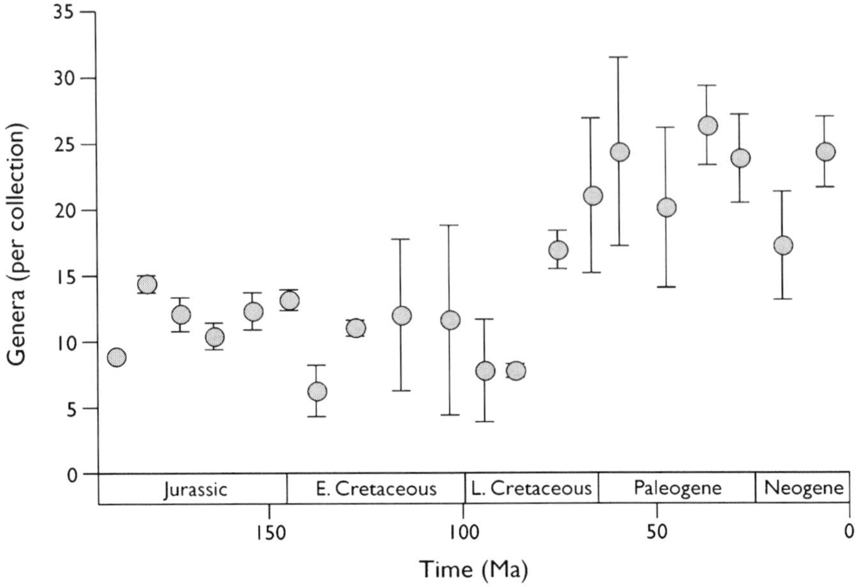

Fig. 3 Variation in mean richness of collections from the Jurassic through Cenozoic

necessarily restrict the assembly of particularly large collections, more often than not the collectors of fossil material will be limited by the volume of sample that can be removed from the field, or limited by the surface area of rock from which specimens can be extracted or recorded. Collectors may also deliberately obtain especially large samples, where faunas are particularly well preserved, rich in apparent taxonomic richness, or of particular paleontologic importance.

With this in mind, it is intuitive to expect that temporal trends in sample size will have a similar affect on taxonomic richness. The mean richness of collections through the Jurassic-Cenozoic (Fig. 3). remains relatively low through Jurassic and Early Cretaceous, ranging from between 7 and 14 genera per collection. A notable increase in richness is observed between the early Late Cretaceous stages and the latest Cretaceous (from 7 to 21 genera), increasing further into the Paleocene. Exceptionally large Campanian and Maastrichtian age collections from the Gulf Coastal Plain of North America contribute to this jump in richness; the extensive faunal lists of Sohl (1960, 1964a, b), and Sohl and Koch (1983, 1984, 1987) contribute disproportionately to Latest Cretaceous occurrence data for this interval. Richness remains relatively high from the Paleocene through Late Neogene (averaging around 24 genera). These results point to an increase in apparent marine benthic community richness of around two- to threefold between the Jurassic-Early Cretaceous and Cenozoic.

In Fig. 4, however, these data are again divided among the three categories of lithification state (lithified, poorly lithified, unlithified) to determine if these different states have an affect on values of mean richness. Assemblages from lithified collections have relatively low richness through the Jurassic and Early Cretaceous, ranging from 7 to 14 genera per list, although there is no significant

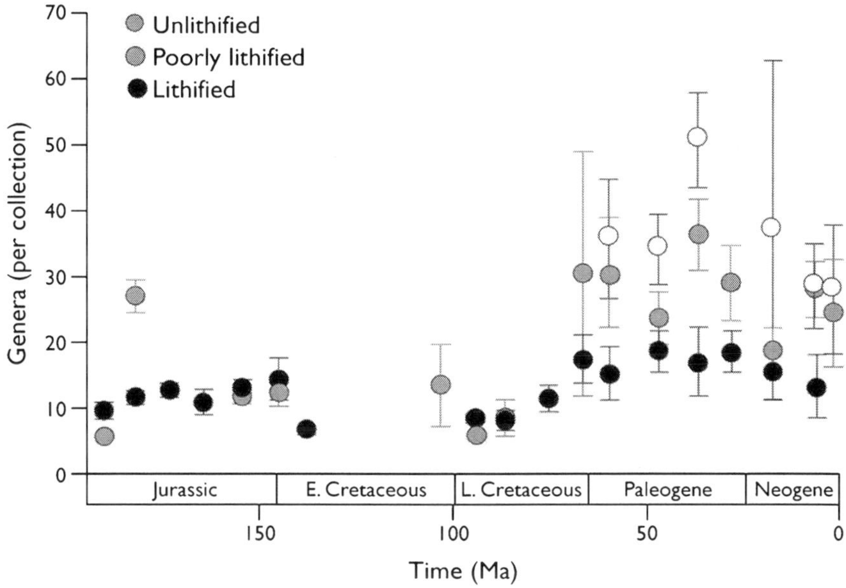

Fig. 4 Variation in mean richness of collections derived from unlithified, poorly lithified, and lithified sediments from the Jurassic through Cenozoic

difference in the richness of assemblages from contemporaneous poorly lithified sediments (where sampled adequately) though this time interval. The richness of lithified collections climbs steadily through the Late Cretaceous fossil record from 9 to 18 genera and remains steady throughout the Paleogene and Neogene, fluctuating between 13 and 19 genera. Poorly lithified assemblages show distinctly higher diversity from the latest Cretaceous and throughout the Paleogene and Neogene where sample sizes are large enough to permit meaningful averages. Notably, the richness of unlithified assemblages is consistently higher than those of poorly lithified assemblages, and statistically distinct to those lithified collections (at 95% confidence interval). There are difficulties in accurately measuring alpha diversity, however. Most notable is the need for abundance data to standardize for variable sample size. At a global scale, such comparisons are additionally hindered by latitudinal and environmental heterogeneity among the assemblages that the dataset comprises.

Koch and Sohl (1983) had previously noted a dramatic disparity in richness between well-preserved and poorly preserved Late Cretaceous assemblages. They categorized their dataset of Maastrichtian Gulf Coastal Plain assemblages into one of six preservation types, ranging from those in which calcite and aragonite shells were well preserved to those having only calcite shells preserved. Collections in which both aragonite and calcite were well preserved consistently had more taxa than those of poorer preservation quality (Fig. 5) and in addition contain many taxa not found in other collections (Fig. 6). In one of the earlier efforts to standardize diversity for variations in the number of available fossils (Fig. 7), Koch and Sohl

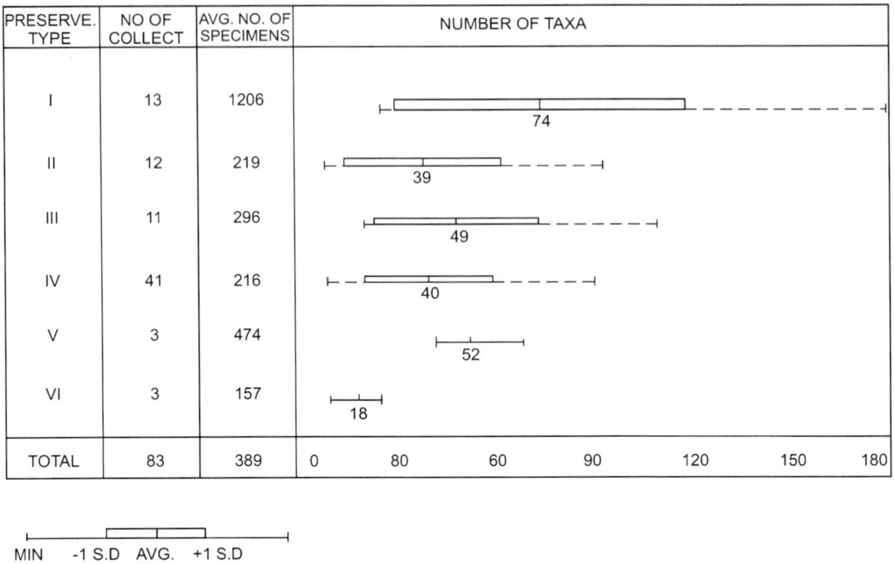

Fig. 5 Chart showing the average number of taxa recovered for the six preservation categories along with variance about the mean and maximum and minimum values. Also shown is the number of specimens for each category. Preservational types: I, aragonite and calcite well preserved; II, calcite well preserved, aragonite poorly preserved; III, aragonite preserved as molds; IV, calcite fossils and molds; V, only molds; VI, calcite only, no molds (From Koch and Sohl 1983)

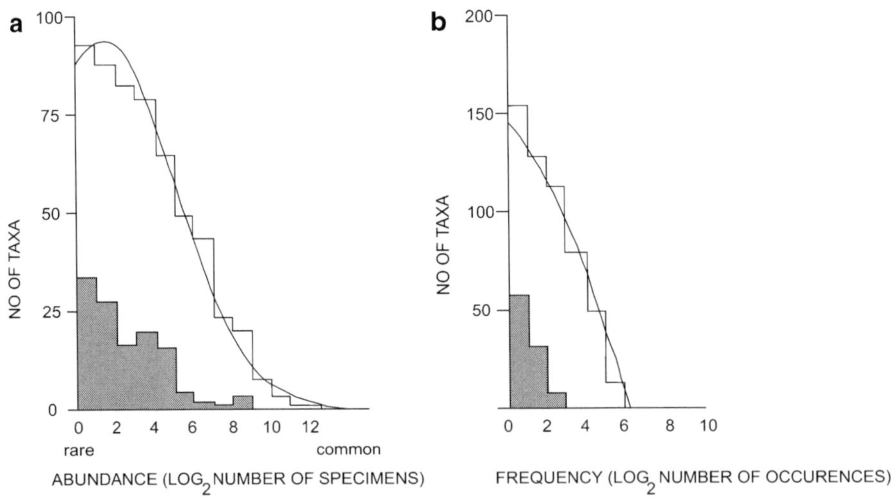

Fig. 6 Histograms showing abundance (**a**) and occurrence frequency (**b**) of 643 taxa used in Koch and Sohl (1983). Also shown is the distribution for taxa limited to collections in which aragonite and calcite are well preserved (Type I; see Fig. 5) (*stippled*). Curve in **a** is log-normal fit to histogram; curve in **b** is log-series fit to histogram (From Koch and Sohl 1983)

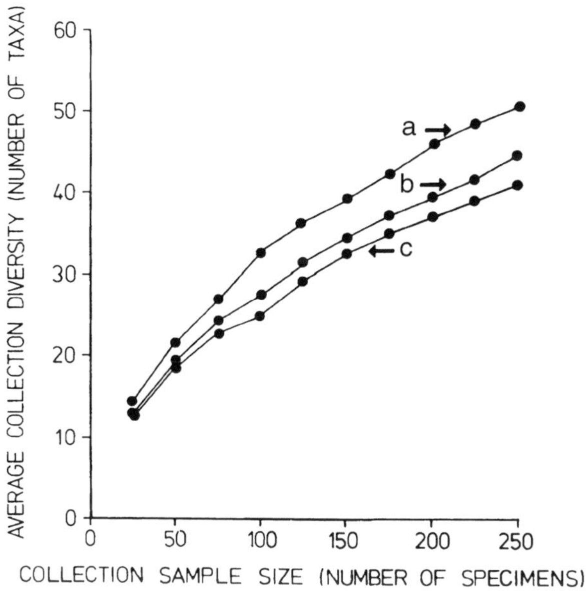

Fig. 7 Plot of average collection diversity (number of taxa) vs. collection sample size (number of specimens) after rarefaction of collections; (**a**) collections with aragonite and calcite well preserved (type 1), (**b**) collections without aragonite and calcite well preserved but with molds and calcite preserved (types II, III, and IV) and (**c**) collections from silty fine sands but without well preserved aragonite. See Fig. 5 for explanation to assemblage types (From Koch and Sohl 1983)

(1983) rarefied specimens from each of three broader preservation categories further demonstrating the influence that preservational quality, diagenetic degradation, and matrix lithology have on sample-level richness. The analysis indicates, for this example, a modest 25–30% increase in richness between poorly preserved and well preserved collections at similar sample size. These results also hint at taphonomic controls on the evenness of abundance distributions, complicating straightforward interpretation of Phanerozoic trends in community evenness (e.g., Powell and Kowalewski 2002; Bush and Bambach 2004).

Because the process of lithification commonly involves the cementation of matrix by precipitation of dissolved carbonate, a likely cause for the genus richness decline in lithified sediments is the preferential dissolution of aragonitic skeletal hardparts. Important factors negatively affecting preservation include small size, fragility and shell composition (Schopf 1978; Koch and Sohl 1983; Paul 1998; Valentine 1989; Glover and Kidwell 1993; Kidwell and Flessa 1995; Jablonski and Sepkoski 1996; Cherns and Wright 2000; Wright et al. 2003; Valentine et al. 2006), although recent analysis (e.g., Kidwell 2005) has suggested that biases that act against skeletal composition have little net impact on diversity patterns. Nevertheless, it is argued here that size and mineralogy are significant factors in the preservation potential of skeletal components of marine benthic communities through the Mesozoic and Cenozoic. Additionally, small fossils might also be more readily overlooked by collectors in the field because of difficulty in extracting them from lithified sediments (Cooper et al. 2006). The process of fossil extraction from lithified fossil assemblages in the field, or preparation in the lab, inherently involves the splitting of hardened slabs or the fragmentation of larger blocks, processes during which small and fragile specimens are more likely to be damaged or destroyed.

Data for the analysis of body size composition of taxa were collected from the literature on Cenozoic and extant Mollusca. Each genus is assigned to one of four size categories based on the average maximum linear dimension of specimens belonging to the species of each genus. These sizes, where possible, are based on multiple species and on multiple specimens of each species where possible, generally representing type or figured specimens. Each species in the dataset (including those not contributing measurements) is allotted a size category (maximum linear dimension) representing their genus; very small (<5 mm), small (6–15 mm), medium (16–50 mm), and large (>50 mm). Analysis of mineralogical composition used data derived from the general literature (primarily Coan et al. 2000; Mikkelson and Bieler 2008). As mineralogy is highly conserved among species and genera, composition was assumed to be consistent within each family (Taylor et al. 1969; Kidwell 2005). Taxa are here classified as being of dominantly aragonitic, calcitic, or mixed calcite-aragonite skeletal mineralogies.

The affects of lithification on retrieval of taxa representing various size classes and mineralogical composition can be assessed a number of ways (Fig. 8). Differences in composition of lithified and unlithified collections can be estimated using the mean percentage of taxa in each collection (Fig. 8a), the mean percent of specimens in each collection (Fig. 8b) (where abundance data are available), the percent of all occurrences from a particular time interval (Fig. 8c), and the percent of all taxa from a particular time interval (Fig. 8d) (summarized in Table 1). Each of the metrics for composition based on richness or occurrences (Fig. 8a,c,d) reveal a consistent pattern, a near lack of very small taxa, and reduced numbers of small taxa, among lithified collections, relative to their unlithified counterparts. Correspondingly, large and medium-sized taxa tend to contribute a far greater percentage of occurrences among lithified collections than unlithified ones. Abundance data (Fig. 8b) yields the more conservative pattern, and although very small taxa and large taxa differ as anticipated in proportion between lithified and unlithified collections, the differences are minor. This may be an artefact of the small number of collections available from this time interval with information on both abundance and lithification state (Table 2).

Any of the above occurrence or richness metrics (Fig. 8) could be used to monitor the affects of lithification through geological time, although an appropriate choice of measure is probably determined by the scale and resolution of investigation. For instance, changes in taxic composition measured using abundance data are probably best limited to investigations of local scale, in which sufficient census counts are available, and where environmental and taphonomic heterogeneity is controlled. At a global scale, changes in taxic composition are probably best-determined using occurrence or richness data, which are available for a considerably larger number of collections that fairly represent a broad range of geographic, environmental settings and taphonomic conditions.

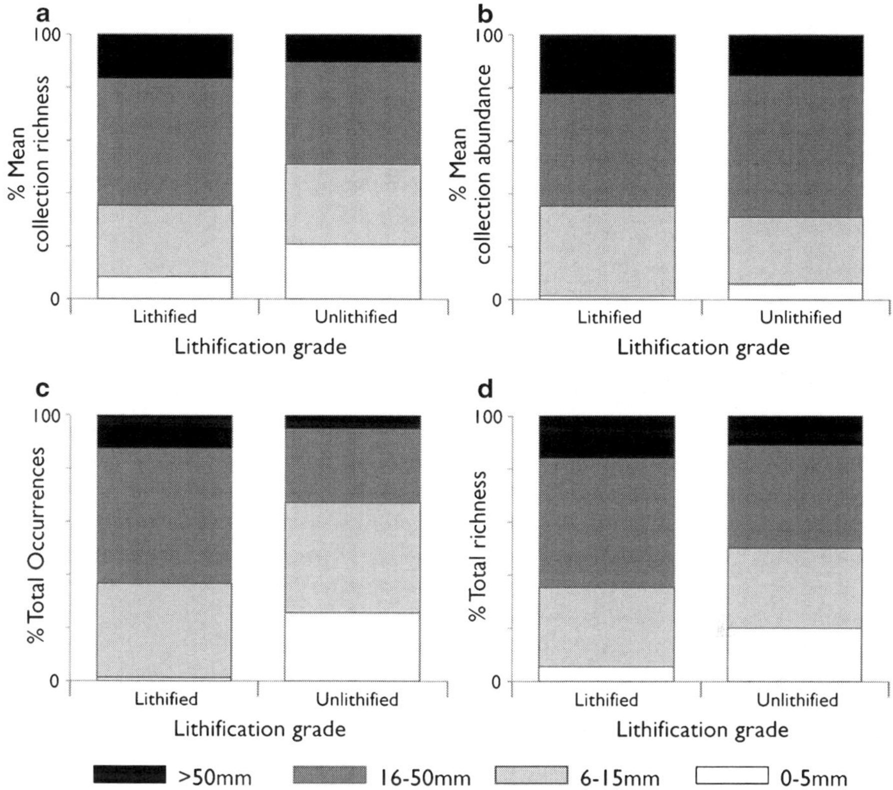

Fig. 8 Variation in size distribution among fossil data from lithified and unlithified sediments from the Late Miocene-Pleistocene. (**a**) Mean percentage of taxa within individual collections. (**b**) Mean percentage of specimens within individual collections. (**c**) Percentage of all Late Miocene-Pleistocene occurrences. (**d**) Percentage of all Late Miocene-Pleistocene taxa

Table 1 Estimates of the effect of lithification on small (<15 mm) molluscs using different protocols. The magnitude by which unlithified sediments contain more specimens, occurrences, higher mean richness and total diversity than comparable lithified sediments (herein referred to as the increase factor) is presented for four different data types

	Total occurrences	Total richness	Collection richness	Collection abundance*
Late Miocene-Pleistocene	2.36	1.91	1.50	
Early-Middle Miocene	2.61	1.72	1.47	
Neogene				1.28
Late Eocene	1.86	1.40	1.41	
Paleogene				2.17

* = The increase factor calculated with collection abundance is based on samples pooled from the Neogene and Paleogene to increase sample size

Table 2 Availability of occurrence and abundance data for which ithification state is known for three time intervals investigated

	Total collections	With abundance
Late Miocene-Pleistocene	757	22
Early-Middle Miocene	562	37
Late Eocene	226	23

Figure 9 presents an analysis of differences in composition of lithified and unlithified collections using the percentage of all occurrences that belong to various size or mineralogical categories for three time-intervals, the Late Miocene-Pleistocene (Fig. 9a,b), the Early-Middle Miocene (Fig. 9c,d), and the Late Eocene (Fig. 9e,f). There is a clear difference between the representation of very small and small size classes in lithified and unlithified collections. Additionally, there is a considerable increase (10–20%) in the retrieval of aragonitic molluscs in unlithified sediments, relative to their lithified counterparts for all time intervals. This further strengthens the claim that lithification has a considerable impact on the quality of paleontological data beyond simply influencing the availability of fossil material. Additionally, this analysis indicates that the pattern is not merely a function of processes influencing the Late Neogene, but one that is prevalent throughout the interval for which a non-lithified fossil record in preserved.

2.2 *Time-Series Analysis of Lithification and Alpha Diversity: A Regional Perspective*

2.2.1 Cenozoic of New Zealand

A dataset using Eocene-Pleistocene occurrence data from New Zealand's Fossil Record Electronic Database (FRED; www.fred.org.nz), is used herein to evaluate the extent that lithification through time, within a single region, impacts the relationship between regional and alpha diversity (Fig. 10). In the absence of abundance data spanning the entire time interval of interest, mean collection richness is used as a proxy for alpha diversity. The revealed pattern is remarkably similar to that of sampling-standardized total diversity, except in the latest Pliocene and Pleistocene (Crampton et al. 2006; Hendy 2007). Mean collection richness declines slightly from the Late Eocene to the Early Oligocene before increasing to an early peak in earliest Miocene age assemblages. Following the Early Miocene there is a steady decline in mean richness until the latest Miocene, which apparently represents the most depauperate stage of the Neogene. Collection richness then increases first to a middle Pliocene peak, declining in the Late Pliocene before climbing further to a Cenozoic high from Pleistocene assemblages. The nearly threefold increase in mean collection richness between the latest Miocene (~10 genera) and Pleistocene (27 genera) is remarkable given the time span of only 4–5 Ma between the two intervals. Even in localized studies, however, difficulties in interpretation of collection richness are posed by variability in the sample size of assemblages, environmental heterogeneity, biostratinomic and diagenetic effects (including lithification), and time averaging.

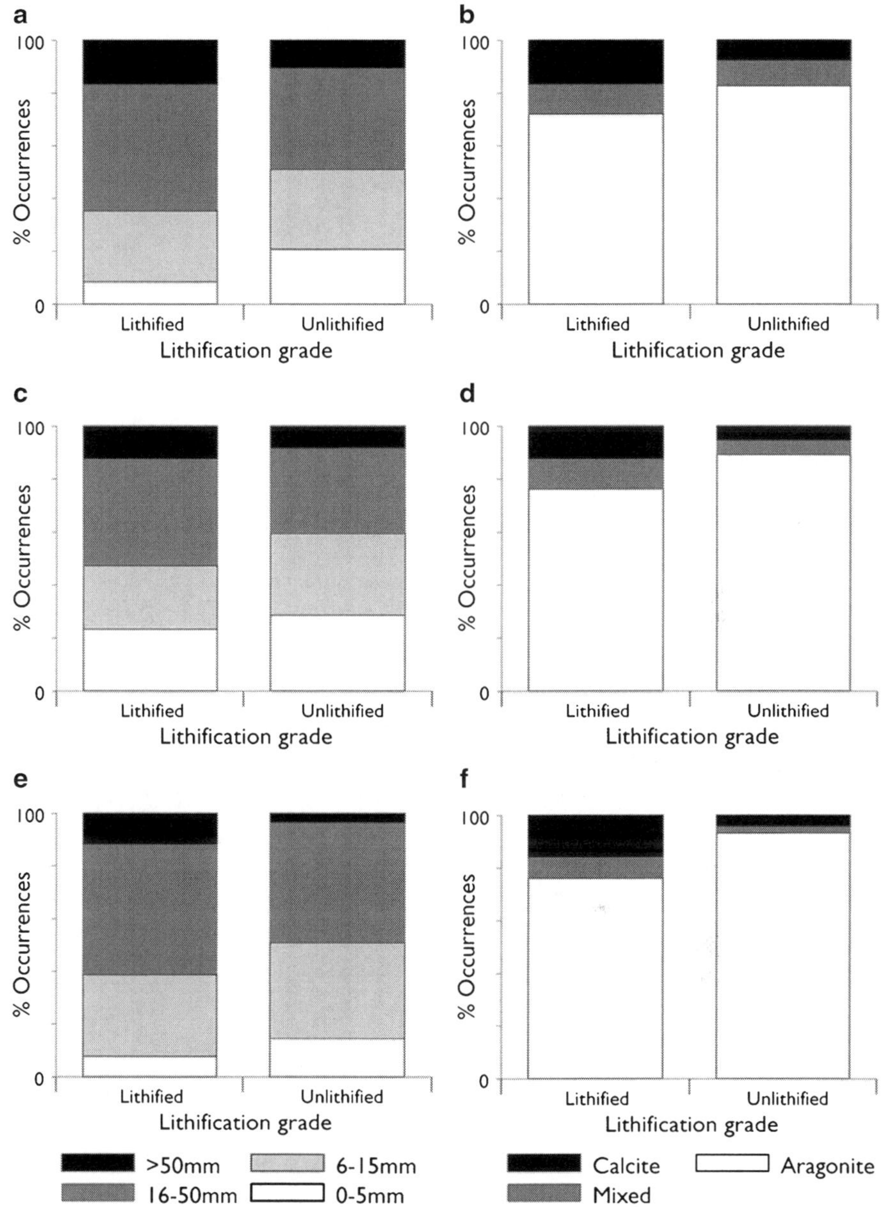

Fig. 9 Variation in size distribution and mineralogical composition among fossil occurrences from lithified and unlithified sediments from Late Miocene-Pleistocene (**a**, **b**), Early-Middle Miocene (**c**, **d**), and the Late Eocene (**e**, **f**). Percentage was calculated on composition of all occurrences in each time interval

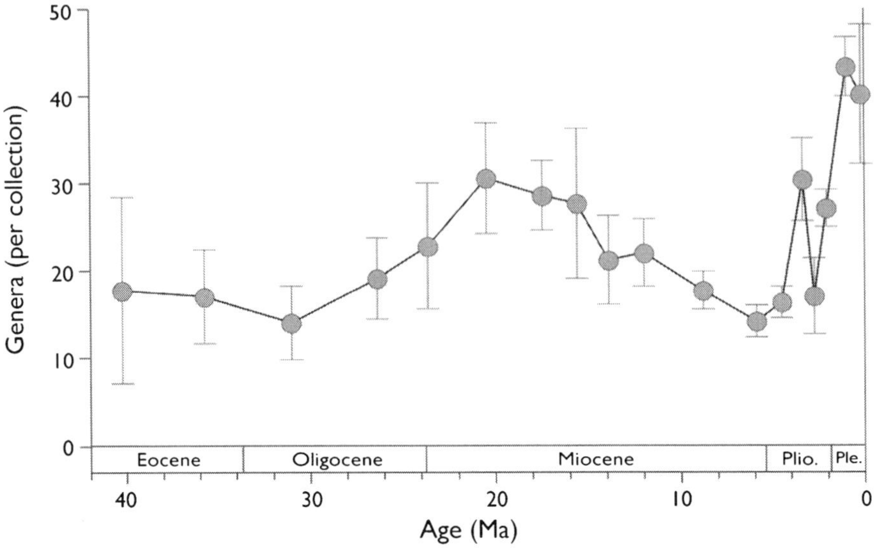

Fig. 10 Mean richness of collections of New Zealand Cenozoic Mollusca. Restricted to collections >2 occurrences and <100 occurrences. *Error bars* represent 95% confidence intervals

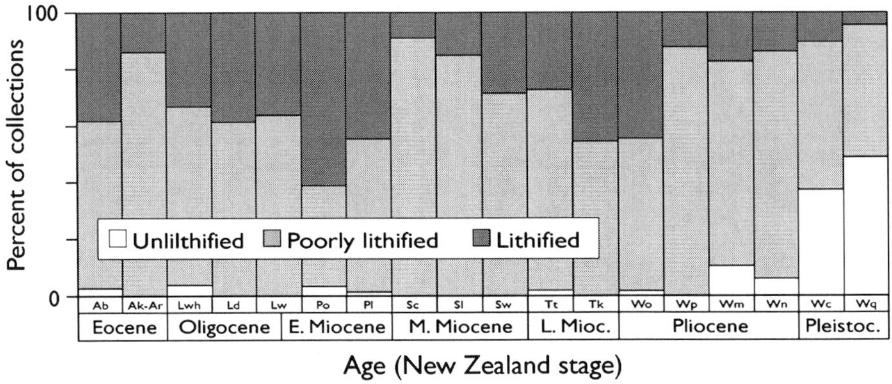

Fig. 11 Proportional distribution of collection data among three categories of lithification

Though degree of lithification is not often quantified in the literature, the semi-quantitative category used to describe "hardness" of matrix for collections reposited in FRED, in addition to anecdotal evidence associated with collections, permit assessment of how lithification influences richness of individual collections. This is not without some difficulty given that only about a third of collections are assigned hardness values (Fig. 11). Two broad scale trends are apparent, the significant increase in unlithified samples representing the final two stages of the Cenozoic, and the reduction in lithified assemblages following the Opoitian.

Because of sample size limitations no estimates can be made for a number of stages during the late Eocene, early Oligocene, and middle Miocene. Indeed, there is

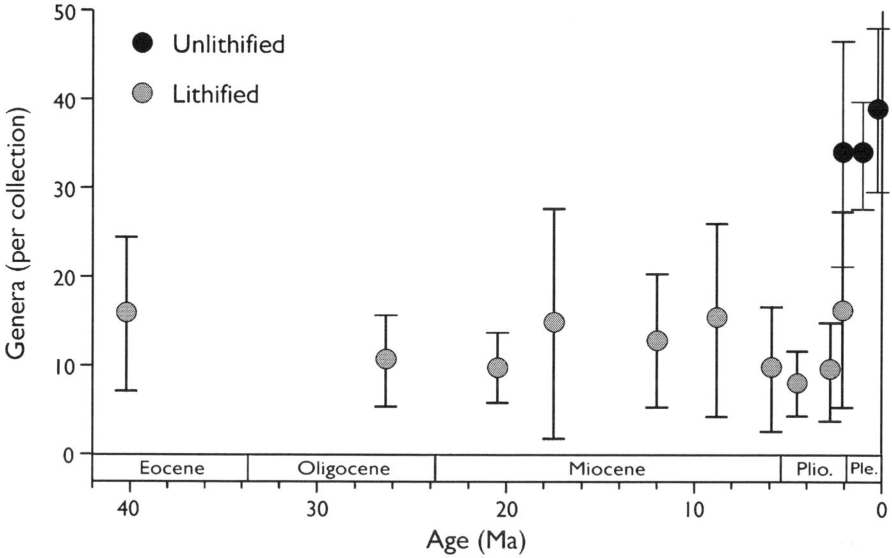

Fig. 12 Mean richness of lithified and unlithified collections of New Zealand Cenozoic Mollusca. Restricted to collections >2 occurrences and <50 occurrences; means calculated for stages with >10 collections and whose lithification is accurately determined. *Error bars* represent 95% confidence intervals

only sufficient sample size to analyze unlithified collections from the latest Pliocene-Pleistocene, and no collections determined to be lithified are available from the Pleistocene. Nevertheless, the only stage in which both categories can be assessed is the Nukumaruan (Fig. 12). The difference in mean richness of Eocene-Early Pliocene and the latest Late Pliocene-Pleistocene is considerable (approximately two- to three-fold). While error bars (95% confidence intervals) overlap for the Nukumaruan, the difference in comparison of diversity in all other stages is significant.

Assuming that the largely lithified Eocene-Early Pliocene fossil record is free of other secular trends in preservation biases, it appears that there is a strong pattern of community scale change in biological diversity. If this was indeed a biological signal then it would suggest that, much of the Eocene-Early Pliocene variation in New Zealand's regional biodiversity owes its origin to changing within-community richness, rather than changes in beta or gamma diversity. This is indeed a significant result and encourages focused investigation on environmental factors that could play a role in influencing community-scale diversity, for example, sea-surface temperature and productivity. The apparently rapid increase in community-richness at the conclusion of the Neogene is, however, overprinted by a lithification bias and any biological interpretation of this trend should proceed with caution (Hendy 2009a).

2.2.2 Paleogene of the Gulf Coastal Plain

Sessa et al. (2009) studied the impact of lithification on mollusc-dominated assemblages of Early Paleogene age from the Gulf Coastal Plain (Texas through Alabama). This region is renowned for its well-preserved and diverse record of Paleogene marine invertebrates, and correspondingly has received considerable taxonomic and biostratigraphic investigation. They assembled abundance data from field-collected bulk samples, and a number of previously published datasets (Toulmin 1977; Hansen et al. 1993a, b; C. Garvie 2008) that spanned the early to late Paleocene. Using sampling standardization procedures they showed a dramatic difference in sample size and diversity between lithified and unlithified fossiliferous deposits; the latter become the increasingly dominant mode of preservation through the Paleocene and Eocene in this region. On average, unlithified samples have 2.4 times the diversity of lithified samples of comparable sample size. Significantly these authors demonstrated that one effect of this bias was to extend the perceived duration of the recovery period following the Cretaceous-Paleogene mass-extinction by as much as 7 my. (Fig. 13). An important implication is that observations of the fate of particular taxonomic or ecologic groups and investigations into the duration and dynamics of recovery faunas need to be evaluated with respect to taphonomic processes, such as the lithification bias.

An additional dataset, derived from measurements of museum-reposited specimens from a similar lithology and geography was used to contrast the apparent disparity in size of taxa recovered from lithified and unlithified units (Fig. 14). Sessa et al. (2009)

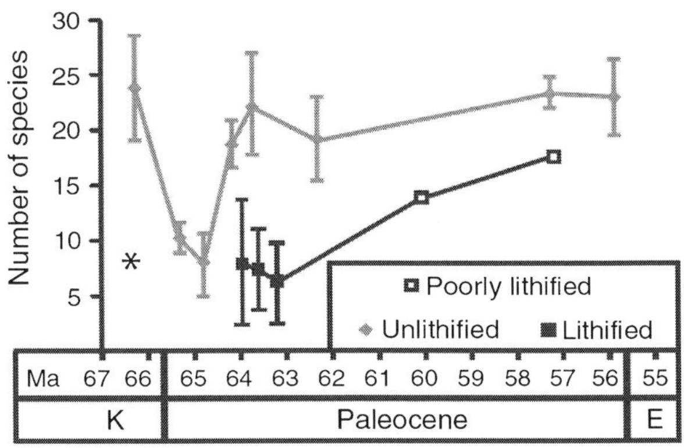

Fig. 13 Averaged sample-level species richness of bulk samples from the Gulf Coastal Plain, rarefied to 70 individuals and with standard deviations, from the latest Cretaceous through Paleocene. *, lack of abundance data for lithified Cretaceous units (From Sessa et al. 2009)

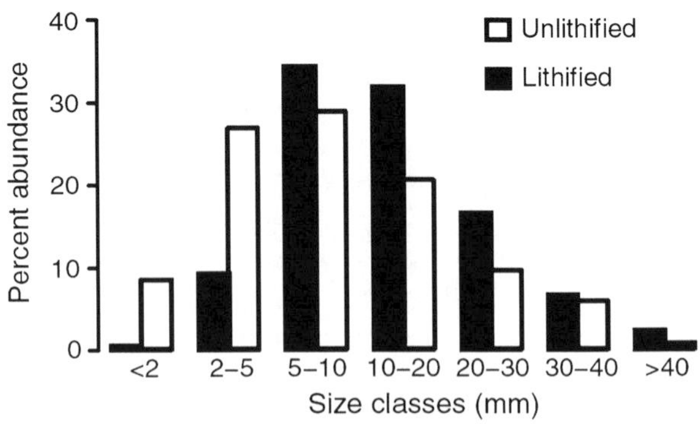

Fig. 14 Size frequency distribution for museum reposited specimens indicating lack of small specimens from lithified sediments (n = 1,001) relative to those from unlithified sediments (n = 729) (From Sessa et al. 2009)

showed that lithified samples contained individuals with a median size of 11.3 mm (median of 16.1 mm for unique genera), while unlithified samples possessed individuals with a median size of 7.1 mm (median of 6.2 mm). They found that lithification concealed considerable diversity among small taxa, reduced taxonomic resolution, and caused the undersampling of already rare taxa. Their study suggested that a size threshold of 5 mm exists, below which specimens were more easily dissolved or more difficult to identify. Sessa et al. (2009) suggest that while the organisms of particular interest and preparation techniques will contribute to observed size distributions, specimens of smaller size typically dominate assemblages. Therefore overabundance of larger specimens in paleontological samples should be cause for concern.

2.3 *Within-Interval Analysis of Lithification and Alpha Diversity: A Local Perspective*

An ideal approach to rigorously unraveling the potential effects of this important transition in preservation should include an attempt to constrain variations in the depositional environment, latitudinal position, time-averaging, and temporal variations in biodiversity itself (Kowalewski et al. 2006). A large dataset of bulk-sampled fossil assemblages in the late Neogene of New Zealand (Hendy 2009a) provides just such an opportunity to estimate the loss of taxonomic information associated with lithification bias among contemporaneous assemblages. The primary data for this investigation were mollusc-dominated assemblages, that ranged in age from Late Miocene to Pleistocene, collected from a narrow range of sedimentary facies in two sedimentary basins (Wanganui and East Coast) of New Zealand. The extensive and continuous late Neogene succession in these basins exhibits a strong lithification gradient between its oldest and youngest sedimentary components. Sampling was restricted to transgressive shell bed facies (Hendy et al. 2006) to control as much as possible for between-sample variation in time averaging and to allow the comparison of relatively consistent environments through the time series. These samples represented lower shoreface to mid-shelf bathymetric settings, from sandy or mixed sandy silty substrates, and exhibited characteristics consistent with within-habitat time-averaging. Additionally, Hendy (2009a) applied consistent methods of collection (stratigraphic and spatial integrity of samples), preparation, counting and identification, although sample treatment varied from assemblage to assemblage because of the nature of enclosing sediments (e.g., weathered or fresh outcrops, lithified or unlithified bedding planes).

If earlier examples of the lithification bias were related simply to the size of the sample collected from individual localities, then techniques that standardize for variations in sampling intensity, such as rarefaction, should mitigate this bias (e.g., Bush and Bambach 2004). Figure 15a shows rarefaction curves for 169 field-collected bulk samples of Late Miocene-Early Pleistocene age, representing 37 unlithified, 66 poorly lithified, and 66 lithified fossil assemblages. At comparable levels of sampling, most unlithified samples yield considerably higher richness than those from lithified sediments, with poorly lithified assemblages showing an intermediate position, a pattern that is further amplified by the mean curves for each lithification category (Fig. 15b).

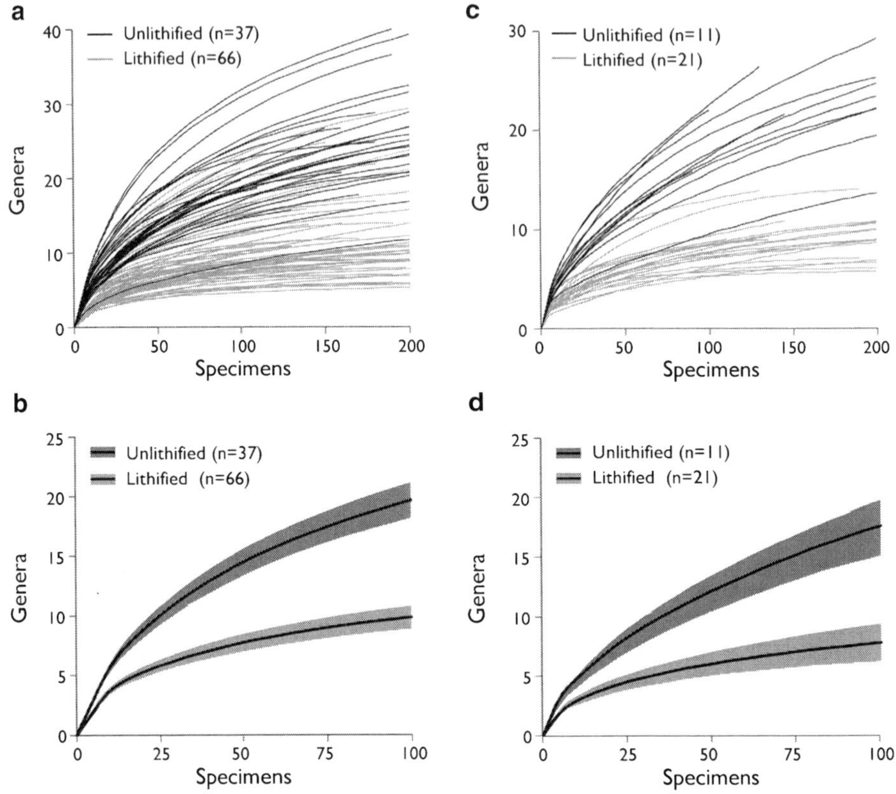

Fig. 15 Rarefaction of census counts from bulk samples of varying lithification from Late Miocene-Pleistocene sediments of Wanganui Basin, New Zealand. (**a**) Rarefaction curves for individual samples coded by lithification category (poorly lithified samples excluded for clarity). (**b**) Means of individual curves in (**a**) within each lithification category with shaded 95% confidence intervals. (**c**) Rarefaction curves for individual samples dominated by *Tawera*. (**d**) Means of individual curves in **c** within each lithification categories with shaded 95% confidence intervals

At a quota of 100 specimens, unlithified sediments yield on average close to 20 genera, whereas lithified sediments produce slightly fewer than 10 genera for the same sampling intensity. The disparity was even greater at larger quotas (Table 3).

Hendy (2009a) further constrained environmental heterogeneities in this dataset by restricting the analyses to a subset of these samples that were dominated by a single ubiquitous infaunal bivalve, *Tawera*, which is present throughout late Neogene fossiliferous deposits in New Zealand (Beu and Maxwell 1990). This subset of samples represents a single paleocommunity from the shelfal transgressive environmental gradient through the time series (Hendy and Kamp 2004; Hendy et al. 2006; Hendy and Kamp 2007). Rarefaction (Fig. 15c, d) of samples dominated by *Tawera* indicate again that at comparable levels of sampling most unlithified samples show considerably higher richness than those from lithified sediments, with poorly lithified assemblages occupying an intermediate position. Mean curves for each lithification category confirm this pattern. At a quota of 100 specimens mean richness of unlithified

Table 3 Genus richness in lithified, poorly lithified, and unlithified sediments of the late Neogene New Zealand from the FRED and from field-collected bulk samples

Data set	Subset	Unlithified	Poorly lithified	Lithified
FRED	Mean	25.1	9.6	6.0
	Max	88	54	19
Field samples	All	19.7 (25.1)	15.9 (20.5)	9.9 (10.4)
	Pleistocene	20.6	19.6	–
	Late Pliocene	17.4	14.8	12.6
	Early Pliocene	20.9	16.0	8.8
	Late Miocene	–	–	8.6
Tawera association		17.5 (22.5)	13.4 (14.0)	7.8 (8.7)

Mean genus richness for FRF data is unstandardized; genus richness for field samples and Tawera samples rarefied to 100 specimens (and to 200 specimens, in parentheses)

assemblages was approximately two and a half times that of lithified sediments (Table 3). Unlithified sediments yield on average close to 19 genera, whereas lithified sediments produce slightly more than seven genera for the same sampling intensity.

A further analysis, reported by Hendy (2009a), restricted comparisons to individual time intervals in order to minimize the possibility that temporal variation in composition of faunas affected the patterns illustrated in Fig. 15. Although unlithified and lithified sediments were lacking from Late Miocene and Pleistocene successions, respectively, the pattern of increasing diversity with decreasing degree of lithification is evident for each time interval analyzed independently (Table 3), but not through time within any single lithification category. These results demonstrate that sampling standardization techniques alone cannot reconcile the high diversities yielded from the easier recovery of fossils from unlithified samples with the lower diversities of lithified samples, indicating a fundamental difference in the recoverable taxonomic composition of lithified and unlithified samples.

The results presented in Fig. 16 suggest that skeletal size and mineralogy, indeed, account for at least part of the difference in taxonomic content between lithified and unlithified sediment. There is an observable, albeit small, decrease in the proportion of taxa (and occurrences) with predominantly aragonitic skeletons in lithified sediment (Fig. 15a). Likewise there is an increase in the proportion of observed diversity contributed by the smallest and medium sized classes of invertebrates in poorly lithified and unlithified sediments (Fig. 15b). The difference, while slight, corroborates an independent analysis of the removal of small size classes on sample-level diversity (Kowalewski et al. 2006).

2.4 Influence of Lithification and Diagenesis on Preservational Quality: Implications for Taxonomy

Biases in preservational quality at the scale of individual specimens and populations play an important, yet often overlooked, role in perceptions of biodiversity and paleoecology. After all, fossil specimens are the raw material in which the

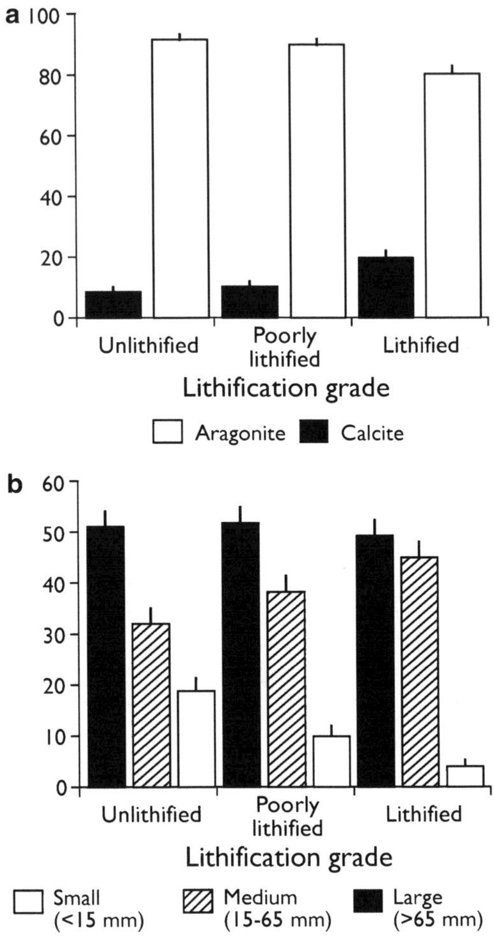

Fig. 16 Influence of lithification on taphonomic features of skeletal assemblages. (**a**) Relative composition of aragonitic and calcitic skeletal types. (**b**) Relative composition of various size classes. *Error bars* indicate 95% confidence intervals

fossil record is based. The term preservational quality is used broadly to include the nature of skeletal material, such as preservation in its original form, replacement by another mineral, complete dissolution (represented by molds), articulation, fragmentation, abrasion, and the affect of encasement in sediment. Influences on the preservation of taxa may take two forms, either by distortion of sampling probability and patterns or relative abundance within assemblages (fossil material may be reduced in frequency or absent due to destructive taphonomic processes), or through influencing taxonomic identifiability (fossil material may be preserved, but not identifiable to a given taxonomic level). Effects on sampling probability and changes to the abundance structure of former communities are widely acknowledged in the literature (e.g., Plotnick and Wagner 2006), although investigations commonly focus on analyses at fairly coarse resolution, for example, phyla and class (e.g., Foote and Sepkoski 1999), or clades with typically similar mineralogical composition (e.g., Kidwell 2005; Valentine et al. 2006). Studies of taxonomic identifiability are, however, greatly lacking from recent taphonomic and biodiversity literature (but see Rofthus 2002, 2005). Such affects are especially important given that preservational quality probably varies fairly predictably through geologic time and across geographic gradients for given taxonomic groups.

Taphonomic processes may profoundly affect the known fossil record of a taxonomic group, resulting in skewed perceptions of diversity trends and evolutionary relationships. The diversity history of any group as deduced from fossil data has a distinct taphonomic overprint (Greenstein 1992), although this may vary from group to group, depending on the morphological complexity of their body plan, preservation potential, and their geological age. Additionally, fossil assemblages themselves are often a mixture of well-specimens with varying degrees of taphonomic alteration (fragmentation, bioerosion, surface alteration). The post mortem alteration of specimens has the potential to introduce bias into paleoecological data by preventing taxonomic identification of some portion of the assemblage.

2.4.1 Direct Observation of Fossil Specimens

Assessment of the impact that taphonomy has on taxonomic identification and the alteration of morphological data cannot be derived from traditional or even occurrence-based fossil databases. Rather, data must be derived from direct observation of fossil specimens or whole assemblages. Using data derived from observations of specimens reposited in major natural history museums (collected between 2005 and 2007), the relationships between preservation and facies characteristics are explored. Variations in preservational quality are determined through observations of morphological detail for eight families (or superfamilies) of Bivalvia. These families were chosen for data collection because of their long geological record (in some cases, Ordovician-Recent), and the diversity of mineralogical composition and skeletal durability that they comprise (Table 4). Figure 17 presents examples of specimens of each family from lithified, poorly lithified, and unlithified host-sediments.

Figure 18 illustrates the preservational quality of the eight families (or superfamilies) of bivalves through the Phanerozoic. Preservational quality is presented as the mean percentage of the five or six key types of characters chosen for each family, potentially observable across all specimens in each interval. Plotted additionally are the mean values for lithification of host sediment per interval for each family. While a number of taphonomic conditions could be investigated for their relationship to preservation of observed morphological details, lithification is explored in this case study. Table 4 summarizes these character groupings, which will vary from family to family depending on their inherent morphological characteristics. The character types chosen in each family are considered essential to the diagnosis of genus-level taxa in those families, and do not need to be present (i.e. not all Nuculidae possess crenulation of their inner ventral margin), but must be potentially observable (i.e. inner ventral margin is exposed for inspection) to be scored as preserved.

Two trends are apparent in each of the family plots in Fig. 18. First, there is a general increase in observable characters from the Paleozoic to the Cenozoic, culminating

Table 4 Summary of variations in preservational quality relative to lithification state and geologic time for several families of Bivalvia

Taxonomic group	Character groups observed (generalized)	Preserved characters L (%)	Preserved characters UL (%)	Observations (N) L	Observations (N) UL	Skeletal characteristics and age range
Nuculidae	Shape, sculpture, inner ventral margin, hinge plate, resilifer	39	96	121	62	Aragonitic, nacreous, sculpture fine or absent; Silurian-Recent
Nuculanoidea	Shape, sculpture, rostrum, escutcheon, hinge plate	53	98	78	54	Aragonitic, sculpture fine or absent; Devonian-Recent
Mytiloidea	Shape, sculpture, beak, inner ventral margin, hinge plate	54	99	181	31	Aragonitic and calcitic, nacreous, sculpture absent-coarse; Devonian-Recent
Pinnidae	Sculpture, ornamentation, longitudinal sulcus, adductor muscle scar, hinge plate	40	81	40	29	Aragonitic and calcitic, nacreous, sculpture coarse; Carboniferous-Recent
Limidae	Shape, sculpture, auricle, byssal gape, hinge plate	47	100	123	41	Aragonitic and calcitic, sculpture fine or coarse; Carboniferous-Recent
Anomioidea	Shape, presence of right valve, muscle scars, foramen, crura	30	72	30	75	Calcite and aragonite, subnacreous, sculpture irregular; Permian?, Jurassic-Recent

Lucinoidea	Shape, sculpture, lunule, cardinal dentition, pallial line	43	89	112	3	Aragonite, sculpture fine or absent; Ordovician-Recent
Astartidae	Shape, sculpture, lunule, inner ventral margin, cardinal dentition, lateral dentition	52	98	102	73	Aragonitic, sculpture fine; Devonian-Recent
Crassatelloidea	Shape, sculpture, umbonal ridge, cardinal dentition, lateral dentition, posterior truncation	36	97	47	41	Aragonitic, sculpture fine or absent; Permian-Recent

L, lithified; UL, unlithified. Skeletal characteristics and choice of character groups based on Cox (1969), Coan et al. (2000), and Mikkelson and Bieler (2008)

Fig. 17 Examples of the preservation of specimens from unlithified, poorly lithified and lithified lithification states from several groups of Bivalvia: (**a**) Nuculanoidea, (**b**) Nuculanoidea, (**c**) Anomiidae, (**d**) Limidae, (**e**) Mytiloidea, (**f**) Astartidae, (**g**) Lucinoidea, (**h**) Crassatelloidea. Identifications and specimen numbers listed in Appendix

in the highest values for the Neogene and Recent (in cases where data were collected). Second, in many families, peaks may be observable in either the Carboniferous or Permian intervals. For the Carboniferous, this reflects inclusion in the dataset of either well-preserved calcite-replaced aragonite taxa from the Mississippian of North America, or silicified specimens from the type Visean of Belgium. For the Permian, uncommon, but well-preserved silicified aragonite and calcitic taxa from west Texas

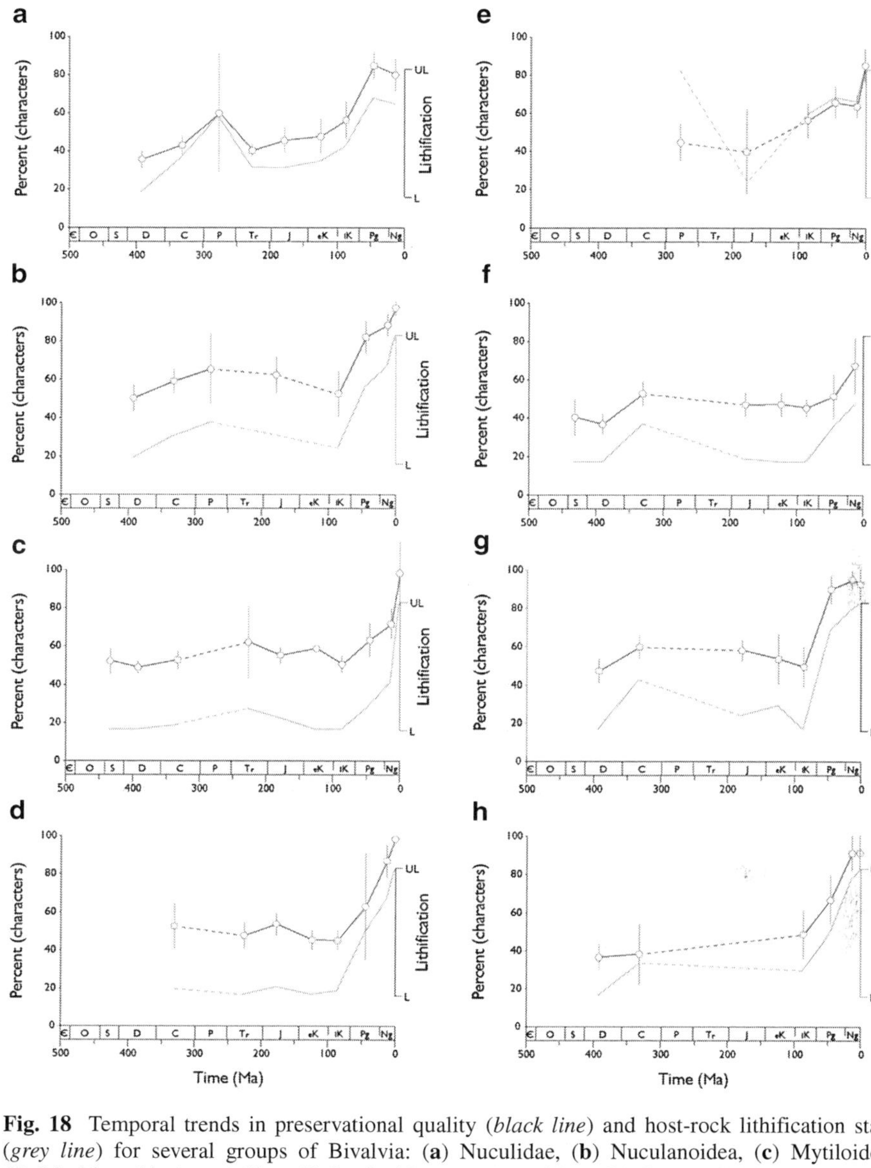

Fig. 18 Temporal trends in preservational quality (*black line*) and host-rock lithification state (*grey line*) for several groups of Bivalvia: (**a**) Nuculidae, (**b**) Nuculanoidea, (**c**) Mytiloidea, (**d**) Limidae, (**e**) Anomioidea, (**f**) Lucinoidea, (**g**) Astartidae, (**h**) Crassatelloidea. *Error bars* indicate 95% confidence intervals; dashed lines indicate unsampled intervals

provide a source for well-preserved specimens. Generally Devonian specimens, which are numerous in museum repositories of North America, are judged to be poorly preserved (commonly as external molds) and therefore lacking internal skeletal features such as dentition or muscle scars. In most cases, a significant increase in preservational quality occurs between commonly lithified specimens of Cretaceous age, and those from the Paleogene interval. Most Neogene specimens in the dataset are from non-lithified host sediment and hence permit observation of most of the key diagnostic characters that are typically available from Recent material.

In Fig. 19, the preservation quality (mean percent of characters observed) of the eight taxonomic groups is plotted against degree of lithification. It is clear from each family that a strong negative trend exists between preservational quality and degree of lithification. Again, families containing genera that are largely defined on the basis of internal character groups (e.g., nature of dentition, presence of resilifer, position of muscle scars, sculpture of interior ventral margin) tend to show highest decrease (steepest curves) in preservational quality between unlithified and poorly lithified host sediment categories (i.e. Nuculidae, Nuculanoidea). Carboniferous and Permian specimens that have undergone silica replacement are not typically derived from originally unlithified sediments. Nevertheless they indicate that the process of silicification and subsequent sample preparation techniques (which free skeletonized specimens from their enclosing matrix), have similar implications for the preservation and observation of morphological features.

Figure 20a provides more detail on how diagenetic alteration of specimens and their host sediment can affect morphological details. Data are presented on the five key character groups required for definition of genus and subgenus-level groups in the family Nuculidae. Internal features such as the nature of the interior ventral margin, hinge plate and associated dentition, and resilifer are rarely observable in specimens from lithified host sediment. Evidence of the latter two character groups are rarely, if ever, observed on internal molds, whereas evidence of the interior ventral margin, which is often crenulated, is sometimes preserved by molds. The shell shape is perhaps the easiest feature (and potentially the least diagnostic for genus or subgenus level taxonomy) to observe, even among specimens that are preserved as shells embedded in lithified sediment or as internal and external molds. Shell surficial sculpture occupies an intermediate position. Being an exterior feature, it is more readily recognized among embedded specimens and external molds, although processes of delamination, abrasion and bioerosion often degrade its appearance even when original shell material remains. Each of the five key character groups for Nuculidae show improvement in observation probability between lithified, poorly lithified, and unlithified host sediments. Likewise, when the mean observation probability is plotted by time interval, a similar increase is noted from the Paleozoic to the Mesozoic, and the Cenozoic, matching increased representation among the dataset by specimens from unlithified host sediments.

Figure 20b, presents an independent dataset on the taxonomic resolution of published occurrences of the Nuculidae, from the Paleobiology Database. These data represent a random and large sampling of recorded occurrences from the available literature. The taxonomy largely reflects original published nomenclature and specifically reported occurrences are not subject to more recent revision, although nomenclature is corrected for changes in taxonomic rank and synonymy (following Wagner et al. 2007). The percentage of occurrences that are identified to species level (Fig. 20b) among recognized genera in this family (e.g., *Nucula proxima, Acila divaricata*) increases steadily through the Paleozoic to intermediate levels in

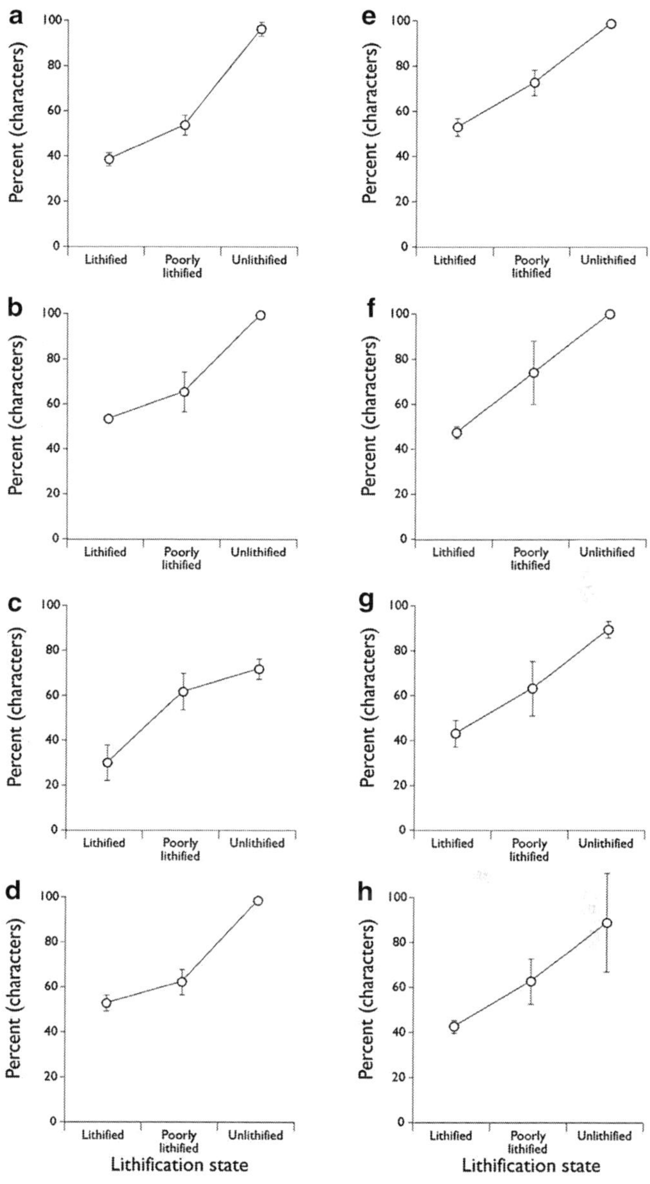

Fig. 19 Relationship between preservational quality and host-rock lithification state for several groups of Bivalvia: (**a**) Nuculidae, (**b**) Nuculanoidea, (**c**) Mytiloidea, (**d**) Limidae, (**e**) Anomioidea, (**f**) Lucinoidea, (**g**) Astartidae, (**h**) Crassatelloidea. *Error bars* indicate 95% confidence intervals; dashed lines indicate unsampled intervals

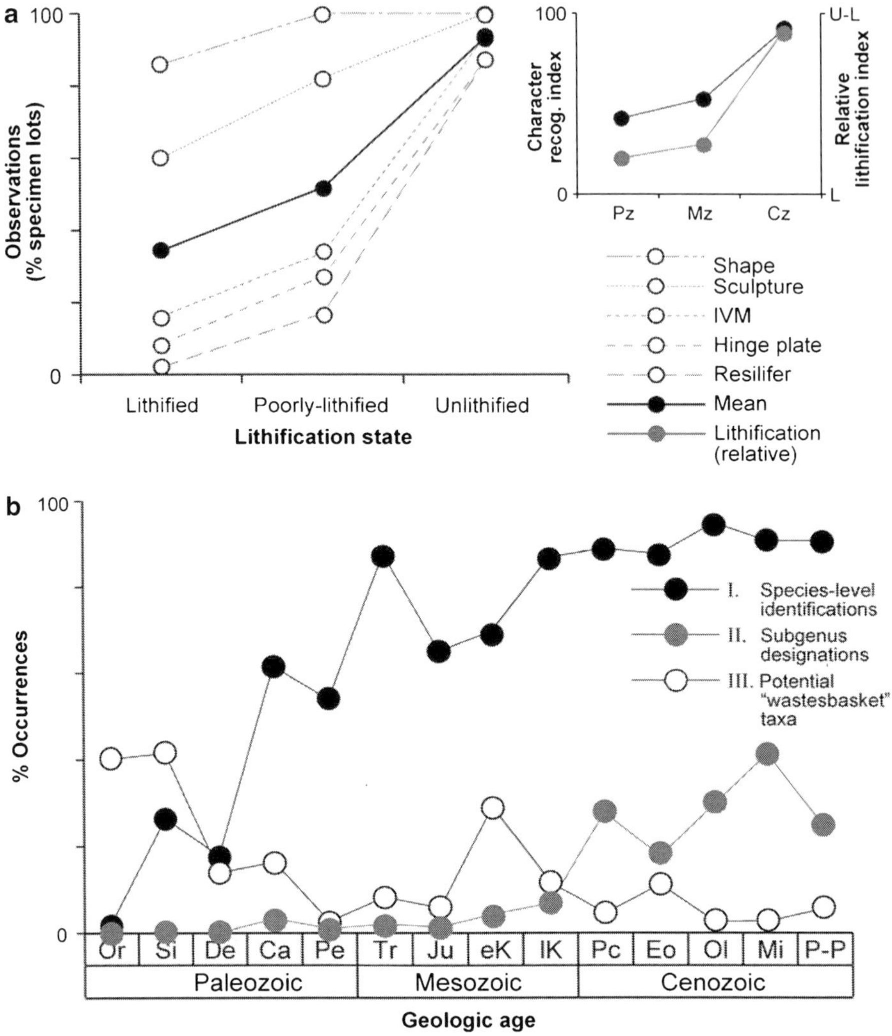

Fig. 20 Preservational quality (measured as percent of specimens bearing key characters), lithification state of sediments from which specimens were derived, and nomenclatural characteristics for Nuculidae (Bivalvia). (**a**) Proportion of specimens preserving individual characters and mean value plotted against lithification state. *Inset* shows temporal trends in the mean value and relative lithification state of those specimens; (**b**) Proportion of occurrences carrying (I) species-level identifications, (II) subgenus designations, and (III) identified as *Nucula* sp. (a potential "wastebasket taxon")

the Mesozoic, reaching a plateau near 90% in the Cenozoic. The percentage of occurrences carrying a subgenus-resolution designation (e.g., *Acila* (*Truncacila*)) remains low for Paleozoic and early Mesozoic records, only increasing substantially with Cenozoic occurrences. Unidentified occurrences of the type genus (Fig. 20b) for Nuculidae (i.e. *Nucula* sp.), here regarded as potential "wastebasket taxa", are remarkably frequent in rocks of early Paleozoic age, but are fairly uncommon among records of younger age.

These data series are not necessarily controlled by the degree of lithification of the sediments from which fossil occurrences are derived, but do reflect quality of preservation of those fossils, which co-varies with lithification. A lack of species-level identifications among greater than 70% of early Paleozoic occurrences suggests that while authors can recognize (correctly or incorrectly) the fossils as members of the Nuculidae, sufficient diagnostic characters are lacking to permit species recognition or encourage the description of new taxa at species-resolution. The increasing use (largely post-Mesozoic) of the subgenus rank for identification of Nuculidae appears to reflect a bias imposed by systematists working with living specimens. The definitions of recent subgenera commonly incorporate anatomical features that are only available from living material. Systematists working with Cenozoic-age taxa attempt to conform to the taxonomic framework established for the recent fauna, and relate fossil taxa to their recent counterparts on the basis of similarity in shell form or assumed ancestor–descendent relationships. The identification of so many early Paleozoic occurrences as *Nucula* sp. is quite troubling, given that the type for the genus is recent *Nucula nucleus* (Linné 1758), and although *Nucula* has a very good Cenozoic fossil record it probably originated no earlier than the Late Mesozoic (Cox 1969; Wingard and Sohl 1990). The assignment of specimens to *Nucula* sp. probably reflects a combination of bad taxonomy on the part of authors, and poor preservation that would limit subsequent reclassification to true Paleozoic members of the Nuculidae (e.g., *Nuculoidea, Nuculopsis,* or *Palaeonucula*).

2.4.2 Other Studies

Greenstein's (1992) rigorous study, showed that taphonomic bias affected the diversity history of cidaroid echinoderms, in a systematic, non-random fashion through time. Greenstein hypothesized that the Jurassic diversification of cidaroids was the result of evolutionary changes that permitted expansion into new ecospace. The increased diversity of preservation style noted by Greenstein could have resulted from occupation of a greater range of environments, which offer differing modes of preservation, and evolutionary changes that permitted development of more robust skeletons. An additional observation was that the wider range of preservational styles following the Jurassic provided additional material for the description of new taxa, based on the larger number of morphological characters that became available for description (e.g., lantern muscle-attachment structures). Smith (1990) suggested that these features are a crucial characteristic in defining and recognizing major taxonomic groups among Echinoidea.

Rofthus (2002, 2005) investigated the relationship between preservation state and taxonomic identifiability for bivalves and brachiopods from Silurian and modern environments. Rothfus found that shell modification by fragmentation (modifying shell shape characteristics) and surface alteration (modifying sculptural features) were among the most important taphonomic variables in reducing taxonomic identifiability. These processes, and their causes are thought to have either increased (e.g., through durophagous predation or bioerosion) or remained constant (e.g., wave energy, abrasion) over the course of the Phanerozoic. This is an interesting counter to other biases, such as increased sampling intensity and reduced lithification, which favour the preservation of younger fossil material (Rofthus 2005). Nevertheless, documentation of taphonomic characteristics, such as fragmentation frequency, in fossil assemblages may yield additional useful criteria for grading preservational quality and identifying taxonomic bias in paleobiological studies.

3 Exploring Other Taphonomic Trends in the Quality of the Phanerozoic Fossil Record

Lithification is just one of a range of secular taphonomic biases. Other potential biases, which are discussed to varying degrees elsewhere in this volume (e.g., Butts and Briggs, Dornbos, Kidwell this issue), include complete aragonite dissolution (as evidenced by preservation of molds), silicification, phosphatization, and preservation of konservat-lagerstätten. The following section summarizes the presence of these preservation styles in the published fossil record, as recorded by the Paleobiology Database (data downloaded 9/4/2007). The nature of any observed trends in their occurrence in the fossil record and the ways in which they may influence measures of diversity are described, but more thorough discussion of bio- and geo-chemical processes responsible for these taphonomic and sedimentologic conditions is left for other authoritative contributions. Nevertheless, their description here provides an initial assessment of whether large-scale occurrence-based datasets, such as the Paleobiology Database, adequately capture important taphonomic/sedimentologic trends.

3.1 Preservation as Casts and Molds

The dissolution of mineralized skeletons has long been considered to be a significant bias on the fossilized record of marine invertebrate organisms (Aller 1982; Flessa and Brown 1983; Davies et al. 1989). The chemical equilibrium of calcium carbonate in seawater is balanced between dissolution and precipitation. Dissolution converts solid $CaCO_3$ into component ions, CO_3^{2-} and Ca^{2+}, thereby removing skeletal material. Inorganic precipitation then creates new solid $CaCO_3$ from ions in sea or pore water. This precipitation often forms cement that binds together existing skeletal and non-skeletal grains (lithification). Early diagenetic dissolution may also completely remove aragonite material, resulting in a “dissolution fauna” where susceptible groups are entirely missing (e.g., Beu et al. 1972). More commonly recognized in the fossil record, however, are cases where aragonitic material survives dissolution at the sea floor and become enclosed within host-sediments that experience processes of cementation and lithification. Dissolution of shell material and concurrent cementation of surrounding sediments by shell-derived cement preserves traces of former skeletal hardparts as biomolds (Smith and Nelson 2003; Cherns et al. 2008). Surface marine waters are less saturated with respect to aragonite than calcite, and aragonite has a lower thermodynamic stability. Thus aragonite is more likely to dissolve than calcite (Chave et al. 1962). Furthermore, taxa dominated by organic-rich microstructures (e.g., nacre) are known to be commonly preserved as molds in the Paleozoic fossil record, even when co-occurring with taxa bearing organic-poor microstructures, and preserved with original shell material (Taylor et al. 1969, 1973; Glover and Kidwell 1993). Given that seawater saturation states of calcite and aragonite have varied throughout the Phanerozoic (Sandberg 1983; Holland 1984; Wilkinson and Given 1986) it is appropriate to investigate the fossil record for trends of increased dissolution (preservation as casts and molds) and to determine the consequences of skeletal dissolution on perceptions of biodiversity and community composition (see Cherns and Wright in review, for thorough analysis).

Interpreting trends in the presence of fossil biota preserved as molds through the geological record is rather complicated, given the secular variation in mineralogy of skeletal hardparts and multitude of geochemical pathways by which the dissolution of former skeletons can undergo. Nevertheless, Fig. 21a presents a preliminary analysis of biotas reported in the Paleobiology Database as lacking body fossil preservation. A significant percentage (75%) of Early Cambrian biotas are reported as being preserved as internal or external molds. This percentage decreases significantly through the remainder of the Cambrian, reaching a low of less than 5% through the Ordovician. A steady increase in representation of fossils as molds is observed from the Ordovician to the end of the Carboniferous, where values reach around 25–30%. The Permian is characterized by especially low values (<10%), but a significant increase in records of cast and mold preservation in observed in the Triassic and early Jurassic, with values ranching between 20% and 60%. For much of the remainder of the Mesozoic and Cenozoic, casts and molds contributed to less than 20% of reported occurrences.

Almost certainly, the complete dissolution of skeletal hardparts has consequences on community-level richness, as shown by comparisons with contemporaneous silicified biota by Cherns and Wright (2000), Wright et al (2003), and Cherns and Wright (2009). Figure 21b compares the mean sample richness of those moldic assemblages plotted in Fig. 20a with background (non-moldic assemblages), suggesting that this bias is prevalent throughout the Phanerozoic. Sixteen of 21 intervals yield lower mean richness from moldic assemblages than their background counterparts, although the measured difference in richness is never greater than 40%. It should be noted that this analysis takes no account of the sample size, taxonomic composition, environmental or geographic origin, and any additional taphonomic overprints affecting either moldic or background assemblages. The comparison therefore lacks rigorous control, but provides a starting point for further detailed investigation.

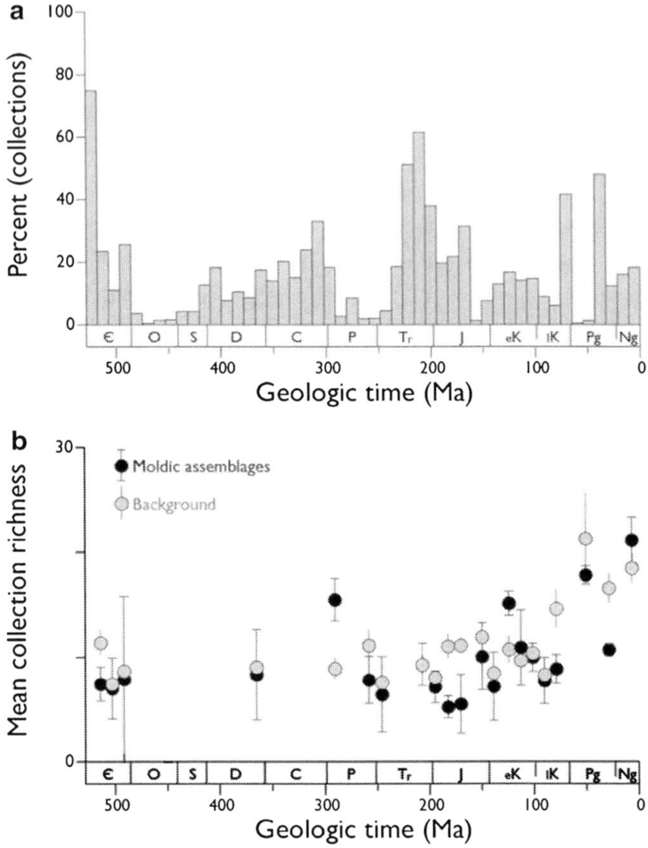

Fig. 21 Preservation of marine invertebrate fossils as molds through the Phanerozoic based on data reposited in the Paleobiology Database. (**a**) Proportion of collections preserved as molds. (**b**) Alpha diversity of contemporaneous assemblages preserved as either molds or as body fossils; collections with fewer than four taxa (potentially incomplete) are excluded from analyses, and only intervals with >15 collections with moldic preservation are compared to background assemblages

3.2 Lagerstätten and the Preservation of Soft-Bodied Fossils

Preservation of entirely soft-bodied organisms requires unusual environmental conditions that are geologically rare, such as anoxia or catastrophic burial with rapid mineral replacement by specialist microbial communities (Allison and Briggs 1991, 1993; Briggs and Crowther 2001). When it occurs, such preservation provides valuable windows into the anatomy and habitats of these groups and can be important simply by virtue of being the earliest record of taxa and morphological characters. However, stratigraphic horizons with comparable preservation are generally so dispersed through the stratigraphic record that any time series and evolutionary conclusions for these groups are highly incomplete (Kidwell and Holland 2002).

Examples of soft-bodied biotas are not well reported or appear not to have been targeted by the Paleobiology Database, but occur uncommonly during the Early through Late Cambrian, Carboniferous and then during the Late Triassic through Late Jurassic (Fig. 22). These include some of the classic konservat-lagerstätten, such as the Chengiang biota, Burgess Shale, and Alum Shale ('Orsten' biota), but others, including the Hunsrück Slate, Holzmaden and Solnhofen Limestone are insufficiently represented in the database. Additional soft-bodied assemblages of Cretaceous and Cenozoic age include preserved remains of non-shelled cephalopods. While low in frequency, collections exhibiting soft-body preservation comprise a greater range of taxonomic groups (particularly arthropods and annelids) and body compositions (including soft-bodied, and chitinous or phosphatic skeletons) than background (contemporaneous collections with normal preservation).

Allison and Briggs (1993) determined that the Cambrian and Jurassic (Fig. 23) show significantly higher concentrations of exceptional faunas than predicted by chance. Those of the Cambrian accumulated in both deep marine (e.g., Kinzers Fm, Burgess Shale, Wheeler Fm) and shallow water (e.g., Chengjiang biota) settings. Jurassic faunas, however, tend to be limited to marine environments with restricted circulation and a stratified water column (Seilacher et al. 1985). The Paleobiology Database appears not to have captured data from a large number of assemblages. However, the gross number of exceptionally preserved faunas tabulated by Allison and Briggs (1993) is itself quite low, with a maximum of only nine reported for the Cambrian (Fig. 23); by their tabulation an "exceptional fauna" denotes a single "formation"-scale unit from which konservat-lagerstätten are recorded. Although the database attempts to assemble data on multiple collections from individual

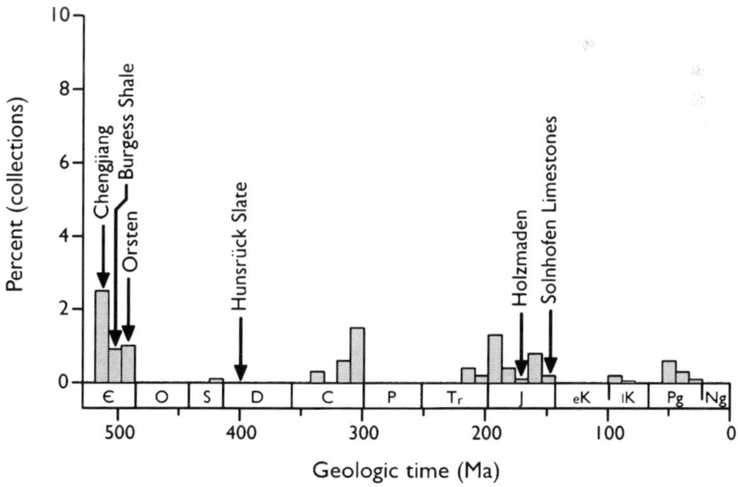

Fig. 22 Preservation of marine invertebrate fossils in konservat-lagerstätten through the Phanerozoic (proportion of total collections) based on data reposited in the Paleobiology Database. *Arrows* indicated particularly important marine invertebrate lagerstätten

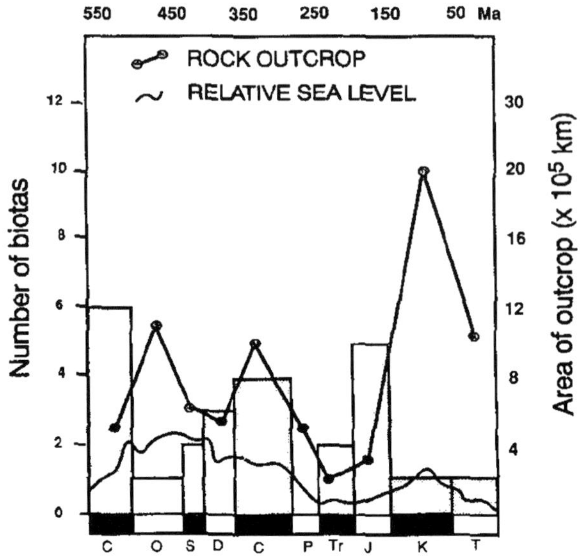

Fig. 23 Distribution of exceptional faunas in relation to sedimentary outcrop and sea level (sea level after Hallam 1984) (From Allison and Briggs 1993)

lithostratigraphic units and local sections, it appears as though the few collections derived from konservat-lagerstätten are overwhelmed by the proportion of faunal data yielded from fossiliferous sediments with normal preservation. Allison and Briggs (1993) reported that exceptionally preserved faunas are often omitted from analyses of evolutionary patterns because of their infrequent occurrences (hence they can distort diversity metrics) and because their environment of accumulation is often atypical of the bulk of the fossil record (e.g., deep-water, reduced circulation). Nevertheless, the taxonomic data from such accumulations does contribute to global biodiversity estimates (e.g., Sepkoski 2002; Alroy et al. 2008) and their propensity for recording the first appearances for many higher taxonomic groups with low preservation potential has relevant implications for paleobiological analyses.

3.3 *Concentrations of Fossils*

Shell beds (densely packed concentrations of skeletal remains) are a conspicuous feature of the Phanerozoic (Simões et al. 2000). A number of recent papers have presented quantitative analyses of changes in the taphonomic quality of the shelly marine record through the Phanerozoic (Kidwell 1990; Kidwell and Brenchley 1994, 1996; Kidwell this issue). In particular, these studies have demonstrated an increase in internal complexity and thickness of shell beds through this interval. These patterns have been suggested to result from Phanerozoic-scale increases in macrofaunal diversity, average body-size of benthic taxa, the depth and intensity of infaunalization, the durability of biomineralized skeletons, and the occupation of high-energy habitats. While not the primary focus of their investigations, the authors also suggested an increase in the overall abundance of shell beds, independent of thickness-frequency distributions. As shell beds tend to be a focus for paleontological collection effort they are potentially a rich source of data on biodiversity. Fluctuations in their thickness and frequency in the sedimentary record therefore might have significant implications on perceptions of past biodiversity.

Records of skeletal concentrations (e.g., shell beds, bioclastic limestones) are not well documented in the Paleobiology Database due to the lack of specific descriptive lithology fields concerning such deposits. The time-series of Fig. 24a was derived from both semi-quantitative lithology descriptor fields (shelly/skeletal) and informal comments in the lithology comments field, including such terms as shell bed and coquina. It is unfortunate that objective data (e.g.. dimensions, packing density) are not readily available to confidently classify collections following the definitions of Kidwell (1990) and others. Nevertheless, "skeletal assemblages", as defined above, comprise between 5% and 18% of collections for most of the Paleozoic, although their abundance is low (<7%) during the Cambrian, and again in the late Carboniferous through Late Cretaceous. A significant increase in skeletal assemblages, reported in the Paleobiology Database, takes place following the Late Cretaceous, with values ranging between 10% and 25% for most of the Cenozoic. The mean richness of assemblages derived from skeletal concentrations was not consistently greater than that assumed to be derived from background sediments (Fig. 24b), although five of six of the Cenozoic data points did yield higher collection richness. Given lack of certainty regarding the classification of collections as shell beds, and the limited availability of abundance data to standardize sample size, it is difficult to place much emphasis on these results.

3.4 Silicification

Silicified fossils are not only an important component of the Paleozoic fossil record (Schubert et al. 1997), but preserve data of tremendous significance to paleontology through their retention of exquisite morphologic detail. Silicification can take place in a variety of paleoenvironments, including hypersaline and evaporitic shallow marine settings, oxic through anoxic deep-water sediments, and near hydrothermal vents (Carson 1991). In addition, the factors involved in silicification vary in each environment or facies through geologic time. Processes such as evaporation, meteoric interaction, changes in redox and pH reactions, and depositional saturation (e.g., influx of volcanic ash, hydrothermal discharge, and dissolution of biogenic silica) can contribute, sometimes together, towards silica replacement. The nature of skeletal material (i.e. carbonate mineralogy), the timing relative to burial history of the material and the presence of other elements in an organism's hard-parts or depositional environment can influence the pathway by which silica replacement

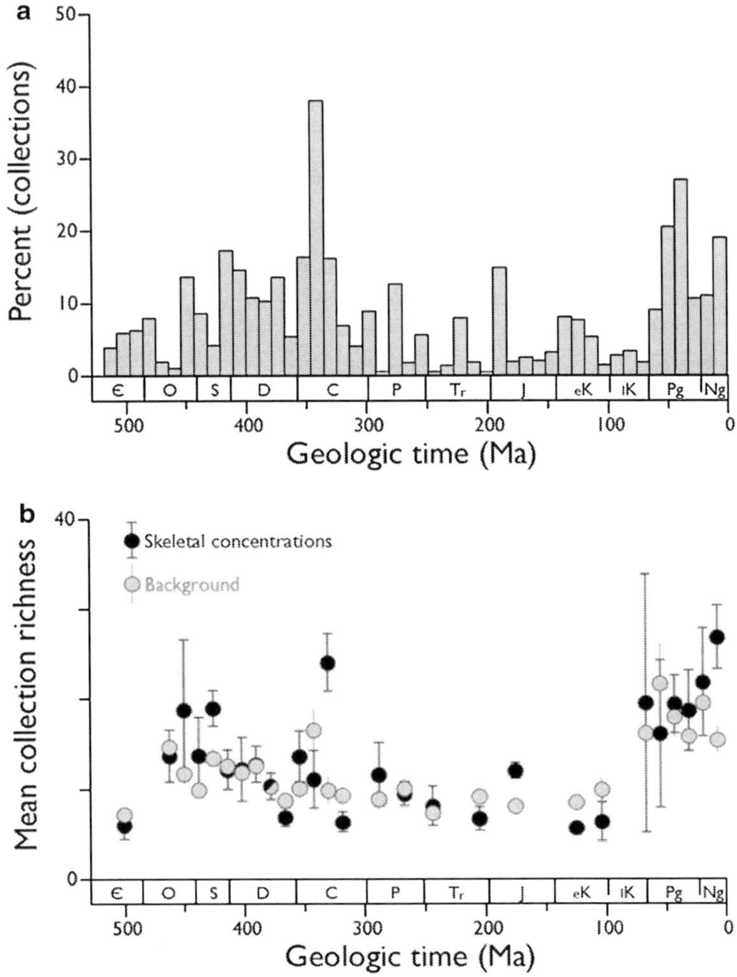

Fig. 24 Preservation of marine invertebrate fossils in shelly/skeletal rich sediments through the Phanerozoic based on data reposited in the Paleobiology Database. (**a**) Proportion of collections derived from shelly/skeletal rich sediments. (**b**) Alpha diversity of contemporaneous assemblages preserved in shelly/skeletal rich sediments and the background fossil record; collections with fewer than four taxa (potentially incomplete) are excluded from analyses, and only intervals with >15 collections from shelly/skeletal rich sediments are compared to background assemblages

proceeds. Schubert et al. (1997) found that silicified faunas compose about 20% of the published record during the Paleozoic, with an almost negligible record silicified during the Mesozoic and Cenozoic (Fig. 25). The post-Paleozoic decline in silicification has been suggested by a number of authors (Finks 1960, 1970; Schubert et al. 1997) to correspond to changes in siliceous sponge and radiolarian abundance or diversity. These authors also found that the dominant locus of silica

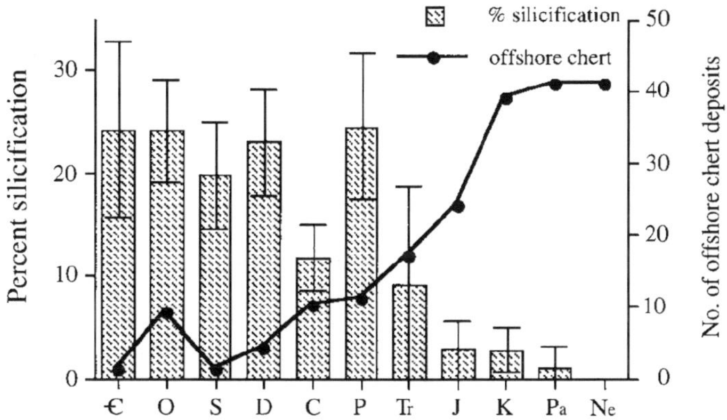

Fig. 25 Percentage of analyzed paleontological literature describing benthic marine macrofossils replaced with silica is shown in histogram. Scale is to left. Line marks number of offshore bedded cherts, based on data from Hein and Parrish (1987); scale is to right. Confidence intervals 95% for percentage silicification were calculated following Raup (1991) (From Schubert et al. 1997)

deposition moved to deep-marine environments at the end of Paleozoic, concurrent with this transition. If silicification is enhanced by the presence of organisms producing siliceous skeletons, namely sponges, then the non-random distribution of these organisms through time and across environmental gradients (Finks 1960; Brunton and Dixon 1994) may produce another temporal mega-bias in the fossil record (Schubert et al. 1997).

Analysis of data from the Paleobiology Database in Fig. 26a, indicates that silica replacement of macrofossils occurs with reasonable frequently throughout Paleozoic. As much as 15% of Cambrian through Carboniferous fossil occurrences are noted as showing evidence of silica replacement. A major peak, however, is clear during the latest Carboniferous and Early Permian, for which greater than 30% of fossil occurrences are silicified. Post-Paleozoic silicified occurrences are considerably less frequent, but a clear peak is observed during the Late Triassic with as much as 10% of occurrences noted as silicified. Fewer than 3% of post-Triassic fossil occurrences are silicified.

The frequency distribution of occurrences with silica replacement derived from data reposited in the Paleobiology Database (Fig. 26a) agrees with previously published trends by Schubert et al. (1997) (Fig. 25), in exhibiting a significant post-Paleozoic decline in silicification. Nevertheless, the relatively flat-lying trend observed through the Paleozoic by Schubert is not replicated by the higher-resolution results in Fig. 26a. This distinction might be related to greater and more representative sampling of the paleontological literature in the Paleobiology Database. More likely, however, it is the result of the focused efforts by Schubert et al. to identify evidence of silicification among their own literary dataset, and the frequent omission of ancillary taphonomic data for collections when entered in the Paleobiology Database. It stands to be tested whether the dramatic increase in the Permian is truly above Paleozoic levels as

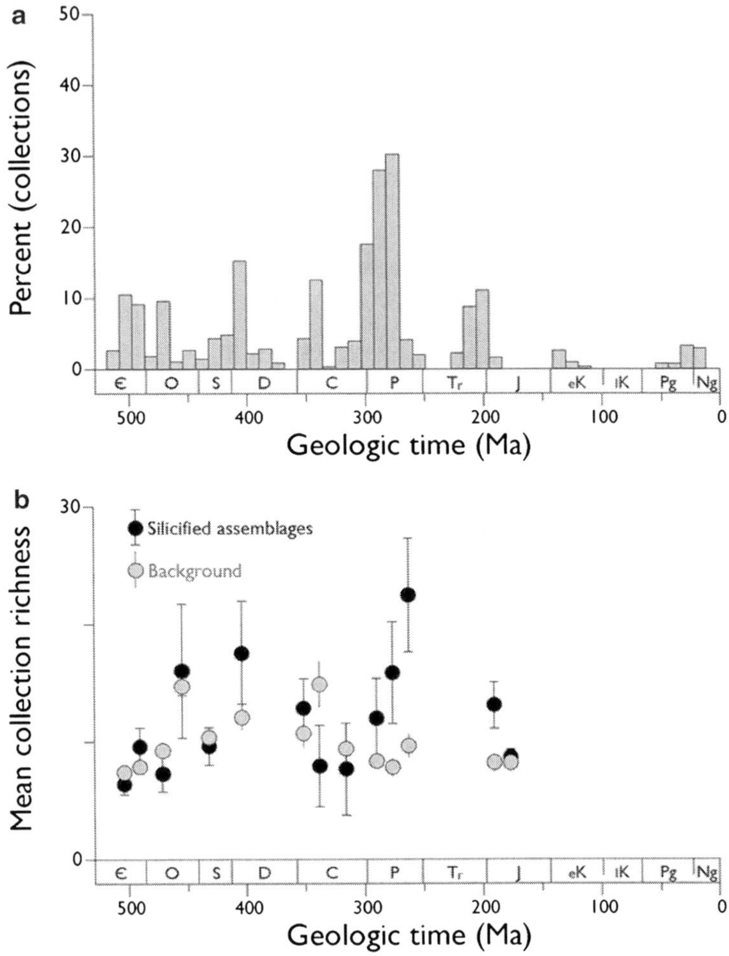

Fig. 26 Preservation of marine invertebrate fossils with silica replacement through the Phanerozoic based on data reposited in the Paleobiology Database. (**a**) Proportion of collections preserved with silica replacement of fossils. (**b**) Alpha diversity of contemporaneous assemblages preserved with silica replacement; collections with fewer than four taxa (potentially incomplete) are excluded from analyses, and only intervals with >15 collections with silica replacement are compared to background assemblages

suggested with these new data, or an artefact of data entry bias towards well known rich silicified assemblages (e.g., Cooper and Grant 1972). Figure 26b compares the mean richness for samples from exhibiting silica replacement with those from contemporaneous background (normal preservation) assemblages. In ten of fourteen cases the silicified assemblages yield higher collection richness. Early Devonian and the Permian background assemblages were depleted as much as 30–60% in richness relative to contemporaneous silicified assemblages. This raises the possibility that estimates of diversity, particularly those of Permian stages where 20–30% of collections are silicified, could be significantly inflated relative to adjacent intervals where silicification is less frequent. Again, it is worth noting that this comparison does not rigorously ensure that collections are derived from similar environmental and geographic provenance, reflect similar sampling effort, have consistent preservational modes, or ensure compatibility in taxonomic composition.

3.5 Phosphatization

Phosphate may preserve calcareous and siliceous skeletons either by replacing the primary shell mineral or by forming internal or external molds. This type of preservation is usually rare because of low background phosphate concentrations. Occurrences of secondarily phosphatized skeletons are therefore predominantly associated with phosphate beds, or hardgrounds where conditions for apatite precipitation are enhanced (Prévôt and Lucas 1986, 1990; Lucas and Prévôt 1991). Additionally, it is possible for soft tissues to be preserved by diagenetic apatite in natural phosphorites. Replication of soft tissues is rare and requires extraordinary conditions (Allison 1988a), typically occurring in argillaceous sediments deposited in oxygen-depleted environments (Prévôt and Lucas 1990; Lucas and Prévôt 1991). Phosphatized soft remains exhibit exceptional preservation, including three-dimensional preservation of soft-parts and the retention of cellular morphology (Wilby et al. 1996; Wilby and Briggs 1997). While phosphorite deposits serve as a probable source of phosphorous involved in replacement of skeletal hard parts, Allison and Briggs (1993) showed no evidence for global controls on the phosphatization of soft tissue.

Figure 27 presents trends captured by the Paleobiology Database in occurrence of fossils that have undergone phosphatic replacement. This preservational mode is recorded at its peak in the Cambrians (ranging between 5% and 9% of recorded occurrences). Reported occurrences of phosphatic replacement are patchy throughout the remainder of the Phanerozoic, although another small peak in occurrences is noted during the Early Cretaceous. It should be noted that this analysis does not make rigorous estimates of the nature of the fossils that are preserved through phosphatic replacement (i.e. original mineralogy) or the type of replacement itself (i.e. replication of soft parts, replacement of skeletal hard parts, or as internal molds), and therefore should be regarded as a preliminary attempt to characterize temporal trends in this mode of fossilization.

The observed pattern of Cambrian and Early Cretaceous peaks in phosphatization are corroborated by other authors. Numerous recent investigations have detailed secondary phosphatization among konservat-lagerstätten (e.g., Müller 1985; Butterfield 1990; Zhu et al. 2005), and small shelly faunas (e.g., Bengston et al. 1990; Brasier 1990; Dzik 1994) of the Cambrian. These fossil occurrences almost certainly skew perceptions of early Paleozoic biodiversity (Brasier 1990). Small shelly fossils, for example, dominate Early Cambrian diversity, but suffer a decline by the Middle Cambrian (Porter 2004), primarily owing to a significant reduction in phosphogenesis (Cook and McElphinny 1979; Cook 1992). Porter (2004) suggests that the lack of abundant middle and Late Cambrian small shelly faunas is

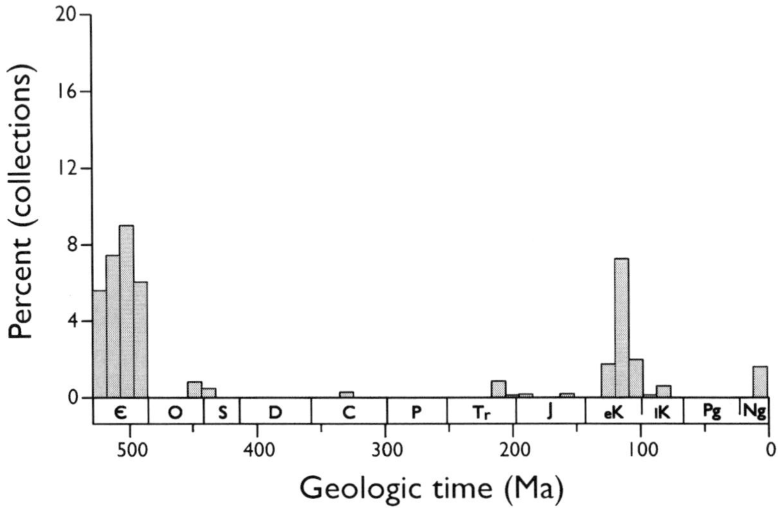

Fig. 27 Preservation of marine invertebrate fossils with phosphatic replacement (proportion of total collections) through the Phanerozoic based on data reposited in the Paleobiology Database

attributable in part to the closure of this phosphatization window. In addition to phosphatized Cambrian konservat-lagerstätten, examples of phosphatized marine invertebrate soft parts are known from the Devonian (Briggs and Rolfe 1983), and are particularly prevalent in Jurassic (Donovan 1983; Pinna 1985; Allison 1988b), and Cretaceous (as reported by Allison and Briggs 1993).

4 Discussion

4.1 Evaluation of the Paleobiology Database in Capturing Taphonomic Trends

It is clear from the examples presented here that particular taphonomic biases can significantly distort biological trends in biodiversity trajectories. While the focus is on describing the lithification bias (and associated diagenetic effects), it is clear that a number of other taphonomic and sedimentology characteristics of fossil assemblages show secular variation during the course of the Phanerozoic. The Paleobiology Database provided the data used to support these claims. Given that the database is presently regarded as the primary source of data on biodiversity, and geographic, stratigraphic, and environmental distribution of fossil occurrences in the Phanerozoic, it is appropriate to assess whether the database has adequately captured notable taphonomic/sedimentologic trends. Furthermore does the variation observed between previously noted patterns and those observed from the database truly reflect the fossil record, features inherent in published paleontologic data, or simply inadequacies in database structure?

Data used in the global Phanerozoic-scale analyses presented here (lithification, moldic preservation, silicification, phosphatization, preservation as concretions, and lagerstätten) were downloaded at a stage in the development of the Paleobiology Database during which temporal coverage of fossil assemblage data was considered to have become fairly complete and evenly distributed. Subsequent improvements to the composition of the database have primarily focused on adding to data from a few poorly sampled Phanerozoic stages and improving geographic coverage and evenness of geographic data distribution. A concern, however, remains the taxonomic composition of the database, which tends to be dominated by especially well-documented groups (e.g., brachiopods, trilobites, corals, bivalves and gastropods) during particular intervals of the Phanerozoic (see Fig. 1). This reflects not only their ecological dominance (e.g., early Paleozoic collections are dominated by brachiopod and trilobite occurrences; Cenozoic collections are dominated by bivalve and gastropod occurrences), but also factors such as paleontological interest (i.e. biostratigraphic utility), and preservation potential (e.g., echinoids are not well suited to frequent preservation in the fossil record). With these patterns in mind, any similar use of the database should attempt to mitigate biases caused by secular taxonomic variations by limiting investigations to particular taxonomic groups, or combinations of groups with similar sampling and taphonomic characteristics, particularly for the purposes of temporal comparisons.

The Paleobiology Database was initially designed (see Alroy et al. 2001, 2008) to collect data in support of analyses of biodiversity change over geological time, and not necessarily to assemble data appropriate for analyses in a taphonomic framework. Nevertheless, the database does attempt to format taxonomic, geographic, stratigraphic, sedimentologic, environmental, and taphonomic data in a standardized framework, by using descriptive values from pulldown lists for respective fields. Hence, considerable taphonomic and sedimentologic data are associated with each fossil occurrence, when entered. Fields that have particular potential for future taphonomic analyses are listed in Table 5, and include those that consider modes of preservation (e.g., as body fossils, casts, molds, as traces, in concretions), the nature of original or replaced skeletal mineralogy (e.g., as original aragonite or calcite, or replaced through silicification or phosphatization), preservation as konservat-lagerstätten, the degree of concentration, the spatial orientation of fossils, the preservation of anatomical data, temporal and spatial resolution, and biostratinomic damage (e.g., articulation, sorting, fragmentation, bioerosion and encrustation). Taphonomic data are also contained within qualitative comments for each collection, or can be gleaned from sedimentologic fields (see Table 4). Additional taphonomic analyses can be undertaken using data pertaining to taxonomic units (e.g., species, genera, orders), such as information on original skeletal composition, body size, relative thickness, and the presence of skeletal reinforcement and characteristics of skeletal architecture. This source of data supported recent analyses on the relationship

Table 5 Data entry fields in the paleobiology database with applications to taphonomic analysis. Complete table structure and field definitions can be found at http://paleodb.org/public/tips/tips.html

Field	Example	Availability (%)
Sedimentology		
Lithification	Metamorphosed, lithified, poorly lithified, unlithified	59
Lithology modifier	Bioturbated, concretionary, nodular, pyritic, shelly/skeletal, siliceous	42
Lithology	Dolomite, limestone, sandstone, siltstone	93
Taphonomy		
Modes of preservation	Body, cast, mold, impression	57
Original biominerals	Aragonite, calcite, chitin	9
Replacement minerals	Calcite, pyrite, silica	9
Lagerstätten type	Conservation, concentration	<1
Concentration of fossils	Dispersed, concentrated, lag, hiatal	<1
Spatial orientation	Life position, random, preferred	<1
Anatomical detail	Excellent, good, medium, poor	10
Degree of articulation	Articulated, associated, disassociated	<1
Size sorting	Poor, medium, good	<1
Fragmentation	None, occasional, frequent, extreme	2
Bioerosion	None, occasional, frequent, extreme	<1
Encrustation	None, occasional, frequent, extreme	<1
Feeding/pred. traces	Drill holes, repair scars, fractures	<1
Temporal resolution	Snapshot, time-averaged, condensed	5
Spatial resolution	Autochthonous, parautochthonous, allochthonous	5

Availability, percent of all collections that contain field entries (based on data downloaded 9/4/2007)

between skeletal durability and frequency of occurrence in the fossil record (Behrensmeyer et al. 2005).

A prerequisite to the analyses shown here has been the fundamental shift in the nature of fossil databases away from compendia that depict only the first and last known global occurrences of taxa, towards more comprehensive compilations that list occurrences of taxa wherever they are found in the world, regardless of whether they are the first or last occurrences. Each occurrence can be accompanied by a range of supplementary data that can be used not only to assess the overprint of preservational biases on diversity trajectories, but also potentially significant paleoenvironmental or paleobiogeographic selectivity in diversification and extinction. The Paleobiology Database and numerous other databases now typically include this kind of information, thereby permitting the kind of comprehensive assessments of diversity and biases outlined in this chapter. Nevertheless, the database remains poorly populated with many kinds of data appropriate for the analyses of taphonomic trends (Table 5), although an increasing proportion of newly entered fossil assemblages include semi-quantitative taphonomic data (Fig. 28a). Some of this shift is the result of data entry (e.g., availability of fields, interest and skill of the data enterer), but it may also reflect the quality of the published literature (Fig. 28b). Modern paleontologic literature increasingly focuses on the paleoecological or paleoenvironmental interpretation of fossil data, for which taphonomic information plays a critical role. Much of the assemblage-level data published through the nineteenth and early twentieth century, however, were the product of taxonomic treatments or biostratigraphic reports for particular fossil groups, where aspects of preservation were lacking or not easily converted into the semi-quantitative data fields that the Paleobiology Database accepts. As the database continues its growth it should be able to support broad-scale analyses (where appropriate) of many of the variables listed in Table 5.

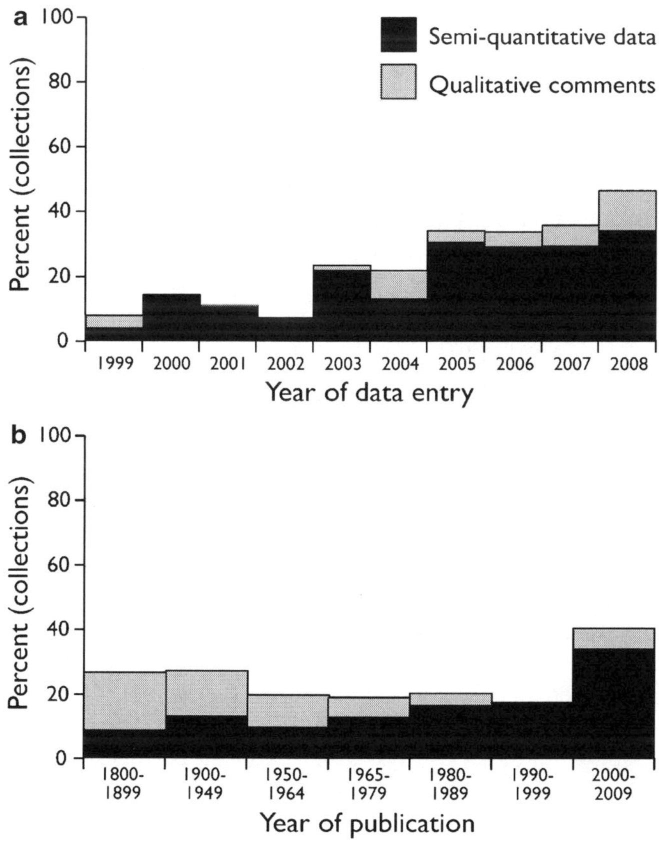

Fig. 28 Growth of semi-quantitative (pulldown fields) and qualitative taphonomic data (keystroked comments) reposited in the Paleobiology Database. (**a**) Historical trends in quality of entered data; (**b**) trends in quality of published data

4.2 Research Opportunities and the Mitigation of Taphonomic Biases

4.2.1 Taphonomic Biases and the Biodiversity Record

It is clear that for temporal analyses of paleontological data we need to understand and mitigate for potential biases affecting the fossil record. This has required a fundamental change in the nature of fossil databases used for large-scale and coarse-resolution analyses. At the broadest-scale, databases of taxonomic data must provide data on more than their first and last appearances, but also the nature of their occurrences and abundance with respect to depositional environments and past paleogeography (Smith 2001, 2003). Electronic databases now allow high levels of stratigraphic and environmental dissection of paleontological data that was not possible in the past. Sampling-standardization techniques are critical in removing many of the effects (some of which are associated with lithification and taphonomic biases) on variations in sample size, although these may reveal further issues, such as variation in heterogeneity of geographic and environmental sampling (see Alroy et al. 2008). Despite major advances in our analytical approach to reading biodiversity and paleoecological data from the fossil record (e.g., Alroy et al. 2001; Bush et al, 2004), concerns still remain over the biological veracity of observed results. Databases are only as good as the data they contain (Markwick and Lupia 2002). Indeed the quantitative assessment presented here demonstrates the potential of large-scale occurrence-based datasets in taphonomic investigation, but also reveals a number of shortcomings, such as heterogeneity in data quantity and quality and difficulty in establishing causation. Furthermore, the kind of information required for analysis of biodiversity trends, for instance, can only reveal limited clues about contemporaneous taphonomic patterns and processes. While large-scale assessments can indicate the temporal scope and magnitude of biases, they will struggle to fully understand their causal mechanisms without corroboration from small-scale studies or experiments in which the potential overprint of environmental and geographic variation is controlled.

Focused taphonomic research on the fossil record needs to continue to collect quantitative or semi-quantitative (i.e. rank or relative assessment) data on the preservation potential of taxa. Improvements in our knowledge of these patterns among taxa, environments, and through time will lead to better modelling and empirical analytical approaches. Such research needs to operate at both the global scale to refine the timing and magnitude of potential biases, but also at the local scale (field-based) to establish better constrained magnitudes of taphonomic bias and effective methods to avoid them either during collecting or data analysis. Experimental (e.g., Flessa and Brown 1983; Hof and Briggs 1997; Martin et al. 2002; Gupta et al. 2006) and actualistic (e.g., Davies et al. 1989; Walker et al. 1998; Parsons-Hubbard et al. 1999; Powell et al. 2002) studies over the past decade have considerably improved our understanding of the mechanisms involved in some taphonomic processes, and their rates, at least over short durations (ecological snapshots rather than geologic time) and in present day conditions of sea water geochemistry and biological activity.

In developing research strategies, it is important to realize that no single bias applies to all scales of evolutionary analyses or to all taxonomic groups and that no scale or type of evolutionary analysis or taxonomic group is free of taphonomic bias. Taphonomic biases may not affect paleontological analyses conducted at coarse spatial scale and temporal resolution, and indeed, this is one recognized approach for mitigating potential deficiencies in the fossil record. Nevertheless, recent basin-scale and high-resolution assessments of the lithification (e.g., Hendy 2009a; Sessa et al. 2009) and aragonite-dissolution (e.g., Cherns and Wright 2000, in review; Cherns et al. 2008; Wright et al. 2003) biases show such significant change in community composition and richness that it is difficult not to expect their manifestation among globally aggregated data. Future effort is therefore required to better quantify not only the temporal, geographic and taxonomic range of particular taphonomic biases, but also the ways in which they alter diversity at a range of spatial scales (i.e. local through global) and the paleoecological properties of individual assemblages.

The non-random nature of the above biases raises concern regarding the analysis of diversity or ecological complexity over the course of the Phanerozoic or comparison among contemporaneous faunal groups. Furthermore, a number of the biases have tremendous potential to affect community-scale patterns, either degrading (e.g., lithification, aragonite dissolution) or enhancing (e.g., silicification, phosphatization) the relative quality of fossil data. Several of approaches can be undertaken to minimize these biases, including the selective filtering of datasets to remove taphonomically vulnerable groups or the use of taphonomic control taxa that indicate the consistent preservation state of fossil assemblages (e.g., Jablonski et al. 1997). In light of recent results indicating that lithification has the potential to alter our perception of faunal recovery to a mass-extinction event, other recognized extinction and faunal turnover events should also be revisited to establish their veracity. Sessa et al. (2009) demonstrated that lithification had a marked effect on perceptions of recovery dynamics following the Cretaceous-Paleogene mass extinction in one of the most well known and statigraphically complete regional sections (Gulf Coastal Plain). With this in mind it is possible that a similar bias influences global compilations of diversity during this interval. Other intervals of extinction, in which both lithified and non-lithified fossiliferous facies are represented, should also be treated with some caution. If other, taphonomic biases (e.g., phosphatization, silicification) are shown to have similar consequences on community level richness and evenness, and selective destruction of particular size cohorts or mineralogical components, then such effects could be widespread throughout the Phanerozoic, and influence many globally and regionally significant extinction and turnover events.

4.2.2 Implications for Taxonomic and Morphologic Analyses

A number of taphonomic processes have the ability to degrade as well as enhance the relative quality of fossil material, and in doing so can introduce error into taxonomic practice, skew morphological patterns, and influence evolutionary inference. Phosphatization and silicification of carbonate skeletons, for instance, plays an important role in enhancing the fossil record of the Early Cambrian, preserving skeletons of thin and small (<40 mm) early metazoans, which were vulnerable to diagenetic loss (Brasier 1990). These secondarily mineralized specimens are routinely obtained by acid digestion of limestones, rendering exquisitely preserved specimens, free of matrix. There is a growing suspicion that this methodology could be introducing a preservational bias towards small fossils (Runnegar and Pojeta 1985; Dzik 1991; Mus et al. 2008). Mus et al. (2008) suggested that this has resulted in the evolutionary model that the earliest molluscs (e.g., helcionellids) were millimetre-scale animals, despite occasionally reported occurrences of much larger molluscs in non-carbonate lithologies (e.g., Sundukov and Fedorov 1986). Most reports of large Cambrian molluscs predate the development of acid-based extraction techniques and scanning electron microscopy. It appears therefore, at least in the Early Cambrian, that the excellent preservation of smaller, more easily phosphatized fossils (phosphatization preferentially occurs in small cavities) has driven taxonomic effort away from larger poorly preserved material (usually preserved as molds or through recrystallization), and shifted ideas on the evolution of body-size among early Paleozoic marine invertebrates. A systematic investigation of body-size trends in relation to taphonomic properties and preparation techniques for Neoproterozoic and early Paleozoic organisms would be a valuable prerequisite to further interpretation of early evolutionary trends in diversity or morphology.

Elsewhere in the Phanerozoic fossil record, changes in quality and preservation tend to bias against the preservation of small specimens. Indeed few taphonomic factors act to distort the preservation of larger specimens without also influencing the smaller cohort. Certainly, the analyses of size composition presented in this study would suggest that in lithified rocks (Figs. 8, 9, 18) and in cases of complete aragonite dissolution, small taxa are less likely to be preserved in fossil assemblages. Cooper et al. (2006) showed convincingly that the size bias is responsible for removing as much as 36% of a well-studied, and apparently well-preserved regional fauna of Cenozoic age. In an independent study, Sessa et al. (2009) also indicated the selective loss of small specimens, demonstrating that the median size of specimens from lithified assemblages was nearly 1.5 times as large as those from unlithified sediments; the median size of unique, lithified genera was over twice that of unique, unlithified genera. Analyses of Phanerozoic trends in body-size (Hendy 2009b) reveal a similar pattern for gastropods, with lithified sediments yielding consistently larger specimens than unlithified sediments or assemblages having undergone silica replacement. Payne (2005) acknowledged that the size distribution of silicified species contrasts strongly with other contemporaneous material, clearly indicating the importance of preservation and preparation for sampling the small end of the size spectrum. It is likely that smaller taxa, particularly those with less stable carbonate mineralogy (i.e. organic-rich aragonite), are missing from other time intervals because they are less well preserved. This presents a particularly problematic bias on estimates of body size or biomass through the Phanerozoic (e.g., Payne 2005; Payne and Finnegan 2006; Payne et al. 2009).

Useful approaches for mitigating the bias on body-size include choice of proxies that are independent of the smaller end of the size spectrum (i.e. size maxima; see Payne et al. 2009), although this greatly restricts the degree to which observed patterns can be generalized to whole faunas, and indeed the interpretation of their causes. Future studies should dissect trends in body-size among their taxonomic, geographic, environmental and taphonomic components to ensure correct interpretation of evolutionary and/or paleoenvironmental controls.

Nevertheless, this bias also serves as a reminder that much is lost with respect to taxonomic composition, thereby influencing assessments of ecological structure and biodiversity (Hendy 2009a; Sessa et al. 2009). A commonly adopted strategy to filter the size-frequency distribution of paleontological samples is through the physical sieving of fossil material, although this is only feasible with unconsolidated sediments. While this preparation method is often viewed as degrading neontological as well as paleontological data (Bush et al. 2007), and hence is termed the size-filtering bias, it offers a useful approach to even the playing field between unlithified Mesozoic and Cenozoic strata (where small taxa are easy to collect and identify) and lithified rocks of the Paleozoic (in which small taxa are difficult to extract, poorly preserved, or missing). In other words, it is possible to exclude a size-range that has a geologic history of variable preservation. This approach to standardizing paleontological data need not be restricted to the physical preparation of material; rather it can also be achieved through analytical sieving (see Kowalewski and Hoffmeister 2003; Bush et al. 2007), whereby measurements of sampled specimens are used to filter the dataset to appropriate size-ranges. Proxy data (e.g., type specimen or mean dimensions) can be used to provide the information necessary to include or exclude particular size ranges from paleobiological analyses (e.g., Allmon et al. 1993). With the growth of taxonomic (e.g., Rosenberg 2005) and occurrence-based datasets it is becoming easier to filter body size as a routine step during paleobiological analyses. The Paleobiological Database now has the ability to collect not only semi-quantitative data (i.e. the dominant size category for each taxon) but also the dimensions of individual specimens (i.e. type specimens or population-scale datasets).

Heterogeneity in preservation can also place constraints on the amount of morphological complexity that is available to taxonomists, although the taphonomic nature of studied specimens is seldom reported. Lack of diagnostic characters not only increases the likelihood of misidentification or error in the classification of new taxa, but they can provide misleading information on phylogenetic relationships (Grantham 2004). Gastropods, for example, have an extensive fossil record that stretches back to the Cambrian. However, determining the relationship of early taxa to one another, and especially to extant forms present major challenges owing to the limited number of characters preserved and architectural restrictions apparent among large clades (Frýda et al. 2008). Features that might be of additional use in diagnosing these clades, such as shell mineralogy and protoconch morphology are preserved only infrequently in early fossils. Additionally, the poor preservation of type material can present a barrier to establishing relationships between species and genera within contemporaneous assemblages. Malinky (1990) in a review of Early and Middle Cambrian hyolithids described how the poor preservation of many type specimens rendered their generic identification uncertain and in some cases suggested that their classification to Hyolitha was quite unlikely. In his systematic treatment of the group Malinky (1990), therefore, makes a great effort to describe the preservational state of type and other available specimens. Taphonomic characterization of type material has a role in the future assessments of taxon validity and such data should be routinely included in systematic descriptions using standard terminology. The Paleobiology Database recently added fields to collect similar data where available for type specimens.

5 Conclusions

The quality of the fossil record as an archive of evolutionary and ecological development has almost certainly changed over the Phanerozoic. Much of the variation in the fossil record is due to secular taphonomic biases, including those involving diagenetic alteration of fossils or their host sediments, concentration of skeletal hard-parts, and the exceptional preservation of soft-bodied organisms. These mega-biases not only introduce heterogeneity into the quality of the fossil record, but skew paleontologists' perception of trends in biodiversity and paleoecology.

The lithification of most Paleozoic and Mesozoic fossiliferous sediments, typically by carbonate cements, has had a significant influence on perceptions of within-community diversity and paleoecological composition. Independent investigations of global and regional fossil occurrence datasets reveals that the increasing availability of unlithified sediments in rocks of late Mesozoic through Cenozoic age coincides with a two- to threefold increase in local-scale diversity. Previously such a relationship has been related to the ease of sampling in unlithified sediments; in other words, a sample-size bias. However, this discrepancy remains even after using sampling-standardization on a high-quality set of abundance data in which other taphonomic, environmental, and biogeographic effects have been controlled for. A significant variation in the body-size and mineralogical characteristics of biota from lithified and unlithified sediments suggests that assemblages undergo considerable compositional change during the diagenesis associated with lithification. Compositional differences may also result from preparation difficulties and decreased taxonomic identification.

A specimen-based analysis indicates that lithified and generally poorly-preserved specimens, typical of Paleozoic and early Mesozoic sediments, yield fewer diagnostic characters necessary for precise generic identification than their unlithified and well-preserved Cenozoic counterparts. Not only are fewer potential characters observable, but occurrence-based datasets also indicate that late Mesozoic and Cenozoic rocks are more likely to yield taxa identified to species-resolution, subgenera, and a more even distribution of species among the constituent genera of families. This analysis indicates a largely ignored mechanism by which taphonomic damage could hinder taxonomic practice and estimates of diversity at the global-scale.

Lithification is one of many taphonomic mega-biases that have the potential to affect biodiversity trends. Occurrence- and specimen-based datasets, in which taxonomic data is tied to information pertaining to the environmental, lithologic, and taphonomic context of faunal occurrences, are potentially effective tools in highlighting intervals and particular faunal groups that are susceptible to bias. Silica replacement, which is thought to enhance diversity among groups composed of less stable skeletal composition, appears most frequently among Permian fossil assemblages. Phosphatic replacement, which plays a key role in the preservation of soft-bodied and small-shelly faunas, appears commonly in assemblages of Cambrian age. Konservat-lagerstätte, while providing a rich source of information on the rarely preserved soft-bodied biota, are infrequent in the fossil record, but perhaps are most notable from rocks of Cambrian age. Shell beds are well known as sources of tremendous diversity in Neogene sedimentary basins. Although they are not easily defined these beds appear to show increased frequency in middle Paleozoic and Cenozoic age successions. Fossil molds, unlike previously mentioned biases, suggest lost diversity, and are most frequent in rocks of early Cambrian and early Mesozoic age.

The considerable value of paleontological data in evolutionary analyses provides motivation for thorough understanding of the deficiencies in the fossil record and developing appropriate protocols to gain the maximum quality and quantity of data from the fossil record. Taphonomic biases may not affect paleontological analyses conducted at coarse spatial scale and temporal resolution. However, the non-random nature of the above biases raises concerns regarding the comparison of diversity or ecological complexity over the course of the Phanerozoic or between contemporaneous faunal groups. Furthermore, a number of the biases have tremendous potential to affect community-scale patterns, either degrading (e.g., lithification, aragonite dissolution) or enhancing (e.g., silicification, phosphatization) the relative quality of fossil data. A number of approaches can be undertaken to minimize these biases, including the selective filtering of datasets to remove taphonomically vulnerable groups or the use of taphonomic control taxa that indicate the appropriate preservation state of fossil assemblages.

Future effort is required to better quantify not only the temporal, geographic and taxonomic range of particular taphonomic biases, but also the ways in which they alter diversity at a range of spatial scales (i.e. local through global) and the paleoecological properties of individual assemblages. In the light of recent results indicating that lithification has the potential to alter our perception of faunal recovery to a mass-extinction event, other recognized extinction and faunal turnover events should also be revisited to establish their veracity. The quantitative assessment presented here has in part relied on large-scale occurrence-based datasets. In doing so, their potential in taphonomic investigation has been highlighted, but also their shortcomings (e.g., data quality, objective definitions for taphonomic conditions) are apparent. Additionally, it is clear that large-scale assessments will struggle to establish the true effect of taphonomic biases and their causal mechanisms without small-scale studies in which the potential overprint of environmental and geographic variation is controlled.

Acknowledgments The following institutions and staff are acknowledged for permitting and facilitating access to museum collections used in this study for analysis of lithification and diagenesis on specimens: American Museum of Natural History (Bushra Hussaini), Cincinnati Museum Center (Brenda Hanke), Paleontological Research Institution (Warren Allmon, Greg Dietel, Ursula Smith), and the Yale Peabody Museum of Natural History (Susan Butts, Cope MacClintock). Jessica Bazeley and Penny Benson are thanked for photography and digital archiving of fossil collections used in this study. Initial research was undertaken while AH was supported by a University Dean Distinguished Dissertation Award from the University of Cincinnati. A Gaylord Donnelley Environmental Fellowship from the Yale Institute of Biospheric Studies allowed final completion of this investigation and funded museum visits. Carlton Brett is thanked for reading an earlier version of this manuscript, while Arnie Miller, Devin Buick, Katherine Bulinski and Chad Fergusson, are acknowledged for helpful discussions during its infancy. Lastly, I appreciate the support of John Alroy in maintaining the Paleobiology Database. This is Paleobiology Database publication number 99.

6 Appendix

Specimens illustrated in Fig. 19; read from left to right (lithified to unlithified) in figure (a) Nuculanoidea: *Phestia* sp. (x0.5), YPM 507054, Mississippian, Indiana; *Phestia bellistriata* (x0.5), YPM 507043, Pennsylvanian, Oklahoma; *Hilgardia brogniarti* (x.05), YPM 325744A, Eocene, Alabama. (b) Nuculidae: *Nuculoidea corbuliformis* (x0.5), YPM 507052–507053, Devonian, New York; *Nucula ovata* (x0.5), YPM 383450, Early Cretaceous, England; *Nucula percrassa* (x0.3), YPM 507066, Late Cretaceous, Tennessee. (c) Anomiidae: "*Anomia*" sp. (x0.8), YPM 507055, Late Cretaceous, Mississippi; *Anomia anomialis* (x0.8), YPM 507039, Eocene, England; *Anomia argentaria* (x0.8), YPM 507014–507015, Late Cretaceous, England. (d) Limidae: *Lima dichotoma* (x0.3), YPM 507051, Late Cretaceous, Czechoslovakia; *Limatula gibbosa* (x0.8), YPM 507028, Jurassic, France; *Ctenoides spatula* (x0.8), YPM 507012–507013, Eocene, France. (e) Mytiloidea: *Phthonia nodicostata* (x0.5), YPM 500843, Devonian, New York; *Promytilus swallovi* (x0.5), YPM 507056, Pennsylvanian, Kansas; *Modiolus modiolus* (x0.3), YPM 309185, Pleistocene, Scotland. (f) Astartiidae: *Neocrassina elegans* (x0.3), YPM 507031, Jurassic, France; *Astarte incrassata* (x.0.5), YPM 507023, Pliocene, England; *Iocrassina omalii* (x0.8), YPM 507065, Pliocene, England. (g) Lucinoidea: Lucinidae indet. (x0.8), YPM 507048, Late Jurassic, England; *Eophysema ozarkana* (x0.5), YPM 6937, Eocene, Alabama; *Epilucina concentrica* (x0.5), YPM 507036, Eocene, France. (h) Crassatelloidea: *Cypricardella bellastriata* (x1.0), YPM 507050, Devonian, New York; *Bathytormus* sp. (x1.2), YPM 507034, Oligocene, Mississippi; *Hybolophus speciosa* (x0.5), YPM 507030, Pliocene, North Carolina.

References

Aller, R. C. (1982). Carbonate dissolution in nearshore terriginous muds: The role of physical and biological reworking. *Journal of Geology, 90*, 79–95.

Allison, P. A. (1988a). Konservat-Lagerstätten: Cause and classification. *Paleobiology, 14*, 331–344.

Allison, P. A. (1988b). Phosphatized soft-bodied squids from the Jurassic Oxford Clay. *Lethaia, 21*, 403–410.

Allison, P. A., & Briggs, D. E. G. (1991). The taphonomy of soft-bodied animals. In S. K. Donovan (Ed.), *The processes of fossilization*. Columbia University Press: New York.

Allison, P. A., & Briggs, D. E. G. (1993). Exceptional fossil record: Distribution of soft-tissue preservation through the Phanerozoic. *Geology, 21*, 527–530.

Allmon, W. D., Rosenberg, G., Portell, R. W., & Schindler, K. S. (1993). Diversity of Atlantic coastal plain mollusks since the Pliocene. *Science, 260*, 1626–1629.

Alroy, J., Marshall, C. R., Bambach, R. K., Bezusko, K., Foote, M., Fürsich, F. T., et al. (2001). Effects of sampling standardization on estimates of Phanerozoic marine diversification. *Proceedings of the National Academic Sciences of the United States of America, 98*, 6261–6266.

Alroy, J., Aberhan, M., Bottjer, D. K., Foote, M., Fürsich, F. T., Harries, P. J., et al. (2008). Phanerozoic trends in the global diversity of marine invertebrates. *Science, 321*, 97–100.

Bambach, R. K. (1977). Species richness in marine benthic habitats through the Phanerozoic. *Paleobiology, 3*, 152–167.

Behrensmeyer, A. K., & Kidwell, S. M. (1985). Taphonomy's contribution to paleobiology. *Paleobiology, 11*, 105–119.

Behrensmeyer, A. K., Kidwell, S. M., Gastaldo, R. A. (2000). Taphonomy and paleobiology. In D. H. Erwin, S. L. Wing (Eds.), *Deep time: Paleobiology's perspective. Paleobiology 26*(Suppl. 4), 103–147.

Behrensmeyer, A. K., Fürsich, F. T., Gastaldo, R. A., Kidwell, S. M., Kosnik, M. A., Kowalewski, M., et al. (2005). Are the most durable shelly taxa also the most common in the marine fossil record? *Paleobiology, 31*, 607–623.

Bengston, S., Conway Morris, S., Cooper, B. J., Jell, P. A., & Runnegar, B. N. (1990). Early Cambrian fossils from South Australia. *Member Association of Australian Paleontology, 9*, 1–364.

Benton, M. J. (1995). Diversification and extinction in the history of life. *Science, 268*, 52–58.

Beu, A. G., & Maxwell, P. A. (1990). Cenozoic Mollusca of New Zealand. *New Zealand Geological Survey Paleontological Bulletin, 58*, 1–518.

Beu, A. G., Henderson, R. A., & Nelson, C. S. (1972). Notes on the taphonomy and paleoecology of New Zealand Tertiary Spatangoida. *New Zealand Journal of Geology and Geophysics, 15*, 275–286.

Bottjer, D. J., Clapham, M. E., Fraiser, M. L., & Powers, C. M. (2008). Understanding mechanisms for the end-Permian mass extinction and the protracted Early Triassic aftermath/recovery. *GSA Today, 18*, 4–10.

Brasier, M. D. (1990). Phosphogenic events and skeletal preservation across the Precambrian-Cambrian boundary interval. In A. J. G. Notholt & I. Jarvis (Eds.), *Phosphorite Research and Development. Geological Society of London, Special Publication 52* (pp. 289–303).

Briggs, D. E. G., & Crowther, P. R. (2001). *Palaeobiology: A synthesis*. Oxford: Blackwell.

Briggs, D. E. G., & Rolfe, W. D. I. (1983). New Concavocarida (new order? Crustacea) from the Upper Devonian of Gogo, Western Australia, and the paleoecology and affinities of the group. *Special Paper Paleontology, 30*, 249–276.

Brunton, F. R., & Dixon, O. A. (1994). Siliceous sponge-microbe biotic associations and their recurrence through the Phanerozoic as reef mound constructors. *Palaios, 9*, 370–387.

Bush, A. M., & Bambach, R. K. (2004). Did alpha diversity increase through the Phanerozoic? Lifting the veils of taphonomic, latitudinal, and environmental biases. *Journal of Geology, 112*, 625–642.

Bush, A. M., Kowalewski, M., Hoffmeister, A. P., Bambach, R. K., & Daley, G. W. (2007). Potential paleoecological biases from size-filtering of fossils: Strategies for sieving. *Palaios, 22*, 612–622.

Bush, A.M., Markey, M. J., & Marshall, C. R. (2004). Removing bias from diversity curves: the effects of spatially organized biodiversity on sampling-standardization. *Paleobiology, 30*, 666–686.

Butterfield, N. (1990). Organic preservation of non-mineralizing organisms and the taphonomy of the Burgess Shale. *Paleobiology, 16*, 272–286.

Butts, S. H., & Briggs, D.E.G. (this issue). Silicification through time. In P. A. Allison, D. J. Bottjer (Eds.), *Taphonomy: Process and bias through time.* Berlin: Springer.

Carson, G. A. (1991). Silicification of fossils. In P. A. Allison & D. E. G. Briggs (Eds.), *Taphonomy: Releasing the data locked in the fossil record.* New York: Plenum.

Chave, K. E., Deffeys, K. S., Weyl, P. K., Garrels, R. M., & Thompson, M. E. (1962). Observations on the solubility of skeletal carbonates in aqueous solutions. *Science, 137*, 33–34.

Cherns, L., & Wright, V.P. (2009). Quantifying the impacts of early diagenetic aragonite dissolution on the fossil record. *Palaios, 24,* 756–771.

Cherns, L., & Wright, V. P. (2000). Missing molluscs as evidence of large-scale, early skeletal aragonite dissolution in a Silurian sea. *Geology, 28*, 791–794.

Cherns, L., Wheeley, J. R., & Wright, V. P. (2008). Taphonomic windows and molluscan preservation. *Palaeogeography, Palaeoclimatology, Palaeoecology, 270*, 220–229.

Coan, E. V., Scott, P. V., & Bernard, F. R. (2000). *Bivalve seashells of western North America.* Santa Barbara, CA: Santa Barbara Museum of Natural History.

Cook, P. J. (1992). Phosphogenesis around the Proterozoic-Phanerozoic transition. *Journal of the Geological Society of London, 149*, 615–620.

Cook, P. J., & McElphinny, M. W. (1979). A re-evaluation of the spatial and temporal distribution of sedimentary phosphate deposits in the light of plate tectonics. *Economic Geology, 74*, 315–330.

Cooper, G. A., & Grant, R. E. (1972). Permian brachiopods of West Texas, I. *Smithson Contributions to Paleobiology, 14*, 1–231.

Cooper, R. A., Beu, A. G., Crampton, J. S., Jones, C. M., Marshall, B., & Maxwell, P. A. (2006). Completeness of the fossil record: Estimating losses due to small body size. *Geology, 34*, 241–244.

Crampton, J. S., Beu, A. G., Cooper, R. A., Foote, M., Jones, C. M., Marshall, B., et al. (2006). The ark was full! Constant to declining Cenozoic shallow marine biodiversity on an isolated mid-latitude continent. *Paleobiology, 32*, 509–532.

Davies, D. J., Powell, E. N., & Stanton, R. J., Jr. (1989). Relative rates of shell dissolution and net sediment accumulation – a commentary: Can shell beds from by the gradual accumulation of biogenic debris on the sea floor? *Lethaia, 22*, 207–212.

Donovan, D. T. (1983). *Mastigophora* Owen 1856, a little known genus of Jurassic coleoids. *Neues Jahrbuch für Geologie und Paläontologie Abhandlungen, 165*, 484–495.

Donovan, S. K., & Paul, C. R. C. (1998). *The adequacy of the fossil record.* New York: Wiley.

Dornbos, S. Q. (this issue). Phosphatization through the Phanerozoic. In P. A. Allison & D. J. Bottjer (Eds.), *Taphonomy: Process and bias through time*. Berlin: Springer.

Dzik, J. (1991). Is fossil evidence consistent with traditional views of the early metazoan phylogeny? In A. M. Simonetta & S. Conway Morris (Eds.), *The early evolution of Metazoa and the significance of problematic taxa.* Cambridge: Cambridge University Press.

Dzik, J. (1994). Evolution of 'small shelly fossils' assemblages of the Early Paleozoic. *Acta Paleontologica Polonica, 39*, 247–313.

Finks, R. M. (1960). Late Paleozoic sponge faunas of the Texas region. *American Museum of Natural History Bulletin, 120*, 1–160.

Finks, R. M. (1970). The evolution and ecologic history of sponges during Palaeozoic times. *Zoological Society of London, Symposium, 25*, 3–22.

Flessa, K. W., & Brown, T. J. (1983). Selective solution of macroinvertebrate calcareous hard parts – a laboratory study. *Lethaia, 16*, 193–205.

Foote, M. (2006). Substrate affinity and diversity dynamics of Paleozoic marine animals. *Paleobiology, 32*, 345–366.

Foote, M., & Sepkoski, J. J., Jr. (1999). Absolute measures of the completeness of the fossil record. *Nature, 398*, 415–417.

Forey, P. L., Fortey, R. A., Kenrick, P., & Smith, P. (2004). Taxonomy and fossils: A critical appraisal. *Philosophical Transactions of the Royal Society of London, B, 359*, 639–653.

Frýda, J., Nützel, A., & Wagner, P. J. (2008). Paleozoic Gastropoda. In W. F. Ponder & D. R. Lindberg (Eds.), *Phylogeny and evolution of the Mollusca.* Berkeley: University of California Press.

Glover, C. P., & Kidwell, S. M. (1993). Influence of organic matrix on the post-mortem destruction of molluscan shells. *Journal of Geology, 101*, 729–747.

Grantham, T. (2004). The role of fossils in phylogeny reconstruction: Why is it so difficult to integrate paleobiological and neontological evolutionary biology? *Biology and Philosophy, 19*, 687–720.

Greenstein, B. J. (1992). Taphonomic bias and the evolutionary history of the family Cidaridae (Echinodermata: Echinoidea). *Paleobiology, 18*, 50–79.

Gupta, N. S., Michels, R., Briggs, D. E. G., Evershed, R. P., & Pancost, R. D. (2006). The organic preservation of fossil arthropods: an experimental study. *Proceedings of Royal Society, London B, 273*, 2777–2783.

Hallam, A. (1984). Pre-quaternary sea-level changes. *Annual Review of Earth and Planetary Sciences, 12*, 205–243.

Hansen, T. A., Farrell, B. R., & Upshaw, B., III. (1993). The first 2 million years after the Cretaceous-Tertiary boundary in east Texas: Rate and paleoecology of the molluscan recovery. *Paleobiology, 19*, 251–265.

Hansen, T. A., Upshaw, B., III, Kauffman, E. G., & Gose, W. (1993). Patterns of molluscan extinction and recovery across the Cretaceous-Tertiary boundary in east Texas; report on new outcrops. *Cretaceous Research, 14*, 685–706.

Hein, J. R., & Parrish, J. T. (1987). Distribution of siliceous deposits in space and time. In J. R. Hein (Ed.), *Siliceous sedimentary rock-hosted ores and petroleum*. New York: Van Nostrand Reinhold.

Hendy, A. J. W. (2007). Cenozoic molluscan biodiversity: An examination of biodiversity change at global, regional, and local spatial scales. Unpublished PhD thesis, University of Cincinnati, USA.

Hendy, A. J. W. (2009a). Lithification and the measurement of biodiversity. *Paleobiology, 35*, 51–62.

Hendy, A. J. W. (2009b). Taphonomic overprints on Phanerozoic trends in body-size and morphology. North American Paleontological Conference Abstracts and Program.

Hendy, A. J. W., & Kamp, P. J. J. (2004). Late Miocene-Early Pliocene biofacies of Wanganui and Taranaki Basins, New Zealand. *New Zealand Journal of Geology and Geophysics, 47*, 769–785.

Hendy, A. J. W., & Kamp, P. J. J. (2007). Paleoecology of Late Miocene-Early Pliocene sixth-order glacioeustatic sequences in the Manutahi-1 core, Wanganui-Taranaki Basin, New Zealand. *Palaios, 22*, 325–337.

Hendy, A. J. W., Kamp, P. J. J., & Vonk, A. J. (2006). Cool-water shell bed taphofacies from Miocene-Pliocene shelf sequences: Utility in sequence stratigraphic analyses. *Geological Society of London, Special Publication, 255*, 285–307.

Hof, C. H. J., & Briggs, D. E. G. (1997). Decay and mineralization of mantis shrimps (Stomatopoda: Crustacea) – A key to their fossil record. *Palaios, 12*, 420–438.

Holland, H. D. (1984). *The chemical evolution of the atmosphere and oceans*. Princeton, NJ: Princeton University Press.

Jablonski, D., & Sepkoski, J. J., Jr. (1996). Paleobiology, community ecology, and scales of ecological pattern. *Ecology, 77*, 1367–1378.

Jablonski, D., Lidgard, S., & Taylor, P. D. (1997). Comparative ecology of bryozoan radiations: Origin of novelties in cyclostomes and cheilostomes. *Palaios, 12*, 505–523.

Kidwell, S. M. (1990). Phanerozoic evolution of macroinvertebrate shell accumulations: Preliminary data from the Jurassic of Britain. In W. Miller III (Ed.), *Paleocommunity temporal dynamics: The long-term development of multispecies assemblies. Paleontological Society, Special Publication 5* (pp. 309–327)

Kidwell, S. M. (2005). Shell composition has no net impact on large-scale evolutionary patterns in mollusks. *Science, 307*, 914–917.

Kidwell, S. M. (this issue). Taphonomy of shell beds through time. In P. A. Allison & D. J. Bottjer (Eds.), *Taphonomy: Process and bias through time*. Berlin: Springer.

Kidwell, S. M., & Brenchley, P. J. (1994). Patterns in bioclastic accumulation through the Phanerozoic: Changes in input or in destruction. *Geology, 22*, 1139–1143.

Kidwell, S. M., & Brenchley, P. J. (1996). Evolution of the fossil record: Thickness trends in marine skeletal accumulations and their implications. In D. Jablonski, D. H. Erwin, & J. H. Lipps (Eds.), *Evolutionary paleobiology*. Chicago: University of Chicago Press.

Kidwell, S. M., & Flessa, F. W. (1995). The quality of the fossil record: Populations, species, and communities. *Annual Review of Ecology and Systematics, 24*, 433–464.

Kidwell, S. M., & Holland, S. M. (2002). The quality of the fossil record: Implications for evolutionary analyses. *Annual Review of Ecology and Systematics, 33*, 561–588.

Kiessling, W., & Aberhan, M. (2007). Environmental determinants of marine benthic biodiversity dynamics through Triassic-Jurassic times. *Paleobiology, 33*, 414–434.

King, P. R., Naish, T. R., Browne, G. H., Field, B. D., & Edbrooke, S. W. (1999). Cretaceous to recent sedimentary patterns in New Zealand. *Institute of Geological and Nuclear Science folio series 1*, Lower Hutt, New Zealand.

Koch, C. F., & Sohl, N. F. (1983). Preservational effects in paleoecological studies: Cretaceous mollusc examples. *Paleobiology, 9*, 26–34.

Kowalewski, M. K., & Hoffmeister, A. P. (2003). Sieves and fossils: Effects of mesh size on paleontological measures. *Palaios, 16*, 566–579.

Kowalewski, M., Kiessling, W., Aberhan, M., Fürsich, F. T., Scarponi, D., Barbour Wood, S. L., et al. (2006). Ecological, taxonomic, and taphonomic components of the post-Paleozoic increase in sample-level species diversity of marine benthos. *Paleobiology, 32*, 533–561.

Lucas, J., & Prévôt, L. E. (1991). Phosphates and fossil preservation. In P. A. Allison & D. E. G. Briggs (Eds.), *Taphonomy: Releasing the data locked in the fossil record*. New York: Plenum.

Malinky, J. M. (1990). Early and Middle Cambrian Hyolitha (Mollusca) from Northweastern China. *Journal of Paleontology, 64*, 228–240.

Markwick, P. J., & Lupia, R. (2002). Palaeontological databases for palaeobiography, palaeoecology and diversity: A question of scale. *Geological Society of London. Special Publication, 194*, 169–178.

Martin, D., Briggs, D. E. G., & Parkes, R. J. (2002). Experimental mineralization of invertebrate eggs and the preservation of Neoproterozoic embryos. *Geology, 31*, 39–42.

McGowran, B., Holdgate, G. R., Li, Q., & Gallagher, S. J. (2004). Cenozoic stratigraphic succession in southeastern Australia. *Australian Earth Science, 51*, 459–496.

McKinney, M. L. (1991). Completeness of the fossil record: An overview. In S. K. Donovan (Ed.), *The processes of fossilization*. Columbia University Press: New York.

Mikkelson, P. M., & Bieler, R. (2008). *Seashells of Southern Florida: Living marine mollusks of the Florida Keys and adjacent regions – Bivalvia*. Princeton, NJ: Princeton University Press.

Miller, A. I. (1997). Dissecting global diversity patterns: Examples from the Ordovician radiation. *Annual Review of Ecology and Systematics, 28*, 85–104.

Miller, A. I. (2000). Conversations about Phanerozoic global diversity. In D. H. Erwin & S. L. Wing (Eds.), *Deep time: paleoecology's perspective. Paleobiology 26*(Suppl. 4), 53–73.

Moore, R. C. (1969). Treatise on invertebrate paleontology. Part N. Mollusca 6, Bivalvia. Geological Society of America and the University of Kansas Press, 1 and 2, 952 pp. and 953–1224 pp.

Müller, K. J. (1985). Exceptional preservation in calcareous nodules. *Philosophical Transactions of the Royal Society of London, B, 311*, 1–18.

Mus, M. M., Palacios, T., & Jensen, S. (2008). Size of the earliest molluscs: Did small helcionellids grow to become large adults. *Geology, 36*, 175–178.

Nichol, D. (1977). The number of living animal species likely to be fossilized. *Florida Scientist, 40*, 135–139.

Parsons-Hubbard, K. M., Callender, W. R., Powell, E. N., Brett, C. E., Walker, S. E., Raymond, A. L., et al. (1999). Rates of burial and disturbance of experimentally-deployed Molluscs: Implications for preservation potential. *Palaios, 14*, 337–351.

Paul, C. R. C. (1998). An overview of the completeness of the fossil record. In S. K. Donovan & C. R. C. Paul (Eds.), *The adequacy of the fossil record*. New York: Wiley.

Payne, J. L. (2005). Evolutionary dynamics of gastropod size across the end-Permian extinction and through the Triassic recovery interval. *Paleobiology, 31*, 269–290.

Payne, J. L., & Finnegan, S. (2006). Controls on animal biomass through geological time. *Geobiology, 4*, 1–10.

Payne, J. L., Boyer, A. G., Brown, J. H., Finnegan, S., Kowalewski, M., Krause, R. A., Jr., et al. (2009). Two-phase increase in the maximum size of life over 3.5 billion years reflects biological innovation and environmental opportunity. *Proceedings of the National Academy of Sciences of the United States of America, 106*, 24–27.

Pinna, G. (1985). Exceptional preservation in the Jurassic of Osteno. *Philosophical Transactions of the Royal Society of London, B, 311*, 170–180.

Plotnick, R. E., & Wagner, P. J. (2006). Round up the usual suspects: common genera in the fossil record and the nature of wastebasket taxa. *Paleobiology, 32*, 126–146.

Porter, S. M. (2004). Closing the phosphatization window: Testing for the influence of taphonomic megabias on the pattern of small shelly fossil decline. *Palaios, 19*, 178–183.

Powell, M. G., & Kowalewski, M. (2002). Increase in evenness and sampled alpha diversity through the Phanerozoic: Comparison of early Paleozoic and Cenozoic marine fossil assemblages. *Geology, 30*, 331–334.

Powell, E. N., Parsons-Hubbard, K. M., Callender, W. R., Staff, G. M., Rowe, G. T., Brett, C. E., et al. (2002). Taphonomy on the continental shelf and slope: Two-year trends – Gulf of Mexico and Bahamas. *Palaeogeography, Palaeoclimatology, Palaeoecology, 184*, 1–35.

Prévôt, L., & Lucas, J. (1986). Microstructure of apatite replacing carbonate in synthesized and natural samples. *Journal of Sedimentary Petrology, 56*, 153–159.

Prévôt, L., & Lucas, J. (1990). Phosphate. In D. E. G. Briggs & P. R. Crowther (Eds.), *Palaeobiology: A synthesis*. Oxford: Blackwell.

Raup, D. M. (1976). Species diversity in the Phanerozoic: An interpretation. *Paleobiology, 2*, 289–297.

Raup, D. M. (1991). The future of analytical paleontology. In N. L. Gilinsky & P. W. Signor (Eds.), *Analytical paleontology. Short courses in paleontology 4*, The Paleontol Society.

Rofthus, T. A. (2002). Relationship between taphonomic damage and taxonomic identifiability: Implications for paleoecologic analysis. *Geological Society of American Abstract Programs, 34*, 36.

Rofthus, T. A. (2005). Taphonomic damage and taxonomic identifiability: Implications for secular bias in the fossil record. *Geological Society of American Abstract Programs, 37*, 118.

Rosenberg, G. (2005). Malacolog 4.1.0: A database of western Atlantic marine Mollusca. http://www.malacolog.org/. Accessed 19 May 2009.

Runnegar, B., & Pojeta, J. (1985). Origin and diversification of the Mollusca. In E. R. Trueman & M. R. Clarke (Eds.), *The Mollusca, volume 10, evolution*. New York: Academic.

Sandberg, P. A. (1983). An oscillating trend in Phanerozoic non-skeletal carbonate mineralogy. *Nature, 205*, 19–22.

Schopf, T. (1978). Fossilization potential of an intertidal fauna: Friday Harbor, Washington. *Paleobiology, 4*, 261–270.

Schubert, J. K., Kidder, D. L., & Erwin, D. H. (1997). Silica-replaced fossils through the Phanerozoic. *Geology, 25*, 1031–1034.

Seilacher, A., Reif, W.-E., Westphal, F. (1985). Sedimentological, ecological and temporal patterns pf fossil Lagerstätten. In H. B. Whittington & S. Conway-Morris (Eds.), *Extraordinary fossil biotas: Their ecological and evolutionary significance. Philosophical Transactions of Royal Society of London, B 311* (pp. 5–23)

Sepkoski, J. J., Jr. (1982). *A compendium of fossil marine families*. Milwaukee, WI: Milwaukee Publication Museum.

Sepkoski, J. J., Jr. (1997). Biodiversity: Past, present, and future. *Journal of Paleontology, 71*, 533–539.

Sepkoski, J. J., Jr. (2002). A compendium of fossil marine animal genera. *Bulletins of American Paleontology, 363*, 1–563.

Sessa, J. A., Patzkowsky, M. E., & Bralower, T. J. (2009). The impact of lithification on the diversity, size distribution, and recovery dynamics of marine invertebrate assemblages. *Geology, 37*, 115–118.

Simões, M. G., Kowalewski, K., de Freutas, T. F., Ghilardi, R. P., & de Mello, L. H. C. (2000). Early onset of modern-style shell beds in the Permian sequences of the Paraná Basin:

Implications for the Phanerozoic trend in bioclastic accumulations. *Revista Brasileira de Geociencias, 30*, 496–499.

Smith, A. B. (1990). Echinoid evolutions from the Triassic to Lower Liassic. *Cahiérs de l'Institut Catholique de Luon, Séries Sci, 3*, 79–115.

Smith, A. B. (2001). Large-scale heterogeneity in the fossil record: Implications for Phanerozoic biodiversity studies. *Philosophical Transactions of Royal Society of London, B, 356*, 351–367.

Smith, A. B. (2003). Getting the measure of diversity. *Paleobiology, 29*, 34–36.

Smith, A. M., & Nelson, C. S. (2003). Effects of early sea-floor processes on the taphonomy of temperate shelf skeletal carbonate deposits. *Earth Science Reviews, 63*, 1–31.

Sohl, N. F. (1960). Archeogastropoda, Mesogastropoda and stratigraphy of the Ripley, Owl Creek, and Prairie Bluff Formations. *United States Geological Survey. Professional Paper, 331A*, 1–152.

Sohl, N. F. (1964a). Gastropods from the coffee sand (Upper Cretaceous) of Mississippi. *United States Geological Survey. Professional Paper, 331C*, 345–394.

Sohl, N. F. (1964b). Neogastropoda, Opisthobranchia, and Basommatophora from the Ripley, Owl Creek, and Prairie Bluff Formations. *United States Geological Survey. Professional Paper, 331B*, 153–344.

Sohl, N. F., & Koch, C. F. (1983). Upper Cretaceous (Maestrichtian) Mollusca from the *Haustator bilira* Assemblage Zone in the East Gulf coastal Plain. *U.S. Geological Survey Open File Report, 83–451*, 1–239.

Sohl, N. F., & Koch, C. F. (1984). Upper Cretaceous (Maestrichtian) larger invertebrate Fossils from the *Haustator bilira* Assemblage Zone in the west gulf coastal Plain. *U.S. Geological Survey Open File Report, 84–687*, 1–282.

Sohl, N. F., & Koch, C. F. (1987). Upper Cretaceous (Maestrichtian) larger invertebrate from the *Haustator bilira* Assemblage Zone in the Atlantic Coastal Plain with further data for the East Gulf. *U.S. Geological Survey Open File Report, 87–194*, 1–172.

Stanley, S. M. (1968). Post-Paleozoic adaptive radiation of infaunal bivalve molluscs: A consequence of mantle fusion and siphon formation. *Journal of Paleontology, 42*, 214–229.

Sundukov, V. M., & Fedorov, A. B. (1986). Palaeontology and age of the beds containing algal-archaeocyathan bioherms of the Medvezhya-River. In I. T. Zhuravleva (Ed.), *Biostratigrafi ya i paleontologiya kembriya severnoj Azii*, Trudy Akademiya Nauk SSSR Sibirskoe Otdelenie.

Taylor, J. D., Kennedy, W. J., & Hall, A. (1969). The shell structure and mineralogy of the Bivalvia, Introduction, Nuculacea-Trigonacea. *Bulletin of British Museum (Natural History) Zoology Supplement, 3*, 1–125.

Taylor, J. D., Kennedy, W. J., & Hall, A. (1973). The shell structure and mineralogy of the Bivalvia II, Lucinacea-Clavagellacea, conclusions. *Bulletin of British Museum (Natural History) Zoology, 22*, 255–294.

Thayer, C. W. (1983). Sediment-mediated biological disturbance and the evolution of marine benthos. In M. J. S. Tevesz & P. L. McCall (Eds.), *Biotic interactions in recent and fossil benthic communities*. New York: Plenum.

Toulmin, D. (1977). Stratigraphic distribution of Paleocene and Eocene fossils in the eastern Gulf coast region. *Geological Survey of Alabama, Monograph, 13*, 1–602.

Valentine, J. W. (1989). How good was the fossil record? Clues from the Californian Pleistocene. *Paleobiology, 15*, 83–94.

Valentine, J. W., Jablonski, D., Kidwell, S., & Roy, K. (2006). Assessing the fidelity of the fossil record by using marine bivalves. *Proceedings of the National Academy of Sciences of the United States of America, 103*, 659–6604.

Vermeij, G. J. (1987). *Evolution and escalation: An ecological history of life*. Princeton, NJ: Princeton University Press.

Wagner, P. J., Aberhan, M., Hendy, A. J. W., & Kiessling, W. (2007). The effects of taxonomic standardization on sampling-standardized estimates of historical diversity. *Proceedings of the Royal Society B: Biological Sciences, 274*, 439–444.

Walker, S. E., Parsons-Hubbard, K., Powell, E. N., & Brett, C. E. (1998). Bioerosion or bioaccumulation? Shelf-slope trends for epi-and endobionts on experimentally deployed gastropod shells. *Historical Biology, 13*, 61–72.

Wall, P. D., Ivany, L. C., & Wilkinson, B. H. (2009). Revisiting Raup: Exploring the influence of outcrop area on diversity in light of modern sample-standardization techniques. *Paleobiology, 35*, 146–167.

Wilby, P. R., & Briggs, D. E. G. (1997). Taxonomic trends in the resolution of detail preserved in fossil phosphatized soft tissues. *Geobios, 20*, 493–502.

Wilby, P. R., Briggs, D. E. G., & Riou, B. (1996). Mineralization of soft-bodied invertebrates in a Jurassic metalliferous deposit. *Geology, 24*, 847–850.

Wilkinson, B. H., & Given, R. K. (1986). Secular variation in abiotic marine carbonates: Constraints on Phanerozoic atmospheric carbon dioxide contents and oceanic Mg/Ca ratios. *Journal of Geology, 94*, 321–333.

Wingard, G. L., & Sohl, N. F. (1990). Revision of the Nucula percrassa Conrad, 1858 group in the Upper Cretaceous of the Gulf and Mid-Atlantic Coastal Plains: An example of bias in the nomenclature. *U.S. Geological Survey of Bulletin, 1881*, D1–D25.

Wright, V. P., Cherns, L., & Hodges, P. (2003). Missing molluscs: Field testing taphonomic loss in the Mesozoic through early large-scale aragonite dissolution. *Geology, 31*, 211–214.

Zhu, M., Babcock, L. E., & Steiner, M. (2005). Fossilization modes in the Chengjiang Lagerstätte (Cambrian of China): Testing the roles of organic preservation and diagenetic alteration in exceptional preservation. *Palaeogeography, Palaeoclimatology, Palaeoecology, 220*, 31–46.

Chapter 3
Taphonomic Bias in Shelly Faunas Through Time: Early Aragonitic Dissolution and Its Implications for the Fossil Record

Lesley Cherns, James R. Wheeley, and V. Paul Wright

Contents

Abstract Early diagenetic dissolution of skeletal carbonate in environments from seafloor to shallow burial has the potential to skew the marine fossil record of aragonitic shells, particularly molluscs. Taphonomic windows leading to the preservation of labile skeletal components include relatively rare occurrences of early mineral replacement by silica (skeletal lagerstätten). Another, much more frequent process is event deposition where dissolution is halted by rapid burial of shells. Shell plasters form in basinal mud or low energy lagoonal environments during temporary dysoxic episodes, such as are caused by algal blooms. Preservation potential for

L. Cherns (✉)
School of Earth and Ocean Sciences, Cardiff University, Park Place,
Cardiff CF10 3YE, UK
e-mail: cherns@cardiff.ac.uk

J.R. Wheeley
School of Geography, Earth and Environmental Sciences, University of Birmingham,
Edgbaston, Birmingham B15 2TT, UK

V.P. Wright
BG-Group, 100 Thames Valley Park, Reading RG6 1PT, UK

P.A. Allison and D.J. Bottjer (eds.), *Taphonomy: Process and Bias Through Time*,
Topics in Geobiology 32, DOI 10.1007/978-90-481-8643-3_3,

aragonitic fossils may be enhanced by early cementation during shallow burial (hardgrounds) that protects the delicate dissolution moulds from destruction by bioturbation, or in high energy shoal environments where the drive for microbial dissolution is reduced. A data-based environmental model summarizes the main taphonomic zones, and illustrates significant taphonomic bias against aragonitic shells in lower energy settings of platform interiors and mid-outer ramps/shelves. The temporal distribution of various taphonomic windows shows the limited occurrence of silicified faunas, while the nature and extent of shell beds also change, but there is no obvious correlation with periods of 'calcite' and 'aragonite seas'.

1 Introduction

A growing weight of evidence points to dissolution of biogenic carbonate as a significant process in modern carbonate sea-floors in shallow water settings (Morse et al. 1985; Rude and Aller 1991; Walter et al. 1993). This process, which preferentially affects the less stable polymorphs, aragonite and high-Mg calcite, is also significant in ancient shelf environments (e.g. Hendry et al. 1995, 1996; Munnecke and Samtleben 1996; Munnecke et al. 1997; Sanders 2003). There is now increasing recognition of its potential to skew the marine fossil record of aragonite shells, particularly molluscs (Cherns and Wright 2000; Wright et al. 2003; Bush and Bambach 2004; James et al. 2005; Knoerich and Mutti 2006). Yet the mollusc fossil record is good (Harper et al. 1997; Harper 1998), and Kidwell (2005) concluded that taphonomic bias was unimportant in the macroevolutionary record of bivalves. The skeletal lagerstätten represented by early silicified faunas that have been used to quantify the loss of molluscs (Cherns and Wright 2000; Wright et al. 2003) remain relatively rare in the fossil record. Hence, it appears that other processes that increased the preservation potential of aragonitic faunas, such as storms, temporary sea floor anoxia, and synsedimentary cementation, are also important in capturing fossil biodiversity (Cherns et al. 2008).

Molluscs biologically mediate the precipitation of their calcareous shells directly as aragonite and/or calcite layers composed of a variety of crystalline structures (Lowenstam and Weiner 1989; Falini et al. 1996). The earliest mollusc shells in the Cambrian were aragonite (Runnegar and Bentley 1983; Porter 2007). As molluscs evolved and diversified, calcite was incorporated in some groups as the outer shell layer (pteriomorphs), or forming all but a thin hypostracum (oysters). The functional significance of the evolution of calcitic layers or shells for epifaunal habits relates to advantages of lower solubility and density (Carter 1980; Carter et al. 1998). In some bivalves, environmental factors such as seawater temperature affect the secretion of calcite (Carter et al. 1998). There is some evidence that ambient seawater may have exerted control in the evolution of calcitic shelled groups during 'calcite sea' intervals (Harper et al. 1997). However, the wider pattern of mollusc shell mineralogy over time does not appear to correlate with alternations of 'aragonite' and 'calcite seas' (Stanley and Hardie 1998; Stanley 2006).

Shell mineralogy and microstructure tend to be conservative among species, genera and families (Taylor et al. 1969), which makes it possible to extrapolate to fossil taxa even where the shell is poorly or not preserved. Shell microstructures include prismatic calcite (simple/fibrillar) or aragonite (simple/complex), foliated calcite, nacreous aragonite, homogeneous aragonite, crossed-lamellar and complex crossed lamellar aragonite (Carter 1980, 1990). The surface area to volume ratios of crystallites (SAV), and the content of organic matrix in the shell are both important in determining resistance to dissolution through chemical reactivity (Glover and Kidwell 1993; Kidwell 2005; Valentine et al. 2006). Shells with high SAV and high organic content, such as nuculoids, are most vulnerable to early dissolution (Flessa and Brown 1983). Size, shell thickness and ornament are further measures of robustness (e.g. Behrensmeyer et al. 2005). Juveniles whose shells are thin and weakly calcified are especially prone to dissolution (Green et al. 2004).

There is a strong correlation between life mode and shell composition. Infaunal bivalve groups are aragonitic-shelled, while the major epifaunal group, the pteriomorphs, are typically calcitic or bimineralic. Therefore, those compositionally more labile shells, if remaining buried in the sediment during post mortem decay of soft parts, are already in the taphonomically active zone (TAZ) where the potential for dissolution is extensive (see below). This contrasts with a presumption of better preservation potential for infauna because when they die in the sediment they are protected from seafloor processes of destruction and diagenesis (e.g. Kier 1977; Cummins et al. 1986). It is however notable that burrowers typically come up to the surface when stressed (e.g. Oschmann 1988, 1991); in situ preservation of infaunal bivalves commonly represents rapid burial.

2 Environments of Dissolution

Immediately post mortem, in marine oxygenated settings molluscan remains are altered by chemical, physical and biological processes. Scavenging and microbial decay rapidly remove organic tissue, and intraskeletal organics can produce acids that are corrosive to mineralized skeletal elements. Three diagenetic environments are important in dissolving aragonite: the seafloor, the TAZ and shallow sub-TAZ burial (Fig. 1).

2.1 Seafloor Diagenesis

Shells at the seafloor are subject to transport, abrasion, sorting, bioerosion/encrustation, dissolution and cementation. Physical breakdown into constituent components increases susceptibility to attrition, abrasion and dissolution, and increases surface area for boring micro-organisms. Molluscan and other shells are typically destroyed before they can be buried; if this were not the case shell beds would form through general accumulation (Davies et al. 1989). Seafloor dissolution of aragonite is particularly well known from Recent carbonate undersaturated cool-water settings (e.g. Alexandersson 1978, 1979; Freiwald 1995). On the other hand, over

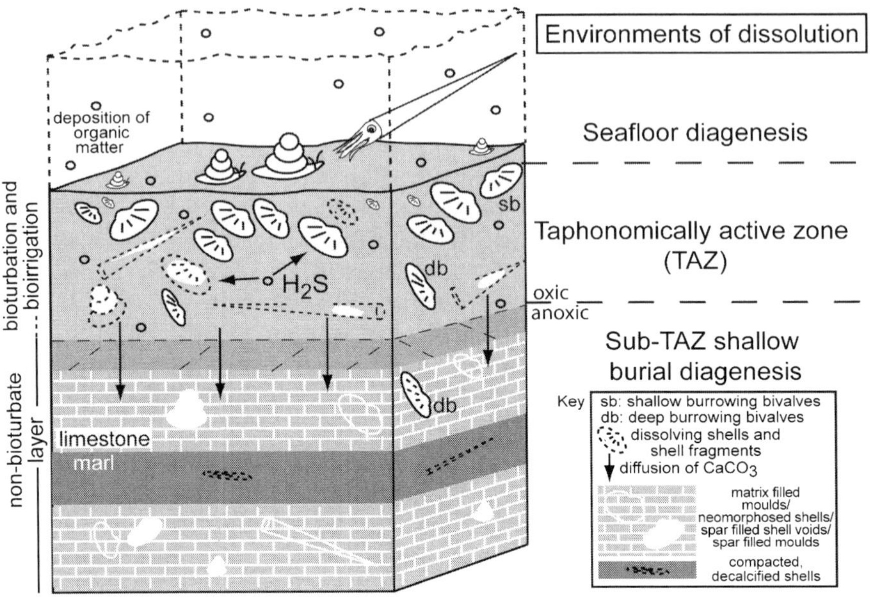

Fig. 1 Environments of dissolution close to a carbonate sea floor. Shells are potentially destroyed through physical processes, bioerosion and by dissolution already on the seafloor. If entering the bioturbated sediment of the TAZ, acidity raised through decay of organic matter and re-oxidation, for example producing H_2S, leads to differential dissolution of aragonite and high-Mg calcite shells, further filtering the shell assemblage. Only minor amounts of aragonitic shell material (e.g. in situ deep burrowing bivalves) survive into the shallow sub-TAZ burial diagenetic environment. Carbonate liberated from early dissolution, at least partly contributed by shells in TAZ and shallow sub-TAZ environments, provides cement through diffusive transfer for limestone beds in LMA successions (Munnecke and Samtleben 1996; Wheeley et al. 2008)

a 13-year time span, experimental studies in the warm seas of the Gulf of Mexico and Bahamas show little corrosion of aragonite shells (Powell et al. 2008).

2.2 *Taphonomically Active Zone (TAZ)*

The TAZ is defined as the upper, bioturbated, mixed zone extending down from the sediment–water interface (Davies et al. 1989). This is equivalent to the 'active layer' of Sanders (2004) where dissolution of aragonite occurs due to acidity resulting from the degradation of organic matter that is exacerbated by burrowing (e.g. Aller 1982; Aller and Aller 1998; Sanders 2001). Most important are dissolution-precipitation reactions that result from changes to pore-water saturation through bacterially-mediated oxidation of organic matter and re-oxidation of reaction by-products such as solid phase sulfides (producing H_2S), as well as the build-up of CO_2 from aerobic oxidation. The TAZ, however, refers to more than the extent and depth of bioturbation. Aragonitic

Table 1 Net reactions of organic matter oxidation with regard to calcium carbonate saturation in pore-waters of pure carbonate sediment. Vertical arrangement of reactions 1–4 reflects prevalent redox reactions down section, or from a burrow wall into the adjacent sediment. In bioturbated/bio-irrigated shallow-water sediments, reactions 1–4 may proceed adjacent to each other in a patchy 'redox-mosaic' (After Sanders 2004)

Reactions of organic matter oxidation	$CaCO_3$ saturation state
1. Aerobic oxidation $CH_2O + O_2 = H_2O + CO_2$	Down
2. Aerobic oxidation of hydrogen sulfide produced by sulfate reduction $HS^- + 2O_2 = SO_4^{2-} + H^+$; $H_2S + 2O_2 = SO_4^{2-} + 2H^+$	Down
3a. Sulfate reduction $2\ CH_2O + SO_4^{2-} = HS^- + HCO_3^- + H_2O + CO_2$	Down-up
3b. Sulfate reduction by methane oxidation (anaerobic methane oxidation) $CH_4 + SO_4^{2-} = H_2S + H_2O + CO_3^{2-}$	Up
4. Methanogenesis (two step process) *4.1.* $2CH_2O + 2H_2O = 2CO_2 + 4H_2$ *4.2.* $4H_2 + CO_2 = CH_4 + 2H_2O$	Constant-up
5. Dissolution of calcium carbonate due to production of carbon dioxide (reactions 1–3a) $CaCO_3 + CO_2 + H_2O = Ca^{2+} + 2HCO_3^-$	The role of carbonic acid
6. Dissolution of calcium carbonate due to production of H+ (reactions 1–3a) $CaCO_3 + H^+ = Ca^{2+} + HCO_3^-$	Control on pH
7. Net equation of stoichiometric coupling of carbonate dissolution to organic matter oxidation (Ku et al. 1999, their equation 7) $2CaCO_3 + 2CH_2O + 2O_2 = 2Ca^{2+} + 4HCO_3^-$	n/a

molluscan bioclasts that survive the burrowed zone pass into the zone of accumulation, or 'non-bioturbate' (i.e., no longer actively bioturbated) layer of Sanders (2004), where anoxic redox processes and pore-water chemistry also influence their taphonomy. Detailed reviews of the geochemical processes relating to syn-depositional to early post-depositional carbonate dissolution are given by Sanders (2003, 2004; also Morse and Mackensie 1990; Canfield and Raiswell 1991a; Martin 1999; see Table 1).

Bioturbation forces the movement of both solid and dissolved components through redox mosaics. In Recent carbonate sediments of the Bahamas and South Florida, the upper 50–350 cm are completely turned over within 100–600 years (Tedesco and Aller 1997). Organic matter in the active layer of the TAZ is typically driven hundreds to thousands of times through a cycle of aerobic oxidation to methanogenesis (Sanders 2004), fuelling dissolution of vulnerable carbonate components (i.e., aragonite and high-Mg calcite). Residence times for carbonate bioclasts can vary greatly depending on the depositional setting. For carbonate settings, components in cool-water regimes typically reside longer in the TAZ because of lower accumulation rates than in tropical settings (Nelson et al. 2003). In rapidly deposited and buried shell beds, vulnerable bioclasts may escape the zone of bioturbation and with close-packing there is increased potential for 'self buffering' (e.g. Kidwell 1991). Burrowing organisms bio-irrigate sediments by actively pumping and channelling water through burrows (Martin 1999); this process has been interpreted for burrow networks as far back as the Ordovician (Cherns et al. 2006). This facilitates oxygenation of sediments leading to aerobic decay of organic matter, and allows back-flux of dissolved calcium carbonate from pore-waters into the overlying seawater, also increasing sulfate supply and interstitial water exchange (Sanders 2004). Both the rate and depth of bioturbation have increased through the Phanerozoic, which may have enhanced the potential for syn-depositional aragonite dissolution. Early Paleozoic (Ordovician) burrowing depths are generally ≤10 cm (Droser and Bottjer 1989), although Orr (2003) reported bioturbation depth of 4 m in the Arkansas Blakely Sandstone (Middle Ordovician). Deep burrowing (≤1m) becomes common only from Late Paleozoic times (e.g. Bottjer and Ausich 1986).

2.3 Shallow Sub-TAZ Burial Diagenesis

This is distinct from both well documented near-surface seafloor diagenesis (e.g. Smith and Nelson 2003), and deeper burial diagenesis characterized by compaction, pressure solution and late cements. It starts with undersaturation of seawater-filled pores with respect to aragonite and high-Mg calcite, and thus overlaps partially with the TAZ, and continues until the supply of aragonite is consumed. Deeper marine pore fluids are likely to stabilize any vulnerable bioclasts or partially dissolved bioclasts that have survived processes nearer to the surface (Melim et al. 2002, 2004).

3 Taphonomic Windows

Processes leading to the preservation of the labile skeletal component of fossil assemblages include early mineral replacement, as in silicified 'skeletal lagerstätten', and event deposition involving rapid burial of shell beds. Other taphonomic windows are produced by dysoxic episodes leading to shell plasters, and by hardgrounds (Cherns et al. 2008). Here we review the temporal distribution of such faunas. Reference is made where appropriate to the taphonomic and preservational model presented in the following section (Section 4).

3.1 'Skeletal Lagerstätten'

'Skeletal lagerstätten' refer to rare, unusually aragonite-rich fossil faunas preserved when early aragonite replacement arrests the dissolutional destruction of these unstable shells. Silicification is one process that can preserve carbonate shells and fine skeletal detail, and silicified faunas feature prominently among systematic studies particularly of Paleozoic fossils (Schulbert et al. 1997). Both the primary skeletal structure, and the relative rates of silica supply and carbonate dissolution, are important in determining the replacement fabric (Carson 1991). Very early replacement is critical to preservation of aragonite shells. Holdaway and Clayton (1982) interpreted pre-lithification replacement of bioclasts and trace fossils by silica as postdating aragonite dissolution and occurring within the sulfate reduction zone at shallow depths of 5–10 m. Aragonitic faunas are preserved (if rarely) in cherts forming in this diagenetic zone; Upper Cretaceous ammonite moulds in flints associated with chalk hardgrounds may indicate a raised sulfate reduction zone in response to reduced bottom-water circulation and oxygenation (Carson 1991). Through the Phanerozoic the source of silica for marine chert and silica-replaced fossils is predominantly biogenic, derived from sponges, radiolarians and diatoms, but with input also from volcanic deposits and terrestrial weathering (Maliva et al. 1989; Kidder and Erwin 2001; Fig. 4a). A decline in shelf silicified fossil assemblages in post-Paleozoic times corresponds to an offshore shift in biogenic silica and bedded chert deposition (Kidder and Erwin 2001; also Schulbert et al. 1997; Fig. 4c).

The case studies referred to below range from Ordovician – Jurassic age, all bar one representing silicified faunas and all indicative of early lithification (Cherns and Wright 2009). For the Silurian example from Gotland, Sweden, a volcanic source of silica is indicated by close association with bentonite horizons (Laufeld and Jeppsson 1976; Cherns and Wright 2000). Upper Ordovician silicified faunas from Kentucky, USA also come from sequences with frequent bentonite horizons (Hoare and Pojeta 2006). The source of silicification in the two other cases is more equivocal. The Lower Devonian case study from SE Australia has regional association with volcanic rocks (Johnston 1993). In the Jurassic example from South Wales, UK (Wright et al. 2003), Mississippi Valley-type metallic mineralization (Fletcher 1988; Bevins and Mason 1997) affecting underlying rocks may indicate hot spring environments, with evidence for very shallow burial mineral replacement coming from silicified burrow systems. The remaining case, from the Lower Carboniferous of the Mendips, UK, is represented by faunas with carbonate preservation (Mitchell 1987), but where very early lithification is indicated by the fine preservation of abundant small and thin shells, including colour banding on some (Batten 1966).

The faunas are summarized briefly here to illustrate the potential skewing of fossil biodiversity. All are characterized by aragonitic, mollusc dominated assemblages, which contrast with those characteristically calcitic, mollusc-poor Paleozoic faunas and the calcitic/bimineralic Mesozoic bivalve faunas that are typical of associated shelf successions (Sepkoski, 1984; Paleozoic cf. Modern Evolutionary Faunas).

Extensive Upper Ordovician silicified faunas from limestones in Kentucky, USA have formed the basis for many systematic studies of fossil groups (>55,000 specimens; Pojeta 1971; review in Hoare and Pojeta 2006; Fig. 4c). From Middle Ordovician times, brachiopods became prevalent among shelly faunas (e.g. Li and Droser 1999; Droser 2002). For example, in the middle Upper Ordovician of Kentucky, brachiopods dominate faunas from offshore shelf to inshore sand shoals (Holland and Patzkowsky 2004). In the Cincinnati Arch region, molluscs mostly comprise <10% in faunas dominated by brachiopods and bryozoans (Novack-Gottshall and Miller 2003). However, among the silicified faunas, both molluscs and brachiopods are very numerous. In three faunas across an offshore-onshore environmental gradient (Hoare and Pojeta 2006), species diversity is dominated by molluscs (58–84%), with aragonite shells representing up to 71% (mean 47%). By abundance, aragonitic molluscs dominate in all three faunas (Fig. 4b-1, -2). The nearshore epifaunal community is a very diverse, almost entirely molluscan assemblage dominated by gastropods and pteriomorph (bimineralic) bivalves. The lagoonal community is dominated by nuculoid bivalves, and chitons are also common. The open shelf community is characterized by pteriomorphs, gastropods and brachiopods.

Silurian brachiopod-dominated shelly assemblages of level bottom shelf environments have been extensively studied, and applied widely in paleoenvironmental interpretations (e.g., Johnson 1996; Boucot and Lawson 1999). Faunas from Gotland analysed in detail through a logged Lower Silurian section in inter-reef carbonates appear typical (Vattenfallet Project; Jaanusson et al. 1979). Brachiopods are overwhelmingly prevalent (72%), with corals (11%) and trilobites (12%), and relatively infrequent molluscs (5%) (Fig. 4b-2). A marked contrast in the quality of preservation between originally calcitic and aragonitic shells contributes to taxonomic bias. A very diverse brachiopod fauna preserved as calcite shells comprises >70 species, but includes limited new taxa (two new species, with several less well represented left in open nomenclature; Bassett in Jaanusson et al. 1979). This reflects how well known the calcitic faunas are as a consequence of typically good preservation throughout the Gotland succession. Mollusc preservation is mouldic and poor, except for bimineralic shells (e.g. some gastropods) where the thin outer calcitic shell layer is preserved; as a result, much of the taxonomy was left in open nomenclature (Jaanusson et al. 1979). Lower Silurian silicified material from Gotland comes from comparable inter-reef carbonate facies, with a brachiopod assemblage characteristic of the same subtidal environmental zone, Benthic Assemblage 2 that allows direct comparison with part of the Vattenfallet section (Högklint b, c; Boucot 1975; Boucot and Lawson 1999; Fig. 4c). However, although brachiopods are still common they form a subsidiary component among a mollusc dominated assemblage. Since both faunas here have been studied extensively and comprise >5,000 specimens, far above the rarefaction 'plateau' for Silurian (and Gotland) shelly faunas (Watkins 1996, 2000), they can reasonably be compared based on raw diversity data. The silicified preservation reveals morphological detail also of aragonite shells. Among a bivalve assemblage of 11 species, seven species (four genera) were new (Liljedahl 1985), and shell composition by abundance was 96% aragonitic. An equally abundant gastropod assemblage, which is at least predominantly aragonitic, still awaits systematic description but was reported as including 20 species (Liljedahl 1985). A significant and rare, aragonitic polyplacophoran (chiton) assemblage includes seven new species (six new genera; Cherns 1998a, b). Although brachiopods (15%) are common and diverse in this fauna, they are hugely outnumbered by molluscs and include no new taxa (M.G. Bassett, N.M.W. Cardiff, personal communication, 2007). A ternary diagram of shell mineralogy by abundance illustrates the contrast between the 'relict' yet rich calcitic faunas represented at Vattenfallet, and the silicified assemblage Fig. 4b-2). The silicified bivalves (mostly nuculoids) and gastropods represent a deposit feeding assemblage (Liljedahl 1984). Many small shells and articulated bivalves (20–30%) among the fauna indicate that this is an in situ to neighbourhood assemblage (Liljedahl 1983, 1984, 1985).

In the Lower Devonian of SE Australia, brachiopod faunas are again generally abundant and dominate most level bottom communities, while molluscs are sparse (Talent et al. 2000). However, silicified limestones yield rich and diverse faunas among which molluscs are common and greatly increase the known mollusc biodiversity. Johnston (1993) described a silicified bivalve fauna comprising 39 species, of which 22 species (three genera) were new. Along an environmental gradient from inshore to deeper shelf (Fig. 4c), aragonitic bivalves (nuculoids and heteroconchs) dominate the inner ramp and deeper subtidal assemblages, while bimineralic pteriomorphs are dominant in an intermediate, shallow subtidal setting. Gastropods are also common and diverse among the silicified faunas (Tassell 1982). Brachiopods have similar species diversity to bivalves in the two shallower silicified faunas, but are more diverse in the deeper shelf setting (Chatterton 1973).

The pattern identified from silicified faunas of under-representation of molluscs in many shelf faunas continues in the Lower Carboniferous Hotwells Limestone fauna from the Mendips, SW England (Mitchell 1987; Fig. 4c). In the Carboniferous Limestone of the South Wales-Mendip shelf, coral-brachiopod shelly assemblages are characteristic (Cossey et al. 2004). A very diverse, small-shell fauna was obtained through extensive individual study (by Mr Cliff Salter) of one thin 'rubbly-weathering bed...with masses of corals' (Green and Welch 1965, p. 27) in the massive crinoidal-oolitic limestones in Cliff Quarry, Compton Martin. As well as the typical coral-brachiopod assemblage, there is an exceptionally abundant small-sized molluscan fauna. This is dominated numerically by gastropods (>3,000), which include 98 species (46 genera), of which 18 are new (Batten 1966). Although not a silicified fauna, the abundant small shells, with colour banding preserved on some (Batten 1966), indicate a rapidly stabilized fauna. The less numerous bivalve assemblage (~730 specimens) includes seven species, and is dominated by aragonitic shells. The aragonitic shelled rostroconch *Conocardium* is also common, and the fauna also includes a small chiton assemblage.

In the Mesozoic, bivalves are by far the dominant shelly benthos. Wright et al.'s (2003) silicified fauna came from the typically bivalve-rich Blue Lias facies (Porthkerry Member) of the Lower Jurassic of South Wales, UK (Fig. 4c). Large, thick-shelled calcitic and bimineralic pteriomorphs (*Gryphaea*, *Plagiostoma*, *Pinna*) dominate shelly faunas in the alternating nodular limestone and shale facies that is developed widely across southern Britain (Cox et al. 1999). In such assemblages, only anomalodesmatan bivalves represent originally aragonitic shells. Their mouldic, commonly in situ preservation contrasts with the bedding assemblages of pteriomorph shells, and they comprise no more than a minor component by abundance (<2%; Wright et al. 2003; Fig. 4b-3). By contrast, in the silicified assemblage, where bivalves are still dominant, the characteristic pteriomorph genera form only part of a far more diverse and well preserved assemblage (Fig. 2). Gastropods

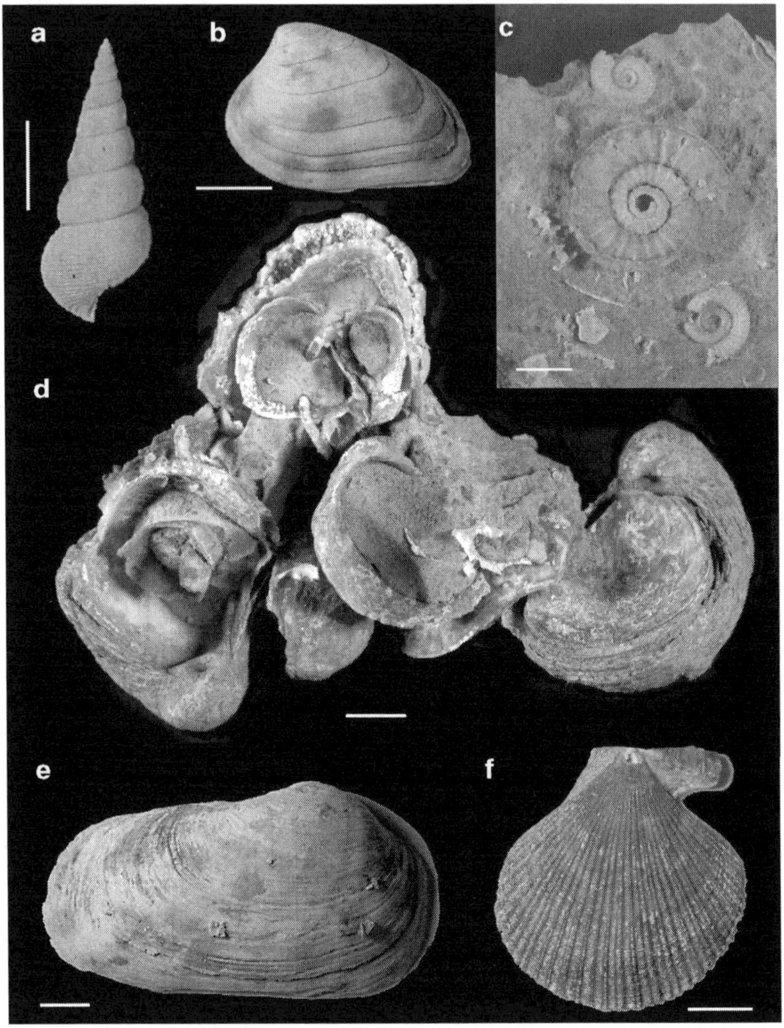

Fig. 2 Lower Jurassic silicified fauna, Bridgend, South Wales, UK showing good preservation of aragonitic (**a–c**, **e**), bimineralic (**f**) and calcitic (**d**) shells. (**a**) *Allocosmia*; (**b**) *Cardinia* – note beekite rings; (**c**) ammonites; (**d**) *Gryphaea* cluster including articulated shells, single and imbricated shells – note silicification from outer shell layer into inner void at top of cluster; **e**, *Pholadomya*; **f**, *Chlamys*. Scale bars 1 cm (Images **a**, **b**, **e**, **f** courtesy of P. Hodges)

are also very common, and 77% of the mostly small shelled assemblage is aragonitic (Fig. 4b-3). The most abundant bivalve genera are a heteroconch and arcoid, neither of which is present in the typical carbonate fauna. Lower Jurassic bivalves are well known and this fauna lacks new species (Hodges 1987); the gastropods include as yet undescribed taxa.

3.2 Other Deposits Capturing Biodiversity

Skeletal lagerstätten remain rare in the fossil record, but all cases described above are characterized by higher diversity and/or abundance of molluscs than their comparator faunas. All show aragonite dominated assemblages, whose depletion from faunas typical of associated successions is represented by a shift toward calcitic/bimineralic faunas (Fig. 4b). The role of other, more frequent taphonomic processes that increase the preservation potential of shells vulnerable to early dissolution is reviewed below.

3.2.1 Storm and Shell Beds

Storm beds and shell beds can halt dissolution of vulnerable aragonite shells through rapid burial below the TAZ (or zone of bioturbation; Cherns et al. 2008; Fig. 4a). Storm events are commonly represented by concentrations of epifaunal shells and exhumed shallow infauna (e.g. Kidwell and Brenchley 1994, 1996). These have been considered proxies for the dominant faunal components, and hence as records of paleocommunities (e.g. Li and Droser 1997, 1999; Boyer et al. 2004). Storm beds and shell coquinas occur abundantly from the early Paleozoic, increasing in frequency and thickness in late Cenozoic rocks (Kidwell and Brenchley 1994, 1996; Fig. 4c). In the Cenozoic, molluscs are abundant in shelf faunas, and storm shell beds in a variety of environmental settings commonly comprise scoured, transported aragonitic bivalves (Hendy et al. 2006). Molluscan-rich storm event beds in older rocks are also commonly dominated by such bivalves even when those are otherwise uncommon in the fossil record (Fursich and Pandey 2003).

Many processes associated with the high energy of storm events, such as shell disarticulation, fragmentation and abrasion during transport, are destructive for fossils (Kidwell and Bosence 1991). Sediment erosion and reworking is likely to bias the record of aragonitic shells through loss of partly dissolved, pitted shells and early formed moulds. However, some processes increase the potential to preserve aragonitic faunas. Removal of fine grained sediment and organic matter by winnowing and transport reduces the drive for microbially mediated dissolution (Wright et al. 2003). Rapid burial of shells reduces or avoids residence time in the TAZ, and anaerobic decay of organic material within the sulfate reduction zone is associated with increased alkalinity, favouring shell preservation (Canfield and Raiswell 1991a; Table 1 (3a and 3b)). In storm concentrations of close-packed shells, buffering of the pore water through dissolution of fragmented shells can lead to early firmground matrix in which moulds of aragonitic shells can be preserved (e.g. Sanders 2003, 2004).

Storm beds and other shell beds are discussed here selectively to illustrate the record of aragonitic biodiversity and abundance represented by these faunas through the Phanerozoic by comparison with surrounding sediments (Fig. 4c). A fuller account is given in Cherns et al. (2008).

Trimerellacean brachiopod faunas in storm beds are reported widely from brachiopod-coral-stromatoporoid carbonate facies of Ordovician-Silurian age (e.g.

Webby and Percival 1983; Bassett 2005). These large, thick-shelled brachiopods are interpreted unusually as aragonitic shelled because of their consistently mouldic preservation (Jaanusson 1966). Thick storm beds of mostly disarticulated valves form high density monospecific concentrations in shallow intertidal-subtidal carbonate deposits. In situ shell banks are preserved in the shoal zone (Webby and Percival 1983), where high wave energy ensures a low drive for dissolution (cf. Wright et al. 2003) because of the removal of fine particulate organic matter.

Storm-winnowed shell beds in the brachiopod-dominated Khuff Formation (Permian) of Oman were deposited in outer carbonate ramp environments (Angiolini et al. 2003; Fig. 4b). These coquinas have notably molluscan-rich faunas of bivalves (Dickins 1999), gastropods and scaphopods, among successions with brachiopod dominated faunas. From the slightly younger Al Jil Formation (early Triassic) of Oman, winnowed storm beds of shallow subtidal environments yield mollusc-dominated faunas that are partly silicified (Krystyn et al. 2003; Fig. 4c). Shell beds dominated by aragonitic bivalves (*Promyalina*) represent more proximal, very shallow subtidal storm beds, while slightly deeper subtidal storm beds have more pteriomorph bivalves (aviculopectinids); interbedded sediments are poor in bivalves (Krystyn et al. 2003).

Triassic silicified faunas from shallow subtidal storm beds include diverse microgastropod assemblages from Oman (Wheeley and Twitchett 2005). Other Triassic silicified microgastropod faunas are also from storm induced shell beds, e.g. the Sinbad Limestone (Nammalian) of the western USA, which includes scaphopods and bivalves (Fraiser and Bottjer 2004) and the Gastropod Oolite of northern Italy (Nützel and Schulbert 2005). Both these represent inner ramp carbonate deposits (Fig. 4c).

The Saltford Shale of the Lower Jurassic (Hettangian) of Worcestershire, UK includes outer ramp storm-influenced shell beds that contain a diverse and abundant molluscan assemblage, including >15 bivalve and several gastropod species (Fig. 4c). Many of the originally aragonitic bivalves (e.g. *Cardinia, Grammatodon, Mactromya, Neocrassina, Pholadomya, Modiolus*) are well preserved (Fig. 3). These shell beds most likely formed on oyster-encrusted, exhumed sandy hardgrounds; they demonstrate taphonomic feedback/ecological succession with evidence for storm and current winnowing and reworking (e.g. Kidwell 1986; Tsujita et al. 2006).

From the Upper Jurassic-Lower Cretaceous of Kachchh, western India, bivalve-rich storm beds occur among otherwise unfossiliferous siliciclastic successions (Fursich and Pandey 2003; Fig. 4c). The storm beds are dominated by large bivalves, all of which were originally aragonitic (heteroconchs including trigonioids, and arcoids; *Seebachia, Indotrigonia, Pisotrigonia, Megacucullaea*). Thin shell lenses of abraded, robust shells of calcitic *Gryphaea* accumulated during periods of sediment starvation, but a notable lack of small, thin and originally aragonitic shells was ascribed to diagenetic dissolution by Fursich and Pandey (2003).

In the Cenozoic (Miocene-Pliocene) of New Zealand, bivalve-rich shell beds on a storm-wave dominated siliciclastic shelf comprise mostly aragonitic bivalves from innermost to outer shelf (Hendy et al. 2006; Fig. 4c). Storm beds dominated by aragonitic bivalves interrupt sequences with calcitic barnacle-bivalve-bryozoan

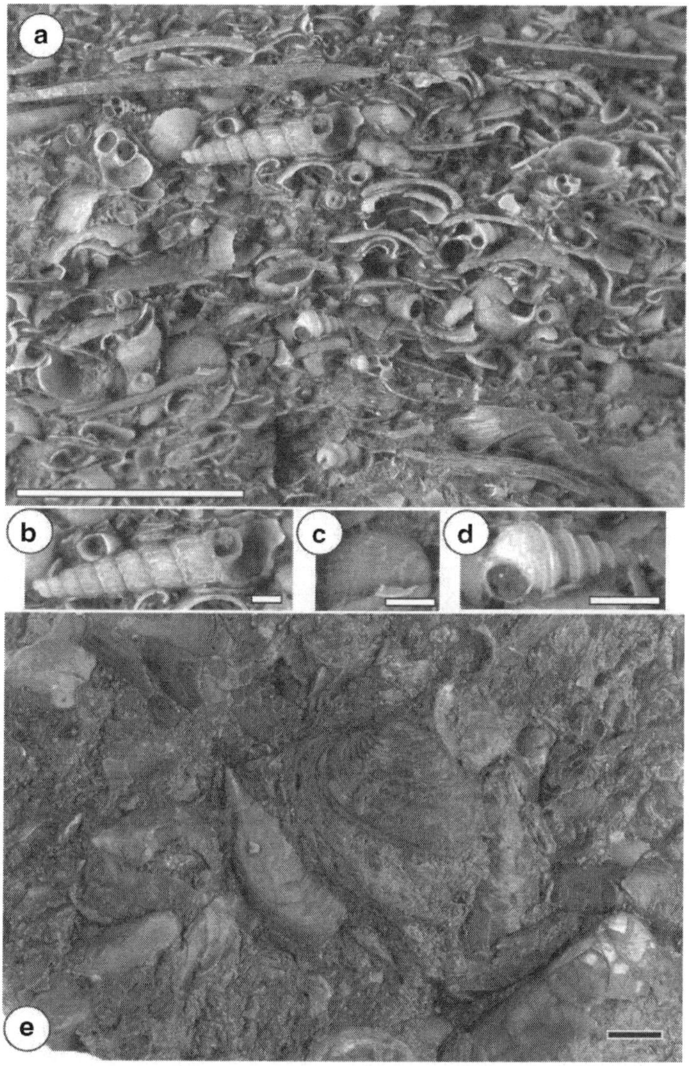

Fig. 3 Lower Jurassic Saltford Shale shell bed, Worcestershire, UK. (**a**) Section through a winnowed shell bed with convex-up, thin aragonitic bivalve shells (cf. thick oyster shell bottom right) and abundant microgastropods. (**b–d**) Details of exquisitely preserved micro-molluscs in shell bed (**a**). (**e**) Plan view of bivalve dominated shell bed showing convex-up orientation of several originally aragonitic shelled taxa. Scale bars (**a**), (**e**) 1 cm, (**b–d**) 1 mm

faunas (Nelson et al. 2003). Taphofacies range from amalgamated shell beds of densely packed, disarticulated and broken shells to rapidly buried deposits with in situ, articulated shallow burrowers (Hendy et al. 2006). By contrast to the shelf-wide distribution of aragonitic shells, the robust calcitic shells of oysters dominate only in inner-mid shelf environments in sediment starved beds and proximal tempestites.

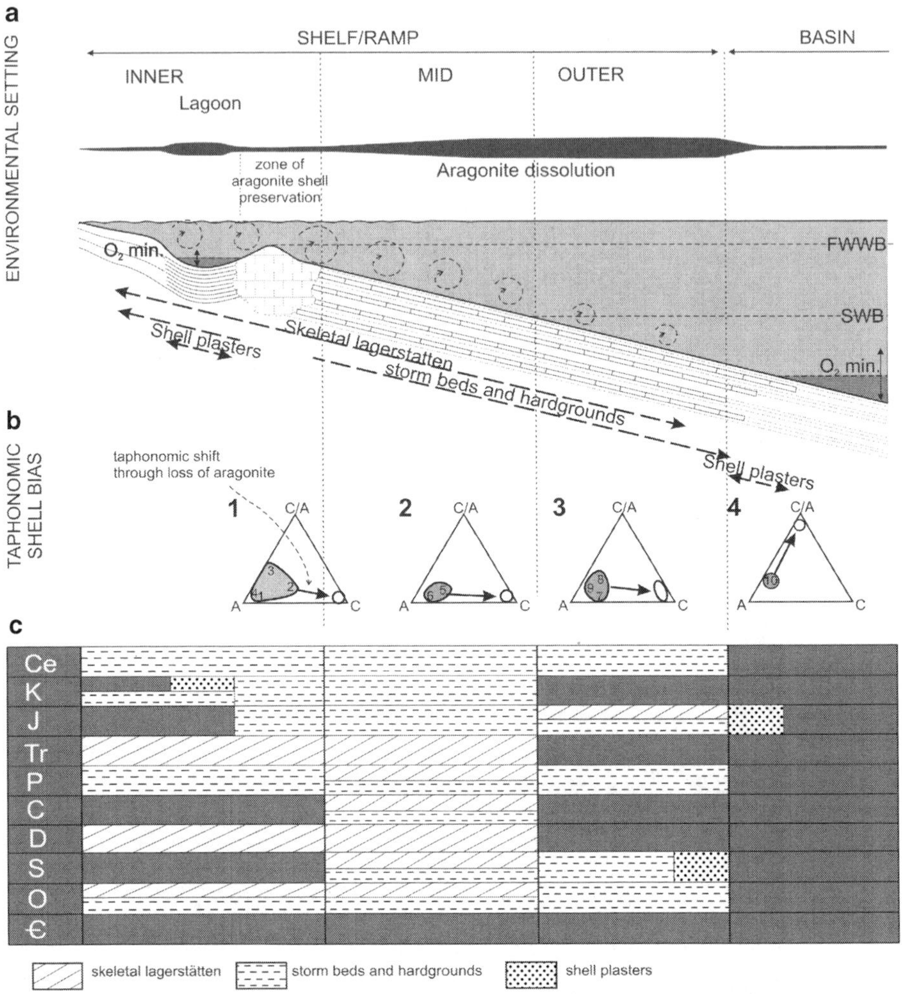

Fig. 4 Model for taphonomic gradients and fossil preservation. (**a**) Environmental profile from shelf/ramp to basin for siliciclastic and carbonate settings (terminology for carbonate ramps following Burchette and Wright 1992), showing the major controls of water energy, environments where temporary dysoxia/anoxia may develop, sedimentary facies distribution (nearshore and shoal grainstones, mid-outer ramp LMA, outer ramp–basin mudstones/shales), and distribution of taphonomic windows. The drive for aragonite dissolution is higher in organic-rich muddy sediments of mid-outer shelf/ramp to basin settings and inner shelf/ramp lagoons. (**b**) Taphonomic bias resulting from original shell mineralogy of shelly faunas across shelf/ramp to basin zones, comparing faunas of taphonomic windows (*shaded zones*) to typical, or'relict', faunas (*open zones*). 1, Inner ramp shoals and lagoons, e.g. Ordovician; 1–1, Lower Ordovician (Cope 1996); 1–2, Nearshore and 1–3, Lagoonal silicified Upper Ordovician (Hoare and Pojeta 2006); 1–4, trimerellid banks (Webby and Percival 1983); cf. typical, brachiopod-rich faunas of inner-mid ramp (Williams et al. 1981; Holland and Patzkowsky 2004). 2, Mid ramp; 2–5, Upper Ordovician open shelf silicified (Hoare and Pojeta 2006); 2–6, Silurian silicified (Cherns and Wright 2000); cf. typical brachiopod-rich faunas of Upper Ordovician (Holland and Patzkowsky 2004) and Silurian

3.2.2 Shell Plasters

Shell plasters refer to thin layers or pavements of aragonitic and bimineralic shells preserved in dark, organic-rich mudstones representing environments prone to temporary anoxia/dysoxia, such as lagoons and deep ramp or basin settings (e.g. Noe-Nygaard et al. 1987; Oschmann 1988, 1991; Martill and Hudson 1991; Martill et al. 1994; Fig. 4a, b-4, c). Temporary anoxic/dysoxic events such as those caused by algal blooms draw infaunal bivalves up onto the sediment surface where they die (Oschmann 1991). Residence time for aragonitic shells in the acidic TAZ remains critical, but raising of the redox boundary and the sulfide oxidation zone removes them from this zone even at shallow depths. Without bioturbation to disturb the sediment, local alkalinity may build up around the shell layers (e.g. Canfield and Raiswell 1991b; Sanders 2003). Diagenetic factors such as organic matter content, sulfate availability and iron concentration control the style of shell preservation (Hudson in Canfield and Raiswell 1991b). In the dark, sulfide-rich shales of the Lower Oxford Clay, originally aragonite fossils are preserved as aragonite or as pyritic moulds (Martill and Hudson 1991). However, in the anoxic/dysoxic shale environments of the Posidonia Shale, shell dissolution at very shallow depths is evident from the reduction of mollusc shells to periostracal (organic) films (e.g. Allison et al. 1995). Similarly, aragonite dissolution is indicated by shell plasters in black shales restricted to bimineralic pteriomorphs such as *Bositra* and *Meleagrinella,* where very thin shell preservation indicates loss of the aragonitic inner shell layer (Oschmann 1993; Wilby et al. 2004).

Taphofacies contrasts between fossil preservation in dark grey and black shales, comparable to those described above, are evident in the Middle Devonian of New York, USA (Brett et al. 1991). Mass concentrations of nuculoid bivalves are preserved as pyritized moulds in dark grey shales, while fossils in black shales are decalcified, even the calcitic brachiopods (Dick and Brett 1986; Speyer and Brett 1991). The nuculoid beds, with many articulated or splayed individuals, likely represent death assemblages of burrowers drawn to the sediment surface by stressed conditions associated with the onset of a severe storm event, and then buried by the later effects of that same event (Speyer and Brett 1991).

3.2.3 Hardgrounds

James et al. (2005) predicted that aragonitic shells should be preferentially preserved at submarine limestone hardgrounds through rapid cementation of

Fig. 4 (continued) (Jaanusson et al. 1979). 3, Mid-outer ramp; 3–7, Ordovician storm bed (Wilcox and Lockley 1981); 3–8 Permian storm bed (Dickins 1999); 3–9, Lower Jurassic silicified (Wright et al. 2003); cf. background faunas, respectively. 4, Outer ramp–basin; 4–10, Jurassic shell plasters (Oschmann 1991), cf. restricted pterioid faunas (Duff 1975). **(c)** Temporal distribution of skeletal lagerstätten, storm bed and hardground faunas, and shell plasters across the environmental zones (data from text, and Cherns et al. 2008, Table 1) the host sediment (also Walker and Diehl 1985). This protects the delicate solution moulds of molluscs from destruction through bioturbation. For example, in hardgrounds of the Ordovician cephalopod limestones of Sweden, large, originally aragonitic orthocone cephalopods are commonly preserved as moulds (e.g. King 1990). In Ordovician and Jurassic hardgrounds, encrusting organisms inside the moulds of molluscs were taken to indicate that shell dissolution took place very early in diagenesis (Palmer et al. 1988; Palmer and Wilson 2004). It is not clear whether those moulds formed in direct contact with overlying seawater, or if dissolution occurred in contact with modified pore-waters in the near-surface sediment and the moulds were subsequently exhumed and encrusted. The differential taphonomic vulnerability of aragonite is evidenced by loss of the inner aragonitic shell layer of bimineralic bivalves before encrustation of the calcitic inner shell faces (Palmer and Wilson 2004). Intraformational conglomerates associated with hardgrounds can also preserve evidence of aragonitic mollusc faunas, e.g. bivalve, cephalopod and gastropod bioclasts abundant in the Ordovician Kanosh Formation, Utah USA (Palmer and Wilson 2004).

3.2.4 Shoal Deposits

In high energy inner ramp, shoreface settings, well oxygenated waters and limited organic matter reduce the drive for microbial dissolution, and thus the preservation potential for aragonite shells is increased (Fig. 4a). In the coarse conglomeratic limestones of the Lower Jurassic Sutton Member of South Wales, UK, formerly aragonitic shells are preserved as open moulds or recrystallized to calcite, by contrast to their dissolutional loss in outer ramp Blue Lias facies (Fig. 4b-3; Wright et al. 2003). Ordovician trimerellid brachiopods from carbonate shoal environments are preserved as open moulds in situ forming shell banks (Webby and Percival 1983; see Section 3.2.1; Fig. 4c).

In the siliciclastic settings of the early Paleozoic Welsh Basin, rare inshore shelf faunas are dominated by diverse assemblages of aragonitic bivalves preserved as moulds, but molluscs are otherwise sparse among the brachiopod-rich shelf faunas (Cope 1996, 1999; Ratter and Cope 1998). The Middle Devonian Hamilton Group, and many Upper Devonian deposits, of the Appalachian Basin have both bivalves and brachiopods very well preserved as external moulds in the coarser siltstones and sandstones of nearshore to mid-shelf environments (Brett et al. 1991; Speyer and Brett 1991). Preservation of surface ornament and epibionts on moulds is consistent with relatively late shell dissolution in environments where bottom circulation and aeration ensured that acid products of decay were flushed through sediments. The moulds are less compacted and deformed than fossils in more offshore mudstones. It is notable that many of the diverse bivalves illustrated by Hall and Conrad came from such facies (e.g. Bailey 1983), as well as other molluscs (e.g. Rollins et al. 1971).

4 Discussion

4.1 Taphonomic Gradients and Molluscan Preservation: A Model

The taphonomic windows outlined above (Fig. 4) define a bias against aragonitic skeletons and illustrate the potential for early dissolution to skew the fossil record, both in terms of mollusc biodiversity and of paleocommunity structure. Even in relatively shallow tropical settings, extensive dissolution of aragonite (and high-Mg calcite) takes place during very early burial (Walter et al. 1993). Some facies differential is likely through higher rates of shell loss for bivalves in carbonate sediments than in siliciclastics through bioerosion on the seafloor and TAZ dissolution (Kidwell et al. 2005). In low energy, offshore settings, much of the molluscan fauna is likely to have been removed by shallow, early, microbially-mediated aragonite dissolution, even in Mesozoic communities where fossil faunas remain mollusc dominated.

In inner ramp, shoreface settings above fair-weather wave base, where low organic matter content and lack of shallow anoxia reduce the drive for microbial dissolution in the TAZ, aragonitic shells can be buried until replaced by calcite or preserved as open moulds (Section 3.2.4). Early cementation forming hardground layers prior to dissolution allows ghosts of the labile aragonitic fauna to be captured (Section 3.2.3). Early silicification in inner ramp environments preserves small shell, gastropod-rich, diverse molluscan assemblages (Hoare and Pojeta 2006; Fig. 4b-1; Nützel and Schulbert 2005). In the Cenozoic, storm beds increase in thickness and frequency across inner-mid shelf/ramp environments, and preserve thick molluscan deposits (Kidwell and Brenchley 1994, 1996; Fig. 4c). In the inner ramp environmental zone, the preservation potential of aragonitic faunas is probably enhanced by reduced drive for dissolution compared with quieter and more offshore settings.

In more protected, lower energy lagoonal areas, such as Florida Bay and the interior of the Great Bahama Bank (Walter and Burton 1990; Walter et al. 1993; Hendry 1993), organic rich sediment increases the drive for synsedimentary loss of aragonite through microbially mediated dissolution. A silicified Ordovician fauna of molluscs and brachiopods (Hoare and Pojeta 2006; Fig. 4b-1) demonstrates the potential skewing of the fossil record. In such protected areas, more complete molluscan faunas can be preserved by storm events (e.g. Radley and Barker 2000; Fig. 4c). Also, thin shell plasters that formed as a consequence of dysoxic episodes, such as seasonal algal blooms, comprise aragonitic faunas (Noe-Nygaard et al. 1987; Section 3.2.2; Fig. 4c).

Across the mid to outer ramp zones, current energy becomes generally lower, and the frequency of storm reworking decreases (Fig. 4a). In bioturbated sediment that becomes increasingly muddy towards the outer ramp, silicified lagerstätten faunas suggest that many level bottom shelly faunas are likely to have been selectively affected by aragonite dissolution (Section 3.1; Fig. 4c; Wright and Cherns 2004). In the shallow subtidal mid-ramp, Silurian and Ordovician silicified faunas provide evidence for significant loss of aragonitic shells (e.g. Cherns and Wright 2000; Hoare and Pojeta 2006; Fig. 4b-2). In the muddier outer ramp setting, the diverse

Lower Jurassic silicified fauna (Wright et al. 2003) is dominated numerically by two aragonitic bivalve taxa unrepresented in the typical Blue Lias fauna (Porthkerry Member; Section 3.1; Fig. 4b-3). The potency of the aragonite dissolution process is clearly evidenced by the abundance of limestone-marl alternations (LMA) in mid-outer ramp settings through the Phanerozoic record (Westphal and Munnecke 2003; Fig. 4a). Many LMA, and especially those formed during 'calcite sea' intervals and in cool-water carbonate environments, were likely sourced from molluscan aragonite (Wheeley et al. 2008).

In subtidal settings, storm deposits associated with rapid burial that removed shells from the TAZ can preserve aragonitic molluscs among a more complete fauna (Section 3.2.1). Storm bed faunas are illustrated (Fig. 4b-3) from the siliciclastic Ordovician succession of South Wales, UK (Wilcox and Lockley 1981) and Permian carbonates of Oman (Khuff Formation; Dickins 1999; Angiolini et al. 2003). The fidelity of such faunas varies with the magnitude of storm scouring, competency of the storm current and thickness of the storm bed (e.g. Finnegan and Droser 2008). Some selectivity may also result from physical destruction during transport. Thus it is relatively rare that a snapshot of the community is captured. Single storm events are typically thin (≤10 cm) beds, and are most likely to show relative enrichment of aragonitic shells. Thick (≥50 cm) beds are mostly amalgamated, multiple event concentrations (Kidwell and Brenchley 1994), and those formed during periods of sediment starvation may be depleted in aragonitic shells (e.g. Jurassic *Gryphaea* beds of Fursich and Pandey 2003; Section 3.2.1). In the late Ordovician (Cincinnatian) of Ohio and Kentucky, USA, only the very thin, commonly lenticular shell beds are rich in molluscs relative to brachiopods, while the widely traceable, major shell beds have brachiopod–bryozoan–crinoid, calcitic faunas with few, bimineralic (pterioid) bivalves (Brett et al. 2008; Dattilo et al. 2008).

Some areas of the sea floor are prone to surface dysoxia or anoxia, such as lagoons where lack of tidal exchange leads to stratification, and in deeper water, outer ramp to basin settings. In such muddy environments, a typically molluscan fauna may become preserved as thin shell plasters (Section 3.2.3; Fig. 4b-4). These represent a limited benthic fauna of small, thin shelled, aragonitic and bimineralic bivalves, killed by temporary anoxic episodes and preserved as original, recrystallized or pyritic shells (e.g. Oschmann 1988, 1991). Periodic dysaerobic sea floor conditions in outer carbonate ramp environments are represented by shell beds in the widespread 'cephalopod limestone facies' of early Paleozoic Gondwana (e.g. Upper Silurian Kopanina Formation, Czech Republic; Gnoli 2003; Fig. 4c). Shell pavements of large cephalopod shells were colonized by aragonitic bivalve faunas (Kříž 1992), with all shells preserved recrystallized to calcite.

Temporal trends are evident from the distribution of several types of taphonomic windows through the Phanerozoic (Fig. 4c). The rare, early silicified, mollusc-rich skeletal lagerstätten are recorded from Ordovician through Jurassic (Schulbert et al. 1997; Section 3.1). Shell beds are considerably more frequent, but show changes in thickness and extent; thin (mostly <20 cm), single event storm beds form the majority among early Paleozoic shell beds, while by Neogene times thick (≥50 cm), multiple event beds are also very common (Kidwell and Brenchley 1994). In dysaerobic mud environments, the typically sparse benthic faunas are largely infaunal, aragonitic molluscs, whose preservation potential may be enhanced by episodic raising of the redox boundary and sulfate oxidation zone (Section 3.2.2).

4.2 Molluscan Preservation During 'Calcite' and 'Aragonite Seas'

An important consideration with regard to differential taphonomy through time is the influence of 'calcite' and 'aragonite seas'. The dominant abiogenic carbonate seafloor precipitates are controlled by atmospheric pCO_2 and/or Mg:Ca ratios in seawater (Hardie 1996). In Recent oceans, aragonite, and to a lesser extent high-Mg calcite, are the main precipitates at the seafloor at least in the tropical realm. Through the Phanerozoic, similar periods of 'aragonite seas' (Precambrian – end Cambrian; end mid Carboniferous – end Triassic; end Cretaceous – present day) have oscillated with 'calcite seas' (Ordovician – end early Carboniferous; Jurassic – Cretaceous; e.g. Sandberg 1983). These periodic changes in inorganic carbonate precipitation broadly correlate with patterns of calcification in some invertebrates and algae (Stanley and Hardie 1998). It may at first seem that periods with more aragonitic taxa (i.e. 'aragonite seas') were intervals in geological history when taphonomic loss was potentially higher. However, in an assessment of syndepositional dissolution through time, Sanders (2003) found no correlation between 'calcite' and 'aragonite seas' and extent of syndepositional dissolution, suggesting that site-specific conditions are more important. The temporal occurrence of the taphonomic windows discussed above concentrates on 'calcite sea' intervals, which cover the main distribution of silicified fossil assemblages (Schulbert et al. 1997; Fig. 4). The effects of 'calcite' or 'aragonite seas' seem to have been more important in influencing the calcification of some groups (e.g. Ries 2005) rather than on shallow burial dissolution processes.

For the Ordovician, Palmer et al. (1988) and Palmer and Wilson (2004) proposed that ambient 'calcite seas' dissolved aragonite shells directly at the sediment–water interface (SWI) to form open moulds that became encrusted by contemporaneous organisms. With recent modelling that predicts those 'calcite seas' were in fact supersaturated with respect to aragonite (Riding and Liang 2005, cf. Locklair and Lerman 2005), it may be that biomoulds formed through dissolution in the shallow sediment, either through interaction with modified marine pore-waters or through meteoric leaching, and were later exhumed and encrusted (e.g. Kenyon-Roberts 1995).

5 Conclusions

The taphonomic effects of differential, early aragonite dissolution on the marine shelly fossil record and biodiversity from carbonate and siliciclastic settings are highlighted by relatively infrequent faunas capturing aragonitic shells. As well, early shell stabilization forming skeletal lagerstätten, the processes leading to storm and shell beds, shell plasters, hardgrounds and shoal deposits all enhance the preservation potential of originally aragonitic shells. A data-based model for taphonomic gradients and molluscan preservation through the Phanerozoic indicates that the early aragonite dissolution in lower energy settings such as platform interiors (including lagoons), mid and outer ramps/shelves and basins was a major process skewing the fossil record. Dissolution at relatively shallow depths in the TAZ is due to undersaturation caused by microbially-mediated decay of organic matter and related reactions. A substantial mineralogical shift in the shelly faunas, quantified from examples across a shelf/ramp to basin profile, is away from aragonite-dominated molluscan assemblages towards calcite or bimineralic-dominated, diagenetically filtered 'relicts'. Taphonomic windows record aragonitic, molluscan-rich faunas from Ordovician times onwards. Although the temporal distribution of the several types of deposit discussed is variable (e.g. silicified faunas), there is no apparent correlation with the intervals of 'calcite' and 'aragonite seas'.

Acknowledgments We thank Carl Brett (University of Cincinnati) and Peter Allison (Imperial College, London) for constructive reviews of the original manuscript. JRW is grateful to Robert Raine (University of Birmingham) for help in collecting Jurassic Saltford Shale material, and for discussion on the origins of these shell beds.

References

Alexandersson, E. T. (1978). Destructive diagenesis of carbonate sediments in eastern Skagerrak, North Sea. *Geology, 6*, 324–327.

Alexandersson, E. T. (1979). Marine maceration of skeletal carbonates in the Skagerrak, North Sea. *Sedimentology, 26*, 845–852.

Aller, R. C. (1982). Carbonate dissolution in nearshore terrigenous muds: The role of physical and biological reworking. *Journal of Geology, 90*, 79–95.

Aller, R. C., & Aller, J. Y. (1998). The effect of biogenic irrigation intensity and solute exchange on diagenetic reaction rates in marine sediments. *Journal of Marine Research, 56*, 905–936.

Allison, P. A., Wignall, P. B., & Brett, C. E. (1995). Palaeo-oxygenation: Effects and recognition. In D. W. J. Bosence & P. A. Allison (Eds.), *Marine palaeoenvironmental analysis from fossils. Geological Society of London, Special Publication 83*, 97–112.

Angiolini, L., Balini, M., Garzanti, E., Nicora, A., Tintori, A., Crasquin, S., et al. (2003). Permian climatic and paleogeographic changes in Northern Gondwana: The Khuff Formation of Interior Oman. *Palaeogeography, Palaeoclimatology, Palaeoecology, 191*, 269–300.

Bailey, J. B. (1983). Middle Devonian Bivalvia from the Solsville Member (Marcellus Formation), central New York State, USA. *Bulletin of the American Museum of Natural History, 174*(3), 196–325.

Bassett, M. G. (2005). *Silurian brachiopods and brachiopod biofacies of Gotland, Sweden: An excursion guide*. Fifth International Brachiopod Congress, Copenhagen. National Museums and Galleries of Wales, Cardiff.

Batten, R. L. (1966). The Lower Carboniferous gastropod fauna from the Hotwells Limestone of Compton Martin, Somerset. *Palaeontographical Society Monographs 509*, 1–52, pl. 51–55 (Part 1) and *513*, 53–109, pl. 106–110 (Part 2).

Behrensmeyer, A. K., Fursich, F. T., Gastaldo, R. A., Kidwell, S. M., Kosnik, M. A., Kowalewski, M., et al. (2005). Are the most durable shelly taxa also the most common in the marine fossil record? *Paleobiology, 31*(4), 607–623.

Bevins, R. E., & Mason, J. S. (1997). Welsh metallophyte and metallogenic evaluation project: Results of a minesite survey of Dyfed and Powys. CCW Contract Science Report 156. National Museums & Galleries of Wales, Cardiff.

Bottjer, D. J., & Ausich, W. I. (1986). Phanerozoic development of tiering in soft substrata suspension-feeding communities. *Paleobiology, 12*, 400–420.

Boucot, A. J. (1975). *Evolution and extinction rate controls*. Amsterdam: Elsevier.

Boucot, A. J., & Lawson, J. D. (1999). *Paleocommunities: A case study from the Silurian and Lower Devonian*. Cambridge: Cambridge University Press.

Boyer, D. L., Bottjer, D. J., & Droser, M. L. (2004). Ecological signature of Lower Triassic shell beds of the western United States. *Palaios, 19*(4), 372–380.

Brett, C. E., Dick, V. B., & Baird, G. C. (1991). Comparative taphonomy and paleoecology of Middle Devonian dark gray and black shales facies from western New York. In E. Landing & C. E. Brett (Eds.), *Dynamic stratigraphy and depositional environments of the Hamilton Group (Middle Devonian) in New York State,* Part II. *New York State Museum Bulletin 469*, 5–36.

Brett, C. E., Kirchner, B. T., Tsujita, C. J., & Dattilo, B. F. (2008). Depositional dynamics recorded in mixed siliciclastic-carbonate marine successions: Insights from the Upper Ordovician Kope Formation of Ohio and Kentucky, USA. In B. R. Pratt & C. Holmden (Eds.), *Dynamics of epeiric seas. Geological Association Canada, Special Papers, 48,* 73–102.

Burchette, T. P., & Wright, V. P. (1992). Carbonate ramp depositional systems. *Sedimentary Geology, 79*, 3–57.

Bush, A. M., & Bambach, R. K. (2004). Did alpha diversity increase during the Phanerozoic? Lifting the veils of taphonomic, latitudinal, and environmental biases. *Journal of Geology, 112*(6), 625–642.

Canfield, D. C., & Raiswell, R. (1991a). Carbonate precipitation and dissolution; its relevance to fossil preservation. In P. A. Allison & D. E. G. Briggs (Eds.), *Taphonomy: Releasing the data locked in the fossil record*. New York: Plenum.

Canfield, D. C., & Raiswell, R. (1991b). Pyrite formation and fossil preservation. In P. A. Allison & D. E. G. Briggs (Eds.), *Taphonomy: Releasing the data locked in the fossil record*. New York: Plenum.

Carson, G. A. (1991). Silicification of fossils. In P. A. Allison & D. E. G. Briggs (Eds.), *Taphonomy: Releasing the data locked in the fossil record*. New York: Plenum.

Carter, J. G. (1980). Environmental and biological controls of bivalve shell mineralogy and microstructure. In D. C. Rhoads & R. A. Morse (Eds.), *Skeletal growth of aquatic organisms*. New York: Plenum.

Carter, J. G. (1990). *Skeletal biomineralization patterns, processes and evolutionary trends*. New York: Van Nostrand and Reinhold.

Carter, J. G., Barrera, E., & Tevesz, M. J. S. (1998). Thermal potentiation and mineralogical evolution in the Bivalvia (Mollusca). *Journal of Paleontology, 72*, 991–1010.

Chatterton, B. D. E. (1973). Brachiopods of the Murrumbidgee Group, Taemas, New South Wales. *Australian Bureau of Mineral Resources, Geology and Geophysics Bulletin, 137*, 146. + 135 pl.

Cherns, L. (1998a). *Chelodes* and closely related Polyplacophora (Mollusca) from the Silurian of Gotland, Sweden. *Palaeontology, 41*, 545–573.

Cherns, L. (1998b). Silurian polyplacophoran molluscs from Gotland, Sweden. *Palaeontology, 41*, 939–974.

Cherns, L., & Wright, V. P. (2000). Missing molluscs as evidence of large-scale, early skeletal aragonite dissolution in a Silurian sea. *Geology, 28*(9), 791–794.

Cherns, L., & Wright, V. P. (2009) Quantifying the impacts of early diagenetic aragonite dissolution on the fossil record. *Palaios, 24*, 756–771.

Cherns, L., Wheeley, J. R., & Karis, L. (2006). Tunneling trilobites: Habitual infaunalism in an Ordovician carbonate seafloor. *Geology, 34*, 657–660.

Cherns, L., Wheeley, J. R., & Wright, V. P. (2008). Taphonomic windows and molluscan preservation. *Palaeogeography, Palaeoclimatology, Palaeoecology, 270*(3–4), 220–229.

Cope, J. C. W. (1996). Early Ordovician (Arenig) bivalves from the Llangynog Inlier, south Wales. *Palaeontology, 39*, 979–1025.

Cope, J. C. W. (1999). Middle Ordovician bivalves from mid-Wales and the Welsh Borderland. *Palaeontology, 42*, 467–499.

Cossey, P., Adams, A. E., Wright, V. P., Whiteley, M. J., Whyte, M., & Purnell, M. (2004). British Lower Carboniferous stratigraphy. *JNCC Geological Conservation Review, 29*, xix +617 pp.

Cox, M., Sumbler, M.G., & Ivimey-Cook, H.C. (1999). A formational framework for the Lower Jurassic of England and Wales (onshore area). *British Geological Survey Research Report RR/99/01.*

Cummins, H., Powell, E. N., Stanton, R. J., & Staff, G. M. (1986). The rate of taphonomic loss in modern benthic habitats: How much of the potentially preservable community is preserved? *Palaeogeography, Palaeoclimatology, Palaeoecology, 52*, 291–320.

Dattilo, B. F., Brett, C. E., Tsujita, C. J., & Fairhurst, R. (2008). Sediment supply versus storm winnowing in the development of muddy and shelly interbeds from the Upper Ordovician of the Cincinnati region, USA. *Canadian Journal of Earth Sciences, 45*(2), 243–265.

Davies, D. J., Powell, E. N., & Staff, G. M. (1989). Relative rates of shell dissolution and net sedimentary accumulation – A commentary: Can shell beds form by the gradual accumulation of biogenic debris on the sea floor? *Lethaia, 22*, 207–212.

Dick, V. B., & Brett, C. E. (1986). Petrology, taphonomy, and sedimentary environments of pyritic fossil beds from the Hamilton Group, Middle Devonian of Western New York USA. *New York State Museum Bulletin, 457*, 102–128.

Dickins, J. M. (1999). Mid-Permian (Kubergandian-Murgabian) bivalves from the Khuff formation, Oman: Implications for world events and correlation. *Rivista Italiana di Paleontologia e Stratigrafia, 105*(1), 23–35.

Droser, M. L. (2002). Ecological changes through geological time. In D. E. G. Briggs & P. R. Crowther (Eds.), *Palaeobiology II.* UK: Blackwell.

Droser, M. L., & Bottjer, D. J. (1989). Ordovician increase in extent and depth of bioturbation - implications for understanding early Paleozoic ecospace utilization. *Geology, 17*(9), 850–852.

Duff, K. L. (1975). Palaeoecology of a bituminous shale – The Lower Oxford Clay of central England. *Palaeontology, 18*, 443–482.

Falini, G., Albeck, S., Weiner, S., & Addadi, L. (1996). Control of aragonite or calcite polymorphism by mollusk shell macromolecules. *Science, 271*(5245), 67–69.

Finnegan, S., & Droser, M. L. (2008). Reworking diversity: Effects of storm deposition on evenness and sampled richness, Ordovician of the Basin and Range, Utah and Nevada, USA. *Palaios, 23*, 87–96.

Flessa, K. W., & Brown, T. J. (1983). Selective solution of macroinvertebrate calcareous hard parts: A laboratory study. *Lethaia, 16*, 193–205.

Fletcher, C. J. N. (1988). Tidal erosion, solution cavities and exhalative mineralization associated with the Jurassic unconformity at Ogmore, South Glamorgan. *Proceedings of the Geologists' Association, 99*, 1–14.

Fraiser, M. L., & Bottjer, D. J. (2004). The non-actualistic Early Triassic gastropod fauna; a case study of the Lower Triassic Sinbad Limestone Member. *Palaios, 19*, 259–275.

Freiwald, A. (1995). Bacteria-induced carbonate degradation – A taphonomic case- study of *Cibicides Lobatulus* from a high-boreal carbonate setting. *Palaios, 10*(4), 337–346.

Fursich, F. T., & Pandey, D. K. (2003). Sequence stratigraphic significance of sedimentary cycles and shell concentrations in the Upper Jurassic-Lower Cretaceous of Kachchh, western India. *Palaeogeography, Palaeoclimatology, Palaeoecology, 193*(2), 285–309.

Glover, C. P., & Kidwell, S. M. (1993). Influence of organic matrix on the postmortem destruction of molluscan shells. *Journal of Geology, 101*(6), 729–747.

Gnoli, M. (2003). Northern Gondwanan Siluro-Devonian palaeogeography assessed by cephalopods. *Palaeontologia Electronica, 5*(2), 1–19.

Green, G. W., & Welch, F. B. A. (1965). *Geology of the country around Wells and Cheddar.* London: HMSO.

Green, M. A., JM, E., Boudreau, C. L., Moore, R. L., & Westman, B. A. (2004). Dissolution mortality of juvenile bivalves in coastal marine deposits. *Limnology and Oceanography, 49*, 727–734.

Hardie, L. A. (1996). Secular variation in seawater chemistry: An explanation for the coupled secular variation in the mineralogies of marine limestones and potash evaporites over the past 600 my. *Geology, 24*(3), 279–283.

Harper, E. M. (1998). The fossil record of bivalve molluscs. In S. K. Donovan & C. R. C. Paul (Eds.), *The adequacy of the fossil record*. Chichester: Wiley.

Harper, E. M., Palmer, T. J., & Alphey, J. R. (1997). Evolutionary response by bivalves to changing Phanerozoic sea-water chemistry. *Geological Magazine, 134*(3), 403–407.

Hendry, J. P. (1993). Calcite cementation during bacterial manganese, iron and sulphate reduction in Jurassic shallow marine carbonates. *Sedimentology, 40*, 87–106.

Hendry, J. P., Ditchfield, P. W., & Marshall, J. D. (1995). Two-stage neomorphism of Jurassic aragonitic bivalves: Implications for early diagenesis. *Journal of Sedimentary Geology, A65*(1), 214–224.

Hendry, J. P., Trewin, N. H., & Fallick, A. E. (1996). Low Mg-calcite marine cement in Cretaceous turbidites: Origin, spatial distribution and relationship to sea-water chemistry. *Sedimentology, 43*, 877–900.

Hendy, A. J. W., Kamp, P. J. J., & Vonk, A. J. (2006). Cool-water shell bed taphofacies from Miocene-Pliocene shelf sequences in New Zealand: Utility of taphofacies in sequence stratigraphic analysis. In H. M. Pedley & G. Carannante (Eds.), *Cool-water carbonates. Geological Society Special Publication, 255*, 283–305.

Hoare, R. D., & Pojeta, J., Jr. (2006). Ordovician Polyplacophora (Mollusca) from North America. *Journal of Paleontology, 80*(3, Suppl II), 1–28.

Hodges, P. (1987). Lower Lias (Lower Jurassic) Bivalvia from South Wales and adjacent areas. Unpublished Ph.D. thesis, University of Wales, Swansea.

Holdaway, H. K., & Clayton, C. J. (1982). Preservation of shell microstructure in silicified brachiopods from the Upper Cretaceous Wilmington Sands of Devon. *Geological Magazine, 119*(4), 371–382.

Holland, S. M., & Patzkowsky, M. E. (2004). Ecosystem structure and stability: Middle Upper Ordovician of central Kentucky, USA. *Palaios, 19*, 316–331.

Jaanusson, V. (1966). Fossil brachiopods with probable aragonitic shell. *Geologiska Föreningens i Stockholm Förhandlingar, 88*, 279–281.

Jaanusson, V., Skoglund, R., Laufeld, S. (1979). Lower Wenlock faunal and floral dynamics – Vattenfallet section, Gotland. *Sveriges Geologiska Undersökning, Serie C, 762*, 294 pp.

James, N. P., Bone, Y., & Kyser, T. K. (2005). Where has all the aragonite gone? – Mineralogy of Holocene neritic cool-water carbonates, southern Australia. *Journal of Sedimentary Research, 75*(3), 454–463.

Johnson, M. E. (1996). Stable cratonic sequences and a standard for Silurian eustasy. In B. E. Witzke, G. A. Ludvigsen, & J. Day (Eds.), *Palaeozoic sequence stratigraphy: Views from the North American craton. Geological Society of America Special Papers 306*, 203–211

Johnston, P. A. (1993). Lower Devonian Pelecypoda from southeastern Australia. *Association of Australasian Palaeontologists, Memoir, 14*, 134 pp.

Kenyon-Roberts, S. M. (1995). The petrography and distribution of some calcite sea hardgrounds. Unpubl Ph.D. thesis, University of Reading.

Kidder, D. L., & Erwin, D. H. (2001). Secular distribution of biogenic silica through the Phanerozoic: comparison of silica-replaced fossils and bedded cherts at the series level. *Journal of Geology, 109*, 509–522.

Kidwell, S. M. (1986). Taphonomic feedback in Miocene assemblages: testing the role of dead hard parts in benthic communities. *Palaios, 1*, 239–255.

Kidwell, S. M. (1991). The stratigraphy of shell concentrations. In P. A. Allison & D. E. G. Briggs (Eds.), *Taphonomy: Releasing the data locked in the fossil record*. New York: Plenum.

Kidwell, S. M. (2005). Shell composition has no net impact on large-scale evolutionary patterns in mollusks. *Science, 307*(5711), 914–917.

Kidwell, S. M., & Bosence, D. W. J. (1991). Taphonomy and time-averaging of marine shelly faunas. In P. A. Allison & D. E. G. Briggs (Eds.), *Taphonomy: Releasing the data locked in the fossil record*. New York: Plenum.

Kidwell, S. M., & Brenchley, P. J. (1994). Patterns in bioclastic accumulation through the Phanerozoic – changes in input or in destruction. *Geology, 22*(12), 1139–1143.

Kidwell, S. M., & Brenchley, P. J. (1996). Evolution of the fossil record; thickness trends in marine skeletal accumulations and their implications. In D. Jablonski, D. H. Erwin, & J. H. Lipps (Eds.), *Evolutionary paleobiology*. Chicago: University of Chicago Press.

Kidwell, S. M., Best, M. M. R., & Kaufman, D. S. (2005). Taphonomic trade-offs in tropical marine death assemblages: Differential time averaging, shell loss, and probable bias in siliciclastic vs. carbonate facies. *Geology, 33*(9), 729–732.

Kier, P. (1977). The poor fossil record of the regular echinoid. *Paleobiology, 3*, 168–174.

King, A. H. (1990). Lower and Middle Ordovician Cephalopoda of Baltoscandia. Unpubl. Ph.D. thesis, University of Wales, Swansea.

Knoerich, A. C., & Mutti, M. (2006). Missing aragonitic biota and the diagenetic evolution of heterozoan carbonates: A case study from the Oligo-Miocene of the central Mediterranean. *Journal of Sedimentary Research, 76*(5–6), 871–888.

Kříž, J. (1992). Silurian field excursions: Prague Basin (Barrandian), Bohemia. *National Museum Wales, Geology Series, 13*, 1–111.

Krystyn, L., Richoz, S., Baud, A., & Twitchett, R. J. (2003). A unique Permian–Triassic boundary section from the Neotethyan Hawasina Basin, Central Oman Mountains. *Palaeogeography, Palaeoclimatology, Palaeoecology, 191*, 329–344.

Ku, T. C. W., Walter, L. M., Coleman, M. L., Blake, R. E., & Martini, A. M. (1999). Coupling between sulfur recycling and syndepositional carbonate dissolution: Evidence from oxygen and sulfur isotope composition of pore water sulfate, South Florida Platform, USA. *Geochimica et Cosmochimica Acta, 63*, 2529–2546.

Laufeld, S., & Jeppsson, L. (1976). Silicification and bentonites in the Silurian of Gotland. *Geologiska Föreningens i Stockholm Förhandlingar, 98*(1), 31–44.

Li, X., & Droser, M. L. (1997). Nature and distribution of Cambrian shell concentrations: Evidence from the Basin and Range Province of the Western United States (California, Nevada, and Utah). *Palaios, 12*(2), 111–126.

Li, X., & Droser, M. L. (1999). Lower and Middle Ordovician shell beds from the Basin and Range Province of the Western United States (California, Nevada, and Utah). *Palaios, 14*(3), 215–233.

Liljedahl, L. (1983). Two silicified bivalves from Gotland. *Sveriges Geologiska Undersökning, Serie C, 799*, 51 pp.

Liljedahl, L. (1984). Silurian silicified bivalves from Gotland. *Sveriges Geologiska Undersökning, Serie C, 804*, 1–82.

Liljedahl, L. (1985). Ecological aspects of a silicified bivalve fauna from the Silurian of Gotland. *Lethaia, 18*(1), 53–66.

Locklair, R. E., & Lerman, A. (2005). A model of Phanerozoic cycles of carbon and calcium in the global ocean: Evaluation and constraints on ocean chemistry and input fluxes. *Chemical Geology, 217*(1–2), 113–126.

Lowenstam, H. A., & Weiner, S. (1989). *On biomineralization*. New York: Oxford University Press.

Maliva, R. G., Knoll, A. H., & Siever, R. (1989). Secular change in chert distribution: A reflection of evolving biological participation in the silica cycle. *Palaios, 4*, 519–532.

Martill, D.M., & Hudson, J.D. (1991). Fossils of the Oxford Clay. *Palaeontological Association Field Guide to Fossils,* 4.

Martill, D. M., Taylor, M. A., & Duff, K. L. (1994). The trophic structure of the Peterborough Member, Oxford Clay Formation (Jurassic), UK. *Journal of the Geological Society of London, 151*(1), 173–194.

Martin, R. E. (1999). *Taphonomy: A process approach. Cambridge Paleobiology Series 4.* Cambridge: Cambridge University Press.

Melim, L. A., Westphal, H., Swart, P. K., Eberli, G. P., & Munnecke, A. (2002). Questioning carbonate diagenetic paradigms: evidence from the Neogene of the Bahamas. *Marine Geology, 185*(1–2), 27–53.

Melim, L. A., Swart, P. K., & Eberli, G. P. (2004). Mixing-zone diagenesis in the subsurface of Florida and The Bahamas. *Journal of Sedimentary Research, 74*(6), 904–913.

Mitchell, M. (1987). The fossil collection of C.B. Salter from Cliff Quarry, Compton Martin, Mendip Hills. *Geological Curator, 4*(8), 487–491.

Morse, J. W., & MacKensie, F. T. (1990). *Geochemistry of sedimentary carbonates. Developments in sedimentology 48*. New York: Elsevier.

Morse, J. W., Zullig, J. J., Bernstein, L. D., Millero, F. J., Milne, P., Mucci, A., et al. (1985). Chemistry of calcium-rich shallow water sediments in the Bahamas. *American Journal of Science, 285*, 147–185.

Munnecke, A., & Samtleben, C. (1996). The formation of micritic limestones and the development of limestone-marl alternations in the Silurian of Gotland, Sweden. *Facies, 34*, 159–176.

Munnecke, A., Westphal, H., Reijmer, J. J. G., & Samtleben, C. (1997). Microspar development during early marine burial diagenesis: a comparison of Pliocene carbonates from the Bahamas with Silurian limestones from Gotland (Sweden). *Sedimentology, 44*(6), 977–990.

Nelson, C. S., Winefield, P. R., Hood, S. D., Caron, V., Pallentin, A., & Kamp, P. J. J. (2003). Pliocene Te Aute limestones, New Zealand: expanding concepts for cool-water shelf carbonates. *New Zealand Journal of Geology and Geophysics, 46*, 407–424.

Noe-Nygaard, N., Surlyk, F., & Piasecki, S. (1987). Bivalve mass mortality caused by toxic dinoflagellate blooms in a Berriasian-Valanginian lagoon, *Bornholm Denmark, Palaios, 2*, 263–273.

Novack-Gottshall, P. M., & Miller, A. I. (2003). Comparative geographic and environmental diversity dynamics of gastropods and bivalves during the Ordovician Radiation. *Paleobiology, 29*(4), 576–604.

Nützel, A., & Schulbert, C. (2005). Facies of two important Early Triassic gastropod lagerstatten: implications for diversity patterns in the aftermath of the end-Permian mass extinction. *Facies, 51*, 495–515.

Orr, P. J. (2003). Ecospace utilization in early Phanerozoic deep-marine environments: Deep bioturbation in the Blakely Sandstone (Middle Ordovician), Arkansas, USA. *Lethaia, 36*(2), 97–106.

Oschmann, W. (1988). Kimmeridge clay sedimentation – A new cyclic model. *Palaeogeography, Palaeoclimatology, Palaeoecology, 65*(3–4), 217–251.

Oschmann, W. (1991). Distribution, dynamics and palaeoecology of Kimmeridgian (Upper Jurassic) shelf anoxia in western Europe. In R. V. Tyson & T. H. Pearson (Eds.), *Modern and ancient continental shelf anoxia. Geological Society London, Special Publication, 58*, 381–395.

Oschmann, W. (1993). Environmental oxygen fluctuations and the adaptive response of marine benthic organisms. *Journal of the Geological Society of London, 150*, 187–191.

Palmer, T. J., & Wilson, M. A. (2004). Calcite precipitation and dissolution of biogenic aragonite in shallow Ordovician calcite seas. *Lethaia, 37*(4), 417–427.

Palmer, T. J., Hudson, J. D., & Wilson, M. A. (1988). Paleoecological evidence for early aragonite dissolution in ancient calcite seas. *Nature, 335*(6193), 809–810.

Pojeta, J., Jr. (1971). Review of Ordovician pelecypods. *U.S. Geological Survey Professional Papers* 695.

Porter, S. M. (2007). Seawater chemistry and early carbonate biomineralization. *Science, 316*(5829), 1302.

Powell, E. N., Callender, W. R., Staff, G. M., Parsons-Hubbard, K. M., Brett, C. E., Walker, S. E., et al. (2008). Molluscan shell condition after eight years on the sea floor - taphonomy in the Gulf of Mexico and Bahamas. *Journal of Shellfish Research, 27*(1), 191–225.

Radley, J. D., & Barker, M. J. (2000). Palaeoenvironmental significance of storm coquinas in a Lower Cretaceous coastal lagoonal succession, Vectis Formation, Isle of Wight, southern England. *Geological Magazine, 137*, 193–205.

Ratter, V. A., & Cope, J. C. W. (1998). New Silurian neotaxodont bivalves from South Wales and their phylogenetic significance. *Palaeontology, 41*, 975–991.

Riding, R., & Liang, L. (2005). Geobiology of microbial carbonates: metazoan and seawater saturation state influences on secular trends during the Phanerozoic. *Palaeogeography, Palaeoclimatology, Palaeoecology, 219*(1–2), 101–115.

Ries, J. B. (2005). Aragonite production in calcite seas: Effect of seawater Mg/Ca ratio on the calcification and growth of the calcareous alga *Penicillus capitatus*. *Paleobiology, 31*, 449–462.

Rollins, H. B., Eldredge, N., & Spiller, J. (1971). Gastropoda and Monoplacophora of the Solsville Member, Middle Devonian Marcellus Formation in the Chenango valley, NewYork State. *Bulletin of the American Museum of Natural History, 144*(2), 133–170.

Rude, P. D., & Aller, R. C. (1991). Fluorine mobility during early diagenesis of carbonate sediment: An indicator of mineral transformations. *Geochimica et Cosmochimica Acta, 55*, 2491–2509.

Runnegar, B., & Bentley, C. (1983). Anatomy, ecology and affinities of the Australian Early Cambrian bivalve *Pojetaia runnegari* Jell. *Journal of Paleontology, 57*(1), 73–92.

Sandberg, P. A. (1983). An oscillating trend in Phanerozoic non-skeletal carbonate mineralogy. *Nature, 305*(5929), 19–22.

Sanders, D. (2001). Burrow-mediated carbonate dissolution in rudist biostromes (Aurisina, Italy): Implications for taphonomy in tropical, shallow subtidal carbonate environments. *Palaeogeography, Palaeoclimatology, Palaeoecology, 168*(1–2), 39–74.

Sanders, D. (2003). Syndepositional dissolution of calcium carbonate in neritic carbonate environments: Geological recognition, processes, potential significance. *Journal of African Earth Sciences, 36*, 99–134.

Sanders, D. (2004). Potential significance of syndepositional carbonate dissolution for platform banktop aggradation and sediment texture: A graphic modeling approach. *Austrian Journal of Earth Science, 95*(96), 71–79.

Schulbert, J. K., Kidder, D. L., & Erwin, D. H. (1997). Silica-replaced fossils through the Phanerozoic. *Geology, 25*(11), 1031–1034.

Sepkoski, J. J. (1984). A kinetic-model of Phanerozoic taxonomic diversity. 3. Post-Paleozoic families and mass extinctions. *Paleobiology, 10*(2), 246–267.

Smith, A. M., & Nelson, C. S. (2003). Effects of early sea-floor processes on the taphonomy of temperate shelf skeletal carbonate deposits. *Earth Science Reviews, 63*(1–2), 1–31.

Speyer, S. E., & Brett, C. E. (1991). Taphofacies controls: Background and episodic processes in fossil assemblage preservation. In P. A. Allison & D. E. G. Briggs (Eds.), *Taphonomy: Releasing the data locked in the fossil record*. New York: Plenum.

Stanley, S. M. (2006). Influence of seawater chemistry on biomineralization throughout Phanerozoic time: Paleontological and experimental evidence. *Palaeogeography, Palaeoclimatology, Palaeoecology, 232*, 214–236.

Stanley, S. M., & Hardie, L. A. (1998). Secular oscillations in the carbonate mineralogy of reef-building and sediment-producing organisms driven by tectonically forced shifts in seawater chemistry. *Palaeogeography, Palaeoclimatology, Palaeoecology, 144*, 3–19.

Talent, J. A., Mawson, R., Aitchison, J. C., Becker, R. T., Bell, K. N., Bradshaw, M. A., Burrow, C. J., Cook, A. G., Dargan, G. M., Douglas, J. G., Edgecombe, G. D., Feist, M., Jones, P. J., Long, J. A., Phillips-Ross, J. R., Pickett, J. W., Playford, G., Rickards, R. B., Webby, B. D., Winchester-Seeto, T., Wright, A. J., Young, G. C., & Zhen, Y.-Y. (2000). Devonian palaeobiogeography of Australia and adjoining regions. In A. J. Wright, G. C. Young, J. A. Talent , & J. R. Laurie (Eds.), *Palaeobiogeography of Australasian faunas and floras. Association of Australasian Palaeontologists, Memoir, 23*, 167–258.

Tassell, C. B. (1982). Gastropods from the Early Devonian 'Receptaculites' Limestone, Taemas, New South Wales. *Records of the Queen Victoria Museum, Launceston, 77*, 1–59.

Taylor, J. D., Kennedy, W. J., & Hall, A. (1969). The shell structure and mineralogy of the Bivalvia: Introduction, Nuculacea–Trigonacea. *Bulletin of British Museum Natural History London, 3*, 1–125.

Tedesco, L. P., & Aller, R. C. (1997). ^{210}Pb chronology of sequences affected by burrow excavation and infilling: Examples from shallow marine carbonate sediment sequences, Holocene South Florida and Caicos Platform, British West Indies. *Journal of Sedimentary Research, 67*, 36–46.

Tsujita, C. J., Brett, C. E., Topor, M., & Topor, J. (2006). Evidence of high-frequency storm disturbance in the Middle Devonian Arkona Shale, southwestern Ontario. *Journal of Taphonomy, 4*(2), 49–68.

Valentine, J. W., Jablonski, D. S., & Roy, K. (2006). Assessing the fidelity of the fossil record by using marine bivalves. *Proceedings of the National Academy of Sciences OF the United States of America, 103*, 6599–6604.

Walker, K. R., & Diehl, W. W. (1985). The role of marine cementation in the preservation of Lower Palaeozoic assemblages. *Philosophical Transactions of the Royal Society London B, 311*, 143–153.

Walter, L. M., & Burton, E. A. (1990). Dissolution of platform carbonate sediments in marine pore fluids. *American Journal of Science, 290*, 601–643.

Walter, L. M., Bischof, S. A., Patterson, W. P., & Lyons, T. W. (1993). Dissolution and recrystallization in modern shelf carbonates – Evidence from pore-water and solid-phase chemistry. *Philosophical Transactions of the Royal Society London B, 344*(1670), 27–36.

Watkins, R. (1996). Skeletal composition of Silurian benthic marine faunas. *Palaios, 11*(6), 550–558.

Watkins, R. (2000). Silurian reef-dwelling brachiopods and their ecologic implications. *Palaios, 15*, 112–119.

Webby, B. D., & Percival, I. G. (1983). Ordovician trimerellacean brachiopod shell beds. *Lethaia, 16*, 215–232.

Westphal, H., & Munnecke, A. (2003). Limestone-marl alternations: A warm-water phenomenon? *Geology, 31*(3), 263–266.

Wheeley, J. R., & Twitchett, R. J. (2005). Palaeoecological significance of a new Griesbachian (Early Triassic) gastropod assemblage from Oman. *Lethaia, 38*(1), 37–45.

Wheeley, J. R., Cherns, L., & Wright, V. P. (2008). Provenance of microcrystalline carbonate cement in limestone–marl alternations (LMA): Aragonite mud or molluscs? *Journal of the Geological Society London, 165*(1), 395–403.

Wilby, P. R., Hudson, J. D., Clements, R. G., & Hollingworth, N. T. J. (2004). Taphonomy and origin of an accumulate of soft-bodied cephalopods in the Oxford Clay Formation (Jurassic, England). *Palaeontology, 47*(5), 1159–1180.

Wilcox, C. J., & Lockley, M. G. (1981). A reassessment of facies and faunas in the type Llandeilo (Ordovician), Wales. *Palaeogeography, Palaeoclimatology, Palaeoecology, 34*(3–4), 285–314.

Williams, A., Lockley, M. G., & Hurst, J. M. (1981). Benthic palaeocommunities represented in the Ffairfach Group and coeval Ordovician successions of Wales. *Palaeontology, 24*(4), 661–694.

Wright, V. P., & Cherns, L. (2004). Are there 'black holes' in carbonate deposystems? *Geologica Acta, 2*, 285–290.

Wright, V. P., Cherns, L., & Hodges, P. (2003). Missing molluscs: Field testing taphonomic loss in the Mesozoic through early large-scale aragonite dissolution. *Geology, 31*(3), 211–214.

Chapter 4
Comparative Taphonomy and Sedimentology of Small-Scale Mixed Carbonate/Siliciclastic Cycles: Synopsis of Phanerozoic Examples

Carlton E. Brett, Peter A. Allison, and Austin J.W. Hendy

Contents

C.E. Brett (✉)
Department of Geology, University of Cincinnati, Cincinnati, OH 45221, USA
e-mail: C. Brett@uc.edu

P.A. Allison
Department of Earth Science and Engineering, Imperial College, London, UK
e-mail: P.A.Allison@imperial.ac.uk

A.J.W. Hendy
Center for Tropical Palaeoecology and Archaeology, Smithsonian Tropical Research Institute, Panamá, República de Panamá, USA
and
Department of Geology and Geophysics, Yale University, New Haven, CT 06510, USA
e-mail: hendyaj@si.edu

P.A. Allison and D.J. Bottjer (eds.), *Taphonomy: Process and Bias Through Time*,
Topics in Geobiology 32, DOI 10.1007/978-90-481-8643-3_4,

Abstract Small scale cycles deposited over 10–100 kyr are a common component of Phanerozoic shelfal deposits. A combination of detailed outcrop analysis and data-mining from published literature of cycles largely deposited in greenhouse regimes reveals a series of recurring sedimentological, paleoecological, and taphonomic motifs. In general, each cycle is composed of three to four components: (a) a basal skeleton-rich bed with evidence of condensation and, in some cases mineralization; (b) a medium-dark gray siliciclastic mudstone/shale interval; (c) a calcareous and/or silty mudstone interval with common concretionary, diagenetic overprint. A series of exemplars are highlighted from proximal and distal shelf settings and described using a depositional sequence approach. The cycles studied include examples deposited under greenhouse (Cambrian, Ordovician, Devonian, Jurassic and Cretaceous) and, for comparison purposes, icehouse (Neogene) conditions. The fact that repetitive patterns can characterize deposits that formed over a 500 million year interval is striking. The primary taphonomic moderator in these cycles is rate of sedimentation, which varies exponentially from sediment-starved concentrations to obrutionary deposits. The occurrence of a persistent motif over this time scale suggests that biological innovations, which might be expected to impact upon fossil preservation, have in fact been overprinted by the extremes of sedimentation preserved in these small-scale cycles. Having a skeleton, which is twice as resistant to abrasion, is of little import when sedimentation is dominated by the extremes: instant obrution or condensation.

1 Introduction

Small-scale cyclicity, typically of meter-scale, is pervasive through geologic time (Fig. 1). The advent of cyclostratigraphy recognizes this ubiquity of cyclic sedimentation as a critical tool not only in stratigraphic correlation (Vail 1987; Vail et al. 1991; Coe 2003; Catuneanu 2006), but also for understanding the temporal context of physical and biotic processes (House 1985, 1995). The inference that many cycles are periodic and related to Milankovitch orbital variations potentially enables them to be used as a geologic "metronome" (Gilbert 1895; Berger et al. 1992; deBoer and Smith 1994; House and Gale 1995; Hinnov 2000, 2004). This regular "pulse" is expressed to greater or lesser extent at different times during the Phanerozoic as a result of long-term secular variations in climate and environmental regime (Fig. 1). Cycles may be amplified during global icehouse climatic phases as the result of large amplitude glacioeustatic fluctuations. Calcite vs. aragonite seas also may play a role in skeletal carbonate preservation (see Wilson and Palmer 1992; Stanley and Hardie 1998). A general sincrease in deep bioturbation has no doubt had the effect of homogenizing thin beds and producing distinctive fabrics

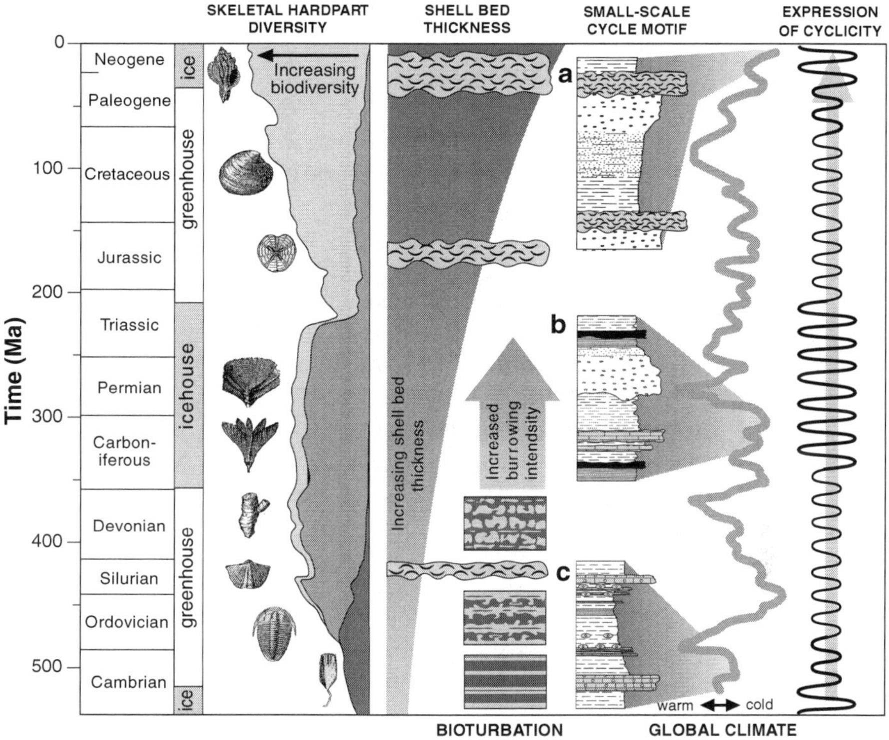

Fig. 1 Schematic of the Phanerozoic time scale illustrating the interplay between the regular beat of climatic/sea-level cycles and long term, secular changes in climate, the greenhouse-icehouse super-cycles, changes in biodiversity of skeletonized marine benthic faunas (note: black, medium gray, and light gray shaded areas represent the biodiversity of "Cambrian". "Paleozoic", and "Modern" faunas, respectively) and bioturbation. Small-scale cycle motifs include: (**a**) late Neogene decameter-scale cycle; (**b**) Carboniferous coal-centered cyclothem, 3–5 m thick; (**c**) Cambrian meter-scale cycle. Note amplification of cycles during icehouse times. Diversity curve modified from Sepkoski (1997), fossil images from Sepkoski (1984); see text for further discussion of the schematic

(Droser and Bottjer 1988). Moreover, the abundance and type of skeletal material available for formation of shell beds has varied greatly from Cambrian to Neogene times (Fig. 1). A secular increase in skeletal production and durability may have led to a general increase in thickness of skeletal limestones, a key component of cyclic sedimentation (Kidwell and Brenchley 1994; Kidwell et al. 1996).

Milankovitch band oscillations of climate, sea level and ocean chemistry produce distinctive lithologic, taphonomic and paleoecological signatures that may be modulated by the prevailing climatic regime of greenhouse vs. icehouse megacycles (Fischer 1980, 1984). A number of parameters may co-vary in cycles, including relative water depth, sedimentation rate, geochemistry, and benthic oxygenation. Milankovitch-related climatic variations may produce transgressive-regressive, redox, or dissolution cycles, cyclic variations in sedimentation, or combinations thereof (see Fischer 1980, 1984; Schwarzacher and Fischer 1982; LaFerriere et al. 1987; Einsele and Ricken 1991; Ricken 1991, 1994). These parameters are among the most important in controlling marine biofacies and their preservation as fossil assemblages (Bennett 1990; DiMichele et al. 2004). For example, water depth oscillations may cause lateral tracking of distinctive benthic communities (Brett et al. 2007b) leading to variation in the type and abundance of skeletal accumulations or trace fossil assemblages (Savrda and Bottjer 1991, 1994). Predictable variations in sedimentation rate, related to sea level or climate oscillations, in turn, may have a very significant influence on the concentration or dilution of skeletal material as predicted by the R-sediment model for shell bed accumulation of Kidwell (1985, 1986, 1989, 1991a). In addition, variation in burial rates will have an important influence on the taphonomy of skeletal material. Modelling predicts that the degree of taphonomic degradation should be positively correlated with extent of skeletal concentration, if rates of burial are indeed responsible for skeletal concentration. Alternatively, skeletal concentration may be related to variations in skeletal production (Tomasovych et al. 2006). In turn, correlated variations in sediment geochemistry may leave a strong imprint in preservation of fossils. For instance, the development of a persistent zone of sulfate reduction during times of lowered sedimentation may lead to formation of carbonate concretions that encapsulate fossils.

Herein, examples of small-scale cycles through the Phanerozoic (Figs. 1 and 2) are reviewed and compared, a generalized model of physical–chemical change through an idealized cycle is proposed, and a standardized approach to describing cycles and their litho-, bio- and taphofacies is established. This paper begins with an overview of well-described late Neogene examples of short-term cycles (Fig. 3). This facilitates a comparison of cycle architecture in ancient examples to those

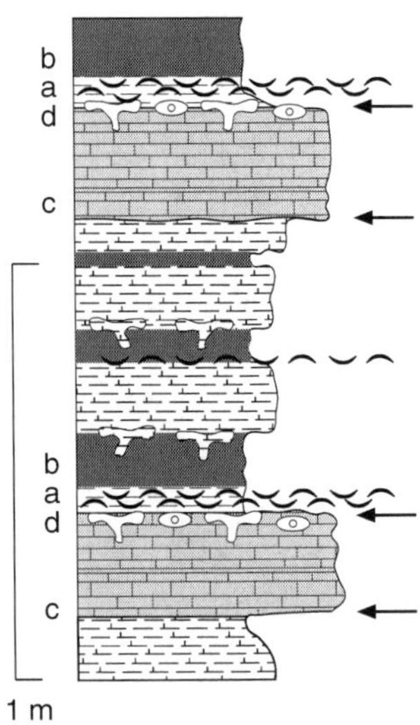

Fig. 2 Idealized cycles modeled on the Jurassic Blue Lias, showing possible positions for placing cycle boundaries. (**a**) At base of thin shell hash; (**b**) at base of black shale; (**c**) at base of concretionary carbonate; (**d**) at top of concretionary carbonate. Note that the sharp base and top of the concretionary carbonates (*marked by arrows*) probably do not represent useful cycle boundaries because they typically represent diagenetic rather than primary sedimentary contacts

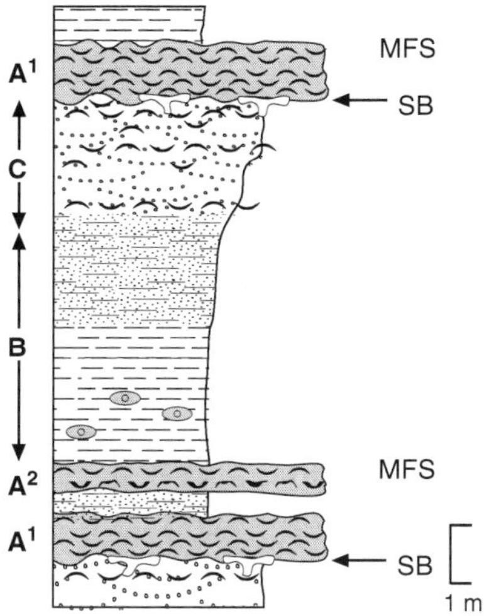

Fig. 3 Idealized, 40-thousand year cyclothem from the Pleistocene Matemateaonga Formation in New Zealand. Sequence interpretation following Hendy et al. (2006). SB, sequence boundary; MFS, maximum flooding surface. Note the following portions of this cycle. (**A**-1) Basal transgressive shell bed; (**A**-2) secondary condensed "backlap" shell bed; (**B**) thick, fine-grained highstand portion of cycle; may show concretions; (**C**) regressive or shallowing-upward portion of cycle showing increasing abundance of graded storm deposited shell hash and sandstone beds

accumulated in an interval from which changes in global sea level oscillation, basin subsidence, and sediment flux is well understood. This summary is followed by a review of examples of taphofacies in small-scale cycles from Cambrian to Paleogene time in order to determine common or unifying features of cycles.

2 Small-Scale Sedimentary Cycles

2.1 Defining Cycles

Herein, small-scale cycles are defined as those that encompass tens to a few hundreds of thousands of years (kyr). Thickness varies substantially within such sequences and may vary by an order of magnitude across their outcrop area.

We specifically focus on offshore mixed fine-grained siliciclastic/carbonate cycles, particularly those that show concentrations of skeletal material in portions of the cycle. Particular focus is applied to evaluating the impact of climatic and sea-level oscillation on taphonomy and sedimentation. To provide a degree of control on the motifs of cycles for comparative purposes we concentrate on cycles formed during greenhouse phases of the Phanerozoic Eon, i.e. early to middle Paleozoic, the Jurassic-Cretaceous, and Paleogene (Figs. 1 and 2). Future work will be directed towards elucidation and comparison of cyclic patterns of a similar scale developed during icehouse phases, including those from the Carboniferous-Permian and Neogene.

An important question that arises is: how are the cycle boundaries to be defined? If sedimentation is continuous and the cycle is symmetrical, one could arbitrarily choose to "begin" a cycle at any particular contact of recurring facies, provided that this point is picked consistently. In many cases, however, sedimentation is discontinuous, and the pattern is more obvious. A sharp surface separates one microfacies from another forming a logical place to divide cycles. It is important in comparative taphonomic studies to recognize analogous portions of cycles and to describe them consistently and systematically with respect to cycle base and top (Fig. 2).

In certain cases the most obvious lithological contacts may not actually represent the best boundary for cycles (Fig. 2). An excellent example is the well-known Blue Lias cycles from the Lower Jurassic of southern England (Hallam 1957, 1960, 1964, 1986; Elliott 1996; see below). These cycles are comprised of sub-tabular limestone and shale alternations. At first glance it would seem obvious to place the cycle base either at the top or base of the limestone beds (Fig. 2). However, these bed surfaces are irregular and actually may cut across primary sedimentary boundaries. They clearly represent diagenetic boundaries, the positions of fronts of concretionary cementation within preexisting sediments. In actuality, sharp primary sedimentary surfaces occur slightly above the tops of the limestones at the bases of black or dark gray shales, or at thin skeletal lags along the boundaries between such dark shales and underlying gray marls. The latter clearly represent the primary sediment from which the concretionary carbonates have formed by diagenetic enhancement (Hallam 1964, 1986; Weedon 1985; House 1986). Moreover, not every cycle shows the diagenetic cementation (Fig. 2).

Herein, cycle bases are defined as the bases of skeletal lag deposits that sharply overlie gray, frequently diagenetically enhanced carbonates. Thus, the concretionary beds, while probably connected in terms of diagenetic history with overlying sharp surface and/or skeletal lag deposits, occur in sediments that are portions of the underlying cycles. The most obvious prominently weathering diagenetic limestones of the Blue Lias and Collingwood Formations lie completely within cycles, not at their boundaries.

2.2 *Identifying Analogous Phases of Cycles*

Another problem encountered in a comparative taphonomic approach is to recognize analogous phases of cycles in terms of pattern and inferred process. Having defined an arbitrary, but objective, lower boundary, we can examine attributes of basal, middle, and upper parts of the cycles. Frequently, in fact, three such divisions can be recognized and we suggest that they are small-scale analogs of the transgressive, highstand, and falling stage (regressive) portion of larger scale (smaller order) cycles. That such sequence-like attributes can occur in cycles of just a few 10 s of kyr duration is strongly suggested by a series of studies on thick successions of mixed siliciclastic and shell bed carbonates from the Neogene of New Zealand (Abbott and Carter 1994; Abbott 1997; Naish and Kamp 1997; McIntyre and Kamp 1998; Hendy et al. 2006), Japan (Kitamura and Kondo 1990; Ito 1992; Kitamura et al. 1994; Kamataki and Kondo 1997; Kondo et al. 1998), Italy (Rio et al. 1996; Dominici 2001) California (Carter et al. 2002), and South America (Del Rio et al. 2001;

Di Celma et al. 2002; Cantalamessa et al. 2005). These age-constrained cycles were influenced by 41,000–100,000 year. Milankovitch oscillation; in some cases they can be linked with oxygen isotope stages and even reasonable estimates of magnitudes of sea level fluctuation. They therefore provide an excellent guide for interpreting more ancient and more distal cycles for which age and depth constraints are typically poorer. Although the Neogene cycles were developed during an icehouse phase we argue that they supply a model for the general pattern and frequency, though not the amplitude of sedimentary cycles seen throughout the Phanerozoic.

Despite differences in details, all of these Neogene studies show a common motif for the cycles: (a) one to two shell-rich beds overlying sharp, erosive surfaces, the lower bed typically referred to as an "onlap shell bed" (Naish and Kamp 1997), Type-1 shell bed (Abbott and Carter 1994) or "base of parasequence shell bed" (Banerjee and Kidwell 1991) with mechanical processing and amalgamation of shells, at least in proximal areas. The upper bed ("backlap shell bed" of Naish and Kamp 1997; "type-2 shell bed" of Abbott and Carter 1994; "mid parasequence bed" of Banerjee and Kidwell 1991) represents gradual accumulation during times of sedimentary condensation; (b) a thick, fine-grained succession with scattered fossils sometimes concentrated as patchy to lenticular obrutionary, tempestitic, or lag accumulations; (c) an upper vaguely coarsening-upward part typified by more sparsely fossiliferous coarser sediments with increased numbers of siltstone, calcisilitite or sandstone layers; fossils are typically sparse and scattered but may also be concentrated in lenses or at the bases of graded tempestite beds (Fig. 3). In distal examples the upper portions of the cycles may occur in thicker and more sparsely fossiliferous, silty mudstone or siltstone; in rare cases, concretions may occur in this portion of the section and its top may be marked by heavy bioturbation.

Not all sedimentary cycles relate to sea level; other possibilities include climatic variations and changes in storm frequency and intensity (Fig. 3). Nonetheless, many cycles preserve features that are at least analogous to those expected from sea level oscillations. These three portions of the cycle could be interpreted, respectively, as (a) early to late transgressive systems tract TST), (b) highstand (HST) and (c) regressive (RST; Naish and Kamp 1997) or falling stage systems tracts (FSST; Plint and Nummedal 2000), in some cases with (d) a modified silty-calcareous and diagenetically altered uppermost division. These lettered divisions (a–c and d, when present) for analogous phases of cycles will be used throughout this chapter. In some cases, particularly with distal representatives of cycles, the lower or a) portion of the cycle may be nearly or completely missing such that a sharp surface (flooding surface) separates the coarsest (typically shallowest) facies of the lower cycle from the finer grained (deeper) portion of the overlying one. In such cases, the small-scale cycles have what may be termed a "parasequence motif".

Nearly all of the Neogene shell bed studies have dealt with proximal shelf facies; there is not a similarly detailed paradigm for distal mixed siliciclastic-carbonate cycles. There are, however, still many aspects of these cycles that can be interpreted within the framework of the small-scale sequence paradigm. Complicating factors arise from the typically much thinner representation or absence of the TST portion, the addition of dark, organic-rich shale facies, and various early diagenetic phenomena, especially concretionary cementation of the upper HST or FFST sediments that is typically lacking or subdued in more proximal cycles.

Proximal-distal cycle comparison is complicated by the differing thickness proportions of analogous portions of cycles that are recorded in different areas. The Late Ordovician (Caradocian) cycles of the Kope Formation and Collingwood Member (Lindsay Formation; Brett et al. 2006; see below) are a good example of this problem. In the Kope cycles the most prominent portions (in terms of outcrop weathering) are ledges of skeletal limestone that define cycle bases (see Figs. 7–9). In the Collingwood the analogous unit (see Figs. 10 and 11), both in terms of taphofacies and inferred sedimentational history, are very thin, lenticular shell hashes that are not obvious on casual inspection. In several cycles these thin lags may be absent such that the overlying dark shales rest directly and sharply on underlying gray mudstones. This clear distinction may relate to rates of skeletal production; in proximal shelf areas the relatively higher production of robust skeletal items may lead to development of relatively thick skeletal accumulations during a given interlude of reduced sedimentation. The same or perhaps longer interval of sediment starvation may lead to a thin veneer of material in deepwater or proximal, high sedimentation areas, where small, often thin-shelled organisms produce very minor amounts of debris (Fig. 2). In such cases the cycle could be defined as having a "parasequence motif", whereas, in cases of high skeletal production, the relatively thick skeletal limestone forms an important component of the cycle, analogous to the transgressive systems tract of a larger scale depositional sequence.

The most prominent beds in a cycle may not be these basal shell-rich lags. In the Collingwood Member, concretionary limestone ledges, analogous to those of the Blue Lias, are very prominent. These could easily be confused with the ledge-forming limestones of the coeval Kope but their taphofacies and inferred origin are completely different. The concretionary limestone ledges are diagenetically cemented muds that actually show evidence of having been rapidly deposited. They are analogous to thin concretionary limestones or discrete concretions present below ledge-forming skeletal limestones in some distal Kope cycles.

We define the bases of small-scale siliciclastic/carbonate cycles as the sharp bases of skeletal limestones, or of overlying mudrocks where these skeletal concentrations are lacking. This is analogous with depositional sequences; in many cases the sharp bases of condensed skeletal rich deposits appear to represent an aggrading to retrograding trend toward more distal facies. However, a key question arises as to whether these surfaces are merely analogous to or "homologous" to sequence boundaries in the sense of being generated by the same type of process. It is conceivable, for example, that sharp surfaces of small-scale cycles may be generated by a combination of sediment starvation and/or intermittent, storm-generated erosion that requires no actual change of sea level. Nonetheless, many small-scale cycles do show a condensed basal shelly carbonate lag with taphonomic evidence for substantial reworking and processing of sediments; it is the analog of a transgressive lag. This compact interval is followed, often sharply, by a mudrock (shale, mudstone, marl) interval that is typically thicker than the shelly basal lag and commonly shows evidence of increased frequency, thickness and episodicity of mud or silt layers upward. Again, this portion of the small-scale cycle is somewhat analogous to a larger scale highstand to falling stage systems tract. Early diagenetic features, such as the above-noted tendency for upper portions of the mudstone to become cemented into concretionary or tabular layers, complicate this portion of many small-scale cycles. This cementation actually provides a unique taphonomic window because contained fossils may be buttressed against compaction.

3 Examples of Small-Scale Cycles in the Phanerozoic

Small-scale cyclicity occurs throughout the Phanerozoic sedimentary record. These recurrent facies are often inferred to relate to periods of climatic and/or sea level oscillation, although the forcing mechanisms and frequency of such changes are difficult to determine. Small-scale cycles from distal and proximal shelf settings not only provide favorable environmental settings for benthic communities, but also their preservation, through the interplay of cyclic sedimentation rate and episodicity (through storm action).

In the following sections we review examples of small-scale cycles in a consistent way that begins with the base of the most prominent skeletal concentrations, treated as the "shell-rich" hemicycle, and progresses upward through the mudstone/siltstone hemicycle. This paper focuses on offshore mid- to deep-ramp/basinal representatives of small-scale cycles, rather than nearshore end-members of cycles that have been described extensively in much previous sedimentological literature (Pattison 1995; Van Wagoner and Bertram 1995; Hampson et al. 1999; Miall and Mohamud 2001). To improve comparability of examples, we have confined our study to the relatively thin cycles typical of gently dipping ramps of epicontinental seas and distal foreland basins. For each exemplar, we discuss shallower and deeper examples, where possible these are correlative units; for simplicity we refer to these as "proximal" and "distal" cycles although it should be understood that in no case do we consider examples shallower than lower shoreface, so that "proximal" examples mainly record medial ramp settings.

We discuss proximal and distal examples of cycles in a systematic way that includes: (a) age and geographic location; (b) the thickness and variation of cycles; (c) their regional persistence and variation; (d) characteristics of the cycle base and top, and how they are defined; (e) variation in lithofacies within cycles; (f) the variation of ichnofacies and the relative abundance of ichnotaxa; (g) paleoecological characteristics of assemblages within cycle components; (h) the variation in taphofacies, for instance the articulation, fragmentation, and diagenetic alteration of skeletal hardparts; and, finally (i) variation in diagenetic facies, for instance the occurrence of concretions, pyrite, siderite, and organic matter.

3.1 Middle Cambrian: Great Basin USA

Meter and decameter-scale cycles from the Middle Cambrian of the Great Basin, USA (Liddell et al. 1997; Elrick and Snider 2002; Gaines and Droser 2005; Gaines et al. 2005; Brett et al. 2009; Figs. 4 and 5) accumulated on a distally steepened ramp

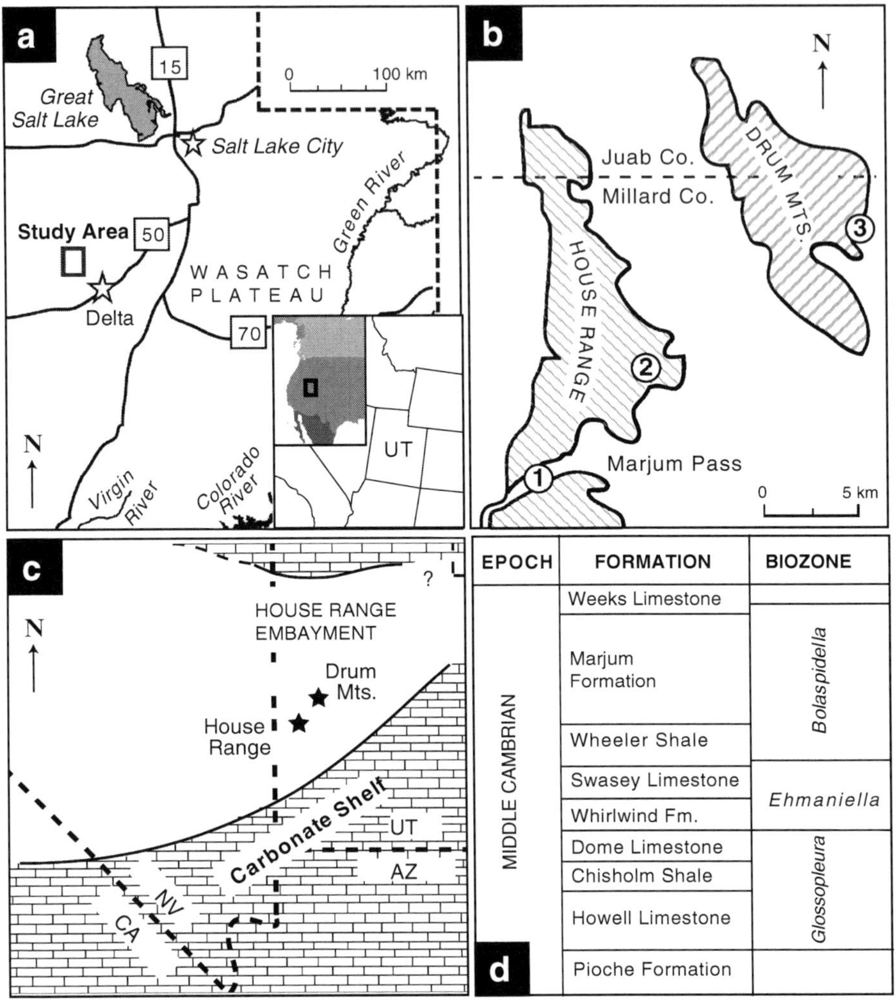

Fig. 4 Location map of study area in south-central Utah and generalized stratigraphic column. (**a**) Position of state of Utah in North America and of Delta, Utah and study area within Utah. (**b**) House Range and Drum Mountains locations, as follows: (1) Marjum Pass; (2) Wheeler Amphitheater; (3) Drum Mountains. (**c**) General paleogeography of the study area showing the position of the House Range Embayment, a fault-bounded intra-shelf basin in which muddy sediments of the Wheeler and Marjum formations accumulated. (**d**) Generalized stratigraphic column showing position of the Wheeler Shale in the Middle Cambrian (Modified from Hintze and Robison 1975; Hintze and Davis 2003)

into an intrashelf basin, the House Range Embayment (Rees 1986; Elrick and Snider 2002). Recurrent taphofacies, including both concentrations of disarticulated, fragmented debris as well as extraordinarily well preserved trilobites and rare soft bodied organisms, occur predictably within a spectrum of facies in sequences in the Middle Cambrian Wheeler, and Marjum Formations of the Drum Mountains and House Range in west-central Utah. The durations of the cycles are somewhat

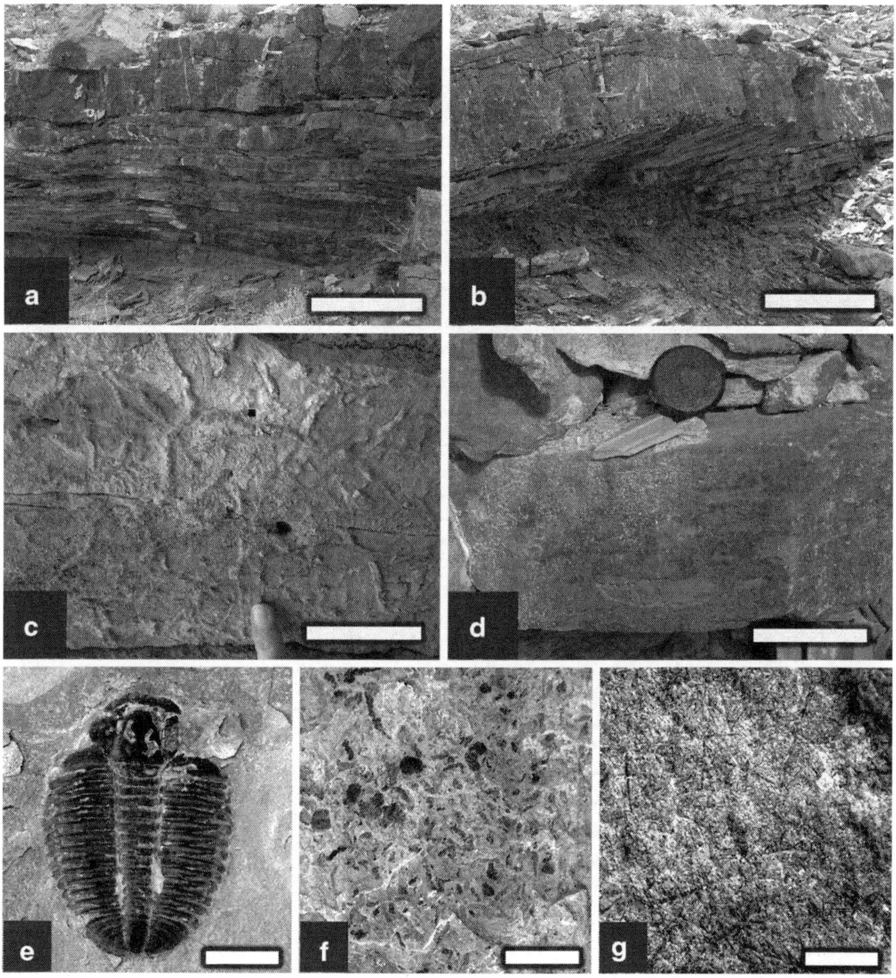

Fig. 5 Aspects of small-scale cycles in the Wheeler Shale, east flank of Sawooth Ridge, Drum Mountains. (**a**) T-1 marker bed, compact oolitic-oncolitic, ledge-forming limestone, sharply overlying a bundle of calcisiltites and fine calcarenites at top of underlying cycle; scale bar 50 cm. (**b**) T-2 limestone ledge sharply overhanging shales and calcisiltites of third upper Wheeler cycle; scale bar 50 cm. (**c**) Detail of the base of limestone ledge showing distinctly incised *Planolites* trace fossils indicative of firmground condition of the underlying muds; scale bar 15 cm. (**d**) Close-up of basal limestone ledge of T-3 showing intraclasts in oolitic packstone; scale bar 10 cm. Typical fossils of Wheeler gray, calcareous shale facies. (**e**) Articulated molted exoskeleton of *Elrathia kingi*; scale-bar is 1 cm; Wheeler Shale, Wheeler Amphitheater, central House Range, Millard County, Utah (**f**) carbonized algal disks; Wheeler Shale, Wheeler Amphitheater, central House Range, Millard County, Utah; scale bar is 5 mm. (**g**) Pink sponge spicule mudstone from condensed bed in lower Wheeler Fm., Stratotype Ridge, Drum Mountains. Mllard Co., Utah; bar scale is 5 mm (From Brett et al. 2009)

uncertain, but Elrick and Snider (2002) relate them to contemporaneous peritidal cycles that are inferred to represent precessional and short term eccentricity forcing of minor sea-level oscillation.

3.1.1 Proximal Cycles

Proximal cycles are 3–20 m thick and comprise alternations of thin carbonates and calcareous shales (Figs. 5 and 6). Small-scale cycles in the Marjum have been correlated over about 10 km within the House Range (Elrick and Snider 2002). Correlation across ranges is difficult because of facies changes and lack of data from intervening graben basins.

(a) Each cycle commences at the sharp base of a ledge-forming limestone bed (Fig. 5a–c). These basal beds are compact, typically 5–50 cm thick, oncolitic to oolitic and skeletal pack- and grainstones that include minor cross stratification. In contrast to many later examples, skeletal fragments, including echinoderm grains, trilobite sclerites, and rare articulate brachiopods, are a minor fraction of these limestones. Lithoclasts eroded from underlying mudstones are common within the beds. Beds show sharply-defined, corroded and mineralized upper contacts. *Thalassinoides*-burrowed firmgrounds are typical of upper surfaces of these beds and in some cases these burrow galleries have been unroofed, leading to irregular depressions in the bed tops. Thrombolitic to stromatolitic mounds may occur at the top surfaces of these beds and extend up into overlying shales.

(b) Basal limestones are overlain by calcareous, medium gray shale, and thin-bedded wacke- to packstones with abundant, largely disarticulated polymeroid and agnostoid trilobites represent late TSTs; cone-in-cone calcite may occur on trilobites and may form small discoid concretions. These beds are commonly followed by pale gray to lavender, siliceous mudstones rich in sponge spicules and comminuted fossil debris that reflect even more condensed intervals (Fig. 5g).

(c) The overlying thin dark gray to black, fissile shales are typically barren, except for carbonized discs of putative algae (Fig. 5f), and include rare soft-bodied animal remains. Interbedded dark gray shales include abundant articulated agnostoid trilobites and diminutive polymeroids.

(d) Overlying platy, calcareous shales that may make up more than half of the cycle thickness include bedding planes covered with articulated bodies and molt ensembles (groupings of associated molted exoskeletal elements) of polymeroid trilobites, especially *Elrathia* from rapid blanketing of undisturbed seafloors by calcareous mud layers (Fig. 5e). The presence of abundant molt ensembles (i.e., associated articulated thoracopygidia and free cheeks discarded in molting) provides excellent evidence of in situ activity and an absence of transport.

(e) The *Elrathia*-rich beds pass upward successively into interbedded sparsely fossiliferous platy to flaggy shale and thin, pale gray weathering calcisiltites and burrow-mottled to nodular limestones. Lower surfaces of siltstone beds display hypichnial burrows. Most beds are barren except for rare basal lags of calcisiltites including trilobite sclerites. Rarely, articulated polymeroid trilobites, notably *Asaphiscus* and the eocrinoid *Gogia*, are preserved at bed bases; these represent obrutionary deposits.

The compact oolitic to oncolitic pack- and grainstones are considered to be early transgressive lag deposits. Their sharp basal contacts are small-scale sequence boundaries, while sharp, mineralized hardground tops record drowning discontinuities and early transgressive systems tracts (TSTs). The overlying early highstand

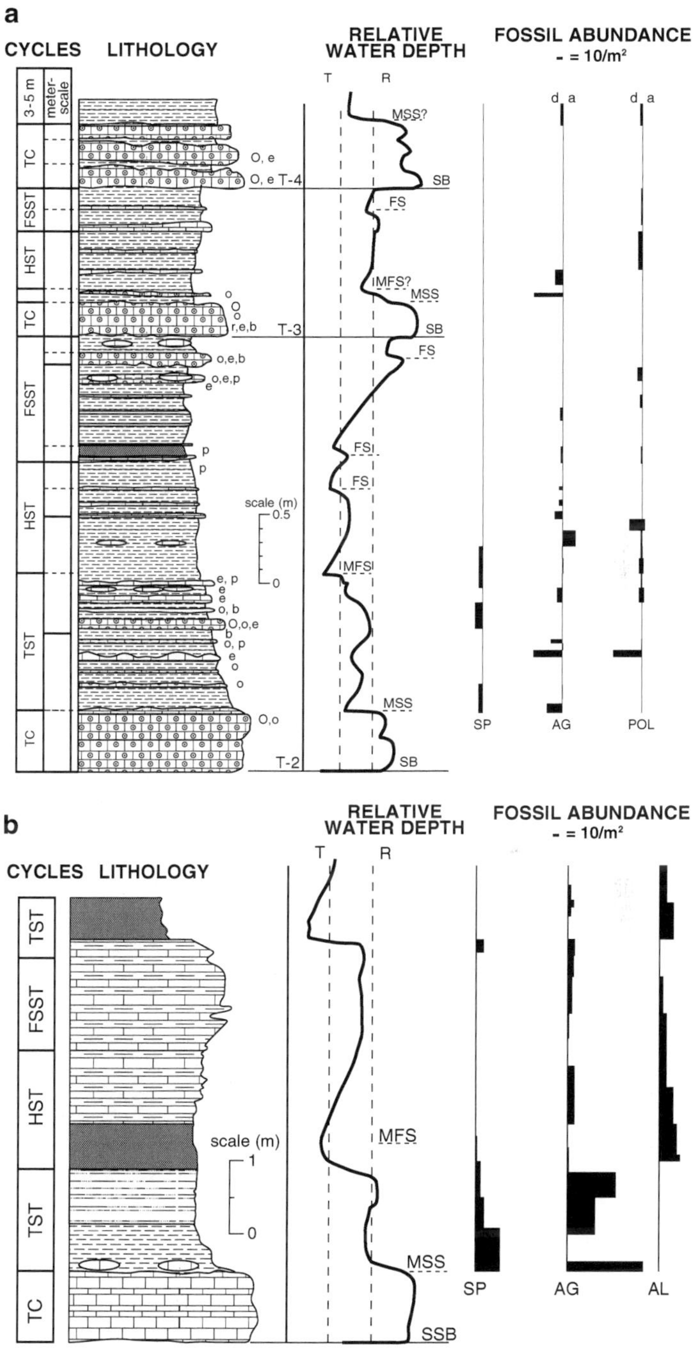

Fig. 6 Detailed log of proximal, medial, and distal small-scale (fifth order) cycles in Middle Cambrian Wheeler Formation with sequence interpretation, inferred relative water depth curve, and fossil distribution; the schematic focuses on the meter-scale cycles. Minor cycles are submeter-scale

(HST) intervals are recorded in black to dark gray laminated shales. A combination of lower dysoxic-anoxic conditions with a fluctuating oxycline and relatively rapid episodic influx of fine-grained detrital sediment favored repeated burial and preservation of abundant organic detritus and rare soft-bodied animals. Laminated calcisiltites record shedding from nearby proximal carbonate banks during late highstand to falling stage. Regression triggered a down-ramp progradation of the carbonate bank and concomitant influx of peloidal carbonate silt and mud. These rapid inputs of calcareous mud occasionally entombed in situ trilobite molts and carcasses and *Gogia*, forming obrution deposits.

3.1.2 Distal Cycles

Distal representatives of meter-cycles in the Middle Cambrian occur in the down-ramp Wheeler Shale section at Marjum Pass in the House Range (Fig. 6b). In this case the cycles commence with: (a) thin (0.5–20 cm) lag deposits of skeletal debris, including sponge spicules and disarticulated agnostoid and small polymeroid trilobites, with rare oncolites. These beds are typically dark gray packstones that may display yellowish-weathering, dolosilt-filled *Thalassinoides* burrows. Upper surfaces may show evidence of corrosion, commonly are ferruginous and may show thin rinds of pyrite. The basal hash beds may abruptly pass upwards into (b) light, purplish-gray spicular platy, siliceous mudstones. Agnostoid trilobites, typically as disarticulated sclerites, occur in minor concentrations. (c) The major lower portion of each cycle comprises fissile to friable barren, black shales. In some cycles these shales lack skeletal fossils, but may include bedding planes covered with carbonized algae. (d) Overlying, medium-dark gray, platy, calcareous shales that comprise the bulk of the cycle thickness are mainly barren, but individual bedding planes may include abundant acrotretid brachiopods and/or agnostoids. The majority of the trilobites are articulated and dorsal side upward (see tables in Brett et al. 2009). In some cycles, especially in the upper part of the Wheeler Shale, the polymeroid *Elrathia* is present as complete, disarticulated sclerites or as articulated specimens, including molt ensembles. The upper quarter to third of the cycle may feature: (e) bundles of thin tabular to nodular, bioturbated calcisiltites. A notable addition to the motif seen in proximal sections is the occasional presence of horizons of concretionary limestone that lie just slightly below the lag deposits of trilobite skeletal debris.

Fig. 6 (continued) variations within fifth order cycles. (**a**) Two proximal ramp cycles; upper Wheeler Fm.; east flank of Sawtooth Ridge, Drum Mountains, Millard County, Utah. (**b**) Distal cycles; Wheeler Shale; Marjum Pass, Millard County, Utah. Sequence Stratigraphy: FSST, falling stage systems tract; FS, flooding surface; MSS, maximum starvation surface; SB, sequence boundary; SSB, sub-sequence boundary; TC, transgressive condensed beds; TST, transgressive systems tract; note TC is the lower portion of the TST. Relative water depth: T, transgression; R, regression. Fossils: SP, sponge spicules; AG, agnostoid trilobites; POL, polymeroid trilobites; IN, inarticulate brachiopods, mainly acrotretids; AL, algae. Taphonomy: d, disarticulated, a, articulated; note bar widths indicate abundance. Lettered symbols include: b, large burrows; e, echinoderm debris; O, oncolites; o, ooids; p, phosphatic nodules; r, rip-up clasts (Adapted from Brett et al. 2009)

Differences between proximal and distal cycles include the absence of thicker basal lag beds of oolitic and oncolitic carbonate and their apparent replacement by thin (mm-scale) beds of trilobite debris. In only one instance were scattered oncolites found in this debris. A second key difference lies in the lesser development of ribbon bedded calcisiltites in upper portions of the cycles. Finally, early diagenetic features, particularly carbonate concretions, may be developed preferentially in upper parts of cycles. Indeed, in places where thin skeletal lag beds are absent, concretions may represent the primary evidence for cycle tops.

The absence of thicker oolitic/oncolitic limestones and a reduction in frequency and thickness of "ribbon" limestone facies suggests increasing distance from the shallow carbonate factory. We infer that concretionary carbonate cementation occurred during intervals of relative sediment starvation associated with overlying skeletal debris beds. During pauses in sedimentation cements developed within muds below the sediment–water interface possibly in the zone of sulfate reduction or near the upper boundary of the methanogenic zone.

Comparable cycles have been described from the somewhat older Middle Cambrian Spence Formation in the Wellsville Mountains of Utah and at Oneida Narrows, Idaho (Liddell et al. 1997).

3.2 *Late Ordovician; Eastern North America*

Alternating mudstone and carbonate-dominated successions are common in cratonic settings in the Upper Ordovician (Caradocian/Cincinnatian; Edenian Stage) of eastern Laurentia (Fig. 7). These intervals consist of distinct, correlatable alternations of thick (meter- to decameter-scale) mudrock-dominated intervals and thinner (decimeter- to meter-scale) shell bed-dominated units. These cycles have been studied extensively from the standpoint of sedimentology (Tobin and Pryor 1981; Jennette and Pryor 1993; Holland et al. 1997; Brett and Algeo 2001a), paleontology/quantitative paleoecology (Holland et al. 2001b; Miller et al. 2001; Webber 2002) and regional stratigraphy (Holland et al. 1997, 2001a; Brett and Algeo 2001b; Brett et al. 2003, 2006, 2008; Kirchner and Brett 2008). The total duration of the Edenian stage is debated, but is estimated at between 2 and 3 million years (Sadler and Cooper 2004). Interpolation suggests that the meter-scale cycles may represent bundles on the order of a few tens of kyr (possibly precessional or obliquity cycles), while decameter scale cycles represent 100–400 kyr eccentricity cycles (Ellwood et al. 2007). Thus, they form a model for early Paleozoic cycles.

3.2.1 Proximal Cycles

The Edenian Kope Formation (Fig. 7b) and its lateral correlatives, the Clays Ferry and Garrard Formations, are subdivisible into about 40 meter-scale cycles which, in turn, comprise portions of some 10 decameter-scale cycles that show increased bundling of limestone beds upward from a thick shale.

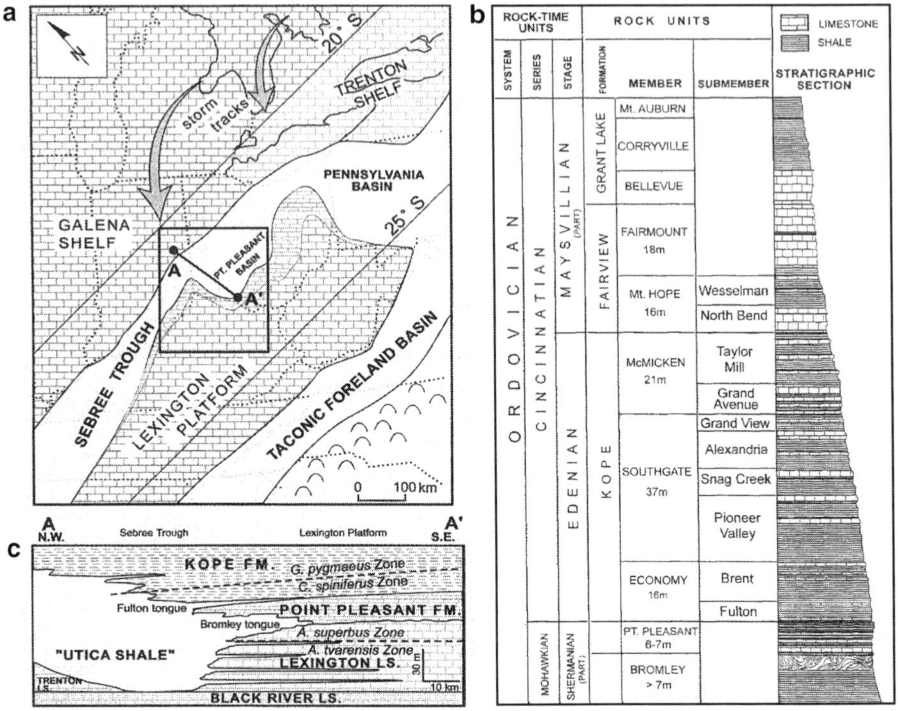

Fig. 7 Geologic setting of Cincinnati Arch, Ohio, Kentucky, and Indiana. (**a**) Paleogeography of the Late Ordovician Edenian Age in eastern Laurentia: Taconic Highlands (From Brett et al. 2003 modified from Mitchell and Bergström 1991). (**b**) General stratigraphy of the Upper Ordovician Cincinnatian Series in the Cincinnati, Ohio area. Kope Formation has been subdivided into submembers on the basis of decameter-scale cycles of thick mudstone to bundled limestones. (**c**) Cross section A-A′ shows transect from shallow Lexington Platform into the Sebree trough (Modified from Caster et al. 1955)

Paleontological gradient analysis using detrended correspondence analysis (DCA) of whole faunal relative abundance data shows patterns of deepening and shallowing within the decameter scale cycles (Holland et al. 2001a) and a crude ("third order") overall shallowing pattern within the Kope to Fairview formations (Fig. 7b). However, detailed faunal studies of meter-scale cycles failed to find consistent patterns of shallowing or deepening on the basis of DCA scores. A general – but not complete – similarity exists between faunas of mudstone versus skeletal limestone portions of any given cycle. However, fossils are extremely concentrated in the limestone portions relative to the shales and include subtle differences in composition.

A typical Kope cycle (Fig. 9) is comprised of several components (Brett and Algeo 2001b). (a) Bases are delimited by a sharp and typically erosionally-based bundle of skeletal pack- to grainstones, in some cases expressed as a single compact amalgamated bed with occasional rip-up clasts of mudstone (Fig. 8c). The limestones are composed of variably disarticulated, fragmented, and in some cases, abraded brachiopod shells, bryozoans and crinoid ossicles. In a basinward direction these beds thin and become more compact skeletal grainstones just a few centimeters

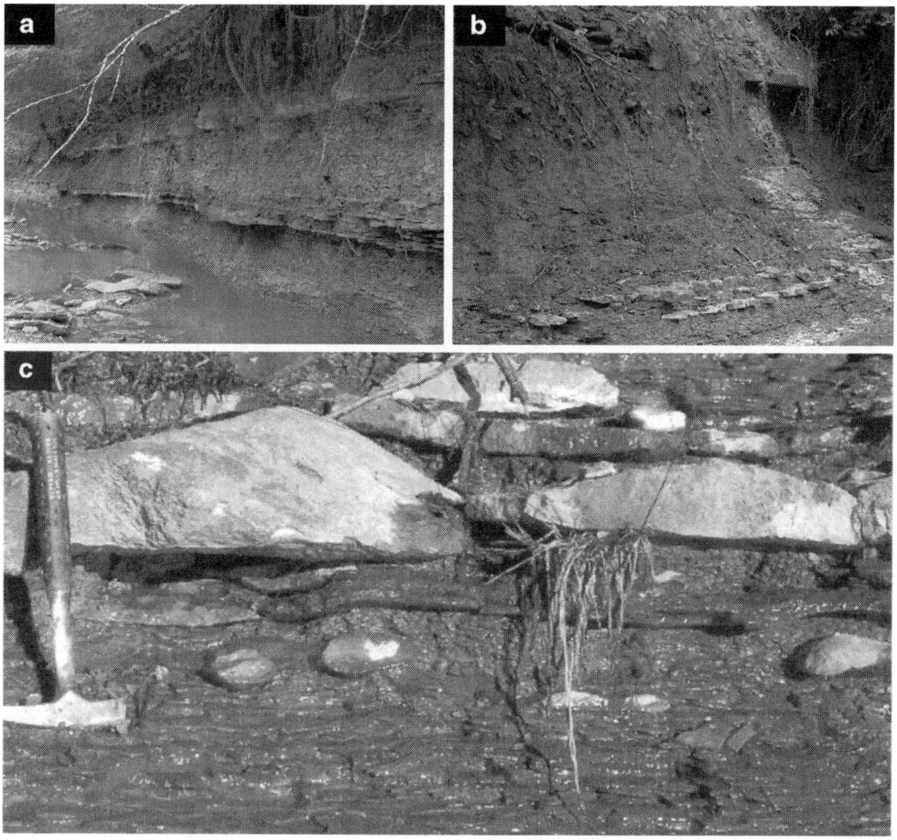

Fig. 8 Limestone-mudstone cycles in Kope Formation, Fowler Creek, near Independence, Kentucky. (**a**) Alexandria Submember, cycle 28, showing mudstone interval with only minor thin beds bracketed between bundles of limestone beds. (**b**) Base of shale overlying limestone bundles 25 and 26; note limestones of bed 26 thinning upward into base of 1.5 m thick mudstone interval. (**c**) View of bed 6 (Brent Submember), showing sharp, nearly planar base and prominently rippled top with foreset beds in cross section; also series of small concretions in shale underlying this bed (below an additional thin siltstone); Industrial Park off Rte. 17, Covington, KY

thick. Skeletal limestones preserve crude grading and rippled tops; firmground burrows, including *Thalassinoides* are typical of bed tops. Hardgrounds with borings (*Trypanites*) and encrusting organisms also occur on tops of a few of the skeletal limestones and are indicative of sediment starvation. (b) The sharp, frequently mounded to rippled tops of the major limestones are commonly overlain by sparsely fossiliferous olive to dark gray mudstone, typically with thin stringers or lenses of shelly debris, including highly comminuted and in some cases phosphatized or pyritized fossil steinkerns. In several instances biostromes of whole bryozoan colonies occur preferentially in the first few centimeters of shale overlying the limestone beds. This interval is followed by (c) the main shale-dominated component of the cycle, comprised of alternating packages of medium to dark gray shale and mudstone. Fresh surfaces have sharp, scoured contacts between shales of subtly different

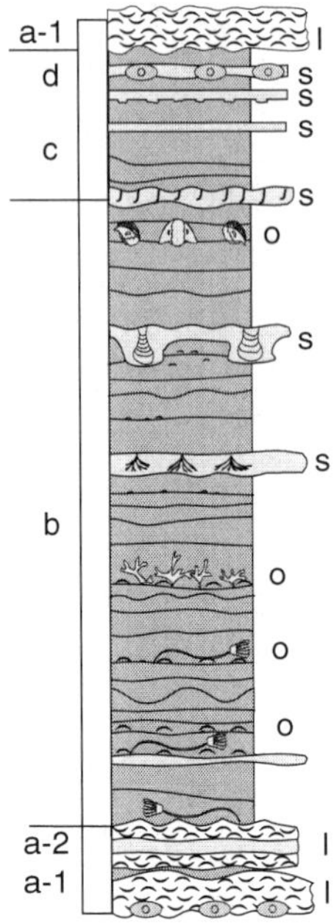

Fig. 9 Representative meter-scale cycle from Upper Ordovician Kope Formation. Lettered divisions include **a**-1, **a**-2; stacked skeletal limestones and thin shales; **a**-1 shows reworked concretions at erosion surface at base of skeletal limestones (l). (**a**-2) Ripple-topped bed with hardground, overlain by obrution deposit. (**b**) bulk of cycle thickness showing a stacking of event mud beds, many of which show obrution beds (o) of crinoids or trilobites, or sharply based siltstones/calcisiltites (s) with burrows, including *Diplocraterion*. (**c**) cluster of siltstones (s) near top of cycle, may show increasing input of silts during minor falling stage; (**d**) diagenetically altered siltstone with concretions formed in older sediment during time of sediment starvation associated with overlying shell bed. Note that the limestones at base and top are condensed relative to the main thickness of cycle (see Fig. 32 for interpretation)

coloration or silt or fossil content. Bedding planes covered with shelly debris may separate the shale packages. Thin and/or lenticular muddy packstones within this interval are primarily confined to up-ramp facies of the cycles. A number of these bedding planes preserve obrutionary deposits including articulated trilobites and crinoid columns, the latter as densely packed, current aligned "logjams". Most cycles also include thin beds of parallel to small-scale hummocky cross-laminated siltstone or calcisiltite. These have sharp bases, typically with distinct tool marks and very minor lags of skeletal debris. Gutter casts occur associated with specific widespread siltstone beds. The tops of the beds are hummocky to planar and frequently heavily burrowed with *Chondrites*, *Diplocraterion* and other traces (Fig. 9). In most proximal sections of the Garrard Siltstone, the upper parts of the cycles are comprised of beds of planar to hummocky laminated siltstone and fine-grained calcareous sandstones that typically show soft sediment deformation interpreted as seismites (Seilacher 1982a; McLaughlin and Brett 2004). Lags of bivalves and concavo-convex brachiopods (*Rafinesquina*, *Strophomena*) occur at bases of siltstone beds, in some cases as edgewise-stacked shell coquinas.

Kope cycles have been correlated in bed-by-bed detail along nearly 100 km of the outcrop belt near Cincinnati, Ohio and into the subsurface (Fig. 10). These detailed correlations demonstrate that not only are the condensed shell-rich beds of allocyclic origin, but also that many of the subtle obrutionary event beds record mud deposition of a regional scale.

3.2.2 Distal Cycles

The distal Kope Formation is dominantly dark gray shale but subtle cycles are picked out by thin (0.2–5 cm) beds of skeletal pack- to grainstones composed of crinoidal and trilobitic debris and small brachiopods (Kirchner and Brett 2008). These thin limestones are overlain by medium gray to dark brownish gray shale, which is largely barren except for fine thread-like pyritic burrows. A few bedding planes are covered with current aligned graptolites; others display an abundance of small inarticulate brachiopods (*Leptobolus*) and disarticulated trilobite sclerites, including small *Triarthrus* and less commonly *Cyptolithus* and *Isotelus*.

An additional component seen in many distal Kope cycles is one or more horizons of ellipsoidal concretions within the upper 5–15 cm of the mudstone underlying the sharp base of the directly overlying skeletal limestone bundle (Fig. 8c). These concretions are typically nucleated on pyritic burrows. In some cases, the concretions are laterally amalgamated to form a semi-continuous layer. Kope concretions rarely contain exceptionally preserved, uncompressed fossils. An excellent example is provided by elongate concretions that underlie bed 23 which formed around fillings of shallow scours or gutters. Their silty infill contains uncompressed current aligned complete rhabdosomes of graptolites (Kirchner and Brett 2008).

Concretions are commonly re-worked, as evidenced encrustation by bryozoans and crinoid holdfasts, and are occasionally re-worked into the bases of overlying skeletal limestone beds. Concretion horizons can be traced from outcrops in which they occur well below skeletal limestones to others in which they become adnate to the limestones and/or fully incorporated. Such horizons indicate regional erosion, typically accentuated in down-ramp directions, along which seafloor currents may have removed several centimeters of mud (see Figs. 9 and 31).

The Upper Ordovician (Edenian to Maysvillian) Collingwood Member of the Lindsay Formation, southern Ontario, Canada is a strikingly cyclic package of shales and carbonates (Figs. 11 and 12; Brett et al. 2006). This unit is approximately coeval with the upper Kope Formation and provides an outcrop illustration of the down-ramp expression of similar scale cycles. About 10–12 cycles are 50–150 cm thick and comprise four major components: (a) very thin lag deposits or in some cases lenses of skeletal debris, including brachiopod shells, crinoid ossicles and trilobite fragments; this unit is absent in some cycles, (b) dark gray to black, organic-rich, laminated shales that grade up-section into, (c) medium to light gray, calcareous

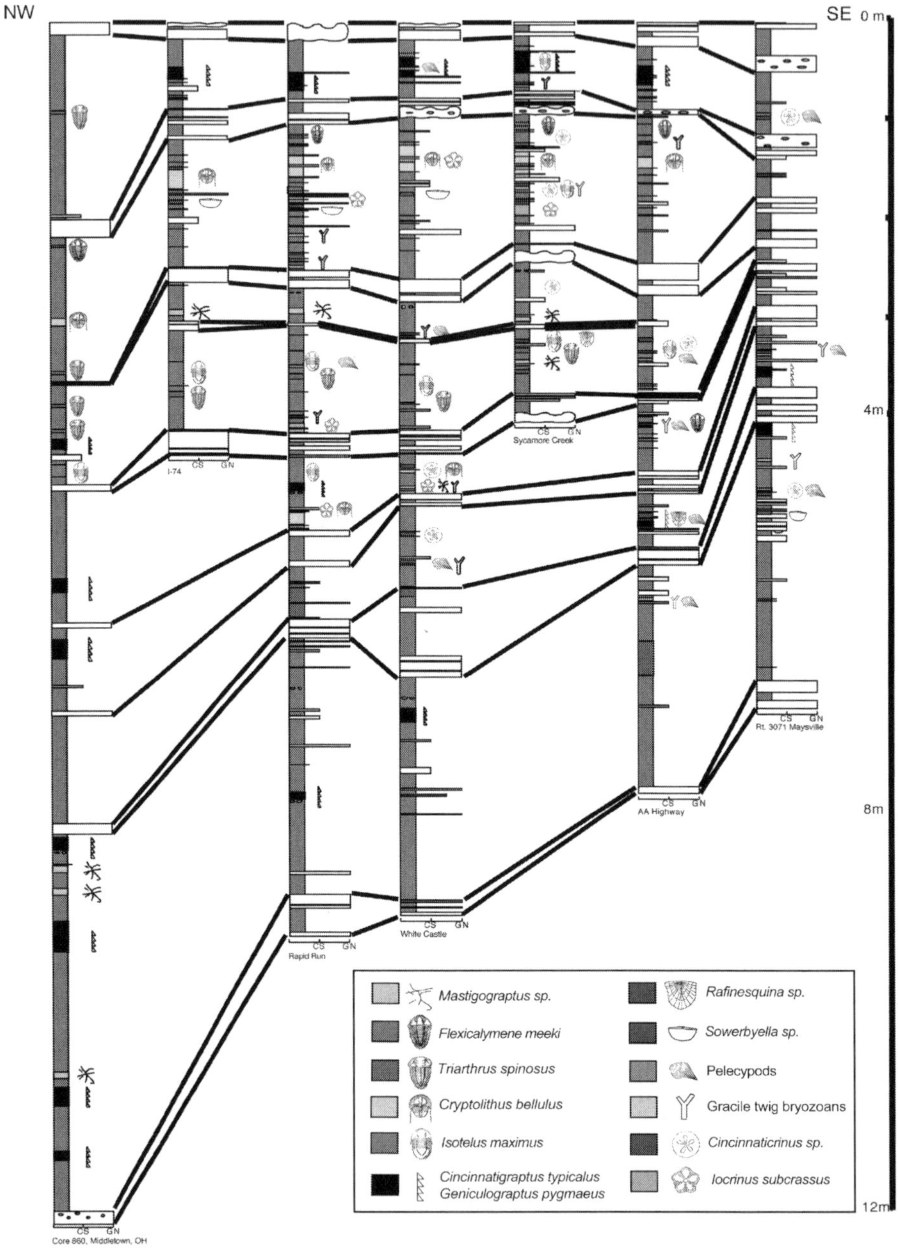

Fig. 10 Correlated stratigraphic sections of the Alexandria sub-member (cycles 25–30) of the Kope Formation from the vicinity of Cincinnati, Ohio/northern Kentucky. Note persistence of proportional spacing of limestone bundles and a series of distinctive fossil event beds. Fossil occurrences and abundances are shown to the left of the column. Black bars denote intervals where a particular fossil was found at six or more of the seven sites. Intervals marked by dark gray or black bars and which occur in relatively thin beds are considered to be marker beds (From Brett et al. 2008)

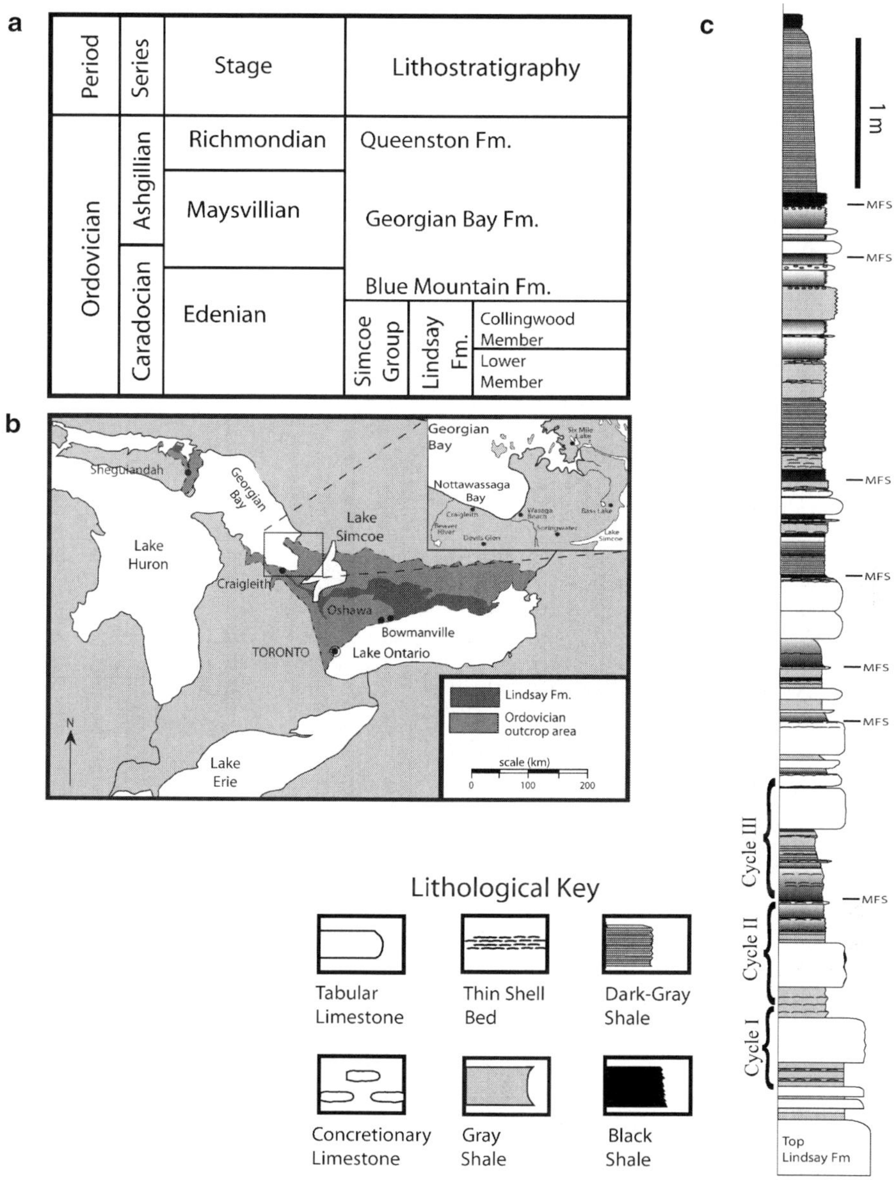

Fig. 11 (**a**) General time relationships and lithostratigraphy of Upper Ordovician strata in southern Ontario, Canada. Note position of Collingwood Member. (**b**) Locality map for Collingwood Member outcrops, inset map shows location of Craigleith Creek and foreshore, located along the southern coast of the Georgian Bay (Nottawasaga Bay) near Craigleith Provincial Park. (**c**) Stratigraphic section of the lower Collingwood Member at Craigleith, maximum flooding surfaces (MFS) and cycles are indicated. See Fig. 12 for detailed log of Cycle III (Adapted from Brett et al. 2006)

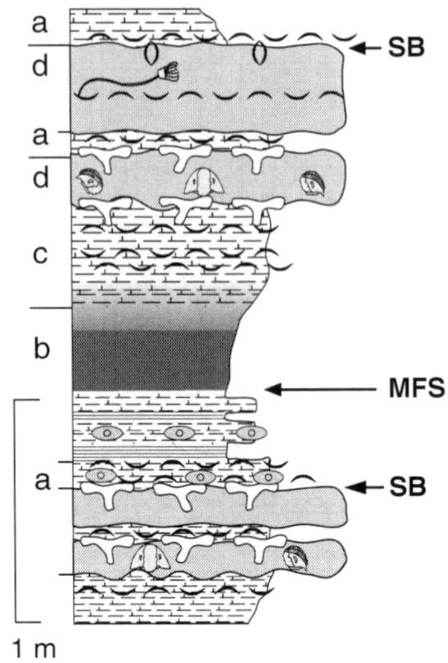

Fig. 12 Typical cycle of Collingwood Formation based on Cycle III (see Fig. 11) at Craigleith, Ontario, showing positions of obrutionary beds and shell hash beds. Note occurrence of well-preserved, non-compacted fossils in concretionary limestones. Lettered divisions are portions of cycle analogous to those shown in Fig. 9. (**a**) Thin skeletal debris layer; (**b**) dark shales in lower part of cycle; (**c**) gray silty mudstone of falling stage of cycle; (**d**) thick concretionary bed developed by early diagenetic cementation of upper obrutionary mudstone (Modified from Brett et al. 2006)

shales or mudstones, and (d) lenticular to tabular concretionary argillaceous limestones and light gray calcareous, fossiliferous mudstones or shales and marls.

Black shale units have a characteristically sharp basal contact and may rest on a condensed shelly pavement. Fossils within shales are preserved as pavements or stringers of trilobite, ostracode and brachiopod debris, with strong taphonomic bias as a result of prolonged exposure at the sediment–water interface. Bioturbated gray mudstones and marls include numerous low diversity orthid brachiopod pavements. Persistent diagenetic tabular limestone bands include shelly beds that alternate with less fossiliferous, calcareous mudstones containing non-compressed, spar-filled burrows and articulated, sometimes in situ fossils.Collingwood cycles involve up-section changes including: (1) benthic oxygenation from lower dysoxic to fully oxic biofacies, (2) increased frequency and episodicity of sedimentation, (3) higher net sedimentation rate within gray mudstone to carbonate intervals, (4) increased environmental energy level, and (5) diagenetic cementation of muds a few centimeters below cycle tops.

A key paradox associated with these cycles is that the best-preserved fossils occur within the sparsely fossiliferous concretionary carbonates near the tops of the cycles (Fig. 12). These parts of the cycles record long periods of low sedimentation that enabled diagenetic redistribution of carbonates. Yet the occurrence of extremely well preserved trilobites and other fossils in the limestones clearly records episodes of rapid influx of calcareous mudstone and engulfing of organism carcasses. The in situ lingulids and spar-filled burrows within these beds reflect colonization of stabilized, perhaps incipiently cemented mudstones. The latter mark the initiation of interludes of low sedimentation following episodic deposition of up to several centimeters of sediment.

During these times, cementation of the muds ensued while thin veneers of marly, disarticulated skeletal hash appear to have accumulated a few centimeters higher.

Variation in fossil abundance in these beds is the result of cyclic variation in sedimentation, ranging from periods of condensation to rapid burial. Consistency of these variations, suggests an allocyclic mechanism for the Collingwood cycles related to short-term eustatic sea level or climatic fluctuation.

3.3 *Early Devonian; Mdaouer-el-Kbir and Khebchia Formations, SW Morocco*

3.3.1 Proximal Cycles

The middle Emsian Mdaouer-el-Kbir Formation (formerly Rich 3, Hollard 1967) of the northeastern Draa Valley in the vicinity of Foum Zguid in south central Morocco (Fig. 13) includes excellent proximal cycles (1.5–10 m thick) of upward coarsening mudstone, siltstone and fine-grained sandstone. Cycles are clustered

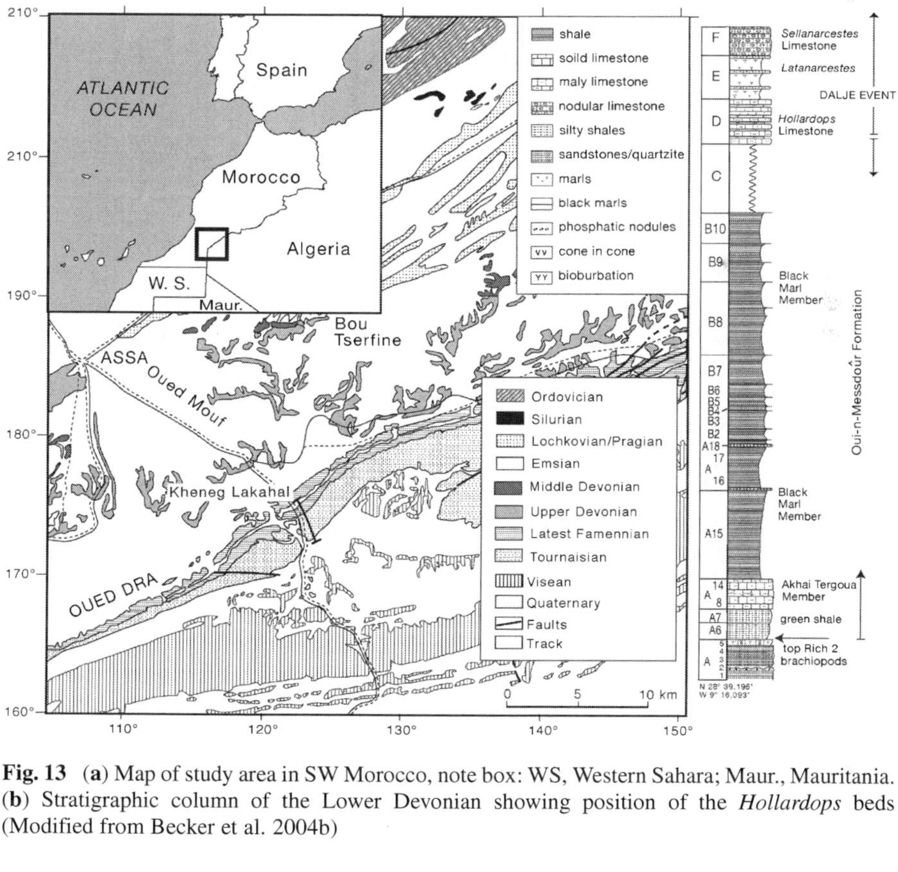

Fig. 13 (**a**) Map of study area in SW Morocco, note box: WS, Western Sahara; Maur., Mauritania. (**b**) Stratigraphic column of the Lower Devonian showing position of the *Hollardops* beds (Modified from Becker et al. 2004b)

into groups that are separated by thicker intervals in siltstone and mudstone, analogous to the "big shales" in the Kope cycles described above. These cycles are presently under study and have not been fully characterized, but a brief description can be given here (Jansen et al. 2004).

Each cycle commences with a sharp contact with underlying sandstone; hence these cycles have a parasequence motif. Nonetheless many of the cycles are based at distinctive shell rich bioturbated calcareous sandstone beds (a) that form the "caps" of the sandy beds. The shell rich beds are comprised of disarticulated shells of spiriferids (e.g. *Euryspirifer*) and strophomenid brachiopods, crinoid debris, bryozoans, and homalonotid trilobites and include phosphatic nodules. They have sharp to abruptly gradational contacts with underlying sandstones and overlying beds. The latter beds are comprised of heavily bioturbated silty, calcareous mudstones, commonly with *Zoophycos* and show stringers and pods of shelly debris. Shells are mainly disarticulated and convex upward, but in some instances articulated specimens occur.

The main portion of each cycle comprises mudstones, siltstones, and muddy sandstones (b). These are typically barren to sparsely fossiliferous, but include thin accumulations of moldic brachiopod and bivalve shells. Upper portions of the cycles (c) consist of hummocky cross-bedded siltstone and sandstone. Symmetrical and interference ripple marks are present on some bedding surfaces. Nodular cementation (d) occurs in the upper parts of sandstones in at least two of the 15 cycles. Nests of articulated, in situ terebratulid brachiopods (*Meganteris*) in the sandstones immediately below caps in two cycles are obrution beds comparable to the well-preserved fossil beds noted at the tops of Ordovician cycles.

Overall, each cycle clearly records a period of sediment starvation and winnowing, associated with development of fossiliferous, and commonly ferruginous or phosphatic lag beds. Overlying sediments are characterized by an influx of increasingly coarse-grained sediments with storm current and wave deposition.

3.3.2 Distal Cycles

The mid Emsian Hollardops Member of the Khebchia Formation is a 5–15 m thick interval of rhythmically bedded limestones and shales that are well exposed in a series of anticlines and synclines in the Draa Valley of southwestern Morocco, particularly at Bou Tserfine, near Assa (Figs. 13–15). It is slightly younger than the proximal facies of Foum Zguid discussed above but grades upslope into comparable facies.

These beds preserve a rich and well preserved trilobite fauna, including the trilobites *Hollardops mesocristata*, *Phacops saberensis, Psychopyge*, *Leonaspis*, and *Scutellum* (see Morzadec 1980; Schraut 2000; Chatterton et al. 2006 for description of similar but slightly younger trilobite faunas from southern Morocco). Other fossils are scattered, but include abundant large orthoconic nautiloids, small athyrid and ambocoeliid brachiopods, and small solitary rugose corals (Becker et al. 2004a, b).

The Hollardops Member and overlying Brachiopod Marl and Sellanarcestes Members interval at its type-section at Bou Tserfine, near Assa, Morocco includes about 160 cycles (238 beds) comprised of calcareous, medium to dark gray shales/

mudstones and decimeter scale concretionary limestones (Figs. 14 and 15). Thicker shale intervals are not homogenous, but are composed of alternating dark gray, laminated calcareous shale and medium gray marly mudstone, minor skeletal hash beds include disarticulated and fragmented trilobite segments, cephala and pygidia, crinoid ossicles, small brachiopods, and flattened molluscan shells (bivalves, gastropods and orthoconic cephalopods). Thus, a general cycle, analogous to those described above comprises: (a) thin skeletal lag, (b) medium dark gray to black shales, (c) calcareous

Fig. 14 (**a**) Typical outcrop of Lower Devonian Hollardops Limestone at Bou Tserfine; note alternating thin-bedded limestones and dark shales. (**b**) Close-up of trilobite bearing limestones. (**c**) Inset shows specimen of *Hollardops* in near vertical orientation; scale bar 1 cm (**c** is adapted from Chatterton et al. 2006)

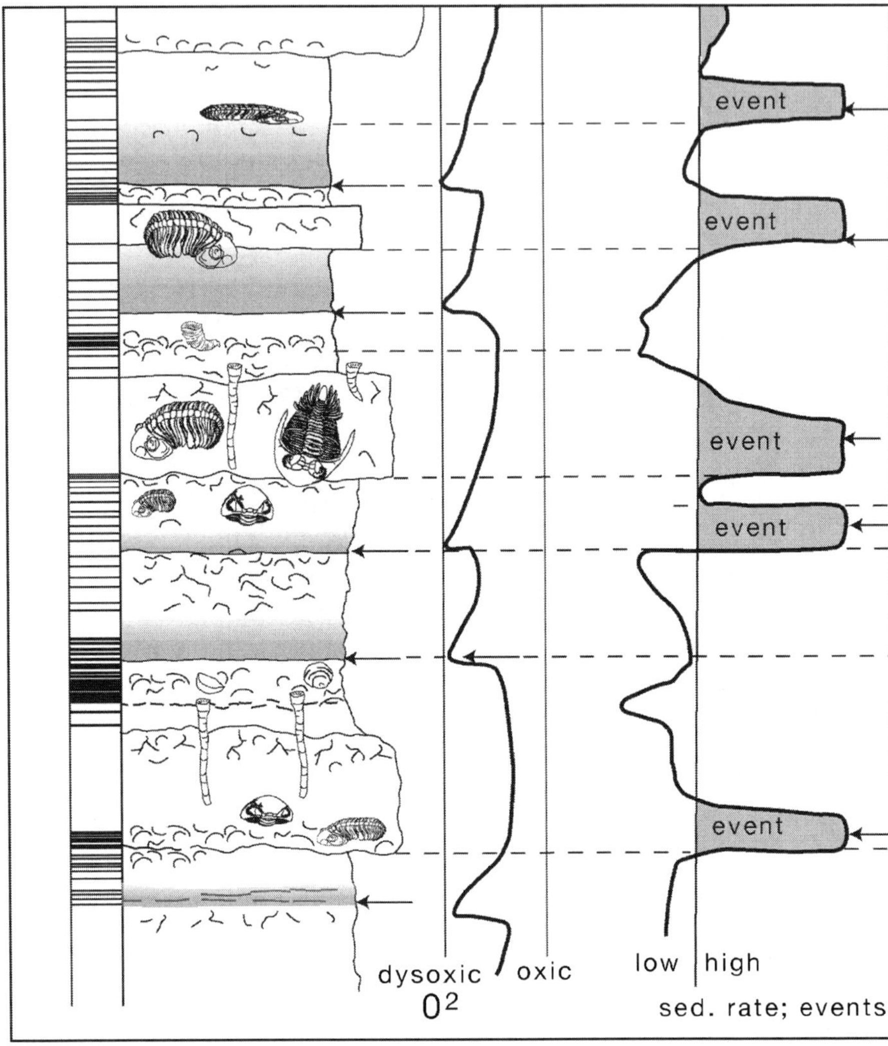

Fig. 15 Schematic of cycles in the Lower Devonian Hollardops beds at Bou Tserfine near Assa, SW Morocco. *Left column* indicates inferred relative time-richness using hypothetical bars of equal time increments; stratigraphic profile shows a series of cycles, each commences with thin skeletal debris beds with trilobite fragments, small brachiopods and small solitary rugose corals, that are sharply overlain by dark, sub-laminated shale recording minor dysoxic episodes (*arrows*), and calcareous mudstones with minor shelly debris and articulated trilobites; note 10–15 cm thick diagenetic limestone beds within gray marls yield well preserved trilobites in unusual orientations and pyritic tubes that may extend from overlying soft mudstone into the cemented layer. Columns to the right show relative benthic oxygenation, sedimentation rate (note pulses of sediment marked "event"); stratigraphic thickness is partially dependent upon degree of diagenetic cementation, as noted by arrows in far right column, which may prop open sediments to approximate original thickness

medium gray mudstone, (d) lenticular to tabular concretionary argillaceous limestones and light gray calcareous, fossiliferous mudstones or shales and marls.

Trilobite remains occurred in 128 (54%) of the beds, with articulated remains in 42 (18%). Limestones were more likely to include obrution beds (27.5%), and in the most trilobite-rich middle portion of the section (50 alternating shales and limestones) about 40% of beds and nearly 60% of limestones had articulated remains. Their highly weathered character undoubtedly biases data collection from shales. Nonetheless, articulated phacopid trilobites occur in several of the shale intervals and thin skeletal debris layers were commonly present in shales just above the contacts with underlying limestones (Fig. 15).

Limestone beds include articulated trilobite remains, including both complete outstretched, gently reflexed, incompletely and completely enrolled specimens as well as molt ensembles. The abundance of molt ensembles indicates a near absence of transport and thus supports the in situ nature of the trilobites. In many cases, articulated trilobites occur in attitudes perpendicular to bedding and single blocks incorporate variably oriented specimens (bed parallel, inverted, upright, and vertical; Fig. 14). All of this evidence strongly suggests that the trilobite and other fossil remains were physically reoriented within viscous uniform muds, perhaps as mudflows that moved carcasses and other skeletal parts very slightly from their living sites in some cases lifting the bodies or skeletal parts upward within the sediments.

Limestones containing well-preserved trilobites are up to 10 cm thick with articulated remains running through the entire thickness (Fig. 15). These are essentially structure-less argillaceous, concretionary micrites. There are no signs of lamination in most although minor hash stringers are present near the bases and/or tops of some beds. Cementation of these beds obviously occurred prior to compaction as the delicate fossils are preserved without distortion. Trilobite bearing limestones commonly show vertical pyritic burrows that may penetrate the entire thickness of the bed. The limestone beds are bioturbated and, in a few instances, burrows that partially disrupted the skeletons have intercepted the buried trilobite carcasses.

Intervening shales do not preserve fossils to the same spectactular extent although some individuals, including articulated trilobites, were buried in varied attitudes. The strong compaction of the muds, however, has greatly distorted these fossils. Very thin skeletal debris layers appear near the bases of many shale intervals, immediately above the concretionary limestone.

Very similar findings are reported from slightly younger beds at Foum Zguid (Chatterton et al. 2006). Similar calcareous rhythmite beds occur in the Lower Devonian Haragan Formation of Oklahoma (Campbell 1977; Ramsköld and Werdelin 1991) as well as the Schoharie or Needmore Formation in Pennsylvania-western Maryland (C. E. Brett, unpublished data).

Trilobite bed cycles involve up-section changes comparable in many ways to those of the Ordovician cycles described above, including: (1) benthic oxygenation from lower dysoxic to fully oxic biofacies, (2) increased frequency and episodicity of sedimentation, (3) higher net sedimentation rate within gray mudstone to carbonate intervals, and (4) diagenetic cementation of muds a few centimeters below cycle tops (Fig. 15).

It is particularly striking that the rhythmic, concretionary limestones appear to transcend facies, occurring in both dysoxic facies with very low diversity assemblages near the base of the Hollardops member to more abundantly fossiliferous sections near the top. Hence, these beds record a regular, recurring cycle superimposed upon an overall shallowing trend. Time series analysis of magnetic susceptibility data from Bou Tserfine shows that these smallest cycles are superimposed upon larger trends that may encompass 100–400 kyr (B. Ellwood, unpublished data). The number and scale of cycles in relation to dating of the early Late Emsian suggests that these cycles record overall durations of 10 s of kyr durations, possibly precessional cycles. The concretionary limestone bands probably formed by carbonate redistribution over several millennia. Nonetheless, more than a third of the concretionary limestones (in some intervals nearly 60% of 50 beds) contain well-preserved, articulated trilobites. Thus, in another sense, they reflect single event deposits of up to several centimeters of mud within no more than days. The paradox is the same as that noted in the Collingwood, and to a lesser extent in concretions of the Kope Formation. The non-random representation of obrutionary muds in the cemented beds appears to reflect the input of thick, perhaps carbonate rich mudflows in the later portions of short term cycles, followed by periods of sediment starvation.

3.4 Middle Devonian; Hamilton Group of New York

Typical cycles are well documented from the Middle Devonian (Givetian) Hamilton Group of the Appalachian basin in eastern North America (Brett and Baird 1985, 1986a, 1996). Proximal cycles are well developed in the upper Hamilton group of central New York State and Pennsylvania, whereas more distal examples occur in western New York (Figs. 16 and 17).

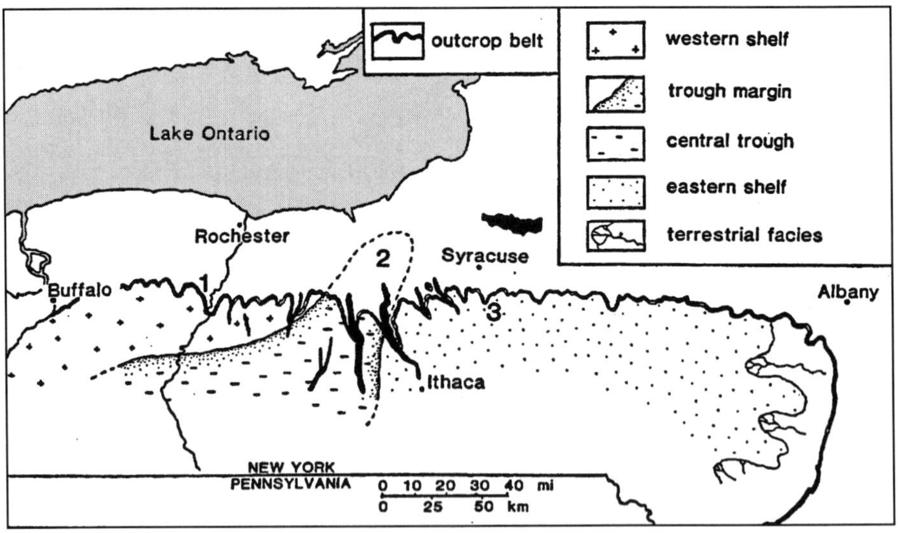

Fig. 16 Map of Middle Devonian Hamilton Group outcrop belt in New York State with superimposed inferred paleogeography. Numbers: (1) western New York shelf; (2) Finger Lakes trough; (3) central New York sandy shelf (Modified from Brett and Baird 1986a)

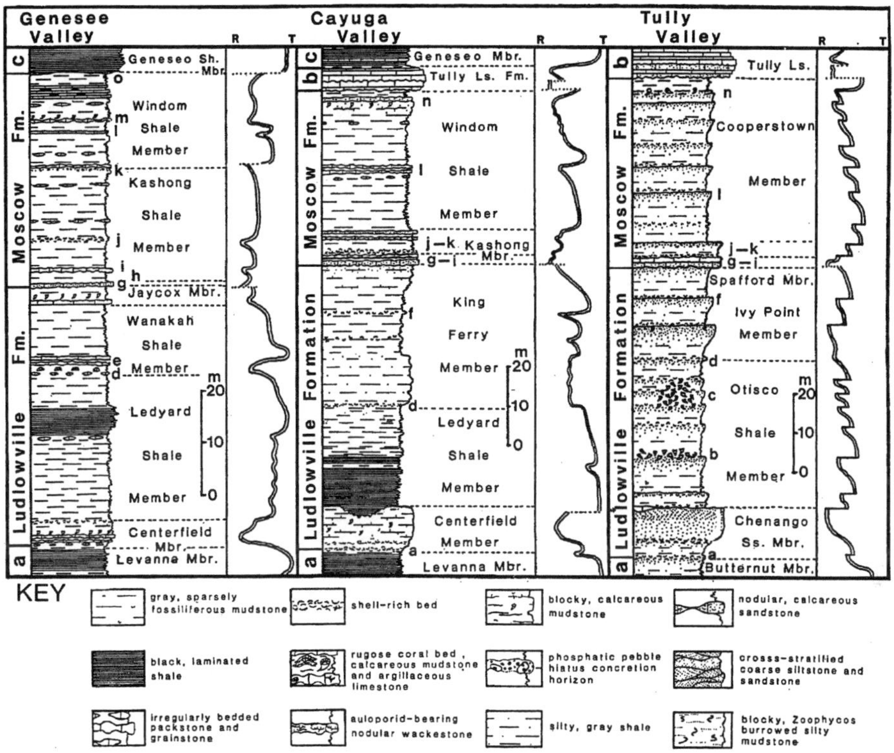

Fig. 17 Decameter-scale cycles in the Middle Devonian upper Hamilton Group in western to central New York State. Genesee Valley shows distal cycles in thin calcareous successions. Cayuga Valley basinal cycles, central Finger Lakes. Tully proximal, coarsening-upward cycles of central New York State. Note relative water depth curves showing the more symmetrical successions of the western sections as opposed to distinctly asymmetrical, shallowing upward, parasequence-type cycles in central sections more proximal to the source area (Modified from Brett and Baird 1986a; see that paper for details and names of marker horizons shown at letters)

3.4.1 Proximal Cycles

Mudstone-siltstone cycles are well represented in central New York where more than 30 such cycles have been examined in detail. These are 1- to 5-m thick packages that commence with relatively thin (up to 0.5 m) shell hash beds and pass upward into mudstones and or siltstones (Figs. 17 and 18). The typical cycle begins with (a) a sharply based skeletal debris bed ranging from decimeters upward to about a meter in thickness. These beds range from silty and muddy packstones to grainstones, with abundant full valves to highly fragmented brachiopods, crinoidal debris, bryozoans, and corals. Internal fabric ranges from moderately to densely packed. Normal grading of shell material is observed in some locations. Admixed with skeletal material near the tops of beds are lithoclasts and diaclasts in the form of reworked concretions and bored phosphatic nodules.

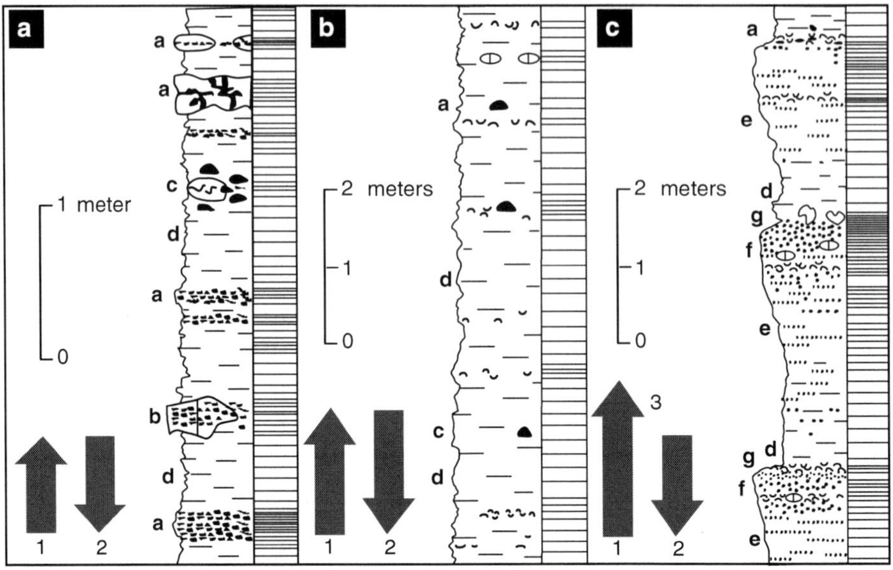

Fig. 18 Schematic of Hamilton Group distal and proximal meter-scale cycles; based on Wanakah Shale, "trilobite beds. (**a**) Distal, calcareous cycles at Lake Erie shore near Eighteenmile Creek. (**b**) Basinal cycles, central Finger Lakes. (**c**) Proximal cycles central New York State. *Bars* to the right of each column show spacing of hypothetical time increments; *close spacing* indicates relative condensation of that portion of the section owing to winnowing and bypass or sediment starvation. *Large arrows* indicate relative vectors of sedimentation (*arrow 1*) and subsidence (*arrow 2*). Note that in the central New York shelf sediment supply may exceed subsidence resulting in winnowing and bypass near cycle tops. Concretions and shell beds of western sections (**a**) and minor coral bearing layers in the central basin (**b**), and thin shell hash beds in the central New York shelf (**c**) reflect intervals of sediment starvation associated with base-level rise. Facies shown by letters include: (a) calcareous to concretionary shell-rich beds; (b) shelly limestone beds; (c) coral-rich beds; (d) gray mudstone; (e) silty, bioturbated mudstone; (f) shell-bearing silt- and sandstone; (g) reworked phosphatic shell hash associated with sediment starvation (Modified from Brett and Baird 1986a)

Upper contacts of the beds are planar to rippled. In some cases they are mantled with a layer of graded silt or calcisiltite. Other examples preserve corrosional discontinuities or hardgrounds. Biostromes of rugose and small aulaporid tabulate corals may occur in these positions (Brett and Baird 1996; Brett et al. 2007a).

Basal shell rich facies pass abruptly into (b) dark gray shales with dysoxic faunas and skeletal beds are sharply overlain by dark gray to black shales with scattered small and brachiopods, especially thin shelled rhynchonellids ("*Leiorhynchus*", *Eumetabolotoechia*), ambocoeliids, and chonetids, as well as small mollusks. Most fossils are disarticulated and may be fragmentary and partially decalcified. Medium gray mudstones that follow typically show sparse, but slightly more diverse assemblages of small ambocoeliid and chonetid brachiopods and infaunal deposit-feeding bivalves (e.g., nuculids, *Palaeoneilo*) and gastropods. Fossils and burrow tubes may show pyritic internal molds.

These facies shallow upward into increasingly silty to sandy mudstones (c), which are typically sparsely fossiliferous, but include lenses of skeletal debris, primarily of brachiopods and disarticulated bivalves plus crinoid ossicles. Intercalated silty beds exhibit hummocky cross-stratification and may be locally amalgamated.

Upper portions of cycles (d) are heavily bioturbated, especially by *Zoophycos*, and pass gradationally into amalgamated silty to sandy mudstones. The upper 0.5–1 m of the cycles may exhibit horizons of rusty weathering ellipsoidal to pipe-like concretions, typically nucleated around pyritized vertical shafts of *Zoophycos* burrows. These intervals are abruptly overlain by shell, coral biostromes and other skeletal rich beds associated with the base of the next cycle.

3.4.2 Distal Cycles

More distal cycles from the Middle Devonian are represented by the "Grabau trilobite beds" of the Wanakah Shale Member and the Smoke Creek bed of the Windom Shale (Speyer and Brett 1986, 1988; Brett and Baird 1986a, 1996; Batt 1996; Fig. 18). These cycles are comparable in many ways to the Hollardops beds of the Lower Devonian in Morocco (discussed above). However, the number of successive rhythms is restricted to three to ten successive beds, as opposed to the 140+ cycles seen in the Moroccan section. Evidently, conditions conducive for diagenetically enhancing the environmental oscillation did not persist nearly as long as in the Moroccan section, although Batt (1996) recognized the trilobite beds in the Wanakah Shale Member as being a special case of minor shell beds that he traced throughout the member. These thin shell hash beds probably record the same scale of cyclicity as the Moroccan rhythms.

Cycles range from about 0.3–0.5 m thick and alternate with medium grey fossiliferous shales. The concretionary beds with distinctively preserved trilobites have each been traced for 80–120 km in outcrops (Speyer and Brett 1985, 1986). The thin debris layers of brachiopods, small rugose and/or auloporid corals that overlie these beds appear to grade laterally eastward into proximal shell and coral-rich beds that occur above minor coarsening upward shale, siltstone cycles in central New York State. Hence, these beds are associated with persistent discontinuities in sedimentation. The concretionary diagenetic beds evidently lie beneath these caps and pass eastward into the upper parts of the mudstone-siltstone cycles discussed above. The shell-rich beds are more persistent and correlate laterally, eastward with complex shell beds in thicker and more clastic-dominated successions discussed above. For example, the "Trilobite beds" pass eastward into complex brachiopod-shell and bivalve-rich shell beds with bioturbated calcareous silty mudstone matrix.

3.5 Lower Jurassic: Lias UK

3.5.1 Proximal Cycles

Proximal examples of meter-scale cycles are recorded by the 5–10 m thick, coarsening-upward successions in the Lower Jurassic (Pleinsbachian) Cleveland Ironstone Formation of Staithes on the Yorkshire Coast of England (Figs. 19–21;

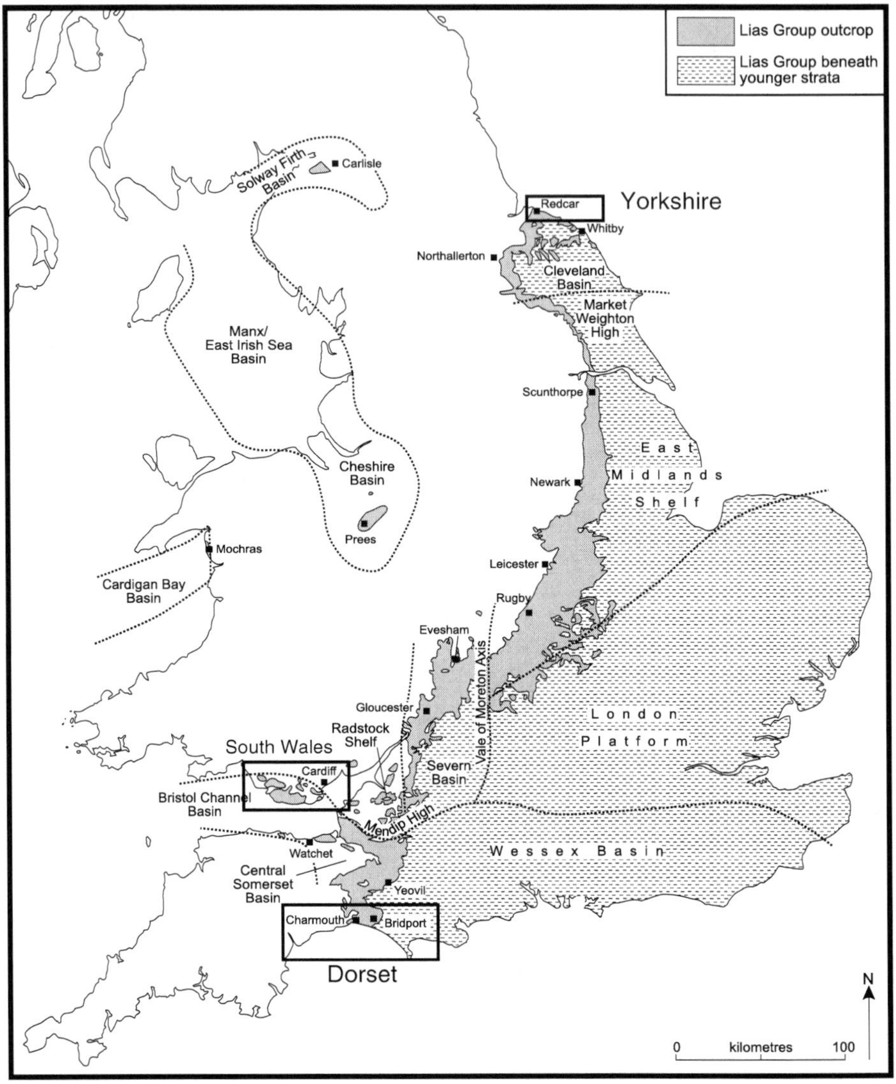

Fig. 19 Outcrop and subcrop map for the Lias Group in England and Wales showing the location of the main sedimentary basins. Outcrops in Dorset, South Wales and Yorkshire (discussed in text) are indicated (Modified slightly after Cox et al. 1999; Simms et al. 2004)

Hemingway 1934; Whitehead et al. 1952; Sellwood and Jenkyns 1975; Smith 1989; Rawson and Wright 1995; Simms et al. 2004; West 2007a, b). The cycles begin with: (a) thin (mm to cm) bioturbated to cross-bedded, shell rich lags, some of which are ferruginous, with siderite and oolitic berthierine (e.g., Anderton et al. 1979, p. 206). These beds contain abundant fragmented and sorted shell debris as well as whole shells including rhynchonellid brachiopods, pectinid bivalves,

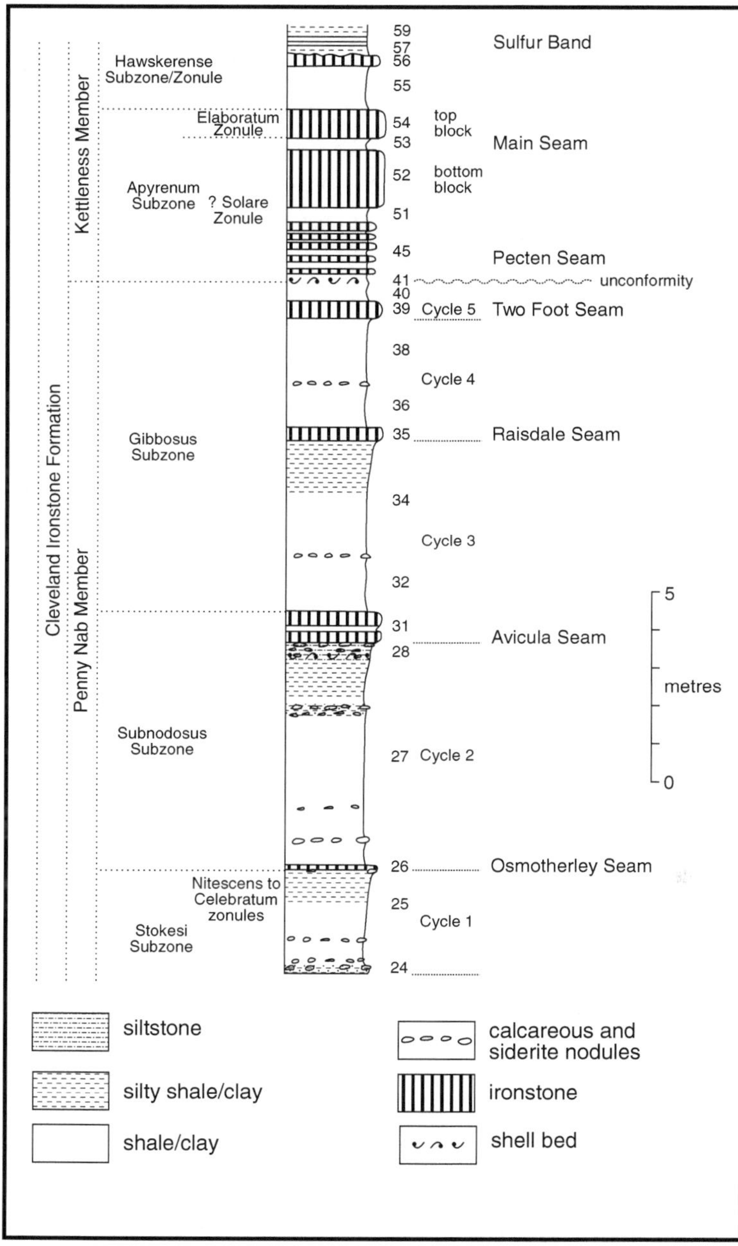

Fig. 20 Section through the Cleveland Ironstone formation between Cowbar Nab, Staithes, and Rosedale Wyke, Port Mulgrave. After Rawson and Wright (1995) and Simms et al. (2004). Bed numbers and cycles are from Howarth (1955) and Howard (1985) respectively

Fig. 21 Cleveland Ironstone Formation at Penny Nab, south of Staithes. The Two Foot Seam (**a**), the five thin ironstones of the Pecten Seam (**b**), and components of the Main Seam (**c**) are indicated (Photo: M.J. Simms, after Simms et al. 2004)

belemnites, and ammonites (Hemingway 1951). (b) The majority of the cycle is represented by 1–7 m of dark gray shale, mudstone with thin siltstone. (c) Higher portions of the cycles, characterized as "striped" facies, comprise interbedded dark gray silty shale and thin siltstone and sandstone beds showing sharp bases and small-scale hummocky cross-stratification, gutter casts (Greensmith et al. 1980), and occasional scattered shells. Prominent, condensed sideritic beds mark the "caps" of several small-scale cycles in the Cleveland Ironstone (Figs. 20 and 21; Hemingway 1951, 1974).

Upper portions of cycles, (d), are amalgamated siltstone beds up to 0.5 m thick with abundant burrows, especially *Rhizocorallium*, suggesting firm sand conditions. In some instances calcitic or sideritic concretions occur in the upper portions of the cycles below the sharp top (Hallam 1967). The latter nucleated on pipe-like mineralized burrows. These minor concretions are inferred to be the sedimentologic analog of the concretionary and tabular cemented limestone beds of more distal cycles; conversely the main body of the cycle comprising upward coarsening siliciclastic is correlative with thin dark shale and gray marl of the distal carbonate-shale rhythmic successions.

The Lower Jurassic (Hettangian to lower Sinemurian) Blue Lias, exposed on the Welsh Coast of the Severn Estuary near Nash Point (Fig. 19), provides a slightly older and somewhat more distal example (Hallam 1960; Shepard et al. 2006). Here the cycles range from 0.5 to 2 m in thickness with sharp bases at the bottoms of shell-rich beds. The latter contain abundant fragmentary bivalve shells, including gryphaeids, pinnids, and pectinids, crinoid debris and rare, small, solitary scleractinian corals. These beds are up to 20 cm thick.

Shell beds are overlain by medium to dark gray shale and mudstone. Upper surfaces of these shell rich beds have gradational to abrupt contacts with dark gray to black shales. Concretionary limestone beds may occur in the upper portions of the shell beds but tend to be more lenticular in form compared with the thick tabular beds of the distal Blue Lias. A high proportion of these concretionary beds contain articulated and closed *Pleuromya,* as well as pinnid bivalve shells, vertically to obliquely positioned in the otherwise sparsely fossiliferous mudstone in apparent life positions.

3.5.2 Distal Cycles

The condensed Jurassic Blue Lias Formation of Dorset, UK (House 1993) contains numerous fossil-rich shale and concretionary limestone cycles (Figs. 19, 22–25; Hesselbo and Jenkyns 1996; Moghadam and Paul 2000; West 2007a, b; Paul et al. 2008; Allison et al. in press). About 45 major concretionary limestone beds have been identified and traced over 10 s km (Lang 1924). Not all cycles show concretionary limestone components. The cyclic pattern is striking because of the alternation of ledge-forming pale gray weathering concretionary limestones and dark gray to black shales and marls (Fig. 22). The beds have been a classic source of fossil Lagerstätten, including ammonites and marine reptiles and fish (Fig. 23).

A typical Blue Lias cycle is 50–150 cm thick and comprises up to five component divisions (Fig. 24: Moghadam and Paul 2000). (a) A thin lag (centimeters) of shelly debris occurs at the tops of the gray marls immediately underlying many dark shales and seems to be the only representation of the transgressive lag. (b) Black and dark gray laminated, organic rich shale, typically with *Chondrites* burrows pass upward into pale-medium gray mudstone. (c) Medium gray calcareous

Fig. 22 Lower Jurassic Blue Lias Formation, Church Bay, east of Lyme Regis. Cycles composed of light gray limestone beds and darker shale and marl beds

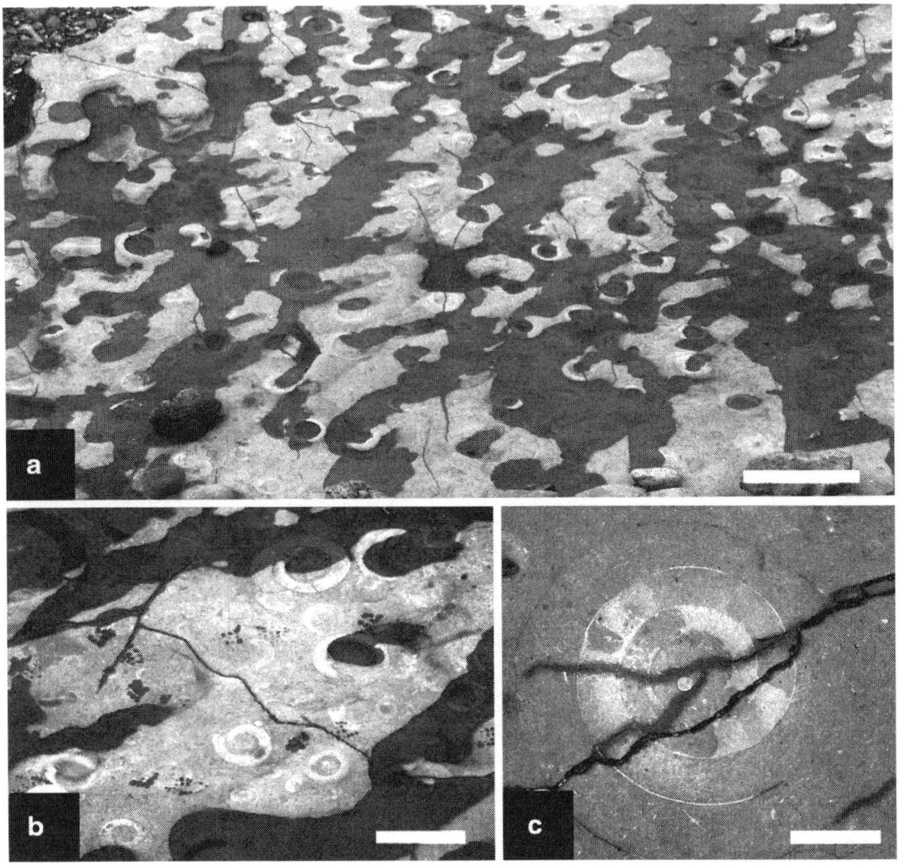

Fig. 23 Ammonites of bed 29 (Top Tape Bed of quarrymen) of the Blue Lias at Monmouth Beach, Lyme. (**a**) Bedding plane from above showing multiple large specimens of the ammonite *Coroniceras*; scale 1.5 m. (**b**) Close-up of ammonites; scale 60 cm (**c**) chamber fillings showing different generations of sediment infill; scale: 8 cm (From Allison et al. in press)

shales (marls) may be extensively bioturbated and include scattered fossils. (d) The prominent 10–25 cm thick ledged-forming, concretionary micritic limestones, when present, occur within the gray mudstone and are typically overlain by (e) a few centimeters of shell-rich marl. As such, the Blue Lias shows a distinctive meter-scale cyclic motif comparable to cycles discussed above for the Upper Ordovician Collingwood Formation and the Devonian Hollardops member, as well as classic Cretaceous cycles discussed below (Fig. 25).

These beds have a long history of study from the standpoint of cycle stratigraphy (Hallam 1960, 1964, 1986; House 1985, 1986, 1993; Weedon 1985, 1986; Weedon et al. 1999; Moghadam and Paul 2000; Paul et al. 2008; Allison et al. in press). Interpretations have ranged from a purely primary alternation of calcareous and clay rich sediment to a diagenetically-induced, rhythmic un-mixing (compare Hallam 1964 with Hallam 1986; Bottrell and Raiswell 1990; Moghadam and Paul 2000; Shepard

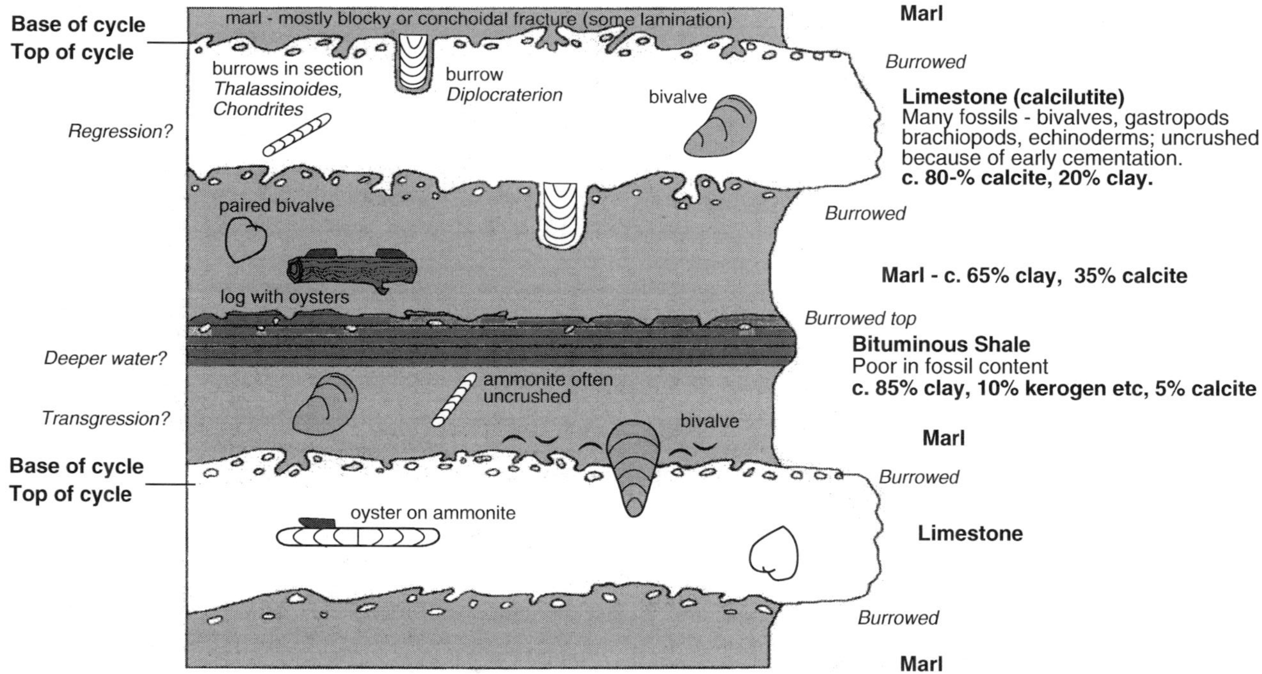

Fig. 24 Idealized Blue Lias rhythms showing principal lithofacies. In rhythms **a** and **b** the pale marl horizon has been diagenetically cemented into a limestone bed and is overlain by a dark marl bed. Adapted from figure in website Lym Regis-West, Blue Lias, Lower Jurassic; http://www.soton.ac.uk/~imw/lyme.htm;by Ian West

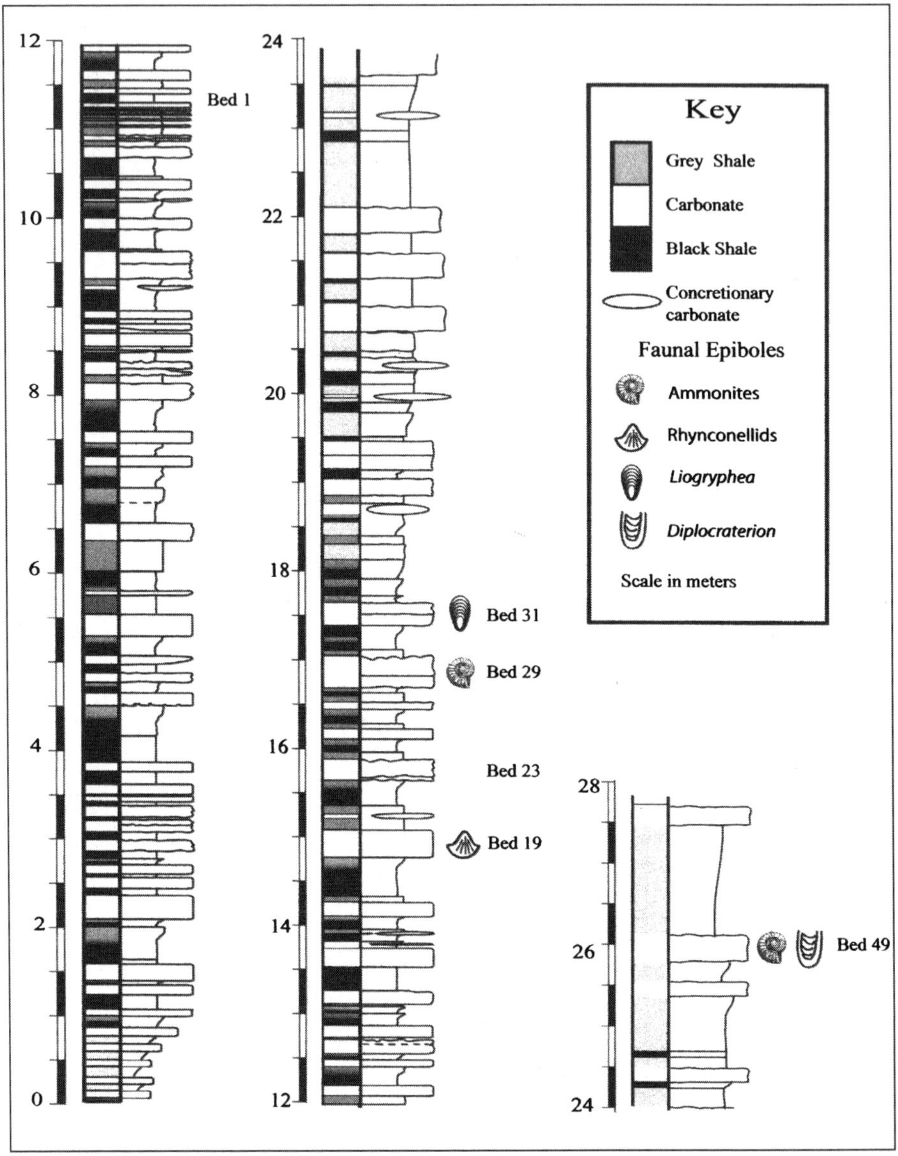

Fig. 25 Stratigraphic section of lower Blue Lias beds at Chippel Bay, Lyme Regis, Dorset, England, showing recurrent pattern of cycles (From Paul et al. 2008)

et al. 2006). A reasonable alternative is Weedon's (1985) proposal that the Blue Lias limestones are actually diagenetically altered marls, i.e. a diagenetic enhancement of a primary difference in sediment composition. Intervals with an originally higher content of fossil material and/or carbonate mud were preferentially cemented, probably in the zone of sulfate reduction (Moghadam and Paul 2000). The overlying fossil-rich horizons may mark the upper portion of a zone of diagenesis within the light marls. Both sets of limestone components show signs of early diagenetic alteration. Fossils within the carbonate units are more widely spaced although articulated specimens, including in situ fauna, are preserved in their original three-dimensional form.

Large ammonites (up to 80 cm) are abundant in some beds covering up to 40% of bedding-surface area (Fig. 23). Ammonites are preserved in 3D in nodular limestones and at variable orientation to bedding. Ichnological, sedimentological, and taphonomic evidence indicates condensation as the primary agent of accumulation (Allison et al. in press; Paul et al. 2008). However, the occurrence of equally condensed Blue Lias limestones lacking ammonite concentrations indicates that this alone is insufficient to account for concentration. Accumulation likely resulted from a combination of condensation and environmental conditions favoring the establishment of ammonites.

Stable isotope values suggest that cements were derived from pore-waters of a similar composition to contemporary seawater (Moghadam and Paul 2000). A clear diagenetic trend exists, with limestones having least, and laminated black shales most, modified stable isotope values. Contrast between trace fossil fills and host sediment demonstrates that Blue Lias rhythms are primary, but limestone beds are diagenetically enhanced. Condensation resulted from the interaction of climatic and sea level fluctuations. Marly skeletal rich sediments formed during minor transgressions. The sharp contacts of the overlying black shales denote maximum flooding surfaces. These times of prolonged sediment starvation favored concretionary cementation of underlying marly sediments and the formation of diagenetic limestones. Black shales record stratification under a stable halocline during maximum high-stands and humid climates. The thicker lower dark to medium gray mudstones evidence increased flux of mud associated with stable to slightly falling base level, and/or increased weathering and run-off.

Similar cycles occur in the Upper Jurassic Kimmeridge Formation of east Dorset Coast (Oschmann 1988; Gallois 2000) and the Lower to Middle Jurassic (Pleinsbachian) of Spain (Fernandez-López 2007; Fernandez-López et al. 2000, 2002).

3.6 Upper Jurassic to Lower Cretaceous; India

The Upper Jurassic to Lower Cretaceous sedimentary cycles of the Kachchh Basin, western India (Fürsich and Oschmann 1993; Fürsich and Pandey 2003) are strongly asymmetric, coarsening upward, and bounded by transgressive surfaces and overlying skeletal lag deposits (Fürsich and Pandey 2003; see Fig. 29). In proximal settings, the skeletons of basal shell beds (a) appear to have been reworked and locally transported, and are moderately time-averaged, with nearly total disarticulation of shells and preferred convex-up orientation. The fauna generally comprises low-diversity benthic communities, dominated by bivalves and corals. Fürsich and Pandey (2003) in contrast to other examples, however, interpret these deposits as the reworked residues of regressive (RST) sediments. The upper surface of these basal skeletal deposits, Maximum Flooding Zone (MFZ) shell concentrations of Fürsich and Pandey (2003) are higher in diversity, poorly-sorted, highly bioturbated, and dominated by mollusks, brachiopods, and randomly oriented, biogenically altered nektonic cephalopods. Taphonomic feedback (Kidwell and Jablonski 1983) favored colonization of the seafloor by epifaunal bivalves and brachiopods. Evidence for low rates of sedimentation include a concentration of iron minerals, with ferruginous ooids and glauconitic grains.

The overlying early highstand sediments (b) of cycles in the Kachchh Basin are typically bioturbated (sometimes planar laminated) claystones and siltstones largely devoid of skeletal accumulations. Fürsich and Pandey (2003) describe well-sorted shell pavements and lenses from late highstand parts of cycles (c), indicating paleoenvironments above fair-weather wave base. These skeletal accumulations are of low diversity, and represent parautochthonous to allochthonous reworked and transported relicts of benthic communities. Unlike skeletal accumulations of the TST, those of the RST in the Kachchh Basin (d) are laterally discontinuous and commonly form lenses, and pods, or form pavements on foresets of large cross-beds.

The most distal cycles of the Kachchh Basin, for instance the lower portion of the Bharodia section, are thinner and are dominated by fine-grained siliciclastic sediments. A well-developed concentration of ammonites, belemnites and bivalves forms the base of the sequence, which is high ferruginous and bioturbated (a). Fürsich and Oschmann (1993) report reworked bored concretions from many shell-rich beds, interpreted as minor transgressive lags, in the Callovian to Oxfordian Chari Formation. Ferruginous siltstone to silty fine-sandstone (b) overlies the ~1 m thick MFZ, and is devoid of skeletal accumulations. The overlying sequence (c), which is more proximal, in contrast to the distal example, is thicker, has a transgressive erosive surface, thicker (~1 m) transgressive lag and thinner MFZ, and is overlain by a thin package of bioturbated fine sandstone, and thick succession of trough cross-bedded sandstone.

3.7 Upper Cretaceous: Greenhorn Formation, Western Interior, USA

Small cycles of the Upper Cretaceous of the Western Interior are similar in scale and motif to limestone-shale cycles of the Lower Jurassic (Lias) of England (see above). Using spectral analysis of carbonate content, gray-scale variation, bioturbation index, and geochemical proxies, Sageman et al. (1997, 1998) detected bundled periodic signatures inferred to represent 20–100 kyr Milankovitch oscillations in climate and/or minor sea level variation (Figs. 26 and 27). The cycles of the Greenhorn and Niobrara Formations have been correlated in bed-by-bed detail from platform facies of Kansas to basinal sections near Pueblo, Colorado (Hattin

Fig. 26 View of Cretaceous Bridge Creek Limestone Member of Greenhorn Formation along railway cut at Rock Springs anticline, Pueblo, Colorado, showing rhythmic alternation of 10–20 cm thick, white-weathering, concretionary limestones and dark shales

1971, 1977a, 1982a, b; Kauffman 1982), to proximal and shoreface settings in central Utah and northern Arizona (Elder et al. 1994).

3.7.1 Proximal Cycles

In the Albian to Cenomanian Greenhorn Formation of southwestern Utah and Arizona skeletal beds occur consistently at the bases of shallowing-upwards 10–20 m thick packages of mudstones, siltstones and sandstones referred to as 'parasequences' by Elder et al. (1994). The shell-rich deposits (a) comprise disarticulated gryphaeid shells as well as other bivalves, baculitid ammonoids and gastropods; they show minor cross stratification and sorting of shells. Concretions are generally absent. Shell hash beds overlie bioturbated siltstones and sandstones with *Thalassinoides* and *Planolites* burrows (b, c).

Further shoreward, shell hashes were thin to absent but their position was marked by sharp contacts of bioturbated sandstone and siltstone beds with overlying shales that were long interpreted as flooding surfaces and were referred to as "lowstand" shell beds by Sageman (1996) by analogy with widespread thin, shelly limestone discussed by Brett and Baird (1986a) from the Appalachian basin Devonian. However, in both cases, these shell beds are better interpreted to represent lags at the bases of shallowing upward cycles and thus, we interpret them as

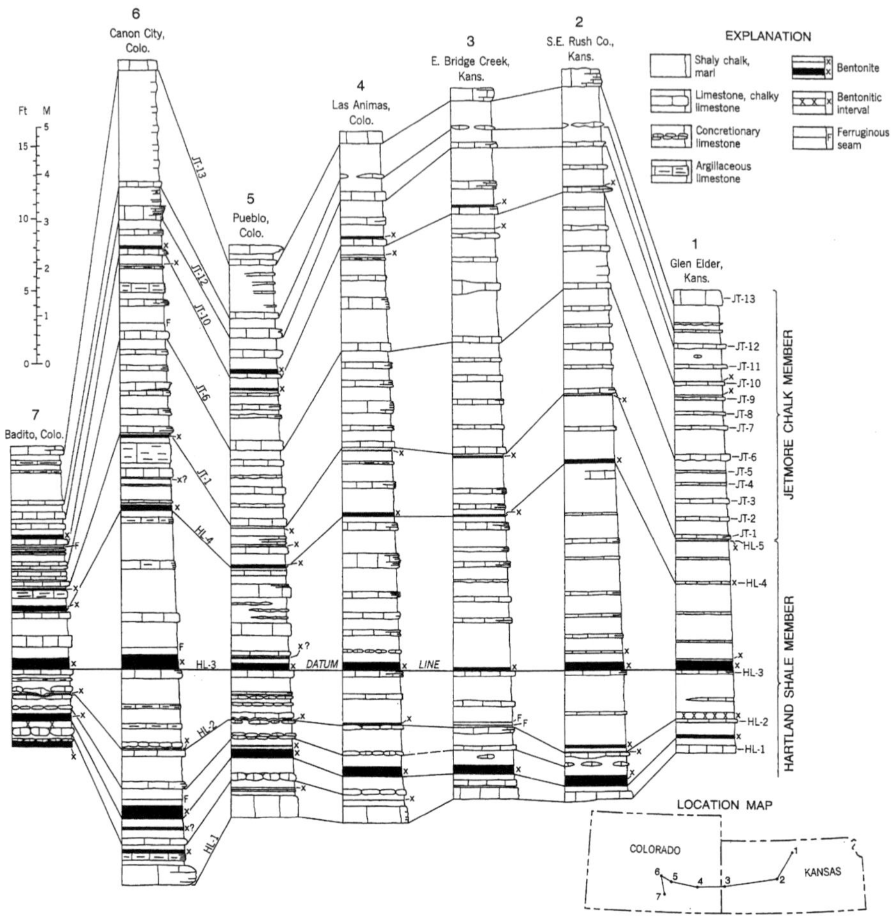

Fig. 27 Correlated sections of Cretaceous Greenhorn cycles from Kansas to Colorado, showing persistence of small-scale cycles (From Hattin 1971)

reworked, time-averaged assemblages that accumulated during intervals of relative siliciclastic starvation (Brett and Baird 1996).

Such transgressions could be invoked to explain the shell beds as trangressive lags and would explain the presence of overlying distal dark shale facies. Progradation of siliciclastics during the alternate highstand/wet interval produced the coarsening upward motif of the parasequences. Alternatively, the shelley, sediment starved intervals were associated either with periods of dry climatic conditions during which offshore transport of terrigenous sediments was greatly reduced during small base level rises.

The offshore cycles were typically attributed to Milankovitch-band climatic oscillations, whereas the parasequences were ascribed to local progradation and subsidence of deltaic lobes, perhaps in response to minor base-level oscillations or to autocyclic processes (Elder et al. 1994). The careful correlation of the cycles across proximal-distal profiles provides strong evidence that the typical 10 m-scale cycles of proximal regions and decimeter scale marl-limestone cycles of distal areas were products of a common, allocyclic forcing function. These Cretaceous cycles parallel observations on the Middle Devonian of the Appalachian Basin and in both cases there is now strong evidence for basin-wide, if not interbasinal oscillations in sea level and/or climate.

Depositional processes responsible for the limestone shale cycles include sediment bypassing during transgressive erosion and onlap in shoreface to inner shelf depths and sediment starvation during deepening in inner to mid-shelf depths. Deposition of entrained relict skeletal material following storm events from shelf depths produced thin, lenticular shell beds. Rapid sedimentation terminated accumulation of sediment-starved fauna in mid-outer shelf depths. Storm erosion and re-deposition of shells occurred above storm wave base in inner shelf and shoreface depths. Reworking and condensation of shells above fair-weather wave base in inner shelf to shoreface depths led to deposition of thin time-averaged shell beds comparable to those of the Jurassic Kachchh basin (Fürsich and Pandey 2003; Fig. 29).

3.7.2 Distal Cycles

The distal facies of the Greenhorn Formation in Kansas and central Colorado are recorded in chalk facies as very thin shelly lags (a), marly chalk (b) to tabular limestones (c), especially well displayed in the Bridge Creek Formation (Hattin 1971, 1977a, b; Kauffman 1982). The tabular bands are approximately 10–25 cm thick and alternate with slightly thicker marls. These bands generally display sharp lower and upper contacts and are overlain by thin lags of inoceramid prisms and other shell debris. Bed tops are bioturbated by *Thalassinoides*, *Planolites* and other large burrow systems (Hattin 1971). Hattin (1977b) described well-preserved, articulated barnacles from the tops of some of these beds that indicate pulses of rapid burial; in situ bivalves may also be present.

Spectral analysis of these rhythmically banded chalky marls and limestones suggest precessional, obliquity, and eccentricity effects (Sageman et al. 1997, 1998). The latter have been attributed to Milankovitch forced climatic and/or minor sea-level oscillations.

In basinal sections, cycles comprise thin concretionary shell hash deposits (c) alternating with calcareous, dark gray shales (b). Shell material is largely disarticulated and fragmented. Rarely, complete, double-valved shells and well-preserved ammonite shells, primarily *Baculites* occur within the concretionary limestone bands. These concretionary intervals thus appear to represent rapidly buried pods of shells and ammonites accumulated on the seafloor perhaps in low hollows (cf. Tsujita 1995). These beds are the sedimentological and diagenetic analogs of concretionary beds with articulated chaotically oriented trilobites in Ordovician and Devonian samples cited above. As in the latter cases, the overlying hashes of broken and disarticulated shells (a) are thin, lenticular and subtle. But it is these surfaces that reflect interludes of sediment starvation during which the concretions developed in underlying sediment packages, perhaps nucleating around clusters of rapidly buried shells.

Stratigraphic correlations (Elder et al. 1994) demonstrate that offshore concretionary beds correlate approximately with the limestone bands of the eastern rhythmically bedded chalk, and with proximal shell hash beds. However, we infer that the concretions reflect diagenesis beneath thin shell lags that formed during interludes of low sedimentation.

Comparative examples include decimeter- and meter-scale Milankovitch cycles of soft and more indurated beds also have been documented in Cretaceous chalks from Great Britain and France (Kennedy and Garrison 1975; Hancock 1975). As with the Greenhorn and Niobrara cycles, and others discussed herein, the compact limestones show extensive firmground burrows of *Thalassinoides, Rhizocorallium, Planolites*, and others; in some cases, *Trypanite*s bored- and oyster-encrusted hardgrounds also occur at the tops of these cycles (Bromley 1967, 1968; Hancock 1975).

3.8 Cenozoic: Ashiya Group, Japan, and Punta Judas Formation, Costa Rica

3.8.1 Proximal Cycles

Relative to other intervals little is known of small-scale sedimentary cycles from the Paleogene. Examples have been noted, however, from the Ashiya Group (Oligocene) of Japan (Hayasaka 1991; Sakakura 2002), and the San Julián Formation (Oligocene) of Argentina (Parras and Casadio 2005). Cycles from the Ashiya Group are reviewed here.

The upper part of the Ashiya Group consists of coarsening-upward sedimentary cycles of 30–100 m in thickness and comprises a range of sandstone and mudstone lithologies (Fig. 2 of Hayasaka 1991). Within cycles of the Ashiya Group, a transgressive basal medium sandstone rests on a distinctive erosional surface that truncates the upper part of the underlying cycle, and fines upward to very fine sandstone (Fig. 28). This basal facies ranges in thickness from 5–20 m in thickness, and is characterized by coarse lithic granules, smectite, glauconite, and occasional dense imbricated shelly lenses that overlie minor erosional surfaces. The upper part of the fining up basal sandstone may contain patches of articulated bivalves. The base of the cycle (a) is often intensely bioturbated by *Thalassinoides* and *Ophiomorpha*. This upper part of this basal sandstone is rich in glauconite, and is conformably capped with mudstone that represents the base of a progradational coarsening upward interval. Most sequences comprise in successive order (b) dark gray laminated or bioturbated mudstone (5–40 m thick), mudstone interbedded with very thin sandstone beds (3–10 m thick), (c) silty fine sandstone or alternating hummocky cross-stratified sandstone (HCS) (3–20 m thick), and mudstone and (d) amalgamated HCS (10 m thick). In some cases tabular cross-stratified sandstone may immediately overlie the coarsening upward mudstone, in place of HCS (10–15 m thick).

The shell beds of the basal sandstone facies (a) are densely concentrated as a parautochthonous accumulation 20–50 cm thick. The erosional surface exhibits

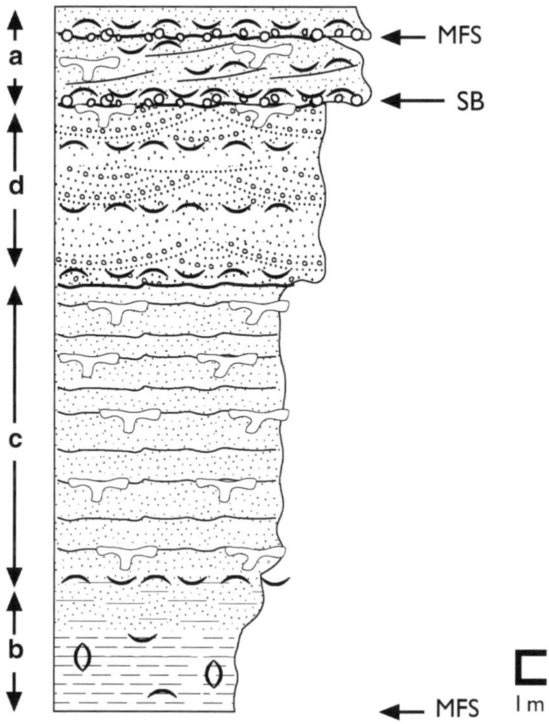

Fig. 28 Typical cycle of the Ashiya Group based on section 7 (Waita Formation), exposed on the coastline between Sakamizu and Waita, Kitakyushu-Ashiya area, southwest Japan. Lettered divisions are portions of cycle analogous to those shown in Fig. 9. (**a**) Poorly-sorted pebbly coarse to fine sand or gravel with an erosional surface and gravel or shell lag at its base; (**b**) massive, dark gray siltstone in lower part of cycle; (**c**) silty very fine sandstone of falling stage of cycle; (**d**) well-sorted fine sandstone, amalgamated type HCS, with lags of heavy mineral or shells (Modified from Hayasaka 1991)

wavy undulation of up to 20 cm in relief. The shell bed comprises typically disarticulated and fragmented bivalve shells, which may be piled up and imbricated bidirectionally along bedform slopes. The upper part of the transgressive unit in the Ashiya Group includes both allochthonous shell beds of approximately 20 cm thick, which rest on wavy erosional surfaces, dispersed assemblages of articulated bivalves, and shell beds that lack an erosional base. The matrix of these shell beds consists of shell fragments showing imbrication, articulated shells filled by geopetal, shells which exhibit high abrasion, and common epifaunal forms (e.g. pectinid and anomiid bivalves), and widespread encrustation by barnacles.

The mudstone unit that represents highstand conditions in Ashiya Group cycles (b) yields autochthonous and scattered bivalve shells. Many of these shells are articulated and in life position. In general the facies lack evidence of bottom erosional or reworking. Occasional shell stringers accumulated along bedding planes in the mudstone as it coarsens up, and show preferred orientations that result from unidirectional currents. Primary sedimentary structures in highstand and regressive parts of Ashiya Group sequences are disturbed by bioturbation. Ichnofossils including *Phycosiphon, Planolites, Palaeophycus, Rosselia* and *Skolithos* are observed in these facies.

The succession of taphofacies in Ashiya Group cycles suggests an upward decreasing influence of wave and current energy. The strong wave influence declines from an erosional and winnowing phase to a quiet muddy phase during transgressions.

3.8.2 Distal Cycles

Although cycles analogous in scale and pattern to those discussed for the Paleozoic and Mesozoic are uncommon in the Cenozoic because of changes in depositional setting (i.e. epicontinental sea vs oceanic shelf), one example described from the Miocene Punta Judas Formation of Pacific Costa Rica (Krawinkel and Seyfried 1996) provides a number of parallels. A brief synopsis is included for sake of comparison. This youngest example comes from a time of transition from greenhouse to icehouse global climates, but it is derived from a subtropical setting.

Two types of open marine cycles are described by Krawinkel and Seyfried (1996). Open shelf cycles consist of meter thick cycles that commence with (a) tempestitic skeletal accumulations and pass upward into mudstones with storm-reworked parautochthonous to allochthonous shell hashes (b). A few concretionary beds (d), present within the mudstones host in situ articulated bivalves. Siltstones and mudstones are heavily bioturbated, with *Thalassinoides*, *Ophiomorpha* and *Skolithus* burrows in the sands. Upper parts of cycles consist of hummocky cross-stratified sandstones.

Distal estuarine delta front cycles are also on the order of a meter thick and may be identified by concretion beds in otherwise monotonous, bioturbated sediments. Most of these beds show autochthonous bivalves in life position. In most cases the autochthonous bivalves are associated with concentrations of disarticulated shells of bivalves, as well as gastropods, echinoids and fish teeth, and in some cases driftwood. We conclude that these concretionary horizons record episodes of rapid burial during times of general sediment starvation associated with abrupt sea level rise. Thus, they are excellent analogs of concretionary shell beds seen throughout the Phanerozoic.

4 Discussion: Synopsis of Examples

Although the age-range of examples considered herein spans nearly 500 million years, the cycles share striking sedimentologic, paleontologic and taphonomic characteristics (Tables 1–5; Figs. 29 and 30). Cycles exhibit the same general sequence of lithofacies. Comparison of the major components of the cycles provides a basis for understanding meter-scale cycles in general.

In this concluding section we first outline common features displayed by all cycles in terms of litho-, bio- tapho-, and ichnofacies, and highlight unique features of small-scale cycles of particular ages. We then consider the underlying causes of these common features and finally, we briefly outline secular changes and their influence on cycle appearance.

4.1 Basal Condensed Shell Bed Taphofacies

4.1.1 Base of Cycle Shell Debris Beds

A basal compact bed, yielding evidence of condensation represented by skeletal and or diagenetic clasts is present in virtually all of the examples of proximal cycles presented herein (Table 1; Figs. 29 and 30). The prominence of this skeletal bed varies substantially both from one time to another and with proximality of individual cycles, yet this bed is the most persistent feature of the cycles; this bed in varied guises can be traced from near basinal center to upramp positions, and, in some cases, to both siliciclastic and carbonate margins of basins.

The extent of development of the condensed shell bed is a complex function of several factors, including: (a) shell productivity; (b) durability of skeletal remains; (c) duration of sediment starvation; and/or (d) extent of winnowing and processing of skeletal material by storm waves and/or bottom currents (see McLaughlin et al. 2008 for review). The basal shell beds of the Ordovician to Cenozoic examples have all been noted as having sharply erosive bases and abrupt, sometimes rippled tops with hardgrounds (Fürsich 1978; Fürsich and Pandey 2003). Examples of phosphatic and hematitic ooids are also present in examples from the Ordovician (Dabard et al. 2007), Silurian (Van Houten and Bhattacharrya 1982; Cotter and Link 1993), and Jurassic (Whitehead et al. 1952; Hallam 1966; Hallam and Bradshaw 1979; Van Houten and Bhattacharrya 1982). Evidence of firmground and/or hardground features is typical of the upper surfaces of the bed, and in many instances a coating or impregnation of iron oxide occurs along this sharp contact (Fig. 29).

In distal facies (Fig. 30) this bed is typically subtle and consists simply of a single layer or pavements of shelly debris. It may be represented by a thin interval of shell-rich marls or mudstones and may directly drape concretionary beds at the top of subjacent cycles.

Despite differences in thickness the basal shell beds of small-scale cycles show comparable taphonomic patterns. Key features of the proximal shell bed taphofacies include: (a) a high degree of concentration and varied preservation of skeletal clasts, ranging from intact skeletal elements and rare articulated material to variably fragmented, abraded and/or corroded and coated grains and prefossilized steinkerns; (b) presence of reworked material including shale clasts, reworked concretions, prefossilized steinkerns, coated and mineralized grains, phosphatic, hematized and or pyritic clasts; (c) evidence of reworking and winnowing; typically packstone to grainstone fabrics; (d) for thicker beds, complex internal amalgamation of varied lenses and pods of skeletal debris; (e) shell beds are dominated by single valves,

Table 1 Comparative taphonomic and paleoecologic aspects of basal (transgressive) shell-rich beds

Age Formation Location	References	# Cycles Duration	Lithology Bedding	Common taxa Guilds	Trace Fossils	Taphonomic features
Cretaceous	Elder et al. (1994)	10–30 cm	Sharply based, sharp	Inoceramids, gryphaeids	*Thalassinoides*	Moderately to densely packed
Greenhorn	Sageman (1996)	20 Cycles	firmground top	ammonites, belemnites	*Rhizocorallium*	disarticulated, fragmented shells
Utah/Colorado		20–40 kyr	sandy packstone		*Glossifungites*	inoceramid prisms; corrosion
			tabular, cross bedding			encrusting and boring of skeletons
Early Jurassic	Allison et al. (in press)	1–20 cm	Sharply based, sharp	Gryphaeids, belemnites	*Thalassinoides*	Moderately to densely packed
Blue Lias	Paul et al. (2008)	40 cycles	firmground top	ammonites, crinoid debris	*Rhizocorallium*	disarticulated, fragmented shells and
Dorset, Gt. Britain		40–100 kyr	sandy packstone	small corals	*Diplocraterion*	crinoid ossicles; corrosion; encrusting and
					Glossifungites	boring of skeletons
Mid-Devonian	Brett and Baird (1985,1996)	1–20 cm	Sharply based, sharp	Crinoid, rugose/tabulate	"Megaburrowed"	Moderately to densely packed,
Hamilton Group	Landing and Brett (1987)	40 cycles	firm- and hardground top	corals, thick shelled	firm-/hardgrounds	disarticulated, fragmented
New York, Ontario		40–100 kyr	packstone-grainstone	brachiopods; rare	*Glossifungites*	abraded and corroded shells, crinoids

			phosphatic nodules	mollusks, trilobites	*Trypanites*	encrusting and boring of skeletons
Lower Devonian	Botquelen et al. (2006)	1–20 cm	Sharply based, sharp	Crinoid, rugose/tabulate	*Thalassinoides*	Moderately to densely packed
"Type B" shell beds		30 cycles	firmground top	corals, thick shelled	*Glossifungites*	disarticulated, fragmented, abraded
Spain		20–100 kyr	sandy pack--grainstone	brachiopods; rare		and corroded shells
				mollusks, trilobites		encrusting and boring of skeletons
Late Ordovician	Brett et al. (2003)	5–60 cm	Sharply based, sharp	Strophomenid/orthid brach	*Planolites*	Moderately to densely packed
Kope Fm.		40 cycles	rippled, firmground tops	ramose bryozoans	*Cruziana*	disarticulated, fragmented, abraded
Ohio/KY		20–100 kyr	sandy pack--grainstone	crinoids, trilobites	*Trypanites*	and corroded shells
						encrusting and boring of skeletons
Late Ordovician	Botquelen et al. (2006)	2–20 cm	Silty packstone	Bryozoans, brachiopods,	Sharp *Planolites*	Densely packed
"Type B" shell beds		~40 cycles	sharp bases; phosphatic	cystoids, crinoids	*Thalassinoides*	disarticulated; all fragmented shells
Sardinia		20–100 kyr	crusts on tops			bioeroded and encrusted skeletons
Mid-Ordovician	Boyer and Droser (2003)	5–20 cm	Pack- to grainstone	Orthid brachiopods	Sharp *Planolites*	Loosely-densely packed

(continued)

Table 1 (continued)

Age Formation Location	References	# Cycles Duration	Lithology Bedding	Common taxa Guilds	Trace Fossils	Taphonomic features
Kanosh Fm.		~140 cycles	sharp bases; rippled	echinoderms, gastropods	*Thalassinoides*	disarticulated; highly fragmented shells
W. Utah			firm- and hardgrounds	ostracodes	*Trypanites*	
Mid-Cambrian	Brett et al. (2009)	10–50 cm	Peloidal to oolitic, and	Orthid brachiopods	*Planolites*	Loosely packed fragmented
Wheeler Fm.		~25 cycles	bioclastic pack- to grain	echinoderms; sponge	*Thallasinoides*	skeletal material in oolitic matrix
W. Utah		?20–100 kyr	stone; sharp bases sharp hardground tops	spicules; trilobite frags.	(hypichniia casts)	

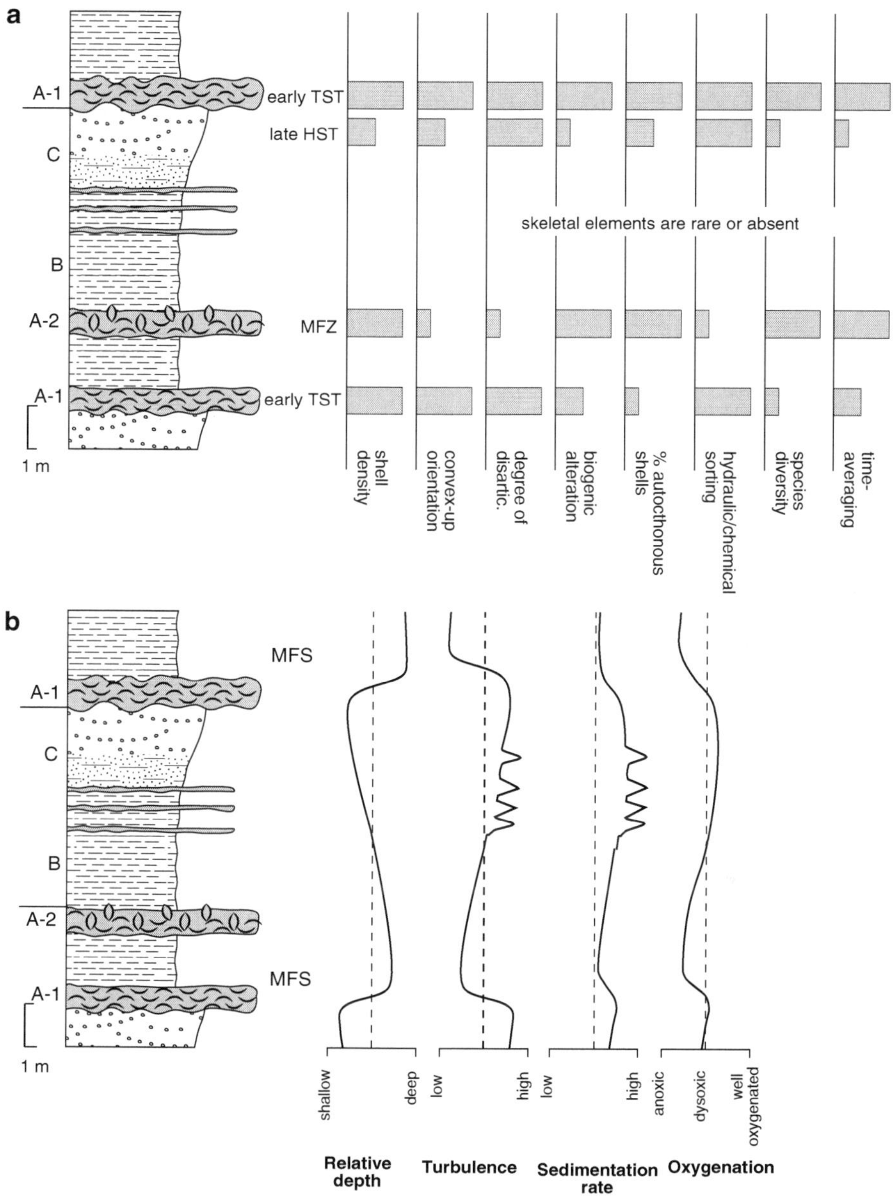

Fig. 29 Generalized proximal small-scale cycle. Note basal shell debris beds, upper bed may be amalgamated to the lower bed followed by shale with thin event beds and capped by cross laminated silt- and sandstone. Bars show relative frequency. (**a**) Distribution of taphonomic features within cycle; note high degree of concentration and orientation in basal bed and very high degree of condensation of upper shell bed. (**b**) Distribution of inferred paleoenvironmental parameters through the cycle. Note spikes on turbulence and sedimentation curves, reflecting instantaneous event beds. Symbols: SB, sequence boundary; MFZ, late TST; MFS, maximum flooding surface; HST, highstand systems tract; TST, transgressive systems tract (Adapted from Fürsich and Pandey 1999)

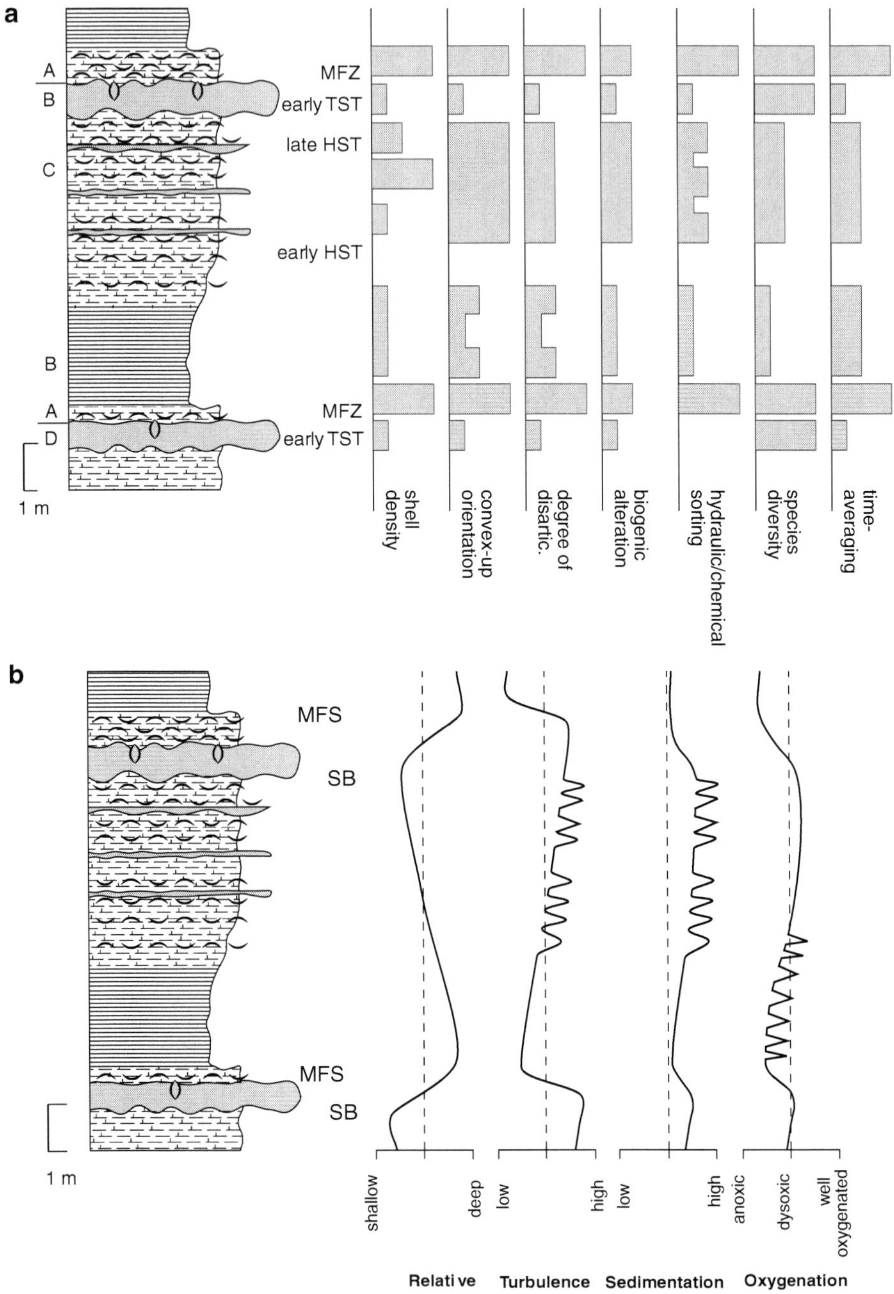

Fig. 30 Generalized distal small-scale cycle. Note basal shell debris beds followed by dark, laminated shale and overlain by gray marly mudstone to siltstone with thin shell hash beds and laminated storm siltstones; also note sparsely fossiliferous (*propped open*) concretionary diagenetic limestone with in situ articulated fossils just below basal shell beds at tops of cycles. (**a**) Distribution of taphonomic features within cycle; note high degree of concentration, orientation in basal bed and very high degree of condensation of upper shell bed. (**b**) Distribution of inferred

predominantly in convex upward positions (see Kidwell and Bosence 1991; Fürsich and Pandey 2003; Hendy et al. 2006).

4.1.2 Gray Marl Beds with Thin Condensed Hashes

The main skeletal beds are typically followed by millimeter to decimeter scale gray calcareous claystone/marl with thin stringers of comminuted shelly debris. In some instances these gray claystones mantle sharp upper surfaces of the shell hash bed preserving articulated fossils at the base as obrution beds (Figs. 29 and 30).

As noted above, the trilobite and diminutive phosphatic brachiopod faunas of deeper water facies did not build thick shell beds. The thin veneers of skeletal hash that typify the Collingwood Shale, as well as the Lower Devonian Hollardops beds are believed to occupy a position analogous to the condensed shell beds in upramp areas. Extreme condensation is recorded by thin veneers of fish bones, conodonts, and teeth that floor black shales in some Devonian as well as Mesozoic cycles (Martill 1985; Baird and Brett 1991). Here there is little or no preservation of shell, if indeed such shells were ever present, and the residue is typically strongly reworked and comminuted. Thus, these distal dysoxic facies still produced "Cambrian style" shell beds in middle Paleozoic (McKerrow 1979; Li and Droser 1997). Much the same can be said for thin debris layers that appear at cycle bases in the distal facies of Jurassic and Cretaceous cycles. The bases of cycles in much of the Blue Lias are marked by thin veneers of fragmentary ammonoids and bivalves; in many dysoxic facies the cycles commence abruptly with a sharp based black shale that may be floored a sprinkling of bones and teeth, or nothing at all.

4.1.3 Biostromes-Bioherms

In some cases the top of the skeletal debris bed is terminated by a more or less in situ accumulation of skeletal remains and/or microbial carbonates. Such examples include small thrombolitic mounds, bryozoan-algal mud mounds and biostromes if intact to fragmental bryozoans and corals. These buildups are present as barren microbial mounds in the Cambrian (e.g. Elrick and Snider 2002), and as thrombolites as well as laterally correlative sponge, tabulate, rugosan coral biostromes and bioherms in the middle Paleozoic. For example, there are numerous examples of small patch reefs and bioherms from the Ordovician to Devonian based upon hardgrounds or stabilized crinoidal shoal deposits recording early TST deposits (Walker and Alberstadt 1975). Sponge, algal mounds have been reported on flooding

Fig. 30 (continued) paleoenvironmental parameters through the cycle. Note spikes on turbulence and sedimentation curves, reflecting instantaneous event beds; also, spikes on oxygenation curve that reflect brief interludes of more oxic conditions that permit temporary seafloor colonization. Symbols: SB sequence boundary; MFZ, late TST; MFS, maximum flooding surface; HST, highstand systems tract; TST, transgressive systems tract surfaces of small-scale cycles from the Middle Jurassic (Bajocian) of the Atlas Mountains and Europe (Addi 2006); here too the buildups are related to relative sea-level rise owing to a combination of tectonics and eustasy. Probably analogous rudistid bivalve patch reefs occur on flooding surfaces in the Albian of Mexico (Lehmann et al. 2000).

4.2 Dark Mudrocks

4.2.1 Dark Laminated Shales

Distal facies of many cycles, especially in ramp settings (e.g. Collingwood Shale, Blue Lias of Dorset and Watchet) include dark gray to black, organic-rich, laminated shales (Table 2; Fig. 30). Such beds are invariably sharp based with a slightly erosive contact with the underlying gray shale or shelly marl and include a very thin basal lag deposit of reworked phosphatic, pyritic and resistant skeletal grains (Baird and Brett 1986, 1991).

Although some shales are barren, many have bedding planes covered with mono- or paucispecific fossil assemblages. Middle Cambrian cycles show a dominance of a few species of agnostoid trilobites, possibly chemosymbiotic olenid polymeroids, such as *Elrathia* (Gaines and Droser 2003), and phosphatic brachiopods. The Ordovician dark shale biofacies, exemplified by the Collingwood Shale, are primarily comprised of inarticulate brachiopods and small orthids, such as dalmanellids, and a few species of trilobites (e.g., *Pseudogygites* and *Triarthrus*). Lingulids and small nuculid bivalves are also relatively common and represent the only infaunal organisms within the assemblage. Pelagic forms are well represented on some bedding planes by graptolites. Devonian dark shales typically include low diversity or monospecific assemblages of inarticulate brachiopods, as well as leiorhynchid and ambocoeliid brachiopods, small nuculid bivalves, and gastropods (Boyer and Droser 2007). Pelagic forms include thin-shelled tentaculitids (e.g. styliolinids), nautiloids, and the oldest goniatitic ammonoids. Mesozoic examples such as the Jurassic Blue Lias shale similarly contain a very low diversity assemblage of small nektobenthic and pelagic organisms. The assemblage is dominated by the small bivalve *Bositra*, ostracodes, small ammonites, and fish scales. In addition, carbonized logs and woody plant debris are present in many examples, and the logs may be encrusted by organisms including bivalves and long stalked crinoids (Seilacher and Hauff 2004).

Low diversity, high dominance fossil assemblages, are typically preserved as hash-rich pavements and thin stringers. Fossils are generally disarticulated, commonly fragmented, and dominated by decalcified, and benthic organisms (e.g. small bivalves, or brachiopods) and/or carbonized and phosphatic remains (e.g. orbiculoid brachiopods, graptolites, coalified wood and other plant remains). In a few Mesozoic examples aragonite is preserved as thin films on highly compressed shells, but in other cases even calcitic fossils are decalcified.

Table 2 Comparative taphonomic and paleoecologic aspects of highstand dark shales

Age Formation Location	References	Thickness Number Duration	Lithology Bedding	Common taxa Guilds	Trace Fossils	Taphonomic features
Mid-Cretaceous	Kauffman (1982)	20–100 cm	Dark gray to black	Small inoceramid	Minor small	Barren to densely packed bedding
Greenhorn Fm.	Sageman (1996)	20–40 kyr	laminated shale	pterioid bivalves	*Chondrites*	plane assemblages, disarticulated
Utah/Colo., USA			disseminated pyrite	small ammonites		fragmented; decalcified, compressed
						minor pyrite patinas
Early Jurassic	Kauffman (1978, 1981)	50–100 cm	Black, laminated shale	*Bositra*, pectenid bivalves	Small *Chondrites*	Barren to densely packed bedding
Posidonienschiefer	Seilacher (1982b)	?	disseminated pyite	ammonites	in some bedding	planes; disarticulated;fragmented
S. Germany		?		crinoids on logs	planes	decalcified, compressed; minor
						pyrite patinas; horions of articulated
						vertebratres and crinoids
Early Jurassic	Mogadam and Paul (Moghadam and Paul 2000)	30–120 cm	Dark gray to black	Small "paper pectens"	Minor small	Barren to densely packed bedding
Blue Lias	Allison et al. (in press)	40 cycles	laminated bituminous	pterioid, bivalves	*Chondrites*	plane assemblages, disarticulated

(continued)

Table 2 (continued)

Age Formation Location	References	Thickness Number Duration	Lithology Bedding	Common taxa Guilds	Trace Fossils	Taphonomic features
Dorset, Great Britain	Paul et al. (2008)	38 kyr	shale	small ammonites		fragmented; decalcified, compressed
				fish scales, fish,		minor pyrite patinas
				marine reptiles		horizons of articulated vertebrates
Mid-Devonian	Brett et al. (1991)	20–500 cm	Dark gray to black	Thin shelled rhynchonellid	Pyritic thread-like	Barren to densely packed bedding
Hamilton Group	Boyer and Droser (2007)	~30 cycles	laminated shale	orbiculoid brachiopods	burrows	plane assemblages, disarticulated
New York, USA		20–100 kyr	disseminated pyrite	small pterioid bivalves		fragmented; decalcified, compressed
						minor pyrite patinas
Lower Devonian	Becker et al. (2004a), b)	20–80 cm	Dark gray to black	Small athyrid, ambocoeliid	Pyritic thread-like	Barren to densely packed bedding
Hollardops beds	Brett, unpub. data	~30 cycles	laminated shale	& orbiculoid brachiopods	burrows; small	plane assemblages, disarticulated
SW Morocco		? 20–40 kyr	disseminated pyrite	small pterioid bivalves	*Chondrites*	fragmented; decalcified, compressed
						minor pyrite patinas
Late Ordovician	Brett et al. (2006)	20–100 cm	Dark gray to black	Orbiculoid, lingulid and	None	Barren to densely packed bedding

Collingwood Shale		15 cycles	laminated oil shale	small orthid brachiopods,		plane assemblages, disarticulated
Ontario, Canada		20–40 kyr	disseminated pyrite	*Triarthrus* trilobites		fragmented; decalcified, compressed
						beds of articulated trilobites; molts
Mid-Cambrian	Gaines et al. (2005)	20–500 cm	Dark gray to black,	Acrotretid brachiopods	None	Barren to densely packed bedding
Wheeler Formation	Gaines and Droser (2005)	25 cycles	laminated shale	agnostoid and polymeroid		plane assemblages, disarticulated
Utah	Brett et al. 2009	20–100 kyr	disseminated pyrite	trilobites (*Elrathia*)		fragmented; decalcified, compressed
						beds of articulated trilobites; molts

A high degree of skeletal fragmentation is typical of dark shales (Fig. 30). Although the cause of this breakage in low energy, low oxygen environments is enigmatic, it points to prolonged exposure and may include episodic disturbance of diagenetically weakened shells by deep currents.

Although fossil debris appears randomly distributed on some bedding planes, prominent alignment of elongated fossils is observed in many examples of black shales, as well exemplified by aligned graptolites in Ordovician-Silurian examples and aligned tentaculitids and orthocones in the mid Paleozoic, or baculitids in the Cretaceous. This evidence, together with minor stringers or gutter-like accumulations of debris points to current activity even in these dysoxic settings.

In contrast to the typical poor preservation, a few bedding planes in many dark shales yield well articulated fossils and even, especially in the Cambrian examples, soft or weakly sclerotized organisms. Well preserved trilobite exoskeletons, both as carcasses and as molt ensembles, are noted on certain bedding planes in Cambrian to Devonian examples, providing direct evidence for episodic rapid deposition. The molt ensembles are particularly significant in pointing to in situ life activity on the seafloor and an absence of lateral transport in obrution deposits. Mesozoic examples typically lack abundant multi-element skeletons making recognition of burial horizons more difficult. However, beds of articulated crinoids-frequently associated with carbonized logs and vertebrates, such as fish and marine reptiles, provide dramatic evidence, for rapid episodic burial (see Kauffman 1978, 1981; Seilacher 1982b).

Shells are typically very strongly compacted, decalcified, and lack mineralization with the rare exception of minor pyritic patinas. Robust pyritized fossils and burrows are nearly lacking in these facies. However, the poor preservation of fossils within the shales indicates a relatively low sedimentation rate (long exposure time to seafloor conditions) and a relatively low pH environment. Some fracturing of shells results from the early compaction of the units. In both cases, aragonitic fossils are decalcified and highly compressed but most aragonitic shells are preserved as plastically deformed molds indicative of early dissolution (see Kauffman 1978, 1981, Seilacher 1982b).

4.2.2 Gray Mudstones and Siltstones

The remainder of simple shelf cycles may comprise gray shales, mudstones and siltstones or calcisiltites (Table 3; Figs. 29 and 30). The mudstone components of the middle Paleozoic and Mesozoic examples both include increased faunal abundance and diversity over the dark shales. In addition, there is a significant increase in bioturbation. Nearly all examples contain numerous burrows and the sediment may, in fact be heavily bioturbated.

Mudstone intervals of distal environmental settings, in fresh samples and cores, reveal discrete, sharply-based packages of relatively uniform muds, up to several centimeters thick. Thin siltstones, calcisiltites and silty mudstones, when present, are sharp based with minor lags of disarticulated skeletal debris on the basal surfaces; these

Table 3 Comparative taphonomic and paleoecologic aspects of highstand mudstones

Age Formation Location	References	Thickness Number Duration	Lithology Bedding	Common taxa Guilds	Trace Fossils	Taphonomic features
Cretaceous	Elder et al. (1994)	20–100 cm	Medium gray, calcareous	*Gryphaea,* inoceramids	*Planolites*	Dispersed to patchy; disarticulated
Greenhorn		~25 cycles	silty mudstone; shelly	burrowing bivalves	pyritic burrows	and articulated shells; minor fragments
Utah/Colorado		20–40 kyr	stringers; concretions	echinoids, ammonites		pods and stringers of shell hash
						concretions with fossils
Early Jurassic	Allison et al. (in press)	30–120 cm	Medium gray, calcareous	*Gryphaea,* burrowing	*Planolites,*	Dispersed to patchy; disarticulated
Blue Lias	Paul et al. (2008)	40 cycles	silty mudstone; shelly	bivalves; crinoids	*Chondrites*	and articulated shells; minor fragments
Gt. Britain		38 kyr	stringers; concretions	echinoids, ammonites	pyritic burrws	pods and stringers of shell hash
						concretions, pyritic burrows/fossils
Mid Devonian	Brett and Baird (1986a)	20–500 cm	Medium gray, calcareous	Ambocoeliid and chonetid	Pyritic burrows	Dispersed to patchy; disarticulated
Hamilton Group	Speyer and Brett (1985)	~30 cycles	silty mudstone; shelly	brachiopods, nuculid and		and articulated shells; minor fragment
New York	Brett et al. (1991)	20–100 kyr	stringers; concretions	modiomorphoid bivalves		pods and stringers of shell hash

(continued)

Table 3 (continued)

Age Formation Location	References	Thickness Number Duration	Lithology Bedding	Common taxa Guilds	Trace Fossils	Taphonomic features
				phacopid trilobites		concretions, pyritic burrows, fossils
Lower Devonian	Botquelen et al. (2006)	20–80 cm	Gray, marly shale	Chonetid brachiopods,	Small burrows	Dispersed to patchy; disarticulated
Amoricain Massif,		~30 cycles	concretions	nuculid and modiomorphid		and articulated shells; minor fragments
France		20–100 kyr		bivalves, phacopid		pods and stringers of shell hash
				trilobites		
Lower Devonian	Becker et al. (2004a, b)	20–100 cm	Gray, marly shale	Small athyrid and chonetid	Minor small	Dispersed to patchy; disarticulated
Hollardops beds	Chatterton et al. (2006)	15 cycles	concretions	brachiopods nuculid and	burrows	and articulated shells; minor fragment
SW Morocco		20–40 kyr		modiomorphoid bivalves		pods and stringers of shell hash
				phacopid trilobites		concretions, pyritic burrows, fossils
						articulated trilobites
Late Ordovician	Brett and Algeo (2001b)	20–500 cm	Medium gray, calcareous	Orthid brachiopods	Small *Planolites*	Dispersed to patchy; disarticulated
Kope Fm.	Brett et al. (2003)	25 cycles	silty mudstone; shelly	bryozoans, small	trilobite traces	and articulated shells; minor fragment

Ohio/KY		20–100 kyr	stringers; concretions	crinoids, bivalves		pods and stringers of shell hash
				calymenid/asaphid trilos		beds with articulated triloibtes, crinoids
Mid Ordovician	Boyer and Droser (2003)	20–120 cm	Medium-olive gray	Small brachiopods	Small Planolites	Dispersed to patchy; disarticulated
Kanosh Fm.		~140 cyc.s	calcareous shale	ostracods, molluscs		and articulated shells; minor fragments
		20–100 kyr				pods and stringers of shell hash
Mid-Cambrian	Gaines et al. (2005)	20–500 cm	Medium, dark gray	Acrotretid brachiopods;	Minor *Planolites*	Dispersed to patchy; disarticulated
Wheeler Fm.	Gaines and Droser (2005)	25 cycles	calcareous mudsone	polymerid/agnostoid		and articulated shells; minor fragments
	Brett et al. (2009)	20–100 kyr	nodular micritic ls.	eocrinoids		pods and stringers of shell hash beds with articulated trilobites, eocrinoids

may feature tool marks, scours, gutter and pot casts etc. They are typically strongly burrowed from the top downward by *Chondrites, Planolites,* and *Diplocraterion*. In some Devonian and later examples more extensive bioturbation by larger traces, such as *Zoophycos*, may homogenize these thin beds with intervening muds. This faunally and ichnofaunally rich assemblage is suggestive of more oxygenated conditions.

Body fossils may be scattered throughout the mudstone but they are typically concentrated on certain bedding planes. These include minor lenses and stringers of skeletal debris, typically including remains of varied soft substrate epifaunal and infaunal benthic fossils. In Cambrian deposits these are mainly trilobites, calcitic brachiopods, and eocrinoids (McKerrow 1979; Li and Droser 1997). For the middle Paleozoic these skeletal layers include abundant strophomenid and spiriferid brachiopods, bryozoans, pelmatozoans, such as crinoids, and trilobites (McKerrow 1979). Jurassic and Cretaceous analogues include gryphaeid bivalves, small bivalves and certain ammonoids. Cenozoic examples feature primarily bivalves and gastropods as well as shark teeth and occasional crustaceans.

Taphonomic conditions of fossils within all examples are similar, with more numerous whole valves and, less commonly, articulated specimens. Overall, the taphonomic grade of these mudstones is better than in the black, laminated shales. The mudstones clearly represent an increase in sedimentation rate. Shells in a majority of beds are dominantly disarticulated, but rarely fragmented, representing reworked slightly time-averaged assemblages. Although this material generally shows little evidence of abrasion or bioerosion, the shells may show encrustation by bryozoans and worm tubes.

Obrution beds are characterized by the preservation of articulated bivalved and multi-element skeletons, such as trilobites, crustaceans, echinoids, crinoids and small vertebrates. They were commonly buried in varied orientations but may have subsequently been crushed; see for example the Hollardops beds. Articulated lingulid brachiopods and bivalves may occur in burrow position or in orientations that suggest escape behavior. Not surprisingly, many obrution beds also contain an abundance of time-averaged, disarticulated and even fragmented material that records earlier generations of organisms that underwent normal mortality and taphonomic degradation prior to the final community (Simoes et al. 1998).

Paleozoic examples include clusters or patches of crinoids, as well as brachiopods and trilobites (Brett and Seilacher 1991; Simoes et al. 1998). Mesozoic and Cenozoic rapid-burial beds include infaunal, shallow-burrowing bivalves typical of offshore, low-energy conditions and clumps of in situ bivalves (Kondo 1997; Fürsich and Pandey 2003; Krawinkel and Seyfried 1996; Hendy et al. 2006).

4.3 Proximal Siltstones and Sandstones

Proximal representatives of small-scale cycles, such as those described above typically have a strongly coarsening-upward parasequence motif (Table 4; Fig. 29). The cycle bases are silty mudstone and siltstone; in many cases, the mudstone-siltstone portion of the cycle may be very nearly barren of fossils (e.g., Lower Devonian Rich 3

Table 4 Comparative taphonomic and paleoecologic aspects of falling stage shell beds

Age Formation Location	References	Thickness Number Duration	Lithology Bedding	Common taxa Guilds	Trace Fossils	Taphonomic features
Cretaceous	Sageman (1996)	20–100 cm	Mudstones and HCS	Inoceramids, gryphaeids	*Zoophycos*	Complete and some fragmented
Greenhorn		25 cycles	Siltstones, sandstone	Burrowing bivalves	*Thalassinoides*	valves; rare articulated fossils; occur
Utah/Colorado		20–100 kyr	gutters; shell lenses, stringers	belemnites	escape traces	as graded lenses and pods
Early Jurassic	Allison et al. (in press)	30–120 cm	Mudstones and HCS	Gryphaeids, belemnites	*Zoophycos*	Complete and some fragmented
Blue Lias	Paul et al. (2008)	40 cycles	Siltstones, sandstone	Burrowing myid bivalves	*Teichichnus*	valves; rare articulated fossils; occur
Gt. Britain		38 kyr	gutters; shell lenses	crinoids		as graded lenses and pods
						rare articulated, e.g., ophiuroids
Mid-Devonian	Brett and Baird (1986)	20–500 cm	Silty mudstone to fine	Orthid, spiriferid and	*Zoophycos*	Complete and some fragmented
Hamilton Group		~30 cycles	grained sandstone; HCS	rhynchonellid brach	*Teichichnus*	valves; rare articulated fossils; occur
New York		20–100 kyr	shelly lenses	Pterioid & modiomorphid	escape traces	as graded lenses and pods
				bivalves, crinoid debris		rare articulated crinoids
Lower Devonian	Botquelen et al. (2006)	20–80 cm	Sandy mudstone and	Spiriferid brachiopods	? *Planolites*	Loosely packed; lenses and pavements

(continued)

Table 4 (continued)

Age Formation Location	References	Thickness Number Duration	Lithology Bedding	Common taxa Guilds	Trace Fossils	Taphonomic features
"Type A" shell beds		~30 cycles	sandstone; HCS, gutters	Bivalves, gastropods	*Thalassinoides*	Disarticulated; slight to complete
Spain		20–100 kyr	packstone lenses			fragmentation; not corroded
						no bioerosion or encrsting
Late Ordovician	Kreisa and Bambach (1982)	20–100 cm	Silty mudstone and	Orthid and strophomenid	*Diplocraterion*	Loosely packed lenses and pavement
Martinsburg		15 cycles	siltstone with graded	Brachiopods; bivalves	*Thalassinoides*	Disarticulated; slight to complete
Virginia		20–40 kyr	shell beds, sharp erosi			fragmentation; not corroded
			basesl HCS			some articulated trilobites, crinoids
Mid-Cambrian	Brett et al. (2009)	20–50 cm	Rhythmic calcisiltites	Orthid brachiopods	*Planolites*	Loosely packed; stringers pavement
Wheeler Fm.		25 cycles	shales; thin skeletal	Acrotretids; poymerid	*Thalassinoides*	Disarticulated; rare articulated fragile
		~20–100 kyr	lag beds	trilobites; eocrinoids		fossils including trilobites and
						eocrinoids

cycles in Morocco; Jurassic cycles in Yorkshire). This may be a function of the combined stresses of low oxygen and rapid sedimentation (Wignall 1993) or may, in part reflect dissolution in undersaturated sediments (Aller 1982). In some instances the siltstone and sandstone portion of the cycle (often 1 to more than 10 m in thickness) may commence abruptly at an erosive surface that may even show evidence of channeling. The soles of the basal bed and subsequent silt and sandstone beds may yield gutter casts and tool marks. Cambrian to Devonian examples may display excellent trilobite trace fossils (*Cruziana* and *Rusphycus*) on lower surfaces (Table 4; Seilacher 2007). Beds are typically coarse silt and sandstones up to 10 s of centimeters thick that may show grading, with shell debris lags on basal surfaces and hummocky lamination, ripple bedding or, rarely, flaser structure. Younger examples may display intense bioturbation of upper surfaces by *Planolites*, *Teichichnus* ("lam-scram" fabrics; Seilacher 2007). Near cycle tops siltstone or sandstone may be more heavily bioturbated, especially by *Rhizocorallium*, *Thalassinoides*, or *Zoophycos*. In some instances pyritic pipe-like burrow replacements, rarely with concretionary overgrowths may be present up to 0.5 m below the sharp cycle top. Such evidence of more intensive bioturbation and mineralization indicates alteration of older sediments during periods of relative sediment starvation associated with ensuing transgressions.

Shell beds associated with these regressive deposits tend to be thin and discontinuous or lenticular (Table 4). Fossil material is typically concentrated at the bases of graded beds and may form discontinuous lags or parautochthonous coquinas up to a few centimeters thick. Fossils are typically whole un-abraded skeletal elements and, if fragmented, show sharp, non-rounded breakage. Shells may occur as edgewise, imbricated, or nested accumulations suggesting storm processing (Seilacher and Meischner 1964). Shells generally show only minor encrustation by bryozoans and small corals, and boring by parasitic forms (e.g. V*ermiformichnus*), which may have been on shells of live organisms. In Paleozoic examples the most typical fossils in these coquinites are disarticulated valves of brachiopods and clams and crinoid ossicles and pluricolumnals; articulated shells may also occur. Current aligned *Tentaculites* tend to be abundant on some sandstone bedding planes in the Late Ordovician to Middle Devonian. Woody plant debris is also typical of Devonian and younger examples and in post-Jurassic examples this wood may contain borings of teredinid bivalves.

Such bedding planes may include well-articulated fragile forms such as complete crinoids or ophiuroids. Excellent examples are provided by the famed ophiuroid bed from hummocky cross-bedded sandstones of Lower Jurassic Eype Clay of the Dorset Coast (House 1989).

4.4 Diagenetic Carbonates

One recurring aspect of many small-scale cycles in distal ramp to basinal settings, is the occurrence of are carbonate cemented concretionary limestones (Table 5; Fig. 29). These may incorporate one or more thin shell hash beds, but are usually nearly

Table 5 Comparative taphonomic and paleoecologic aspects of calcareous/concretionary mudstones

Age Formation Location	References	Thickness Number Duration	Lithology Bedding	Common taxa Guilds	Trace Fossils	Taphonomic features
Mid-Cretaceous	Elder et al. (1994)	10–30 cm	Chalky, micritic shale	Ostreids, inoceramids	*Thallasinoides*	Dispersed fossils; fragmentary
Greenhorn		30 cycles	concretionary limestone	infaunal bivalves	*Zoophycos*	and complete shells; in situ
Utah/Colo., USA		20–100 kyr		rare ammonites		*Gryphaea* and inoceramids
Early Jurassic	Moghadam and Paul (2000)	10–30 cm	Medium gray calcareous	*Gryphaeaa,* pinnids,	*Thallasinoides*	Dispersed, articulated, disart shells
Blue Lias	Paul et al. (2008)	~30 cycles	mudstone, concretions	burrowing bivalves	*Planolites*	articulated, closed bivalves in situ
Dorset, Gt. Britain	Allison et al. (in press)	40–100 kyr		ammonites, crinoids	large *Chondrites*	uncompressed burrows
				rhynchonellid brachs		
Mid-Devonian	Speyer and Brett (1985)	10–30 cm	Calcareous, tabular	Small rugosans	Pyritic burrows	Dispersed to loosely packed
Hamilton Group	Brett et al. (1991)	~10 cycles	concretionary	chonetid and ambocoeliid	small *Zoophycos*	disarticulated and articulated
New York		20–40 kyr	argillaceous limestone	brachiopods	*Chondrites*	clusters of trilobites, including
				phacopid trilobites		enrolled and molts
Lower Devonian	Chatterton et al. (2006)	5–20 cm	Calcareous, tabular	Small rugosans,	Pyritic burrows	Dispersed disarticulated and
Hollardops beds		~80 cycles	concretionary	chonetid and ambocoeliid	rare, small	articulated trilobites; random

SW Morocco		20–40 kyr	argillaceous limestone	brachiopods	*Zoophycos*	orientations
				trilobites		
Late Ordovician	Brett et al. (2006)	10–20 cm	Calcareous, tabular	Orthid, strophomenid, and	*Planolites*	Dispersed disarticulated and
Collingwood		~10 cycles	concretionary	lingulid brachiopods	*Chondrites*	articulated trilobites; random
S. Ontario, Canada		20–40 kyr	micritic limestone	bryozoans, cheirurid and calymenid trilobites		orientations; in situ *Lingula*

structureless argillaceous fine-grained limestone. These horizons range from diffuse, isolated concretions, centered at approximately the same level in the sediment, to ramifying and interlocking concretions, to concretionary limestone and sub-tabular "micritic" bands (Kauffman 2003). In some instances, these may be the most prominent beds in outcrops, masking their true significance as diagenetically cemented layers of background sediment. Many of these beds are nearly barren except for horizontal pyritic (frequently weathered to rusty limonite) thread-like burrows and vertical, pyritic, tubular burrows and highly scattered fossils. In other cases, such as the Collingwood Formation and the Hollardops beds of Morocco, they contain extraordinarily well preserved fossils preserved in varied orientations. Likewise, proximal examples of the Blue Lias show abundant burrowing bivalves preserved in life positions.

In more proximal siliciclastic dominated settings the diagenetic interval is minimal or absent. As noted above, however, the position of the stationary zone of sulfate reduction may be marked by a horizon of pyrite-replaced burrows and/or calcitic to sideritic concretions that commonly are nucleated on discrete pyritic tubes representing mineralized burrow linings. Such rusty concretionary zones are common beneath the tops of small-scale cycles in the Devonian of the Appalachian basin (Brett and Baird 1986a, 1996). Similar cycle top concretionary horizons are known in other settings for example, the Lower Jurassic of Yorkshire (Hallam 1967; Anderton et al. 1979); the Upper Jurassic of Spain (Oloriz et al. 2002); and the Cretaceous of the Western Interior of North America (Kauffman 2003).

5 Inferred Environmental Changes Through Small-Scale Cycles: Implications for Cycle Genesis

Cyclic variation results in repetitive patterns of taphonomic and biotic response. Types of variations include: (a) changes in sedimentation rates; (b) changes in frequency of episodic sedimentation; (c) changes in energy regime; (d) changes in oxygenation; (e) changes in sediment characteristics, particularly TOC values, and (f) varying degrees of diagenetic alteration (Figs. 31–33).

5.1 Environmental Energy

A progressive increase in energy level (favoring carbonate-rich units) is evidenced by the upward increase in current-processed sedimentologic features in many cycles. Within distal cycles black shale bedding surfaces show only minor orientation of lightweight shell material but variations in shell convex up-down orientation in different beds suggest subtle fluctuations in energy level. Narrow gutters and unidirectional alignment of graptolites demonstrate the existence of minor currents. In the mudstones, larger gutters (up to 10 cm across) are present and more numerous,

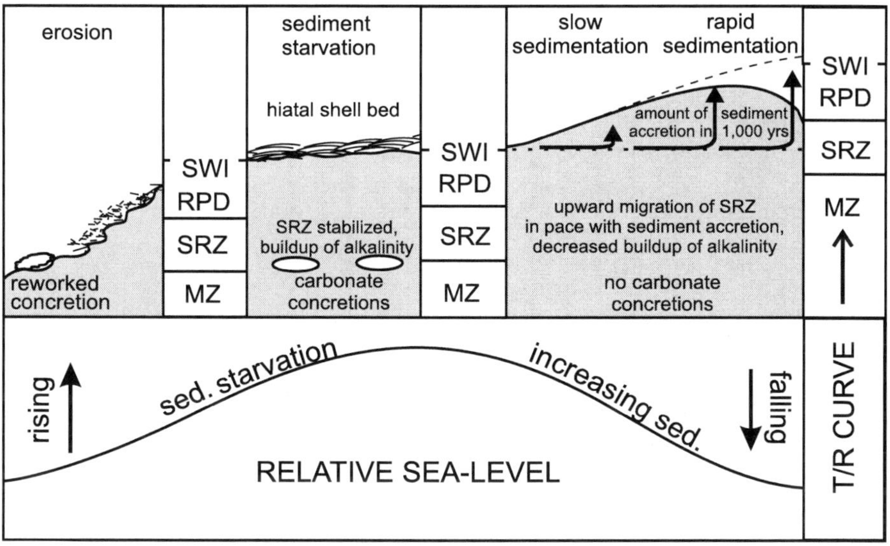

Fig. 31 Relationship of sediment accumulation rates to formation and reworking of concretions. Bar on right shows position of the MZ (methanogenic zone), SRZ (sulfate reduction zone), RPD (redox potential discontinuity) and SWI (sediment–water interface) at outset of interval. Note that these zones remain static during period of sediment starvation but migrate upward as sediments accrete to differing extents. Carbonate concretions will nucleate and accrete during conditions of sediment starvation in the stable SRZ; however, under conditions of sediment buildup; these zones will migrate upward in the sediment and no concretions will form; during sediment starved periods extreme storm erosion may cut down to the SRZ, exhuming concretions. Bar at the bottom shows possible relationship to sea level oscillations, with starvation/erosion and concretion formation during transgression (T) (sediment sequestering in source areas) and increasing rates of sediment aggradation during regression (R) (Modified from Brett et al. 2008)

as are thin beds of convex-up brachiopods. The shelly coquinas, which cap the cycles may contain imbricated and/or nested shells, and occur as parts of graded, hummocky cross-stratified beds indicative of storm wave processing.

5.2 *Oxygenation and Geochemistry*

Organic content within lithofacies decreases up-section within many cycles (see for example Moghadam and Paul 2000). Changes in oxygenation are evidenced by progressive increases in bioturbation and faunal diversity/abundance favoring carbonate-rich units. The increase in ichnofauna and benthic fauna observed in shales up through limestones suggests a transition from low dysoxic to well-oxygenated conditions.

The lower shales and mudstones show little to no diagenetic enhancement. Highly compressed molds of originally aragonitic shells suggests early dissolution and strong compaction of sediments. Fractured molds indicate that shells remained intact up to the time of initial compaction.

Conversely, the limestones clearly underwent early diagenetic cementation. Evidence of synsedimentary cementation includes excellent preservation of fossils (including three-dimensional preservation of material), lack of distortion of burrows, along with the relatively wide spacing of fossil debris horizons, and the concretionary nature of some of the units. Concretionary limestones representing packages of uncompacted carbonate-rich mud are often located below dense hash-rich pavements, indicative of surfaces of maximum sediment starvation (Fig. 31).

5.3 Sedimentation Rates and Time-Averaging

A strong degree of taphonomic alteration of bioclasts in basal skeletal beds indicates low rates of sedimentation coupled with strong reworking (Figs. 32 and 33). Such thick shelly pavements, result from condensation associated with minor sea-level rise and minimal sediment input. Bases of the beds indeed suggest minor to strong erosion of underlying sediments whereas their tops typically evidence sediment starvation. Overlying dark shales have sharp basal contacts, suggestive of strong sediment starvation or minor erosion. Fossil assemblages from dark, organic-rich shales near the bases of cycles are characterized by poor preservation and strong taphonomic biases (e.g. strongly biased pygidia: cranidia ratios). Although very rare articulated remains occur, fossils within the shales are typically fully disarticulated to fragmentary, indicative of long-term exposure to taphonomic processes (Kidwell and Bosence 1991).

The upper mudstone to siltstone portions of cycles contain numerous hash pavements but include greater portions of less fossiliferous matrix in which well-preserved, spar-filled, articulated, and in situ fossils are present. This indicates increasing rates of sedimentation. However, concretion beds within the tops of cycles indicate periods of cementation of older sediment during the time of sediment starvation associated with overlying shell beds.

5.4 Episodicity and Dynamics of Sedimentation

Episodicity is evident throughout the cycles as relatively barren mud layers over shelly layers (Fig. 33) and the presence of obrutionary layers within the dark shales. Obrution beds are more numerous and thicker in the upper mudstone and siltstone

Fig. 32 (continued) mud clasts may be torn up; muds are removed and shell debris is further stacked and concentrated; a thin silt layer may accumulate on top of shell hash. (**f**) Recolonization; in MDP:opportunistic burrowers colonize storm silts producing *Diplocraterion*, *Chondrites*, and other traces; in SDP: recolonization of hard substrate adapted taxa, including bryozoans and crinoids; exhumed concretions may be encrusted or bored with *Trypanites* (Modified from Brett et al. 2008)

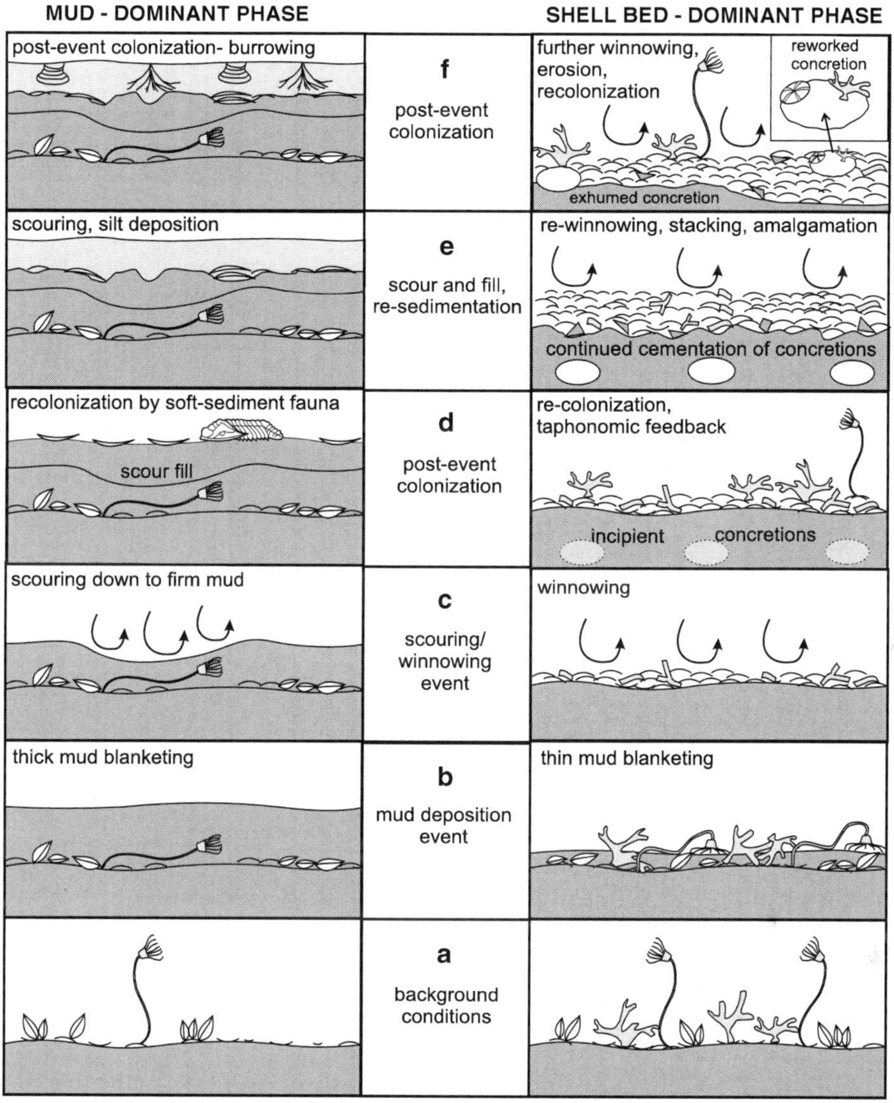

Fig. 32 Analogous seafloor processes during mudstone dominant phase (MDP; *left column*), and shell-bed dominant phase (SDP; *right column*). Note that both sediment accumulation and erosion may occur in either phase, but to different effect. (**a**) Background conditions at outset of interval; note general accumulation of skeletal debris during pause in sedimentation. (**b**) Mud-blanketing; thick layer in the MDP, note obrution deposit with buried intact crinoid; thin mud layer in SDP. (**c**) Seafloor erosion by storm currents; in MDP: scouring is effective in cutting down to firm muds, but does not erode through relatively thick mud blanket; little or no shell lag formed; in SDP: winnowing removes thin mud blankets aggregating shelly debris buried by several previous events. (**d**) Post-event re-deposition and colonization; in MDP relatively thick silt/mud buries scoured surface, muds colonized by "snowshoe strategist" brachiopods and vagrant trilobites; in SDP: minimal mud accumulation; re-colonization involves taphonomic feedback with exposed shell-ground. (**e**) Scour and re-sedimentation; in MDP storm erosion produces irregular scoured surface with gutters buried by silt layer; in SDP scouring creates irregular erosion surface and firm

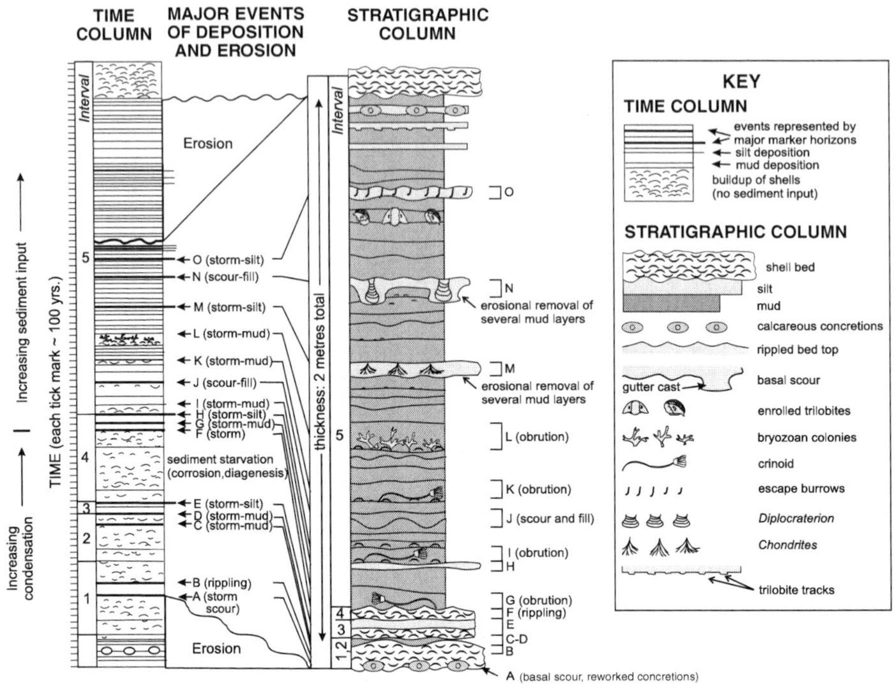

Fig. 33 Schematic of generalized, small-scale cycle typical of cratonic successions, based on Upper Ordovician Kope Formation, showing key components recognized in descriptions in this paper. Individual beds shown in stratigraphic column are coded by letters A–O to levels in the time column. Time intervals 1 and 2 represent intervals of generally low sediment input; intervals 3 and 4 more extreme sediment starvation; note that these intervals have relatively few mud or silt accumulation events; skeletal debris builds up in "background" times. These intervals, comprising about a third of the total time, are represented by thin complex of shell beds. Time interval 5 encompasses a time of increasing sediment aggradation; note deposition of a series of mud and silt layers (including obrution deposits); also note that many upper layers are subsequently removed in an erosional interval, preceding and contemporaneous with, next shell hash accumulation. Portion of preserved mudstone interval comprising about a third of the time occupies the majority of thickness of the preserved cycle (From Brett et al. 2008)

components. Mudflow type events are sufficiently common toward the ends of certain small-scale distal cycles that in some instances a majority of the mudstones-siltstones preserved at cycle tops show evidence of being obrution deposits. This is especially obvious where early diagnesis has stabilized the upper obrutionary layers, such as in the famed Moroccan Devonian trilobite beds where cycle cap beds preserve articulated trilobites (Chatterton et al. 2006; herein). In situ preservation and greater spacing of fossils within the upper (regressive) portions of cycles suggests an increased frequency of burial events, more episodic sedimentation, and higher net sedimentation rate.

The sedimentologic and taphonomic characteristics of both mudrock- and shell bed-dominated divisions of small-scale cycles are heavily influenced by the relative amounts of time represented in mudrock- versus shell bed-dominant phases of the succession (Fig. 31). Both phases record the episodic interruption of low-energy background sedimentation by high-energy events of storm-related scouring and sediment reworking.

Evidence for event deposition is much easier to recognize in bioclastic strata, particularly shell beds, which can preserve bedforms and display obvious heterogeneities in sedimentary fabric and structure such as cross-bedding, grading, edgewise stacking of shells, and rip-up clasts. Evidence for storm activity, although detectable in mudrocks in some instances, is often subtle and is less likely to be recognized than in bioclastic units due to their tendency to lack significant variations in mineralogy, grain size, color and fabric.

Relative to the mudrock-dominated intervals, the thinner bioclastic units are condensed; taphonomic evidence indicates long-term time averaging of skeletal remains within the beds (Figs. 32 and 33). The mixture of skeletons derived from hard and soft substrate adapted organisms suggests a complex succession of minor mud deposition and removal by winnowing (as in the "Jeram" model of Seilacher 1985). The presence of concretions in muds below skeletal hash layers suggests intervals of carbonate cementation in the underlying muds as a result of prolonged stability of the sulfate reduction zone as well as infusion of carbonate saturated waters resulting from dissolution of aragonitic debris in the accumulating skeletal hash beds above (Krawinkel and Seyfried 1996; Brett et al. 2003; Fig. 30). Conversely, the formation of the mudrock dominated hemicycle involved episodic deposition of mud layers up to several centimeters thick. The latter formed by the long-term accumulation of thin, mud-starved veneers of winnowed/reworked parautochthonous skeletal debris.

5.5 *Overview*

Consistent variations in these parameters suggest similar underlying causes of environmental change (Figs. 32 and 33). The oxygen cycle appears coupled with the sediment cycle. Oxygen levels progressively increase as siliciclastic to carbonate proportions decrease. Increased sedimentation rate in association with an increase in episodic formation of $CaCO_3$-rich muds suggest a shallowing-deepening relationship between cycle components. Evidence of sediment starvation low in the cycle seems associated with a deepening trend, with sequestration of siliciclastics and cutback on local carbonate input. Increases in sedimentation rate, episodicity and winnowed shell layers up-section within the carbonate-rich units supports a shallowing trend, as more numerous event beds and winnowed units represent an increased proximity to average storm wave base.

Three models have been proposed to explain the origin of mudrock-shell bed cycles (see Brett et al. 2008 for review). The first two assume that the mudrock intervals reflect low energy/deep-water conditions and that shell beds are primarily the product of increased storm wave energy and winnowing that, in turn, resulted from either (a) relative sea level fall, during which storm wave-base was lowered closer to the seafloor, or (b) increased intensity/frequency of storms without significant change in water depth. A third model proposes that, the shell beds reflect not only storm winnowing, but also the accumulation of time-averaged skeletal debris during prolonged periods of siliciclastic sediment starvation, possibly associated with minor base-level rise. Decimeter-scale shell beds formed during millennial-scale periods of siliciclastic sediment starvation combined with episodes of storm-related reworking and winnowing. This constitutes an alternative interpretation of shell bed genesis that is more in accord with taphonomic, sedimentologic and paleontologic evidence.

6 Long-Term Trends in Cyclic Taphofacies

Comparison of litho-, tapho- and biofacies of small-scale cycles through time highlights commonalities resulting from the constancy of physical processes and differences arising from secular changes in the abundance and diversity of skeleton-producing organisms (Fig. 34). Some differences are obvious particularly between Cambrian and later cycle taphofacies.

Early to Middle Cambrian shallow shelf deposits rarely contain thick skeletal accumulations greater than 1–2 cm thick (Li and Droser 1997). Nonetheless, some basal skeletal lags, of mixed biotic and abiotic carbonate grainstones, particularly ooids, are comparable in thickness and complexity to those of later time. Such mixed oolitic, phosphatic and intraclastic and skeletal beds, ranging up to 50 cm thick exist in shallower facies for example. Comparable shell beds of mixed lingulid brachiopods and ooidal phosphate and hematite beds are well described from the Middle Ordovician of the Armorican Massif (Dabard et al. 2007).

By the Middle Ordovician, condensed shell beds attained thicknesses of 10–30 cm as evidenced by the shelly limestones of the Kanosh and Lehman Formation in the Great Basin (Boyer and Droser 2003). These beds are among the earliest examples of what may be considered relatively typical Paleozoic style shell beds, they are composed of monospecific to moderate diversity assemblages, dominated in offshore facies by orthid brachiopods. Similar shell beds are well documented in the Silurian and Devonian of many areas. For example, brachiopod shell layers up to 50 cm thick occur in the Silurian of Arisaig, the Appalachian Basin and the Welsh Basin (Ziegler et al. 1968). Devonian examples are also widely cited, e.g., the Lower Devonian of Iberia (Botquelen et al. 2006) describe shell beds comparable to their A-type beds in the Ordovician up to several centimeters thick.

Jurassic and Cretaceous shell beds are composed mainly of epifaunal bivalves, together with gastropods and some crinoid material, are typically several centimeters to nearly 2 m in thickness (e.g. Fürsich and Pandey 1999, 2003; Fürsich and Aberhan 1990).

Increased thickness of transgressive shell beds, in part, reflects increased production rates of skeletons relative to the mid Paleozoic, while the absence of internal discontinuities, in contrast to many obviously stacked, amalgamated beds in the Paleozoic, reflects bioturbational mixing of shells. Kidwell and Brenchley

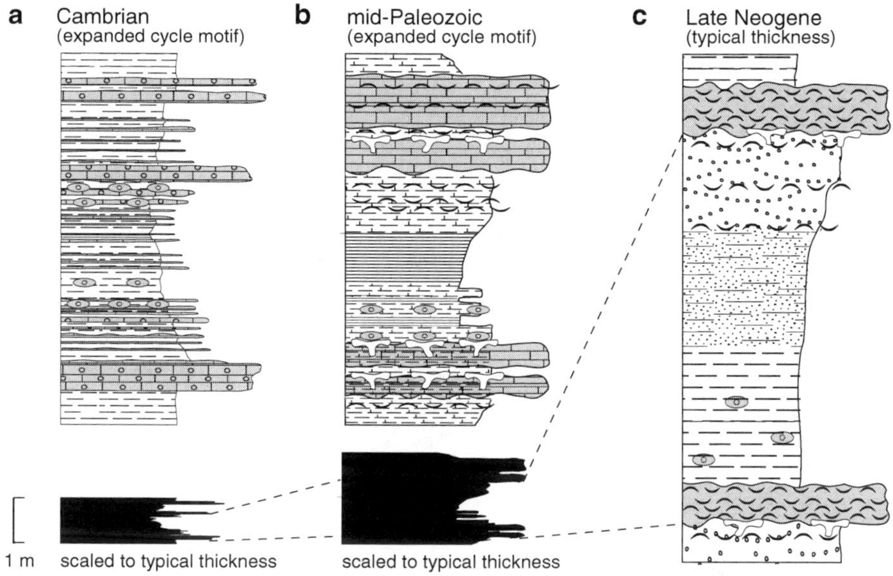

Fig. 34 Small-scale cycles through time. (**a**) Cambrian cycle with thin skeletal lags at bases composed of trilobites with limited bioturbation. (**b**) Middle Paleozoic cycle with better-developed shell beds and more intense bioturbation. (**c**) Cenozoic example showing basal shell rich interval several meters thick, with strongly bioturbated shell debris. Note also the greater thickness of this cycle; this may be an artifact of the site of deposition: most Paleozoic and Mesozoic cycles accumulated in low subsidence epicontinental sea environments, whereas a majority of Cenozoic examples occur in actively subsiding outer continental shelf to forearc basins

(1994, 1996) have systematically documented thickness trends in shell beds and note a substantial change in the mid-Cenozoic that they attribute to the faster production and greater robustness of molluscan shells (Fig. 34).

Examples of Middle Cambrian transgressive shell beds include large branching burrow galleries of "*Thalassinoides*" type within transgressive shell beds. Not surprisingly, however, the degree of bioturbation with the facies steps up from Cambrian to Middle Ordovician (Droser and Bottjer 1988; Manguno and Droser 2004). Larson and Rhodes (1983) and Thayer (1985) also reported a substantial increase in shell bed thickness from Ordovician to Devonian, which they attributed to increased depth and intensity of burrowing. Burrowed firmgrounds are typical of the bases of these beds (e.g., Landing and Brett 1987).

A diverse guild association can be observed within the more carbonate-rich transgressive components of cycles (see discussion in McKerrow 1979; Brett 1995, 1998; and below). Both the calcareous mudstone and limestone intervals contain diverse and abundant assemblages of benthic organisms, in addition to pelagic organisms. In the Cambrian cycles skeletal components include debris of pelmatozoans (eocrinoids), articulate brachiopods, monoplacophorans, and fragments of larger trilobites, such as *Asaphiscus* and *Olenoides*.

In the Middle to Late Ordovician the condensed transgressive lags are more skeleton-rich intervals that can be characterized as shell beds. These include abundant valves of brachiopods, especially concavo-convex strophomenids, dalmanellids and other orthids, trilobites, nuculid bivalves, crinoid ossicles, lingulids, nautiloids, and graptolites. Mid Paleozoic examples include small brachiopods (strophomenids, athyrids, atrypids), small solitary rugose corals, small tabulates (auloporids, *Pleurodictyum*) and nautiloids. Taphonomic feedback became progressively more important in enhancing diversity in shell-rich sediments (Kidwell and Jablonski 1983).

Mid Mesozoic transgressive shell beds are similar in thickness to those of the later Paleozoic and are dominated by the bivalve *Gryphaea*, small rhynchonellid brachiopods (e.g., *Calcirhynchia*), crinoids, small ammonites and rare small, thin flat bivalves. In both mid Paleozoic and Jurassic cases, guild associations are similar in the limestones to those of the mudstones, with increased diversity and abundance of benthic fauna. Noteworthy is the addition of small gastropods and echinoids in Blue Lias limestone biofacies.

Cambrian dark shales are typically fully laminated and show, at most, discrete minute burrows (Gaines and Droser 2005; Gaines et al. 2005). Post-Cambrian examples of dark shales show signs of minor bioturbation at the shale bases. Ordovician Collingwood shales and Devonian black shales display small *Chondrites* (Boyer and Droser 2007). Several of the upper mudstone intervals within the Devonian cycles in New York, and those of the Jurassic Blue Lias and Cretaceous Greenhorn cyclothems include dark burrows piping from the overlying shales (this type of piping is unknown in the Ordovician). This evidence of bioturbation, associated with the laminated shales, suggests that low oxygen levels, associated with these facies were sufficient to support the small tracemakers by the mid Paleozoic and Mesozoic examples, but notably less so in the Cambrian, and perhaps during much of the Ordovician.

7 Summary: Toward General Cyclic Taphofacies Models

Sea-level, climate, and sediment-supply fluctuation have strong biological impacts, controlling the environmental and spatial distribution of organisms at a variety of scales (Bennett 1990; Brett 1998; Brett et al. 2007a, b). Bathymetric and sedimentologic factors, in particular, exert major controls on the distribution patterns of shallow marine benthic organism (Ziegler 1965; Ziegler et al. 1968; McKerrow 1979; Boucot 1975, 1982; Brett 1995, 1998; Boucot et al. 1999; Abbott and Carter 1997; Scarponi and Kowalewski 2004; Brett et al. 2007b; Hendy and Kamp 2007). Water depth/turbulence, sedimentation rate/turbidity, substrate consistency, oxygenation, and other parameters force regular and predictable change in small-scale Milankovitch driven cycles (Bennett 1990; DiMichele et al. 2004; Brett et al. 2007b). In turn these fluctuations result in predictable changes in species composition and preservation of preserved biotas. For example, cyclic variations in relative water depth may lead to recurrent biotic replacement series of bathymetrically zoned organisms (Holland et al. 2001b; Brett et al. 2007b). Redox cycles likewise will show regular variations from low diversity dysoxic adapted assemblages with a dominance of small epibenthic forms to more diverse epi- and endobenthic species; bioturbation fabrics will also reflect changing benthic oxygen levels. Moreover, the predicted variation in siliciclastic sediment supply will vary from relatively low during relative sea-level rise or arid climatic phases to perhaps an order of magnitude higher during sea-level fall or a shift to humid climates with active run-off of sediment from source areas. Such changes in sedimentation patterns will strongly affect benthic substrates, rates of sediment accumulation, and water turbidity (Brett 1998). For instance, high diversity assemblages of stenotopic epibenthic forms are favored by the development of clean, stabilized substrates and shell accumulations during transgressions, whereas limited endofaunal generalized assemblages dominate more rapidly accumulated, food rich, soft substrates during falling stages.

Fossils can be regarded as sedimentary particles (sensu Seilacher 1973), and hence provide sensitive gauges of paleoenvironmental processes, particularly those that relate to dynamics of sedimentation, and geochemistry. Variations in fossil preservation (taphofacies) can be related to variations in burial of organisms or buildup of shell beds, turbulence (degree of disarticulation), redox conditions, pH, and early diagenetic microenvironments. This regular variation leads to predictable variations in fossil assemblage preservation or taphofacies (Brett and Baird 1986b; Speyer and Brett 1986, 1988).

It is striking that similar patterns of fossil preservation should persist in cycles deposited over half a billion years biological evolution and diversification. The varied examples of small scales cycles given here contain a similar variety of fossil accumulations, ranging from thin hash pavements to well preserved in situ assemblages. The primary taphonomic moderator in these cycles is rate of sedimentation, which varies exponentially from sediment-starved concentrations to obrutionary deposits. The occurrence of a persistent motif over this time scale suggests that biological innovations, which might be expected to impact upon fossil preservation, have in fact been overprinted by the extremes of sedimentation preserved in these small-scale cycles. A skeleton, that is twice as resistant to abrasion, is of little import when sedimentation is dominated by the extremes: instant obrution or condensation.

During lowstand or early transgressive phases of sea-level cycles skeletons accumulate in deposits that represent rather shallow water, and accordingly "high" energy conditions. Additionally, sedimentation rates in shallow water settings tend to be low during times of sea level lowstand as a result of winnowing and bypassing of fine-grained sediment. Transgressive phases of sea level change are characterized by low sedimentation across the entire shelf and deeper water settings. Rates of skeletal destruction tend to be high during these times of reduced burial and increased energy, and hence, skeletal lag deposits of highly broken and abraded coquina develop across shelf areas (Kidwell 1991b; Brett 1995; Coe 2003; Catuneanu 2002, 2006).

At times of maximum rate of sea level rise, sedimentation rates in offshore areas are at their lowest due to the entrapment of siliciclastic sediments in coastal alluvial areas and the reduced carbonate production of shelf areas now below the euphotic zone (Vail et al. 1991). Biogenic skeletons therefore tend to be exposed for long intervals of time on the sea floor. Not only are skeletal accumulations subject to reduced wave and current energy, they may also experience intense bio-geochemical corrosion. Hence, the thin-lag deposits that often accumulate during this interval of sea level change are characterized by geochemically resistant particles, such as calcitic (e.g., echinoderm plates) or phosphatic skeletons (e.g., conodont elements, vertebrate bone). These bioclasts may experience reduced abrasion and physical breakage relative to early transgressive beds. However, bioerosion (e.g., microboring), corrosion pitting, and other evidence of biogeochemical dissolution processes may be more prevalent.

During highstand conditions deeper and frequently more dysoxic water masses are established over broad areas of the shelf even as fine-grained siliciclastic sediments begin prograding offshore (Vail et al. 1991; Brett 1995; Van Wagoner and Bertram 1995; Coe 2003). This sets the stage for distinctive types of preservation. Typically, dark shales with fragmented, corroded and decalcified shell material may occur above the condensed; biogenic disturbance may be minimal if benthic oxygen levels are low and variable. This factor favors millimetre scale bedding plane assemblages separated by laminae of barren shale. In some cases this combination of dysoxic conditions and rapid sediment aggradation may favor nearly barren gray mudstone facies.

The later highstand deposits may be expected to show evidence of increasing rates of siliciclastic input, together with improved bottom water oxygen. This may favor a shift to better-preserved benthic epi- and endofauna. These fossils may be expected to be more widely dispersed in the sediment and to show preservation of at least whole valves and some articulated, butterflied, or even closed bivalves. Pulses of sedimentation may be marked by obrution deposits of well-preserved, articulated fossils. Because of the peculiar conditions of moderately oxygenated substrate and bioturbated, organic-poor sediment the episodic burial of organism bodies these facies may be appropriate for the formation of pyritic molds and coatings (Hudson 1982; Brett and Baird 1986b; Brett et al. 1991).

Falling stage – or regressive – deposits (Naish and Kamp 1997; Plint and Nummedal 2000) are characterized by increasing input of coarser siliciclastics, as well as the effects of more numerous turbulence and sedimentation events. Rapid burial should result in dispersed, well preserved fossils (Brett 1995); however, in proximal areas, discrete turbulence events associated with storms may be expected to winnow and concentrate skeletal remains. Fragmentation may be expected, but fossils should generally lack signs of corrosion, abrasion or bioerosion.

In some cases a particular depositional condition may have an effect on more than one aspect of taphofacies, For instance, interludes of sediment starvation, commonly associated with abrupt transgressions, may promote accumulations of variably reworked skeletal debris at the seafloor and enhance cementation of subjacent sediment in the zone of sulfate reduction. This would also enhance preservation of previously buried organism remains.

Although it is beyond the scope of the present paper, further detailed comparative studies may elucidate distinctive time-related features that may relate to large-scale cyclic changes in earth's climate and geochemistry, such as (a) greenhouse-icehouse cycles (Fischer 1980, 1984), (b) calcite vs. aragonite oceans or (c) other secular trends in paleoecology or water mass properties (e.g. increased nutrient run-off following land plant evolution (Algeo et al. 1995, 2001). One may anticipate, in particular, that during icehouse phases of Earth's climatic history, cycles of similar temporal duration to those noted herein will have much larger amplitude, leading to stronger gradients of facies within relatively thin successions. For example, cycles may more frequently show changes from offshore marine facies to marginal marine sediments, coals, and paleosols. Subaerial exposure in many cycles may impart a higher degree of diagenetic alteration and vadose solution to offshore marine cycles. Because of the more rapid shifts in water depth one may anticipate that various portions of the cycles will be thinner and more closely stacked. Improved oceanic circulation during icehouse times may have led to a lesser tendency toward stagnation during highstands and decreased frequency of dark, organic rich facies in small-scale cycles.

Differences between calcite and aragonite oceans may track greenhouse- icehouse conditions because of strong correlations between both climate and oceanic chemistry and rates of ocean floor spreading (Stanley and Hardie 1998, 1999). Times of calcite or aragonite oceans, typically correlated with greenhouse or icehouse megacycle phases, will be characterized by differences in shell bed accumulations. For example, during times of aragonite oceans we anticipate a higher proportion of aragonitic fossils in shell beds. This may permit preservation of thicker skeletal accumulations, at least in settings favoring molluscs. Even in dark shales and mudstones we anticipate that aragonitic (and calcitic) fossils will show a lesser degree of decalcification and compaction than in comparable facies under calcitic oceans. In addition, evidence for early cementation may be substantially less than during greenhouse settings; thus firm- and hardgrounds reported in many examples herein, may be far less typical of the sediment starved transgressive portions of cycles during intervals characterized by aragonitic oceans (Wilson and Palmer 1992; Palmer and Wilson 2004). In addition, we predict that concretionary beds and diagenetic limestone underbeds that have been emphasized in this work may be far less well developed during aragonitic ocean phases and associated icehouse conditions. The lesser degree of early diagenetic cementation may negatively affect the preservation of fossils by inhibiting early cementation of internal shell fillings (leading to more strongly compressed fossils with lesser amounts of internal sparry calcites).

The long-term increase in nutrient content of oceans has also had potential cascading effects on the appearance of cycles. Eutrophication not only promotes increased production of shelly remains that inevitably led to thicker shell beds (Kidwell and Brenchley 1994; Bambach 2006; Bush et al. 2007) but may have led to increased tendency toward production of organic rich laminated facies. This latter trend, however, has inevitably been countered to a degree by the increase in deep bioturbating infauna, especially in low oxygen and sulfide rich environmennts that may have been largely devoid of infauna during the early Paleozoic.

Finally, a general change in depositional settings of marine sediments from a preponderance of epicontinental seas and distal foreland basins in the Paleozoic to continental margins in the later Mesozoic and Cenozoic has inevitably had a strong influence on many aspects of small-scale cycle paleoecology and taphonomy (e.g. see Allison and Wells 2006; Wells et al. 2007). Most Cenozoic examples of cycles are thicker by an order of magnitude than Paleozoic-Mesozoic ones, reflecting the much higher rates of subsidence and sediment input on many narrow continental shelves, especially in areas of active tectonism. Inevitably, this greater rate of sediment accumulation may also have an influence of taphonomy and time averaging, again countered somewhat by increased burrowing rates.

A future program of comparative taphonomy and paleoecology of small-scale cycles should investigate and test these hypotheses rigorously by compiling consistently collected data on many aspects of the litho-, tapho-and biofacies of cycles of comparable time-scale. Icehouse-greenhouse, aragonite-calcite, and other mega-cycles need to be evaluated by comparing deposits laid down in broadly similar oceanographic and paleogeographic settings. Only in this way can the hypotheses listed above be rigorously tested. In turn, variations in taphofacies involving changes in bioturbation, skeletal production and preservation, lithification and other aspects of cycles may have critical implications for long-term trends in taphonomic bias and resolution of the stratigraphic record (see Hendy, this volume).

Acknowledgements This project is an outgrowth of cooperative research between PAA and CB, initially funded by grants from NATO and the Royal Society. CB expresses appreciation to the Donors to the Petroleum Research Fund, American Chemical Society, NSF Grants EAR 0518511 (to W. Huff and C. Brett); and the National Geographic Society for supporting research on the Devonian of Morocco. We have benefited from hours of discussion of ideas with many colleagues and students, but especially Gordon Baird, Alex Bartholomew, Sean Cornell, Patrick McLaughlin, David Meyer, Arnie Miller, Cam Tsujita. AJWH acknowledges funding from the American Museum of Natural History Lerner-Gray Fund, Geological Society of America, Palaeontological Society, the American Association of Petroleum Geologists, and the Department of Geology, University of Cincinnati.

References

Abbott, S. T. (1997). Mid-cycle condensed shell beds from mid-Pleistocene cyclothems, New Zealand: Implications for sequence architecture. *Sedimentology, 44*, 805–824.

Abbott, S. T., & Carter, R. A. (1994). The sequence architecture of mid-Pleistocene (c. 1.1—0.4 Ma) cyclothems from New Zealand: Facies development during a period of orbital control on sea-level cyclicity. *American Association of Petroleum Geological Special Publication, 19*, 367–394.

Abbott, S. T., & Carter, R. A. (1997). Macrofossil associations from mid-Pleistocene cyclothems, Castlecliff Section, New Zealand: Implications for sequence stratigraphy. *Palaios, 12*, 188–210.

Addi, A. A. (2006). The dogger reef horizons of the Moroccan central high atlas: New data on their development. *Journal of African Earth Sciences, 45*, 162–172.

Algeo, T. J., Berner, R. A., Maynard, J. B., & Scheckler, S. E. (1995). Late Devonian oceanic anoxic events and biotic crises: "Rooted" in the evolution of land plants? *Geological Society of America Today, 5*, 64–66.

Algeo, T. J., Scheckler, S. E., & Maynard, J. B. (2001). Effects of the middle to Late Devonian spread of vascular land plants on weathering regimes, marine biota and global climate. In P. Gensel & D. Edwards (Eds.), *Plants invade the land, evolutionary and environmental approaches* (pp. 213–236). New York: Columbia University Press.

Aller, R. C. (1982). Carbonate dissolution in nearshore terrigenous muds. The role of physical and biological reworking. *Journal of Geology, 90*, 79–95.

Allison, P. A., Brett, C. E., Paul, C. R. C., Bilton, J. (in press) Taphonomy of ammonite Konservat Lagerstättten in the early Jurassic blue lias formation of Dorset, UK. *Journal of the Geological Society of London*

Allison, P. A., & Wells, M. R. (2006). Circulation in large ancient epicontinental seas: What was different and why? *Palaios, 21*, 513–515.

Anderton, R., Bridges, P. H., Leeder, M. R., & Sellwood, B. W. (1979). *A dynamic stratigraphy of the British Isles*. London: George Allen & Unwin. 301 pp.

Bambach, R. K. (2006). Phanerozoic biodiversity mass extinctions. In R. Jeanloz, A. L. Albee, K. C. Burke, & K. C. Freeman (Eds.), *Annual Review of Earth and Planetary Sciences, 34* (pp. 127–155).

Banerjee, I., & Kidwell, S. M. (1991). Significance of molluscan shell beds in sequence stratigraphy: An example from the Lower Cretaceous Mannville Group of Canada. *Sedimentology, 38*, 913–934.

Baird, G. C., & Brett, C. E. (1986). Erosion on an anaerobic seafloor: Significance of reworked pyrite deposits from the Devonian of New York State. *Palaeogeography Palaeoclimatology Palaeoecology, 57*, 157–193.

Baird, G. C., & Brett, C. E. (1991). Submarine erosion on the anoxic seafloor: Stratinomic, palaeoenvironmental, and temporal significance of reworked pyrite-bone deposits. In R. V. Tyson & T. H. Pearson (Eds.), *Modern and ancient continental shelf anoxia, Geological Society of America, Special Publication* (pp. 233–257).

Batt, R. (1996). faunal and lithological evidence for small-scale cyclicity in the Wanakah Shale (Middle Devonian) of western New York. *Palaios, 11*, 230–243.

Becker, R. T., Aboussalam, Z. S., Bockwinkel, J., Ebbinghausen, V., El Hassani, A., & Nübel, H. (2004a). Upper Emsian stratigraphy at Rich Tamelougou near Torkoz SW Dra Valley, Morocco. In R. T. Becker & El A. Hassani (Eds.), *Devonian of the western anti atlas: Correlation and events* (pp. 85–89). Documente Institute de Sciences de Rabat 19.

Becker, R. T., Bockwinkel, J., Ebbinghausen, V., Aboussalam, Z. S., El Hassani, A., & Nübel, H. (2004b). Lower and Middle Devonian stratigraphy and faunas at Bou Tserfine near Assa (Dra Valley, SW Morocco). In R. T. Becker & A. El Hassani (Eds.), *Devonian of the western anti atlas: Correlation and events* (pp. 90–100). Documente Institute de Sciences de Rabat 19.

Bennett, K. D. (1990). Milankovitch cycles and their effect on species in ecological and evolutionary time. *Paleobiology, 16*, 11–21.

Berger, A., Loutre, M. F., & Laskar, J. (1992). Stability of the astronomical frequencies over Earth's history for paleoclimatic studies. *Science, 255*, 560–566.

Botquelen, A., Gourvennec, R., Loi, A., Pilola, G. L., & Leone, F. (2006). Replacements of benthic associations in a sequence stratigraphic framework, examples from Upper Ordovician of Sardina and Lower Devonian of the Massif Amoricain. *Palaeogeography Palaeoclimatology Palaeoecology, 239*, 286–310.

Bottrell, S., & Raiswell, R. (1990). Primary versus diagenetic origin of Blue Lias rhythms (Dorset, UK): Evidence from sulphur geochemistry. *Terra Nova, 1*, 451–456.

Boucot, A. J. (1975). *Evolution and extinction rate controls*. Amsterdam: Elsevier. 427p.

Boucot, A. J. (1982). Ecostratigraphic framework for the Lower Devonian of the North American Appohimchi Subprovince. *Neuse Jahrbuch für Geologie und Paläontologie, Abhandlungen, 163*, 81–121.

Boucot, A. J., Lawson, J. D., & Eds. (1999). *Paleocommunities – A case study from the Silurian and Lower Devonian. World and regional geology series no. 10*. New York: Cambridge University Press. 912 pp.

Boyer, D. L., & Droser, M. (2003). Shell beds of the Kanosh and Lehman Formations of western Utah; paleoecological and paleoenvironmental interpretations. *Geology Studies, Brigham Young University, 47*, 1–15.

Boyer, D. L., & Droser, M. L. (2007). Devonian monospecific assemblages: New insights into the ecology of reduced-oxygen depositional settings. *Lethaia, 40*, 321–333.

Brett, C. E. (1995). Sequence stratigraphy, biostratigraphy, and taphonomy in shallow marine environments. *Palaios, 10*, 597–616.

Brett, C. E. (1998). Sequence stratigraphy, paleoecology, and evolution: Biotic clues and responses to sea-level fluctuations. *Palaios, 13*, 241–262.

Brett, C. E., & Baird, G. C. (1985). Carbonate shale cycles in the Middle Devonian of New York: An evaluation of models for the origin of limestones in terrigenous shelf sequences. *Geology, 13*, 324–327.

Brett, C. E., & Baird, G. C. (1986a). Symmetrical and upward shallowing cycles in the Middle Devonian of New York: Implications for the punctuated aggradational cycle hypothesis. *Paleoceanography, 1*, 431–447.

Brett, C. E., & Baird, G. C. (1986b). Comparative taphonomy: A key to paleoenvironmental interpretation based on fossil preservation. *Palaios, 1*, 207–227.

Brett, C. E., & Baird, G. C. (1996). Middle Devonian sedimentary cycles and sequences in the northern Appalachian basin. In B. J. Witzke, G. A. Ludvigson, & J. Day (Eds.), *Paleozoic sequence stratigraphy: Views from the North American Craton. Geological Society of America Special Paper* 306 (pp. 213–241).

Brett, C. E., & Algeo, T. J. (2001a). Event beds and small-scale cycles in Edenian to lower Maysvillian strata (Upper Ordovician) of northern Kentucky: Identification, origin, and temporal constraints. In T. A. Algeo & C. E. Brett (Eds.), *Sequence, cycle, and event stratigraphy of upper Ordovician and Silurian strata of the Cincinnati arch region, Kentucky geological survey field trip guidebook 1, Series XII* (pp. 65–86).

Brett, C. E., & Algeo, T. J. (2001b). Sequence stratigraphy of the upper Ordovician Kope formation in its type area, Northern Kentucky, including a revised nomenclature. In T. A. Algeo & C. E. Brett (Eds.), *Sequence, cycle, and event stratigraphy of upper Ordovician and Silurian strata of the Cincinnati arch region, Kentucky geological survey field trip guidebook 1, Series XII* (pp. 47–64).

Brett, C. E., & Seilacher, A. (1991). Fossil Lagerstätten: A taphonomic consequence of event sedimentation. In G. Einsele, W. Ricken, & A. Seilacher (Eds.), *Cycles and events in stratigraphy* (pp. 284–297) Berlin: Springer.

Brett, C. E., Algeo, T. J., & McLaughlin, P. I. (2003). The use of event beds and sedimentary cycles in high-resolution stratigraphic correlation of lithologically repetitive successions: The upper Ordovician Kope formation of northern Kentucky and southern Ohio. In P. Harries & D. Geary (Eds.), *High-resolution stratigraphic approaches to paleobiology* (pp. 315–351). Boston: Plenum.

Brett, C. E., Allison, P. A., DeSantis, M., Liddell, W. D., & Kramer, T. (2009). Sequence stratigraphy, cyclic facies, and lagerstätten in the Middle Cambrian Wheeler and Marjum Formations, Great Basin, Utah. *Palaeogeography Palaeoclimatology Palaeoecology, 277*, 9–33.

Brett, C. E., Allison, P. A., Tsuijita, C., Soldani, D., & Moffat, H. (2006). Sedimentology, taphonomy and paleoecology of meter-scale cycles from the upper Ordovician of Ontario. *Palaios, 21*, 530–547.

Brett, C. E., Bartholomew, A. J., & Baird, G. C. (2007a). Biofacies recurrence in the Middle Devonian of New York state: An example with implications for habitat tracking. *Palaios, 22*, 306–324.

Brett, C. E., Dick, V. B., & Baird, G. C. (1991). Comparative taphonomy and paleoecology of Middle Devonian dark gray and black shale facies from western New York. In E. Landing &

C. E. Brett (Eds.), *Dynamic stratigraphy and depositional environments of the Hamilton Group in New York Pt. II. NY State Museum Bulletin, 469* (pp. 5–36).

Brett, C. E., Hendy, A. J. W., Bartholomew, A., Bonelli, J., & McLaughlin, P. (2007b). Response of shallow marine biotas to sea level fluctuations: Faunal replacement and the process of habitat tracking. *Palaios, 22*, 230–246.

Brett, C. E., Kohrs, R., & Kirchner, B. (2008). Paleontological event beds from the upper Ordovician Kope formation of Ohio and northern Kentucky and the promise of high-resolution event stratigraphy. In P. I. McLaughlin, C. E. Brett, & S. M. Holland (Eds.), *Stratigraphic renaissance in the Cincinnati arch: Implications for upper Ordovician paleontology and paleoecology. Cincinnati Museum Center Special Publication 2* (pp. 64–87).

Bromley, R. G. (1967). Some observations on burrows of thalassinidean Crustaceans on chalk hardgrounds. *Quarterly Journal Geological Society of London, 123*, 157–182.

Bromley, R. G. (1968). Burrows and borings in hardgrounds. *Meddeleser Dansk Geologisk Forening, 18*, 247–250.

Bush, A. M., Bambach, R. K., & Daley, G. M. (2007). Changes in theoretical ecospace utilization in marine fossil assemblages between the mid-Paleozoic and late Cenozoic. *Paleobiology, 33*, 76–97.

Campbell, K. S. W. (1977). Trilobites of the Haragan, Bois d'Arc and Frisco Formations (Early Devonian), Arbuckle Mountains region, Oklahoma. *Bulletin of Oklahoma Geological Survey, 123*, 1–227.

Cantalamessa, G., Di Celma, C., & Ragaini, L. (2005). Sequence stratigraphy of the Punta Ballena member of the Jama formation (early Pleistocene, Ecuador): Insights from integrated sedimentologic, taphonomic and paleoecologic analysis of molluscan shell concentrations. *Palaeogeography Palaeoclimatology Palaeoecology, 216*, 1–25.

Carter, R. M., Abbott, S. T., Graham, I. J., Naish, T. R., & Gammon, P. R. (2002). The middle Pleistocene Merced-2 and -3 sequences from Ocean Beach, San Francisco. *Sedimentary Geology, 153*, 23–41.

Caster, K. E., Dalvé, E. A., & Pope, J. K. (1955). *Elementary guide to the fossils and strata of the Ordovician in the vicinity of Cincinnati* (47 p). Ohio: Cincinnati Museum of Natural History, Cincinnati.

Catuneanu, O. (2002). Sequence stratigraphy of clastic systems: Concepts, merits, and pitfalls. *Journal of African Earth Science, 35*, 1–43.

Catuneanu, O. (2006). *Principles of sequence stratigraphy*. New York: Elsevier. 386 p.

Chatterton, B. D. E., Fortey, R., Brett, K., Gibb, S., & McKellar, R. (2006). Trilobites from the upper lower to middle Devonian Timrhanrhart formation, Jbel gara el Zguilma, southern Morocco. *Palaeontographica Canadiana, 25*, 1–177.

Coe, A. L. (2003). *The sedimentary record of sea-level change*. Cambridge: Cambridge University Press.

Cotter, E., & Link, J. E. (1993). Deposition and diagnesis of Clinton ironstones (Silurian) in the Appalachian foreland basin of Pennsylvania. *Geological Society of America Bulletin, 105*(7), 911–922.

Cox, B. M., Sumbler, M. G., & Ivimey-Cook, H. C. (1999). A formational framework for the Lower Jurassic of England and Wales (onshore area). British Geological Survey Research Report RR/99/01.

Dabard, M.-P., Loi, A., & Paris, F. (2007). Relationship between phosphogenesis and sequence architecture: Sequence stratigraphy and biostratigraphy in the Middle Ordovician of the Armorican Massif (NW France). *Palaeogeography Palaeoclimatology Palaeoecology, 248*, 339–356.

deBoer, P. L., & Smith, D. G. (Eds.). (1994). Orbital forcing and cyclic sequences. *International Association of Sedimentologists Special Publication 19* (pp. 219–225).

Di Celma, C., Ragaini, L., Cantalamessa, G., & Curzio, P. (2002). Shell concentrations as tools in characterizing sedimentary dynamics at sequence-bounding unconformities: Examples from the lower unit of the Canoa Formation (late Pliocene Ecuador). *Géobios Mémoire Spécial, 24*, 72–85.

Del Rio, C. J., Mártinez, S. A., & Scasso, R. A. (2001). Nature and origin of spectacular marine Miocene shell beds of northeastern Patagonia (Argentina): Paleoecological and bathymetric significance. *Palaios, 16*, 3–25.

DiMichele, W. A., Behrensmeyer, A. K., Olszewski, T. D., Labandeira, C. C., Pandolfi, J. M., Wing, S. L., et al. (2004). Long-term stasis in ecological assemblages: Evidence from the fossil record. *Annual Reviews of EcologyEvolution, and Systematics, 2004*, 285–322.

Dominici, S. (2001). Taphonomy and paleoecology of shallow marine macrofossil assemblages in a collisional setting (late Pliocene-early Pleistocene, Western Emilia, Italy). *Palaios, 16*, 336–353.

Droser, M. L., & Bottjer, D. J. (1988). Trends in depth and extent of bioturbation in carbonate carbonate marine environments. *Geology, 16*, 233–236.

Einsele, G., & Ricken, W. (1991). Limestone-marl alternations – An overview. In G. Einsele, W. Ricken, & A. Seilacher (Eds.), *Cycles and events in stratigraphy* (pp. 23–47). Berlin: Springer.

Elder, W. P., Gustason, E. R., & Sageman, B. B. (1994). Correlation of basinal carbonate cycles to nearshore parasequences in the Late Cretaceous Greenhorn seaway, Western interior U.S.A. *Geological Society of America Bulletin, 106*, 892–902.

Elliott, C. (1996). Recognition and stratigraphic use of orbitally-forced limestone-shale cycles from the Lower Jurassic Blue Lias Formation of Britain: A taphonomic approach. *Geological Society of America (Abstracts with Programs), 28*(7), 308.

Ellwood, B., Brett, C. E., Tomkin, J., & MacDonald, W. D. (2007). Magnetostratigraphy susceptibility of the Upper Ordovician Kope Formation, northern Kentucky. *Palaeogeography Palaeoclimatology Palaeoecology, 210*, 295–329.

Elrick, M., & Snider, A. C. (2002). Deep-water stratigraphic cyclicity and carbonate mud mound development in the Middle Cambrian Marjum formation, House Range, Utah, USA. *Sedimentology, 49*, 1021–1047.

Fernandez-López, S. R. (2007). Ammonoid taphonomy, paleoenvironments and sequence stratigraphy at the Bajocian/Bathonian boundary on the Bas Auran area (Subapline Basin, south-eastern France. *Lethaia, 40*, 377–391.

Fernandez-López, S. R., Duarte, L. V., & Henriques, M. H. P. (2000). Ammonites from lumpy limestones (Lower Pleinsbachian, Portugal). Taphonomic analysis and paleoenvironmental implications. *Revista Sociedad Geologica de España, 13*, 3–15.

Fernandez-López, S. R., Henriques, M. H. P., & Duarte, L. V. (2002). Taphonomy of ammonite condensed associations: Jurassic examples from carbonate platforms of Iberia. *Abhandlungen der Geologischen Bundesanstalt, 57*, 423–450.

Fischer, A. G. (1980). Gilbert-type bedding rhythms and geochronology. In E. I. Yochelson (Ed.), *The scientific ideas of G.K. Gilbert. Geological Society of America Special Paper 11833* (pp. 93–104).

Fischer, A. G. (1984). The two Phanerozoic supercycles. In W. A. Berggren & J. A. Van Couvering (Eds.), *Catastrophes in earth history* (pp. 129–150). Princeton, NJ: Princeton University Press.

Fürsich, F. T. (1978). The influence of faunal condensation and mixing on the preservation of fossil benthic communities. *Lethaia, 11*, 243–250.

Fürsich, F. T., & Aberhan, M. (1990). Significance of time-averaging for paleocommunity analysis. *Lethaia, 23*, 143–152.

Fürsich, F. T., & Oschmann, W. (1993). Shell beds as tools in basin analysis: The Jurassic of Kachchh, western India. *Journal of the Geological Society of London, 150*, 169–185.

Fürsich, F. T., & Pandey, D. K. (1999). Genesis and environmental significance of Upper Cretaceous shell concentrations from the Cauvery Basin southern India. *Palaeogeography Palaeoclimatology Palaeoecology, 145*, 119–139.

Fürsich, F. T., & Pandey, D. K. (2003). Sequence stratigraphic significance of sedimentary cycles and shell concentrations in the Upper Jurassic-Lower Cretaceous of Kachchh, western India. *Palaeogeography Palaeoclimatology Palaeoecology, 193*, 285–309.

Gaines, R. R., & Droser, M. I. (2003). Paleoecology of the familiar trilobite Elrathia kingii: An early exaerobic zone inhabitant. *Geology, 31*, 941–944.

Gaines, R. R., & Droser, M. L. (2005). New approaches to understanding the mechanics of Burgess Shale-type deposits: From the micron scale to the global picture. *The Sedimentary Record, 3*, 4–8.

Gaines, R. R., Kennedy, M. J., & Droser, M. L. (2005). A new hypothesis for organic preservation of Burgess Shale taxa in the Middle Cambrian Wheeler Formation, House Range, Utah. *Palaeogeography Palaeoclimatology Palaeoecology, 220*, 193–205.

Gallois, R. W. (2000). The stratigraphy of the Kimmeridge Clay (Upper Jurassic) in the RGGE Project boreholes at Swanworth Quarry and Metherhills, south Dorset. *Proceedings of the Geologists' Association, I11*, 265–280.

Gilbert, G. K. (1895). Sedimentary measurement of geologic time. *Geology, 3*, 121–127.

Greensmith, J. T., Rawson, P. F., & Shalaby, S. E. (1980). An association of minor fining upward cycles and aligned gutter marks in the Middle Lias (Lower Jurassic) of the Yorkshire Coast. *Proceedings of Yorkshire Geological Society, 42*, 525–538.

Hallam, A. (1957). Primary origin of the limestone-shale rhythm in the British Lower Lias. *Geological Magazine, 94*, 175–176.

Hallam, A. (1960). A sedimentary and faunal study of the Blue Lias of Dorset and Glamorgan. *Philosophical Transactions of the Royal Society London, B243*, 1–44.

Hallam, A. (1964). Origin of the limestone-shale rhythm in the Blue Lias of England: a composite theory. *Journal of Geology, 72*, 157–169.

Hallam, A. (1966). Depositional environment of British Liassic ironstones considered in the context of their facies relationships. *Nature, 209*, 1306–1309.

Hallam, A. (1967). Siderite- and calcite-bearing concretionary nodules in the Lias of Yorkshire. *Geological Magazine, 104*, 222–227.

Hallam, A. (1986). Origin of minor limestone-shale cycles: Climatically induced or diagenetic. *Geology, 14*, 609–612.

Hallam, A., & Bradshaw, M. J. (1979). Bituminous shales and oolitic ironstones as indicators of transgressions and regressions. *Journal of Geological Society, 136*, 157–164.

Hampson, G. J., Howell, J. A., & Flint, S. S. (1999). A Sedimentological and sequence stratigraphic re-interpretation of the Upper Cretaceous Prairie Canyon member ("Mancos B") and associated strata, Book Cliffs Area, Utah, U.S.A. *Journal of Sedimentory Research: Stratigraphic and Global Studies, 69*, 414–433.

Hancock, J. L. (1975). Petrology of the chalk. *Proceedings of the Geologists' Association, 86*, 499–553.

Hattin, D. E. (1971). Widespread, synchronously deposited, burrow mottled limestone beds in Greenhorn Limestone (Upper Cretaceous of Kansas and southeastern Colorado). *American Association of Petroleum Geologists Bulletin, 55*, 412–451.

Hattin, D. E. (1977a). Upper Cretaceous stratigraphy, paleontology and paleoecology of western Kansas. *Mountain Geologist, 144*, 176–217.

Hattin, D. E. (1977b). Articulated lepadomorph cirripeds from the Upper Cretaceous of Kansas: family Stramentidae. *Journal of Paleontology, 51*, 797–825.

Hattin, D. E. (1982a). Stratigraphy and depositional environments pf Smoky Hill Chalk Member, Niobrara Chalk (Upper Cretaceous) of the type area, western Kansas. *Kansas State Geological Survey Bulletin, 225*, 1–85.

Hattin, D. E. (1982b). Distribution, and significance of widespread time-parallel pelagic limestone beds in Greenhorn Limestone (Upper Cretaceous) of the central Great Plains and southern Rocky Mountain. In E. G. Kauffman, L. M. Pratt, & F. B. Zelt (Eds.), *Fine-grained deposits and biofacies of the Cretaceous western Interior Seaway: Evidence of cyclic sedimentary processes. SEPM guidebook 4* (pp. 28–37). Oklahoma: Tulsa.

Hayasaka, R. (1991). Sedimentary facies and environments of the Oligocene Ashiya Group in the Kitakyushu-Ashiya area, Southwest Japan. *Journal of the Geological Society of Japan, 97*, 607–619.

Hemingway, J. E. (1934). The Lias of the Yorkshire coast. *Proceedings of the Geologists' Association, 45*, 250–260.

Hemingway, J. E. (1951). Cyclic sedimentation and the deposition of ironstone in the Yorkshire Lias. *Proceedings of the Yorkshire Geological Society, 28*, 67–74.

Hemingway, J. E. (1974). Ironstone. In D. H. Rayner & J. E. Hemingway (Eds.), *The Geology and Mineral Resources of Yorkshire, Yorkshire Geological Society* (pp. 329-335).

Hendy, A. J. W. (this volume). *Lithification and other taphonomic overprints on Phanerozoic trends in biodiversity. Taphonomy: Process and bias through time*. Springer Berlin.

Hendy, A. J. W., & Kamp, P. J. J. (2007). Paleoecology and sequence stratigraphy of Late Miocene-Early Pliocene sixth-order glacioeustatic cycles in the Manutahi-1 core, Wanganui Basin, New Zealand. *Palaios, 23*, 25–42.

Hendy, A. J. W., Kamp, P. J. J., & Vonk, A. (2006). Cool-water shell bed taphofacies from Miocene-Pliocene shelf sequences in New Zealand: Utility in sequence stratigraphic analysis. In H. M. Pedley & G. Carannate (Eds.), *Cool-water carbonates: Depositional systems and palaeoenvironmental control* (pp. 285–307). London: Geological Society. Special Publications 255.

Hesselbo, S. P., & Jenkyns, H. C. (1996). A comparison of the Hettangian to Bajocian successions of Dorset and Yorkshire. In P. D. Taylor (Ed.), *Field geology of the British Jurassic* (pp. 105–150). London: Geological Society.

Hinnov, L. A. (2000). New perspectives on orbitally forced stratigraphy. *Annual Review of Earth and Planetary Sciences, 28*, 419–475.

Hinnov, L. A. (2004). Earth's orbital parameters and cycle stratigraphy. In F. M. Gradstein, J. Ogg, & A. G. Smith (Eds.), *A geologic time scale* (pp. 142–165). Cambridge: Cambridge University Press.

Hintze, L. F., & Robison, R. A. (1975). Middle Cambrian stratigraphy of the house, Wah Wah, and adjacent ranges in western Utah. *Geological Society of America Bulletin, 86*, 881–891.

Hintze, L. F., & Davis, F. D. (2003) Geology of Millard County, Utah. *Utah Geological Survey Bulletin no. 133* (305 p).

Holland, S. M., Miller, A. I., Dattilo, B. F., Meyer, D. L., & Diekmeyer, S. L. (1997). Cycle anatomy and variability in the storm-dominated type Cincinnatian (Upper Ordovician): Coming to grips with cycle delineation and genesis. *Journal of Geology, 105*, 135–152.

Holland, S. M., Miller, A. I., & Meyer, D. L. (2001a). Sequence stratigraphy of the Kope-fairview interval (Upper Ordovician, Cincinnati, Ohio area). In T. A. Algeo & C. E. Brett (Eds.), *Sequence, cycle, and event stratigraphy of Upper Ordovician and Silurian strata of the Cincinnati arch region, Kentucky Geological Survey Field Trip Guidebook 1, Series XII* (pp. 93–102).

Holland, S. M., Miller, A. I., Meyer, D. L., & Datillo, B. F. (2001b). The detection and importance of subtle biofacies within a single lithofacies: The Upper Ordovician Kope formation of the Cincinnati, Ohio region. *Palaios, 16*, 205–217.

Hollard, H. (1967). Le Dévonien du Maroc et du Sahara nord-occidantal. *Proceedings of International Symposium on Devonian System, 1*, Alberta Society of Petroleum Geologists, Calgary, 203–244.

House, M. R. (1985). A new approach to an absolute timescale from measurements of orbital cyclicities from measurements of orbital cycles and sedimentary microrhythms. *Nature, 316*, 721–725.

House, M. R. (1986). Are Jurassic sedimentary microrhythms due to orbital forcing? *Proceedings of the Ussher Society, 6*, 299–311.

House, M. R. (1989). *Geology of the Dorset coast. Geological Association* (168 p.).

House, M. R. (1993). *Geology of the Dorset coast* (2nd ed.) *Geologists' Association Guide No. 22* (164 p. & plates.) Burlington House, London: Piccadilly. Paperback. ISBN 07073 0485 7.

House, M. R. (1995). Orbital forcing timescales: An introduction. In M. R. House & A. S. Gale (Eds.), *Orbital forcing timescales and cyclostratigraphy. Geological Society Special Publication 85* (pp. 1–18).

House, M. R., & Gale, A. S. (Eds.) (1995). *Orbital forcing timescales and cyclostratigraphy. Geological Society Special Publication 85* (204 p.)

Howarth, M. K. (1955). Domerian of the Yorkshire coast. *Proceedings. Yorkshire Geological Society, 30*, 147–175.

Howard, A. S. (1985). Lithostratigraphy of the Staithes Sandstone and Cleveland Ironstone formations (Lower Jurassic) of north-east Yorkshire. *Proceedings of Yorkshire Geological Society, 45*, 261–275.

Hudson, J. D. (1982). Pyrite in ammonite-bearing shales from the Jurassic of England and Germany. *Sedimentology, 29*, 639–667.

Ito, M. (1992). High-frequency depositional sequences of the upper part of the Kazusa Group, a middle Pleistocene forearc basin fill in Boso Peninsula, Japan. *Sedimentary Geology, 76*, 155–175.

Jennette, D. C., & Pryor, W. A. (1993). Cyclic alternation of proximal and distal storm facies: Kope and fairview formations (Upper Ordovician), Ohio and Kentucky. *Journal of Sedimentary Petrology, 63*, 183–202.

Jansen, U., Becker, G., Plodowski, G., Schindler, E., Vogel, O., & Weddige, K. (2004). The Emsian to Eifelian near Foum Zguid (NE Dra Valley, Morocco). *Document Institute Sciences, Rabat, 19*, 21–35.

Kamataki, T., & Kondo, Y. (1997). 20, 000 or 40, 000-year depositional sequences caused by glacio-eustatic sea-level fluctuation in the middle Pleistocene Jizodo Formation, Boso Peninsula, central Japan. *Journal of the Geological Society of Japan, 103*, 747–762.

Kauffman, E. G. (1978). Benthic environments and paleoecology of the Posidonienschefer (Toarcian). *Neues Jahrbuch für Geologie und Paläontologie–Abhandlungen, 157*, 18–36.

Kauffman, E. G. (1981). Ecological reappraisal of the German Posidonienschefer (Toarcian) and the stagnant basin model. In J. Gray, A. J. Boucot, & W. B. N. Berry (Eds.), *Communities of the past*. Stroudsburg, PA: Dowden, Hutchinson, and Ross.

Kauffman, E. G. (1982). Ecology and depositional environments of chalk-marl and limestone-shale rhythms in the Cretaceous of North America. In G. Ensele & A. Seilacher (Eds.), *Cyclic and event stratification*. Springer: New York.

Kauffman, E. G. (2003). Limestone concretions as near-isochronous surfaces: An example from the Cretaceous of the Western interior of North America. In P. J. Harries (Ed.), *High resolution stratigraphic paleontology*. The Netherlands: Springer.

Kennedy, W. J., & Garrison, R. E. (1975). Morphology and genesis of nodular chalks and hardgrounds in the Upper Cretaceous of southern England. *Sedimentology, 22*, 311–86.

Kidwell, S. M. (1985). Paleobiological and sedimentological implications of skeletal concentrations. *Nature, 318*, 457–460.

Kidwell, S. M. (1986). Models for fossil concentrations: Paleobiologic implications. *Paleobiology, 12*, 6–24.

Kidwell, S. M. (1989). Stratigraphic condensation of marine transgressive records: Origin of major shell deposits in the Miocene of Maryland. *Journal of Geology, 97*, 1–24.

Kidwell, S. M. (1991a). Condensed deposits in siliciclastic sequences: Expected and observed features. In G. Einsele, W. Ricken, & A. Seilacher (Eds.), *Cycles and events in stratigraphy* (pp. 682–695). Berlin: Springer.

Kidwell, S. M. (1991b). The stratigraphy of shell concentrations. In D. E. G. Briggs & P. A. Allison (Eds.), *Taphonomy, releasing information from the fossil record* (pp. 211–290). New York: Plenum.

Kidwell, S. M., & Bosence, D. W. J. (1991). Taphonomy and time-averaging of marine shelly faunas. In P. A. Allison & D. E. G. Briggs (Eds.), *Taphonomy, releasing information from the fossil record* (pp. 116–211). New York: Plenum.

Kidwell, S. M., & Brenchley, P. J. (1994). Patterns in bioclastic accumulation through the Phanerozoic. *Geology, 22*, 1139–1143.

Kidwell, S. M., Brenchley, P. J., & Lipps, J. H. (1996). Evolution of the fossil record: Thickness trends in marine skeletal accumulations and their implications. In D. Jablonski & D. H. Erwin (Eds.), *Evolutionary paleobiology* (pp. 290–336). Chicago: University of Chicago Press.

Kidwell, S. M., & Jablonski, D. (1983). Taphonomic feedback: Ecological consequences of shell accumulation. In P. J. Tevesz & P. J. McCall (Eds.), *Biotic interactions in recent and fossil benthic communities* (pp. 195–248). New York: Plenum.

Kirchner, B., & Brett, C. E. (2008). Subsurface correlation and paleogeography of a mixed siliciclastic-carbonate unit using distinctive faunal horizons: Toward a new methodology. *Palaios, 23*, 173–184.

Kitamura, A., & Kondo, Y. (1990). Cyclic changes of sediments and fossil associations caused by glacio-eustatic sea-level changes during the early Pleistocene – A case study of the middle part of the Omma Formation at the type locality. *Journal of the Geological Society of Japan, 96*, 19–36.

Kitamura, A., Kondo, Y., Sakai, H., & Horii, M. (1994). 41,000–year orbital obliquity expressed as cyclic changes in lithofacies and molluscan content, Early Pleistocene Omma Formation, Central Japan. *Palaeogeography Palaeoclimatology Palaeoecology, 112*, 345–361.

Kondo, Y. (1997). Inferred bivalve response to rapid burial in a Pleistocene shallow-marine deposit from New Zealand. *Palaeogeography, Palaeoclimatology, Palaeoecology, 128*, 87–100.

Kondo, Y., Abbott, S. T., Kitamura, A., Kamp, P. J. J., Naish, T. R., Kamataki, T., et al. (1998). The relationship between shell bed type and sequence architecture; examples from Japan and New Zealand. *Sedimentary Geology, 122*, 109–127.

Krawinkel, H., & Seyfried, H. (1996). Sedimentlogic, paleoecologic, taphonomic and ichnologic criteria for high-resolution sequence analysis: a practical guide for the sequence analysis: A practical guide for the identification. *Sedimentary Geology, 102*, 79–110.

Kreisa, R. D., & Bambach, R. K. (1982). The role of storm processes in generating shell beds in Paleozoic shelf environments. In G. Einsele, & A. Seilacher (Eds.), *Cyclic and Event Stratification* (pp. 200–207). Springer-Verlag, Berlin.

LaFerriere, A., Hattin, D., & Archer, A. W. (1987). Effects of climate, tectonics, and sea-level changes on rhythmic bedding patterns in the Niobrara Formation (Upper Cretaceous), U.S. Western Interior. *Geology, 15*, 233–236.

Landing, E., & Brett, C. E. (1987). Trace fossils and regional significance of a Middle Devonian (Givetian) disconformity in southwestern Ontario. *Journal of Paleontology, 61*, 205–230.

Lang, W. D. (1924). The Blue Lias of the Devon and Dorset coasts. *Proceedings of the Geologists' Association, 35*, 169–185.

Larson, R. L., & Rhodes, D. C. (1983). The evolution of infaunal communities and sedimentary fabrics. In M. J. S. Tevesz & P. L. McCall (Eds.), *Biotic interactions in recent and fossil benthic communities* (pp. 627–648). New York: Plenum.

Li, X., & Droser, M. L. (1997). Nature and distribution of Cambrian shell concentrations: Evidence from the basin and range province of the western United States (California, Nevada, and Utah). *Palaios, 12*, 111–126.

Lehmann, C., Osleger, D. A., & Montañez, I. (2000). Sequence stratigraphy of lower cretaceous (Barremian-Albian) carbonate platforms of northeastern Mexico: Regional and global correlations. *Journal of Sedimentary Research, 70*, 373–391.

Liddell, W. D., Wright, S. H., & Brett, C. E. (1997). Sequence stratigraphy and paleoecology of the middle Cambrian spence shale in northern Utah and southern Idaho. In P. K. Link & B. J. Kowalis (Eds.), *Proterozoic to recent stratigraphy, tectonics, and volcanology, Utah, Nevada, southern Idaho and central Mexico. Brigham Young University, Geology Studies, 42*(1) (pp. 59–78).

Manguno, M. G., & Droser, M. L. (2004). The ichnologic record of the Ordovician radiation. In B. D. Webby, F. Paris, M. L. Droser, & I. G. Percival (Eds.), *The great Ordovician biodiversification event* (pp. 369–379). Columbia University Press: New York.

Martill, D. M. (1985). The preservation of marine vertebrates in the Lower Oxford Clay (Jurassic) of central England. *Philosophical Transactions Royal Society London, Series, B311*, 155–165.

McIntyre, A. P., & Kamp, P. J. J. (1998). Late Pliocene (2.8–2.4 Ma) cyclothemic shelf deposits, Parikino, Wanganui Basin, New Zealand: lithostratigraphy and correlation of cycles. *New Zealand Journal of Geology and Geophysics, 41*, 69–84.

McKerrow, W. S. (1979). *The ecology of fossils*. Cambridge: MIT Press. 400 pp.

McLaughlin, P. I., & Brett, C. E. (2004). Eustatic and tectonic controls on the distribution of marine seismites: Examples from the Upper Ordovician of Kentucky, USA. *Sedimentary Geology, 168*, 165–192.

McLaughlin, P. I., Brett, C. E., & Wilson, M. A. (2008). Hierarchy of sedimentary surfaces and condensed beds from the middle Paleozoic of eastern North America: Implications for cratonic sequence stratigraphy. In B. R. Pratt, C. Holmden (Eds.), *Dynamics of Epeiric Seas. Geological Association of Canada, Special Paper 48* (pp. 175–200).

Miall, A. D., & Mohamud, A. (2001). The castlegate sandstone of the book cliffs, Utah: Sequence stratigraphy, paleogeography, and tectonic controls. *Journal of Sedimentary Research, Section B: Stratigraphy and Global Studies, 71*, 537–548.

Miller, A. I., Holland, S. M., Meyer, D. L., & Dattilo, B. F. (2001). The use of faunal gradient analysis for high-resolution correlation and assessment of seafloor topography in the type Cincinnatian. *Journal of Geology, 109*, 603–613.

Mitchell, C. E., & Bergström, S. M. (1991). New graptolite and lithostratigraphic evidence from the Cincinnati region, U.S.A., for the definition and correlation of the base of the Cincinnatian series (Upper Ordovician), In C. R. Barnes & S. H. Williams (Eds.). *Advances in Ordovician Geology. Geological Survey, Canada Paper* (pp. 90–9, 59–77).

Moghadam, H. V., & Paul, C. R. C. (2000). Trace fossils of the Jurassic, Blue Lias, Lyme Regis, southern England. *Ichnos, 7*, 283–306.

Morzadec, H. (1980). Les Trilobites Asteropyginae du Devonien de l'Anti-Atlas. *Palaeontographica Abteilung A, 262*, 53–85.

Naish, T. R., & Kamp, P. J. J. (1997). High-resolution sequence stratigraphy of 6th order (41 ka) Plio-Pleistocene cyclothems, Wanganui Basin, New Zealand. *Bulletin of the Geological Society of America, 109*, 978–999.

Oloriz, F., Reolid, M., & Rodriguez-Tovar, F. J. (2002). Fossil assemblages, lithofacies, taphofacies and interpreting depositional dynamics in the epicontinental Oxfordian of the Prebetic Zone, Betic Cordillera, southern Spain. *Palaeogeography Palaeoclimatology Palaeoecology, 185*, 53–75.

Oschmann, W. (1988). Kimmeridge Clay sedimentation—a new cyclic model. *Palaeogeography, Palaeoclimatology, Palaeoecology, 65*, 217–251.

Palmer, T. J., & Wilson, M. A. (2004). Calcite precipitation and dissolution of biogenic aragonite in shallow Ordovician calcite seas. *Lethaia, 37*, 417–427.

Parras, A., & Casadio, S. (2005). Taphonomy and sequence stratigraphic significance of oyster-dominated concentrations from the San Julian formation, Oligocene of Patagonia, Argentina. *Palaeogeography Palaeoclimatology Palaeoecology, 217*, 47–66.

Pattison, S. A. J. (1995). Sequence stratigraphic significance of sharp-based lowstand shoreface deposits, Kenilworth Member, Book Cliffs, Utah. *American Association Petroleum Geologists Bulletin, 79*, 444–462.

Paul, C. R. C., Allison, P. A., & Brett, C. E. (2008). The occurrence and preservation of ammonites in the Blue Lias Formation (Lower Jurassic) of Devon and Dorset, England and their palaeoecological, sedimentological, diagenetic significance. *Palaeogeography Palaeoclimatology Palaeoecology, 270*, 258–272.

Plint, A. G., & Nummedal, D. (2000). The falling stage systems tract: Recognition and importance in sequence stratigraphic analysis. In D. Hunt & R. Gawthorpe (Eds.), *Sedimentary responses to forced regressions. Geological Society of London, Special Publications 172* (pp. 1–17).

Ramsköld, L., & Werdelin, L. (1991). The phylogeny and evolution of some phacopid trilobites. *Cladistics, 7*, 29–74.

Rawson, P. F., & Wright, J. K. (1995). Jurassic of the Cleveland Basin, North Yorkshire. In P. D. Taylor (Ed.), *Field geology of the British Jurassic. Geological Society, London* (pp. 173–208).

Rees, M. (1986). A fault-controlled trough through a carbonate platform: The Middle Cambrian House Range embayment. *Geological Society of America Bulletin, 97*, 1057–1069.

Ricken, W. (1991). Variation in sedimentation rates in rhythmically bedded sediments – Distinction between depositional types. In G. Einsele, W. Ricken, & A. Seilacher (Eds.), *Cycles and events in stratigraphy* (pp. 167–187). Berlin: Springer.

Ricken, W. (1994). Complex rhythmic sedimentation related to third-order sea-level variations: Upper Cretaceous, Western Interior Basin, USA. In P. L. deBoer & D. G. Smith (Eds.), *Orbital forcing and cyclic sequences. International association of sedimentology special publication 19* (pp. 167–193). Oxford: Blackwell.

Rio, D., Channell, J. E. T., Masserei, F., Poli, M. S., Sgavetti, M., D'Alessandro, A., et al. (1996). Reading Pleistocene eustasy in a tectonically active shelf setting (Crotone Peninsula, southern Italy). *Geology, 24*, 743–746.

Sadler, D. H., & Cooper, R. A. (2004). Calibration of the Ordovician timescale. In B. D. Webby, F. Paris, M. L. Droser, & I. G. Percival (Eds.), *The great Ordovician biodiversification event* (pp. 48–51). New York: Columbia University Press.

Sageman, B. B. (1996). Lowstand tempestites: Depositional model for Cretaceous skeletal limestones. *Geology, 24*, 888–892.

Sageman, B. B., Rich, J., Arthur, M. A., Burchfield, D. E., & Dean, W. E. (1997). Evidence for Milankovitch periodicities in the Cenomanian-Turonian lithologic and geochemical cycles, Western Interior, US. *Journal of Sedimentary Research, 67*, 285–301.

Sageman, B. B., Rich, J., Arthur, M. A., Dean, W. E., Savrda, C. E., & Bralower, T. J. (1998). *Multiple Milankovitch cycles in the Bridge Creek Limestone (Cenomanian-Turonian), Western Interior Basin. Stratigraphy and Paleoenvironments of the Cretaceous Western Interior Seaway, USA SEPM Concepts in Sedimentology and Paleontology 6* (pp. 153–171).

Sakakura, N. (2002). Taphonomy of the bivalve assemblages in the upper part of the Paleogene Ashiya Group, southwestern Japan. *Paleontological Research, 6*, 101–120.

Savrda, C., & Bottjer, D. J. (1991). Oxygen-related biofacies in marine strata: An overview and update. In R. V. Tyson & T. H. Pearson (Eds.), *Modern and ancient shelf anoxia. Geological Society of London Special Publication 26* (pp. 371–399).

Savrda, C., & Bottjer, D. J. (1994). Ichnofossils and ichnofabrics in rhythmically bedded pelagic/hemipelagic carbonates: Recognition and evaluation of benthic redox and scour cycles. In P. L. deBoer & D. G. Smith (Eds.), *Orbital forcing and cyclic sequences. International association of sedimentology special publication 19* (pp. 219–225). Oxford: Blackwell.

Scarponi, D., & Kowalewski, M. (2004). Stratigraphic paleoecology: Bathymetric signatures and sequence overprint of mollusk associations from late Quaternary sequences of the Po Plain, Italy. *Geology, 32*, 989–992.

Schraut, G. (2000). Trilobiten aus dem Unterdevon des südöstlichen Anti-atlas, Süd Morokko. *Senckenbergiana Lethaea, 7*, 361–433.

Schwarzacher, W., & Fischer, A. G. (1982). Limestone-shale bedding and perturbations of the earth's orbit. In G. Einsele, W. Ricken, & A. Seilacher (Eds.), *Cyclic and event stratification* (pp. 72–95). Berlin: Springer.

Seilacher, A. (1973). Biostratinomy. The sedimentology of biologically standardized particles. In R. N. Ginsberg (Ed.), *Evolving concepts in sedimentology* (pp. 159–177). Baltimore, MD: The Johns Hopkins University Press.

Seilacher, A. (1982a). General remarks about event deposits. In G. Einsele, W. Ricken, & A. Seilacher (Eds.), *Cyclic and event stratification*. Berlin: Springer.

Seilacher, A. (1982b). Ammonite shells as habitats in the Posidonia Shales of Holzmaden: Floats or benthic islands. *Neues Jahrbuch fur Geologie und Palaeontologie Monatshefte, 1982*, 98–114.

Seilacher, A. (1985). The Jeram model: Event condensation in a modern intertidal environment. In U. Bayer & A. Seilacher (Eds.), *Sedimentary and evolutionary cycles. Lecture notes in earth sciences* (pp. 336–342). Berlin: Springer.

Seilacher, A. (2007). *Trace fossil analysis*. New York: Springer. 226 p.

Seilacher, A., & Hauff, R. B. (2004). Constructional morphology of pelagic crinoids. *Palaios, 19*, 3–16.

Seilacher, A., & Meischner, D. (1964). Fazies analyse im Paläozoikum des Oslo-Gebietes. *Geologische Rundschau, 54*, 596–619.

Sellwood, B. W., & Jenkyns, H. C. (1975). Basins and swells and the evolution of an epeiric sea (Pliensbachian – Bajocian of Great Britain). *Journal of the Geological Society of London, 131*, 373–388.

Sepkoski, J. (1984). A kenetic-model of Phanerozoic taxonomic diversity: III Post-Paleozoic families and mass extinctions. *Paleobiology, 10*, 246–267.

Sepkoski, J. (1997). Biodiversity: Past, present, and future. *Journal of Paleontology, 71*, 533–539.

Shepard, T. H., Houghton, R. D., & Swan, A. R. H. (2006). Bedding and pseudo-bedding in the Early Jurassic of Glamorgan: Deposition and diagenesis in the Blue Lias of South Wales. *Proceedings of the Geologists' Association, 117*, 249–264.

Simoes, M. G., Anelli, L. E., Kowalewski, M., & Torello, F. (1998). Long-term time-averaging despite abrupt burial: Paleozoic obrution deposits from epeiric sea settings of the Parana Basin. *Geological Society of America, Abstracts with Programs, 30*(7), 384.

Simms, M. J., Chidlaw, N., Morton, N., & Page, K. N. (2004). *British lower Jurassic stratigraphy. Geological conservation review series No. 30.* Peterborough: Joint Nature Conservation Committee.

Smith, D. (1989). Stratigraphic correlation of presumed Milankovitch cycles in the Blue Lias (Hettangian to earliest Sinemurian). *England Terra Nova, 1*(5), 457–460.

Speyer, S. E., & Brett, C. E. (1985). Clustered trilobite assemblages in the Middle Devonian Hamilton Group. *Lethaia, 18*, 85–103.

Speyer, S. E., & Brett, C. E. (1986). Trilobite taphonomy and Middle Devonian taphofacies. *Palaios, 1*, 312–327.

Speyer, S. E., & Brett, C. E. (1988). Taphofacies models for epeiric sea environments: Middle Paleozoic examples. *Palaeogeography Palaeoclimatology Palaeoecology, 63*, 225–262.

Stanley, S. M., & Hardie, L. A. (1998). Secular oscillations in the carbonate mineralogy of reef-building and sediment-producing organisms driven by tectonically forced shifts in seawater chemistry. *Palaeogeography Palaeoclimatology Palaeoecology, 144*, 3–19.

Stanley, S. M., & Hardie, L. A. (1999). Hypercalcification; paleontology links plate tectonics and geochemistry to sedimentology. *GSA Today, 9*, 1–7.

Thayer, C. W. (1985). Sediment-mediated disturbance and the evolution of marine benthos. In M. J. S. Tevesz & P. L. McCall (Eds.), *Biotic interactions in recent and fossil benthic communities* (pp. 480–626). New York: Plenium.

Tobin, R. C., & Pryor, W. A. (1981). Sedimentological interpretation of an Upper Ordovician carbonate-shale vertical sequence in northern Kentucky. In T. G. Roberts (Ed.), *Geological society of America Cincinnati '81, Field trip guidebooks, v.1, stratigraphy, sedimentology* (pp. 49–57). Falls Church, Virginia, American Geological Institute.

Tomasovych, A., Fürsich, F. T., & Olszewski, T. D. (2006). Modelling shelliness and alteration in shell beds: variation in hardpart input and burial rates leads to opposing predictions. *Paleobiology, 32*, 278–298.

Tsujita, C. J. (1995). Origin of concretion-hosted shell clusters in the Late Cretaceous Bearpaw Formation, southern Alberta, Canada. *Palaios, 10*, 408–423.

Vail, P. R. (1987). Seismic stratigraphy interpretation using sequence stratigraphy, Part 1. Sequence stratigraphy interpretation procedure. In W. Bally (Ed.), *Atlas of seismic stratigraphy, American Association Petroleum Geologists Studies in Geology 27* (pp. 1–10).

Vail, P. R., Audemard, F. E., Eisner, P. N., & Perez-Cruz, C. (1991). The stratigraphic signatures of tectonics, eustasy and sedimentology: An overview. In S. A. Bowman, G. Einsele, W. Ricken, & A. Seilacher (Eds.), *Cycles and events in stratigraphy* (pp. 617–659). Berlin: Springer.

Van Houten, F. B., & Bhattacharrya, D. P. (1982). Phanerozoic oolitic ironstones – geologic record and facies model. *Annual Review of Earth and Planetary Sciences, 10*, 441–457.

Van Wagoner, J. C., & Bertram, G. T. (1995). Sequence stratigraphy and marine to nonmarine facies architecture of foreland basin strata, Book Cliffs, Utah, U.S.A. In J. C. Van Wagoner & G.T. Bertram (Eds.), *Sequence stratigraphy of foreland basin deposits. American Association Petroleum Geologists Memoir 64* (pp. 137 – 223).

Walker, K. R., & Alberstadt, L. P. (1975). Ecological succession as an aspect of structure in fossil communities. *Paleobiology, 1*, 238–257.

Webber, A. (2002). High-resolution faunal gradient analysis and assessment of the causes of meter-scale cyclicity in the Type Cincinnatian series (Upper Ordovician). *Palaios, 17*, 545–555.

Weedon, G. P. (1985). Hemipelagic shelf sedimentation and climatic cycles: The basal Jurassic (Blue Lias) of South Britain. *Earth and Planetary Science Letters, 76*, 321–335.

Weedon, G. P. (1986). Comment on "Origin of cycles, climatically induced or diagenetic?" by Hallam, A. *Geology, 15*, 92–94.

Weedon, G. P., Jenkyns, H. C., Coe, A. L., & Hesselbo, S. P. (1999). Astronomical calibration of the Jurassic time-scale from cyclostratigraphy in British mudrock formations. *Philosophical Transactions of the Royal Society London Series, A357*, 1787–1813.

Wells, M. R., Allison, P. A., Piggott, M. D., Gorman, G. J., Hampson, G. J., Pain, C. C., et al. (2007). Numerical modelling of tides in the Late Pennsylvanian Midcontinent Seaway of North America with implications for hydrography and sedimentation. *Journal of Sedimentary Research, 77*, 943–865.

West, I. M. (2007a). *Staithes, middle Jurassic – geological field guide, Appendix to geology of the Wessex coast.* Internet site: http:\\www.soton.ac.uk/~imw/staithes.htm. National Oceanography Centre, Southampton University. Version: 4th November 2007.

West, I. M. (2007b). *Lyme Regis, West – Blue Lias: Geology of the Wessex Coast.* National Oceanography Centre, Southampton, Southampton University. Internet geological field guide. http://www.soton.ac.uk/~imw/lyme.htm. Version: 27 January 2007

Whitehead, T. H., Anderson, W., Wilson, V., & Wray, D. A. (1952). The Liassic ironstones. The mesozoic ironstones of England. Memoirs of the geological survey of Great Britain, Department of Scientific and Industrial Research, Her Majesty's Stationery Office, 211 pp. With contributions on petrography by Dunham, K. C. See Chapter 2: Petrography of the Liassic ironstones, pp. 16–31, and Chapter 3: Liassic iron ores of the Cleveland District (pp 35–67).

Wignall, P. B. (1993). Distinguishing between oxygen and substrate control in fossil benthic assemblages. *Journal of Geological Society, 150*, 193–196.

Wilson, M. A., & Palmer, T. J. (1992). Hardgrounds and hardground faunas. *University of Wales, Aberystwyth, Institute of Earth Studies Publications, 9*, 1–131.

Ziegler, A. M. (1965). Silurian marine communities and their environmental significance. *Nature, 207*, 270–272.

Ziegler, A. M., Cocks, R. M., & Bambach, R. K. (1968). The composition and structure of Lower Silurian marine communities. *Lethaia, 1*, 1–27.

Chapter 5
Taphonomy of Animal Organic Skeletons Through Time

Neal S. Gupta and Derek E.G. Briggs

Contents

Abstract Investigations of organically preserved invertebrate fossils have focused on abundant taxa such as graptolites and arthropods. Analyses have shown that their composition cannot be explained either as a result of decay resistance, or the introduction of macromolecular material from surrounding sediment. The fossilization of organic materials is a result of the diagenetic transformation of lipids in the organism itself by a process of in situ polymerization which generates a composition with a

N.S. Gupta (✉)
Department of Geology and Geophysics, Yale University, P.O. Box 208109, New Haven, CT 06520-8109, USA
and
Geophysical Laboratory, 5251 Broad Branch Road NW, Washington, DC 20015, USA

D.E.G. Briggs
Department of Geology and Geophysics, Yale University, P.O. Box 208109, New Haven, CT 06520-8109, USA
and
Peabody Museum of Natural History, Yale University, P.O. Box 208118, New Haven, CT 06520-8118, USA

P.A. Allison and D.J. Bottjer (eds.), *Taphonomy: Process and Bias Through Time*, Topics in Geobiology 32, DOI 10.1007/978-90-481-8643-3_5,

significant aliphatic component. While this process causes the fossilized remains of different taxa, even plants and animals, to converge in composition they may still retain differences following diagenesis. Such chemosystematic signatures have the potential to be used in the identification of organic materials that lack diagnostic morphology. The diagenetic transformation of organic materials in macrofossils is similar to the formation of kerogen – the final composition depends on original chemistry, decay and diagenesis. A better understanding of rates and controls on this process will require more experimental investigation of decay and maturation, as well as analyses of fossils of different ages and from different environmental settings.

1 Introduction

The fossil record of animals contrasts with that of plants, in being dominated by organisms with mineralized skeletons: shells, bones and teeth. Such hard parts account for the extensive representation of the familiar shelly invertebrates: sponges, corals, trilobites, ostracodes, brachiopods, molluscs, and echinoderms, as well as the vertebrates. Although they are normally broken down by decay and oxidation, soft-bodied fossils also provide critical data on the history of life. The most decay prone tissues, such as muscle, are preserved by replication in authigenic minerals as a result of bacterial activity (Briggs 2003). More resistant tissues, such as cuticle, however, may survive as organic remains (see Briggs 1999 for a review). They account for the fossil record of a number of important groups including graptolites, chelicerates (eurypterids, horseshoe crabs, scorpions and spiders) and insects. A number of minor groups (e.g. chitinozoans) also have a non-mineralized organic cuticle, and others occasionally preserve organic elements in addition to the mineralized skeleton (e.g., ammonite beaks and some fish scales).

Attempts to understand the organic preservation of animal fossils have focused inevitably on the more abundant remains, those of graptolites and arthropods. It was long assumed that their preservation was a product of a skeleton organically strengthened for rigidity and protection, and consequently more decay resistant than the rest of the animal, and burial in environments where bacterial breakdown and scavenging were inhibited (e.g. low oxygen black shales in the case of graptolites, high salinity in the case of eurypterids). Research in the 1990s, however, showed that this model, commonly referred to as 'selective preservation', is inadequate. This followed the discovery that the composition of the fossilized material (e.g., graptolite periderm, arthropod cuticle) is very different to that of its living precursor.

Graptolites are, of course, extinct. Investigations of well preserved examples using electron microscopy showed that the ultrastructure of the periderm (the organic skeleton) is characteristic of the protein collagen (Towe and Urbanek 1972; Crowther 1981). Confirmation of this interpretation was provided by comparisons with the living pterobranchs, encrusting benthic colonial organisms which are the nearest living relatives of graptolites. Their structure and mode of growth are essentially identical to that of graptolites, and the periderm is composed of collagen (Fig. 1a). Surprisingly, however, analyses of fossil graptolites (Briggs et al. 1995) revealed no evidence of

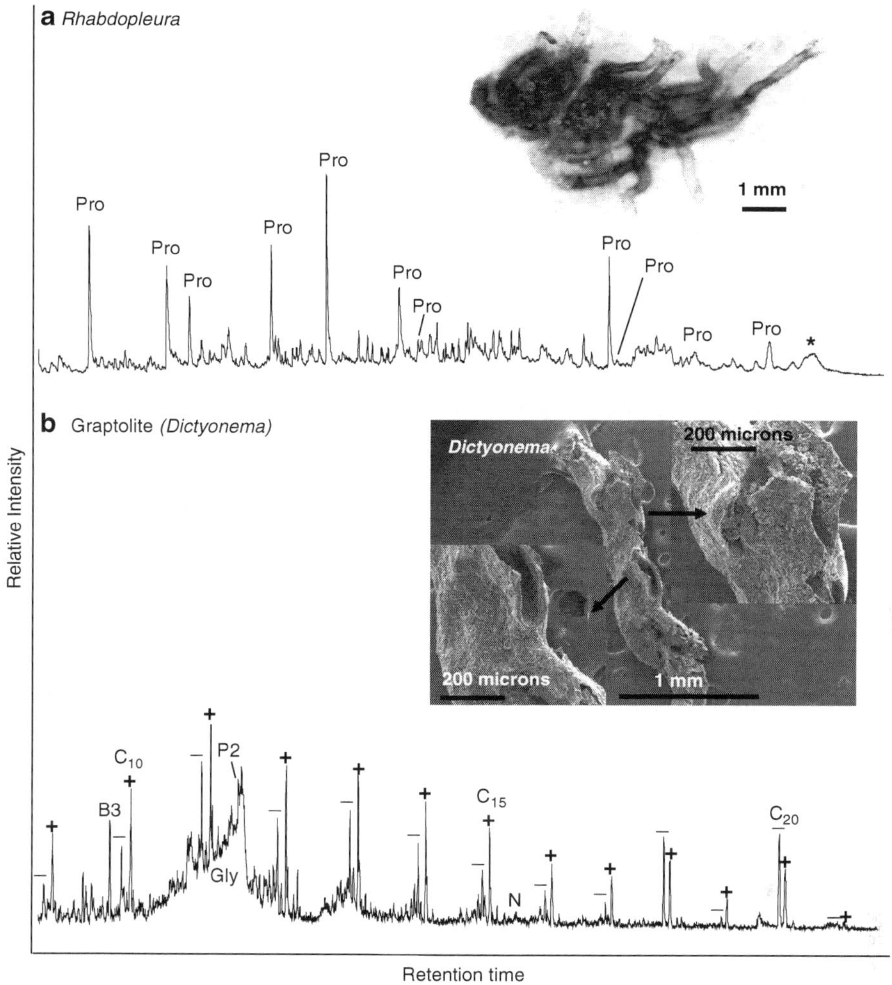

Fig. 1 Partial py-GC-MS chromatogram revealing the chemistry of (**a**) modern pterobranch *Rhabdopleura* and (**b**) graptolite *Dictyonema peltatum* (YPM 202222), Wisby, Sweden, Silurian. Pro, protein pyrolysis products; B3, C_3 benzene; +, *n*-alk-1-ane; –, *n*-alk-e-ene; C_n, where n refers to the carbon chain length; Gly, Glycerine; N, Napthalene; P3, dimethyl phenol/ethyl phenol; *, contaminant. Inset (**a**) branching tubular colony of *Rhabdopleura*; (**b**) *Dictyonema* stipe and close-up of thecae

protein; the composition of the graptolite periderm is aliphatic, composed of long chain hydrocarbons (Fig. 1b). This anomaly was explained at the time by postulating that the macromolecular material had been introduced into the organic skeleton of the graptolite from the surrounding sediment (Briggs et al. 1995).

The history of research on organically preserved arthropod fossils parallels that of graptolites. Until the mid-1990s fossilized cuticles were assumed to be composed of a chitin–protein complex like that of their living relatives (Fig. 2a). The exoskeleton is more decay resistant than the rest of the animal and it survived to become

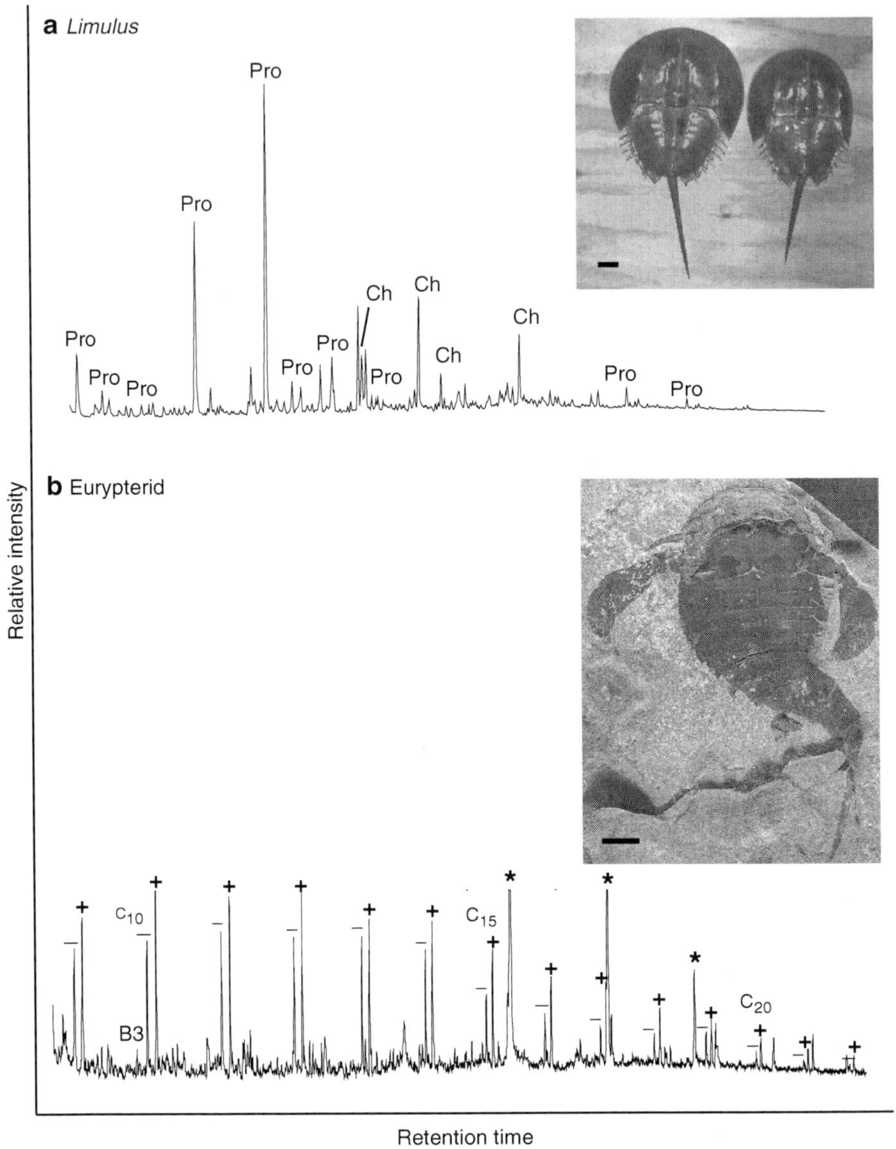

Fig. 2 Partial py-GC-MS chromatogram revealing the chemistry of (**a**) modern *Limulus* cuticle and (**b**) fossil eurypterid cuticle *Eurypterus dekayi* (YPM 209619), Ridgemount Quarry, Ontario, Canada, Williamsville Formation, Pridolian. Note the striking contrast in composition between the modern and the fossil. Pro, protein pyrolysis products; Ch, Chitin; B3, C_3 benzene; +, *n*-alk-1-ane; –, *n*-alk-e-ene; C_n, where n refers to the carbon chain length; *, contaminant. Scale bars: 1 cm

incorporated into sedimentary rock. However, analyses of fossil examples, using a range of techniques, showed that they consist mainly of an aliphatic hydrocarbon very different in composition to the cuticle of living arthropods (Figs. 2b and 3). Indeed, the composition of fossil arthropod cuticle converges on that of graptolite

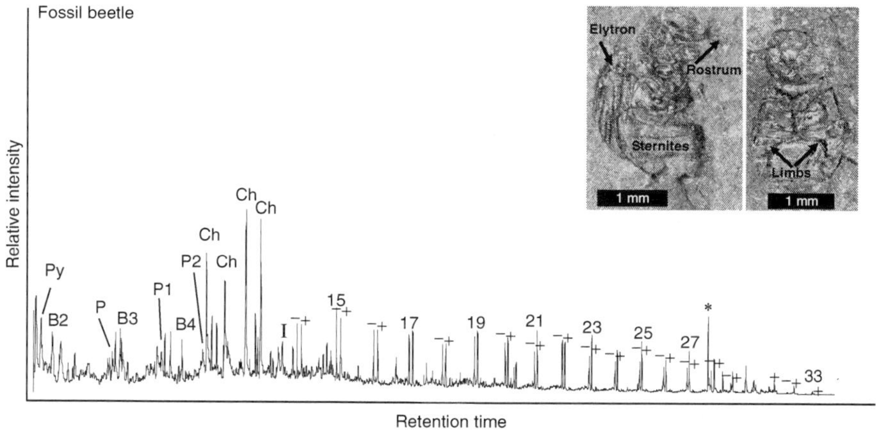

Fig. 3 Partial py-GC-MS chromatogram revealing the chemistry of a fossil curculionid beetle from Enspel, Germany, Oligocene, with preservation of chitin. I, indole (derived from amino acid); Ch, Chitin; B_n, benzene derivative were n refers to the carbon number of the alkyl substitution; +, *n*-alk-1-ane; –, *n*-alk-e-ene; Pn, Phenols numbers refer to the carbon chain length; *, contaminant (This chromatogram focuses on the incorporated aliphatics in contrast to that in Stankiewicz et al. (1997a), which emphasized the chitin markers.)

periderm. As in the case of graptolites, this anomaly was explained by arguing that aliphatic components from the surrounding sediment replaced the original chemistry on a molecular scale (Baas et al. 1995; Briggs et al. 1995).

A new model for the preservation of fossil organic material like that in graptolites and arthropods has now emerged, which does not involve the incorporation of components from the surrounding sediment. It has become clear that diagenetic transformation of lipids in the cuticle and other tissues of the organism itself, by a process of in situ polymerization (Briggs 1999; Stankiewicz et al. 2000), is sufficient to account for the composition of the fossil. Some of the most compelling evidence that fossilization does not involve the introduction of components from an external source comes from situations where such a process is prevented. Insect (and plant) inclusions trapped in natural resins are sealed within a natural chamber, and analyses of progressively older amber fossils showed that their original chemistry is gradually transformed to a more aliphatic composition over time (Stankiewicz et al. 1998b). Artificial maturation experiments, where arthropods were sealed in gold bombs and subjected to high temperatures in the laboratory (350°C/700 bars/ 24 h), resulted in an aliphatic composition that could only have been generated from components within the organism itself (see discussion on experimental maturation: Section 4). Such experiments do not, of course, replicate the conditions of fossilization, but they allow an exploration of which starting components are necessary to generate a particular diagenetic product.

Diagenesis in sediments occurs over millions of years (Briggs et al. 2000). Nucleic acids (DNA and RNA) are very vulnerable to decay and oxidation and do not contribute significantly to the bulk composition of fossils; nor do they yield extensive sequences of base pairs in samples more than 100,000 years old. Traces of proteins and polysaccharides, in contrast, have been detected in much older examples, although they tend to be altered significantly in pre-Tertiary fossils, if present at all (Table 1).

Table 1 Distribution of biomolecules in organically preserved animal and plant fossils through time (Updated from Briggs et al. 2000)

Biomolecule	Source organism	Archeological record	Tertiary record	Mesozoic-Paleozoic record
DNA/RNA	All organisms	Possibly up to 10^5 year. Physical protection (such as in bone) may enhance preservation	None confirmed	None
Proteins	All organisms	10^3–10^6 year in shell and bones	Present in Oligocene beetles	Detected in *T. rex* bone fossils (68 ma) and in kerogen (140 ma); not known in Paleozoic examples
Cellulose	Vascular plants and some fungi	Present	Present in Eocene *Metasequoia*. However these could reflect melanoidins	Not reported so far
Chitin	Arthropods and fungi	In variable amount. Detected in Quaternary beetles	Present in Oligocene beetles	Not reported so far
Lipids	All organisms	Present	Present, significant proportion bound to macromolecule	Present, significant proportion bound to macromolecule
Algaenans	Algae	Present	Present, with greater crosslinking	Present, with greater crosslinking
Resins	Vascular plants	Present	Present	Diagenetically modified
Lignins	Vascular plants	Present	Present, but may be diagenetically modified	Diagenetically modified
Sporopollenin	Vascular plants	Present	Present in altered state	Diagenetically modified
Cutan	Vascular plants	Occurrence in modern taxa very limited	Not reported	Not reported

Cellulose and lignins (aromatic alkoxy phenols), the most important macromolecular constituents of plants, have not been detected in a biopolymeric form in pre-Tertiary fossils (see Collinson chapter 6, this volume). Lignin, however, may be present in a defunctionalized state as part of the aromatic fraction in older plant remains. This diagenetic alteration of organic fossils is a result of polymerization over time, which ultimately converts all of them, regardless of their original chemistry, to a composition converging on kerogen (Gupta et al. 2007b).

2 Organic Skeletons

The periderm of graptolites consists of two layers, an inner fusellar layer, secreted incrementally during growth, and an overlying cortical layer, which is laid down externally by the zooids and varies in thickness on different parts of the colony (Crowther 1981). The periderm of pterobranchs (e.g. *Rhabdopleura*), the closest living relatives of graptolites, consists primarily of collagen (Fig. 1a). Additional components include lipids (a series of n-alkanes and n-alkenes soluble in organic solvents), and fatty acyl moieties. Decay experiments on *Rhabdopleura* showed that this periderm, particularly in the older parts of the colony, is much more decay resistant than the zooids, which become unrecognizable in normal marine conditions within a week of death (Briggs et al. 1995). The periderm of graptolites (Fig. 1b) has transformed to a resistant aliphatic polymer and benzene, phenol and naphthalene derivatives in the fossil (Gupta et al. 2006c). The collagenous jaws of polychaetes are similarly transformed in fossil examples, which are termed scolecodonts (Fig. 4).

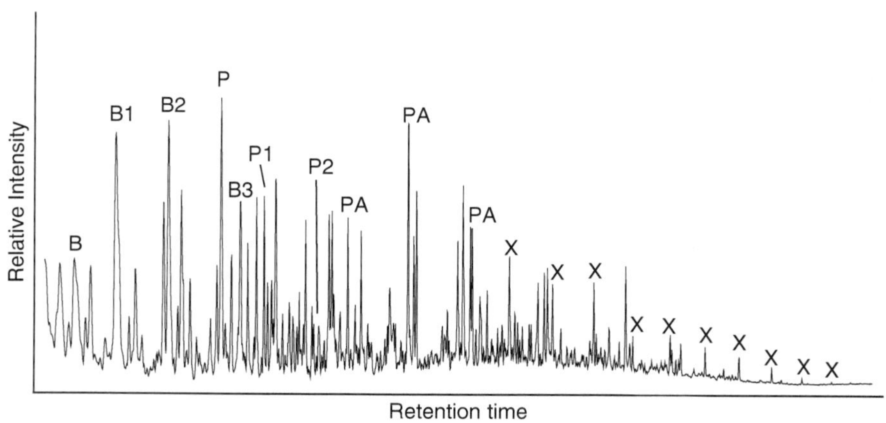

Fig. 4 Partial py-GC-MS chromatogram revealing the chemistry of a scolecodont (YPM 1112997), Cincinnati, Ohio, Upper Ordovician, Cincinnatian, with dominant aromatic composition. B, Benzene; Bn, Benzene derived product where n refers to the number of Carbon atoms attached to the benzene ring; P, Phenol; P1, methyl phenol; P2, dimethyl phenol; PA, polyaromatic compounds; X, alkane/alkene pairs

Arthropod cuticle consists of three layers – an outermost epicuticle and a much thicker exocuticle and endocuticle. The epicuticle of terrestrial arthropods is waxy in order to prevent desiccation; the waxes are composed mainly of hydrocarbons, esters, fatty acids and alcohols (Lockey 1988). The remainder of the cuticle consists of chitin fibers embedded in a protein matrix, cross-linked by catechol, aspartate and histidyl moieties (Schaefer et al. 1987). The cuticle may be strengthened by mineralization, usually in the form of calcium carbonate, but this is a feature of only a small portion of the range of different arthopod taxa: it occurs in trilobites and a number of crustacean groups.

There is extensive loss of the protein in arthropod cuticle during the early stages of decay. Laboratory experiments on shrimps have shown a significant reduction within 2 weeks, whereas the structure of chitin remained largely intact for 8 weeks (Baas et al. 1995). Proteins are clearly more decay-prone than chitin (Tegelaar et al. 1989; de Leeuw and Largeau 1993; but see Nguyen and Harvey 1998) but they are much more resistant where they have undergone cross-linking, as in the collagen in graptolite periderm (Briggs et al. 1995) and the jaws of polychaetes (Briggs and Kear 1993). Protein remnants (Table 1) have been detected in insect remains from archeological sites (McCobb et al. 2004) and in beetles (weevils) in lacustrine sediments as old as 24.7 my from the Oligocene of Enspel, Germany (Fig. 3) (Stankiewicz et al. 1997a; Gupta et al. 2007a). Older traces of proteins have been reported in other taxa: e.g. in the bone of a 68 m.y. old dinosaur (Schweitzer et al. 2007; Asara et al. 2007), and even in sedimentary organic matter 140 my old (Mongenot et al. 2001). Protein fragments in these older examples may have been 'protected' chemically: more reactive functional groups and the bonds that link the carbon chain may be shielded and protected by less charged species such as alkyl chains (Knicker et al. 2001; Mongenot et al. 2001; Riboulleau et al. 2001).

Chitin is also present in young fossils. Traces occur routinely in Pleistocene beetles (Stankiewicz et al. 1997c) but the oldest evidence of chitin is only Oligocene in age (Table 1). This earliest known chitin was found in a weevil from Enspel (Fig. 3) but only remnants were detected in the fossil. The composition of the weevil cuticle is dominated by an aliphatic polymer up to C_{33} in chain length (indicated by *n*-alkane/alkene peaks and consisting mainly of C_{14}, C_{16} and C_{18} fatty acids) with additional aromatic compounds (Stankiewicz et al. 1997a; Gupta et al. 2007a). Neither chitin nor protein are present in other insects analysed from Enspel, or in older arthropod fossils. The fossilized cuticle of Paleozoic examples, such as scorpions and their extinct relatives the eurypterids (Gupta et al. 2007c), is largely aliphatic, consisting of alkane/alkene homologues with chain lengths ranging from C_9 to C_{22} (Fig. 2b). Eurypterids from some sequences yield a more aromatic signature (i.e. with phenols and polyaromatic compounds), which may reflect diagenesis at higher temperatures (thermal metamorphism). Assuming that the original chemistry of the eurypterid cuticle was similar to that of its living relatives, the chitin–protein complex has undergone transformation, including incorporation of the lipid fraction, to a composition with a significant aliphatic component (Gupta et al. 2007c). A similar transformation occurs in fossil crustaceans (e.g. shrimps, Stankiewicz et al. 1997b; Gupta et al. 2008a).

3 Chemosystematics

Organisms and their component tissues vary in composition. At the simplest level the cuticle of terrestrial arthropods like insects, with their waxy outer layer, differs chemically from that of marine taxa like horseshoe crabs. Even though the chemistry is altered dramatically during diagenesis, the composition of a fossil may still reflect its original composition, and therefore its biological affinities. Such chemical differences (chemosystematic signatures) include variation in the chain length of the aliphatic components. These signatures have the potential to help in the identification of fossil fragments that lack diagnostic morphology. They may also illuminate the nature of the diagenetic process by revealing how different starting compositions result in different fossil chemistry.

Research on the preservation of chemosystematic signatures is at an early stage, particularly as applied to animal remains. An example is provided by the Cretaceous of Las Hoyas, Spain, where the composition of fossils differs depending on their identity (Gupta et al. 2008a). The cuticle of beetles from Las Hoyas is composed of aliphatics with carbon chain lengths ranging from C_8 up to C_{31}, presumably reflecting the contribution of long chain waxes (greater than C_{30}) in the epicuticle. Shrimp cuticles from the same locality, in contrast, are composed of alkyl benzene and phenols (aromatics); aliphatics with additional sulfur compounds are also present but they do not range beyond C_{21}. The composition of fish scales is dominated by aliphatics ranging up to $C_{21/22}$ and the plant *Montsechia* is likewise aliphatic, but with chain lengths ranging from C_9 to C_{25} with very little aromatic content. Thus the composition of the Las Hoyas fossils provides a potential means of identifying fragmentary fossil material to at least a major taxonomic group. Compositional differences between different taxa have also been reported from other localities: the cuticles of arthropods and plants from the Carboniferous of North America (Stankiewicz et al. 1998a), for example, differ mainly in the distribution of alkenes/alkanes rather than total chain length or the proportion of aromatics.

The relationship of starting chemistry to final composition in fossils is complicated by factors other than chemistry. Diagenetic history varies with environmental setting and time (Briggs 1999; Briggs et al. 2000), and thermal metamorphism may be reflected in the production of aromatic compounds (Gupta et al. 2007c). In order to investigate fossil chemosystematics further, a comprehensive analysis of modern taxa and related fossils from different ages and environments is required. The development of isotope techniques that target the aliphatic component in the macromolecule, and specifically the incorporated lipids, will provide valuable paleoenvironmental, paleodietary and potentially chemosystematic information.

4 Diagenesis

4.1 Molecules Are Not Introduced from Sediment

The resistant aliphatic components that dominate the chemistry of fossil arthropods are not present in the organic skeleton of living examples. Thus their presence in fossils cannot be explained simply by the survival of a decay resistant biopolymer in the cuticle. A number of lines of evidence indicate that the introduction of aliphatics from sediment can also be ruled out. Experiments on materials sealed in gold bombs and analyses of specimens in amber have shown that input from an external source is not necessary to explain the presence of an aliphatic component in organic fossils (Gupta et al. 2007a). The lipids that occur in leaves from the Miocene of Clarkia, Idaho, are confined to the fossils from which they originate, and were not detected 'leaking' into the sediment (Logan et al. 1995). Aliphatic polymers are even less mobile in sediment due to their insoluble nature (Briggs 1999). Chemosystematic differences between the aliphatic signatures found in insect and associated plant fossils in lithologies in the Cretaceous of Las Hoyas, Spain (Gupta et al. 2008b) and in the Upper Carboniferous of North America (Stankiewicz et al. 1998a) show that diagenesis did not involve replacement with components from a common source such as the sediment. The possibility of migration from sediment was also eliminated in an investigation of samples from the Oligocene of Enspel: the fatty acyl components of insects, plants and the sedimentary matrix are all different (Gupta et al. 2007a).

4.2 Components Contributing to the Composition of the Fossil

It is clear that the composition of a fossil is derived primarily from the chemistry of the living organism. Initial analyses and experiments indicated, surprisingly, that it is not only the decay resistant components that contribute; some much more labile decay-prone components, such as lipids, are also essential to this diagenetic process. Lipids include a diversity of molecules united by their solubility in organic solvents. It is not a straightforward matter to determine which lipids contributed to the composition of fossil cuticle; the fossil material becomes insoluble during diagenesis, and it is necessary to break it down chemically to determine its constituents. The major technique used to analyse inert materials like fossil cuticles is pyrolysis, which vaporizes the sample very rapidly at high temperatures before passing it through a mass spectrometer (i.e. pyrolysis – gas chromatography/mass spectrometry: Py-GC-MS). However, such routine pyrolysis breaks down the macromolecule in an unconstrained fashion and the spectra generated provide little detail of the structure of the aliphatic component or the identity of the molecules that contributed to it.

More detail can be obtained by using techniques that cleave the fossil macromolecule into fragments by breaking specific bonds. The cuticle of fossil eurypterids, for example, consists largely of long-chain ($<C_9$ to C_{22}) aliphatic components. Like other fossil arthropod cuticles, it is very resistant to degradation. It is not broken down by base hydrolysis, for example, suggesting that any ester linkages in the macromolecule are protected by alkyl chains that resist chemical attack on the ester bond. Pyrolysis in the presence of tetramethylammonium hydroxide (a technique known as thermochemolysis or structural analysis: de Leeuw and Baas 1993) cleaves ester linkages, thereby providing evidence of the compounds that were combined during diagenesis. Thermochemolysis of eurypterid cuticle yielded fatty acyl moieties, which are derived from lipids, of chain lengths C_7 to C_{18}. C_{16} and C_{18} are the most abundant (Fig. 5b) indicating that lipids of

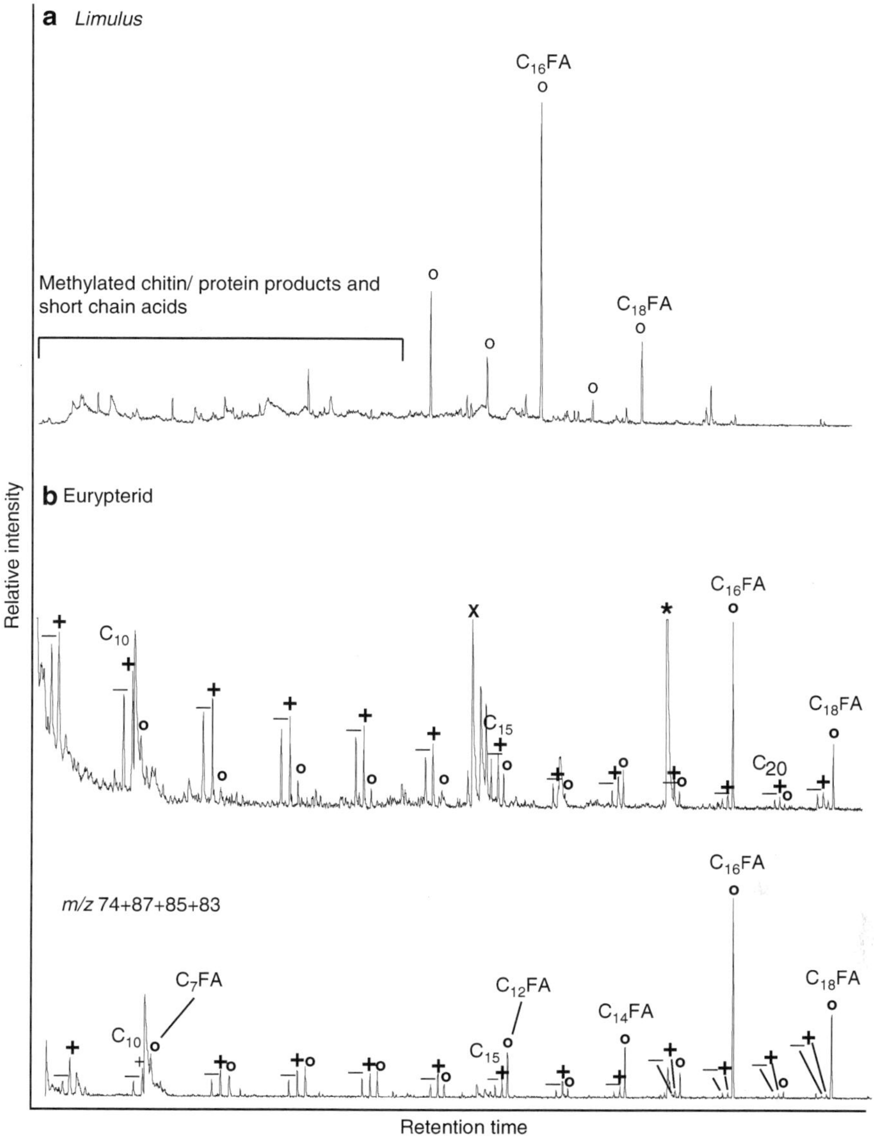

Fig. 5 Partial TMAH-py-GC-MS (thermochemolysis) chromatogram revealing the distribution of fatty acyl moieties in (**a**) modern *Limulus* and (**b**) eurypterid *Eurypterus lacustris*, Ridgemount Quarry, Ontario, Canada, Williamsville Formation, Pridolian. FA(o), fatty acid where C_n refers to the carbon chain length; +, *n*-alk-1-ane; –, *n*-alk-e-ene; C_n, where n refers to the carbon chain length. *M/z* 74+87+85+83 reveals the distribution of fatty acyl moieties relative to alkane/alkene homologues; *, contaminant

this length were the primary contributors to the fossil cuticle. A similar analysis of modern scorpion (Gupta et al. 2007c) and horseshoe crab (Fig. 5a) showed that C_{16} and C_{18} fatty acyl moieties are also dominant in the cuticle of the living relatives of eurypterids. Fossilization must have involved the incorporation of lipids of these chain lengths into the aliphatic polymer. Lipids may also become crosslinked by oxidative reticulation leading to an ether linked macromolecule, e.g. in fossil cephalopod beaks (Gupta et al. 2008b) and fossil algae (Versteegh et al. 2004).

A similar result was obtained from graptolites, even though the starting composition was predominantly collagen as opposed to chitin–protein. The aliphatic polymer (C_9 to C_{21}) in graptolite periderm, like that in arthropod cuticle, is immune to base hydrolysis indicating that here too ester linkages are probably sterically protected. Thermochemolysis released fatty acyl moieties with chain lengths from C_7 to C_{18} (Fig. 6b), dominated by C_{16} and C_{18}. A similar distribution of fatty acyl moieties was detected in the periderm of living *Rhabdopleura* (Fig. 6a) indicating that the fossilization of graptolites involved direct incorporation of lipids mainly of lengths C_{16} and C_{18}. Such a process of in situ polymerization of lipids appears to account for the preservation of organic remains in all those fossil groups so far analysed (Mösle et al. 1998; Briggs 1999; Stankiewicz et al. 2000; Gupta et al. 2006a–c, Gupta et al. 2007a–c, Gupta et al. 2008a, b). Furthermore, electron microscopy shows that the internal structure of fossil cuticles is normally destroyed over an extended period, presumably as a result of the polymerization of the chemical constituents (Stankiewicz et al. 1998a).

The likely effect of original chemistry on the composition of fossil material is illuminated by laboratory experiments. When lipids are removed (by solvent extraction and saponification) from arthropod cuticle prior to thermal maturation in sealed bombs no n-alkyl macromolecular component (aliphatics) are generated (Fig. 7; Gupta et al. 2006b). Thermal maturation of different plant molecular components in gold bombs also results in products that differ, particularly in the length of the carbon chain in the aliphatic polymer obtained (Gupta et al. 2007d).

4.3 *Implications for Kerogen Formation*

Most ancient sedimentary organic matter is a product of the diagenetic alteration of biological material, which typically yields kerogen, a non-hydrolysable macropolymer, insoluble in common organic solvents (Tissot and Welte 1984). The composition and type of kerogen depends on the nature of the biological input, the environment of deposition and the diagenetic pathway (de Leeuw and Largeau 1993). Many kerogens are highly aliphatic (especially Type I/II which is similar in composition to most organic macrofossils) and these are converted to petroleum products during thermal maturation (catagenesis). The process of kerogen formation and preservation is critical to the formation of fossil fuel deposits, impacts the global carbon cycle, and accounts for the preservation of macroscopic and morphologically intact organic remains in the fossil record.

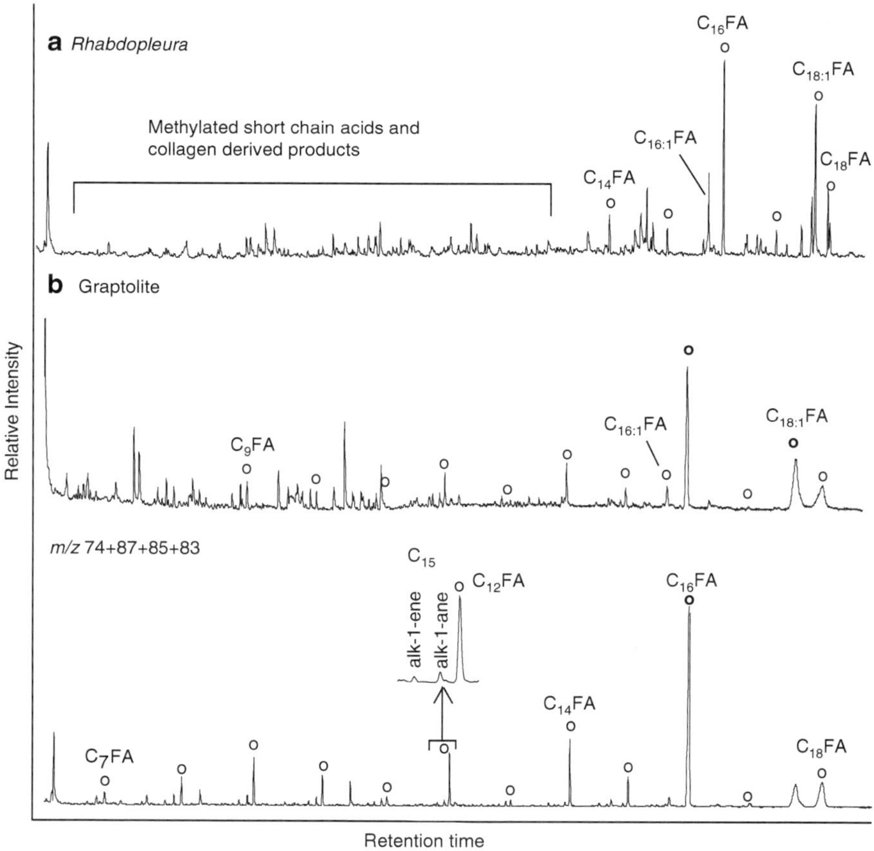

Fig. 6 Partial TMAH-py-GC-MS (thermochemolysis) chromatogram revealing the distribution of fatty acyl moieties in (**a**) modern *Rhabdopleura* and (**b**) graptolite *Dictyonema peltatum* (YPM 202222), Wisby, Sweden, Silurian. FA(o), fatty acid where C_n refers to the carbon chain length; *M/z* 74 + 87 + 85 + 83 reveals the distribution of fatty acyl moieties relative to alkane/alkene homologues

The preservation of organic fossils is a type of kerogen formation. The formation of kerogen has been attributed to a number of processes. Neogenesis (Tissot and Welte 1984) involves random intermolecular polymerization and polycondenzation of sedimentary organic matter. Natural vulcanization involves the reaction between reduced sulfur and various functional groups in organic compounds; it results in the formation of a sulfur-rich macromolecule (Kok et al. 2000). Oxidative reticulation (de Leeuw et al. 2006; Gatellier et al. 1993; Riboulleau et al. 2001; Versteegh et al. 2004; Gupta et al. 2008b) involves crosslinking of lipids with oxygen. Selective preservation involves the preferential survival of highly aliphatic and decay-resistant biopolymers such as algaenan and cutan (Tegelaar et al. 1989; de Leeuw and Largeau 1993).

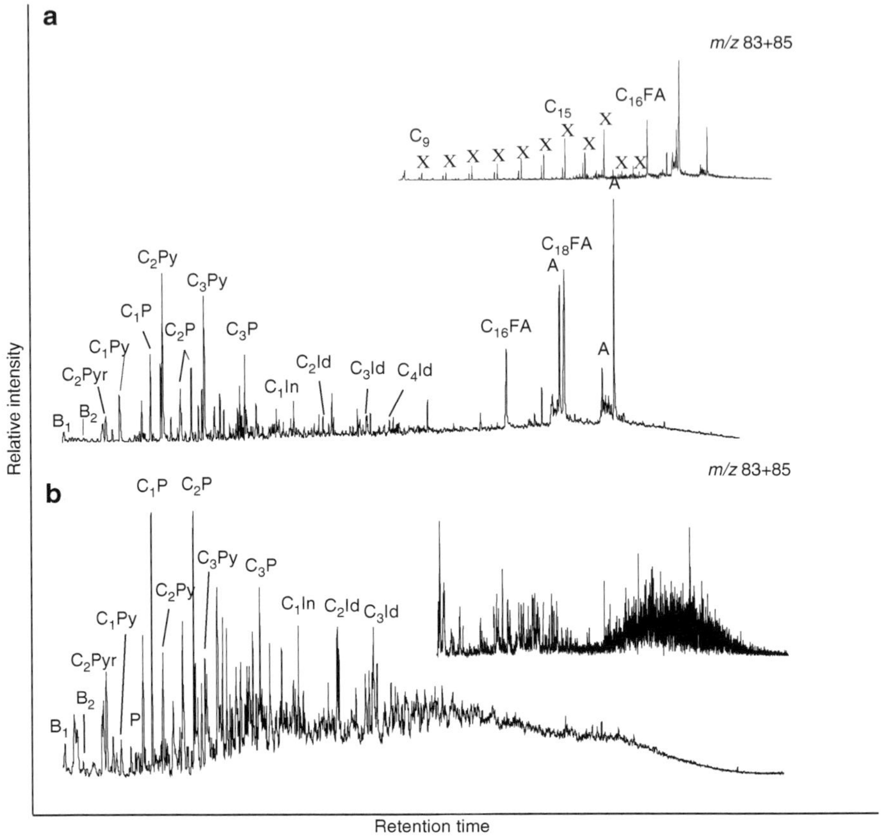

Fig. 7 Partial py-GC-MS chromatogram of cockroach cuticle matured (**a**) without chemical treatment (**b**) after removal of extractable and hydrolysable lipids. B_n, benzene derivatives (n refers to the number of carbon atoms in the alkyl component); C_1In, methyl indole; A, amide derivative (primarily C_{16} and C_{18} derivatives); C_n Py, pyridine pyrolysis products, where n is the number of carbon atoms in the alkyl substituent; P, phenol derivative; Id, alkyl indenes (Inset mass chromatograms *m/z* 83 + 85 reveal the presence of *n*-alkane/alkenes: X.)

The preservation of the organic remains of animals cannot be explained fully by any of the above processes. Fossil arthropods from Pleistocene and Tertiary deposits may preserve traces of the more resistant elements of the chitin–protein complex that make up the cuticle, but older fossils are aliphatic in composition as are many plant fossils. Such a contrast cannot be explained by selective preservation: the cuticles of modern arthropods do not contain decay-resistant aliphatic components and they differ in composition from fossil cuticles (Stankiewicz et al. 2000). Neither does selective preservation explain the composition of most fossil plant material. Most modern leaves lack a resistant aliphatic component such as cutan (Gupta et al. 2006a). The aliphatic signature that dominates the composition of fossil leaves is the result of diagenetic changes (Mösle et al. 1997; Mösle et al. 1998;

Gupta et al. 2007a; b). The convergence in the composition of the organic remains of fossil animals and plants cannot be explained simply by the selective preservation of certain components of the living organisms.

Neither selective preservation nor simple chemical transfer can explain the aliphatic composition of fossil cuticles and it is not the product of random incorporation of components from the surrounding sediment. As clearly demonstrated by the chemical transformation of graptolites, arthropods and other organic structures like cephalopod beaks and fish scales, more labile chemical components, such as free cuticular lipids or hydrolysable lipids (such as fatty acids), are incorporated into the aliphatic dominated fossil cuticle. Such lipid incorporation also has been observed to contribute significantly to the formation of kerogen, where the role of selective preservation is likewise limited (Riboulleau et al. 2001). The major proportion of the fossil record of organic animal remains is the result of some process of in situ polymerization, a process which is also important in the formation of TypeI/II kerogens (Briggs 1999).

The chains of *n*-alkanes/*n*-alkenes in the aliphatic component of marine animal fossils such as cephalopod beaks and eurypterids do not exceed C_{25}; they are not as long as those in terrestrial fossils such as leaves and beetles (Gupta et al. 2007a–c). This presumably reflects the presence of longer chain waxes, which control water loss, in the cuticles of land organisms. Thus the chain lengths in the aliphatic content of macromolecular material in fossils and in sedimentary organic matter may allow marine and terrestrial origins to be distinguished. Neoproterozoic kerogens (e.g. those from the south Oman salt basin), which are exclusively marine in origin, show aliphatic contents mainly up to C_{24} (Höld et al. 1999). Where higher molecular weight aliphatics are present in marine kerogens they may represent the preservation of the biopolymer algaenan which is present in the outer cell walls of some algae and has an aliphatic component $>C_{30}$ (Blokker et al. 2000). In deltaic locations, where marine and terrestrially derived organic matter mix, longer aliphatic components may be sourced from terrestrial input. Although animal cuticles may provide important insight into kerogen formation, kerogen is primarily sourced from plant material, especially algae and terrestrial higher plants, with some bacterial input (Vandenbroucke and Largeau 2007). Only 0.1% to <1% of the organic components produced by autotrophic organisms is incorporated into sediments. The final composition of the kerogen depends on decay prior to sedimentation and diagenesis thereafter. The origin, evolution and structure of kerogen are reviewed by Vandenbroucke and Largeau (2007).

4.4 The Rate of Diagenetic Change

Genetic material (DNA/RNA) can be extracted from fossils up to 10^5 years old, especially where hard parts such as bone provide physical protection from hydrolyzing agents, and sequence data are frequently used in archeological investigations. These nucleic acids are very prone to decay, however, and are unlikely to make any contribution to kerogen. Proteins and carbohydrates, on the other hand, react to form compounds called melanoidins (through the Maillard reaction: Maillard 1912). Melanoidins may cross link with lipids to form protokerogens (Larter and Douglas 1980). Lipids, which are present in all organisms, may make a significant contribution to sedimentary organic matter by becoming incorporated into macromolecules during early diagenesis. Diagenetic changes ensure that animal cuticles older then Tertiary do not contain recognizable traces of their original molecular composition (Table 1). Just how quickly this process of in situ polymerization progresses is unknown. Traces of aliphatics are evident, for example, in *Hymenaea* leaves in 20 Kya resin from Kenya even though they are not present in living *Hymenaea* leaves (Stankiewicz et al. 1998b). Analyses show that resistant aliphatics have formed in Oligocene insect cuticles (flies from Enspel have been transformed even though beetles still reveal traces of chitin and protein: Stankiewicz et al. 1997c). Previous investigations of Tertiary arthropods have focused on the search for the oldest traces of macromolecular components like chitin and may have overlooked the formation of aliphatic components in younger fossils.

5 Future Directions in Molecular Taphonomy

The chemistry of the biopolymeric constituents of invertebrates begins to change in the early stages of decay (Baas et al. 1995). Future experiments are necessary to explore the impact of microbes on the chemistry of a decaying carcass over extended periods of time. Determining the changing structure of biopolymers and associated lipids using high resolution mass spectrometry and ^{13}C NMR spectroscopy will illuminate the reactions involved in the transformation of the living organism to a fossil and its eventual contribution to sedimentary organic matter. Such a coupled analytical approach is important as spectroscopy, which is non-destructive, provides complementary data on functional groups to that obtained from spectrometry (e.g. py-GC-MS and chemolysis).

Preliminary thermal maturation experiments using a hydrothermal apparatus (see Section 6.3) have revealed the importance of lipids in geopolymer formation (Gupta et al. 2006b) but it is essential to track the fate of both the soluble (i.e. lipid) and insoluble fractions to understand the evolution of the macromolecular composition of a fossil. Similar maturation experiments are necessary to evaluate the role of different conditions of pressure and temperature during diagenesis, catagenesis and metagenesis (i.e. the transformation of organic matter with increasing temperature). Such experimentation will help to determine the relative contribution of different organisms and their constituent lipids and biopolymers to sedimentary organic matter and kerogen.

The influence of environmental factors on the preservation of organic fossils is poorly understood. This can be explored by comparing the nature of organic preservation with sedimentological and paleontological evidence for depositional conditions. Lipid biomarkers can also provide data on conditions in the sediment and water column during deposition. Systematic investigations of preservation in different settings are necessary to show how physico-chemical processes generate biases in the record of organically preserved fossils. Investigations of the morphological preservation and chemical composition of organic fossils through time are necessary to determine how one reflects the other. Preliminary results do not reveal any direct correlation (Mösle et al. 1998; Gupta et al. 2006c, Gupta et al. 2007a, b) especially in samples older than Tertiary, which have undergone significant diagenetic alteration.

Accumulation of data on the composition of different animal and plant taxa of different ages and from different settings will lead to an improved understanding of the nature of organic diagenesis. This in turn will lead to an understanding of how different compositions (e.g. relative distribution and chain length of aliphatic components, nature of chemical linkages) are diagnostic of particular taxa or the conditions in which they lived and were diagenetically transformed. The application of new x-ray methods (Section 6.3) will enable chemical characterization of samples at much finer resolutions than are routinely available now, an important consideration when sample size is a limiting factor. Such methods will allow the investigation of the fate of biopolymers and chemical changes during diagenesis, on a subcellular (sub-micron) level where such remains are still preserved. In the case of arthropods that preserve the structure of the cuticle, it will be possible to analyse the composition of the different layers in the fossil (Gupta et al. 2007a) and show how these morphologically specific regions change over time.

6 Appendix: Main Analytical Methods Applied to Organic Remains

6.1 The Soluble Fraction

The soluble fraction (lipids) of organic components in modern and fossil samples can be analysed using *gas chromatography-mass spectrometry (GC-MS)* after solvent extraction in dichloromethane-methanol. The lipids isolated in this way can then be fractionated into different compound classes using column chromatography. This reveals the distribution of lipid carbon chain lengths and associated functional groups.

6.2 The Insoluble Fraction

The molecular and isotopic composition of the insoluble fraction of organic fossils in both modern and fossil specimens is evaluated using a range of analytical techniques. These techniques have specific strengths and biases and it is important to use them together in order to generate complementary data and ensure more robust interpretations. Organic fossil remains generally are not available in abundance and sample size is often limiting. Samples of cuticles can be obtained from sedimentary rock by acid digestion, or by mechanical preparation. *Pyrolysis-gas chromatography/mass spectrometry (py-GC-MS)*, which is destructive, requires a very small sample (200 μg), often all that is available for fossils. It is a rapid and efficient tool that is used routinely for determining the molecular constituents of insoluble organic matter (Larter and Horsfield 1993) in fossils and their modern counterparts. Py-GC-MS involves heating a macromolecule to break chemical bonds thermally. The molecular species generated are then separated using a gas chromatograph (GC) and identified using a mass spectrometer (MS). This technique can be employed to track the changes that take place in the molecular composition of material from living sample, through decay experiments, to progressively older fossils.

Organic fossils often reveal a composition with a dominant aliphatic hydrocarbon-rich component linked by C–C, ester and ether linkages (Figs. 1–4). These functional groups can be targeted specifically using stepwise chemical degradation techniques that cleave molecular linkages in a systematic fashion, thereby yielding structural information beyond that provided by py-GC-MS without such treatments. Such degradation techniques include RuO_4 (ruthenium tetroxide) oxidation treatment and thermochemolysis. RuO_4 oxidation cleaves ether linked carbon components (Blokker et al. 2000) and thermochemolysis (pyrolysis in the presence of tetramethylammonium hydroxide; de Leeuw and Baas 1993) cleaves ester linkages in macromolecules providing evidence of the distribution of fatty acids in fossils (Gupta et al. 2007b). These techniques reveal the distribution of chain lengths and associated functional groups. In marine fossils, such as graptolites and shrimps, molecular components may also be linked through C–S bonds. These can be detected by subjecting the fossils to treatment with Li/NH_3 (Adam et al. 1993; Schaeffer et al. 1995). Thus, following an initial analysis using py-GC-MS, chemolytic methods can be used to cleave specific chemical linkages to provide further structural information.

Spectroscopic tools provide additional structural data on insoluble organic matter. Methods commonly used include *solid state nuclear magnetic resonance spectroscopy (NMR)*, *Fourier transform infrared spectroscopy (FTIR)*, and *Raman spectroscopy*, all of which are non-destructive. Solid state ^{13}C and ^{15}N NMR reveals bulk molecular and structural characteristics that complement the data obtained from pyrolysis but these methods may require around 0.5 g of sample material. FTIR has been used widely to determine the constituent oxygen functional groups (such as carboxyl, carbonyl, ethers) and their relative importance in fossil plant material, providing insights into oxygen-linked bonds that are important in the macromolecule. The application of spectroscopic techniques to animal fossils has been limited so far. Raman spectroscopy allows microscale analysis of carbonaceous samples even when available quantities are very limited. It can be used as an indicator of thermal maturity and has the potential to provide thermal data on organic carbon at temperatures as low as 100°C (Rahl et al. 2005). Current research is also making use of *X-ray absorption spectroscopy of carbon, sulfur and oxygen*

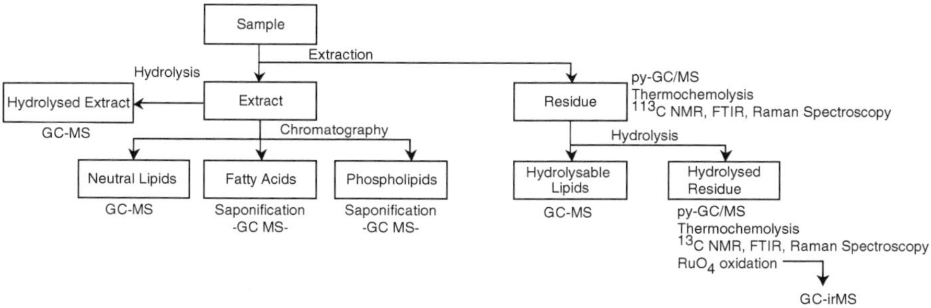

Fig. 8 Schematic analytical protocol used for evaluating the chemistry of modern and fossil samples

(C,S,O-XANES) to provide more information on the valence state and spatial configuration of the molecular species in the fossils. This method allows sub micron level characterization of macromolecules where other methods of chemolysis and pyrolysis are difficult to apply. The presence of lignin, for example, can be mapped at submicron resolution in fossil plant cell walls (Boyce et al. 2002).

Compound specific stable isotope analysis (CSIA), using gas chromatography-isotope ratio-mass spectrometry (GC-IR-MS), provides isotope data on individual molecular components. Such data may be used as a geochemical tracer to analyse the transformation of biomolecules to geomolecules during diagenesis, and to determine the origin of molecular constituents in sedimentary organic matter. The application of compound specific techniques to macromolecules has been problematic due to their insoluble and recalcitrant nature. However, we are currently developing an approach that first employs RuO_4 (ruthenium tetroxide) oxidation treatment (Sharpless et al. 1981) to generate a series of organic acids that provide information on the constituent chain lengths. Applying compound specific techniques to these acids has the potential to identify which lipids are incorporated (Fig. 8).

6.3 Thermal Maturation Experiments

Techniques have been developed recently to explore the chemical transformation of living materials with confined pyrolysis-accelerated maturation techniques using a hydrothermal apparatus (Fig. 9). This approach involves experimental heating of model compounds and modern tissues in a sealed gold cell (within an autoclave) at temperatures from 260°C to 350°C and a confining pressure of 700 atmospheres for a day (or longer at lower temperatures). The end products reveal compositions similar to those of fossils and this provides a means to determine the role of biopolymers in forming geopolymers.

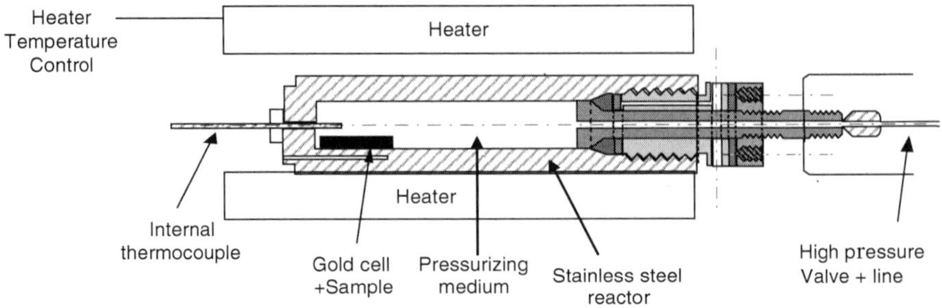

Fig. 9 Confined pyrolysis autoclave and gold cell apparatus used for accelerated maturation of modern tissues and model compounds

6.4 *Investigating Morphology*

Transmission and scanning electron microscopy allow surface morphology and ultrastructural preservation within organic materials to be examined. Surface features may be pristine even if the internal structure of the cuticle has been destroyed. Microscopy in conjunction with rigorous chemical analysis provides the most complete data on the changes that occur in organic fossils though time.

Acknowledgments DEGB's research in this area has been supported by Natural Environment Research Council (UK) grants and mass spectrometry facilities, and by the American Chemical Society Petroleum Research Fund. We are grateful for collaboration and discussions with M.E. Collinson, G. Eglinton, R.P. Evershed, R. Michels, R.D. Pancost, R.J. Parkes, B.A. Stankiewicz and R.E. Summons.

References

Adam, P., Schmid, J. C., Mycke, B., Strazielle, C., Connan, J., Huc, A., et al. (1993). Structural investigation of non-polar sulfur cross-linked macromolecules in petroleum. *Geochimica et Cosmochimica Acta, 57*, 3395–3419.

Asara, J. M., Garavelli, J. S., Slatter, D. A., Schweitzer, M. H., Freimark, L. M., Phillips, M., et al. (2007). Interpreting sequences from mastodon and *T. rex*. *Science, 317*, 1324–1325.

Baas, M., Briggs, D. E. G., van Heemst, J. D. H., Kear, A. J., & de Leeuw, J. W. (1995). Selective preservation of chitin during the decay of shrimp. *Geochimica et Cosmochimica Acta, 59*, 945–951.

Blokker, P., Schouten, S., de Leeuw, J. W., Sinninghe Damsté, J. S., & van den Ende, H. (2000). A comparative study of fossil and extant algaenans using ruthenium tetroxide degradation. *Geochimica et Cosmochimica Acta, 64*, 2055–2065.

Boyce, C. K., Cody, G. D., Feser, M., Jacobsen, C., Knoll, A. H., & Wirick, S. (2002). Preservation of cell wall chemistry and microstructure in plant fossils as old as 400 million years: Detection by carbon X-ray absorption spectromicroscopy. *Geology, 30*, 1039–1042.

Briggs, D. E. G. (1999). Molecular taphonomy of animal and plant cuticles: Selective preservation and diagenesis. *Philosophical Transactions of the Royal Society of London B, 354*, 7–16.

Briggs, D. E. G. (2003). The role of decay and mineralization in the preservation of soft-bodied fossils. *Annual Review of Earth and Planetary Sciences, 31*, 275–301.

Briggs, D. E. G., & Kear, A. J. (1993). Decay and preservation of polychaetes: Taphonomic thresholds in soft-bodied organisms. *Paleobiology, 19*, 107–135.

Briggs, D. E. G., Kear, A. J., Baas, M., De Leeuw, J. W., & Rigby, S. (1995). Decay and composition of the hemichordate *Rhabdopleura*: Implications for the taphonomy of graptolites. *Lethaia, 28*, 15–23.

Briggs, D. E. G., Evershed, R. P., & Lockheart, M. J. (2000). The biomolecular paleontology of continental fossils. *Paleobiology, 26*(Suppl. 4), 169–193.

Crowther, P. R. (1981). The fine structure of graptolite periderm. Special papers in palaeontology 26, 119 pp.

de Leeuw, J. W., & Baas, J. (1993). The behavior of esters in the presence of tetramethylammonium salts at elevated temperatures; flash pyrolysis or flash chemolysis? *Journal of Analytical and Applied Pyrolysis, 26*, 175–184.

de Leeuw, J. W., & Largeau, C. (1993). A review of macromolecular organic compounds that comprise living organisms and their role in kerogen, coal and petroleum formation. In M. H. Engel & S. A. Macko (Eds.), *Organic geochemistry: Principles and applications* (62nd ed., p. 23). New York: Plenum.

de Leeuw, J. W., Versteegh, G. J. M., & van Bergen, P. F. (2006). Biomacromolecules of plants and algae and their fossil analogues. *Plant Ecology, 189*, 209–233.

Gatellier, J.-P. L. A., de Leeuw, J. W., Sinninghe Damsté, J. S., Derenne, S., Largeau, C., & Metzger, P. (1993). A comparative study of macromolecular substances of a Coorongite and cell walls of the extant alga *Botryococcus braunii*. *Geochimica et Cosmochimica Acta, 57*, 2053–2068.

Gupta, N. S., Briggs, D. E. G., & Pancost, R. D. (2006). Molecular taphonomy of graptolites. *Journal of the Geological Society of London, 163*, 897–900.

Gupta, N. S., Collinson, M. E., Briggs, D. E. G., Evershed, R. P., & Pancost, R. D. (2006). Re-investigation of the occurrence of cutan in plants: Implications for the leaf fossil record. *Paleobiology, 32*, 432–449.

Gupta, N. S., Michels, R., Briggs, D. E. G., Evershed, R. P., & Pancost, R. D. (2006). The organic preservation of fossil arthropods: An experimental study. *Proceedings of the Royal Society of London B, 273*, 2777–2783.

Gupta, N. S., Briggs, D. E. G., Collinson, M. E., Evershed, R. P., Michels, R., Jack, K. S., et al. (2007a). Evidence for the *in situ* polymerisation of labile aliphatic organic compounds during the preservation of fossil leaves: Implications for organic matter preservation. *Organic Geochemistry, 38*, 499–522.

Gupta, N. S., Briggs, D. E. G., Collinson, M. E., Evershed, R. P., Michels, R., & Pancost, R. D. (2007b). Molecular preservation of plant and insect cuticles from the Oligocene Enspel Formation, Germany: Evidence against derivation of aliphatic polymer from sediment. *Organic Geochemistry, 38*, 404–418.

Gupta, N. S., Michels, R., Briggs, D. E. G., Collinson, M. E., Evershed, R. P., & Pancost, R. D. (2007c). Experimental evidence for formation of geomacromolecules from plant leaf lipids. *Organic Geochemistry, 38*, 28–36.

Gupta, N. S., Tetlie, O. E., Briggs, D. E. G., & Pancost, R. D. (2007d). The fossilization of eurypterids: A result of molecular transformation. *Palaios, 22*, 399–407.

Gupta, N. S., Briggs, D. E. G., Landman, N. H., Tanabe, K., & Summons, R. E. (2008). Molecular structure of organic components in cephalopods: Evidence for oxidative crosslinking in fossil marine inverstebrates. *Organic Geochemistry, 39*, 1405–1414.

Gupta, N. S., Cambra-Moo, O., Briggs, D. E. G., Love, G. D., Fregenal-Martinez, M. A., & Summons, R. E. (2008). Molecular taphonomy of macrofossils from the Cretaceous Las Hoyas Formation, Spain. *Cretaceous Research, 29*, 1–8.

Höld, I. M., Schouten, S., Jellema, J., & Sinninghe Damsté, J. S. (1999). Origin of free and bound mid-chain methyl alkanes in oils, bitumens and kerogens of the marine, Infracambrian Huqf Formation (Oman). *Organic Geochemistry, 30*, 1411–1428.

Knicker, H., del Rio, J. C., Hatcher, P. G., & Minard, R. D. (2001). Identification of protein remnants in insoluble geopolymers using TMAH thermochemolysis/ GC-MS. *Organic Geochemistry, 32*, 397–409.

Kok, M. D., Schouten, S., & Sinninghe Damsté, J. S. (2000). Formation of insoluble, nonhydrolyzable, sulfur-rich macromolecules via incorporation of inorganic sulfur species into algal carbohydrates. *Geochimica et Cosmochimica Acta, 64*, 2689–2699.

Larter, S. R., & Douglas, A. G. (1980). Melanoidins – kerogen precursors and geochemical lipid sinks: A study using pyrolysis gas chromatography (PGC). *Geochimica et Cosmochimica Acta, 44*, 2087–2095.

Larter, S. R., & Horsfield, B. (1993). Determination of structural components of kerogens by the use of analytical pyrolysis methods. In M. H. Engel & S. A. Macko (Eds.), *Organic geochemistry: Principles and application* (pp. 271–284). New York: Plenum.

Lockey, K. H. (1988). Lipids of the insect cuticle: Origin, composition and function. Comparative biochemistry and physiology Part B. *Biochemistry and Molecular Biology, 89*, 595–645.

Logan, G. A., Smiley, C. J., & Eglinton, G. (1995). Preservation of fossil leaf waxes in association with their source tissues, Clarkia, northern Idaho, USA. *Geochimica et Cosmochimica Acta, 59*, 751–763.

Maillard, L. C. (1912). Action des acides amines sur les sucres: Formation des melanoidines par votre methodiques. *C. R. Academy of Sciences, 154*, 6668.

McCobb, L. M. E., Briggs, D. E. G., Hall, A. R., & Kenward, H. K. (2004). The preservation of invertebrates in 16th century cesspits at St Saviourgate, York. *Archaeometry, 46*, 157–169.

Mongenot, T., Riboulleau, A., Garcette-Lepecq, A., Derenne, S., Pouet, Y., Baudin, F., et al. (2001). Occurrence of proteinaceous moieties in S- and O-rich Late Tithonian kerogen (Kashpir oil Shales, Russia). *Organic Geochemistry, 32*, 199–203.

Mösle, B., Finch, P., Collinson, M. E., & Scott, A. C. (1997). Comparison of modern and fossil plant cuticles by selective chemical extraction monitored by flash pyrolysis-gas chromatography-mass spectrometry and electron microscopy. *Journal of Analytical and Applied Pyrolysis, 40*, 585–597.

Mösle, B., Collinson, M. E., Finch, P., Stankiewicz, B. A., Scott, A. C., & Wilson, R. (1998). Factors influencing the preservation of plant cuticles: A comparison of morphology and chemical comparison of modern and fossil examples. *Organic Geochemistry, 29*, 1369–1380.

Nguyen, R. T., & Harvey, H. R. (1998). Protein preservation during early diagenesis in marine-waters and sediments. In B. A. Stankiewicz & P. F. van Bergen (Eds.), *Nitrogen containing macromolecules in the bio- and geosphere, ACS Symposium Series 707*, American Chemical Society.

Rahl, J. M., Anderson, K. M., Brandon, M. T., & Fassoulas, C. (2005). Raman spectroscopic carbonaceous material thermometry of low-grade metamorphic rocks: Calibration and application to tectonic exhumation in Crete, Greece. *Earth and Planetary Science Letters, 240*, 339–354.

Riboulleau, A., Derenne, S., Largeau, C., & Baudin, F. (2001). Origin of contrasting features and preservation pathways in kerogens from the Kashpir oil shales (Upper Jurassic, Russian Platform). *Organic Geochemistry, 32*, 647–665.

Schaefer, J., Kramer, K. J., Garbow, J. R., Jacob, G. S., Stejskal, E. O., Hopkins, T. L., et al. (1987). Aromatic cross-links in insect cuticle: Detection by solid-state ^{13}C and ^{15}N NMR. *Science, 235*, 1200–1204.

Schaeffer, P., Harrison, B. J., Keely, B. J., & Maxwell, J. R. (1995). Product distributions from chemical degradation of kerogens from a marl from a Miocene evaporitic sequence (Vena del Gesso, N. Italy). *Organic Geochemistry, 23*, 541–54.

Schweitzer, M. H., Suo, Z., Avci, R., Asara, J. M., Allen, M. A., Arce, F. T., et al. (2007). Analyses of soft tissue from *Tyrannosaurus rex* suggest the presence of protein. *Science, 316*, 277–280.

Sharpless, K. B., Carlsen, P. H. J., Katsuki, T., & Martin, V. S. (1981). A greatly improved procedure for ruthenium tetroxide catalysed oxidation of organic compounds. *Journal of Organic Chemistry, 46*, 3936–3938.

Stankiewicz, B. A., Briggs, D. E. G., & Evershed, R. P. (1997a). Chemical composition of Paleozoic and Mesozoic fossil invertebrate cuticles as revealed by Pyrolysis-Gas Chromatography/Mass Spectrometry. *Energy and Fuels, 11*, 515–521.

Stankiewicz, B. A., Briggs, D. E. G., Evershed, R. P., & Duncan, I. J. (1997b). Chemical preservation of insect cuticles from the Pleistocene asphalt deposits of California, USA. *Geochimica et Cosmochimica Acta, 61*, 2247–2252.

Stankiewicz, B. A., Briggs, D. E. G., Evershed, R. P., Flannery, M. B., & Wuttke, M. (1997c). Preservation of chitin in 25-million-year-old fossils. *Science, 276*, 1541–1543.

Stankiewicz, B. A., Mösle, B., Finch, P., Collinson, M. E., Scott, A. C., Briggs, D. E. G., et al. (1998). Molecular taphonomy of arthropod and plant cuticles from the Carboniferous of North America: Implications for the origin of kerogen. *Journal of the Geological Society, 155*, 453–462.

Stankiewicz, B. A., Poinar, H. N., Briggs, D. E. G., Evershed, R. P., & Poinar, G. O., Jr. (1998). Chemical preservation of plants and insects in natural resins. *Proceedings of the Royal Society of London B, 265*, 641–647.

Stankiewicz, B. A., Briggs, D. E. G., Michels, R., Collinson, M. E., & Evershed, R. P. (2000). Alternative origin of aliphatic polymer in kerogen. *Geology, 28*, 559–562.

Tegelaar, E. W., de Leeuw, J. W., Derenne, C., & Largeau, C. (1989). A reappraisal of kerogen formation. *Geochimica et Cosmochimica Acta, 53*, 3103–3106.

Tissot, B., & Welte, D. H. (1984). *Petroleum formation and occurrence* (2nd ed.). Berlin: Springer.

Towe, K. M., & Urbanek, A. (1972). Collagen-like structures in Ordovician graptolite periderm. *Nature, 237*, 443–445.

Vandenbroucke, M., & Largeau, C. (2007). Kerogen origin, evolution and structure. *Organic Geochemistry, 38*, 719–833.

Versteegh, G. J. M., Blokker, P., Wood, G. D., Collinson, M. E., Sinninghe Damsté, J. S., & de Leeuw, J. W. (2004). Oxidative polymerization of unsaturated fatty acids as a preservation pathway for dinoflagellate organic matter. *Organic Geochemistry, 35*, 1129–1139.

Chapter 6
Molecular Taphonomy of Plant Organic Skeletons

Margaret E. Collinson

Contents

Abstract Selective preservation of resistant biomacromolecules, such as cutan in leaf cuticles; lignin in woods, fruit walls and seed coats; sporopollenin in spores and pollen and algaenan in algal cysts, has previously been invoked in survival of these tissues and organs in the fossil record. A growing body of evidence is questioning this paradigm, suggesting that biomacromolecules may provide the structural template for formation of geomacromolecules in fossils which form as the result of (i) polymerization of labile constituents (e.g. in situ polymerization of cutin, waxes and internal lipids in cuticles; oxidative polymerization incorporating an aliphatic component into sporopollenin), (ii) loss (e.g. loss of cellulose from lignin–cellulose complexes), and (iii) transformation (e.g. lignin methoxyphenols to phenols). Recommended future research directions

M.E. Collinson (✉)
Department of Earth Sciences, Royal Holloway University of London,
Egham, Surrey TW20 0EX, UK

P.A. Allison and D.J. Bottjer (eds.), *Taphonomy: Process and Bias Through Time*,
Topics in Geobiology 32, DOI 10.1007/978-90-481-8643-3_6,

include: (a) taphonomic experiments to simulate the molecular alteration sequence in diverse environments, (b) analysis of fossils (time series) from a range of depositional settings, and (c) identifying those modern plant organs that lack an expected fossil record. This will require a combination of microscopical and chemical approaches to monitor alteration and understand specific controls on plant preservation.

1 Introduction

1.1 Aims of This Chapter

Plants are normally represented in the fossil record by a variety of isolated organs such as leaves, fruits and seeds, wood, flowers and pollen. Each has a distinctive combination of organic chemical composition and physical structure and both factors influence their preservation. Other important taphonomic factors include the depositional context, the nature of enclosing sediment, the time available for senescence and decomposition prior to burial and the degree of oxidative exposure prior to and after burial. All of these factors may bias the plant fossil record. This chapter addresses preservation of the plant organic skeleton as a morphologically and structurally recognizable fossil (compression, adpression or mummification) in the absence of entombing or supporting minerals (permineralization). For a recent study on chemical characterization of permineralized plant kerogen see Czaja et al. (2009). The incorporation of plant macromolecular components into sedimentary organic matter (kerogen) is outside the scope of this chapter and the reader is referred to Vandenbroucke and Largeau (2007) for a review. However, it should be noted that that paper was accepted for publication prior to the publication of much of the work cited here, especially on leaf preservation (Section 2) and therefore lacks a lot of more recent information relevant to the origin and structure of kerogen.

To an extent all aspects of applied paleobotany make some degree of assumption that the record of plant fossils is either (i) representative of ancient plants and vegetation or (ii) that the biases in the fossil record are adequately understood. With regard to chemical composition the conventional paradigm has been that the fossil record will be biased in favour of those plants whose organs contain resistant biomacromolecules. Recent research has focussed on a combination of molecular and morphological (including ultrastructural) analyses. These have been used, firstly, to further understand the distribution and location of resistant biomacromolecules within plant groups and plant tissues and, secondly, to establish if plant fossils are indeed composed pre-dominantly of surviving resistant biomacromolecules. This chapter will review recent discoveries, suggest future research directions and reassess the role of resistant biomacromolecules in plant preservation through time.

1.2 Caveats and Barriers to Understanding Resistant Bio- and Geomacromolecules

To establish the presence, physical location and chemical composition of a resistant macromolecule it is necessary to be able to isolate (purify) both the molecule and the structure in which it occurs. Any chemical isolation procedure has the potential to alter the chemistry of the target molecules as well as removing other material.

For modern material this is particularly problematic for most cuticles (some can be physically stripped off the surface but most cannot) and for pollen and spore walls that are multilayered and intimately associated with underlying living tissues. Algal cell walls and cyst walls present similar problems and for both these and pollen and spores the small size of the objects makes physical isolation difficult. The situation is somewhat less problematic for the lignified cell walls of woods and thickened sclerotic layers of seed coats and fruit walls where the physical thickness of the tissue enables parts to be removed more easily. These problems and their consequences are discussed in more detail in van Bergen et al. (2004) and de Leeuw et al. (2006).

Mechanical stripping of cuticles (Collinson et al. 1998) and ultrasonic potter homogenization of algal cells to obtain purified wall samples (Blokker et al. 1998; van Bergen et al. 2004; de Leeuw et al. 2006) are effective physical isolation treatments. However, even physically isolated cuticles do not always consist only of the cuticular membrane as will be discussed in Section 2, showing that monitoring of the structure using transmission electron microscopy (TEM) (or other high resolution methods) is essential.

Fossil material could be regarded as another potential source of purified sample on the basis that the resistant materials survive burial and diagenesis whilst other associated more labile molecules do not. The megaspores and microspore massulae clusters of water ferns have been used as one classic example (Van Bergen et al. 1993a), TEM monitoring of the structure was used to demonstrate that no infill of extraneous organic material was present. However, recent work on sporopollenin suggests that the chemical composition of the fossil spore walls has been altered (de Leeuw et al. 2006) as will be discussed in Section 5, showing that fossil material cannot be simply taken to represent the original composition of a resistant macromolecule. Furthermore, in older fossils, characteristic molecular components may be lost preventing the identification of the original (or altered) macromolecule (see, for example, Section 3).

Recent research, which is reviewed below, focuses on a combination of microscopy and chemistry to closely monitor the effects of isolation procedures and chemical treatments. Where appropriate multiple chemical analytical techniques are employed to provide a variety of evidence on molecular structure. Nevertheless much remains poorly understood and some examples of potential future research directions are indicated in the following sections.

2 Leaves and Cuticles

2.1 Leaf and Cuticle Preservation

The external surfaces of leaves (and other aerial plant organs) are covered by a continuous extracellular membrane named the cuticle. The base material of the cuticle is a cross-linked insoluble biopolyester named cutin, the biosynthesis of which is poorly understood (Suh et al. 2005). The cuticle has many functions including as a permeability barrier and a protective layer (Bargel et al. 2004; Riederer and Müller 2006). Leaves are extremely well-represented as fossils throughout the geological record as are herbaceous axes prior to the evolution of leaves. Dispersed cuticles are also widespread in the fossil record, and are amongst the earliest records of probable land-adapted plants in the Ordovician and Silurian (Raven and Edwards 2004).

The protective role of the cuticle in life might be expected to confer an increased preservation potential, by comparison to other leaf tissues such as mesophyll with cellulosic cell walls. However, this would be an assumption as relatively little is known about the anatomical preservation of leaf tissues. Most paleobotanists study leaf gross morphology (e.g. for paleoclimate analysis) or venation patterns and cuticle details (e.g. for taxonomic purposes) and few have examined the internal anatomy or the ultrastructure of fossil leaves. As Fig. 1a–d shows the structural preservation of leaf compression fossils is extremely variable. A 'fossil leaf' may consist only of a cuticle envelope (Fig. 1c) or it may also exhibit excellent preservation of internal cellular details (Fig. 1d). Both these are conifer leaves and are of the same age and from the same site. Apart from these two extremes 'fossil leaves' may totally lack cuticles (Fig. 1a, b). Furthermore, leaves may exhibit varying degrees of internal morphological preservation ranging from retaining some evidence of their original cellular structure (Fig. 1a) to having internal material which shows absolutely no morphological resemblance whatsoever to that of a leaf (Fig. 1b). Figure 1a, b is from the same horizon at the same site and these flowering plant leaves originally would have possessed cuticles, epidermis and palisade and spongy mesophyll like their modern relatives. This ongoing transmission electron microscope work indicates the potential importance of factors such as degree of senescence, timing of burial, local depositional conditions and subsequent diagenesis for each individual leaf.

The fact that fossil leaves survive after cuticles have been lost calls into question the importance of cuticles in controlling leaf preservation. Equally it is known that cuticles are recalcitrant in microbial decay experiments (e.g. Kelleher et al. 2006) and it is abundantly clear that cuticles survive as fossils both with and without other leaf tissues. It has previously been suggested that cuticles survive as a consequence of the selective preservation of a highly aliphatic resistant biomacromolecule named cutan (Nip et al. 1986a, b; Tegelaar et al. 1991) contained within the cuticle (review in Gupta et al. 2006a and refs cited and de Leeuw et al. 2006). If this were correct the presence/absence of cutan or varying abundances of cutan might control cuticle preservation and hence, in some situations, also control leaf preservation, thus biasing the fossil record in favour of plants whose leaves contain cutan as suggested by Tegelaar et al. (1991).

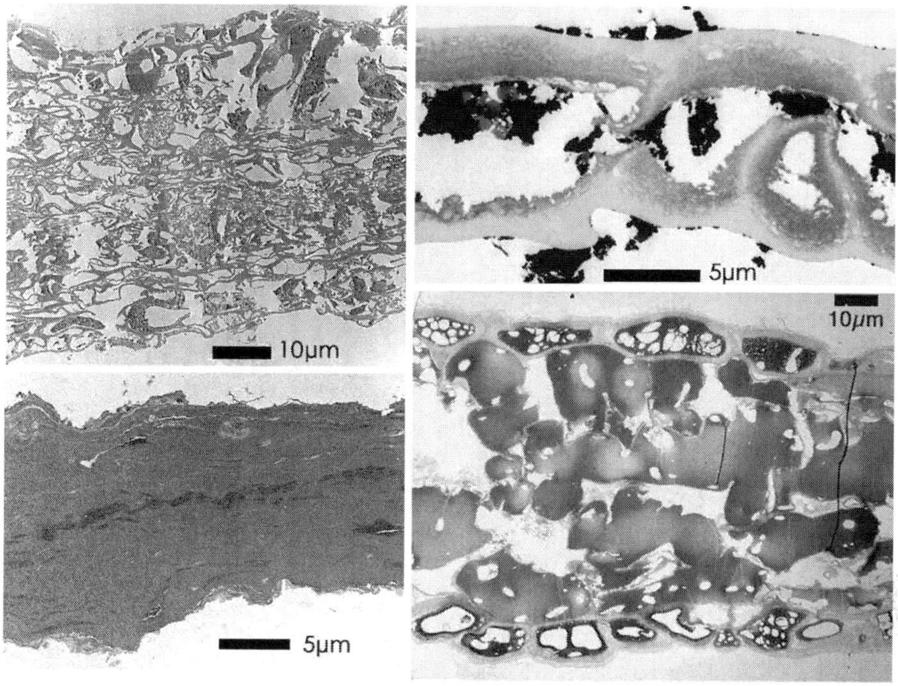

Fig. 1 The range of anatomical preservation in Miocene fossil leaves as shown by vertical sections with transmission electron microscopy (TEM). (**a** and **b**) Ardeche, France **a** (*top left*). *Quercus hispanica* with some evidence of original cellular structure and **b** (*bottom left*) *Castanea vesca* with no recognizable leaf anatomy whatsoever. Neither **a** nor **b** possess cuticles. For further details see Gupta et al. (2007a). (**c** and **d**) Clarkia, USA C (*top right. Amentotaxus*, consisting only of a cuticle envelope and **d** (*bottom right*) *Metasequoia* exhibiting both a cuticle and full anatomical preservation (cell infills probably of diagenetic origin)

It is clear from multiple repeated analyses in different laboratories that the monocotyledon genera *Agave* (Agavaceae) and *Clivia* (Lilliaceae) contain cutan (review in Gupta et al. 2006a and refs there cited). In *Agave*, cutan occupies a considerable proportion of the cuticle and cuticle survives as a recognizable entity after treatment with acetylation and saponification which remove cell walls and cutin respectively (Fig. 2). However, there are no leaves or cuticles of the genera *Agave* and *Clivia* in the fossil record. There are a few other reported occurrences of cutan (summarized in Gupta et al. 2006a) including in one *Podocarpus* species, in *Clusia* and *Prunus laurocerasus* (Boom et al. 2005; Gupta et al. 2006a) and a cutan-like fraction is reported in fruit cuticles of pepper and apple and leaf cuticles of olive on the basis in NMR and FTIR studies (Johnson et al. 2007). Collinson et al. (1998, 2000) studied modern conifer and *Ginkgo* leaves and concluded that they all lacked cutan but this work was criticized by Boom et al. (2005) because chemical treatments used to prepare cuticles might have affected the results. Gupta et al. (2006a) analysed leaves of *Ginkgo* and the conifers *Pinus* and *Metasequoia* without any chemical pretreatments and confirmed the absence of cutan. Gupta et al. (2006a) also analysed leaves from eight families of flowering plants all of which were found to lack cutan whilst *Agave*, as expected, was found to

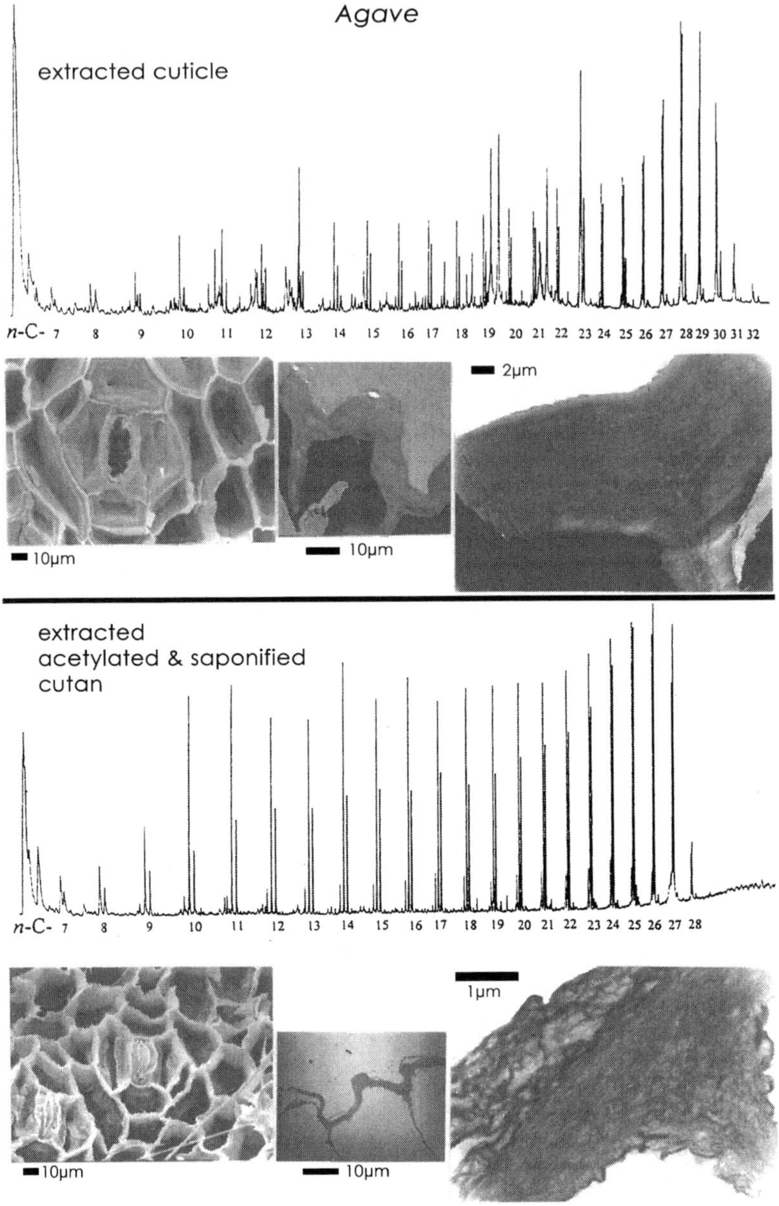

Fig. 2 Chemistry and ultrastructure of *Agave* cuticles showing the importance of the resistant biomacromolecule cutan in this genus. The Py-GC-MS traces of both the extracted cuticle and the extracted, saponified and acetylated cuticle (= cutan only) reveal the characteristic alkene/alkane doublets which dominate in the latter. The EM images show that (i) some cell wall material (electron dense black) remains attached to the extracted cuticle (seen in TEM but not appreciated in SEM) but that this is removed after saponification and acetylation and (ii) that cutan, which remains after a combination of extraction, acetylation and saponification, plays a major structural role in the cuticle and survives as a coherent entity but with much reduced thickness and more open microlamellar organization than the complete cuticle. EM images are from left to right an SEM of the internal surface of a cuticle showing a stoma; a TEM vertical section through the cuticle showing the anticlinal (downward projecting) cuticular flanges which penetrate inwards

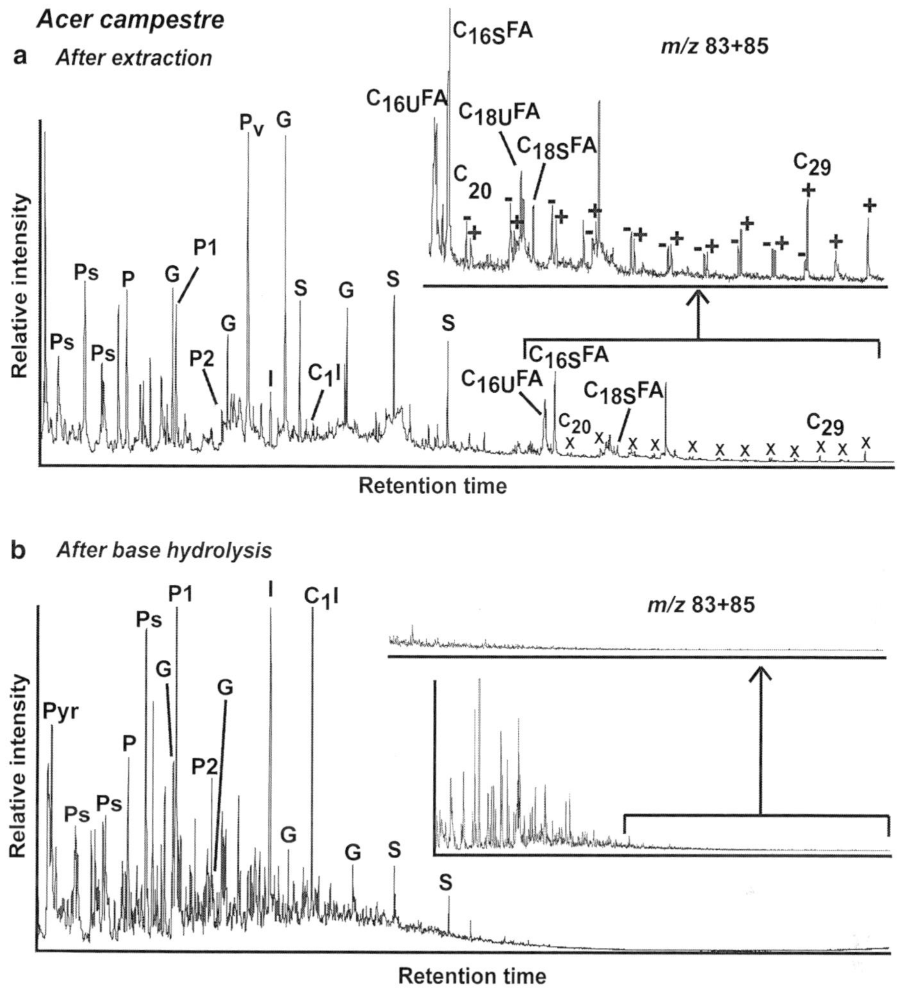

Fig. 3 Absence of cutan in leaves of the flowering plant *Acer* indicated by the absence of alkene/alkane doublets after base hydrolysis (= saponification) which has removed cutin as shown by loss of C16 and C18 fatty acids (C_NFA) (compare to Fig. 4 of *Agave* prepared under exactly the same conditions and without chemical pretreatment and also to Fig. 2). Ps = polysaccharide pyrolysis products, P, S and G indicate presence of lignin derivates as this material has not been subjected to acid hydrolysis (acetylation). For further details of chemical treatments and explanation of chemical annotations see Gupta et al. (2006a). Extracted = chemical extraction with dichloromethane and methanol to remove soluble lipids; Saponified (= base hydrolysis) = chemical extraction with methanolic sodium hydroxide to remove cutin

---

Fig. 2 (continued) into the leaf and into the wall of the epidermal cells; a TEM detail. For further details of the chemistry and treatments of this *Agave* see Mösle et al. (1997). Extracted = chemical extraction with dichloromethane and methanol to remove soluble lipids; Acetylated = chemical extraction with acetyl bromide and acetic acid to remove polysaccharides and lignin; Saponified (= base hydrolysis) = chemical extraction with methanolic potassium hydroxide to remove cutin

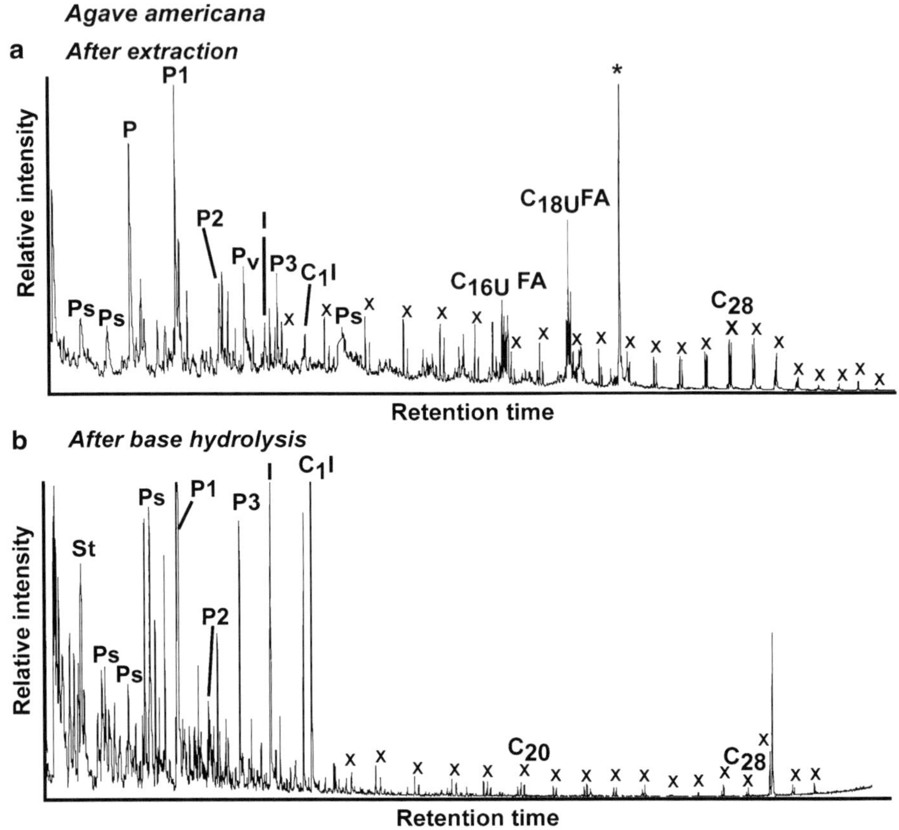

Fig. 4 Presence of cutan in leaves of *Agave* indicated by the presence of alkene/alkane doublets (x) surviving after base hydrolysis which removed cutin. Compare to Fig. 3 prepared under exactly the same conditions and without chemical pre-treatment. Ps = polysaccharide pyrolysis products, P and S indicate presence of lignin derivatives as this material has not been subjected to acid hydrolysis (acetylation). Following base hydrolysis these pyrolysis products dominate the signal but they are removed by acetylation to leave only cutan which yields a chromatogram dominated by alkene/alkane doublets as shown in Fig. 2. For further details of chemical treatments and explanations of chemical annotations see Gupta (2006a)

contain cutan (Compare chemical compositions in Figs. 3, *Acer* and 4, *Agave*). The chemical treatments of acetylation (chemical extraction with acetyl bromide and acetic acid) and saponification (= base hydrolysis, chemical extraction with methanolic potassium or sodium hydroxide) remove cell wall layers and cutin respectively from cuticles of the conifer *Pinus* and the flowering plant *Quercus* (Figs. 5 and 6). If both treatments are applied there is no recoverable residue, in striking contrast to the *Agave* (Fig. 2). Both *Pinus* and *Quercus* (in contrast to *Agave*) have an extensive leaf fossil record both as compression and impression fossils as do many taxa now shown to lack cutan (Gupta et al. 2006a). These data lead to the conclusion that (in spite of its structural role demonstrated in Fig. 2) the presence or absence of cutan does not exert any obvious bias on the preservation of leaves in the fossil record.

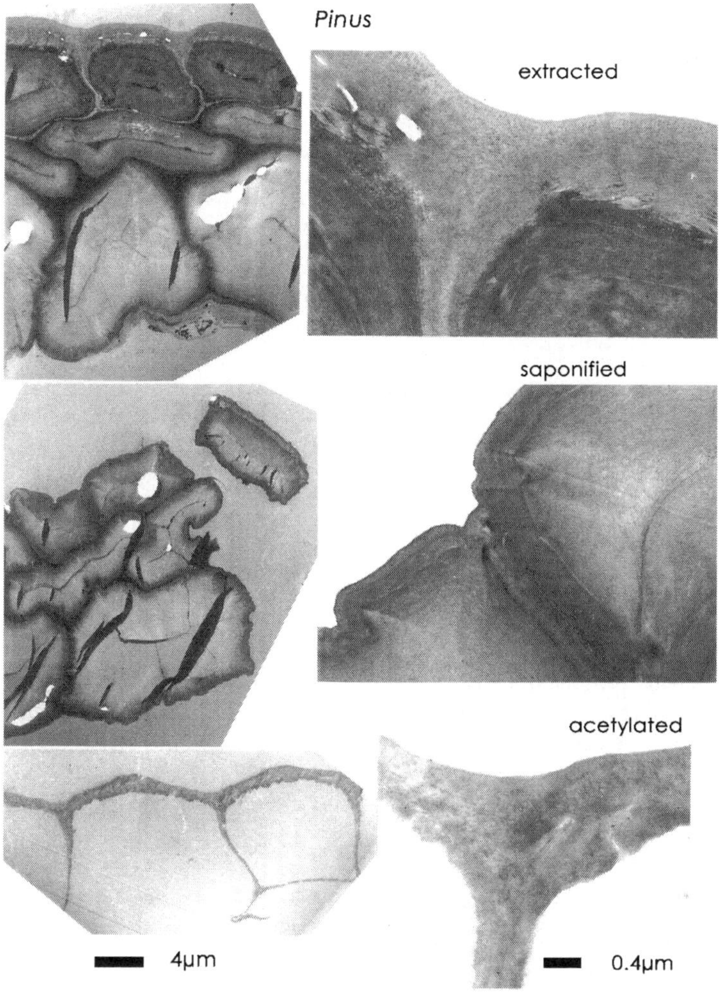

Fig. 5 TEM images of a prepared cuticle of Recent *Pinus*, for comparison with *Agave* in Fig. 2, monitoring the effects of chemical treatments to reveal chemical composition of various layers and an absence of cutan. The prepared and extracted cuticle infact consists not only of the cuticle (paler grey more electron lucent outer layer) but also the cell walls of the epidermal cell and two sub-epidermal cells (*black and pale grey*). Saponification removes the cuticle leaving just cell walls, acetylation removes the cell walls leaving just the cuticle (cutin). A combination of acetylation and saponification yields no recoverable residue, i.e. no cutan, in contrast to *Agave* (Fig. 2). Details of cuticle preparation (removal from the leaf) and chemical treatments are given in Mösle et al. (1997) and Collinson et al. (1998). Extracted = chemical extraction with dichloromethane and methanol to remove soluble lipids; Acetylated = chemical extraction with acetyl bromide and acetic acid to remove polysaccharides and lignin; Saponified (= base hydrolysis) = chemical extraction with methanolic potassium hydroxide to remove cutin

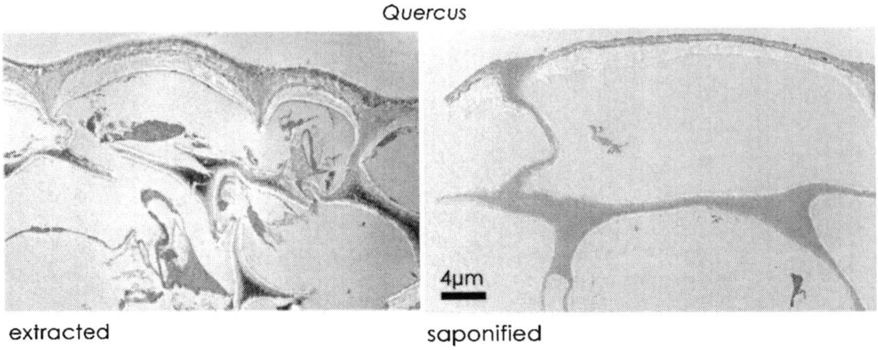

Fig. 6 TEM images of leaf fragments of Recent *Quercus* showing the absence of cuticle (cutin) from the outer surface (top of image) after saponification (base hydrolysis). Chemical analysis of this material is essentially identical to that of *Acer* (Fig. 3), showing an absence of cutan. The material remaining after saponification is essentially lignin moieties as expected from the retention of cell walls seen in the TEM image. See Gupta et al. (2006a) for details.Extracted = chemical extraction with dichloromethane and methanol to remove soluble lipids; Saponified (= base hydrolysis) = chemical extraction with methanolic sodium hydroxide to remove cutin

Combining their studies of modern and fossil leaves Gupta et al. (2006a, 2007a, b, 2009) concluded that, in the absence of cutan, other constituents, including cutin, plant waxes and internal plant lipids contributed to the formation of a highly resistant resistant geomacromolecule in leaf fossils through a process of in situ polymerization. Further support for this hypothesis was derived from experimental maturation which demonstrated that a resistant macromolecule could be generated from plant tissues in the absence of a resistant precursor such as cutan (Gupta et al. 2007c, 2009) and by heat treatment of *Kalanchoe* cuticles where waxy cuticles yielded a chemical signature comparable to that of plant fossils but dewaxed cuticles did not (Finch and Freeman 2001). Yang et al. (2005) found evidence to support the in situ polymerization model in their studies of *Metasequoia* fossils as did Aucour et al. (2009) in their studies of Cretaceous leaf fossils. Polymerization (including oxidative polymerization) of labile components such as lipids has also been recognized as an appropriate mechanism to explain preservation and formation of geopolymers of kerogens, dinoflagellates and animal remains as well as fossil leaves (Kuypers et al. 2002; Versteegh et al. 2004; de Leeuw et al. 2006; De Leeuw 2007; Gupta et al. 2007d). See Stankiewicz et al. (1998), Briggs (1999), Briggs et al. (2000), Gupta et al. (2007d) and Gupta and Briggs (this volume) for further details on animal remains.

2.2 Polymerization and Future Research Directions

In situ polymerization, in particular of labile cell membrane lipids and free fatty acids, could be an important factor in the preservation of any plant materials that contained living cells at the time of death. Even largely dead tissue, such as heartwood, is traversed by rays containing metabolically active parenchyma cells. Therefore, this preservational mechanism may be involved for all plant fossils. De Leeuw et al. (2006) concluded that the aliphatic component of fossil sporopollenin was likely the result of early diagenetic oxidative polymerization of unsaturated lipids whilst Versteegh et al. (2004) suggested a similar process to account for unusually preserved fossil dinoflagellates. Oxidiative exposure is known to increase the aliphatic character of kerogen (Hoefs et al. 1998). This mechanism could have operated since the evolution of cell membrane lipids for which biomarker evidence, in the form of steranes, exists deep in the Proterozoic alongside eukaryotic single cell plant skeletons (Javaux and Marshall 2006). Preservation in the fossil record could thus be strongly influenced by the existence of conditions, including those favouring oxidative cross-linking (as yet not fully understood), that facilitate either in situ polymerization (in the case of cuticles) or incorporation and polymerization of externally sourced labile constituents (or a combination of both). Having recorded the oldest fossil carbohydrates in structurally intact leaves characterised by pyrolysis analysis (polysaccharides in Eocene *Metasequoia* leaves from litter mats) Yang et al. (2005) suggested that the presence, in the leaf litters, of structural polyphenolic compounds as decay products such as tannins might have provided resistance against decomposition. Such possibilities further underscore the importance of the burial context in preservation of the plant organic skeleton.

To further understand controls on preservation one future research direction would be a comparative survey of depositional settings containing impression versus compression fossil leaves (i.e. with and without organic preservation) and a microscale comparison within depositional settings where a range of structural and ultrastructural preservation occurs. One example of the latter is the Miocene of the lacustrine Clarkia site. Here some fossil leaves occur with preservation of chloroplasts containing membrane stacks (Schoenhut et al. 2004; Collinson et al. in progress) and well-preserved cellular anatomy (e.g. Fig. 1d) whilst others consist only of cuticle envelopes (Fig. 1c). Yang et al. (2005) and Gupta et al. (2009) noted very poor preservation of polysaccharides in Clarkia *Metasequoia* compared with *Metasequoia* from other sites. This may be related to poor tissue preservation in specific leaves as the *Metasequoia* from Clarkia shown in their SEM's is stated to have very modified 'amorphous' or 'decayed' internal tissues (in striking contrast to the specimen from the same site illustrated in Fig. 1d herein). Variations in individual leaves may result from different degrees of senescence or from different exposure to oxidative conditions prior to, or after, burial. Combined TEM and chemical studies of fossils are really necessary to fully evaluate molecular preservation.

Carboniferous leaves consisting only of a cuticle envelope (somewhat comparable to the Eocene fossil in Fig. 1c) are well known in concentrated occurrences in what are sometimes termed cuticle coals or paper coals. Based on a variety of chemical and geological evidence Zodrow and Mastalerz (2009) proposed that internal tissues were lost from these fossilized cuticles, probably at a late to post-diagenetic time. This may be the result of oxygen rich groundwater initiating in situ pyritic oxidation and producing sulfuric acid which removed vitrinite (altered internal fossil leaf tissues) resulting in survival of fossil cuticle only.

A second future research direction would be to establish the time frame and conditions over which in situ polymerization occurs. In part this could be accomplished by a series of taphonomic experiments studying chemical and structural alteration of distinctive plant organs and tissues, both fresh and senescent, under varying environmental and depositional conditions (including exposure to oxygen), both pre and post burial. Three studies have investigated both macromolecular chemical composition and morphological changes during decay – Collinson et al. (1998) studied *Ginkgo* cuticles following up to 30 weeks decay, Gupta and Pancost (2004) studied *Arbutus* leaves over 20 days decay and Gupta et al. (2009) studied *Metasequoia* leaves through stages of senescence and from two sedimentary samples inferred to represent 1 year decay. None of these studies found evidence of the formation of a highly aliphatic resistant macromolecule, suggesting that in situ polymerization either had not yet occurred or was insufficient to be recognized by Pyrolysis GC-MS methods. Therefore it would also be appropriate to study a time-series of samples through the Recent, archeological and fossil record from a suite of carefully chosen isotaphonomic contexts. The difficulty with this approach is that the degree of senescence and decomposition prior to burial is likely to be unknown and might be crucial.

A third research direction would be to undertake experiments to artificially simulate chemical alterations. The experimental approach is problematic as it may be impossible to adequately mimic changes that occur over geological time scales. Infact, the purpose of such experiments need not be an attempt to mimic diagenesis but to demonstrate potential sources for highly aliphatic macromolecules from tissues lacking such precursors (Stankiewicz et al. 2000; Finch and Freeman 2001; Gupta et al. 2006b, 2007c, 2009; Gupta and Briggs this volume) and to establish links between physical/visual/structural changes and chemical changes (e.g. in spores – Yule et al. (2000) and cuticles – Stankiewicz et al. (2000)).

Shechter and Chefetz (2008) compared the sorption properties (for hydrophobic organic compounds) of cutin and cutan isolated from tomato (*Lycopersicon*) fruits and *Agave* leaves respectively. This type of study could be extended to include fossil geopolymers such as those in fossil cuticles to provide insights into their chemical interactions in sediments.

3 Xylem (Including Wood), Fruit Walls and Seed Coats

Unequivocal molecular evidence for lignin exists in fossil wood samples from the early Mesozoic onwards (van Bergen et al. 2004). Prior to this time interpretation of the fossil record is complicated by diagenetic transformations with loss of characteristic side chains preventing identification of the lignin biomacromolecule (van Bergen et al. 2004). In Devonian early land plants the presence of lignin is not demonstrable in the earliest vascular thickenings and it is possible that either lignin or other polyphenolic compounds may then have been components of water conducting tissues (Ewbank et al. 1996; Edwards et al. 1997; Boyce et al. 2003; Raven and Edwards 2004; van Bergen et al. 2004). Cell wall lignification may have occurred first in the outer cortex (Boyce et al. 2003) rather than in vascular tissue. Almendros et al. (2005) compared their chemical results for Lower Cretaceous fern leaves to those of Edwards et al. (1997) for Devonian *Psilophyton* and linked the loss of diagnostic lignin signals to an advanced state of maturation of the fossils.

Lignin chemical composition varies in modern plants (Van Bergen et al. 2000b, 2004) and the syringyl units in angiosperm woods are less stable (Fengel 1991) and prone to rapid degradation (Van Bergen et al. 2000b) and preferential removal during decomposition (e.g. Hedges et al. 1985, other references cited in Van Bergen et al. (2000b, p. 71). Other studies also indicate preferential degradation of certain lignins which imply that angiosperm lignin may be less well represented in the fossil record in comparison with gymnosperm lignin (Kim and Singh 2000). In Miocene lignites (brown coals) wood assemblages and collections of large fossil wood specimens are often dominated by coniferous woods. However, the palynological assemblages and small wood fragments and charcoalified wood fragments may indicate a higher proportion of angiosperms in the original plant community (Mosbrugger et al. 1994; Figuerial et al. 1999) thereby suggesting a possible bias against angiosperm woods.

Lignin–cellulose–hemicellulose complexes in sclerotic (= thick-walled, not cuticular) seed coats and fruit walls survive in a modified form in the Cainozoic record typically with loss of polysaccharides and hemicelluloses but retaining lignin markers (Collinson and van Bergen 2004 and refs cited; Stankiewicz et al. 1997). The same is true of woods (Kaelin et al. 2006). In several fruit and seed examples chemosystematically distinctive signatures survive not only in the sclerotesta but also in cuticular layers (van Bergen et al. 1995; Van Bergen et al. 2000a; Collinson and van Bergen 2004; Sawada et al. 2008). In a study of the sclerotesta of the seed coats of *Stratiotes* from the mid Paleogene Hooker et al. (1995) showed that major alteration of the lignin–cellulose–hemicellulose complex can occur (in this case resulting in a very simple polyphenolic macromolecule). Nevertheless, the sclerotesta still remains structurally recognizable and the taxonomic affinity is determinable in the fossil record. This is comparable to the degree of chemical alterations known in leaves (Section 2) where a morphologically recognizable fossil leaf in hand specimen is typically drastically altered chemically. These data suggest little bias to the fossil record of propagules containing lignin–cellulose–hemicellulose complexes of appropriate chemistry. Depositional settings and enclosing lithologies exert some control on chemical preservation of lignin in seed coats (van Bergen et al. 1994; Hooker et al. 1995) and woods (Kaelin et al. 2006). Somewhat counterintuitively preservation of lignin chemistry was better in coarser sediments (sands) than finer sediments such as muds (van Bergen et al. 1993b, 2004) or organic rich 'coaly' layers (Kaelin et al. 2006). Kaelin et al. (2006) suggested that the strong physical compaction in the coaly layers may have been a key driver of geochemical changes as all samples showed very low maturity. These more altered woods also lost some of their structural characteristics due to compaction. In the case of extreme chemical alteration of seed coats some anatomical details maybe lost (Hooker et al. 1995) but nevertheless the taxon remains identifiable and the entity survives as a fossil even when drastic chemical alteration has occurred.

Chemical alterations to woods during fossilizsation have profound implications for applied paleobotany. Poole and van Bergen (2006) emphasise the need for a thorough understanding of the alterations to wood chemistry prior to the application of wood fossils as proxies for paleoclimates. Poole et al. (2006) demonstrated how carbon isotope signatures of bulk wood are influenced by the chemical preservation of the wood and concluded that knowledge of both chemical preservation and taxonomic identity are essential prerequisites for recognition and interpretation of real isotopic shifts which can then be applied in paleoenvironmental analysis. These arguments are strongly underscored by evidence that oxidative degradation alters carbon isotopic composition of kerogens through removal of carbohydrates (Hoefs et al. 1998) and that recalcitrant carbon resulting from microbial decay experiments may be similar despite different plant sources (Kelleher et al. 2006). Richter et al. (2008) drew attention to the risks in assuming that oxygen isotope composition of fossil wood cellulose had not changed during burial. They used XRD to control against mineral contamination and to demonstrate that the crystalline form of fossil cellulose was the same as modern cellulose in some cases. Notably they found that samples from the same site were variable in their preservation and contamination. Lechien et al. (2006) showed that in some morphologically well-preserved Miocene gymnosperm woods cellulose was structurally preserved but was not bioavailable (not degradable e.g. by cellulases) and that lignin had undergone only slight chemical alteration. They suggested that the lignin transformations could have maintained the structural integrity of the wood and led to cellulose preservation through reduced bioavailability. It is important to emphasize that quality of structural preservation need not equate to quality of chemical preservation as was also shown by the absence of assured derivatives of lignin or cellulose in the kerogen of exceptionally structurally preserved permineralized plants (Czaja et al. 2009).

In marked contrast to lignins there is strong evidence suggesting that thick-walled tissues lacking the lignin–cellulose–hemicellulose complex may be biased against in the fossil record. The propagule wall of *Nelumbo* (lotus) is composed of a tannin–polysaccharide complex (Van Bergen et al. 1997b). Although *Nelumbo*-like leaves have a long fossil record and the propagule is extremely physically resistant with high longevity (Shen-Miller et al. 1995) there is no pre-Holocene morphologically recognisable fossil record of the propagules (Van Bergen et al. 1997b). Other propagules are known with distinctive chemistry. Lignin is lacking in peas and wheat grains (Braadbaart et al. 2007) and a polyphenolic macromolecule, along with protein derivatives, was reported in modern and archeological (c. AD 600) seed coats of *Raphanus* (raddish) (Van Bergen et al. 1997a).

The above examples demonstrate that biomacromolecules other than lignin play a role in structural plant tissues. If, as the *Nelumbo* example suggests, some or all of these other biomacromolecules are missing from the fossil record of organic plant skeletons this clearly creates a bias, but the extent of this bias is essentially unknown. Fruits and seeds that lack a sclerotic layer containing a lignin–cellulose–hemicellulose complex and also lack a cuticular layer (see Section 2 for preservation of cuticles), may be strongly biased against in the fossil record (Van Bergen et al. 2000a; Collinson and van Bergen 2004). Obviously the absence of a fossil record is hard to recognize. Future research can be directed in two ways. Firstly further examples like that of *Nelumbo* can be investigated where fossils exist of one organ of a taxon but other organs are lacking. An example may be the fossil record of legumes where the pods and leaves are characteristic widespread fossils (Herendeen and Dilcher 1992) but there appear to be no examples of organically preserved morphologically recognisable legume seed coats as pre-archeological organic fossils inspite of an extensive fossil record of seed assemblages (e.g. Collinson and van Bergen 2004). Secondly, well-calibrated molecular phylogenies may be used to target future chemical studies on the structural entities of taxa with a predicted early origin but no known morphologically recognizable fossil record.

4 Flowers

Flowers consists of a complex mixture of tissues and organs many of which (esp petals and stamens) are typically short-lived and readily shed or detached from the flower. Preservation of the entire flower as a compression fossil will be strongly influenced by conditions which would cause physical break up rather than by chemical composition of the individual organs. Much of the important information on fossils flowers comes from those that have been charcoalified as a result of wildfires (Friis et al. 2006). Discussion of this distinctive preservational pathway is outside the scope of this paper. Compressions fossils of flowers or floral organs are relatively rare and when present are often dominantly impression fossils with little or no organic skeleton remaining, even in settings of exceptional preservation. An exception is the large numbers of flower compression fossils present in the exceptionally preserved biota from the Eocene oil shales of Messel (Schaal and Ziegler 1992; Wilde 2004), an extremely tranquil depositional setting with a very high organic content. The external surfaces of floral organs (e.g. sepals, petals, carpels) are covered by cuticles (Riederer and Müller 2006) and cuticle preservation is discussed in Section 2. Pollen preservation is discussed in Section 5 and fruit walls or seed coats, which will begin to develop in flowers after fertilization, are discussed in Section 3.

5 Spores and Pollen

The outer walls (exines) of pollen and spores of vascular plants contain a highly resistant macromolecule named sporopollenin. For a review and history of study see van Bergen et al. (1993a), Van Bergen et al. (2004) and de Leeuw and Largeau (1993). This is one of the most resistant organic molecules (argueably *the* most resistant) and has been considered to be responsible for the extensive record of land plant pollen and spores in sediments since at least the Ordovician (van Bergen et al. 2004) and which probably extends back into the Cambrian (Raven and Edwards

2004). Spores from at least the latest Ordovician onwards retain a characteristic ultrastructure suggesting comparable development to their modern analogues (Wellman 2004).

At least two chemically different types of sporopollenin (one with aromatic building blocks and the other aliphatic) have been thought to exist, possibly in different plant groups or mixed within the same entity (Reviews in van Bergen et al. 2004; de Leeuw et al. 2006). Most fossil spore and pollen walls analysed consist of both aromatic and aliphatic constituents (de Leeuw et al. 2006). Having taken account of recent analytical results and previous literature de Leeuw et al. (2006) concluded that the aliphatic component in fossils is not part of the original sporopollenin structure but is due to early-diagenetic oxidative polymerization of unsaturated lipids. This hypothesis requires reconsideration of the previous assumption that the presence of sporopollenin is responsible for the preservation of pollen and spore exines in the fossil record. In cuticles (Section 2) it can now be argued that cutan is of little (perhaps no) importance in cuticle preservation which is instead mainly controlled by the in situ polymerization of cutin, plant waxes and internal plant lipids contributing to the formation of a resistant geomacromolecule. Equally it is at least possible that the biomacromolecule sporopollenin plays little role in pollen and spore preservation but that polymerization of lipids and their incorporation into a geomacromolecule (which should be termed fossil sporopollenin, or given a new name) is the controlling factor. The role of sporopollenin in the pollen and spore wall (as also perhaps the role of cutin in the cuticular membrane) could be more as a physical entity providing a structural template for polymerizations.

6 Phytoplankton and Algal Cysts

6.1 Chlorophyta and Prasinophyta

A detailed recent review (de Leeuw et al. 2006) has been published on the resistant macromolecules of algae with an extensive literature survey of their occurrence in fossil and modern examples building on that by Versteegh and Blokker (2004). Relevant references to previous work can be found in de Leeuw et al. (2006) on which the text to follow is based. Only a few Recent algae biosynthesize a resistant cell wall or resting cyst wall containing the highly aliphatic resistant biomacromolecule algaenan – these include Chlorophyta, Eustigmatophyta, Prasinophyta and one member of the Dinophyta. Detailed studies of algaenans (Tegelaar et al. 1989) have been undertaken by Blokker et al. (2000) and by Metzger et al. (2007, 2008), the latter specifically on *Botryococcus*. Algaenan (or an algaenan-like molecule) has been identified in a number of fossil algal cell walls mostly those of the Chlorophyta including *Botryococcus* (a widespread fossil from Permian onwards, well known for hydrocarbon generation potential, Batten and Grenfell 1996), *Tetraedron* (the alga which played a large part in the exceptional preservation of biota in the Eocene Messel oil shales, Schaal and Ziegler 1992) and *Pediastrum* (an important algal indicator taxon for freshwater conditions from the Cretaceous onwards, Batten 1996). The fossil prasinophyte *Tasmanites* also has a similar aliphatic polymer (but with possible additional components). Algaenan in resting cyst walls occurs in modern Chlorophyta such as *Chlamydomonas*, *Spirogyra* and *Dunaniella*. Resting cysts of a variety of Zygnemataceae (Recent analogues *Spirogyra*, *Mougetia*, *Debarya* and *Zygnema*) occur in the fossil record from the Carboniferous onwards (van Geel and Grenfell 1996) but I am not aware of any studies of their chemical composition. Not all algae whose living members produce algaenan have yet been identified in the fossil record but their absence may be due to inadequate morphological characteristics preventing their recognition. Many Recent algae do not produce algaenan, but (with the exception of Dinophyta – see below) it is difficult to name an example where the organic skeleton is strongly diagnostic such that it should be recognizable in the fossil record and so could be demonstrated to be absent. Such cases would be a prerequisite to help to justify an hypothesis that the presence/absence of algaenan exerted a major bias on the algal fossil record. Kodner et al. (2009) showed that algaenan is not widespread ecologically or phylogenetically and argued that it was therefore unlikely to be responsible for a sizeable proportion of refractory organic matter is sediments. As demonstrated for cuticles (Section 2) and as suggested as a possibility for pollen and spores (Section 3) the highly resistant macromolecule (in this case algaenan) may be less important in fossil preservation then previously thought.

6.2 *Dinoflagellates*

A number of modern Dinophyta have been shown to produce acetolysis resistant cell walls and resting cysts (de Leeuw et al. 2006 for a review). The resistant macromolecule in *Lingulodinium* is predominantly aromatic and is often referred to as dinosporin which differs from both algaenan and sporopollenin (de Leeuw et al. 2006). In contrast the cyst walls of *Scrippsiella* also have a substantial aliphatic component (de Leeuw et al. 2006). Most fossil dinoflagellates analysed have both an aromatic and aliphatic component (de Leeuw et al. 2006). A number of modern Dinophyta are also known to lack resistant chemistry in their cell walls (de Leeuw et al. 2006).

It is well known that there are some apparent gaps in the fossil record of dinoflagellates, both the group as a whole and for individual families and genera showing the lazarus taxon phenomenon. One possible explanation is that these gaps are due to phases when resistant cysts were not produced (e.g. Fensome et al. 1996). De Leeuw et al. (2006) suggested, on the basis of chemical analyses of a few modern and fossil dinoflagellates and the differential survival of fossil cyst walls under different extraction procedures, that dinosporin may be variable in its chemical composition. If this hypothesis is supported by further analyses then both the chemical nature and the presence/absence of dinosporin could be exerting a bias on the dinoflagellate fossil record.

Versteegh et al. (2004) argued that oxidative polymerization of lipids was responsible for the preservation of one example of fossil dinoflagellates that lacked a cell wall but were preserved through an infill acting as a 'cast' of the original morphology. This work demonstrates a similar process of polymerization to that discussed above for spores and pollen (Section 5) and cuticles (Section 2) albeit not in situ in the dinoflagellate cell wall or cyst wall. Because the cyst walls of the modern dinoflagellate *Scrippsiella* contain both aromatic and aliphatic components it is possible that the aliphatic/aromatic compositions of fossils cysts are little altered from their modern counterparts. However, the equivalent from the modern genus has not been studied for any of the fossil resting cyst walls listed in the review by de Leeuw et al. (2006). Until this direct comparison is undertaken the relative role in preservation of fossils for (i) the aromatic biomacromolecule dinosporin, or (ii) an original aliphatic component becoming associated with dinosporin through polymerization during diagenesis or (iii) some form of in situ polymerization (see Sections 2 and 5 above) remains unknown.

6.3 Acritarchs

Acritarchs are organic-walled microfossils of unknown biological affinities. Some of these may have been produced by Dinoflagellates (or their close relatives) and some by various algal groups or other groups of life. Acritarch fossils are amongst the oldest organic plant skeletons recovered from the fossil record extending back into the Mesoproterozoic. Recent research has used a combination of transmission electron microscopy and multi-method micro-chemical analysis in an attempt to further elucidate the biological affinities of some very early acritarchs (Javaux and Marshall 2006 for a review). A highly aliphatic signal, similar to that of Recent algaenan, was documented in the wall of the Neoproterozoic *Tanarium*; a potentially new class of biopolymer, with aliphatic, branched aliphatic and saturated/olefinic carbon constituents, in the Neoproterozoic *Leiosphaeridia*; and an aromatic polymer in a Mesoproterozoic *Shuiyousphaeridium* (Marshall et al. 2005; Javaux and Marshall 2006). These authors also noted the occurrence of various different multilayered wall ultrastructures in acritarchs with similar chemical composition. Issues such as those discussed in Sections 2 and 5, in particular the potential for the diagenetic alteration of biopolymers, mean that it cannot be assumed that chemical compositions of these very ancient organic skeletons reflect those of the original material. In situ polymerization or incorporation of external material may well have altered the chemistry of the fossils. As Javaux and Marshall (2006) state a combination of microscopy and chemistry will be important for further characterization of these early eukaryotic fossils.

7 Conclusions and Implications

7.1 *Plant Evolutionary Constraints and Temporal Bias*

The timing of evolution of resistant macromolecules such as lignin and sporopollenin and of the evolution of certain tissues and organs (such as xylem, cuticle, wood and seed coats) is important, but once these had evolved there is no obvious time-related bias. Some obvious bias is taxon-related with the presence/absence of a particular chemistry in a given organ/tissue of a given taxon being an important control on representation in the fossil record. It is predicted that depositional and burial conditions will prove to be extremely important (and possibly the key factor controlling preservation in at least some components of the plant organic skeleton) as these are expected to exert a strong influence on the process(es) of in situ (within organ/tissue) polymerization and oxidative polymerization and have already been shown to influence quality of chemical preservation.

7.2 *Implications for Applied Paleobotany*

7.2.1 Floras and Vegetation Reconstruction, Dating First Occurrences Etc.

It has been shown that preservational biases linked to chemical composition may affect the fossil record of certain organs or tissues of a given taxon. This bias can be overcome by using the fossil record of multiple organs wherever possible. For example, a combination of evidence from pollen and spores, leaves, fruits and seeds and woods can be used to produce a vegetational reconstruction as all are common fossils. There is no evidence that any one organ is less affected by biases than others, the bias affects particular organs of selected taxa. Thus there is no evidence to support preferential selection of particular organ types for any particular application.

7.2.2 Geochemical Applications

Although plant organs/tissues survive as fossils their chemistry may be drastically altered. These alterations vary according to the macromolecular composition of the parent material. The three main categories of alteration involve (i) polymerization (e.g. in situ polymerization of cutin, plant waxes and internal lipids into a geomacromolecule in the cuticle), (ii) loss (e.g. loss of cellulose from lignin–cellulose complexes), and (iii) transformation (e.g. lignin methoxyphenols to phenols). For cuticles in situ polymerization involves little or no migration of material from external sources whilst oxidative polymerization affecting spore and pollen walls may involve incorporation of an external aliphatic component. When geomacromolecules or altered biomacromolecules are being applied as paleoenvironmental proxies it would seem prudent to (i) establish the preservational state of relevant molecules, (ii) use samples of like preservational state for any comparative study, and (iii) monitor the degree of alteration and the contribution of external materials (which may render the proxy invalid) both by microscopy and by chemical analysis. In fossil material, and in modern material which has been physically or chemically isolated for study, it is necessary to establish the presence of wall layers, cells, tissues and organs by microscopy prior to making inferences about the sources of chemical signals. Non-destructive approaches such as Raman microspectroscopy and Scanning Transmission x-ray Microscopy (Bernard et al. 2009; Czaja et al. 2009) have considerable potential to combine microscopical and chemical analyses.

7.2.3 Ultrastructure, Taxonomic Characteristics and Chemotaxonomy

It is to be expected that chemical alteration may result in loss of taxon diagnostic morphology and ultrastructure following polymerization, loss or transformation alterations. In contrast, it is known that taxon diagnostic morphology and ultrastructure may survive even when chemistry is drastically altered. Although chemistry is altered chemical signals of chemosystematic value do sometimes survive and these also provide evidence of the past occurrences of distinctive biosynthetic pathways. Current evidence reveals no obvious overriding factors controlling the survival or loss of taxonomically valuable characters and much more work is needed to establish the circumstances and processes which affects these signals. Further understanding of these factors would offer potential to target particular fossil assemblages for future systematic and evolutionary study.

Exceptional preservation of organelle ultrastructure is known, for example, in some leaf fossils (Section 2). This is variable from specimen to specimen and taxon to taxon within the same sedimentary horizon. In the same horizon some fossil leaves exhibit detailed cellular anatomy and organelle preservation whilst others are composed only of cuticle envelopes. In some other deposits leaves are preserved but without cuticles (Section 2). These facts underscore just how little is understood about the specific controls on plant preservation, warn against wide inferences based on small sample numbers and demonstrate the need for combined microscopical and chemical approaches in studies of plant molecular taphonomy.

Acknowledgments Special thanks are due to Jan de Leeuw and Pim van Bergen for their long term interest, enthusiasm, support, and friendship in our collaborative studies of the organic geochemistry of plant fossils. I would like to thank Ben van Aarsen, Pim van Bergen, Peter Blokker, Tony Brain, Derek Briggs, Richard Evershed, Paul Finch, Neal Gupta, Jan de Leeuw, Raymond Michels, Barbara Mösle, Rich Pancost, Andrew Scott and Gerard Versteegh for their previous, and in many cases ongoing, collaboration; to the chemists amongst them also my thanks for their patience with my inadequate knowledge of chemistry. Any errors in the present work are entirely those of the author. Funding from a Royal Society 1983 University Research Fellowship, the NERC Biomolecular Palaeontology Special Topic, the NERC Ancient Biomolecules Initiative (grants GST/02/1030 and 1390) and from the Petroleum Research Fund, American Chemical Society (the latter to collaborators Pancost, Briggs and Michels) is gratefully acknowledged.

References

Almendros, G., Zancada, M. C., Gonzakez-Vila, F. J., Lesiak, M. A., & Alvarez-Ramis, C. (2005). Molecular features of fossil organic matter in remains of the Lower Cretaceous fern *Weichselia reticulata* from Przenosza basement (Poland). *Organic Geochemistry, 36*, 1108–1115.

Aucour, A.-M., Faure, P., Gomez, B., Hauteville, J., Michels, R., & Thenenard, F. (2009). Insights into preservation of fossil plant cuticles using thermally assisted hydrolysis methylation. *Organic Geochemistry, 40*, 784–794.

Bargel, H., Barthlott, W., Koch, K., Schreiber, L., & Nienhuis, C. (2004). Plant cuticles: Multifunctional interfaces between plant and environment. In A. R. Hemsley & I. Poole (Eds.), *The evolution of plant physiology, Linnean society symposium series no. 21* (pp. 171–194). London: Elsevier.

Batten, D. J. (1996). Green and blue-green algae, 7C colonial chlorococcales. In J. Jansonius & D. C. McGregor (Eds.), *Palynology: principles and applications. American Association of Stragigraphic Palynologists Foundation* (Vol. 1, pp. 191–203).

Batten, D. J. & Grenfell, H. R. (1996). Green and blue-green algae, 7D *Botryococcus*. In J. Jansonius & D. C. McGregor (Eds.), *Palynology: principles and applications. American Association of stragigraphic palynologists foundation* (Vol. 1, pp. 205–214).

Bernard, S., Benzerara, K., Beyssac, O., Brown, G. E., Grauvogel Stamm, L., & Duringer, P. (2009). Ultrastructural and chemical study of modern and fossil sporoderms by scanning transmission X-ray Microscopy (STXM). *Review of Palaeobotany and Palynology, 156*, 248–261.

Blokker, P., Schouten, S., van den Ende, H., de Leeuw, J. W., Hatcher, P. G., & Sinninghe Damsté, J. S. (1998). Chemical structure of algaenans from the fresh water algae *Tetraedron minimum, Scenedesmus communis* and *Pediastrum boryanum*. *Organic Geochemistry, 29*, 1453–1468.

Blokker, P., Schouten, S., de Leeuw, J. W., Sinninghe Damsté, J. S., & van den Ende, H. (2000). A comparative study of fossil and extant algaenans using ruthenium tetroxide degradation. *Geochimica et Cosmochimica Acta, 64*, 2055–2065.

Boom, A., Sinninghe Damsté, J. S., & de Leeuw, J. W. (2005). Cutan, a common aliphatic biopolymer in cuticles of drought-adapted plants. *Organic Geochemistry, 36*, 595–601.

Boyce, C. K., Cody, G. D., Fogel, M. L., Hazen, R. M., Alexander, C. M. O'. D., & Knoll, A. H. (2003). Chemical evidence for cell wall lignification and the evolution of tracheids in Early Devonian plants. *International Journal of Plant Sciences, 164*, 691–702.

Braadbaart, F., Wright, P. J., van der Horst, J., & Boon, J. J. (2007). A laboratory simulation of the carbonization of sunflower achenes and seeds. *Journal of Analytical and Applied Pyrolysis, 78*, 316–327.

Briggs, D. E. G. (1999). Molecular taphonomy of animal and plant cuticles; selective preservation and diagenesis. *Philosophical Transactions of the Royal Society, London, B, 354*, 7–17.

Briggs, D. E. G., Evershed, R. P., & Lockheart, M. J. (2000). The biomolecular paleontology of continental fossils. In D. H. Erwin, & S. L. Wing (Eds.), Deep time: paleobiology's perspective. *Paleobiology, 26* (Suppl. 4), 169–193.

Collinson, M. E., & van Bergen, P. F. (2004). Evolution of angiosperm fruit and seed dispersal biology and ecophysiology: Morphological, anatomical and chemical evidence from fossils. In A. R. Hemsley & I. Poole (Eds.), *The evolution of plant physiology, Linnean society symposium series no. 21* (pp. 343–377). London: Elsevier.

Collinson, M. E., Finch, P. F., Mösle, B., Wilson, R., & Scott, A. C. (1998). The preservation of plant cuticle in the fossil record: A chemical and microscopical investigation. *Ancient Biomolecules, 2*, 251–265.

Collinson, M. E., Finch, P. F., Mösle, B., Wilson, R., & Scott, A. C. (2000). Preservation of plant cuticles. *Acta Palaeobotanica, 1999* (Suppl. 2), 629–632.

Czaja, A. D., Kudryavtsev, A. B., Cody, G. D., & Schopf, J. W. (2009). Characterisation of permineralised kerogen from an Eocene fossil fern. *Organic Geochemistry, 40*, 353–364.

De Leeuw, J. W. (2007). On the origin of sedimentary aliphatic macromolecules, a comment. *Organic Geochemistry, 38*, 1585–1587.

De Leeuw, J. W., & Largeau, C. (1993). A review of macromolecular compounds that comprise living organisms and their role in kerogen, coal and petroleum formation. In M. H. Engel & S. A. Macko (Eds.), *Organic geochemistry. Principles and applications* (pp. 23–72). New York: Plenus.

De Leeuw, J. W., Versteegh, G. J. M., & van Bergen, P. F. (2006). Biomacromolecules of plants and algae and their fossil analogues. *Plant Ecology, 189*, 209–233.

Edwards, D., Ewbank, G., & Abbott, G. D. (1997). Flash pyrolysis of the outer cortical tissues in Lower Devonian Psilophyton. *Botanical Journal of the Linnean Society, 124*, 345–360.

Ewbank, G., Edwards, D., & Abbott, G. D. (1996). Chemical characterization of Lower Devonian vascular plants. *Organic Geochemistry, 25*, 461–473.

Fengel, D. (1991). Aging and fossilisation of wood and its components. *Wood Science and Technology, 25*, 153–177.

Fensome, R. A., Riding, J. B., Taylor, F. J. R. (1996). Dinoflagellates. In J. Jansonius & D. C. McGregor (Eds.), *Palynology: principles and applications. American Association of Stragigraphic Palynologists Foundation* (Vol. 1, pp. 107–169).

Figuerial, I., Mosbrugger, V., Rowe, N. P., Ashraf, A. R., Utescher, T., & Jones, T. P. (1999). The Miocene peat-forming vegetation of northwestern Germany: An analysis of wood remains and comparison with previous palynological interpretations. *Review of Palaeobotany and Palynology, 104*, 239–266.

Finch, P., & Freeman, G. (2001). Simulated diagenesis of plant cuticles – implications for organic fossilisation. *Journal of Analytical and Applied Pyrolysis, 58*, 229–235.

Friis, E. M., Pedersen, K. R., & Crane, P. R. (2006). Cretaceous angiosperm flowers: Innovation and evolution in plant reproduction. *Palaeo, 3*(232), 251–293.

Gupta, N. S., & Briggs, D. E. G. (this volume). Taphonomy of organic animal skeletons through time.

Gupta, N. S., & Pancost, R. D. (2004). Biomolecular and physical taphonomy of angiosperm leaf in early decay: Implications for fossilisation. *Palaios, 19*, 428–440.

Gupta, N. S., Collinson, M. E., Briggs, D. E. G., Evershed, R. P., & Pancost, R. D. (2006). Reinvestigation of the occurrence of cutan in plants: Implications for the leaf fossil record. *Paleobiology, 32*, 432–449.

Gupta, N. S., Michels, R., Briggs, D. E. G., Evershed, R. P., & Pancost, R. D. (2006). The organic preservation of fossil arthropods: An experimental study. *Proceedings of the Royal Society B.* doi:10.1098 rspb 2006.3646.

Gupta, N. S., Briggs, D. E. G., Collinson, M. E., Evershed, R. P., Michels, R., Jack, S. K., et al. (2007a). Evidence for the *in situ* polymerisation of labile aliphatic organic compounds during the preservation of fossil leaves: Implications for organic matter preservation. *Organic Geochemistry, 38*, 499–522.

Gupta, N. S., Briggs, D. E. G., Collinson, M. E., Evershed, R. P., Michels, R., & Pancost, R. D. (2007b). Molecular preservation of plant and insect cuticles from the Oligocene Enspel Formation, Germany: Evidence against derivation of aliphatic polymer from sediment. *Organic Geochemistry, 38*, 404–418.

Gupta, N. S., Michels, R., Briggs, D. E. G., Collinson, M. E., Evershed, R. P., & Pancost, R. D. (2007c). Experimental evidence for the formation of geomacromolecules from plant leaf lipids. *Organic Geochemistry, 38*, 28–36.

Gupta, N. S., Briggs, D. E. G., Collinson, M. E., Evershed, R. P., Michels, R., & Pancost, R. D. (2007d). De Leeuw comment "on the origin of sedimentary aliphatic macromolecules – reply. *Organic Geochemistry, 38*, 1588–1591.

Gupta, N. S., Yangg, H., Leng, Q., Briggs, D. E. G., Cody, G. D., & Summons, R. E. (2009). Diagenesis of plant biopolymers: Decay and macromolecular preservation of *Metasequoia*. *Organic Geochemistry, 40*, 802–809.

Hedges, J. I., Cowie, G. K., Ertel, J. R., Barbour, R. J., & Hatcher, P. G. (1985). Degradation of carbohydrates and lignin in buried woods. *Geochemica et Cosmochimica Acta, 49*, 701–711.

Herendeen, P. S., & Dilcher, D. L. (1992). *Advances in legume systematics, part 4. The fossil record.* Kew: Royal Botanic Gardens. 326 pp.

Hoefs, M. J. L., Sinninghe Damste, J. S., De Lange, G. J. and de Leeuw, J. W. (1998). Changes in kerogen composition across an oxidation front in Madeira abyssal plain turbidites as revealed by pyrolysis GC-MS. *Proceedings of the Ocean Drilling Program, Scientific results, 157* (pp. 591–607).

Hooker, J. J., Collinson, M. E., van Bergen, P. F., Singer, R. L., de Leeuw, J. W., & Jones, T. P. (1995). Reconstruction of land and freshwater palaeoenvironments near the Eocene-Oligocene boundary, southern England. *Journal of the Geological Society, London, 152*, 449–468.

Javaux, E. J., & Marshall, C. P. (2006). A new approach in deciphering early protist paleobiology and evolution: Combined microscopy and microchemistry of single Proterozoic acritarchs. *Review of Palaeobotany and Palynology, 139*, 1–15.

Johnson, E. J., Dorot, O., Liu, J., Cherfetz, B., & Xing, B. (2007). Spectroscopic characterisation of aliphatic moieties in four plant cuticles. *Communications in Soils Science and Plant Analysis, 38*, 2461–2478.

Kaelin, P. E., Huggett, W. W., & Anderson, K. B. (2006). Comparison of vitrified and unvitrified Eocene woody tissues by TMAH thermochemolysis – Implications for the early stages of the formation of vitrinite. *Geochemical transactions, 2006*, 7–9.

Kelleher, B. P., Simpson, M. J., & Simpson, A. J. (2006). Assessing the fate and transformation of plant residues in the terrestrial environment using HR-MAS NMR spectroscopy. *Geochimica et Cosmochimica Acta, 70*, 4080–4094.

Kim, Y. S., & Singh, A. P. (2000). Micromorphological characteristics of wood biodegradation in wet environments: A review. *IAWA Journal, 21*, 135–155.

Kodner, R. B., Summons, R. E., & Knoll, A. H. (2009). Phylogenetic investigation of the aliphatic, non-hydrolyzable biopolymer algaenan, with a focus on green algae. *Organic Geochemistry, 40*, 854–862.

Kuypers, M. M. M., Blokker, P., Hopmans, E. C., Kinkel, H., Pancost, R. D., Schouten, S., et al. (2002). Archaeal remains dominate marine organic matter from the early Albian oceanic anoxic event 1b. *Palaeogeography, Palaeoclimatology, Palaeoecology, 185*, 211–234.

Lechien, V., Rodriguez, C., Ongena, M., Hiligsmann, S., Rulmont, A., & Thonart, P. (2006). Physicochemical and biochemical characterization of non-biodegradable cellulose in Miocene gymnosperm wood from Entre-Sambre-et-Meuse, southern Belgium. *Organic Geochemistry, 37*, 1465–1476.

Marshall, C. P., Javaux, E. J., Knoll, A. H., & Walter, M. R. (2005). Combined micro-Fourier transform infrared (FTIR) spectroscopy and micro-Raman spectroscopy of Proterozoic acritarchs: A new approach to Palaeobiology. *Precambrian Research, 138*, 208–224.

Metzger, P., Rager, M., & Largeau, C. (2007). Polyacetals based on polymethylsqualene diols, precursors of algaenan in *Botryococcus braunii* race B. *Organic Geochemistry, 38*, 566–581.

Metzger, P., Rager, M., & Fosse, C. (2008). Braunicetals: Acetals from condensation of macrocyclic aldehydes and terpene diols in *Botryococcus braunii*. *Phytochemistry, 69*, 2380–2386.

Mosbrugger, V., Gee, C. T., Belz, G., & Ashraf, A. R. (1994). Three-dimensional reconstruction of an in-situ Miocene peat forest from the Lower Rhine Embayment, northwestern Germany – new methods in palaeovegetation analysis. *Palaeogeography, Palaeoclimatology, Palaeoecology, 110*, 295–317.

Mösle, B., Finch, P. F., Collinson, M. E., & Scott, A. C. (1997). Comparison of modern and fossil plant cuticles by selective chemical extraction monitored by flash pyrolysis-gas chromatography-mass spectrometry and electron microscopy. *Journal of Analytical Pyrolysis, 40–41*, 585–597.

Nip, M., Tegelaar, E. W., de Leeuw, J. W., & Schenk, P. A. (1986). A new nonsaponifiable highly aliphatic and resistant biopolymer in plant cuticles. *Naturwissenschaften, 73*, 579–585.

Nip, M., Tegelaar, E. W., Brinkhuis, H., de Leeuw, J. W., Schenk, P. A., & Holloway, P. J. (1986). Analysis of modern and fossil plant cuticles by Curie point Py-GC and Curie point Py-GC-MS: Recognition of a new, highly aliphatic and resistant biopolymer. *Organic Geochemistry, 10*, 769–778.

Poole, I., & van Bergen, P. F. (2006). Physiognomic and chemical characters in wood as palaeoclimate proxies. *Plant Ecology, 182*, 175–195.

Poole, I., Dolezych, M., Kool, J., van der Burgh, J., & van Bergen, P. F. (2006). Do stable carbon isotope compositions of brown coal woods record changes in Lower Miocene palaeoecology? *Palaeo, 3*(236), 345–354.

Raven, J. A., & Edwards, D. (2004). Physiological evolution of lower embryophytes: Adaptations to the terrestrial environment. In A. R. Hemsley & I. Poole (Eds.), *The evolution of plant physiology, Linnean society symposium series no. 21* (pp. 17–41). London: Elsevier.

Richter, S. L., Johnson, A. H., Dranoff, M. M., LePage, B. A., & Williams, C. J. (2008). Oxygen isotope composition of Eocene- to Holocene-aged cellulose. *Geochimica et Cosmochimica Acta, 72*, 2744–2753.

Riederer, M., & Müller, C. (Eds.). (2006). *Biology of the plant cuticle*. Oxford: Blackwell. 438 pp.

Sawada, K., Arai, T., & Tsukagoshi, M. (2008). Compositions of resistant macromolecules in fossil dry fruits of *Liquidambar* and *Nyssa* (Pliocene, central Japan). *Organic Geochemistry, 39*, 919–923.

Schaal, S., & Ziegler, W. (1992). *Messel an insight into the history of life and of the earth*. Oxford: Clarendon.

Schoenhut, K., Vann, D. R., & LePage, B. A. (2004). Cytological and ultrastructural preservation in Eocene Metasequoia leaves from the Canadian High Arctic. *American Journal of Botany, 91*, 816–824.

Shechter, M., & Chefetz, B. (2008). Insights into the sorption properties of cutin and cutan biopolymers. *Environmental Science and Technology, 42*, 1165–1171.

Shen-Miller, J., Mudgett, M. B., Schopf, J. W., et al. (1995). Exceptional seed longevity and robust growth: Ancient sacred lotus from China. *American Journal of Botany, 82*, 1367–1380.

Stankiewicz, B. A., Mastalerz, M., Kruge, M. A., van Bergen, P. F., & Sadowska, A. (1997). A comparative study of modern and fossil cone scale and seeds of conifers: A geochemical approach. *New Phytologist, 135*, 375–393.

Stankiewicz, B. A., Scott, A. C., Collinson, M. E., Finch, P., Mösle, B., Briggs, D. E. G., et al. (1998). The molecular taphonomy of arthropod and plant cuticles from the Carboniferous of North America. *Journal of the Geological Society, London, 155*, 453–462.

Stankiewicz, B. A., Briggs, D. E. G., Michels, R., Collinson, M. E., Flannery, M. B., & Evershed, R. P. (2000). An alternative origin of aliphatic polymer in kerogen. *Geology, 28*, 559–562.

Suh, M. C., Samuels, A. L., Jetter, R., Kunst, L., Pollard, M., Ohlrogge, J., et al. (2005). Cuticular lipid composition, surface structure and gene expression in Arabidopsis stem epidermis. *Plant Physiology, 139*, 1649–1665.

Tegelaar, E. W., de Leeuw, J. W., Derenne, S., & Largeau, C. (1989). A reappraisal of kerogen formation. *Geoochimica et Cosmochimica Acta, 53*, 3103–3106.

Tegelaar, E. W., Kerp, H., Visscher, H., Schenk, P. A., & de Leeuw, J. W. (1991). Bias of the paleobotanical record as a consequence of variations in the chemical composition of higher vascular plant cuticles. *Paleobiology, 17*, 133–144.

Van Bergen, P. F., Collinson, M. E., & de Leeuw, J. W. (1993a). Chemical composition and ultrastructure of fossil and extant salvinialean microspore massulae and megaspores. *Grana, 1993*(Suppl. 1), 18–30.

Van Bergen, P. F., Collinson, M. E., Hatcher, P. G., & de Leeuw, J. W. (1993b). Lithological control on the state of preservation of fossil seed coats of water plants. *Organic geochemistry, 22*, 683–702.

Van Bergen, P. F., Collinson, M. E., Sinninghe Damste, J. S., & de Leeuw, J. W. (1994). Chemical and microscopical characterisation of inner seed coats of fossil water plants. *Geochimica et Cosmochimica Acta., 58*, 231–239.

Van Bergen, P. F., Collinson, M. E., Briggs, D. E. G., de Leeuw, J. W., Scott, A. C., Evershed, R. P., et al. (1995). Resistant biomacromolecules in the fossil record. *Acta Botanica Neerlandica, 44*(4), 319–342.

Van Bergen, P. F., Bland, H. A., Horton, M. C., & Evershed, R. P. (1997). Chemical and morphological changes in archeological seeds and fruits during preservation by desiccation. *Geochimica et Cosmochimica Acta, 61*, 1919–1930.

Van Bergen, P. F., Hatcher, P. G., Boon, J. J., et al. (1997). Macromolecular composition of the propagule wall of *Nelumbo nucifera*. *Phytochemistry, 45*, 601–610.

Van Bergen, P. F., Collinson, M. E., & Stankiewicz, B. A. (2000). The importance of molecular palaeobotany. *Acta Palaeobotanica, 1999*(Suppl. 2), 629–632.

Van Bergen, P. F., Poole, I., Ogilvie, T. M. A., & Evershed, R. P. (2000). Evidence for demethylation of syringyl moieties in archaeological wood using pyrolysis-gas chromatography/mass spectrometry. *Rapid Communications in Mass Spectrometry, 14*, 71–79.

Van Bergen, P. F., Blokker, P., Collinson, M. E., Sinninghe Damste, J. S., & de Leeuw, J. W. (2004). Structural biomacromolecules in plants: What can be learnt from the fossil record? In A. R. Hemsley & I. Poole (Eds.), *The evolution of plant physiology, Linnean society symposium series no. 21* (pp. 133–154). London: Elsevier.

Van Geel, B., & Grenfell, H. R. (1996). Green and blue-green algae, 7A Spores of Zygnemataceae. In J. Jansonius & D. C. McGregor (Eds.), Palynology: *Principles and applications. American Association of Stragigraphic Palynologists Foundation* (Vol. 1, pp. 173–179).

Vandenbroucke, M., & Largeau, C. (2007). Kerogen origin, evolution and structure. *Organic Geochemistry, 38*, 719–833.

Versteegh, G. J. M., & Blokker, P. (2004). Resistant macromolecules of extant and fossil microalgae. *Phycological Research, 52*, 325–339.

Versteegh, G. J. M., Blokker, P., Wood, G. D., Collinson, M. E., Sinninghe Damsté, J. S., & de Leeuw, J. W. (2004). Oxidative polymerization of unsaturated fatty acids as a preservation pathway for dinoflagellate organic matter. *Organic Geochemistry, 35*, 1129–1139.

Wellman, C. H. (2004). Origin, function and development of the spore wall in early land plants. In A. R. Hemsley & I. Poole (Eds.), *The evolution of plant physiology, Linnean society symposium series no. 21* (pp. 43–63). London: Elsevier.

Wilde, V. (2004). Aktuelle Übersicht zur flora aus dem mitteleozänen 'Öilschiefer' der Grube Mesel bei Darmstadt (Hessen, Deutschland). *Courier Forschungsinstitut Senckenberg, 252*, 109–114.

Yang, H., Huang, Y., Leng, Q., LePage, B. A., & Williams, C. J. (2005). Biomolecular preservation of Tertiary *Metasequoia* fossil *lagerstätten* revealed by comparative pyrolysis analysis. *Review of Palaeobotany and Palynology, 134*, 237–256.

Yule, B. L., Roberts, S., & Marshall, J. E. A. (2000). The thermal evolution of sporopollenin. *Organic Geochemistry, 31*, 859–870.

Zodrow, E. L., & Mastalerz, M. (2009). A proposed origin for fossilized Pennsylvanian plant cuticles by pyrite oxidation (Sydney Coalfield, Nova Scotia, Canada). *Bulletin of Geosciences, 84*, 227–240.

Chapter 7
The Relationship Between Continental Landscape Evolution and the Plant-Fossil Record: Long Term Hydrologic Controls on Preservation

Robert A. Gastaldo and Timothy M. Demko

Contents

Abstract Continental depositional environments preserve the majority of the macrofloral record since the advent of land-plant colonization in the mid-Paleozoic, and wetland representatives are encountered more commonly than those that grew under more seasonal conditions. It has been assumed that preservation potential and future recovery of plant debris are high once detritus is introduced into any appropriate environment of deposition (e.g., fluvial-lacustrine or paludal setting), regardless of prevailing associated climate, sediment load, or geochemistry at the time of emplacement or interval thereafter. If a plant fossil is identified in any part of a stratigraphic interval, even if it occurs solely as an impression, it has been presumed that favorable conditions persisted over time to facilitate this record. Conversely, the

R.A. Gastaldo (✉)
Department of Geology, Colby College, Waterville, ME 04901, USA
e-mail: ragastal@colby.edu

T.M. Demko
Department of Geological Sciences, University of Minnesota Duluth, Duluth, MN 55812, USA
and
ExxonMobil Exploration Company, Houston, TX 77210, USA

P.A. Allison and D.J. Bottjer (eds.), *Taphonomy: Process and Bias Through Time*, Topics in Geobiology 32, DOI 10.1007/978-90-481-8643-3_7,

absence of fossil plants in a stratigraphic sequence commonly has been interpreted as the result of catastrophic perturbation across the landscape, rather than the ascribing their absence to taphonomic filters that may have operated millennia after burial. Terrestrial landscapes are affected by aggradational, equilibrium, and degradational processes that control not only the local or regional water table, but also the long-term fossilization potential of organic debris entombed within these sediments. Fossil plants have the highest preservation potential when high water tables are maintained long-term within soils (e.g., histosols, entisols, gleyed soils), or in settings that are maintained below the maximum draw down of the regional water table (e.g., channel barforms, abandoned channels, lakes) of aggradational landscapes. When landscapes reach equilibrium, extensive pedogenesis ensues and the development of deep mature soils (e.g., calcisols) results in the bacterial degradation of any previously buried plant debris due to extreme penetration of atmospheric gases. When sediment is removed during landscape degradation, the local and/or regional water table is reset lower in the unconsolidated stratigraphy, once again promoting rapid decay of previously buried detritus at depth. These processes, operating under time frames of centuries to millennia and longer (lakh), control the ultimate preservational mode of plants recovered from the fossil record.

This chapter reviews the factors influencing the preservation of terrestrial plants in both subaqueous and subaerial environments based on actualistic studies, and develops a conceptual framework for landscape evolution in continental regimes. A model is presented in which preservational mode is related to the taphonomic and sedimentary history of the landscape in which plant detritus is buried. Case studies of the plant-fossil record, ranging from the Triassic to the Eocene, in exclusively aggradational and in aggradational/degradational landscapes are presented.

1 Introduction

Plant communities form the base of terrestrial ecosystems, serving multiple functions including, but not limited to: acting as the primary food resource for life; biogeochemical cycling and carbon storage; development and enrichment of soils; moderation of local and regional temperature; and animal habitats and shelters. Colonization of land may have occurred very early in the Phanerozoic, with evidence of cryptogam and bryophyte-grade plants found within nearshore marine deposits (Strother 2000; Baldwin et al. 2004) earlier than fragmentary debris of multicellular plants preserved in fluvial siliciclastic and associated environments (Pratt et al. 1978; Gensel and Edwards 2001). As plant clades evolved various architectures imparting more robust growth statures beginning in the Silurian, both aerial and subterranean plant parts become more prevalent in the stratigraphic record. Early preservational modes range from adpressions (Shute and Cleal 1987) to pyritization (Grimes et al. 2001) and charcoalification (Glasspool et al. 2004). With the advent of higher vascular plants in the Late Silurian (Edwards and Feehan 1980; Rickards 2000) and the appearance of the seed habit in the Late Devonian

(Fairon-Demaret and Scheckler 1987; Rothwell et al. 1989), plants evolved strategies for successful colonization under wide climatic regimes within various enriched or depauperate soil types. Individual clades developed a broad range of adaptions that allowed taxa to grow under moderate to extreme climates, although fossil assemblages rarely are preserved or encountered in these parts of the landscape for a variety of reasons (DiMichele and Gastaldo 2008). The principal preservational mode where such assemblages are found usually is permineralization (Demko et al. 1998) or authigenic cementation (Schopf 1975), promoted by physico-chemical groundwater conditions interacting with entombed vascular tissues during early diagenesis (e.g., Drum 1968; Allison and Pye 1994).

There is no doubt that the terrestrial plant record consists of an over-representation of wetland assemblages (Greb et al. 2006; Wing and DiMichele 1995). But, it is not unreasonable, and more parsimonious, to hold that deep time landscapes outside of the wetlands also were vegetated to some degree at least as far back as the latest Devonian, when the evolution of the seed habit allowed for biogeographic expansion of clades into more inhospitable regions. Of course, there does exist the possibility that community representatives could be preserved in these extrabasinal areas (e.g., Beraldi-Campsei et al. 2006), but their general absence in the pre-Tertiary stratigraphic record generally is construed to reflect their true absence in the landscape at that moment in geologic time. Lazarus taxa are known, although they are envisioned as having been wetland-centered species (Mamay 1992) and not representative of the remaining coeval landscape. Hence, the prevalence of wetland assemblages appears to have resulted in a prevailing paradigm that when plant-part debris is buried within a suitable depositional regime, early diagenetic processes generally will promote preservation. Conversely, when there is an absence of plant fossils in strata, it is assumed that the landscape was hostile to their successful colonization and they were extirpated from the region (often, in spite of paleopedological evidence) or that this absence marks a major extinction event. But, it is equally probable that the processes that promote preservation also will promote degradation and recycling via fungal and bacterial activity (Gastaldo 1994; Gupta and Pancost 2004). One must remember that the majority of organic matter is recycled for reuse, with a very small proportion of biomass sequestered in the rock record. Hence, concepts tying the presence or absence of the plant-fossil record to the evolution of continental landscapes generally have been overlooked or neglected.

The plant-fossil record plays a major role in understanding and interpreting the response of ecosystems to changes in climate, evolution, and crises in Earth systems. The physical presence (e.g., Gastaldo et al. 2009) or absence (Gastaldo et al. 2005) of terrestrial plants within any particular depositional regime at any specific point in time is integrally tied with the packaging of continental sedimentary successions. Recent workers have suggested methods and frameworks to subdivide continental rocks based upon differences in rates of accommodation and their relationship to allogenic and autogenic factors in basin fill (e.g., Shanley and McCabe 1994; McCarthy and Plint 1998; Etheridge et al. 1998). And with these models in mind, the preservation of terrestrial plant-fossil assemblages may be more a function of longer term processes operating within the landscape than any other factor.

It is essential to understand the relationships between the short-term (facies) and long-term (landscape) controls on the taphonomic biases controlling this record before using these data in paleoecological, paleoclimatic, and macro-evolutionary studies. This contribution will provide a model for continental sedimentary successions constraining the physico-chemical conditions within which terrestrial plant-fossil assemblages, and their preservational mode(s), can be understood and utilized.

2 Factors Influencing Plant-Part Preservation

Vegetated landscapes vary enormously in systematic composition, plant density, and vegetational architecture on continental spatial scales that are controlled by climate, topography, and edaphic conditions. Yet, the prerequisites that allow for preservation of the continental plant-fossil record are met under relatively restricted sedimentological and geochemical conditions (Krasilov 1975; Spicer 1989; Gastaldo 1992, 1994). These prevent incorporation of many types of plant assemblages in the deep-time record. In addition, there are a limited number of potential depositional sites where plant parts accumulate under conditions that may promote their long-term (10^1 to 10^{4-5}) preservation prior to deeper burial within any landscape. Aerial debris must accumulate within a depositional regime where (1) dysoxia and/or anoxia prevails (i.e., at the sediment–water interface in a lake system), (2) micro-environmental geochemical gradients are strong (e.g., fluvial channel-bar troughs; Gastaldo et al. 1995), (3) resistant and diagnostic phytoclasts persist unaltered long after decay has removed all volatiles (e.g., phytoliths – Strömberg 2004), and (4) sedimentary entombment maintains (preventing degradation) or enhances (through pore-water interactions with organic ligands) the geochemical environment within the facies. In the majority of instances, biomass is fated to be reused within various biogeochemical cycles by the living biota, which is the rule rather than the exception.

2.1 *Plant-Part Decay Rates*

The loss of vegetative and reproductive structures, either through physiological or traumatic disarticulation (Gastaldo 1992, 1994), subjects them to decay through a myriad of potential interactions including those with saprotrophs, fungi and bacteria, as well as autocatalytic cellular and subcellular breakdown. Although microbial films may promote preservation under specific environmental conditions (Dunn et al. 1997), phytoclasts have a tendency to degrade instantaneously in a geological sense. Subterranean rooting structures, already pre-entombed, may remain less affected for longer intervals of time, particularly when influenced by a change in pore-water chemistries promoting the precipitation of carbonate of various mineralogical compositions (calcite, siderite, pistomesite, etc.; Retallack 2001).

There is an abundance of actualistic data in the ecological and plant taphonomic literature focused on rates of forest-litter decay across the latitudinal (climate) spectrum (e.g., Bray and Gorham 1964; Gastaldo and Staub 1999 and references therein). In general, the refractory nature of the original biochemical composition of the plant organ, or part thereof, will influence the rate of decomposition which is calculated as the decay constant *k* (Perry 1994; Fig. 1). A leaf with a decay constant of $k = 1$ will be completely degraded in 1 year's time. But, complete decay can proceed within time frames on the order of weeks (e.g., flowers and leaves), months (leaves), or years to several decades (wood, gymnospermous cones, fruits, and seeds; Burnham 1993). It is well documented from neoecological and actuopaleontological studies that decay rates not only differ within taxa of a single clade or between various clades, but also under different climatic conditions (Gastaldo and Staub 1999). Rates even may vary within microhabitats under the same general climate (Bray and Gorham 1964). Hence, all plant assemblages essentially provide a geologically instantaneous (T_0) snapshot of preservable landscape constituents. It is true that certain resistant phytoclasts may be reworked, such as woody debris, heavily lignitized cones, fruits, seeds, charcoal (Scott 2000), and palynomorphs. Labile plant parts, such as leaves, flowers, and less reinforced reproductive structures, will sustain physical abrasion when re-entrained into bedload and reduced to unidentifiable phytoclasts (Gastaldo et al. 1987). But, published criteria allow for recognition of such recycled parts (e.g., wood-clast rounding – Gastaldo 1994; change in palynomorph fluorescence – Traverse 1994) within an allochthonous

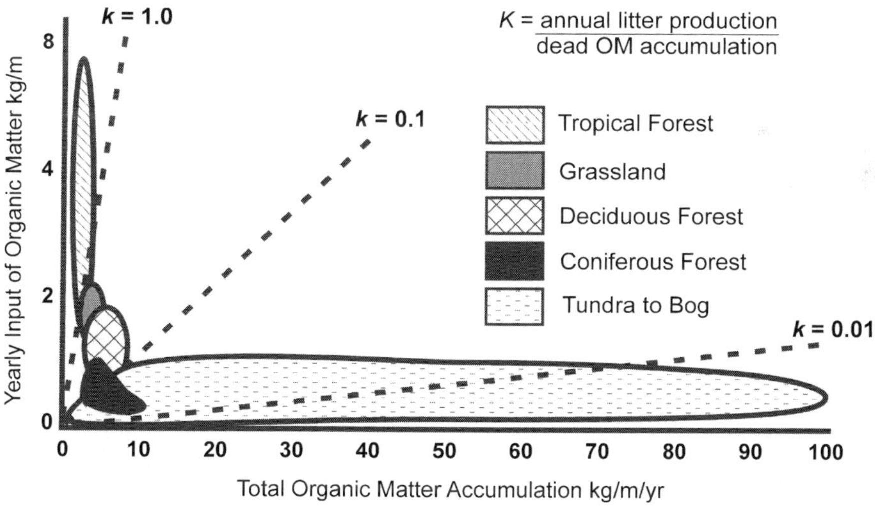

Fig. 1 The relationship between the yearly production of plant biomass and the total organic accumulation in various terrestrial biomes is described by the k constant (Perry 1994). A plant part with a decay constant of $k = 1$ will be completely degraded in 1 year's time. Note that decay constants differ between the most labile (leaves, fruits, flowers, etc.) and refractory (xylem elements, wood, amber/dammar, etc.) plant parts

assemblage. And, in reality, most phytoclasts do not possess structural attributes that allow them to be buried, exhumed, and recycled, if at all, more than once or twice (in the case of lignitized plant parts) before being reduced to palynofacies-grade debris.

2.2 *Relationship Between Rates of Decay and Sedimentation*

Inasmuch as decay rates of the most labile plant parts are, at best, on the order of months to only a few years, this rate exceeds average sedimentation rates in most instances, precluding potential preservation anywhere in the landscape. Sedimentation rate usually is expressed in cm/ka, which is an insufficient rate to promote plant-part preservation. Hence, a convergence must exist in nature where the sedimentation rate at some point in time exceeds the decay rate of plants for there to be any potential preservation of terrestrial vegetation in the stratigraphic record. In addition, the geochemical conditions associated with entombment that promote preservation must be maintained in both the short and long term for that organic debris to be identifiable.

2.2.1 Subaqueous Environments

Plant parts accumulate at the sediment–water interface within discharging and standing water bodies either when their specific gravity exceeds that of water (Gastaldo 1994) or when flow rate is reduced sufficiently to allow for settling from suspension (Spicer and Greer 1986; Spicer 1990). Assemblages within active channels include lag deposits of wood and carpological (fruit and seed) remains (Gee et al. 1997; Gee 2005). Dense accumulations occur within both channel bottoms and various barforms, as well as isolated coarse woody debris (CWD) scattered within the system (e.g., Fielding et al. 1997; Alexander et al. 1999; Gastaldo 2004). In contrast, coarse woody assemblages also have been recognized at the top of fluvial channel fills preserved as log jams (Gastaldo and Degges 2007). Such relationships require an understanding of the contextual taphonomic framework before interpreting Late Devonian (Meyer-Berthaud et al. 1999) to Recent dense woody assemblages in the fluvial stratigraphic record. Troughs within and between fluvial barforms, particularly point bars and lateral barforms, are sites where an admixture of aerial plant parts tend to accumulate when conditions allow for suspension-load settling. This may be in response either to a decrease in discharge velocities following seasonal changes in water supply, the lowering of river stage following a high discharge event (either bankfull or flood stage; e.g., Scheihing and Pfefferkorn 1984), or interactions with meso- to macro-tidal processes transforming a free flowing river to a standing body of water at tidal-bore turnaround (e.g., Gastaldo et al. 1996a). Similarly, organic drapes consisting of various phytoclast components often are found at bounding surfaces separating foreset laminae created by bedform migration. Preservation potential of all these assemblages increases when they are buried by continued bedform migration and maintained below the air–water interface. Geochemical properties inherent within the accumulation, such as the release of organic acids and tannins, may promote long-term preservation in spite of the fact that pore-waters are in chemical equilibrium with the water column. But, when water stage falls to a level below that of the buried organic debris, sediment oxygenation and fluctuating redox conditions promote bacterial and fungal activity that reduces most identifiable plant parts to palynofacies-grade detritus (Gastaldo, 1989). This results in an organic residue in which only the most resistant phytoclasts (e.g., palynomorphs, cuticle, and structured organic matter = mesofossils fraction) may be recovered.

Standing bodies of water have the highest preservation potential for plant-assemblage preservation. These include settings that one envisions as stereotypical lakes and ponds, although the blockade of drainage systems within watersheds either through mass wasting or effects of volcanogenic activity within active tectonic settings also will result in a standing body of water equivalent in scale to lakes (Spicer 1989). In addition, plant parts will accumulate within inactive and abandoned fluvial (oxbow; Gastaldo et al. 1989) and tidal (Gastaldo and Huc 1992) channels. Plant parts transported through feeder channels into lake bodies often are sequestered in shallow water, Gilbert-type deltaic deposits (Spicer and Wolfe 1987) where preservation may result if lake levels are maintained. Lake margins vegetated by aquatic and semi-aquatic plants may act as filters, trapping organic debris in the shallows (Gastaldo 1994). But, once lake level falls and subaerially exposes these areas, buried organic matter shares the same fate as accumulations noted above within barforms. And, if lake levels fall significantly, pedogenesis will overprint these to some degree (e.g., Wing 1984; Gastaldo et al. 1998). Assemblages that accumulate at marked water depth in more distal parts of the water body have a higher probability of preservation if several physical (associated high sedimentation rate) and chemical (i.e., redox conditions operating at and below the sediment–water boundary) conditions are met. Otherwise, debris that settles to the sediment–water interface will be recycled via microbial, invertebrate, or vertebrate activity.

Abandoned (or blocked) channels remaining in connection with an active fluvial (Gastaldo et al. 1996b) or tidal (Gastaldo and Huc 1992) regime provide very localized sites in which primarily parautochthonous assemblages accumulate in association with high sedimentation rates resulting in their entombment. The maintenance of high water levels in a largely confined and restricted setting promotes redox states wherein reducing conditions are not controlled, necessarily, by acidic waters but, rather, by the development of strongly negative Eh values in the sediment. The pH in these water bodies actually may be near neutral (e.g., Gastaldo and Huc 1992), yet the rate of decay is retarded and even labile tissues are conserved in the subsurface. In such settings the promotion of bacterial films, as identified by Dunn et al. (1997), developed at the sediment–water interface in conjunction with sediment influx, may be controlling the preservation potential in these assemblages. Hence, there is a localized geochemistry with its own internal equilibrium very different than the surrounding landscape that controls the taphonomic character of these regimes.

2.2.2 Subaerial Environments

In subaerially exposed floodplain settings, it is generally held that plant-fossil assemblages have the highest preservation potential when buried by overbank flood deposits. High discharge, sediment-laden flood events allow for the spread of fine-grained clastics away from the trunk channel over a short time interval that may last a few weeks before flood waters recede and suspension load settles to what previously was the soil–air interface (Fig. 2). During flooding, the regional water table is elevated to above the soil interface and standing water may promote site-specific dysoxyia and a change in redox conditions. But, once flood waters recede and the regional water table is reestablished at its previous level, normal decay processes proceed within the buried litters as groundwater levels fluctuate in response to meteoric water supply, evaporation, and plant growth (transpiration). Fluctuations in the groundwater table introduce oxygenated waters promoting biochemical and biogeochemical processes, reducing plant parts to meso- and micro-detrital sizes without physical abrasion. In effect, the return to the local or regional equilibrium resets the degradational processes, reducing to null the potential for preservation as

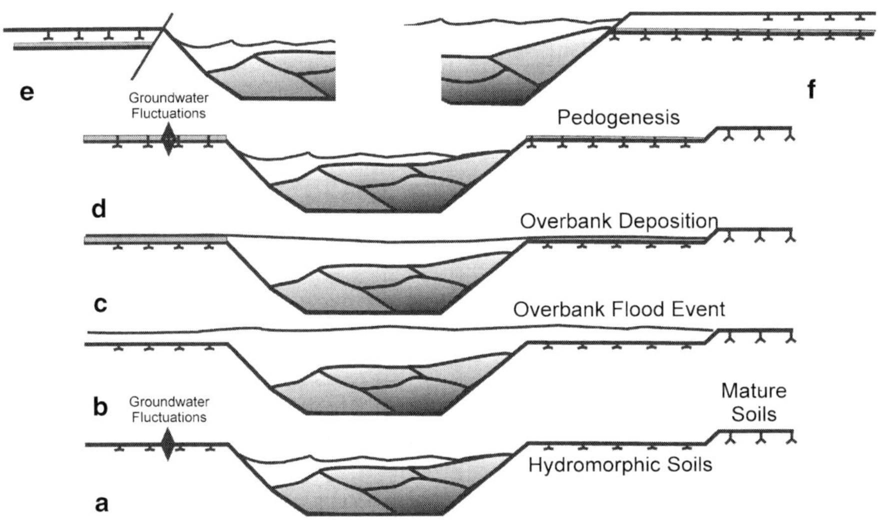

Fig. 2 The fate of organic matter at the soil–air interface within interfluves. (**a**) Groundwater fluctuations in hydromorphic (wetland) soils are responsible, in part, for decay of organic matter at and immediately below ground level. (**b**, **c**) During overbank flooding, fine-grained sediment is transported into the interfluves where it is deposited from suspension load as a thin to thick blanket over the former soil. (**d**) As flood waters subside and return to the original river stage, groundwater table also returns to its previous placement in the landscape. Pedogenic and biological activity, along with yearly-centennial fluctuations in groundwater table, promote decay of buried plant debris. To isolate the buried litter (O–) horizon and increase the probability of preservation, the surface must be placed beneath the regional watertable. This may be done either in response to tectonic subsidence (**e**) or rapid aggradation of the landscape (**f**)

pedogenesis becomes the primary mechanism operating within the landscape. This scenario also holds for landscapes influenced by avulsion unless a perched water-table is established, preventing root penetration, oxygenation, and microbial decay of buried debris. Hence, it is only possible to preserve subaerial plant assemblages that accumulate at the soil–air interface by not only (1) entombing the debris within overbank (and avulsion) deposits that are relatively thick, but also (2) in an area where the post-depositional groundwater table has been elevated above the former soil profile to insure that subsequent pedogenesis and associated biological decay processes are eliminated (Fig. 2e, f).

Plants in volcanogenic regimes (Spicer 1989, 1991) may be encountered in a variety of settings ranging from autochthonous litters, preserved at the contact between the soil and ashfall deposits (Burnham and Spicer 1986; Wing et al. 1993; Opluštil et al. 2007), to allochthonous assemblages encased within lahars and debris flows (Fritz 1980; Fritz and Harrison 1985) or reworked tuffaceous sediments (Jacobs et al. 2005). Besides the possibility for the presence of charcoal, either the result of temperatures experienced during the blast event (Spicer 1989) or following burial (Scott and Glasspool 2005), very early diagenetic interactions as a result of reactive pore-water chemistries may promote sulfide precipitation enhancing preservation potential of more labile parts (Burnham and Spicer 1986). Such rapid mineralization may allow for identification of those plants in the stratigraphic record subsequent to the reestablishment of regional groundwater table and promotion of degradation in the buried assemblages. In other instances where mineral-charged groundwaters are transported through xylary conducting tissues (tracheids in plant steles and tracheids and/or vessel elements in wood), reactions with organic ligands at the boundaries of cell walls may result in silicification (Sigleo 1978, 1979). Subsequently, abiotically mineralized plant parts will be retained in the stratigraphic record even when others may be removed by degradation. Hence, even in explosive volcanogenic regimes where sedimentation rates exceeds decay rates the control on potential preservation is linked with maintaining the assemblage within the phreatic zone.

3 Models of Stratigraphic Frameworks and Landscape Evolution

With the advent of the sequence-stratigraphy paradigm and its primary application to marine-influenced sedimentary successions, extension of sequence boundaries onto continental terrains became necessary to understand how these were expressed in subaerial environments. Initial models for continental regimes relied upon base-level changes tied to eustacy (e.g., Posamentier and Vail 1988; Miall 1991; Shanley and McCabe 1991, 1994), although Wright and Marriott (1993) note that models controlled by simple base-level changes can not be applied to fluvial systems where complexity is controlled, in part, by climate. They also dispute the concept of available accommodation in continental settings by noting that floodplain sedimentation not only is controlled by the elevation of the channel but also by its bankfull depth. More recently, Muto and Steel (2000) extended the concept of accommodation, sensu lato, when they argued that the term "potential accommodation" is equivalent to the maximum possible accommodation that essentially coincides with water-column height at a specific place in time. Hence, the depth and extent to which sediment can accumulate within any continental regime will be controlled by both the water level within fluvial conduits (including avulsion channels) and at over-bank stage when sediment is distributed between channel systems that, in effect, increase the height of confining levees.

3.1 Continental Sequence Stratigraphy

As originally conceptualized, the sequence stratigraphy of continental successions was placed within the context of base-level change within coastal plain and deltaic, marginal marine settings (Posamentier and Vail 1988). Posamentier and Vail (1988) differentiated between fluvial deposition in the coastal plain, at or just above sea level, and in the alluvial plain, above sea level. In the coastal plain, fluvial deposition was constrained to incised valleys during lowstand, and during the latter stages of highstand progradation of the shoreline depositional system. Several authors were critical of this framework because of the fact that fluvial architecture is related to concepts of equilibrium-profile change over time (Miall 1991; Wright and Marriott 1993). With this in mind, Wright and Marriott (1993) modified the model relating base level and accommodation to alluvial architecture and soil development. Although still visualized within the context of eustacy (e.g., Blum and Törnquist 2000), fluvial sediment accumulation was constrained to the coastal plain in transgressive and highstand systems tracts when there is sufficient accommodation to store sediment within the system. But, as Wright and Marriott (1993) note, their model is only one of many possible scenarios.

The range of possible controls on stratigraphic base level within continental successions was discussed by Shanley and McCabe (1994) wherein they noted an interdependency of climate, tectonism, and eustacy that may determine fluvial architectural patterns. Their and subsequent concepts of a continental sequence followed that of Mitchum et al. (1977) in which genetically related strata are bounded at the top and base by unconformities or their correlative conformities. Such boundaries mark a depositional hiatus, the origin of which may have been controlled more by climate and/or tectonic processes within continental sequences than in marine ones, and are reflected in the character of interfluve paleosols developed within coastal plain settings (e.g., McCarthy and Plint 1998, 2003; Plint et al. 2001). Debate continues as to how to confidently identify the boundaries of marine systems tracts within correlative alluvial strata (e.g., Etheridge et al. 1998), as well as the application of terminology devised for marine sequences used in continental settings. General models for continental sequences continue to be proposed (e.g., Boyd et al. 2000) although all models, to date, tend to focus on eustatically influenced coastal plain deposits to a significant degree (Atchley et al., 2004). In this paper, we use the term *continental* to refer to settings in the alluvial plain, in the sense of Posamentier and Vail (1988), which are beyond the reach of marine influences.

3.2 Graded Profiles, Paleosols, and Landscape Evolution

As many authors have noted (e.g., Bull 1991; Quirk 1996; Posamentier and Allen 1999; Muto and Steel 2000), fluvial systems are in a state of disequilibrium when they are doing stratigraphic "work" (depositing or eroding). Fluvial equilibrium can be defined as a state when there is no significant erosion or deposition occurring along the course of a river that would permanently changes that system's overall profile (Machin 1948; although see arguments about theory and reality discussed by Muto and Steel (2000)). This hypothetical graded state results in a rate of bedload transport equivalent to that of sediment supply; hence, there is a balance between the energy required to carry the bedload and the amount of bedload transported (Quirk 1996). Disequilibrium occurs when either there is: (1) insufficient discharge or a decrease in alluvial gradient that results in a decrease in stream power to carry bedload through the system, resulting in aggradation of the alluvial plain (positive fluvial accommodation of Posamentier and Allen 1999), or (2) higher discharge than that required to move bedload, or an increase in alluvial gradient that results in an increase in stream power causing degradation (downcutting) of the alluvial plain by incising channels (negative fluvial accommodation of Posamentier and Allen 1999). Hence, fluvial systems only contribute additional strata or surfaces of erosion to the stratigraphic record when in disequilibrium, although, as Schumm (1993) pointed out, reworking of the landscape occurs due to channel and channel belt migration during times of relative stasis. Disequilibrium is promoted by changes in allogenic forcing factors such as tectonic subsidence or uplift in the basin, climate and precipitation regimes (Cecil and Dulong 2003), relative height of ultimate base level (sea level or, in the case of closed basins, interior lakes), and processes operating in both the provenance area and coastal zones (baselevel change; Schumm 1993). Quirk (1996) remarks that when climate or the position of relative sea-level are influencing factors, large parts of the drainage basin are likely to be affected. And, as such, intervals of aggradation and degradation may provide a means of chronostratigraphic correlation in alluvial strata. With this in mind, Quirk (1996) introduced the concept of base profile, an idealized graded profile of a drainage basin constrained in time to a potential chronostratigraphic datum. Rivers aggrade up to, or degrade down to, this base profile (referred to as the "base level of erosion" by Bull (1991)).

Substantial attention has been focused on sandstone-prone channel facies in alluvial plain stratigraphy due to their economic importance in hydrocarbon and groundwater exploration and exploitation. Yet, interfluve areas (floodplain, overbank, etc.) make up the majority of any alluvial plain, with the trunk drainage channels of major rivers covering a relatively small component of the overall landscape. Hence, the character of paleosols adjacent to, and coeval with, within-channel deposits provides critical information about the prevailing climate (e.g., Sellwood and Price 1993; Retallack 2001) and the relative timing of their formation. Because paleosols that form under a distinctive climatic signature represent an amount of time that is roughly equivalent to, or less than Milanković-scale cycles, they, in effect, allow for chronostratigraphic control in alluvial strata (McCarthy and Plint 1998, 2003; Plint et al. 2001).

When fluvial systems are in equilibrium, paleosols in adjacent areas are not provided with any significant new sediment supply, outside arid and semi-arid areas where aeolian deposition may dominate. During this time, climate is the predominate influence across the landscape and the result is the development of a mature, complexly overprinted, or amalgamated paleosol with features reflecting the physio-chemical processes of formation (e.g. histosols – high water tables and organic production, vertisols – seasonal wetting and drying, aridisols – accumulation of salts, etc.). More mature paleosols record longer durations over which the fluvial regime remained in equilibrium providing for increased depth of pedogenic alteration. When fluvial systems are in disequilibrium, one of two general scenarios will ensue. Landscape degradation will result in the loss of alluvial plain stratigraphy due to increased downcutting and floodplain scavenging. Potential evidence for the conditions that prevailed during equilibrium may be restricted to channel lag deposits (e.g., reworked soil nodules: Pace et al. 2005) or aerially restricted sites that were unaffected by, or resistant to, erosion (e.g., Gastaldo et al. 1998). Conversely, landscape aggradation will result in the rapid buildup of interfluves with a concomitant rise in watertable. These conditions promote the development of immature and wetland paleosols (Kraus 2002).

In summary, continental landscapes can be envisioned as existing within one of three different states at any point in time during their histories – degradation, aggradation, and equilibrium (stasis). During degradation, fluvial systems are downcutting, removing previously deposited material or bedrock because of excess stream power. The regional water table will follow the downcutting, and floodplain sediments that may have been stored under saturated or submerged conditions will be subject to deeper drainage, infiltration, and more oxidizing conditions. During aggradation, fluvial systems deposit material, filling up potential accommodation in order to increase the gradient to a point that is adequate to carry the bedload that is available. Once this gradient is reached, the fluvial system is in a state that will carry the bedload provided. As the system aggrades, the regional watertable will rise with the increased elevation of the fluvial/overbank system. Previously deposited channel-fill and floodplain sediments will be buried below this rising water table and will be cut off from further connection to vadose-zone weathering and/or oxidation. If this aggradation is taking place within a valley cut during a previous period of degradation, the fluvial/overbank system will have to first fill the confined space within the valley before being able to aggrade the regional landscape surface. This initial period of valley filling may produce a landscape that has both well-drained oxidized soils on interfluve areas and poorly-drained areas with relatively wet edaphic conditions within the incised valley. Since only the valley floor is available for aggradation and storage of channel-fill and overbank material, relative sedimentation rates may be high while the valley topography is annealed. The volume of valley accommodation also may increase because of lateral retreat of the valley walls due to cutbank erosion, slope wash, and mass-wasting during aggradation. If the potential equilibrium profile of the system is at a higher elevation than the interfluve areas, the fluvial system will overtop the valley walls and will have a greater area over which to deposit channel and overbank sediments. At this time, relative sedimentation rates will apparently decrease, and there will be evidence of a related increase in floodplain soil maturity. Systems subject to repeated or cyclic changes in the forcing factors that control landscape aggradation and degradation (discharge, bedload sediment supply, tectonically-controlled gradient, etc.) may have successive valley fill and post valley-fill strata that have been subject to superimposed pedogenic and paleohydrologic regimes. The preservation of plant-fossil assemblages with identifiable remains within these successions will be controlled by the magnitude and timing of these depositional and hydrological processes.

4 A Model for Plant-Part Preservation in Continental Landscapes

The potential for encountering fossil plants within a continental framework is related directly not only to whether or not the organic detritus accumulates within a potential preservational site, but also whether or not it is maintained in a burial site where pore-water geochemistry retards or halts degradation. The highest probabilities for preservation exist within bodies of water in which fine-clastic sediments accumulate at rates greater than those for organic-matter decay. Conversely, the lowest probabilities exist within subaerially exposed sites, including wetlands (Gastaldo et al. 1989), where groundwater flux and meteoric water input vary over diverse time scales promoting pedogenesis and carbon recycling. Hence, for plant parts to become recognizable (identifiable) biological remains in the stratigraphic record, they first must be confined to stratigraphic intervals where high A/S (Accommodation/Sediment Supply) ratios prevailed. And, once buried, these assemblages also have to be maintained at a stratigraphic level beneath the maximum draw down of the prevailing watertable to prevent subsequent decay, deterioration, and loss from the sedimentary succession. Envisioned within the framework of continental landscape evolution, there are limited potential instances where these criteria are met in space and time.

During periods when the landscape is static and fluvial systems are in equilibrium, plant-part preservation is confined to: (1) within channel, subaqueous deposits (CWD concentrated in basal lags or barforms, heteromeric [admixture of plant-part types and sizes; Krasilov 1975] hydrodynamically equivalent assemblages in barform troughs or abandoned channels), and (2) lakes where the depth-to-bottom exceeds the chemical and biological depth of reactivity (Fig. 3). Pedogenesis across the interfluves and along river-and-lake margins (aquatic and semi-aquatic colonization)

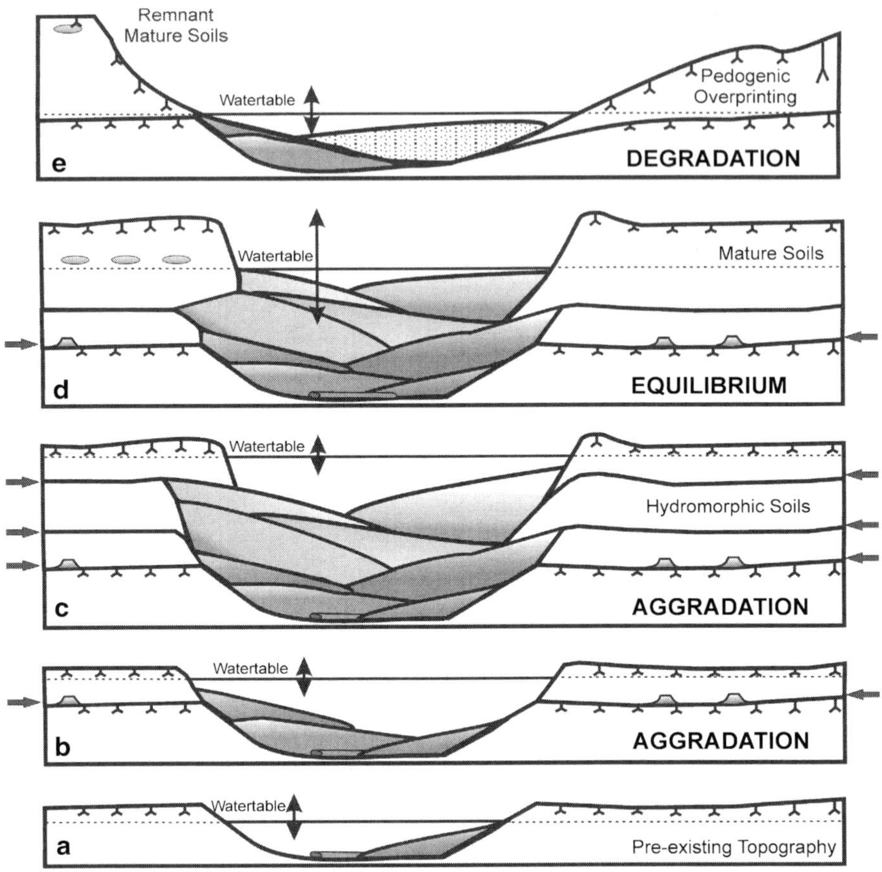

Fig. 3 A model for preservation of plant assemblages in continental regimes. See text for details

promotes decay and carbon recycling, preventing preservation of all but the most resistant aerial plant parts (e.g., palynofacies debris, palynomorphs; Gastaldo et al. 1996a). Evidence for subterranean rooting structures may be preserved depending upon the type of vegetation or soil, and the reactivity of pore waters with entombed organic material (e.g., drab halos, rhizoconcretions, etc.; Retallack 2001).

The onset of fluvial disequilibrium and landscape aggradation promotes sediment accumulation within the interfluves that, in turn, may bury organic litters once residing at the soil–air interface or shallow subsurface. As incremental bedload deposits accumulate within channels in response to discharge rates that are lower than needed to flush the system, CWD lags may occur within various parts of barforms (Gastaldo 2004). Incremental deposition of siliciclastics also occurs along channel margins (levees) which respond to overbank sedimentation or avulsion with an increase in their vertical height and lateral extent. The change in relative position of the river's water surface is reflected in the watertable across the interfluve, and it ascends in the section in response to sediment buildup (Fig. 3b, c). As the watertable continues to reestablish at a higher stratigraphic level in response to aggradation, buried plant debris is maintained within anoxic geochemical conditions that prevent decay past the state of the original organic matter when entombed. Organic matter that accumulates at the soil–air interface within the new (primarily wetland) landscape still is subjected to pedogenic activity and carbon recycling.

A phase of continuous aggradation without landscape degradation, accompanied by a stratigraphic rise in interfluve watertable, will result in a sequence of stacked fine-grained clastic deposits wherein fossil-plant assemblages will be confined to the basal unit(s) immediately above the disconformity (previous soil horizon) if geochemical conditions are met for potential preservation. Where semi-continuous aggradation occurs without landscape degradation, the possibility exists, but rarely is met, for fossil-plant assemblages to be preserved at the contact between each soil horizon and the overlying aggradational, fining upwards sequence if subsequent aggradational intervals accumulate rapidly (Fig. 3). These plant assemblages will be preserved primarily as adpressions (coalified compressions and impressions) with carbon, lipid, resin (if present systematically), and cuticular residua. Depending upon the timing between soft-tissue decay and the rate of sediment influx and final entombment, casts of axes (e.g., pith casts) also may be found (Allen and Gastaldo 2006).

Once equilibrium is reestablished across the landscape, pedogenesis again dominates the interfluves (Fig. 3d). Depending upon the prevailing climate and associated distribution of rainfall over the year, deep soil alteration can occur, buried organic matter recycled, and pedogenic concretions may form. Under more arid or highly seasonal and restricted precipitation regimes, stable-isotope signatures sequestered in carbonate concretions will reflect atmospheric gas concentrations because of formation within a geochemically open system (Tabor et al. 2007). But, carbonate concretions also may form in geochemically closed systems, wherein stable-isotope signatures reflect bacterially-mediated decay of organic matter within saturated soils (Tabor et al. 2007). Hence, the presence, alone, of carbonate concretions at some depth within a paleosol may not be the sole criterion for interpreting prevailing climate at the time of landscape equilibrium and, hence, soil formation (Gastaldo and Rolerson 2008). But, as long as entombed plant-part assemblages are maintained below the stratigraphic level of deepest watertable penetration, their preservation potential remains high (Fig. 3).

The onset of fluvial disequilibrium resulting in landscape degradation, either through tectonic uplift or tilting, increased river discharge and/or decreased sediment supply associated with a shift in climate, or base level change, leads to valley cutting and scavenging of the former landscape due to the downward shift in equilibrium profile. The removal of strata and establishment of a new base profile, resets the regional watertable (Fig. 3e). The outcome of such a regional watertable reset exposes any previously buried plant assemblages to renewed decay processes if they are positioned now (or at any time) in the vadose zone. Depending upon the original grain-size, mineralogical parameters, and degree of compaction of the entombing sediment, a number of different plant-part preservational modes may result. For example, if non-woody plant parts originally were entombed within fluvial sand, sediment porosity and permeability will promote complete deterioration of the most labile organic matrix. The only remnants that may be left behind would be unrecognizable or unidentifiable debris and "poorly" preserved axial remains, which could be assigned to a major plant group if distinguishing morphological characters exist (e.g., sphenopsid axes). In contrast, if the confining sediments were fine-grained in nature, decay will result in loss of the organic matrix while retaining the overall impression of the features of the buried plant part in the mudrock. Impressions will be the result of sediment and organic-matter compaction within the confining pressures of the matrix (Rex and Chaloner 1983). Plant assemblages that remain at a stratigraphic depth below the new watertable have the potential for continued retention in the stratigraphic record, unless landscape degradation in the future again resets the level of regional watertable. Preservation potential for plant parts again increases once aggradational processes begin to fill these incised landscapes, usually resulting in wetland assemblages (Demko et al. 1998).

In summary, plant-fossil assemblages in continental regimes have the highest preservation potential when there is a turnover from landscape degradational to landscape aggradational processes. In systems where aggradation is continuous, the highest preservation potential exists at the contact above the disconformity between the degradational and aggradational landscapes and within standing bodies of water. There are decreasing probabilities for the preservation of plant debris higher in these sections depending upon the A/S ratio in effect at the time of plant-part entry into an aquatic (fluvial or limnic) environment. In systems where aggradation is semi-continuous without landscape degradation, the highest preservation potential exists at the contact between fining upwards intervals, each of which can be considered analogous to a parasequence in marginal-marine settings. In this case, instead of the genetically related strata being bounded by marine flooding surfaces, they are bounded by surfaces that mark an increase in relative accommodation rate filled nearly instantaneously by fluvial and overbank aggradation. As a new, long-term watertable is established at the top of the stratigraphic section in response to the buildup of the landscape, plant assemblages maintained stratigraphically within saturated sediments are buffered from decay and loss from the potential fossil record. When fluvial equilibrium and landscape stasis are attained, thick mature soils form that can be used as chronostratigraphic marker beds across the region.

5 Case Studies

The following case studies are provided as examples illustrating the basic premises between plant-part preservational modes within the landscape model context. By no means do these cover the full spectrum of possible scenarios over space and time. Rather, each sedimentary basin within the range of various tectonic settings must be evaluated independently to determine the relationships between presence or absence of fossiliferous plant beds within stratigraphic and regional context.

5.1 Plant Assemblages in Aggradational/Degradational Landscapes

5.1.1 Paleogene Weißelster Basin, Central Europe

The Paleogene Weißelster basin, Germany, is comprised of interfingering continental and marine sequences, and is paleogeographically in close proximity to the paleo-North Sea (Krutzsch 1992). The Eocene and Oligocene deposits were subdivided by Eissmann (1970) into three stratigraphic units. The Borna beds are predominantly terrestrial and of Middle and Late Eocene in age, the Middle Oligocene Böhlen beds are predominantly brackish-marine, whereas the overlying Middle Oligocene Bitterfelder beds are indicative of terrestrial to brackish-marine settings. The fluvial Borna beds (Halfar et al. 1998) consist of coarse sand and gravel, with intercalated thick clay deposits representing oxbow lakes and wetland paleosols. Several thick economic coals occur within the Eocene sequence, in addition to several well preserved fossil-leaf assemblages that have been used for paleoecologic (Gastaldo et al. 1998) and paleoclimatic (e.g., Mosbrugger et al. 2005) restorations of central Europe.

In the Bockwitz mine near Borna, Germany, a Middle Oligocene flora is preserved in close proximity to a Late Eocene coal – Oberflöz II – and an overlying interfluve paleosol (Gastaldo et al. 1998; Fig. 4), which stratigraphically correlates with sediments below the Late Eocene Oberflöz III coal in the Schleenhain mine several kilometers to the west (Halfar et al. 1998). Taphonomic criteria (Gastaldo et al. 1996b) allow for the recognition of autochthonous floodplain assemblages, parautochthnous channel-fill sequences, and allochthonous accumulations either of coarse woody detritus (Gastaldo 2004) or heteromeric assemblages within trough fills of channel barforms.

Three autochthonous plant assemblages were exposed in the mine highwall in the early 1990s; more recently, mine reclamation has left little surface exposure of any plant-bearing intervals. Besides the lignite bed (histosol of Oberflöz II) of varying regional thickness, two additional paleosols occur in the stratigraphic sequence. The first is found terminating the overlying fluvial complex, and represents the culmination of alluvial plain aggradation following peat cessation. This Late Eocene paleosol is only a remnant of a laterally extensive, well-developed soil that attained a thickness of at least 1.7 m. A prominent A-horizon is underlain by a weathered and stained B interval, beneath which was a well-developed C zone (Gastaldo et al. 1998, Fig. 4). In situ woody tree bases are preserved in the A-horizon, with extensive and deeply penetrating roots found in the B and C horizons. Clay mineralogy and major element analyses show little variance. The second paleosol is a very thick, at least up to 17 m or more (due to missing section at ground level), kaolinitic massive and homogenized silt of Middle Oligocene age. Large diameter woody roots penetrate deeply throughout the interval, accompanied by slickensides and other features that were used to interpret pedogenesis of a vertisol. This soil formed across a middle Oligocene landscape characterized by

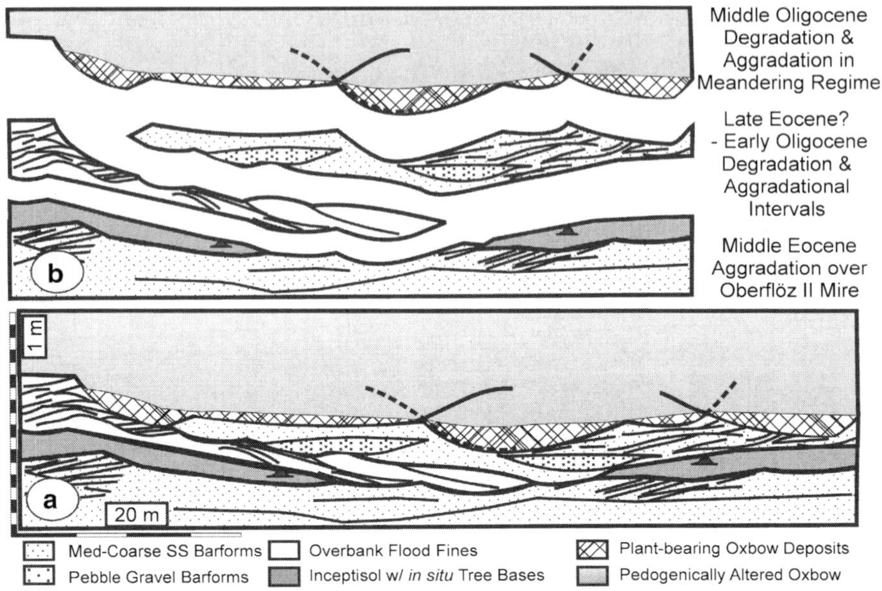

Fig. 4 An example of the effects of an aggradational/degradational landscape on plant-fossil assemblages in Eocene-Oligocene fluvial deposits of the Weißelster Basin, central Europe. (**a**) Illustration of the mine highwall based on photomosaic taken in the Bockwitz mine near Borna, Germany (Gastaldo et al. 1998). The stratigraphic sequence, exposed above a thick Eocene coal (Oberflöz II), consists of aspects of both aggradational and degradational landscapes that encompass approximately 10 MY. (**b**) Aggradational sequences and boundaries marking the base of degradational phases in chronostratigraphic context are isolated in a Wheeler-type diagram depicting the relationship between the type of plant-fossil assemblage and these events. Erect, in situ, woody trees and an organic (O) horizon are preserved at the top of a Middle Eocene landscape, with much of this landform removed by subsequent degradation. Exceptionally well preserved leaf assemblages occur only in the base of Middle Oligocene oxbow lake deposits, the lowest meters of which were not subjected to pedogenic overprinting

the aggradation of one or more meandering fluvial regimes, each of which were abandoned and filled as oxbow lakes. Pedogenic alteration was subsequent to channel fill.

Identifiable parautochthonous plant assemblages only are restricted to the lowermost 1.5 m of Oligocene channel-fill sequences, although these probably were more extensive (Gastaldo et al. 1996b) prior to pedogenic overprinting (Fig. 4). The aquatic fern *Salvinia* is commonly interbedded with aerial leaf, branch, fruit-and-seed remains, representing autochthonous elements within this parautochthonous assemblage. Leaf accumulations occur as isolated leaves or clusters of leaves (mats) on bedding planes without any preferred abiotic orientation, either solely as impressions or inclusive of cuticular remains.

Allochthonous assemblages are restricted to channel lags as CWD above scour contacts (Gastaldo et al. 1998; Gastaldo 2004) and within channel barforms as admixtures of leaf, fruit, and seed debris. Due to the grain size of these channel deposits, ranging from medium and coarse sand to pebble conglomerate, non-woody detritus is restricted to troughs of bedforms where most exhibit a very low quality of preservation.

The quality, quantity, and stratigraphic position of these assemblages in the Bockwitz mine are consistent within the context of the proposed preservation model. Autochthonous assemblages that include erect woody trees, deeply penetrating rooting systems, and differentiated soil horizons, mark intervals of landscape stasis and equilibrium of the fluvial regime. Phases of disequilibrium occurred when river avulsion or fluvial reactivation occurred in response to Alpine tectonic activity, climate change, or both. This resulted in the burial of paleosols and overall continental aggradation in the Late Eocene (Halfar et al. 1998). Oxbow lake deposits and the plant-fossil assemblages preserved therein were coeval with in-channel deposits of meandering fluvial systems elsewhere in the region.

The stratigraphic hiatus recognized between the Late Eocene and Middle Oligocene is a function of disequilibrium that resulted in landscape degradation. The first preserved degradation event (Fig. 4) emplaced fluvially derived fine-grained sediment above an erosional contact that removed the latest Eocene deposits (including Oberflöz III) and much of the Eocene autochthonous woody assemblage. Although the base of the channel form that marks the erosional contact occurs at places below that paleosol, the regional watertable was maintained higher in the landscape profile. A high regional watertable prevented decay and degradation of these buried trees, forest-floor litter, and subterranean rooting structures. If regional watertable had dropped at any time below the stratigraphic position of this paleosol, organic matter decay and pedogenesis would have proceeded in the same manner as that found in the overlying Middle Oligocene.

The middle Oligocene experienced renewed aggradation in the landscape, with the emplacement of at least two stratigraphically stacked fluvial systems (Fig. 4). Plant-fossil assemblages in coarse fluvial bedforms reflect the persistence of refractory CWD in the absence of other plant detritus, which underwent decay in oxygen-rich waters and/or exposure of bedforms during intervals of low discharge. Abandoned trunk channels filled with a mixture of fine-grained clastics and aerial plant parts derived from channel margin, riparian communities (Gastaldo et al. 1989, 1996b). Stratigraphic sequences in modern oxbow lakes show a pattern of interbedded intervals of leaf clusters (mats), representing annual leaf fall, and mud, deposited during overbank flood events from the base of the abandoned channel to the top of the channel fill. The Bockwitz fossil assemblage, though, is restricted to the basalmost 1.5 m of section because of channel-fill sequence overprinting by pedogenetic activity. The depth of pedogenic activity is marked by the contact between the well preserved plant-fossil assemblages and the homogenized, kaolinite-rich and root penetrated paleosol. Hence, the regional watertable at the time of paleosol development dropped to within 1.5 m of the contact between the bottom of the oxbow-lake channel and the underlying fluvial barforms. If regional watertable had dropped below that level, there would have been no evidence of megafloral remains at this stratigraphic horizon.

5.1.2 Upper Triassic Chinle Formation, Southwestern United States

The Chinle Formation in the Colorado Plateau region of the southwestern United States was deposited in a fully continental basin along the western margin of equatorial Pangaea (Dubiel 1994). The bulk of the Chinle Formation consists of a succession of fluvial, lacustrine, and minor aeolian strata that fill a dynamic basin subsiding between the old Permo-Carboniferous Ancestral Rockies Uplift to the east and the Mogollon Highlands continental volcanic arc to the south (Stewart et al. 1986), and separated from the Luning marine basin to the west by a back-arc bulge (Lawton 1994). The lower part of the Chinle is dominated by volcanic-sourced smectitic mudstone and mineralogically immature sandstone and conglomerate with gleysol- and aflisol-type paleosols, while the upper part has more detrital clay, aeolian sandstone with calcisol-, and aridosol-type paleosols (Dubiel et al. 1991, Demko et al. 1998). Paleosol types, paleobiologic and taphonomic data indicate that the paleoclimate during Chinle deposition was highly seasonal with respect to precipitation, characterized as megamonsoonal by Dubiel et al. (1991).

Plant fossils are locally abundant in the lower part of the formation, including the bulk of the described assemblages of adpressions and the cuticular remains of gymnosperms, cycadophytes, ferns, and other plant types (Daugherty 1941; Ash 1970, 1980) and the justly famous permineralized gymnosperm log assemblages (e.g., Petrified Forest National Park – Ash and Creber 2000; Creber and Ash 2004). The upper part of the formation is characterized by rare fossil plant occurrences, mostly all impressions of robust foliar material (Ash 1987).

A locality in Petrified Forest National Park in Arizona illustrates the relationship between plant preservation, paleosol type, and timing of landscape degradation and aggradation in the lower part of the Chinle Formation (Fig. 5). This locality, within the Tepees area of the central portion of the park, has produced a significant number

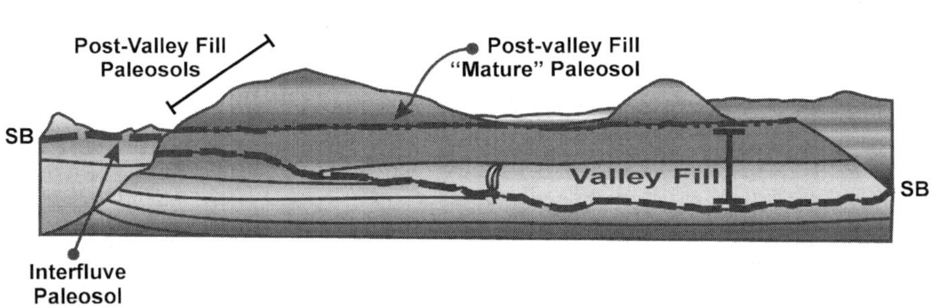

Fig. 5 Strata of the Chinle Formation at the Tepees locality, Petrified Forest National Park, Arizona, were deposited during two periods of fluvial aggradation separated by an episode of landscape degradation, resulting in a 12–13 m deep valley cut into the underlying aggradational succession. Fossil plants are preserved in the initial phases of landscape aggradation that onlap the paleovalley walls. Fossil-plant abundance in the lower Chinle, when compared to a sparse record in the upper part of the formation, is the result of deposition within aggrading fluvial and overbank systems confined to paleovalleys under a hydrologic setting characterized by perennially high watertables. SB = sequence boundary

of the fossil plants that have been described from the Chinle (Ash 1972, 1991, 2001). These strata were deposited by two periods of fluvial aggradation separated by an episode of landscape degradation that cut a valley 12–13 m deep into the first aggradational succession (Demko 1995a). The entire succession at the locality can be divided into three units characterized by different depositional facies, paleosol types, and fossil-plant taphocoenoces: (1) an initial aggadational succession characterized by drab purplish-gray floodplain mudstones with discontinuous coarse-grained sandstone channel-fills; (2) an aggradational valley-fill succession comprised of light greenish-gray mudstone, and light red and brown siltstone with climbing-ripple cross-laminated very fine-grained sandstone, and lateral-accretion beds of fine- to medium grained current ripple cross-laminated and trough cross-bedded sandstone; and (3) a post valley-fill succession of dark reddish brown floodplain mudstones with moderately- to well-developed calcareous and vertic paleosols (Demko 1995a). The light greenish-gray mudstone within the aggradational valley-fill succession preserve abundant parautochthonous and autochthonous cycadophyte, fern, and other leaf material in distinct litter layers at bed boundaries. Autochthonous coalified gymnosperm logs are preserved in the base of, and erect, casted in situ *Neocalamites* trunks and leaf mats are preserved in the upper parts of time-equivalent lateral-accretion units.

The fossil plants at the Tepees locality were preserved in the initial phases of landscape aggradation after a period of fluvial incision and degradation. The purplish-gray floodplain mudstones and coarse-grained sandstones below the surface of incision, a sequence boundary, were deposited by aggradation of an avulsion-dominated fluvial system (Demko 1995a). The floodplain mudstones are characterized by pedogenic features such as drab root haloes and color mottles that indicate groundwater gleyization and fluctuating watertables associated with the aggradational succession. No fossil plants are preserved in either the partially-filled sandstone channel deposits or their associated levee and floodplain deposits. The surface of incision cuts 12–13 m into these deposits. At the base of this valley cut, this degradation surface is overlain by lateral-accretion beds of the Newspaper Rock sandstone and associated levee and overbank wetland deposits. These units onlap the valley walls and preserve the fossil plants noted above. The overbank wetland deposits are characterized by meter-scale coarsening-upwards units comprised of greenish-gray mudstone, siltstone, and climbing-ripple cross-laminated fine-grained sandstone. In areas near the facies change into levee and lateral-accretion beds, the tops of the coarsening-upwards units are pedogenically modified and have a slight reddish-gray coloration. At interfluve areas along the margins of the valley cut, a well-developed dark reddish brown calcareous paleosol is developed on the underlying pre-incision units. A dark red calcareous vertic paleosol is developed over the valley fill and marks the overlap of the incisional topography, a return to unconfined fluvial and overbank conditions, and an apparent slowing of depositional rate.

As discussed by Demko et al. (1998), the distribution of coalified compression-impressions (adpressions) and preserved cuticle of foliar plant material in the lower part of the Chinle Formation was controlled by sedimentological and near-surface geochemical conditions inherent to depositional settings within incised valley-fill successions. The abundance of plant fossils in the lower Chinle, as compared to their rarity in the upper Chinle, is the result of deposition within aggrading fluvial and overbank systems confined to paleovalleys. Evidence from lower Chinle paleosols and trace fossils in areas outside the valley-fills indicates a climatic and hydrologic setting characterized by seasonal fluctuations in recharge and groundwater table (e.g., Dubiel et al. 1991; Hasiotis and Mitchell 1993). However, during aggradation of the valley-fill succession, watertables in these deposits were perennially high, as they are in modern aggrading alluvial valleys (e.g., Gallaher and Price 1966; Runkle 1985). Burial of plant material within depositional environments in these paleovalleys (fluvial channels, crevasse splays, lakes, wetlands, and lacustrine deltas; Demko 1995b) increased the preservation potential because reducing conditions and low biological demand in these areas of high watertables (near and at the surface), and groundwater discharge were conducive to low rates of decay. Confined fluvial systems also have a relatively small area in which to store and deposit aggrading material, contributing to comparatively high rates of vertical aggradation (Posamentier and Allen 1999). Preservation potential for foliar material was dramatically lower in areas outside the paleovalleys during deposition of the lower Chinle, and in upper Chinle strata overall, conditions of fluctuating and low water tables, associated with both fluvial depositional styles (e.g., avulsion-dominated and ephemeral streams) and increasing aridification. Only rare robust (coriaceous) leaves and stems in areas outside channels, and permineralized wood in channels filled with, or laterally-adjacent to, facies with abundant volcanogenic material, can be found (Demko et al. 1998).

5.1.3 Lower Triassic Katberg Formation, South Africa

The Katberg Formation in the Karoo Basin represents deposition within Early Triassic fluvial aggradational and degradational successions (Pace et al. 2009) within a fully continental basin. These rocks are assigned to the Tarkastad Subgroup of the Beaufort Group (Johnson et al. 1997), and were sourced by sediments transported several hundred kilometers from the Cape Fold Belt (Smith et al. 1993). The distance from the provenance and overall grade resulted in fluvial regimes dominated by very fine, feldspathic sand to silt, with overbank interfluves consisting of silt and thin sheet sand bodies. Pedogenic carbonate pisoliths and intraformational mudclasts (Smith and Ward 2001) may be concentrated as lags at the base of each erosional contact marking landscape degradation. Hiller and Stravrakis (1984) interpreted the sandstone bodies as wide, shallow channels of a shifting braided pattern, while Retallack et al. (2003) ascribe several climate attributes to a variety of paleosols across the landscape. Recently, Gastaldo and Rolerson (2008) were unable to apply the criteria proposed by Retallack et al. (2003) to paleosols exhibiting bioturbation and pedogenic nodules, resulting in their recognition of more wetland paleosols in this part of the stratigraphy than previously interpreted.

Plant-fossil assemblages in this part of the sequence are poorly preserved and, in many instances, represent impressions of the most resistant parts including sphenopsid and unidentifiable axial remains. Gastaldo et al. (2005) report new collections from bedform drapes, scour-and-fill structures, and troughs within channel barforms, and from within overbank deposits in aggradational sequences above paleosols. Collections from Bethulie and Carlton Heights, to date, have been preserved in fine- to very fine sandstone matrices.

The Katberg Sandstone exposed along the N9 highway at Carlton Heights (Fig. 6) will serve to illustrate the relationships between plant-assemblage preservation and the role of landscape aggradation and degradation. A sequence of stacked wetland paleosols (Gastaldo and Rolerson 2008) overlie each principal fluvial sandstone body, where not removed through subsequent erosion. Each paleosol is comprised of silt-sized clasts. Thin, very fine-grained and rippled sheet sandstone bodies are interbedded with the siltstone, along with isolated lenses of pedogenic carbonate nodule lags. It is within one of the sandstone bodies that scattered, poorly preserved plant remains occur in a megaripple trough fill. There is no apparent order to the assemblage, with most fragmentary debris on the order of 5 cm in maximum dimension. Impressions of sphenopsid axes, dispersed fern pinnules (identified due to the vascular tissue having been accentuated by iron staining), and reproductive structures are dispersed in the matrix. The poor state of preservation, with resistant plant parts remaining solely as impressions, indicates that all soft tissues were degraded.

The timing when these plant parts underwent degradation is important to interpret the preservational attributes of the assemblage. The plant parts are allochthonous, being emplaced within the bedform trough via suspension-load settling. There is no evidence for physical abrasion of these parts that would be encountered during bedload transport (Gastaldo 1994). If the regional watertable had dropped following deposition, the most delicate plants (i.e., fern pinnules) would have

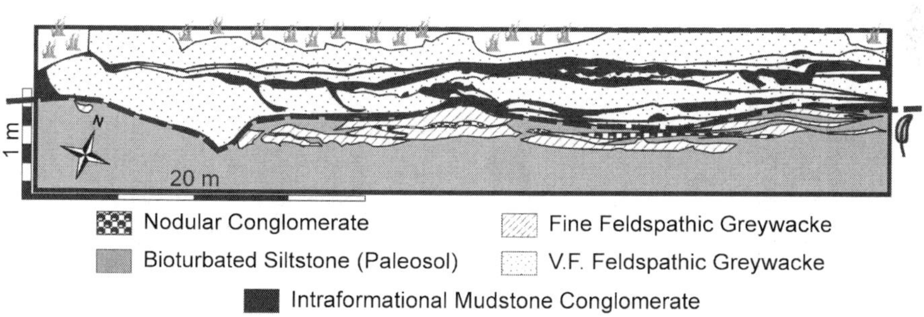

Fig. 6 An Early Triassic example of the effects of aggradational and degradational landscapes on plant-fossil assemblages in the Karoo Basin, South Africa. Only poorly preserved impressions of conducting tissues of fern pinnules, sphenopsid axes, and isolated reproductive structures remain in overbank interfluvial deposits of fine-grained sandstone that were not removed by subsequent landscape degradation (*dashed black line*). The plant debris remained beneath the watertable established following landscape degradation, but was subjected to oxygenated pore waters resulting in loss of the most labile tissues

decayed and been removed from the record without any evidence of having been there. The development of impressions of isolated plant parts in the sediment is not possible immediately following burial, due to the maintenance of pore-water pressures in the matrix. Some degree of compaction and dewatering must occur before the outlines of plant parts are imparted to the entombing matrix. Hence, plant-tissue decay occurred subsequent to continued burial and early compaction.

The Lower Triassic Katberg Formation is characterized by marked degradational events that removed significant portions of the stratigraphic record (Pace et al. 2009). Landscape degradation in the Katberg exposure at Carlton Heights resulted in the removal of the paleosol and overbank sheet sandstones in the north of the outcrop (Fig. 6). A multi-storied sand body exhibiting a braided configuration was emplaced above the erosional contact. Stratigraphically, the preserved plant-fossil assemblage occurs just above the lateral equivalent of this channel base. Landscape degradation and the development of a fluctuating watertable in adjacent sediments promoted decay of any previously buried organic matter (paleosol TOC values are very low, ≤0.5%; Gastaldo and Rolerson 2008). But, because of prior early compaction of the floodplain deposits, impressions of the most resilient plant tissues remained in the fine-grained sand when volatile tissues were lost. It is probable that subsequently introduced pore-water chemistries promoted diagenetic staining of the remaining structural tissues, accentuating their morphologies. The near absence of plant-fossil assemblages in the Early Triassic of South Africa can be explained as the result of changes in regional watertable in response to landscape-degradation events in the basin. Hence, the paucity of a plant-fossil record in this case is a taphonomic artifact of the basin (Gastaldo et al. 2005) rather than the absence of a terrestrial flora in response to terrestrial extinction (Ward et al. 2000).

5.2 *Plant Assemblages in Aggradational Landscapes*

5.2.1 Eocene Willwood Formation, Western United States

The Bighorn Basin in Wyoming, USA, is a large intermontane basin encompassing approximately 10,000 mi^2 of dominantly Cretaceous rock exposures in addition to localized outcrops of Early Mesozoic and Tertiary successions. Part of the Lower Eocene stratigraphy is assigned to the Willwood Formation, characterized by alluvial deposits consisting of vertically stacked floodplain mudrock (moderate to strongly developed paleosols), thick, multi-storied sheet sandstone bodies (meandering trunk channels), and heterolithic deposits consisting of ribbon sandstone surrounded by pedogenically altered mudrock (avulsion deposits associated with weakly developed paleosols; Kraus 1996, 2001; Fig. 7). Poorly developed paleosols within their heterolithic facies indicate high sedimentation rates controlled by avulsion (Kraus and Gwinn 1997). In contrast, better developed and more mature soils are indicative of low sedimentation rates on floodplains controlled by overbank deposition (Kraus 2002). A gleyed, drab-colored hydromorphic paleosol underlies

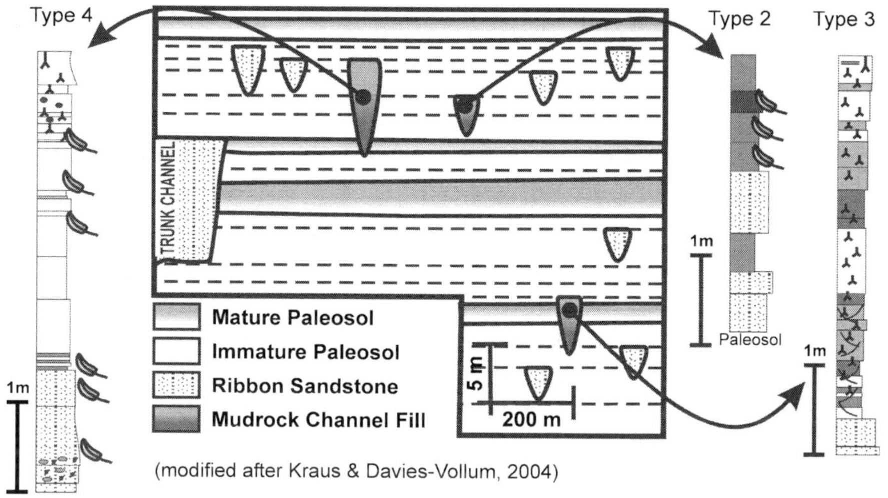

Fig. 7 An Eocene example of aggradational landscapes in the Willwood Formation, Wyoming, USA. Stacked poorly developed (immature) paleosols occur throughout the sequence, with interspersed better developed (more mature) paleosols. Fossil plant assemblages are most common within siltstone channels interpreted as abandoned avulsion channels within the floodplain. Kraus and Davies-Vollum (2004) recognize four different channel-fill types. Channel-fill Types 2 and 4 may contain well-preserved leaf mats and show the least effects of pedogenic activity. Channel Types 1 (*not illustrated*) and 3 rarely preserve plant debris due to pedogenic alteration of sediments

the carbonaceous shale which, in turn, is overlain by the development of a yellow-brown paleosol in the former sequence (Davies-Vollum and Kraus 2001). The better developed soil types range from vertisols to Fersiallitic paleosols (no direct equivalent in the Soil Survey Staff 2006; Kraus 2002). Spatial variation in the more developed paleosols are suspected to be indicative of variations in regional drainage which are linked to grain size and, hence, permeability.

Plant-fossil assemblages are common in the carbonaceous shale facies, and have been used to reconstruct not only the paleoecological mosaic but also the paleoclimate of the Early Eocene (Davies-Vollum and Wing 1998). Carbonaceous shale bodies exhibit both tabular and lenticular geometries (Wing 1984), and the plant accumulations within display the entire spectrum from exceptionally well preserved (Davies-Vollum and Wing 1998) to very poorly preserved assemblages (Kraus and Davies-Vollum 2004). The best preserved material occurs in two facies. These are mudrock that are relatively low in total organic carbon and unaltered by soil-forming processes, and carbonaceous shales that also are the least pedogenically altered. Both facies are interpreted as overbank-wetland deposits (Davies-Vollum and Wing 1998; Davies-Vollum and Kraus 2001) and, more recently, lenticular bodies have been recognized as channel-fill sequences associated with avulsion events. Where plant debris is preserved in carbonaceous shale of a tabular nature, Davies-Vollum and Kraus (2001) interpreted the overlying paleosols as having formed on the floodplain during avulsions. The underlying carbonaceous shale, then, represents forest-floor (in situ) accumulations within the distal floodplain. They note that conditions promoting organic-matter accumulation in this setting require a high water table, at or above the soil surface. Such a state results in waterlogged, anoxic conditions that may prohibit bioturbation and inhibit decay of organic material. Hence, the maintenance of a high local (or regional) water table during the initiation of each aggradational phase, associated with its rise in the section during the emplacement of stacked soil horizons, resulted in preservation of backswamp (autochthonous) debris.

Parautochthonous assemblages are found within mudrock facies interpreted as abandoned crevasse-splay feeder channels (Kraus and Davies-Vollum 2004). These often are overprinted by pedogenic activity in the later stages of, or following, channel fill, as plant colonization occurs and the most recent deposits were amalgamated into the floodplain (Wing 1984). Based on what can be determined from the literature, there does not seem to be evidence for the preservation of coeval litter (O) horizons at the soil–air interface equivalent with channel-fills. Four different categories of channel-fill successions are recognized by Kraus and Davies-Vollum (2004) in which plant part preservation varies. Poorly preserved plant assemblages are found in Types 1 and 3 fills (Fig. 7). In the former, decay of organic matter is attributed to the presence of circulating oxygenated water and low sedimentation rates, whereas pedogenesis overprints Type 3 fills and eliminates any buried organic matter. Type 3 fills seemingly are associated more commonly with more mature paleosol horizons (Kraus and Hasiotis 2006), resulting in the greater depth of pedogenic alteration and soil modification. The depth of root penetration and soil modification ultimately are controlled by variations in local watertable. Type 2 fills (Fig. 7) contain lenticular carbonaceous shale that preserves leaf mats as well as dispersed plant debris. The presence of organic-rich mudrock reflects slower sedimentation rates and a change in prevailing Eh/pH conditions promoting preservation (Gastaldo and Huc 1992). Lastly, Type 4 fills are characterized by conglomerate or sandstone basal deposits with allochthonous plant debris, and intervals of interbedded mudstone and siltstone in which moderate-to-well preserved parautochthonous fossils occur. Kraus and Davies-Vollum (2004) attribute the variety of fill characteristics to the differences in local drainage conditions that followed channel abandonment, as well as how rapidly the channel was filled. Their assemblages display similar taphonomic attributes to those described by Gastaldo et al. (1998) in Germany (see above). Hence, within the aggradational landscape of the Eocene Willwood Formation, fossil-plant assemblages have the highest preservation potential when they are maintained below the prevailing watertable, either in the short term (aggradational phase characterized by poorly developed paleosols) or longer term (the phase during which more mature soils developed). Once the local/regional watertable fell below the level of buried plant detritus in Types 1 and 3 channel fills, pedogenic activity removed most, if not all traces of the remains.

The type and nature of rhizoliths preserved within these aggradational paleosols recently have been used to identify differences in the paleodrainage of the landscape. Kraus and Hasiotis (2006) provide criteria to separate those soil (more mature) types that formed under relatively well-drained from those formed under poorly drained (immature) conditions. Iron (hematite) and manganese staining adjacent to rhizoliths, as well as the presence of calcareous rhizoconcretions are considered to be indicative of better drained soils (albeit without stable isotope geochemical constraint), whereas rhizoliths in poorly drained paleosols commonly are surrounded by goethite rims and may be preserved by jarosite (pyrite oxidation product). Fluxes in the geochemical conditions of the soil, in response to climate, result in the variety of preservational states of rooting structures as well as previously buried litters.

5.2.2 Upper Jurassic Morrison Formation, Western United States

The Morrison Formation was deposited in a broad, continental, retroarc basin in western North America during the initial stages of collision- and subduction-related tectonic mountain building processes that eventually culminated in the Sevier orogeny (DeCelles 2004). Demko et al. (2004) interpret the Morrison to represent two aggradational sequences with an intervening degradational episode. The upper and lower boundaries, and the medial bounding surface separating the two aggradational successions, were interpreted as basinwide sequence-bounding unconformities. These were generated by changes in dynamic uplift and subsidence patterns due to hinterland tectonics and deep crust-mantle reorganization. The initial aggradational episode in the Colorado Plateau region is represented by the restricted-marine to lacustrine facies of the Tidwell Member and the overlying thick package of low-sinuosity fluvial sandstones and associated floodplain deposits of the Salt Wash Member. Overlying a surface marked by a widespread paleosol and minor incision, the second aggradational succession is represented by the mudstone-prone fluvial and lacustrine facies of the Brushy Basin Member. Paleosols, trace fossils, and other sedimentary paleoclimatic proxies indicate that the Morrison was deposited under arid to tropical wet-dry conditions (Demko and Parrish 1998; Demko et al. 2004; Hasiotis 2004; Turner and Peterson 2004).

Even though outcrop of the Morrison Formation covers a very large area of the Western Interior of the United States and is, in most cases, very well exposed in the modern semi-arid climate, plant fossils are typically very rare, with the exception of the very uppermost beds in the northern reaches of the outcrop belt in a coal and organic-rich mudstone interval (Parrish et al. 2004). However, traces of roots, moderately well-developed paleosols, a diverse and abundant soil ichnofauna associated with socialized insects, and the abundant and diverse remains of large herbivorous saurpod dinosaurs for which the formation is famous, indicate that the landscape supported extensive vegetation (Demko et al. 2004; Hasiotis 2004; Engelmann et al. 2004). The plant fossils that do occur include coniferophytes, ginkophytes, cycadophytes, ferns, and other plants (Tidwell 1990; Parrish et al. 2004).

In a taphonomic study of fossil plants in the Morrison, Parrish et al. (2004) concluded that within the bulk of the formation, surface and near-surface conditions during deposition were such that even small fragments of carbonaceous debris were not preserved. They attribute this relative paucity of identifiable plant remains to oxidation in well-drained soil conditions and the relative scarcity of durable woody vegetation on the landscape in general. The few well-preserved foliar assemblages described by Parrish et al. (2004) were associated with wetland and marginal-lacustrine depositional environments where watertable would have remained seasonally high in an otherwise well-drained landscape. However, both wetland and shallow or marginal-lacustrine settings that may have existed in overall net-evaporative climatic settings like the Morrison basin, must have been subject to regular low lake or watertable conditions within the realm of typical water budget fluctuations over tens to thousands of years, reducing the preservation potential of large volumes of organic material. Also, these types of depositional environments can only exist in the absence of high siliciclastic sedimentation rates, also reducing the potential for quick burial before decay or detritivory. Parrish et al. (2004) note that evidence suggests that the uppermost Morrison strata in the northern part of the depositional basin indicate of a less restrictive water budget, reflected not only in the type and amount of plant material preserved, but also in the types of paleosols and other paleoclimatic indicators (Fig. 8). They suggest that this may have been due to a lower overall temperature regime in that area at that time, rather than any relative increase in yearly precipitation as compared to the seemingly drier areas to the south.

The plant-fossil record of the Morrison Formation illustrates that even in overall aggradational settings, preservation of identifiable remains is still intimately linked to a balance between the A/S ratio and the maintenance of a perennially high watertable in the area where plants either are growing or introduced into the depositional environment. Paleosol types, ichnofossils, stable-isotope geochemistry of pedogenic and lacustrine/palustrine carbonates, and other sedimentary paleoclimatic indicators in the Morrison indicate that even though the landscape was dominantly aggrading though time, watertables were low or fluctuating (Demko et al. 2004; Dunagan and Turner 2004). Significant assemblages of identifiable plant material are rare and only present in deposits that accumulated in isolated lows on the landscape, near lake margins and floodplain ponds.

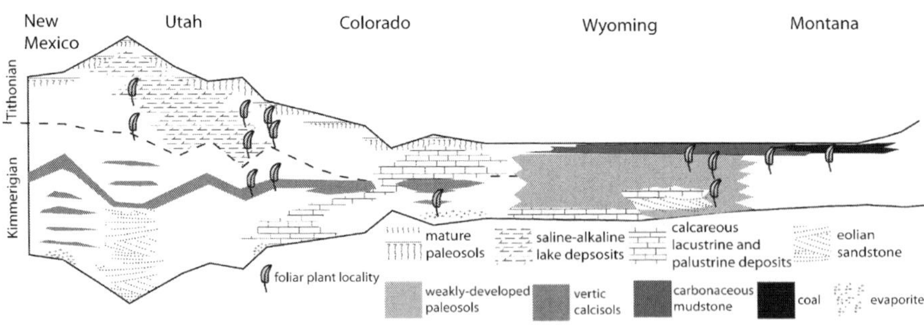

Fig. 8 Schematic north-south cross-section of the Morrison Formation in the western USA showing distribution of sedimentary paleoclimatic indicators and foliar fossil-plant localities (After Parrish et al. 2004)

6 Conclusions

There is no doubt that the general processes that promote preservation of a megafloral fossil record (or any fossil record) are restricted in space and time. Based on actuopaleontological studies, it continues to be presumed that once plant detritus is introduced into an appropriate depositional setting, the chances for its preservation and retention in the stratigraphic record are high (which is not necessarily the case). This assumption is particularly true if the debris is buried rapidly, entombing it in a sedimentary package. Hence, when these environments of deposition are encountered in a basin without evidence for terrestrial plants, catastrophic perturbation often has been invoked to explain their absence. In other words, the lack of a plant-fossil record in a particular stratigraphic interval is considered to be more a function of ecosystem reorganization, extirpation, or extinction than the mere fact that prevailing conditions at the time of sediment accumulation were outside of the preservational window. In reality, without an understanding of the plant taphonomic attributes within the sedimentological and regional/basinal context, the absence of a plant-fossil record may be either a function of (1) sediment supply, attendant geochemistry, and climate preventing preservation at the onset; or (2) the interaction of landscape and climate, over the short (millennial) and long (lakh to millions of years) term, rather than extirpation or outright extinction.

Sediment deposition within continental landscapes, regardless of their geographic position, is a function of disequilibirum within the fluvial (graded) profile across the drainage system(s). Aggradational landscapes that form adjacent to river systems in disequilibrium are characterized by wetland (histosols, entisols, gleyed soils) and moderately drained soil types. Plant-fossil assemblages have the highest preservation potential in settings that are maintained below the maximum draw down of the regional watertable (e.g., channel barforms, abandoned avulsion channels, oxbow lakes, other limnic regimes, etc.). Under normal circumstances, plant debris that accumulates at the air-soil interface is recycled even when buried by overbank flood deposits due to the re-establishment of the regional watertable and exposure to oxygenating interstitial conditions. Preservation potential only increases when this O-horizon is displaced to a stratigraphic position some distance below the prevailing water table, and is maintained under geochemical conditions that retard or eliminate autocatalytic, bacterial, or detritivore activity (Fig. 2). Hence, the highest fidelity plant-fossil assemblages are preserved within aggradational landscapes above each disconformity (marked by a soil profile).

When fluvial systems are in equilibrium, the landscape is static providing sufficient time over which pedogenic features indicative of mature soil types form (e.g., thick O-horizons in histosols, well-developed Bk or K horizons in calcisols, etc.). Similar to aggradational landscapes, the highest preservation potential for plants exists in subaqueous environments (permanent standing bodies of water) when climate variables result in evaporation exceeding precipitation, and within peat mires/bogs when precipitation exceeds evaporation (and a perched watertable prevents or retards drainages; Gastaldo, 2010). In most instances where the landscape is controlled by high evaporation rates under a dry or seasonally dry climate, there is an improbable chance for plant-part preservation anywhere in that landscape (although vertebrate assemblages often are preserved under these conditions). Deep penetration of pedogenic processes and drying by atmospheric gases results in organic matter disintegration. Preservation potential exists only for those materials existing below the level of deepest pedogenic penetration. Similarly, and although it may seem counterintuitive, in most instances where the landscape is controlled by high precipitation rates under a humid or everwet climate, there is a low chance for plant-part preservation in the landscape. This is because sediment that would have been available for transport out of the landscape is sequestered in soils by the rootstocks of dense vegetation. The perception of muddy rivers in Recent tropical regimes is a function of human agricultural and silvicultural activities (e.g., Staub et al. 2000).

Sediment removal within continental landscapes, again regardless of their geographic position, is a function of disequilibrium within the fluvial profile across the drainage system(s). Degradational landscapes are characterized by fluvial incision, sediment excision and re-entrainment of interfluvial deposits, and transport of these materials, along with any previously entombed organics, to a depositional site below the new base level. Fluvial incision resets the relative position of the water table in the "new" landscape in a stratigraphic position below its previous height. As such, watertable fluctuations in response to whatever prevailing climate now operating in the region will result in the re-exposure of once buried organic matter to renewed biogeochemical processes which, in effect, will remove all traces of organic residuum that may have been present originally.

The long-term fate of buried organic matter and the size of the sediment clasts in which it was entombed controls the ultimate preservational mode (Schopf 1975) of fossil-plant material collected in the field. Compressions, where a coalified residuum exists on the bedding-plane surface, are the result of plant debris buried within aggradational landscapes wherein the subfossil assemblage was maintained below the prevailing and subsequent water table(s). Devolatilization and depolymerization leave a residuum of the most refractory chemicals surrounded by matrix. The external impression (mold) of those plant parts is best developed in fine-grained matrices (clay and silt-sized clasts), whereas the fidelity of the impression decreases with an increasing clast size. When both the compression and impression of the original plant remains exist, this preservational mode has been termed an adpression (Shute and Cleal 1987). Impressions, where there is no evidence of an organic residual on the bedding plane, are the result of organic decay following early diagenetic compaction by burial. There would be no transfer of plant morphology to the sediment without a decrease in pore space within the entombing matrix. Hence, for such a transfer of morphology to exist, pore space must have been reduced through burial at some depth less than that required for lithification. But, the removal of all organic residuum from the matrix requires that active decay processes (or thermal heating in response to tectonic activity) were reestablished prior to cementation. Such processes occur when landscapes are static (equilibrium) or when the landscape has been degraded. The larger the size of the entombing clasts, the higher the probability that once buried plant debris, now exposed to oxygenating waters under a reset watertable, will be removed from the stratigraphic record. Other preservational states of plant material, including permineralization, pyritization, replacement, etc. can be explained as a function of changing geochemical groundwater-solute concentrations and conditions, and have the potential to occur under all landscape phases.

Acknowledgments Support for RAG includes: a Forschungspreis from the Alexander von Humboldt Foundation for studies in the Weißelster basin, Germany; NSF EAR 0417317 and a Mellon Foundation grant to Colby, Bates, and Bowdoin Colleges for research efforts in the Karoo Basin, South Africa; and NSF EAR, ACS PRF, NATO, and other agencies for plant-taphonomic investigations in the southeastern U.S., Kalimantan, Indonesia, Sarawak, Malaysia, and central Europe. Support for TMD includes: NSF EAR 9305087, USGS-NPS Interagency Agreement 1443-IA-1200-94-003, Chevron, Colorado State University, and the University of Minnesota Duluth.

References

Alexander, J., Fielding, C. R., & Jenkins, G. (1999). Plant-material deposition in the tropical Burdekin River, Australia; implications for ancient fluvial sediments. *Palaeogeography, Palaeoclimatology, Palaeoecology, 153*, 105–125.

Allen, J. P., & Gastaldo, R. A. (2006). Sedimentology and taphonomy of the Early to Middle Devonian plant-bearing beds of the Trout Valley Formation, Maine. In W. A. DiMichele & S. Greb (Eds.), *Wetlands through time: Geological Society of America, Special Publication* 399 (pp. 57–78).

Allison, P. A., & Pye, K. (1994). Early diagenetic mineralization and fossil preservation in modern carbonate concretions. *Palaios, 9*, 561–575.

Atchley, S. C., Nordt, L. C., & Dworkin, S. I. (2004). Eustatic control on alluvial sequence stratigraphy: A possible example from the Cretaceous-tertiary transition of the Tornillo Basin, Big Bend National Park, West Texas, USA. *Journal of Sedimentary Research, 74*, 391–404.

Ash, S. R. (1970). Ferns from the Chinle Formation (Upper Triassic) in the Fort Wingate area. *New Mexico: US Geological Survey Professional Paper, 613D*, 1–40.

Ash, S. R. (1972). Plant megafossils of the Chinle Formation. In C. S. Breed & W. J. Breed (Eds.), *Investigations of the Triassic Chinle Formation: Museum of Northern Arizona Bulletin, 47* (pp. 23–43).

Ash, S. R. (1980). Upper Triassic floral zones of North America. In D. L. Dilcher & T. M. Taylor (Eds.), *Biostratigraphy of fossil plants* (pp. 153–170). Stroudsburg: Dowden, Hutchinson, Ross.

Ash, S. R. (1987). The Upper Triassic red bed flora of the Colorado Plateau, western United States. *Journal of the Arizona-Nevada Academy of Sciences, 22*, 95–105.

Ash, S. R. (1991). A new pinnate cycad leaf from the Upper Triassic Chinle Formation of Arizona. *Botanical Gazette, 152*, 123–131.

Ash, S. R. (2001). New cycadophytes from the Upper Triassic Chinle Formation of the southwestern United States. *PaleoBios, 21*, 15–28.

Ash, S. R., & Creber, G. T. (2000). The Late Triassic *Araucarioxylon arizonicum* trees of the Petrified Forest National Park, Arizona, USA. *Palaeontology, 43*, 15–28.

Baldwin, C. T., Strother, P. K., Beck, J. H., & Rose, E. (2004). Palaeoecology of the Bright Angel Shale in the eastern Grand Canyon, Arizona, USA, incorporating sedimentological, ichnological and palynological data. *Geological Society Special Publications, 228*, 213–236.

Beraldi-Campsei, H., Cevallos-Ferriz, S. R. S., Centeno-García, E., Arenas-Abad, C., & Fernández, L. P. (2006). Sedimentology and paleoecology of an Eocene-Oligocene alluvial-lacustrine arid system, Southern Mexico. *Sedimentary Geology, 191*, 227–254.

Blum, M. D., & Törnqvist, T. E. (2000). Fluvial responses to climate and sea-level change: A review and look forward. *Sedimentology, 47*, 2–48.

Boyd, R. C., DiEssel, C., Wadsworth, J. A., Leckie, D. A., & Zaitlin, B. A. (2000). Developing a model for non-marine sequence stratigraphy – Application to the western Canada sedimentary basin (abstract): *GeoCanada 2000 Conference Abstracts, CD-ROM* (p. 4).

Bray, J. R., & Gorham, E. (1964). Litter production in forests of the world. In J. B. Cragg (Ed.), *Advances in ecological research.* New York: Academic (Vol. 2, pp. 101–157).

Bull, W. B. (1991). *Geomorphic responses to climatic change.* Oxford, UK: Oxford University Press. 326 p.

Burnham, R. J. (1993). Time resolution in terrestrial macrofloras: Guidelines from modern accumulations: Short courses in paleontology. *Paleontological Society, 6*, 57–78.

Burnham, R. J., & Spicer, R. A. (1986). Forest litter preserved by volcanic activity at El Chichon, Mexico: A potentially accurate record of the pre-eruption vegetation. *Palaios, 1*, 158–161.

Cecil, C. B., & Dulong, F. T. (2003). Precipitation models for sediment supply in warm climates. In C. B. Cecil & N. T. Edgar (Eds.), *Climate controls on stratigraphy, SEPM Special Publication 77* (pp. 21–27).

Creber, G. T., & Ash, S. R. (2004). The Late Triassic *Schilderia adamanica* and *Woodworthia arizonica* trees of the Petrified Forest National Park, Arizona, USA. *Palaeontology, 147*, 21–38.

Daugherty, L. H. (1941). *The Upper Triassic flora of Arizona: Carnegie Institute of Washington Publication 526* (108 p.)

Davies-Vollum, K. S., & Kraus, M. J. (2001). A relationship between alluvial backswamps and avulsion cycles: An example from the Willwood Formation of the Bighorn Basin. *Wyoming: Sedimentary Geology, 140*, 235–249.

Davies-Vollum, K. S., & Wing, S. L. (1998). Sedimentological, taphonomic, and climatic aspects of Eocene swamp deposits (Willwood Formation, Bighorn Basin, Wyoming). *Palaios, 13*, 28–40.

DeCelles, P. G. (2004). Late Jurassic to Eocene evolution of the Cordilleran thrust belt and foreland basin system, western USA. *American Journal of Science, 304*, 105–168.

Demko, T. M. (1995a). Taphonomy of fossil plants in Petrified Forest National Park, Arizona. In *Fossils of Arizona: Proceedings, 1995 Southwest Palaeontological Society and Mesa Southwest Museum Mesa*, Arizona (pp. 37–52).

Demko, T. M. (1995b). *Taphonomy of fossil plants in the Upper Triassic Chinle Formation: Ph.D. dissertation.* Tucson: University of Arizona. 274 p.

Demko, T. M., & Parrish, J. T. (1998). Paleoclimatic setting of the Upper Jurassic Morrison Formation. *Modern Geology, 22*, 283–296.

Demko, T. M., Dubiel, R. F., & Parrish, J. T. (1998). Plant taphonomy in incised valleys: Implications for interpreting paleoclimate from fossil plants. *Geology, 26*, 1119–1122.

Demko, T. M., Currie, B. S., & Nicoll, K. A. (2004). Regional paleoclimatic and stratigraphic implications of paleosols and fluvial-overbank architecture in the Morrison Formation (Upper Jurassic), Western Interior, USA. *Sedimentary Geology, 167*, 117–137.

DiMichele, W. A., & Gastaldo, R. A. (2008). Deep time plant paleoecology. *Annals of the Missouri Botanical Gardens, 95*, 144–198.

Drum, R. W. (1968). Silicification of Betula wood tissue in vitro. *Science, 161*, 175–176.

Dubiel, R. F. (1994). Triassic deposystems, paleogeography, and paleoclimate of the Western Interior. In M. V. Caputo, J. A. Peterson, & K. J. Franczyk (Eds.), *Mesozoic systems of the rocky mountain region, USA: SEPM Rocky Mountain Section, Denver, CO* (pp. 133–168). Tulsa, Oklahoma: SEPM.

Dubiel, R. F., Parrish, J. T., Parrish, J. M., & Good, S. C. (1991). The Pangaean megamonsoon – Evidence from the Upper Triassic Chinle Formation, Colorado Plateau. *Palaios, 6*, 347–370.

Dunagan, S. P., & Turner, C. E. (2004). Regional paleohydrologic and paleoclimatic settings of wetland/lacustrine depositional systems in the Morrison Formation (Upper Jurassic), Western Interior, USA. *Sedimentary Geology, 167*, 269–296.

Dunn, K. A., McLean, R. J. C., Upchurch, G. R., Jr., & Folk, R. L. (1997). Enhancement of leaf fossilization potential by bacterial films. *Geology, 25*, 1119–1122.

Edwards, D., & Feehan, J. (1980). Records of Cooksonia-type sporangia from late Wenlock strata in Ireland. *Nature, 287*, 41–42.

Eissmann, L. (1970). Geologie des Bezirkes Leipzig. *Natura Regionis Lipsiensis, 1/2*, 1–172.

Engelmann, G. F., Chure, D. J., & Fiorillo, A. R. (2004). The implications of a dry climate for the paleoecology of the fauna of the Upper Jurassic Morrison Formation. *Sedimentary Geology, 167*, 297–308.

Etheridge, F. G., Wood, L. J., & Schumm, S. A. (1998). Cyclic variables controlling fluvial sequence development: Problems and perspectives. In K. W. Shanley & P. J. McCabe (Eds.), *Relative role of eustacy, climate and tectonism in continental rocks: SEPM Special Publication 59* (pp. 17–29)

Fairon-Demaret, M., & Scheckler, S. E. (1987). Typification and redescription of *Moresnetia zalesskyi* Stockmans, 1948, an early seed plant from the upper Famennian of Belgium: Bulletin de l'Institut Royal des Sciences Naturelles de Belgique. *Sciences de la Terre, 57*, 183–199.

Fielding, C. R., Alexander, J., & Newman-Sutherland, E. (1997). Preservation of *in situ*, arborescent vegetation and fluvial bar construction in the Burdekin River of North Queensland, Australia. *Palaeogeography, Palaeoclimatology, Palaeoecology, 135*, 123–144.

Fritz, W. J. (1980). Reinterpretation of the depositional environment of the Yellowstone "fossil forests". *Geology, 8*, 309–313.

Fritz, W. J., & Harrison, S. (1985). Transported trees from the 1982 Mount St. Helens sediment flows: Their use as paleo-current indicators. *Sedimentary Geology, 42*, 49–64.

Gastaldo, R. A. (1989). Preliminary observations on phytotaphonomic assemblages in a subtropical/temperate Holocene bayhead delta: Mobile Delta, Gulf Coastal Plain, Alabama. *Review of Palaeobotany and Palynology, 58*, 61–84.

Gastaldo, R. A. (1992). Taphonomic considerations for plant evolutionary investigations. *The Palaeobotanist, 41*, 211–223.

Gastaldo, R. A. (1994). The genesis and sedimentation of phytoclasts with examples from coastal environments. In A. Traverse (Ed.), *Sedimentation of organic particles* (pp. 103–127). Cambridge: Cambridge University Press.

Gastaldo, R. A. (2004). The relationship between bedform and log orientation in a Paleogene fluvial channel, Weißelster basin, Germany: Implications for the use of coarse woody debris for paleocurrent analysis. *Palaios, 19*, 595–606.

Gastaldo, R. A. (2010). Peat or No Peat: Why do the Rajang and Mahakam Deltas Differ?: International Journal of Coal Geology, v. **, p. ***-*** (doi; 10.1016/j.coal.2010.01.005)

Gastaldo, R. A., & Degges, C. W. (2007). Sedimentology and paleontology of a carboniferous Log Jam. *International Journal of Coal Geology, 69*, 103–118.

Gastaldo, R. A., & Huc, A. Y. (1992). Sediment facies, depositional environments, and distribution of phytoclasts in the Recent Mahakam River delta, Kalimantan, Indonesia. *Palaios, 7*, 574–591.

Gastaldo, R. A., & Rolerson, M. W. (2008). *Katbergia* gen. nov., a New Trace Fossil from the Late Permian and Early Triassic of the Karoo Basin: Implications for paleoenvironmental conditions at the P/Tr extinction event. *Palaeontology, 51*, 215–229.

Gastaldo, R. A., & Staub, J. R. (1999). A Mechanism to explain the preservation of leaf litters lenses in coals derived from raised mires. *Palaeogeography, Palaeoclimatology, Palaeoecology, 149*, 1–14.

Gastaldo, R. A., Douglass, D. P., & McCarroll, S. M. (1987). Origin, characteristics and provenance of plant macrodetritus in a Holocene crevasse splay, mobile delta, Alabama. *Palaios, 2*, 229–240.

Gastaldo, R. A., Bearce, S. C., Degges, C., Hunt, R. J., Peebles, M. W., & Violette, D. L. (1989). Biostratinomy of a Holocene oxbow lake: A backswamp to mid-channel transect. *Review of Palaeobotany and Palynology, 58*, 47–60.

Gastaldo, R. A., Allen, G. P., & Huc, A. Y. (1995). The tidal character of fluvial sediments of the Recent Mahakam River delta, Kalimantan, Indonesia. *Special Publications International Association of Sedimentologists, 24*, 171–181.

Gastaldo, R. A., Feng, W., & Staub, J. R. (1996). Palynofacies patterns in channel deposits of the Rajang River and delta, Sarawak, East Malaysia. *PALAIOS, 11*, 266–279.

Gastaldo, R. A., Walther, H., Rabold, J., & Ferguson, D. (1996). Criteria to distinguish parautochthonous leaves in cenophytic alluvial channel-fills. *Review of Palaeobotany and Palynology, 91*, 1–21.

Gastaldo, R. A., Riegel, W., Püttmann, W., Linnemann, U. H., & Zetter, R. (1998). A multidisciplinary approach to reconstruct the Late Oligocene vegetation in central Europe. *Review of Palaeobotany and Palynology, 101*, 71–94.

Gastaldo, R. A., Adendorff, R., Bamford, M. K., Labandeira, Neveling, J., & Sims, H. J. (2005). Taphonomic trends of macrofloral assemblages across the Permian-Triassic boundary, Karoo Basin, South Africa. *Palaios, 20*, 478–497.

Gastaldo, R. A., Purkyňová, E., Šimůnek, Z., & Schmitz, M. D. (2009). Ecological persistence in the Late Mississippian (Serpukhovian – Namurian A) Megafloral Record of the Upper Silesian Basin, Czech Republic. *Palaios, 24*, 336–350.

Gee, C. T. (2005). The genesis of mass carpological deposits (bedload carpodeposits) in the Tertiary of the Lower Rhine Basin, Germany. *Palaios, 20*, 464–479.

Gee, C. T., Abraham, M., & Sander, P. M. (1997). The occurrence of carpofloras in coarse sand fluvial deposits: Comparison of fossil and recent case studies. *Mededelingen Nederlands Instituut voor Toegepaste Geowetenschappen TNO, 58*, 171–178.

Gensel, P. G., & Edwards, D. (2001). *Plants invade the land: Evolutionary and environmental perspectives*. New York: Columbia University Press. 304 p.

Glasspool, I. J., Edwards, D., & Axe, L. (2004). Charcoal in the Silurian as evidence for the earliest wildfire. *Geology, 32*, 381–383.

Greb, S. F., DiMichele, W. D., & Gastaldo, R. A. (2006). Evolution of wetland types and the importance of wetlands in earth history. In W. A. DiMichele & S. Greb (Eds.), *Wetlands through time, Geological Society of America, Special Publication, 399* (pp. 1–40).

Grimes, S. T., Brock, F., Richard, D., Davies, K. L., Edwards, D., Briggs, D. E. G., et al. (2001). Understanding fossilization: Experimental pyritization of plants. *Geology, 29*, 123–126.

Gupta, N. S., & Pancost, R. D. (2004). Biomolecular and physical taphonomy of angiosperm leaf during early decay: Implications for fossilization. *Palaios, 19*, 428–440.

Halfar, J., Riegel, W., & Walther, H. (1998). Facies architecture and sedimentology of a meandering fluvial system: A Palaeogene example from the Weisselster Basin, Germany. *Sedimentology, 45*, 1–17.

Hasiotis, S. T. (2004). Reconnaissance of Upper Jurassic Morrison Formation ichnofossils, Rocky Mountain Region, USA: Paleoenvironmental, stratigraphic, and paleoclimatic significance of terrestrial and freshwater ichnocoenoses. *Sedimentary Geology, 167*, 177–268.

Hasiotis, S. T., & Mitchell, C. E. (1993). A comparison of crayfish burrow morphologies; Triassic and Holocene fossil, paleo- and neo-ichnological evidence, and the identification of their burrows signatures. *Ichnos, 2*, 291–314.

Hiller, N., & Stravrakis, N. (1984). Permo-Triassic fluvial systems in the Southeastern Karoo basin, South Africa. *Palaeogeography, Palaeoclimatology, Palaeoecology, 45*, 1–21.

Jacobs, B. F., Tabor, N., Feseha, M., Pan, A., Kappelman, J., Rasmussen, T., et al. (2005). Oligocene terrestrial strata of northwestern Ethiopia: A preliminary report on paleoenvironments and paleontology. *Palaeontologia Electronica, 8*(1), 19.

Johnson, M. R., Van Vuuren, C. J., Visser, J. N. J., Cole, D. I., Wickens, H., Christie, A. D. M., et al. (1997). The Foreland Karoo Basin, South Africa. In R. C. Selley (Ed.), *Sedimentary basins of the world* (pp. 169–185). New York: Elsevier.

Krasilov, A. (1975). *Paleoecology of terrestrial plants: Basic principles and techniques*. New York: Wiley. 283 p.

Kraus, M. J. (1996). Avulsion deposits in lower Eocene alluvial rocks, Bighorn Basin, Wyoming. *Journal of Sedimentary Research, 66*, 354–363.

Kraus, M. J. (2001). Sedimentology and depositional setting of the Willwood Formation in the Bighorn and Clark's Fork basins. In P. D. Gingerich (Ed.), *Paleocene-Eocene stratigraphy and biotic change in the Bighorn and Clarks Fork basins, Wyoming*: *Papers on Paleontology, 33* (pp. 15–28).

Kraus, M. J. (2002). Basin-scale changes in floodplain paleosols: Implications for interpreting alluvial architecture. *Journal of Sedimentary Research, 72*, 500–509.

Kraus, M. J., & Davies-Vollum, K. S. (2004). Mudrock-dominated fills formed in avulsion splay channels: Examples from the Willwood Formation, Wyoming. *Sedimentology, 51*, 1127–1144.

Kraus, M. J., & Gwinn, B. M. (1997). Facies and facies architecture of Paleogene floodplain deposits, Willwood Formation, Bighorn Basin, Wyoming, USA. *Sedimentary Geology, 114*, 33–54.

Kraus, M. J., & Hasiotis, S. T. (2006). Significance of different modes of rhizolith preservation to interpreting paleoenvironmental and paleohydrologic settings: Examples from Paleogene paleosols, Bighorn Basin, Wyoming, USA. *Journal of Sedimentary Research, 76*, 633–646.

Krutzsch, W. (1992). Paläobotanische Klimagliederung des Alttertiärs (Mitteleozän bis Oberoligozän) in Mitteldeutschland und das Problem der Verknüpfung mariner und kontinentaler Gliederungen (klassische Biostratigraphien-paläobotanisch-ökologische Klimastratigraphie-Evolutionsstratigraphie der Vertebraten. *Neues Jahrbuch für Geologische und Paläontologische Abandelung, 186*, 137–153.

Lawton, T. F. (1994). Tectonic setting of Mesozoic sedimentary basins, Rocky Mountain region, United States. In M. V. Caputo, J. A. Peterson, & K. J. Franczyk (Eds.), *Mesozoic systems of the Rocky Mountain Region, USA: SEPM Rocky Mountain Section, Denver, CO* (pp. 1–25). Tulsa, Oklahoma: SEPM.

Machin, J. J. (1948). Concept of the graded river. *Bulletin of the Geological Society of America, 59*, 463–512.

Mamay, S. H. (1992). *Sphenopteridium* and *Telangiopsis* in a *Diplopteridium*-like Association from the Virgilian (Upper Pennsylvanian) of New Mexico. *American Journal of Botany, 79*, 1092–1101.

McCarthy, P. J., & Plint, A. G. (1998). Recognition of interfluve sequence boundaries: Integrating paleopedology and sequence stratigraphy. *Geology, 26*, 387–390.

McCarthy, P. J., & Plint, A. G. (2003). Spatial variability of palaeosols across Cretaceous interfluves in the Dunvegan Formation, NE British Columbia, Canada: Palaeohydrological, palaeogeomorphological and stratigraphic implications. *Sedimentology, 50*, 1187–1220.

Meyer-Berthaud, B., Scheckler, S. E., & Wendt, J. (1999). Archaeopteris is the earliest known modern Tree. *Nature, 398*, 700–701.

Miall, A. D. (1991). Stratigraphic sequences and their chronostratigraphic correlation. *Journal of Sedimentary Petrology, 61*, 497–505.

Mitchum, Jr., R. M., Vail, P. R., & Thompson III, S. (1977). Seismic stratigraphy and global changes of sea level, part 2: The depositional sequence as a basic unit for stratigraphic analysis. In C. E. Payton (Ed.), *Seismic stratigraphy – Applications to hydrocarbon exploration: AAPG Memoir 26* (pp. 53–62).

Mosbrugger, V., Utescher, T., & Dilcher, D. L. (2005). Cenozoic continental climatic evolution of Central Europe. *Proceedings of the National Academy of Sciences of the United States of America, 102*(42), 14964–14969.

Muto, T., & Steel, R. J. (2000). The accommodation concept in sequence stratigraphy: Some dimensional problems and possible redefinition. *Sedimentary Geology, 130*, 1–10.

Opluštil, S., Pšenička, J., Libertín, M., & Šimůnek, Z. (2007). Vegetation patterns of Westphalian and Lower Stephanian mire assemblages preserved in tuff beds of the continental basins of Czech Republic. *Review of Palaeobotany and Palynology, 143*, 107–154.

Pace, D. W., Gastaldo, R. A., & Neveling, J. (2009). Aggradational and Degradational Landscapes in the Early Triassic of the Karoo Basin and Evidence for Dramatic Climate Shifts Following the P/Tr Event: *Journal of Sedimentary Research, 79*, 276–291.

Parrish, J. T., Peterson, F., & Turner, C. E. (2004). Jurassic "savannah" – Plant taphonomy and climate of the Morrison Formation (Upper Jurassic, Western USA). *Sedimentary Geology, 167*, 137–162.

Perry, D. A. (1994). *Forest ecosystems*. Baltimore: The Johns Hopkins University Press. 649 p.

Plint, A. G., McCarthy, P. J., & Faccini, U. F. (2001). Nonmarine sequence stratigraphy: Updip expression of sequence boundaries and systems tracts in a high-resolution framework, Cenomanian Dunvegan Formation, Alberta foreland basin, Canada. *AAPG Bulletin, 85*, 1967–2001.

Posamentier, H. W., & Allen, G. P. (1999). *Siliciclastic sequence stratigraphy – Concepts and applications: SEPM Concepts in Sedimentology and Paleontology 6* (210 p.).

Posamentier, H. W., & Vail, P. R. (1988). Eustatic controls on clastic deposition. II. Sequence and system tract models. In C. K. Wilgus, B. S. Hastings, C. G. St. C. Kendall, H. W. Posamentier, C. A. Ross, & J. C. Van Wagoner (Eds.), *Sea-level changes: An integrated approach: SEPM Special Publication 42* (pp. 125–154).

Pratt, L. M., Phillips, T. L., & Dennison, J. M. (1978). Evidence of non-vascular land plants from the Early Silurian (Llandoverian) of Virginia, USA. *Review of Palaeobotany and Palynology, 25*, 121–149.

Quirk, D. G. (1996). 'Base profile': A unifying concept in alluvial sequence stratigraphy. In J. A. Howell & J. F. Aitken (Eds.), *High resolution sequence stratigraphy: Innovations and applications: Geological Society of America Special Publication, 104* (pp. 37–49).

Retallack, G. J. (2001). *Soils of the past: An introduction to paleopedology*. Malden, MA: Blackwell. 404 p.

Retallack, G. J., Smith, R. M. H., & Ward, P. D. (2003). Vertebrate extinction across Permian-Triassic boundary in Karoo Basin, South Africa. *Geological Society of America Bulletin, 115*, 1133–1152.

Rex, G. M., & Chaloner, W. G. (1983). The experimental formation of plant compression fossils. *Palaeontology, 26*, 231–252.

Rickards, R. B. (2000). The age of the earliest club mosses: The Silurian *Baragwanathia* flora in Victoria, Australia. *Geological Magazine, 137*, 207–209.

Rothwell, G. W., Scheckler, S. E., & Gillespie, W. H. (1989). *Elkinsia* gen. no, a late Devonian gymnosperm with cupulate ovules. *Botanical Gazette, 150*, 170–189.

Runkle, D. R. (1985). *Hydrology of the alluvial, buried channel, basal Pleistocene and Dakota aquifers in west-central Iowa: USGS Water-Resources Investigations Report 85–4239*, 111 p.

Scheihing, M. H., & Pfefferkorn, H. W. (1984). The taphonomy of land plants in the Orinoco Delta: A model for the incorporation of plant parts in clastic sediments of Late Carboniferous age of Euramerica. *Review of Palaeobotany and Palynology, 41*, 205–240.

Schopf, J. M. (1975). Modes of fossil preservation. *Review of Palaeobotany and Palynology, 20*, 27–53.

Schumm, S. A. (1993). River response to baselevel change: Implications for sequence stratigraphy. *The Journal of Geology, 101*, 279–294.

Scott, A. C. (2000). The pre-quaternary history of fire. *Palaeogeography, Palaeoclimatology, Palaeoecology, 164*, 281–329.

Scott, A. C., & Glasspool, I. J. (2005). Charcoal reflectance as aproxy for the emplacement temperature of pyroclastic flow deposits. *Geology, 33*, 589–592.

Sellwood, B. W., & Price, G. D. (1993). Sedimentary facies as indicators of Mesozoic palaeoclimate. *Philosophical Transactions: Biological Sciences, 341*, 225–233.

Shanley, K. W., & McCabe, P. J. (1991). Predicting facies architecture through sequence stratigraphy – An example from the Kaiparowits Plateau, Utah. *Geology, 19*, 742–745.

Shanley, K. W., & McCabe, P. J. (1994). Perspectives on the sequence stratigraphy of continental strata. *AAPG Bulletin, 78*, 544–568.

Shute, C. H., & Cleal, C. J. (1987). Palaeobotany in museums. *Geological Curator, 4*, 553–559.

Sigleo, A. C. (1978). Organic geochemistry of silicified wood, Petrified Forest National Park, Arizona. *Geochimica et Cosmochimica Acta, 42*, 1397–1405.

Sigleo, A. C. (1979). Geochemistry of silicified wood and associated sediments, Petrified Forest National Park, Arizona. *Chemical Geology, 26*, 151–163.

Smith, R. M. H., & Ward, P. D. (2001). Pattern of vertebrate extinctions across an event bed at the Permian-Triassic boundary in the Karoo Basin of South Africa. *Geology, 28*, 227–230.

Smith, R. M. H., Eriksson, P. G., & Botha, W. J. (1993). A review of the stratigraphy and sedimentary environments of the Karoo-aged basins of South Africa. *Journal of African Earth Science, 16*, 143–169.

Soil Survey Staff. (2006). *Keys to soil taxonomy* (10th ed.). Washington, DC: US Department of Agriculture, Natural Resources Conservation Service. 312 p.

Spicer, R. A. (1989). The formation and interpretation of plant fossil assemblages. *Advances in Botanical Research, 16*, 96–191.

Spicer, R. A. (1990). Transport/hydrodynamics of plant material. In D. E. G. Briggs & P. R. Crowther (Eds.), *Palaeobiology: A synthesis* (pp. 230–232). Oxford: Blackwell.

Spicer, R. A. (1991). Plant taphonomic processes. In D. E. G. Briggs & P. Allison (Eds.), *Taphonomy: Releasing the data locked in the fossil record* (pp. 71–113). New York: Plenum.

Spicer, R. A., & Greer, A. G. (1986). Plant taphonomy in fluvial and lacustrine systems. In T. Broadhead (Ed.), *Land plants: University of Tennessee, Department of Geological Sciences Studies in Geology, 15* (pp. 10–26).

Spicer, R. A., & Wolfe, J. A. (1987). Plant taphonomy of late holocene deposits in trinity (Clair Engle) lake, Northern California. *Paleobiology, 13*, 227–245.

Staub, J. R., Among, H. L., & Gastaldo, R. A. (2000). Seasonal sediment transport and deposition in the Rajang River Delta, Sarawak, East Malaysia. *Sedimentary Geology, 133*, 249–264.

Stewart, J. H., Anderson, T. H., Haxel, G. B., Silver, L. T., & Wright, J. E. (1986). Late Triassic paleogeography of the southern Cordillera; the problem of a source for voluminous volcanic detritus in the Chinle Formation of the Colorado Plateau region. *Geology, 14*, 567–570.

Strömberg, C. A. E. (2004). Using phytolith assemblages to reconstruct the origin and spread of grass-dominated habitats in the Great Plains of North America during the late Eocene to early Miocene. *Palaeogeography, Palaeoclimatology, Palaeoecology, 207*, 239–275.

Strother, P. K. (2000). Cryptospores: The origin and early evolution of the terrestrial flora. In R. A. Gastaldo & and W. A. DiMichele (Eds.), *Phanerozoic Terrestrial Ecosystems: The Paleontological Society Papers, 6* (pp. 3–19).

Tabor, N. J., Montanez, I. P., Steiner, M. B., & Schwindt, D. (2007). δ^{13}C Values of carbonate nodules across the Permian-Triassic boundary in the Karoo Supergroup (South Africa) reflect a stinking sulfurous swamp, not atmospheric CO_2. *Palaegeography, Palaeoclimatology, Palaeoecology, 252*, 370–381.

Tidwell, W. D. (1990). Preliminary report on the megafossil flora of the Upper Jurassic Morrison Formation. *Hunteria, 2*, 12.

Traverse, A. (Ed.). (1994). *Sedimentation of organic particles* (p. 556). Cambridge: Cambridge University Press.

Turner, C. E., & Peterson, F. (2004). Reconstruction of the Upper Jurassic Morrison Formation extinct ecosystem – a synthesis. *Sedimentary Geology, 167*, 309–355.

Ward, P. D., Montgomery, D. R., & Smith, R. M. H. (2000). Altered river morphology in South Africa related to the Permian-Triassic extinction. *Science, 289*, 1740–1743.

Wing, S. L. (1984). Relation of paleovegetation to geometry and cyclicity of some fluvial carbonaceous deposits. *Journal of Sedimentary Petrology, 54*, 52–66.

Wing, S. L., & DiMichele, W. A. (1995). Conflict between local and global changes in plant diversity through geological time. *Palaios, 10*, 551–564.

Wing, S. L., Hickey, L. J., & Swisher, C. C. (1993). Implications of an exceptional fossil flora for Late Cretaceous vegetation. *Nature, 363*, 342–344.

Wright, V. P., & Marriott, S. B. (1993). The sequence stratigraphy of fluvial depositional systems: The role of floodplain sediment storage. *Sedimentary Geology, 86*, 203–210.

Chapter 8
Hierarchical Control of Terrestrial Vertebrate Taphonomy Over Space and Time: Discussion of Mechanisms and Implications for Vertebrate Paleobiology

Christopher R. Noto

Contents

Abstract There is no doubt among paleontologists that the fossil record of terrestrial vertebrates is fragmented and unevenly distributed over space and time. The underlying causes of this patchiness derive from a combination of factors occurring before and after the deposition of vertebrate remains. Large-scale vertebrate fossil distribution patterns present challenges in addressing the effects of small-scale taphonomic processes on distribution patterns, and what, if any, effect they may have on biodiversity reconstructions. This chapter presents a hierarchical model connecting small-scale taphonomic processes and large-scale fossil preservation patterns. Factors acting at

C.R. Noto (✉)
Department of Biomedical Sciences, Grand Valley State University, Allendale, MI 49401, USA
e-mail: crnoto@life.bio.sunysb.edu

P.A. Allison and D.J. Bottjer (eds.), *Taphonomy: Process and Bias Through Time*, Topics in Geobiology 32, DOI 10.1007/978-90-481-8643-3_8,

higher levels in the hierarchy constrain the range of taphonomic processes acting at lower levels, whereas lower level processes are responsible for determining vertebrate preservation and the resulting fossil record for an area. Secular changes in climate, tectonics, sea-level, etc. alter the distribution of both environments and biodiversity over time. These changes in turn may alter the congruence between standing biodiversity and the fraction of that diversity faithfully represented in the fossil record, skewing our understanding of extinct vertebrate ecosystems and their evolution over time.

1 Introduction

The growth of terrestrial taphonomic science requires not only developing new analytical tools and techniques, but expanding the scope of research questions into new theoretical territory. Research conditions are rapidly changing as the development of large online databases allow for the compilation of data from a variety of sources into common, searchable formats (Alroy 2003; Barnosky and Carrasco 2000; Rees and Noto 2005). This development provides an unprecedented resource for studying the taphonomy and paleobiology of terrestrial vertebrates, particularly the ability to analyze regional and global patterns of fossil distribution. The potential for discovering and analyzing large-scale patterns in fossil distribution has been discussed for decades (Behrensmeyer et al. 2000), yet it remains to be explored how taphonomic factors, acting over multiple scales, interact to influence spatio-temporal preservational patterns of vertebrates. Martin (1999) proposed that many taphonomic processes follow a hierarchical organization (Rule 19, p. 391), though uncertainty remains about the strength of interaction between different levels. This concept has yet to be comprehensively explored in terrestrial systems, least of all its effect on vertebrate preservation and interpretations of large-scale patterns in fossil distribution.

1.1 Top-Down Versus Bottom-Up Controls on Terrestrial Taphonomy

In ecology, the processes structuring ecosystems or communities can be described as being exerted from the "bottom-up" or "top-down". Bottom-up refers to lower-level inputs (resources) exerting control over higher-level processes (community dynamics), whereas top-down control is where the structure of lower levels (such as species diversity) depends on processes acting at higher levels (predation or environmental disturbance) (Begon et al. 2006). Taphonomic processes and their effects on the fossil record also can be approached analogously, as suggested by Martin (1999). In this case, top-down processes restrict the range of taphonomic conditions that can act on remains prior to and following fossilization. Bottom-up processes act at the level of an individual carcass or bone in response to higher-order restrictions, producing the fossil record for an area and influencing our view of fossil distribution at that time (Fig. 1).

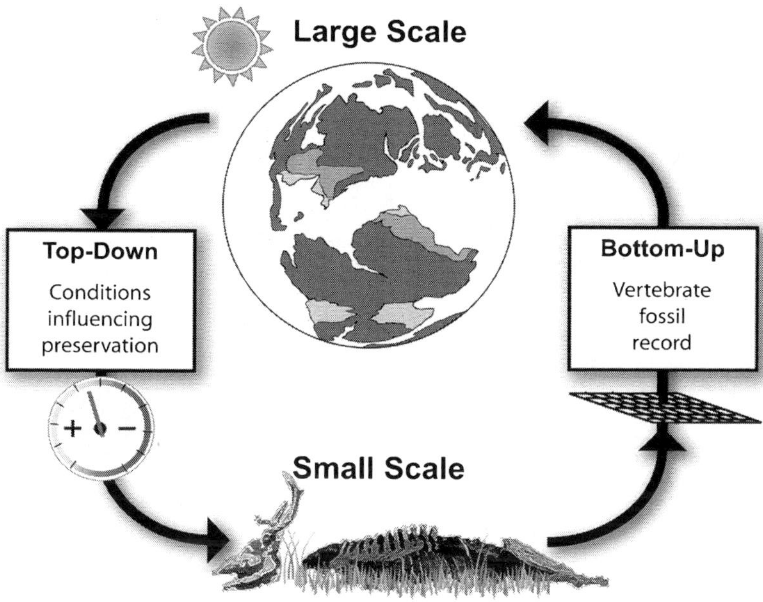

Fig. 1 The role of top-down and bottom-up processes on the terrestrial vertebrate fossil record. Large-scale conditions (*see text*) influence the range of taphonomic modes available in a given local environment. Generally favorable (+) or unfavorable (–) taphonomic modes drive the probability of preservation at small scales. The combination of preservation patterns at smaller scales form the taphonomic filters responsible for creating the fossil assemblage for a given time and place. The various fossil assemblages available for study inform our view of life on Earth during the time period in question

Climate, tectonic activity, and solar energy input exert top-down control over taphonomic processes through driving the distribution of different environments and the conditions that preserve or destroy vertebrate remains. Rogers (1993) suggested that the tectonic regime alone controls vertebrate fossil-accumulation patterns and therefore would affect paleoecological interpretations. This may be true in certain regions, especially in tectonically-active areas that experience aggradational–degradational cycles. However, climate appears to be the more important factor. Fiorillo (1999), in a review of fossil sites from the Late Cretaceous Foreland Basin of western North America, found that while regional tectonism did play a role, climatic influence was paramount in the formation of the area's vertebrate fossil record. Many studies across terrestrial vertebrate taxa have noted relationships between the distribution of fossils at local scales and regional-to-global climatic and biogeographic patterns (Benton 1985; Fastovsky 1987; Graham et al. 1996; Lehman 1997; Markwick 1998; Barnosky et al. 2003; Engelmann et al. 2004; Rees et al. 2004). Preservation patterns also vary over time in response to climate change. Millennial-scale climate changes due to plate tectonic movements and Milankovitch oscillations shift prevailing global climate patterns, altering not only environments and the distribution of species, but the distribution of taphonomic modes. Loope et al. (1998) and Brain (1995) found that periods of fossil assemblage formation in very different environments (eolian and cave, respectively) coincided with distinct climatic shifts towards greater precipitation, which created sedimentation events favoring vertebrate burial and preservation. Outside of these intervals the vertebrate fauna went largely unrecorded. Therefore, large-scale processes determine when and where preservation may occur at smaller scales by constraining local environmental conditions and taphonomic modes.

Bottom-up control is a product of local environmental conditions and includes small-scale processes such as necrolysis, biostratinomy, and diagenesis acting on individual remains within depositional settings, creating the taphonomic modes that drive preservation (Table 1). The sum of these processes over time contributes to

Table 1 Terrestrial depositional environments that contribute to the vertebrate fossil record and some of their important taphonomic attributes. Based on information found in Behrensmeyer and Hook (1992)

Depositional environment	Vertebrate occurrence	Taphonomic characteristics
Poorly-drained floodplain	Present	Heavily vegetated; high water table; reducing soils; sometimes acidic; fine-grained sediments frequently deposited
Well-drained floodplain	Present	Less vegetated; variable water table; well-developed, oxidizing soils; sometimes acidic; infrequent sedimentation; bioturbation
Eolian (dune, interdune, loess)	Uncommon	Fine-to-coarse grained sediments; dry conditions; periods of rapid burial
Lacustrine	Common	Range of productivity, sedimentation, temperature, chemistry, and oxygen content
Fluvial (channel lags, bars)	Common	Low-to-high energy; rapid burial; hydraulic transport and sorting
Abandoned channel fill	Common	Fine-grained sediment; abundant clays; organic-rich
Crevasse splay	Variable	Coarse-grained sediment; rapid burial; hydraulic transport and sorting
Levee	Uncommon	Heavily vegetated; well-drained, fining-upward sediment; soil development; bioturbation
Springs	Common	Fine-grained sediment; vary in temperature and mineral content; bioturbation
Tar seeps	Very common	Excellent preservation; vertical mixing; geologically unstable
Karst (caves, sinkholes, fissures)	Very common	Natural sediment traps; subject to lacustrine and fluvial influence; geologically unstable
Volcanigenic (mudslides, ashfall)	Uncommon	Excellent preservation; mass death; rapid burial; alter chemistry; climate independent

the formation of fossil assemblages, the composition of which may differ because each mode has its own associated set of biases (Behrensmeyer and Hook 1992). Studies of vertebrate decomposition and diagenesis following burial provide evidence that short-term, environment-dependent processes are vital in determining the long-term preservation potential of vertebrate remains (Andrews 1995; Bell et al. 1996; Fernández-Jalvo et al. 2002; Berna et al. 2004; Nielsen-Marsh et al. 2007; Noto 2009). The distribution of environment types determines what part of the original biological signal is preserved, exerting a well-known filter over the fossil record that may persist at larger scales, contributing to large-scale spatiotemporal patterns of vertebrate fossil distribution. Behrensmeyer and Hook (1992) note that the distribution of various taphonomic modes through time likely reflects the sum of environmental variation at regional-to-global scales; referred to here as a taphonomic regime. Such variation in taphonomic regimes likely drives taphonomic megabiases, the existence of which is recognized, although they remain poorly understood (Behrensmeyer and Hook 1992; Behrensmeyer et al. 2000).

1.2 Hierarchical Integration of Terrestrial Vertebrate Taphonomy

In order to aid the development of new tools and techniques in the study of terrestrial vertebrate taphonomy, any hierarchical framework should take into account the complex relationship between Earth system processes (including climatic and tectonic processes), ecological and evolutionary responses of the biosphere, and the resulting vertebrate fossil record.

The following sections explore some of the prominent processes acting in the formation of the terrestrial vertebrate fossil record at different spatial scales. Thus providing a context and timeframe through which these processes may act and the degree to which they may be influenced by factors at other levels of the hierarchy. The purpose is not to examine every possible process that may occur, as many extraordinary examples have been documented that may not represent the typical pathway of fossil formation (Dal Sasso and Signore 1998; Chin et al. 2003; Channing et al. 2005; McNamara et al. 2006; Schweitzer et al. 2007). The focus when discussing taphonomic processes will be on the production of body fossils through bone preservation. Because widespread soft tissue preservation in terrestrial settings is relatively rare, it will not be discussed in detail here. For reviews see Martin (1999) and Schweitzer et al. (2007).

This chapter is organized into five main sections. The first section presents a brief overview of vertebrate bone and how it is affected by taphonomic processes. The second section is an overview of a proposed taphonomic control hierarchy for terrestrial vertebrates. The third section deals with the connections between these hierarchical levels and the major factors constraining processes at each level. Possible effects of the taphonomic control hierarchy on the fossil record over time and paleobiological patterns reconstructed from the fossil record are discussed in the final two sections. Finally a conceptual model is proposed for approaching changes in the fossil record brought about by environmental change.

2 The Structure of Vertebrate Bone

The interplay between osseous tissue properties and taphonomic processes is often underappreciated, even though such knowledge allows for better prediction of a bone's "behavior" following organismal death. Bone is a general term that describes a group of biologically-derived materials that use the mineralized collagen fibril as a fundamental element in its construction (Weiner and Wagner 1998). Bone has three main constituents: a close-packed framework of collagen fibrils, layers of carbonated apatite crystals packed in the spaces between fibrils, and a "cement" consisting of mucopolysaccharides, glycoproteins, lipids, carbonate, citrate, sodium, magnesium, fluoride, and water (Weiner and Wagner 1998).

The carbonated apatite crystals found in bone and teeth (bioapatite) have the general chemical composition of $Ca_{10}(PO_4)_3(OH)_2$, with carbonate making up 5–6 wt% (Pasteris et al. 2004). Often F^- or Cl^- ions substitute for OH^- in the crystal lattice. Various authors refer to bone mineral as hydroxyapatite, hydroxylapatite, frankolite, or dahllite. However compared to the geologic standard, bioapatite demonstrates several characteristics setting it apart as a unique mineral phase, including extremely small crystal size (50 × 25 × 4 nm), poor crystallinity, and low OH^- content in the crystallites (Weiner and Price 1986; Weiner and Wagner 1998; Pasteris et al. 2004). Bioapatite is most stable under homeostatic conditions in the body. Once removed from this environment, the non-stoichiometric nature of bioapatite crystallites makes them highly unstable and prone to alteration.

In mineralogical apatite, OH^- is necessary for maintaining charge balance in channel sites of the crystal lattice; its removal would lead to a local charge imbalance (Pasteris et al. 2004; Wopenka and Pasteris 2005). In bioapatite local charge balance in the channel sites is maintained through ionic bonding between collagen, which contains many OH^- groups (mainly from hydroxyproline), and the bioapatite lattice (Pasteris et al. 2004). Sharing OH^- groups leads to a strong bond that enables simple chemical means for rapid coupling or decoupling of the mineral-collagen bond in response to physiological needs, most likely accomplished by altering pH (Pasteris et al. 2004). Low OH^- content and poor crystallinity leads to the low buffering capacity necessary for bone remodeling; higher OH- and higher crystallinity in tooth enamel leads to better buffering capacity necessary for resisting acids such as those that regularly attack teeth (Pasteris et al. 2004; Pasteris et al. 2008). This may also explain why vertebrate teeth are more readily preserved than bone.

The structural and chemical properties of a juvenile skeleton are inherently different from those of adults. During early stages of bone development amorphous or poorly-crystalline calcium phosphate is laid down and later replaced by crystalline bioapatite (Menczel et al. 1965; Termine et al. 1967; Glimcher 1984; Grynpas and Omelon 2007). Continued bone development involves the incorporation of more carbonate and/or fluorine into the crystal lattice, especially during periods of bone growth, after which it reaches a relatively constant level (Rey et al. 1991; Freeman et al. 2001; Magne et al. 2001; Pasteris et al. 2004). During life this would serve to further stabilize the mineral in much the same way that fluoride added to drinking water prevents tooth decay. Juvenile skeletons are less mineralized; mineral density increases over time as the animal matures (Symmons 2005). Through ontogeny, ultrastructural changes within the bone also occur, representing changing metabolic strategies and physical requirements (such as rapid increases in body size) (Barreto et al. 1993; Horner et al. 2000; Montes et al. 2005). Therefore, the probability of preservation of an individual may change with ontogeny, especially between early and late life stages (Symmons 2005).

The ossified tissues of vertebrates are composed of multiple, hierarchically arranged structures, which vary in chemistry, structure, and organization (Enlow and Brown 1956, 1957, 1958; Weiner and Traub 1992; Aerssens et al. 1998; Dirrigl and Frank 2001). Differences in size, shape, and internal structure among elements exist within a skeleton and between taxa due to age, ecology, and evolutionary history. For example, small, but significant chemical and structural differences exist between cortical and trabecular bone (Bigi et al. 1997; Aerssens et al. 1998). Small changes to the chemical or crystal structure of a mineral can have large effects on its properties, altering how the mineral reacts to external conditions (Bigi et al. 1997). Structurally different regions of the same bone may follow different diagenetic trajectories. For example, the fractionation of various common elements (Mn, Fe, Cu, and Ba) (Carvalho et al. 2004) and rare earth elements (REEs) (Williams and Potts 1988; Trueman and Tuross 2002) differ between cortical and trabecular bone tissue. A similar situation exists in the fractionation of oxygen and carbon isotopes incorporated into bone tissue and dental enamel during early diagenesis (Zazzo et al. 2004). Analysis of dinosaur (Pawlicki and Bolechala 1987; Goodwin et al. 2007) and human (Lambert et al. 1983; Schoeninger et al. 1989) compact bone show that diagenesis, as measured by elemental concentrations of Ca, P, Fe, Mn, and others, proceeds differentially through the vascular canals and compact lamellae of bone due to differences in the porosity and composition of these tissues. Therefore, the diagenetic alteration of bone tissue is not uniform and can vary due to environmental differences and/or the structural and chemical properties of the tissue.

3 The Terrestrial Taphonomic Hierarchy

The description of any hierarchy requires a delineation of specific, inclusive levels (see Martin 1999, p. 11). This model includes three major spatial levels: micro-, meso-, and macroscale (Fig. 2). Each level in the hierarchy contains a set of associated biological, chemical, and/or physical processes that influence the preservation potential of vertebrate remains. Conditions present at an overlying level restrict the range of possible taphonomic processes and biogeochemical conditions acting at

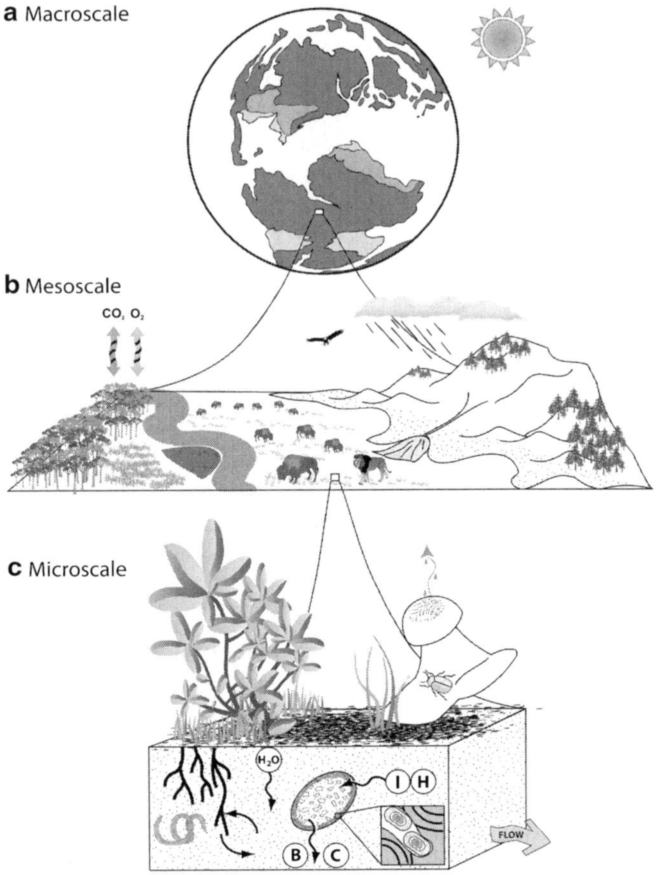

Fig. 2 Hierarchy of terrestrial taphonomic processes and controls. (**a**) Macroscale: distribution of continental landmasses, sea-level, ocean circulation, atmospheric composition and circulation, intensity and distribution of solar radiation on the surface, and biome distribution. (**b**) Mesoscale: landscape characteristics, local weather patterns, species population dynamics, biogeochemical cycles, predation/death, and scavenging of remains. (**c**) Microscale: soft tissue decay, bone exposure, desiccation and cracking from solar radiation, invertebrate utilization, bioturbation, nutrient use and organic acid release by plant roots, leaching of bone mineral (B) and collagen (C), incorporation of exogenous ions (I) and humics (H) into bone matrix, bacterial and fungal degradation (inset), diagenesis, and hydraulic flow of groundwater

lower levels. It should be noted that the temporal extent of some taphonomic processes may cross more than one level of the hierarchy. This model is intended to organize and relate the work of many different researchers and highlight important relationships among taphonomic processes that, when considered collectively and at higher scales, will lend insight into the importance of these factors in the preservation of vertebrate remains. It is also worth noting that certain processes may behave similarly regardless of scale (e.g., temperature).

Processes described at each level are split into two categories: those that act at the surface and those that act in the subsurface, following a similar approach used in marine paleontology (Kidwell 1986; Plotnick et al. 1988; Walker and Goldstein 1999; Behrensmeyer et al. 2000). Here, surface processes refer to those acting at or near the sediment surface. Subsurface processes act beneath the sediment–air or sediment-water interface and may be independent of and/or influenced by conditions at the sediment surface, especially at shallow burial depths. Together these two sets of processes describe a taphonomically active zone (TAZ) for the terrestrial realm [after Davies et al. (1989) and Lyman (1994); see also "highly dynamic mixed zone" of Behrensmeyer et al. (2000)]. Additionally, the spatio-temporal extent of processes and the elements (e.g., molecules, tissues, carcasses, assemblages) interacting at each level are discussed.

3.1 *Microscale Processes*

Microscale processes cover spatial scales in the micrometer to centimeter range and a temporal scale anywhere from ≤1 day to upwards of 100 ky. A major difficulty in assigning specific time estimates to the duration of microscale processes results from an inability to directly observe the processes in action. This creates uncertainty about the amount of time necessary for these reactions to go to completion. Instead, different authors have inferred the amount of time needed, using qualitative terms such as "early" or "late" diagenesis. Still, an attempt is made to estimate the time windows for these reactions when their activity is most prominent during diagenesis based on indications given in the literature. Elements interacting at this scale are molecules, cells, and tissues on both internal and external bone surfaces. This includes the components of bone tissue (collagen, bioapatite, cells, etc.), bacteria, fungi, exogenous ions, and water.

3.1.1 Surface Processes

Diagenetic alteration of bone can begin almost immediately following death. Within hours metabolic processes shut down and body cells undergo autolysis as their structural integrity deteriorates, releasing contents of the cytoplasm (organelles, etc.) into interstitial fluid, leading shortly to soft tissue hydrolysis (Tappen 1994; Andrews 1995). In bone, autolysis is restricted to cells only; collagen and extracellular matrix proteins ("cement") remain unaffected (Child 1995). Within days the collagen-mineral bond weakens, and the bioapatite begins to recrystallize into a more thermodynamically stable form, beginning the transition to a composition closer to the geologic standard and increasing overall crystallite size (Nielsen-Marsh and Hedges 1997; Berna et al. 2004; Trueman et al. 2004). These processes will occur regardless of surface exposure, rapid burial, or submergence under water, although the extent to which reaction rate depends on these conditions is not well understood.

Water is a strong limiting factor in determining the rate at which diagenetic change and microbial degradation occurs. Large-scale and/or rapid diagenesis and microbial attack of the remains becomes possible only when the remains are saturated with water. In submerged bodies microbial attack and diagenetic alteration to bone tissue can take place almost immediately, causing substantial damage to histological structures and altering chemical composition after only a few years (Yoshino et al. 1991; Bell et al. 1996; Davis 1997; Zazzo et al. 2004). At the sediment–air interface this does not occur usually until after burial, although archeological and experimental evidence indicates that loss of both collagen and carbonate ions (in the form of CO_2) can occur in the absence of mineral dissolution in water (Person et al. 1995, 1996). This difference between subaerial and buried bones has been well documented in the archeological literature (Yoshino et al. 1991; Andrews 1995), although the ability to discern between them depends on the difference between prevailing climate and surface conditions during exposure versus those below ground. When surface and subsurface conditions are divergent, different taphonomic signatures will result; when a similar set of conditions occur, such as in moist environments, the taphonomic signatures are indistinguishable (Nicholson 1996).

Bacteria responsible for initiating soft-tissue decomposition are likely of endogenous origin. As tissues hydrolyze and mucous barriers break down, bacteria from the intestinal lumen proliferate and migrate to other parts of the body through the vasculature, continuing aerobic decomposition of soft tissue; this can occur within as little as 24 h post mortem (Dolan et al. 1971; Kellerman et al. 1976). These bacteria similarly could invade the skeleton via the vasculature and spread intracortically through the bone vascular system (Bell et al. 1996). Predator or scavenger action that removes large amounts of viscera and vasculature may retard this type of bacterial invasion, although the removal of flesh exposes bone to potentially destructive abiotic conditions. Some researchers have noted that regions of the skeleton closest to the abdomen, such as the vertebrae and ribs, can be more degraded than distal elements due to the production of organic acids by anaerobic bacterial decomposition of soft tissue (Boddington 1987; Child 1995).

The important abiotic controls at this scale differ between aqueous and subaerial environments. At the sediment–air interface ambient temperature, ultraviolet (UV) radiation exposure, the amount and frequency of precipitation, and the composition of the sediment or substrate where the remains lie, each play an important role. The rate of most chemical reactions depends on temperature, approximately doubling for every 10°C increase (Henderson 1987). This relationship is especially important in determining the rate of soft tissue decay (Shean et al. 1993), collagen hydrolysis and peptide loss in bone (Collins et al. 2002; Hedges 2002). At very high temperatures, small amounts of carbonate are released from the bioapatite crystallites in the form of CO_2 and OH^- ions are incorporated into the lattice (Person et al. 1996; Nielsen-Marsh and Hedges 1997; Pasteris et al. 2004). Temperature and UV levels are positively correlated due to their relation to sunlight: unshaded, exposed surfaces experiencing high temperature are exposed to higher levels of UV radiation. UV radiation is particularly damaging to organic components and is likely a leading cause of damage to subaerially-exposed bone (Andrews 1995; Fernández-Jalvo et al. 2002). UV facilitates the break-down of the collagen matrix by cleaving specific peptide bonds, lowering the denaturation temperature of the collagen molecule (Sionkowska 2005).

Precipitation provides moisture that can have a variable effect on the remains: soft-tissue desiccation (a form of natural mummification) can act to preserve remains, while bone desiccation leads to cracking and loss of structural integrity (Behrensmeyer 1978; Cutler et al. 1999; Trueman et al. 2004). Large diurnal fluctuations in temperature and precipitation have an especially destructive effect on exposed bone (Martin 1999). The overall effect of temperature, UV radiation, and precipitation on exposed remains varies due to differences in latitude and seasonality; the presence of vegetation can help mitigate some of these destructive effects (Behrensmeyer 1978; Tappen 1994; Cutler et al. 1999).

The ground surface itself plays a role in mediating the effects of the above-mentioned factors (Shalaby et al. 2000). It can act as a reflector and/or radiator of incoming solar radiation, increasing the temperature experienced by remains well above that in the air a short distance above the remains. It also may extend the operable time of decomposition processes by releasing stored heat after sundown. Porous sediments with higher hydraulic flow draw moisture downwards, away from the surface, thereby accelerating desiccation. Sediment with low hydraulic flow retains moisture near the remains, retarding water loss. Under particular situations when evaporation at the surface is extreme, ground water may be drawn up through the sediment, leading to the formation of destructive evaporite minerals (e.g., gypsum) in the remains (Trueman et al. 2004).

Temperature plays the largest role in driving aquatic decomposition patterns. Under aqueous conditions, the surrounding water dulls the effect of the above factors on submerged remains by mediating diurnal fluctuations. Higher temperatures are associated with more rapid tissue decomposition, due mainly to increased bacterial activity (Elder 1985; Elder and Smith 1988; Minshall et al. 1991). Oxygen concentration also is important, determining the availability of the remains to macroconsumers and the range of biogeochemical reactions that may occur on the remains (Elder and Smith 1988). The more anoxic an environment, the more closed it becomes to consumers. Water pH will play a role only in cases of very low pH, which is unusual in most terrestrial aquatic environments. Low pH (<5) would create acidic conditions capable of dissolving bioapatite crystallites and eroding osseous tissues (Hare 1980).

The formation of adipocere (hydrogenated fatty acids; "grave soap") is another byproduct of decay commonly found in submerged or saturated environments, but is largely absent as a decay stage in subaerial environments (Mellen et al. 1993; Haglund and Sorg 2002; O'Brien and Kuehner 2007). Adipocere may play a role in fossilization by preserving soft tissues and/or the three-dimensional arrangement of skeletal elements through encouraging early calcium carbonate mineralization (Berner 1968; Martill 1988). Little work has been done to understand these mechanisms, making it an area ripe for experimental work.

3.1.2 Subsurface Processes

The processes that occur after burial have received far less attention, despite the fact that diagenesis only begins at the surface. At least over the short-term (months to years), burial can offer protection from some of the destructive surface processes described above, such as direct UV radiation, significantly hindering their effect (Behrensmeyer 1978; Andrews 1995; Martin 1999; Trueman et al. 2004). Once buried, many of the same processes described for surface exposure continues, however the intensity with which they occur and the factors that control them change.

Living plants affect buried remains through both physical and chemical means. Roots physically infiltrate bone, sometimes causing large cracks and opening the interior to destructive agents. Roots are also capable of chemically digesting bone as the plant seeks out nutrients and minerals, with the type of attack and its extent varying depending on the species and community composition (Henderson 1987; Berner et al. 2004).

Given the highly unstable nature of bioapatite crystallites in the absence of physiological maintenance, it is perhaps more relevant to ask why bioapatite crystals do not spontaneously dissolve upon death (Berna et al. 2004). Loss of the stabilizing presence of a strong mineral-collagen bond opens the crystallites to undergo further alteration by reacting with available pore water. This is accomplished through the processes of dissolution (preferential loss of less thermodynamically stable crystallites) and recrystallization (crystallites defer to a more stable state, usually incorporating exogenous ions) (Nielsen-Marsh and Hedges 1997). These two processes are by no means mutually exclusive, and may be described more accurately as end-members of a continuum in which ions are lost and gained between the crystallites and surrounding pore water. The relative difference between rates of ionic loss or gain determines the prevailing alterations to the bone. This leads to the change in mineral identity from bioapatite to a more stable apatite phase, usually through the uptake of F and CO_3 (Hedges 2002) and Fe and Si (Johnsson 1997). The degree of alteration to bioapatite crystal structure and chemical composition is considered a relative measure of diagenesis in the bone. At this stage the stability of the bone mineral (i.e., its propensity to dissolve and recrystallize) is controlled by the pH, Eh (redox potential), and ionic concentration of the pore water.

It is unknown to what degree the intimate relationship between collagen and bioapatite influences decomposition dynamics, including whether collagen or mineral loss must occur first for diagenesis to proceed. Some authors propose that collagen protects the mineral component from significant alteration until its removal (Person et al. 1996). However, the bulk of empirical data supports the opposite scenario, in which bioapatite crystallites and inorganic matrix protect the collagen from immediate microbial attack (Yoshino et al. 1991; Collins et al. 1995, 2002; Nielsen-Marsh et al. 2000; Pfretzschner 2004). Collins et al. (1995) proposed that the intra-crystalline spaces of the mineral fraction are too small for microbial enzymes to penetrate, effectively forming a barrier to everything but water. Under this model collagen loss is controlled mainly by hydrolysis and temperature, with microbial digestion playing a secondary role until significant mineral loss has occurred.

While the presence and activity of microorganisms (bacteria, fungi, and protozoans) is an accepted tenet of organic decay, the role they play (especially bacteria) in diagenesis is less clear. Several species of collagenase-producing bacteria and fungi are known from archeological bones (Child 1995). Both archeological and experimentally manipulated bones from a variety of species and environments, covering timescales over 1–40,000 years, show clear evidence of bacterial and fungal attack (Hackett 1981; White and Hannus 1983; Piepenbrink 1986; Yoshino et al. 1991; Child 1995; Hedges et al. 1995; Bell et al. 1996; Davis 1997; Trueman and Martill 2002; Jans et al. 2004). Zones of microbial attack, known as microscopic focal destruction (MFD; Hackett 1981), are observed readily within compact bone, usually within and surrounding osteons, and consist of individual tunnels or honeycomb-type structures 0.1–10 μm in diameter (Yoshino et al. 1991; Bell et al. 1996; Jans et al. 2004). These structures hasten collagen and mineral loss by exposing more internal surface area to dissolution and leaching (Jans et al. 2004 and references therein). Avascular bone will be less susceptible to this kind of attack, due to the lack of routes permitting access to the bone interior (Nicholson 1996).

The overall proportion of bone microstructure attacked appears relatively small and the overall number of collagenase-producing microorganisms isolated from bones and surrounding soil is low (Child 1995). Some authors conclude that biodegradation plays only a minor role in collagen loss and bone degradation, the main control instead being abiotic conditions (Child 1995; Collins et al. 1995; Pfretzschner 2004). Alternatively, the incidence of MFD in fossil bones is minimal or nonexistent compared to archeological specimens, where ~50% exhibit extensive bioerosion of histological structures (Hedges et al. 1995; Trueman and Martill 2002; Chinsamy-Turan 2005). This difference between fossil and archeological bone suggests that bioerosion is an important determinant of bone preservation and must be prevented altogether, or halted in its earliest stages, by environmental conditions for the specimen to be fossilized (Trueman and Martill 2002). However, microbial activity alone cannot account for all collagen loss during diagenesis, especially when original histology is preserved, indicating collagen loss in bone is controlled by a separate process (Hedges et al. 1995). Therefore, bioerosion represents an early stage of diagenesis that will lead to rapid deterioration of internal structure unless quickly halted. But, over longer timescales collagen loss is controlled by abiotic factors, such as gelatinization rate (Trueman and Martill 2002; Pfretzschner 2006).

Others workers have suggested that fossilization cannot proceed without bacterial activity. It has been experimentally shown that bacteria are necessary for authigenic mineral deposition within bone (Daniel 2003; Carpenter 2005). Soil-derived bacteria recently have been shown to mediate $CaCO_3$ mineralization as a byproduct of their metabolism and it is thought this property may be common to many soil bacteria (Lian et al. 2006; Barabesi et al. 2007). Archeological bone sometimes is observed to contain additional mineral deposition lining the walls of bone tissue presumably damaged by bacterial activity (Yoshino et al. 1991). These mineral deposits are noted for having different properties from surrounding bone tissue but few have been recognized to have bacterial origins. These most likely represent the initial stages of mineral precipitation (see Daniel 2003).

Given the breadth of environments covered in taphonomic studies, it is more likely that conditions present in the burial environment are regulating which fraction of the osseous tissue – organic or mineral – is degraded and/or removed first. Therefore, as a general precondition, degradation in at least one fraction must occur in order for diagenesis to proceed. As diagenesis proceeds, especially in the early stages, bone porosity initially increases as collagen is lost, with the intervening spaces filled by the mineral as it recrystallizes and increases in size (Hedges and Millard 1995; Hedges 2002). Where pore-size distribution shifts to larger diameters, a greater surface area is opened to reaction, facilitating mineral loss through diffusion to the pore water. This process appears to take place over a timescale of hundreds to a few thousand years. Environmental settings that inhibit this initial precondition will delay the onset of diagenesis and potentially act to stabilize the remains against future destructive change from geochemical and/or biological agents. Binding of collagen with humic substances released from decaying plant matter can stabilize collagen against hydrolytic loss and biodegradation (Collins et al. 2002). Clay minerals may have anti-enzymatic properties that could prohibit many types of biodegradation (Butterfield and Nicholas 1996; see Gaines 2008 for an alternative view of this mechanism).

Sediment hydrology is the single most important factor determining the decay and diagenesis of buried remains, influencing numerous biogeochemical properties of the sediment. Two factors determine the moisture regime: (i) how water moves through a sediment, called the hydraulic conductivity; and (ii) the potential for that water to migrate within the pores of the sediment via osmosis or capillary action, called the hydraulic (matric) potential (Brady 1974; Retallack 1990; Hedges and Millard 1995). Coarse-grained sediments typically have higher conductivity and lower potential, while the opposite is true of fine-grained sediments. These two factors determine the overall solute concentration surrounding a bone and the maximum pore size within a bone that will be occupied by groundwater (Pike et al. 2001). Diagenesis proceeds only when internal bone pores are in contact with water (Hedges and Millard 1995). Osseous tissue appears to undergo the greatest degradation in high conductivity environments, while diagenesis may be favored with increasing matric potential (Hedges and Millard 1995; Nicholson 1996; Pike et al. 2001; Noto 2009).

The hydraulic properties of the sediment control oxygen availability, thereby exerting a strong influence on the nature and depth of the terrestrial TAZ. Water movement influences organismal abundance and activity, as highly conductive sediments provide oxygen to the large proportion of aerobic organisms residing in the sediment, while anoxia resulting from low flow depresses biological activity. Oxygen availability also determines the redox potential, which will determine the range of possible chemical reactions affecting the remains and surrounding sediment (Retallack 1990). Redox conditions, especially in submerged sediments, may be important to the formation of certain authigenic minerals such as pyrite (FeS), which are often associated with fossilized bone (Pfretzschner 2000, 2001a, b; Leng and Yang 2003).

The effect of temperature on decay and diagenesis varies because of differences in latitude, season, and depth of burial. Areas with little seasonal variation in temperature should experience higher rates of degradation, as opposed to those places where the ground remains frozen for part of the year (Henderson 1987). This variation affects not only chemical reactions, but the influence of soil biota as well. Freezing of the ground causes most biotic activity and water movement to cease.

3.2 Mesoscale Processes

While it may be easier to differentiate between micro- and macroscale processes, the "middle ground" between them, mesoscale processes, are more difficult to distinguish. Mesoscale processes form the bridge connecting localized biogeochemical conditions with regional-level distribution patterns of preserved bone. The spatial scale used here encompasses areas on the order of 10^0–10^4 m: a scale that is expected to include areas subject to the same local environmental conditions and regional climatic regime. Temporally, mesoscale processes may be more constrained than at other levels, covering ~1 day up to 10^2–10^3 years. Individual bones up to whole carcasses will be affected at this level, with alterations brought about through interactions with local biota, geomorphological features of the landscape (including river and lake beds), and biogeochemical characteristics of the surrounding sediment.

3.2.1 Surface Processes

Disarticulation, skeletonization, desiccation, and bone utilization by vertebrate, invertebrate, and fungal consumers are among the main forces acting on vertebrate remains at the surface. In aquatic environments, fresh carcasses may go through a well-known process of bloating, tissue decay, bone exposure, and finally disarticulation (Haglund and Sorg 2002; Anderson and Hobischak 2004). The soft tissues of a carcass often are consumed by predators and scavengers and is a well known part of the taphonomic process, with examples reaching back into the fossil record (Rogers et al. 2003; Spencer et al. 2003). The rate and extent of soft tissue removal determines when and how much of the skeleton is exposed to the surface environment, with important implications for bone survival prior to and following burial (Behrensmeyer 1978; Andrews and Cook 1985; Henderson 1987; Weigelt 1989; Andrews 1995).

Subsequent movement by fluid transport will physically abrade and damage exposed bone after deposition. The degree of movement is regulated by flow rate, bed morphology, and skeletal element size, shape, density, and degree of articulation (Voorhies 1969; Boaz and Behrensmeyer 1976; Elder and Smith 1988; Coard and Dennell 1995; Coard 1999). Both physical transport and consumer behavior can then lead to the selective removal and/or concentration of certain elements (e.g., limbs), age classes, and/or species (Norman 1987; Wood et al. 1988; Lyman 1994; Martin 1999).

Following skeletonization various organisms may utilize exposed bone. Many mammals are known to break open bones to consume the energy-rich marrow within or gnaw them to obtain essential nutrients, however there is no direct evidence of this strategy outside of Mammalia (Fiorillo 1991; Van Valkenburgh and Molnar 2002; Reisz and Tsuji 2006). If the environment contains sufficient moisture, fungi will colonize any exposed bone, using the surface as a growth substrate while digesting the composite matrix beneath (Piepenbrink 1986; Nicholson 1996). Various species of insects also utilize bone, either by excavating the surface or boring into the interior, consuming the bone tissue and weakening the ultrastructure (Kaiser 2000; Paik 2000; Roberts et al. 2007). This may also occur following burial. In aquatic environments, bone may be colonized and consumed by vertebrates, macroinvertebrates, and algae (Davis 1997; Haglund and Sorg 2002; Hobischak and Anderson 2002; Haefner et al. 2004; Goffredi et al. 2005). Still, little is known about freshwater decay processes and their consequences for preservation (Hobischak and Anderson 1999).

Species identity, size, age, sex, and health at the time of death affect how the remains respond to surface processes (Martin 1999; Behrensmeyer et al. 2003). Species vary in the chemistry, structure, and organization of their ossified tissues, which affects the size, shape, and density of elements. Body size determines the surface-area-to-volume ratio of the entire carcass, and individual elements available to interact with the environment, with smaller taxa and elements more prone to loss and destruction than larger counterparts (Von Endt and Ortner 1984; Nicholson 1996; Martin 1999; Munoz-Duran and Van Valkenburgh 2006). Under certain conditions large carcasses may decompose faster because they attract more consumers (Hewadikaram and Goff 1991).

The age (ontogenetic stage), sex, and health of the individual at death may be more important than size in determining decay susceptibility because of the close relationship between these factors and bone structure. Juveniles not only tend to be smaller than adults but their skeletons are less mineralized and differ chemically (Symmons 2005). Reproductive status may play an important role, as the females of many vertebrate groups utilize calcium reserves from the skeleton during gestation (e.g., egg production) and parental care (e.g., lactation, brooding, protection) (Randall et al. 1997; Arias and Fernandez 2001; Schweitzer et al. 2005). Many of the factors described above are also affected by individual health, but diseases that affect the mineral density and structure of bone may contribute to loss of bone integrity following death and can occur regardless of age (Henderson 1987).

On the ground surface temperature, precipitation, solar energy input, and sediment/soil type continue to affect vertebrate remains at the mesoscale. In submerged environments, temperature, pH, light availability, nutrient availability, flow regime, and sediment/substrate type play a similar role (Barnes and Mann 1991). Their interactions, together with local geomorphology, determine the suite of flora and fauna that forms the local community of which the remains are a part. Both the diversity and numbers of predators, scavengers, and decomposers influence the extent of soft tissue removal and bone tissue modification. On land surfaces foliage can modify local conditions and provide a certain degree of protection for exposed remains by decreasing diurnal fluctuations in temperature and moisture and/or obscure the remains from detection. This can lead to the long-term survival of bones on the surface in vegetated areas compared to those exposed only a short distance away (Behrensmeyer 1978; Kerbis Peterhans et al. 1993; Shean et al. 1993; Sept 1994; Tappen 1994; Cutler et al. 1999).

The greater habitat heterogeneity a local environment supports, the larger the number of potentially favorable microenvironments that exist to protect the carcass from the most destructive conditions (of course, depending on where the animal comes to rest after death). The frequency and intensity of changes in the local environment determines how long surface conditions last, including seasonal changes (Dubiel et al. 1991; de Carvalho and Linhares 2001; Rogers 2005) and episodic events such as fires, floods, or mudslides, that may occur over cycles from decades to centuries (Watson and Alvin 1996; Loope et al. 1998; Greenwald and Brubaker 2001). The distribution of landscape features (rivers, hills, plains, caves, etc.) interact with local flora to control rates of sediment aggradation and erosion over sediment surfaces, which in turn help drive burial of exposed remains (Lyman 1994).

3.2.2 Subsurface Processes

Mesoscale subsurface processes involve mainly diagenetic alteration of the sedimentary body and the effects this has on the diagenesis and preservation of buried remains (Tucker 1991; Lyman 1994). Both aquatic and ground surface sediments are altered though a suite of physical, chemical, and biological processes that are controlled by land surface morphology (topography), sediment characteristics (composition, grain size) and moisture availability (precipitation, water table). Subaerially-exposed sediments may also undergo weathering, considered a set of specific alterations related to but separate from diagenetic processes, although significant overlap exists (Middleton 2003). Weathering reactions play a large role in both erosion and soil formation (Retallack 1990).

Physical processes acting on sediment bodies include loosening/cracking caused by heating-freezing expansion cycles and the movement of water and gasses. Chemical alterations result from four principal reactions: hydrolysis, oxidation, hydration, and dissolution (Brady 1974; Retallack 1990). Hydrolysis leads to the displacement of cations by hydronium ions, creating a new, insoluble mineral product. Oxidation involves reactions in which electron loss forms new compounds. Hydration includes the addition of water into the mineral structure. Dissolution is the disaggregation of a compound into ions, a classic example being a salt cube (NaCl) dissolving in water into Na^+ and Cl^-. Biotic influences on sediment alteration are profound. Living organisms influence the availability of nutrients through their life processes and alter the physical structure of the sediment profile through bioturbation and the release of various byproducts (Retallack 1990). Many groups of bacteria create biofilms that can significantly slow the movement of water through sediment, altering biogeochemical conditions from those predicted by sediment characteristics alone (Vandevivere and Baveye 1992; Baveye et al. 1998; Battin and Sengschmitt 1999).

Soils and paleosols are common environments of vertebrate preservation and the degree of soil development plays an important role in vertebrate preservation. Soil formation is a complex process involving the interaction of climate, living organisms, the nature of the parent materials, topography of the area, and time that parent materials are subject to alteration (Retallack 1990; Schaetzl and Anderson 2005). Over time a dynamic balance is reached between sedimentary and organic inputs with their biogeochemical modification that may last thousands of years. These prevailing conditions determine bulk soil properties, making them environment-specific. Therefore, different environments can be characterized by their soils, even long after the conditions that created them cease to exist (Retallack 1990). Over the long-term, soil formation is affected by (i) shifts in regional tectonics, which alter basin drainage and sediment aggradation patterns; (ii) volcanic activity, depositing ash and altering atmospheric chemistry; (iii) changes in atmospheric or ocean circulation patterns, altering precipitation patterns; and (iv) alterations in the level of solar radiation reaching the surface (Retallack 1990; Martin et al. 1999). Over time the evolution of the sediment profile changes preservation conditions through alteration of the biogeochemical properties of the sediment, for example the production of clays or iron oxides (see Retallack 1990 for detailed description of processes; Martin 1999).

Changing sediment-moisture content causes many clay minerals, such as montmorillonite, present in the sediment to shrink or swell, which can physically distort or damage bones lying within clay-rich sediment. The shrink-swell cycles of clayey sediments may cause more distortion than sediment compaction can over time because they occur more frequently and result in movement in multiple directions, while sediment compaction is a long-term and unidirectional event (Henderson 1987; Retallack 1990).

Sediment properties along with topography influence local hydrological conditions, including the height and relative flow of the water table. Changing hydrologic conditions over both space and time have been implicated in patterns of bone decay and diagenesis by determining the amount of hydrologic recharge and solute concentration surrounding the remains (Hedges and Millard 1995). Under these conditions, it may not be possible for bones to undergo preservative diagenesis (that is, stabilization of the organic and/or mineral components) and fossilization until they are beneath the water table, where they are buffered against rapidly shifting biogeochemical conditions. The faster remains come to lie beneath the water table, the greater their chances of preservation. A similar taphonomic model is proposed by for plants, in which regional changes in sediment aggradation and accommodation that lead to base level change and subsequent rise in water table are best for plant preservation (see Gastaldo et al. this volume). Even if the biogeochemical requirements for vertebrate and plant preservation differ, the lower the residence time in the terrestrial TAZ above the water table, the more likely the remains will be preserved. This may help explain the relative wealth of vertebrate remains from lacustrine, palustrine, and fluvial environments, and from those settings with relatively high water tables such as wet floodplains in close proximity to water sources that frequently deposit sediment (Noto unpublished data).

3.3 *Macroscale Processes*

The macroscale consists of many large-scale phenomena not often considered in studies of the vertebrate fossil record. These processes occur over spatial scales of 10^1–10^4 km and temporal scales ranging from 10^2–10^6 years. Note that the spatial and temporal boundaries for this level fall within the resolution considered typical for most terrestrial fossil assemblages (Behrensmeyer and Hook 1992). The leading control at this level is the global climate regime, driven ultimately by the interplay of plate tectonics and Milankovitch cycles altering the distribution of land area and arrangement of continents, eustatic sea-level, atmospheric concentrations of carbon dioxide and oxygen, volcanism, mean global temperature, and global atmosphere and ocean circulation patterns (Behrensmeyer and Hook 1992; Martin 1999). This complex interaction determines the distribution of biomes. Biomes represent large regions united by similar climatic and ecological conditions, which produce distinct assemblages of organisms (Begon et al. 2006). Because the range of taphonomic processes acting within a particular area is determined by the environment and local biota, many areas within the biome are united under a similar taphonomic regime. The distribution of biomes and their constituent environments will influence community composition and patterns of biodiversity. When coupled with variation between taphonomic regimes, these will have a direct impact on how different communities and species are preserved in the fossil record. The relative stability of these factors will determine the type of fossil accumulation formed (if any at all) and its fidelity to the original biota.

3.3.1 Surface Processes

Large-scale factors control several key processes at this level. First, weathering processes, while highly variable at small spatial scales, are coordinated regionally, altering the land surface for thousands of square kilometers in a similar way (Simon-Coinçon et al. 1997). Changes in weathering regime can be traced through time and correspond to shifts in sea-level, topography, and climate. Similarly, the distribution and morphology of water bodies and associated features are influenced by geomorphology, tectonics, sea-level, biota, and climate (Prothero and Schwab 1996; Blum and Tornqvist 2000; Leier et al. 2005). Second, the dynamics of plant and animal populations vary with changing abiotic conditions, which can alter the size and location of species' geographic ranges (Dynesius and Jansson 2000). The expansion and contraction of species ranges over time affect the distribution of biodiversity. For example, large-scale climate patterns have been shown to drive population dynamics in groups of caribou and musk oxen, each separated by more than 1,000 km of ice on the coasts of Greenland (Post and Forchhammer 2002). Because individual species can respond differently to climate change, community assemblages change over time as new communities are created through species reorganization (Stone et al. 1996). Third, and perhaps of greatest interest to paleontologists, is the close relationship between development of Earth's abiotic systems and the evolution of its biotic inhabitants. The positions of continents, sea-level, tectonic activity, and climate regime have all played important roles in the history of life, influencing the evolution and extinction of taxa. Orbitally-driven oscillations in climate patterns have been implicated with species-range dynamics at different latitudes, related to the amount of seasonality and environmental heterogeneity (Dynesius and Jansson 2000). These dynamics vary spatio-temporally, affecting speciation rates, speciation mechanisms, and degree of adaptive specialization. This variation leads to differing evolutionary rates over both space, in the form of latitudinal gradients in species richness (Wright et al. 2006), and time, in secular patterns of global species diversity (e.g., Sepkoski 1998).

3.3.2 Subsurface Processes

Changing climate and tectonic conditions over time can effectively alter the subsurface environment and establish biogeochemical conditions that promote the physical and/or chemical decomposition of organic remains. Long-term development of the soil profile in conjunction with a lowering of the regional water table can lead to expansion of the terrestrial TAZ, spreading oxidative conditions, bioturbation, and consumer access to buried remains. This is most common in areas that support high productivity and biodiversity due to relatively stable climatic and tectonic conditions over 10^4–10^5 year timescales. Not all soil development is necessarily destructive. Certain soil orders, such as the aridisols and mollisols, contain calcium carbonate-bearing horizons at relatively shallow depths (~30–60 cm) (Brady 1974), that can provide a ready source for calcite formation within the bone and/or sediment during diagenesis. Several productive fossil formations, mainly from the Mesozoic and Cenozoic, consist primarily of paleosols at low to moderate stages of development that were formed within seasonal or semi-arid environments (Bown and Kraus 1981; Winkler 1983; Maas 1985; Badgley and Gingerich 1988; Downing and Park 1998; Paik 2000; Clyde et al. 2005). As a general rule, environments that support a greater degree of soil development lead to biogeochemical conditions promoting organic decomposition and, in extreme cases, leaving only the most recalcitrant remains behind (see Retallack 1990 for more extensive description).

Water and sediment availability, geomorphology, topography, and tectonic stability influence riverbed geometry and flow dynamics. These factors affect the sinuosity and lateral migration rate of the riverbed, as well as the size and distribution of overbank deposits (Prothero and Schwab 1996; Einsele 2000). Highly sinuous and mobile rivers cut into the surrounding landscape, reworking the sediment and exhuming previously buried remains. Over time this reworking process creates time-averaged assemblages of varying duration and composition (Behrensmeyer 1988; Behrensmeyer and Hook 1992). This process favors the preservation of more resistant skeletal elements, including teeth and large bones or fragments within the most active of fluvial systems.

As sediment accumulates on the surface, underlying layers experience compaction, cementation, and authigenesis as lithification proceeds (Prothero and Schwab 1996).

Since most fossil bones demonstrate some amount of structural deformation, prolonged sediment compaction likely plays a role in this phenomenon. This distortion has the potential to artificially inflate skeletal element variability and confuse the taxonomic assignment of the remains (Wilborn 2007). Grain size, pore space, and bone size and shape determine the extent of compaction and its effect on the interred bone (Prothero and Schwab 1996; Smoke and Stahl 2004). Cementation depends on the hydrology and chemistry of the ground water, but silica or calcite typically forms the cement. Early cementation has been suggested as a pathway for fossil preservation, especially in soft tissues (Briggs and Kear 1993; Sagemann et al. 1999). The presence of calcite within fossil bone has been implicated as a condition favoring preservation (Holz and Schultz 1998; Fernández-Jalvo et al. 2002; Berna et al. 2004; Wings 2004). Diagenetic concretions may even form, preserving small-bodied and delicate taxa that would otherwise be destroyed (Downing and Park 1998). Authigenesis leads to the deposition of new or altered mineral species within and around remains and, due to the environmental-specificity of mineral formation, can provide a detailed diagenetic history (Bao et al. 1998; Clarke 2004).

The extent of diagenetic alteration depends on sedimentary composition, depth of burial, temperature at depth, and the length of time the unit is exposed to these conditions. Depth of burial is particularly important, because certain diagenetic changes – including dewatering, cementation, recrystallization, dissolution, and replacement – can occur at shallow depths of 1–3 km (Prothero and Schwab 1996). The survival of fossils within the unit is contingent upon the extent of these alterations, potentially destroying or rendering unidentifiable fossils that were stable at shallower depths. It appears that plant fossils, which are far more delicate than bone, can withstand moderate to high diagenetic processes at depths of 3.8–5.2 km and temperatures between 150°C and 170°C (Howe and Francis 2005).

While the stability of bone apatite at different pressures and temperatures has yet to be comprehensively studied, the diagenetic behavior of another bioapatite – from conodonts – has been well described and may provide a useful model. Color changes in conodont elements have been linked with an increase in the temperature experienced by the enclosing rock, providing a relatively precise scale from low (pale yellow, <80°C) to high (black, >300°C) temperatures (Epstein et al. 1977; Prothero and Schwab 1996). Conodont elements contain structures homologous to vertebrate bone (Sansom et al. 1992) and undergo a similar change in crystallite size during diagenesis (Noth 1998). The bone of more advanced vertebrates may react similarly and help to explain the wide diversity of colors observed in bone from terrestrial fossil assemblages, providing valuable insight into the diagenetic conditions experienced by the sedimentary unit.

McKean et al. (2007) hypothesized that bone color change results from geothermal alteration of the bone's remaining organic content. They found that bone color (yellowish-white to black) and organic content correlate with depth of burial if one assumes a geothermal gradient of 27.5°C/1,000 m. This represents a promising first step, but more work on this is certainly necessary. This relationship is relevant to those studying rare earth element (REE) signatures in fossil bone for use in paleoenvironmental reconstruction, because the REE signature of conodont elements is altered at high temperatures due to recrystallization (Armstrong et al. 2001). These results could have a significant impact on environmental reconstructions that use REE signatures of fossil bones and merits further investigation.

4 Large-Scale Spatio-Temporal Controls Over Taphonomic Processes

Changes in the global climate regime, the ultimate causes of which are still not fully understood, have a far-reaching impact on not only the history of life by driving extinction and evolution, but also the fossil record, by controlling the distribution of environments and taphonomic modes. Hence, not only do the players change, but the stage changes as well. Within the relatively narrow temporal window provided by many fossil occurrences, these long-term secular changes in taxa and environments play a relatively minor role in fossil assemblage formation, occurring near the minimum resolution recordable by the fossil record (Behrensmeyer et al. 2000). The effect of secular changes on preservation cannot be fully appreciated by studying individual fossil assemblages.

Large-scale environmental changes due to changing tectonic activity, atmospheric CO_2 concentration, and/or insolation patterns may shift the taphonomic window of preservation, altering the biota we are likely to recover in the fossil record. This dynamic may explain why the fossil record can dramatically improve or deteriorate (widen or narrow the taphonomic window) during suspected periods of major environmental change (also see Frasier et al. this volume). Alteration to taphonomic modes can be subtle, such as is often found with changing sedimentation rates brought about by changes in erosion patterns (Behrensmeyer et al. 1997; Martin 1999). Changing environmental conditions may create or remove critical depositional environments and taphonomic modes over time (Smith and Swart 2002; Retallack 2005b; Smith and Botha 2005). On the extreme end, environmental changes behind mass extinction events may severely perturb preservation conditions, leading to unusual, short-term taphonomic modes. For example, during the biotic crisis surrounding the Permian-Triassic transition elevated volcanic activity altered atmospheric chemistry, leading to massive plant die-offs and extensive terrestrial erosion that was rapid and short-lived (Huey and Ward 2005; Retallack 2005a; Sephton et al. 2005; Arche and Lopez-Gomez 2006). In all of these examples, higher-level changes were required before large-scale alterations to environmental distribution and taphonomic modes could proceed.

4.1 Geophysical Dynamics

Above all, vertebrate remains must be buried before they can fossilize. Tectonic activity, primarily uplift and subsidence, is the ultimate control of sediment erosion and deposition. Subsidence increases basin accommodation and allows for rapid burial, even in the absence of high sedimentation rates. Continental accretion and mountain building affects topography and can generate regionally higher rates of sediment accumulation in fluvial, lacustrine, and deltaic environments distal to the uplifted area (Behrensmeyer and Hook 1992; Behrensmeyer et al. 2000). Tectonically active periods also tend to see an increase in volcanism. Volcanoes deposit various silica-rich particulates which, when weathered, alter sediment chemistry and provide an important source of diagenetic materials (Behrensmeyer and Hook 1992; Downing and Park 1998; Martin 1999). Increased volcanism may form rift valleys in zones of continental extension, providing a basin for sediment deposition and altering the local water table, which can create river and lake systems in valley interiors, leading to important fossil accumulation sites (Rogers et al. 2001; Smith and Swart 2002).

Significant alterations of geomorphology that affect burial can occur even in the absence of tectonic activity. For example, the development of continental glacial conditions during "icehouse" periods promotes widespread eolian silt (loess) deposition, alluvial outwash, and lake formation from meltwater over land surfaces in areas proximal to the glacial front (Behrensmeyer and Hook 1992; Prothero and Schwab 1996). In other, more distant areas the onset of glaciations leads to a marked shift in temperature and precipitation patterns. These environments promote burial of remains, especially those influenced by periodic flooding from glacial meltwater. Glacial retreat exposes new depositional basins and topographic sources for weathering, and enhances erosion by altering base level following isostatic rebound. These features create new opportunities for both terrestrial and freshwater preservation following glaciation. Glacial mass compacts any underlying non-consolidated sediments. Over larger spatial scales, glacial formation reduces sea-level, expanding continental area. Larger land areas support an overall greater abundance of species, although the exact mechanisms underlying the pattern are still under scrutiny (Rohde 1992; Chown and Gaston 2000; Storch et al. 2005). The species-area effect has been observed to operate in the late Cenozoic for various taxa and at multiple scales, signifying its importance in understanding paleobiodiversity patterns (Flessa 1975; Marui et al. 2004; Barnosky et al. 2005).

Continental drift transports a land surface, submitting it to changing climatic conditions even when global climate remains stable. The effect of drift vs. large-scale climate change can be difficult to discern from the fossil record of a particular region if widespread fossil localities are unavailable to place the inferred changes into a larger context. New technology, in the form of GIS software, coupled with geophysical models of crustal plate movement,[1] allows paleontologists to reconstruct the probable pathways and extents of regional movement over time, providing an additional comparison between climate change and drift-induced changes (of course these need not be mutually exclusive). If continental drift has a similar effect on environmental distribution as climate change does, then we can expect the prevailing taphonomic regime to change as a result.

[1]Paleogeographic Atlas Project at the University of Chicago (pgap.uchicago.edu) and the Paleomap Project at the University of Texas, Arlington (www.scotese.com).

4.2 Atmospheric Carbon Dioxide

The concentration of atmospheric CO_2 is one of, if not the most, important drivers of global taphonomic patterns because of its role in determining both global temperature and biogeochemical cycles at multiple spatio-temporal scales. Two sets of carbon cycles control CO_2 concentration: (i) the short-term exchange between the biosphere, soil, ocean, and atmosphere, operating on a 10^0–10^4 year timescale; and (ii) a long-term exchange between the atmosphere, rocks, and ocean on a 10^5–10^9 year timescale (Rothman 2001 and references therein).

Periods of elevated atmospheric CO_2 accelerate chemical weathering of silicate rocks, providing calcium, bicarbonate, and silicon ions for the formation of $CaCO_3$ and SiO_2 in marine and terrestrial environments (Martin 1999). Changing atmospheric CO_2 concentrations, due to volcanism or climate change, alter the cycling of critical elements like C, N, Ca, and P. However, the relationship is complex, involving several coupled feedback mechanisms between terrestrial, marine, and atmospheric sources (Igamberdiev and Lea 2006).

Carney et al. (2007) found that experimental doubling of CO_2 in a forest community actually enhanced terrestrial carbon cycling instead of leading to greater plant assimilation by altering the relative abundance and activity of soil microbes. This discrepancy is due to differences between photosynthesis and respiration, with photosynthesis being more temperature sensitive, placing a lower maximum response limit to CO_2 enrichment than respiration (Allen et al. 2005; Igamberdiev and Lea 2006). In other words, an increase in atmospheric CO_2 leads to relatively minor gains in plant productivity (which will scale with size) while supporting a much greater increase in decomposition, labile C cycling, and microbial biomass within soils (Allen et al. 2005).

When metabolically or structurally critical elements are abundant in the burial environment, extraordinary preservation can result, but their absence leads to extensive biogeochemical recycling and subsequent destruction of remains (Behrensmeyer and Hook 1992). Many of the extraordinary fossil Lagerstätten, such as the Early Cretaceous Liaoning deposits (Zhou 2006), were formed during periods of high atmospheric CO_2, when perturbations to major biogeochemical cycles led to periods of exceptional preservation (Retallack 2005b), perhaps driving tissue carbonization as a major mode of preservation. As paleogeographic and paleoclimate models improve, it should become possible to explicitly test changes in biogeochemical cycles due to climate and tectonic change at varying spatio-temporal scales.

4.3 Orbital Cycles in Solar Energy

The geometry of Earth's orbit varies over time, causing changes in the amount of solar energy received on the surface. These Milankovitch oscillations are long-term cyclical changes in eccentricity (100 ky period), obliquity (40 ky period), and precession

(26 ky period) (Prothero and Schwab 1996). Individually, the adjustments in orbital geometry are small, but the combination of these cycles over millennia affect how and where solar energy is distributed, and they have been implicated in causing periods of rapid regional-to-global climate change (Wright and Vanstone 2001; Rial 2004).

Due to the semi-chaotic nature of Earth's orbital geometry it is nearly impossible to accurately project these oscillations into the distant past (>40–50 Ma) (Laskar et al. 2004). However, the large-scale effect of changes in these values on climate is observable in the fossil record. Some examples reach as far back as the Devonian (Yang et al. 1996) and cover climatic responses in both the marine (Yang et al. 1996; Balog et al. 1997; Fenner 2001; Wendler et al. 2002; Gibbs et al. 2004) and terrestrial (Olsen et al. 1991; Miller and Eriksson 1999; Retallack et al. 2004) records. Such responses induced major changes to atmospheric and ocean circulation patterns, sea-level, seasonality, precipitation, and surface temperature. Changes in these patterns have widespread effects on continental weathering and sediment transport (Van der Zwan 2002; and see below), and therefore the number and distribution of depositional systems and fossil-bearing units.

Cycles in solar activity have been found to drive short-term increases (100 ky cycles) in global temperature and CO_2 release into the atmosphere (Rial 2004), although the overall effect on global climate dynamics appears small compared to that of orbital oscillations (Cubasch et al. 2006).

5 Implications for the Terrestrial Vertebrate Fossil Record

5.1 *The Existence of Terrestrial Megabiases*

Megabiases result from the combined effect of multiple, often secular, changes in taphonomic processes controlling the destruction of remains (Behrensmeyer et al. 2000). Megabiases were first recognized and studied in the marine invertebrate fossil record, foremost among them cycles of preservation and biomineralization (Martin 1999). Together changes in these factors over time profoundly affect reconstructions of biodiversity and macroevolutionary patterns in the marine fossil record (Smith et al. 2001; Martin 2003; Bush and Bambach 2004). Although Behrensmeyer and Hook (1992) noted the possibility of a megabias in the terrestrial fossil record and recommended further research, actual research into the existence of terrestrial megabiases similar to those found in the marine realm have yet to infiltrate large-scale studies of the terrestrial fossil record.

When major trends in terrestrial fossil preservation are considered, they are interpreted in light of marine trends, i.e., sea-level change as a driver of terrestrial fossil preservation (e.g., Sereno 1997; Wolfe and Kirkland 1998). Fara (2002), in a study on gaps in the Late Jurassic–Eocene terrestrial fossil record, found no evidence that sea-level change influenced continental fossil preservation. Instead he suggests that the marine and continental records behave independently of each other. However, Van der Zwan (2002) found that sea-level effects on continental sediment supply depend on global climate. Milankovitch cycles are more important during "greenhouse" intervals, whereas sea-level change dominates during "icehouse" intervals. Regardless, examining marine megabiases should lend some insight into potential parallel patterns in the terrestrial record. The marine record demonstrates many processes, acting at similar spatio-temporal scales and falling under the same influences as the continental fossil record.

Long-term changes in sea-level and atmospheric CO_2 lead to multiple, cascading alterations in the habitat and sediment character of marine environments, affecting: (i) the spatial distribution of environments, (ii) species diversity and abundance levels, and (iii) preservation of the biotic assemblage; all these factors changing predictably with depth (Smith et al. 2001; Martin 2003). A similar pattern exists in terrestrial systems, where changes in the sedimentary record often are observed to coincide with noticeable changes in vertebrate fossil accumulations (Maas 1985; Bown and Beard 1990; Martin 1999; Smith and Botha 2005). In terrestrial environments elevation and atmospheric CO_2 (a factor in long-term climate change) control the same set of habitat and sediment character changes. These factors change predictably with elevation: both environmental distribution and taxonomic diversity vary with altitude (Gaston 2000) and the opportunity for burial and fossilization decreases as one moves from low-lying areas of net deposition to elevated areas of net erosion. The poor representation of highland environments is a well-known feature of the terrestrial record. It has yet to be identified as a true megabias, possibly due to the difficulty in distinguishing the effect of elevational changes from other factors (although methods are improving). Regions experiencing increased tectonic activity or isostatic rebound most likely are affected by this megabias.

Marine $CaCO_3$ saturation decreases at higher latitudes as shallow waters become cooler – the "latitudinal lysocline" – causing dissolution in $CaCO_3$ skeletons at higher latitudes (Martin 1999). Atmospheric CO_2, tectonic, and climate changes mediate the strength and steepness of the lysocline over time by shifting water depth and continental shelf area (Martin 1999; Pearson and Palmer 2000). Bone preservation may be similarly controlled by the position of the Intertropical Convergence Zone (ITCZ), a wide belt of precipitation following seasonal north-south movement cycles, which changes in response to continental arrangement and land area (Ziegler et al. 2003). Together these factors influence the geographic range and position of moist, seasonally wet/arid, and arid environments in tropical and temperate regions.

The greatest extent of moist conditions in tropical and temperate regions are supported during times of high sea-level and continental divergence, forming a continuous latitudinal band at the equator and two others at mid to high latitudes (Ziegler et al. 2003). Low sea-level and continental accretion lead to greater seasonality and aridity, reducing the size of the equatorial moist tropical zone while leaving the high latitude moist temperate zone intact, forming a precipitation gradient extending down to the low latitude arid zones. Predominantly moist or arid zones contain a lower diversity of depositional environments and taphonomic modes favorable to bone preservation (sedimentation rates, sediment biogeochemistry, biotic activity, etc.). Intermediate regions that receive enough seasonal precipitation to support a variety of favorable environments for preservation are more likely to form a substantial fossil record because more opportunities for burial exist. Bone preservation in very moist or arid regions does occur but may be more infrequent and subject to seasonal patterns or other intermittent changes in climate that shift conditions towards favoring burial and preservation (Fastovsky et al. 1997; Loope et al. 1998; Rogers 2005). It appears this pattern has remained relatively stable through most of the Phanerozoic because of Hadley cell circulation (Ziegler et al. 2003), with the steepness of moist-arid gradients mediated by sea-level and continental configuration. This taphocline has important consequences for vertebrate paleobiology, because most of what we know about extinct life in the terrestrial fossil record comes from seasonal intermediate regions at mid-latitudes. Such a pattern has been observed in Mesozoic terrestrial ecosystems (Anderson et al. 1998; Rees et al. 2004; Noto, unpublished data) but remains to be studied in other vertebrate groups and other times.

Another potentially widespread cause of megabias is the evolution of vascular land plants, whose morphological and metabolic adaptations have altered biogeochemical cycles over time (Berner et al. 2004). This is especially true in the replacement of gymnosperms by angiosperms as the dominant flora is most terrestrial environments, starting in the Cretaceous. Many studies have found distinct differences between each group in their effect on sediment biogeochemical processes, which influences the type of weathering regime found in the host soil (Kelly et al. 1998; Berner et al. 2004). Significant disagreement exists as to which group promotes greater weathering. The evidence is slightly in favor of gymnosperms, although a great deal of variability exists between taxa (average ratio of angiosperm over gymnosperm weathering rate is 0.5–1.5) within each group, requiring much more work (see Berner et al. 2004).

Geologic measures of weathering may not be as important in bone decomposition as other soil properties under gymnosperm vs. angiosperm cover. Grown in the same conditions, soil under gymnosperms tends to be more acidic, contains less exchangeable Ca (due to lower Ca content of leaf litter), and a lower abundance of heterotrophic consumers (Reich et al. 2005). The low sediment pH will be more destructive to buried bone over time than microbial or fungal activity. Each group alters soil texture and composition differently. Gymnosperm-altered soils are more organic-rich and sandy, while angiosperm-altered soils are more clayey and dense (Andrews et al. 2006), affecting soil hydrologic properties and chemistry. Higher sand content permits greater fluid flow through the sediment and creation of biogeochemical conditions detrimental to the stability of osseous tissue. High clay content reduces fluid flow; slowing bioapatite and collagen loss, while encouraging early diagensis (see Section 3.2.2).

These differences may have lead to greater bone destruction in the gymnosperm-dominated communities of the Paleozoic and Mesozoic than later angiosperm-dominated communities of the Cenozoic, though this may have been offset somewhat by the higher consumer abundance and decomposition supported in angiosperm-derived soils. Furthermore, differences between the soils in gymnosperm vs. angiosperm communities are dependent on external environmental factors and therefore will not act uniformly across environments. Particular combinations of parent sediment, plant taxa (or functional group), and climate are likely more important than plant type alone.

The dramatic rise in terrestrial vertebrate diversity starting in the Cretaceous (Benton 1985) may not only be the result of "pull of the recent" effects or rock volume/exposure area bias, but a long-term change in the biogeochemical conditions of environments where angiosperms proliferated, creating conditions more favorable for bone preservation. Further work on paleosols from the late Mesozoic and early Cenozoic are necessary to track the chemical and physical changes to the soil profiles before, during, and after angiosperms diversified.

5.2 Examples of Changing Taphonomic Regimes Over Time

5.2.1 Paleozoic

The fossil record of terrestrial vertebrates begins in the late Devonian with the evolution of tetrapods and their first forays onto land. Early tetrapods inhabited marginal-marine, deltaic, and fluvial environments in seasonally wet, semi-arid climates during a time of high global mean temperature and elevated CO_2 and low O_2 concentrations in the atmosphere (Fig. 3a) (Berner 2006; Cressler 2006; Scotese 2009). Two major gaps exist in this record: one during the Famennian (374–360 mya) and the other between the late Devonian and early Carboniferous, known as "Romer's Gap". Both gaps have been interpreted as a true decrease is tetrapod diversity resulting from the sharp drop in atmospheric oxygen concentration during this period (Ward et al. 2006). However, a comprehensive examination of the fossil record from this time reveals changes in tetrapod morphology, diversity, and fossil distribution that are more compatible with a taphonomic interpretation (Clack 2007; Coates et al. 2008).

In early tetrapodomorphs ("stem tetrapods") such as *Eusthenopteron*, *Tiktaalik*, and *Panderichthys*, large body size, obligatory aquatic lifestyle, presence of scales, and well-ossified skeletal elements may have aided in their preservation. Decay dynamics of these taxa may have been more like those of large, bony fish such as the alligator gar (*Atractosteus spatula*). Gars may provide a viable model for early tetrapod taphonomy, not only because of their large size, but their ecology and habitat preferences closely match those of the earliest tetrapods. Well-preserved taxa like *Tiktaalik* (Daeschler et al. 2006) and *Eusthenopteron* (Schultze and Cloutier 1996) demonstrate taphonomic features, such as the broad head resting parallel to the bedding plane and dorsoventral flattening of the axial skeleton, similar to those described in carcasses of alligator gar deposited following tropical storms on the gulf coast of Texas (Weigelt 1989).

Later more derived, limbed taxa like *Acanthostega* and *Ichthyostega* inhabited freshwater fluvial and shallow-water environments under semi-arid climate conditions (Long and Gordon 2004). Under these environmental conditions early tetrapod remains were at risk of prolonged subaerial exposure. These taxa also show loss or reduction of scales on the body (Coates et al. 2008). Scale loss may have made the bodies of these animals more prone to disarticulation due to

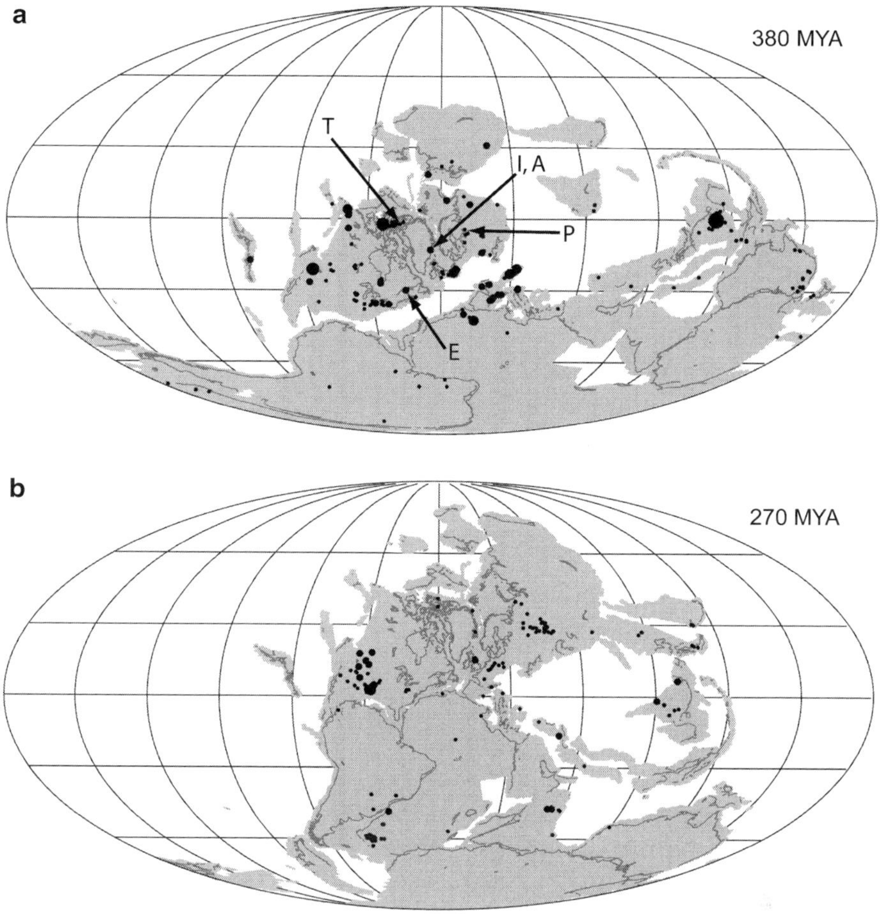

Fig. 3 Distribution of terrestrial vertebrate fossil localities. The size of the marker is proportional to the number of taxa known from a locality: larger markers contain greater vertebrate diversity. (**a**) Devonian, 380 MY reconstruction. Location of tetrapodomorph taxa marked with letters: A=*Acanthostega*, E=*Eusthenopteron*, I=*Ichthyostega*, P=*Panderichthys*, T=*Tiktaalik*. (**b**) Permian, 270 MY reconstruction. Data from the Paleobiology Database. Paleogeographic map software created by John Alroy and Chris Scotese

gas bloating and/or scavenger consumption. These factors likely affected their preservation, as seen in the prevalence of isolated postcranial elements and/or large, robust skulls, contributing to the lack of fossil material during much of the Fammenian.

Starting in the late Devonian terrestrial plants underwent a rapid evolutionary radiation, expanding into a variety of new niches and culminating in the first recognizable forests (Bateman et al. 1998). An overall rise in secondary plant productivity, as evidenced by larger, woody stems with deeper roots, resulted in deeper weathering of the soil profile. This was especially true in warm temperate and tropical everwet environments (Retallack 1997; Algeo et al. 1998; Scheckler and Maynard 2001). During the early Carboniferous the onset of Pangean assembly drove the development of extensive ice sheets in the southern hemisphere (Rygel et al. 2008). The ice sheets would have prevented the spread of vertebrates to the south, but created ecological opportunities in the north. Glaciation altered global climate and supported a differentiation and expansion of tropical and temperate biomes, with retraction of arid regions (Scotese 2009). All of this added up to increased ecological opportunity for tetrapods, who likely began radiating into these new environments. However, low atmospheric oxygen favored smaller body sizes (Clack 2007). Smaller average body sizes and increased weathering of temperate and tropical soils was a likely driver of Romer's Gap during the early radiation of terrestrial tetrapods.

This trend continued until the late Carboniferous when widespread glaciation at high latitudes lead to increasing seasonality and contraction of the tropical biome. Increasing atmospheric oxygen and development of the amniotic egg allowed tetrapods to increase in size and expand into more seasonal environments with higher preservation potential (DiMichele and Hook 1992). Cyclical glacioeustatic sea-level changes occurred frequently from late Carboniferous to early Permian (Rygel et al. 2008), driven by Milankovitch oscillations (Wright and Vanstone 2001). Sea-level changes created repeated cycles of continental sediment erosion and deposition (Van der Zwan 2002) that increased preservation opportunities during specific intervals.

During the Permian, continued assembly of the large Pangean landmass continued the drying trend begun in the late Carboniferous (Fig. 3b). The southern hemisphere ice sheet disappeared. Atmospheric circulation and precipitation distribution (i.e., ITCZ position) was highly monsoonal (Ziegler et al. 2003). With the drying of the continental interior and strongly seasonal precipitation pattern, tropical everwet and warm temperate conditions were rare on the main continental mass. More environments favorable for bone preservation became available, as evidenced by the reasonably good vertebrate record (Sander 1987, 1989; DiMichele and Hook 1992; Smith 1993; Sidor et al. 2005). Extreme aridity in some places led to highly destructive environmental conditions. Evidence for extensive ($\geq$200 K km^2) red bed formation in North America during the mid-Permian shows extremely acidic (~pH 1) groundwater and soils, which would have destroyed any buried bone (Benison et al. 1998). Given these hostile environments, it is likely that vertebrate diversity in these regions was also low.

5.2.2 Mesozoic

A rapid increase in atmospheric CO_2 concentration at the beginning of the Triassic (Berner 2006) relaxed the latitudinal temperature gradient as mean surface temperatures increased, establishing even more widespread aridity – perhaps one of the most arid periods in Earth history. The absence of polar ice caps meant warm temperate conditions extended all the way to the poles for most of the Mesozoic. The symmetrical arrangement of Pangea about the equator created a "megamonsoonal" precipitation regime, with a wide latitudinal excursion for the ITCZ that made climate patterns highly seasonal (Ziegler et al. 2003). Vertebrate preservation was best at low- to mid-latitudes, where the monsoonal precipitation led to seasonal flooding, rapidly burying exposed remains under poorly developed soils with high water tables that promoted early diagenesis (Dubiel et al. 1991; Smith and Swart 2002). Like the Permian, the most hostile desert regions (e.g., much of the continental USA) sport few fossils due to poor preservation and low diversity.

As CO_2 levels declined from the mid-Triassic to mid-Jurassic, tropical and warm temperate biomes once again expanded their ranges and a seasonally wet tropical biome was reestablished at the equator. There is no good evidence of tropical everwet (i.e., rainforest) conditions throughout the Jurassic due to a weakened but still operational monsoonal circulation (Ziegler et al. 2003). As such, biome distribution remains more or less stable during the Jurassic, with an increase in global temperature and humidity as Pangea began to rift apart and CO_2 levels once again peak in the Late Jurassic, before beginning a steady decline in the Cretaceous (Price and Sellwood 1997; Berner 2006; Sellwood and Valdes 2006).

The Late Jurassic contains spectacular assemblages of dinosaurs and other vertebrates, preserved mostly in the seasonally-dry biomes at mid-latitudes (Fig. 4a). The number of vertebrate fossils drops sharply at higher latitudes, coinciding with the transition to the warm temperate biome, where the peak in plant fossil diversity occurs (Rees et al. 2004). This drop in vertebrate preservation is likely due to the relatively poor preservation conditions that existed in high latitude forests. The record for Early and Middle Jurassic vertebrates is not as good but demonstrates the same pattern: the vertebrate fossil peak follows the north-south migration of the biome (Noto, unpublished data). It is interesting that the Middle Jurassic vertebrate record is so poor, considering it contains the same basic arrangement of continental area and biomes. Further work on this time period will be necessary to tease out the cause of this idiosyncrasy.

The Cretaceous was perhaps one of the most equable times in Earth history. Global temperature cooled as CO_2 levels fell, most likely due to the increased weathering and runoff from the diverging continents (Donnadieu et al. 2006). As the continents drifted further apart, the smaller continental interiors became moister and the large desert biomes characteristic of the Triassic and Jurassic were greatly reduced. Instead, widespread seasonally wet savannah-like environments persisted in the tropics at the low to mid latitudes (Upchurch et al. 1999). Small patches of tropical rainforest were restricted to the equator, while warm- and cool-temperate forests reached all the way to the poles. Widespread seasonally wet conditions persisted across much of North America, Europe, and China. Global temperature and humidity were still higher relative to today (Scotese 2009).

Cretaceous vertebrate fossils are relatively widespread, found at some abundance in most biomes and latitudes (Fig. 4b). This is in stark contrast to previous patterns, where the greatest number of vertebrate fossils is restricted to a narrow range of environments. High sea-level, in concert with the seasonal precipitation pattern, promoted broad plains proximal to sea shores (Horner 1989; Horner et al. 1992), creating ideal environments for bone accumulation and burial. In tropical

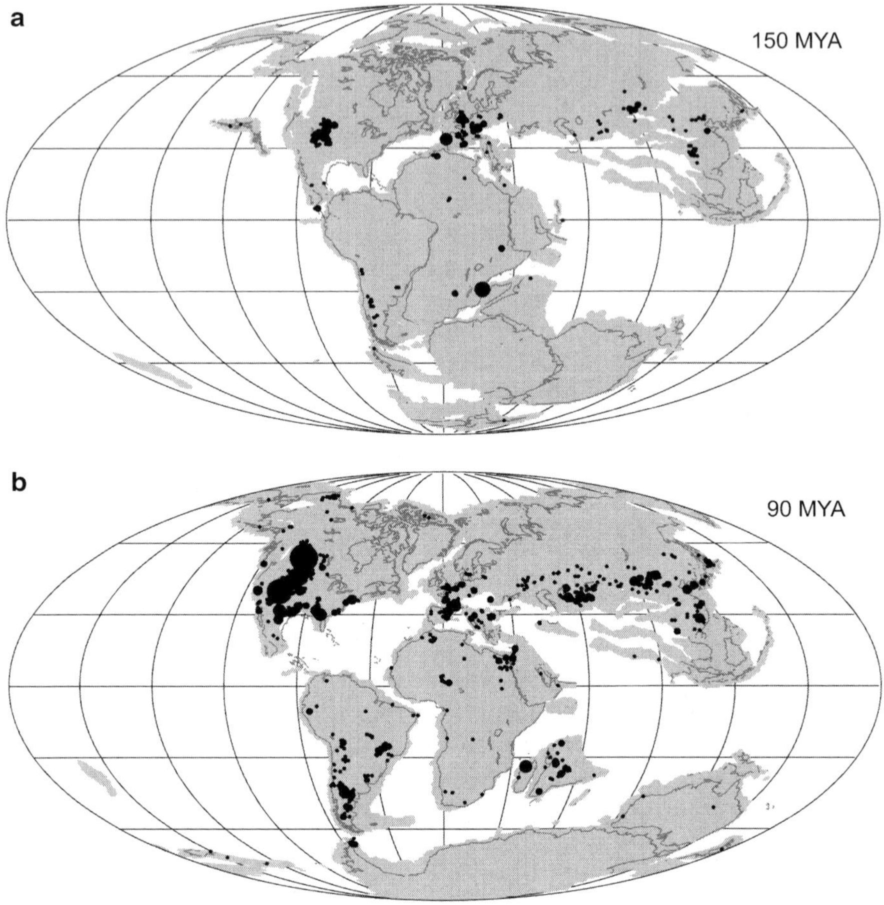

Fig. 4 Distribution of terrestrial vertebrate fossil localities. The size of the marker is proportional to the number of taxa known from a locality: larger markers contain greater vertebrate diversity. (**a**) Late Jurassic, 150 MY reconstruction. (**b**) Late Cretaceous, 90 MY reconstruction. Data from the Paleobiology Database. Paleogeographic map software created by John Alroy and Chris Scotese

environments, these conditions contributed to the formation of lagoons and other brackish water bodies, which provide an outstanding record of soft-tissue preservation (Dal Sasso and Signore 1998; Zhou et al. 2003).

By the Late Cretaceous, angiosperms are important members of the flora, found in every major environment. Their modifications to the burial environment potentially contributed to the apparent rise in diversity by promoting conditions more amenable to bone preservation. Indeed, a survey of vertebrate fossil distribution does appear to show more fossil localities near shorelines. The lack of a fossil record across the K/T boundary may as much be a function of loss of these habitats due to lowering sea-level and increasing aridity, which destroyed the most faithful recorders of the terrestrial biota, as it was widespread extinction (Fastovsky 1990).

6 Implications for Vertebrate Paleobiology

6.1 Changing Patterns of Species Diversity

Cycles of evolutionary and ecological change observed in the fossil record are not surprising, considering the control that the global climate and tectonic regimes exert on both biogeochemical processes and species' population dynamics. These cycles are explicitly linked because they are controlled by the same set of factors, yet the changes they elicit often are treated separately. Given the hierarchical nature of both biotic and abiotic response to climate and tectonic changes, how each level of the hierarchy interacts during these transitions to form the fossil record takes on special importance.

As Behrensmeyer and Hook (1992, p. 88) observe, "[m]any paleoecologic studies have emphasized the role of global climate as the major factor controlling biotic change, but climate pattern alone neither explains nor predicts the composition of subsequent biotas." Climate and tectonic changes that bring about environmental change lead to shifts in not only floral and faunal assemblages but in biogeochemical cycling as well, thus altering the suite of taphonomic processes operating in the area, in effect creating the very real problem of distinguishing between local shifts in species distributions and large-scale biotic events. The use of other time-equivalent deposits for reference is required to understand the scope and nature of the perceived change (Behrensmeyer and Hook 1992). This idea needs further enhancement by extending the phenomenon to a larger scale.

At the global scale, shifting environmental parameters and species ranges are symptomatic of long-term alterations to the properties and distribution of the major biomes, themselves dependent on the state of tectonic activity and insolation. As the biomes change, this will also significantly change the overall nature and distribution of taphonomic regimes, effectively moving the taphonomic window of preservation and our view of life during the periods of time in question. It is not the completeness of the terrestrial fossil record per se that is at issue here but how this sliding taphonomic window affects our ability to reconstruct the ecology and evolution of extinct species, specifically the distribution of regional biodiversity. Much of the research concerning biodiversity patterns has focused on quantifying global patterns, charting the rise and fall of life's diversity through time, while neglecting potentially important dynamics occurring at the regional level (Miller 2003). It is precisely the nature and effect of these dynamics in determining species distributions that continues to be a major source of study for ecologists, the most important pattern being latitudinal gradients in species richness.

The current latitudinal gradient consists of highest diversity in the tropics, which decreases with increasing latitude toward the poles. This pattern has been demonstrated in plants, mammals, reptiles, amphibians, insects, and fish (Rosenzweig 1995). Although well documented, the mechanisms driving this pattern of species distribution still are debated actively, being most strongly related to productivity and/or evolutionary rate (Gaston 2000).

Several examples of latitudinal diversity gradients are also known from the fossil record, found in plants (Silvertown 1985; Crane and Lidgard 1989; Haskell 2001), dinosaurs (Rees et al. 2004), radiolarians (Kiessling 2002), bivalves (Crame 2002), and brachiopods and foraminifera (Stehli et al. 1969). Given the growing interest in the effect of ecological interactions on speciation and evolutionary rates among paleontologists and neontologists alike, the presence of latitudinal gradients in the fossil record suggests that the mechanisms driving these patterns are an ancient phenomenon.

Studying regional patterns of alpha, beta, and gamma diversity in the fossil record enables a better understanding of the effects of environment and ecology on micro- and macroevolutionary processes in ways that may only be addressed with the fossil record (Miller 2003; Jackson and Erwin 2006). Allison and Briggs (1993) found that, since biodiversity varies with latitude, the latitudinal range sampled by the fossil record at a particular time will influence estimates of extinction and origination rates, especially when comparing patterns from times that experienced different global climate regimes.

A global point of view is necessary to place local- and regional-scale fossil distributions within the context of global taxonomic diversity patterns, but few studies have addressed this directly. Applying these principles to the terrestrial fossil record will first require a better understanding of how taphonomic bias due to environmental differences influences large-scale patterns. Once addressed, it will be possible to implement correction factors to the diversity data. Such an approach was applied to Middle Paleozoic and Late Cenozoic marine benthic communities (Bush and Bambach 2004).

6.2 *Model of Diversity Gradients and Climate Change*

Modern continental surfaces can be grouped into broad regions of common climate and ecology to form biomes (see Section 3.3). The latitudinal and longitudinal extent of each biome depends on the distribution of constituent environments and biota, which is driven largely by precipitation and insolation patterns (Begon et al. 2006). These same factors also lead biomes to support different levels of biodiversity; moist, high-energy areas support more species than dry, low-energy areas (Hawkins et al. 2003; Evans et al. 2005; Kreft and Jetz 2007). Specific chemical, physical, and biological processes present within an environment define a taphonomic mode, with particular modes preserving vertebrates more frequently in certain environments over others (Behrensmeyer and Hook 1992).

In the proposed model, vertebrate taxa differ in the taphonomic modes necessary for fossil preservation and vary in abundance due to ecological requirements. The number of different taphonomic modes together describes the taphonomic regime for a given biome. The greater the heterogeneity of taphonomic modes, in particular the proportion of these modes with conditions necessary for preservation of Taxon X, the greater the possibility of it entering the fossil record. Therefore the fossil record of

Taxon X varies between biomes due to both abundance and preservation requirements, creating latitudinal and longitudinal gradients between taphonomic regimes. The collection of all taphonomic regimes, in conjunction with the level of biodiversity supported in each biome, produces the overall fossil record for a given time period. This model then allows one to explore changes to the fossil record over time brought about by changes in climate, biodiversity, and taphonomic processes.

Consider the following scenario shown in Fig. 5. At Time A the prevailing climate regime supports three major biomes: a moist, high-latitude Biome 1; a seasonally-wet Biome 2; and a moist, equatorial Biome 3. Biodiversity is highest in Biome 3, decreasing with increasing latitude towards the poles (3 > 2 > 1). When living biodiversity is surveyed at Time A, the resulting diversity pattern is a unimodal curve, similar to that observed today. However, when forming the fossil record under the conditions during Time A, the probability of vertebrate preservation in Biome 2 is greater than either Biomes 1 or 3 (2 > 3 > 1) due to supporting a greater

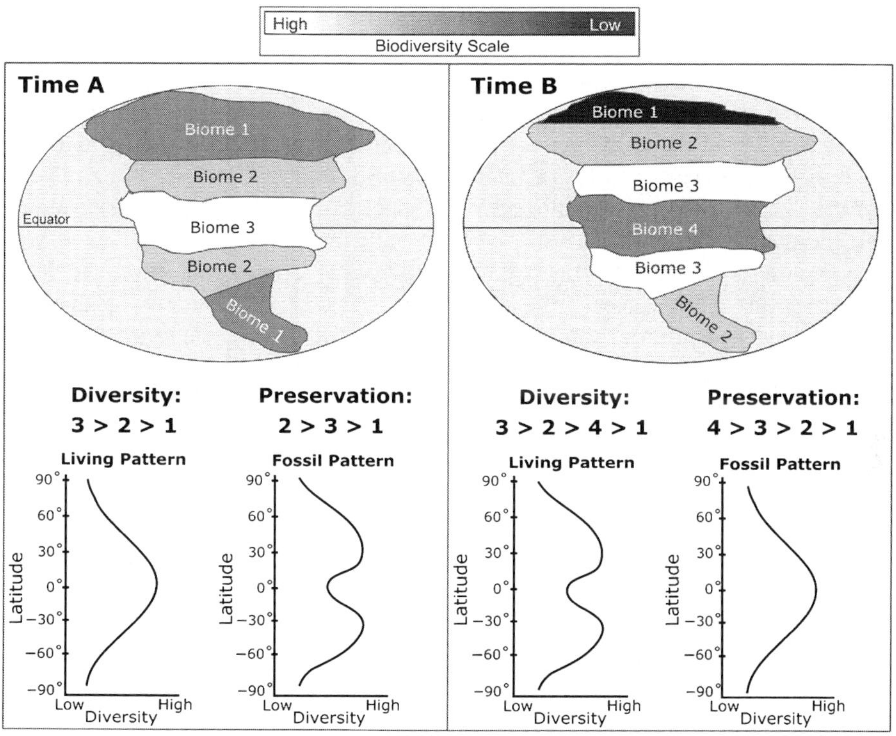

Fig. 5 Conceptual model of diversity patterns and fossil preservation for a hypothetical landmass at two different times: Time A and Time B. Global climate patterns have changed in the transition from Time A to Time B, changing the distribution and types of biomes on the continental surface (*see text*), which has altered both biodiversity and vertebrate preservation patterns. The preserved fossil pattern differs significantly from the living biodiversity curve for each time period. Light shades represent higher and dark shades lower biodiversity levels in each biome

variety of taphonomic modes. The resulting paleolatitudinal diversity curve represented by the fossil record of Time A would appear bimodal instead of the actual unimodal pattern.

When we move forward to Time B, climate patterns have shifted, altering environmental and taxonomic distributions. Thus there is a reorganization of diversity and taphonomic regime within the biomes. The major biomes at Time B include: a cool, high-latitude Biome 1; a moist, temperate Biome 2; a seasonally-wet, tropical Biome 3; and a seasonally-arid, equatorial Biome 4. The diversity pattern is now 3 > 2 > 4 > 1, creating a bimodal diversity curve. Although possessing fewer taxa in this hypothetical world, Biome 4 contains a wide range of taphonomic modes due to a seasonal climate. Increasing moisture and decreasing diversity as one moves away from the equator leads to depressed vertebrate preservation at higher latitudes. The resulting preservation pattern 4 > 3 > 2 > 1 yields a unimodal paleolatitudinal diversity curve from the fossil record of Time B.

Comparing diversity curves between Times A and B, from both the original living distribution and reconstructed fossil distribution, reveals two very different and conflicting patterns. Contrasting interpretations can be drawn regarding ecological properties of the biosphere at each time and evolutionary responses to climate change during the transition from Time A to B. For example, based on the fossil record from both times, Biomes 2 and 4 appear to support the greatest biodiversity, which could lead to erroneous conclusions about productivity patterns and their effect on the biota. Biased fossil diversity patterns could further lead to an underestimate of morphological and/or adaptive space occupied by a taxon or community, because areas of high biodiversity usually contain the greatest ecological diversity due to intense competition for resources (Pfennig et al. 2007 and references therein). From an evolutionary perspective, comparing terrestrial diversity patterns over time, in the manner of Sepkoski and others, may lead to gross over- and underestimations of diversity depending on how the living diversity patterns were filtered by prevailing taphonomic conditions at the time in question. Fossil diversity estimates could, in part, be a function of alignment between living diversity patterns and favorable taphonomic conditions.

While not quantitative, this model offers a predictive framework for hypothesis testing. The reconstruction of extinct communities usually is accomplished by examining the fossil assemblage for patterns, then comparing these patterns with present knowledge derived from community and landscape ecology to create a meaningful picture of the biota and its environment. Awareness of the many filters this information passes through is part of the process. Nevertheless, an incomplete understanding of the factors controlling these filters may be playing a direct role in how one interprets biologically relevant information from the fossil record and consequently how the same patterns are viewed today. If we were to create a fossil record for the present world based on what we know about preservation, what would it look like? Does the distribution of fossil diversity from different times in the past match with what we would predict based on environmental and climatic reconstructions? Such exercises may prove useful when considering the spatiotemporal patterns of biodiversity for different times in the past.

7 Summary and Conclusions

Beyond simply cataloging new specimens, paleontologists desire to understand sparse fossil remains, not only as once living individuals, but within their community and ecosystem context. Understanding the dynamics of past ecosystems has important consequences for how we view biogeographic and evolutionary patterns. To detect ecological and evolutionary patterns from the fossil record, we need a comprehensive taphonomic framework that acknowledges the multiple, hierarchical factors controlling the surface and subsurface destruction of remains in different environments, herein termed the micro-, meso-, and macroscales. Each level represents the spatio-temporal extent over which a set of taphonomic processes act.

Every environment contains a specific set of taphonomic conditions that act on remains above and below the sediment–atmosphere or sediment–water interface, and together represent the combined influence of local conditions (e.g., landscape, precipitation, temperature). Local conditions are controlled by global patterns of climate, tectonics, and insolation. Therefore, the taphonomic processes acting on a set of remains at the microscale reflect the prevailing conditions at the macroscale during exposure and diagenesis.

While particular taphonomic processes (modes) are associated with certain environments, environments, in turn, can be grouped together as biomes. Each biome contains a subset of taphonomic modes and can be referred to collectively as a taphonomic regime. As biomes shift in response to macroscale change, the nature and distribution of taphonomic regimes also changes, creating cascading effects through the lower levels. These lead not only to ecological and evolutionary change but also to changes in taphonomic processes, which directly impact the subsequent fossil record. Distinguishing between fossil patterns formed by these very different processes may be difficult unless hierarchical taphonomic change and initial conditions are explicitly considered.

With a hierarchy of taphonomic control, paleontologists must recognize that preservation bias is passed on to larger spatio-temporal scales, directly impacting ecologic, evolutionary, and biogeographic reconstructions. This perspective provides a powerful tool for analyzing fossil datasets by constraining the range of potential alteration to the original biotic community, allowing for a more comprehensive assessment of information loss (and gain) for different regions of the Earth at different times.

Acknowledgments This chapter is dedicated to Alfred M. Ziegler (University of Chicago, retired), who inspired me to think about big problems at big scales and helped me develop the scholarly tools to approach them. This chapter owes its existence to the intellectual heritage he instilled in me as a lowly undergraduate working in his lab many years ago. Working for Fred opened the opportunity to work with David Weishampel (Johns Hopkins University), who deserves credit for letting me get my hands on the *Dinosauria* distribution data, which helped get me interested in the factors behind fossil distribution patterns. I would like to thank Bob Gastaldo (Colby College) for detailed comments and criticisms on an early draft of the manuscript and to Catherine Forster (The George Washington University), who further helped shape this mass of ideas into a coherent whole through multiple drafts. Thanks also go to Kay Behrensmeyer

(Smithsonian), Tony Fiorillo (Dallas Museum of Nature and Science), Louis Jacobs (Southern Methodist University), and Ray Rogers (Macalester College) for many fruitful discussions and encouragement. I am grateful to David Bottjer and Peter Allison for the opportunity to contribute to this book. Special thanks go to my family and Summer Ostrowski for their continued support in all my endeavors, paleontological and otherwise, throughout the years. Paleogeographic and paleoclimate maps produced by the Paleogeographic Atlas Project (pgap.uchicago.edu), The Paleomap Project (www.scotese.com), and Ron Blakey (jan.ucc.nau.edu/~rcb7/RCB.html) proved invaluable in the preparation of this manuscript. Some of the symbols used in Fig. 2 are courtesy of the Integration and Application Network (ian.umces.edu/symbols/), University of Maryland Center for Environmental Science.

References

Aerssens, J., Boonen, S., Lowet, G., et al. (1998). Interspecies differences in bone composition, density, and quality: Potential implications for in vivo bone research. *Endocrinology, 139*, 663–670.

Algeo, T. J., Scheckler, S. E., & Scott, A. C. (1998). Terrestrial-marine teleconnections in the Devonian: links between the evolution of land plants, weathering processes, and marine anoxic events [and discussion]. *Philosophical Transactions of the Royal Society of London. Series B: Biological Sciences, 353*, 113–130.

Allen, A. P., Gillooly, J. F., & Brown, J. H. (2005). Linking the global carbon cycle to individual metabolism. *Functional Ecology, 19*, 202–213.

Allison, P. A., & Briggs, D. E. G. (1993). Paleolatitudinal sampling bias, phanerozoic species-diversity, and the end-permian extinction. *Geology, 21*, 65–68.

Alroy, J. (2003). Global databases will yield reliable measures of global biodiversity. *Paleobiology, 29*, 26–29.

Anderson, G. S., & Hobischak, N. R. (2004). Decomposition of carrion in the marine environment in British Columbia, Canada. *International Journal of Legal Medicine, 118*, 206–209.

Anderson, J. M., Anderson, H. M., & Cruickshank, A. R. I. (1998). Late Triassic ecosystems of the Molteno Lower Elliot biome of southern Africa. *Palaeontology, 41*, 387–421.

Andrews, P. (1995). Experiments in taphonomy. *Journal of Archaeological Science, 22*, 147–153.

Andrews, P., & Cook, J. (1985). Natural modifications to bones in a temperate setting. *Man, 20*, 675–691.

Andrews, M. Y., Ague, J. J., & Berner, R. A. (2006). Trees and weathering: Using soil petrographic and chemical analyses to compare the relative weathering effects of gymnosperms and angiosperms in the Cascade Mountains of Washington State, USA. Eos: Transactions of American Geophysical Union 87: Fall Meeting supplementary, abstract V53D–1787.

Arche, A., & Lopez-Gomez, J. (2006). Late Permian to Early Triassic transition in central and NE Spain: Biotic and sedimentary characteristics. *Geological Society of London, Special Publication, 265*, 261–280.

Arias, J. L., & Fernandez, M. S. (2001). Role of extracellular matrix molecules in shell formation and structure. *World Poultry Sciences Journal, 57*, 349–357.

Armstrong, H. A., Pearson, D. G., & Griselin, M. (2001). Thermal effects on rare earth element and strontium isotope chemistry in single conodont elements. *Geochimica et Cosmochimica Acta, 65*, 435–441.

Badgley, C., & Gingerich, P. D. (1988). Sampling and faunal turnover in Early Eocene mammals. *Palaeogeography, Palaeoclimatology, Palaeoecology, 63*, 141–157.

Balog, A., Haas, J., Read, J. F., et al. (1997). Shallow marine record of orbitally forced cyclicity in a late Triassic carbonate platform, Hungary. *Journal of Sedimentary Research, 67*, 661–675.

Bao, H. M., Koch, P. L., & Hepple, R. P. (1998). Hematite and calcite coatings on fossil vertebrates. *Journal of Sedimentary Research, 68*, 727–738.

Barabesi, C., Galizzi, A., Mastromei, G., et al. (2007). Bacillus subtilis gene cluster involved in calcium carbonate biomineralization. *Journal of Bacteriology, 189*, 228–235.

Barnes, R. S. K., & Mann, K. H. (1991). *Fundamentals of aquatic ecology*. Oxford: Blackwell.

Barnosky, A. & Carrasco, M. (2000). MIOMAP: a GIS-linked database for assessing effects of environmental perturbations on mammal evolution and biogeography. *Journal of Vertebrate Paleontology, 20*, 3 suppl., 28A.

Barnosky, A. D., Hadly, E. A., & Bell, C. J. (2003). Mammalian response to global warming on varied temporal scales. *Journal of Mammalogy, 84*, 354–368.

Barnosky, A. D., Carrasco, M. A., & Davis, E. B. (2005). The impact of the species-area relationship on estimates of paleodiversity. *PLoS Biology, 3*, e266.

Barreto, C., Albrecht, R. M., Bjorling, D. E., et al. (1993). Evidence of the growth-plate and the growth of long bones in juvenile dinosaurs. *Science, 262*, 2020–2023.

Bateman, R. M., Crane, P. R., DiMichele, W. A., et al. (1998). Early evolution of land plants: Phylogeny, physiology, and ecology of the primary terrestrial radiation. *Annual Review of Ecology and Systematics, 29*, 263.

Battin, T. J., & Sengschmitt, D. (1999). Linking sediment biofilms, hydrodynamics, and river bed clogging: evidence from a large river. *Microbial Ecology, 37*, 185–196.

Baveye, P., Vandevivere, P., Hoyle, B. L., et al. (1998). Environmental impact and mechanisms of the biological clogging of saturated soils and aquifer materials. *Critical Reviews in Environmental Sciences and Technology, 28*, 123–191.

Begon, M., Townshend, C. R., & Harper, J. L. (2006). *Ecology: From individuals to ecosystems*. Malden: Blackwell.

Behrensmeyer, A. K. (1978). Taphonomic and ecologic information from bone weathering. *Paleobiology, 4*, 150–162.

Behrensmeyer, A. K. (1988). Vertebrate preservation in fluvial channels. *Palaeogeography, Palaeoclimatology, Palaeoecology, 63*, 183–199.

Behrensmeyer, A. K., & Hook, R. W. (1992). Paleoenvironmental contexts and taphonomic modes. In A. K. Behrensmeyer, J. D. Damuth, W. A. DiMichele, et al. (Eds.), *Terrestrial ecosystems through time: Evolutionary paleoecology of terrestrial plants and animals*. Chicago: University of Chicago Press.

Behrensmeyer, A. K., Todd, N. E., Potts, R., et al. (1997). Late Pliocene faunal turnover in the Turkana Basin, Kenya and Ethiopia. *Science, 278*, 1589–1594.

Behrensmeyer, A. K., Kidwell, S. M., & Gastaldo, R. A. (2000). Taphonomy and paleobiology. In: Erwin DH, Wing SL, editors. Deep Time: Paleobiology's Perspective. *Paleobiology, 26*, 4 suppl., 103–147.

Behrensmeyer, A. K., Stayton, C. T., & Chapman, R. E. (2003). Taphonomy and ecology of modern avifaunal remains from Amboseli Park, Kenya. *Paleobiology, 29*, 52–70.

Bell, L. S., Skinner, M. F., & Jones, S. J. (1996). The speed of post mortem change to the human skeleton and its taphonomic significance. *Forensic Science International, 82*, 129–140.

Benison, K. C., Goldstein, R. H., Wopenka, B., et al. (1998). Extremely acid Permian lakes and ground waters in North America. *Nature, 392*, 911–914.

Benton, M. J. (1985). Patterns in the diversification of Mesozoic nonmarine tetrapods and problems in historical diversity analysis. *Special Papers in Palaeontology, 33*, 185–202.

Berna, F., Matthews, A., & Weiner, S. (2004). Solubilities of bone mineral from archaeological sites: The recrystallization window. *Journal of Archaeological Science, 31*, 867–882.

Berner, R. A. (1968). Calcium carbonate concretions formed by the decomposition of organic matter. *Science, 159*, 195–197.

Berner, R. A. (2006). GEOCARBSULF: A combined model for Phanerozoic atmospheric O_2 and CO_2. *Geochimica et Cosmochimica Acta, 70*, 5653–5664.

Berner, E. K., Berner, R. A., & Moulton, K. L. (2004). Plants and mineral weathering: Present and past. In J. I. Drever (Ed.), *Treatise on geochemistry (Vol. 5)*. New York: Elsevier.

Bigi, A., Cojazzi, G., Panzavolta, S., et al. (1997). Chemical and structural characterization of the mineral phase from cortical and trabecular bone. *Journal of Inorganic Biochemistry, 68*, 45–51.

Blum, M. D., & Tornqvist, T. E. (2000). Fluvial responses to climate and sea-level change: A review and look forward. *Sedimentology, 47*, 2–48.

Boaz, N. T., & Behrensmeyer, A. K. (1976). Hominid taphonomy – Transport of human skeletal parts in an artificial fluviatile environment. *American Journal of Physical Anthropology, 45*, 53–60.

Boddington, A. (1987). Chaos, disturbance and decay in an Anglo-Saxon cemetery. In A. Boddington, A. N. Garland, & R. C. Janaway (Eds.), *Death, decay and reconstruction*. Manchester: Manchester University Press.

Bown, T. M., & Beard, K. C. (1990). Systematic lateral variation in the distribution of fossil mammals in alluvial paleosols, Lower Eocene Willwood Formation, Wyoming. *Geological Society of America Special Paper, 243*, 135–151.

Bown, T. M., & Kraus, M. J. (1981). Vertebrate fossil-bearing paleosol units (Willwood Formation, Lower Eocene, Northwest Wyoming, USA) – Implications for taphonomy, biostratigraphy, and assemblage analysis. *Palaeogeography, Palaeoclimatology, Palaeoecology, 34*, 31–56.

Brady, N. C. (1974). *The nature and properties of soils*. New York: Macmillan.

Brain, C. K. (1995). The influence of climatic changes on the completeness of the early hominid record in southern African caves, with particular reference to Swartkrans. In E. S. Vrba, G. H. Denton, T. C. Partridge, et al. (Eds.), *Paleoclimate and evolution, with emphasis on human origins*. New Haven: Yale University Press.

Briggs, D. E. G., & Kear, A. J. (1993). Fossilization of soft tissue in the laboratory. *Science, 259*, 1439–1442.

Bush, A. M., & Bambach, R. K. (2004). Did alpha diversity increase during the Phanerozoic? Lifting the veils of taphonomic, latitudinal, and environmental biases. *Journal of Geology, 112*, 625–642.

Butterfield, N. J., & Nicholas, C. J. (1996). Burgess Shale-type preservation of both non-mineralizing and 'shelly' Cambrian organisms from the Mackenzie Mountains, Northwestern Canada. *Journal of Paleontology, 70*, 893–899.

Carney, K. M., Hungate, B. A., Drake, B. G., et al. (2007). Altered soil microbial community at elevated CO_2 leads to loss of soil carbon. *Proceedings of the National Academy of Sciences of the United States of America, 104*, 4990–4995.

Carpenter, K. (2005). Experimental investigation of the role of bacteria in bone fossilization. *Neues Jahrbuch für Geologie und Paläontologie–Monatshefte, 2*, 83–94.

Carvalho, M. L., Marques, A. F., Lima, M. T., et al. (2004). Trace elements distribution and post-mortem intake in human bones from Middle Age by total reflection X-ray fluorescence. *Spectrochimica Acta Part B: Atomic Spectroscopy, 59*, 1251–1257.

Channing, A., Schweitzer, M. H., Horner, J. R., et al. (2005). A silicified bird from Quaternary hot spring deposits. *Proceedings of the Royal Society of London. Series B: Biological Sciences, 272*, 905–911.

Child, A. M. (1995). Towards an understanding of the microbial decomposition of archaeological bone in the burial environment. *Journal of Archaeological Science, 22*, 165–174.

Chin, K., Eberth, D. A., Schweitzer, M. H., et al. (2003). Remarkable preservation of undigested muscle-tissue within a late Cretaceous tyrannosaurid coprolite from Alberta, Canada. *Palaios, 18*, 286–294.

Chinsamy-Turan, A. (2005). *The microstructure of dinosaur bone: Deciphering biology with fine-scale techniques*. Baltimore: Johns Hopkins University Press.

Chown, S. L., & Gaston, K. J. (2000). Areas, cradles and museums: The latitudinal gradient in species richness. *Trends in Ecology and Evolution, 15*, 311–315.

Clack, J. A. (2007). Devonian climate change, breathing, and the origin of the tetrapod stem group. *Integrative and Comparative Biology, 47*, 510–523.

Clarke, J. B. (2004). A mineralogical method to determine the cyclicity in the taphonomic and diagenetic history of fossilized bones. *Lethaia, 37*, 281–284.

Clyde, W. C., Finarelli, J. A., & Christensen, K. E. (2005). Evaluating the relationship between pedofacies and faunal composition: Implications for faunal turnover at the Paleocene-Eocene boundary. *Palaios, 20*, 390–399.

Coard, R. (1999). One bone, two bones, wet bones, dry bones: Transport potentials under experimental conditions. *Journal of Archaeological Science, 26*, 1369–1375.

Coard, R., & Dennell, R. W. (1995). Taphonomy of some articulated skeletal remains: transport potential in an artificial environment. *Journal of Archaeological Science, 22*, 441–448.

Coates, M. I., Ruta, M., & Friedman, M. (2008). Ever since owen: Changing perspectives on the early evolution of tetrapods. *Annual Review of Ecology, Evolution and Systematics, 39*, 571–592.

Collins, M. J., Riley, M. S., Child, A. M., et al. (1995). A basic mathematical simulation of the chemical degradation of ancient collagen. *Journal of Archaeological Science, 22*, 175–183.

Collins, M. J., Nielsen-Marsh, C. M., Hiller, J., et al. (2002). The survival of organic matter in bone: A review. *Archaeometry, 44*, 383–394.

Crame, J. A. (2002). Evolution of taxonomic diversity gradients in the marine realm: A comparison of Late Jurassic and Recent bivalve faunas. *Paleobiology, 28*, 184–207.

Crane, P. R., & Lidgard, S. (1989). Angiosperm diversification and paleolatitudinal gradients in Cretaceous floristic diversity. *Science, 246*, 675–678.

Cressler, W. L. (2006). Plant palaeoecology of the late Devonian red hill locality, north central Pennsylvania, and Archaeopteris-dominated wetland plant community and early tetrapod site. In S. F. Greb and W. A. DiMichele (Eds.), *Wetlands through time. Geological Society of America, Special Papers 399.*

Cubasch, U., Zorita, E., Kaspar, F., et al. (2006). Simulation of the role of solar and orbital forcing on climate. *Advances in Space Research, 37*, 1629–1634.

Cutler, A. H., Behrensmeyer, A. K., & Chapman, R. E. (1999). Environmental information in a recent bone assemblage: Roles of taphonomic processes and ecological change. *Palaeogeography, Palaeoclimatology, Palaeoecology, 149*, 359–372.

Daeschler, E. B., Shubin, N. H., & Jenkins, F. A., Jr. (2006). A Devonian tetrapod-like fish and the evolution of the tetrapod body plan. *Nature, 440*, 757–763.

Dal Sasso, C., & Signore, M. (1998). Exceptional soft-tissue preservation in a theropod dinosaur from Italy. *Nature, 392*, 383–387.

Daniel, J. (2003). The role of bacterially mediated precipitation in the permineralization of bone. Masters thesis. University of Colorado, Boulder.

Davies, D. J., Powell, E. N., & Stanton, R. J. (1989). Taphonomic signature as a function of environmental process – Shells and shell beds in a hurricane-influenced inlet on the Texas coast. *Palaeogeography, Palaeoclimatology, Palaeoecology, 72*, 317–356.

Davis, P. G. (1997). The bioerosion of bird bones. *International Journal of Osteoarchaeology, 7*, 388–401.

de Carvalho, L. M. L., & Linhares, A. X. (2001). Seasonality of insect succession and pig carcass decomposition in a natural forest area in southeastern Brazil. *Journal of Forensic Sciences, 46*, 604–608.

DiMichele, W. A., & Hook, R. W. (1992). Paleozoic terrestrial ecosystems. In A. K. Behrensmeyer, J. D. Damuth, W. A. DiMichele, et al. (Eds.), *Terrestrial ecosystems through time: Evolutionary paleoecology of terrestrial plants and animals*. Chicago: University of Chicago Press.

Dirrigl, J., & Frank, J. (2001). Bone mineral density of wild turkey (*Meleagris gallopavo*) skeletal elements and its effect on differential survivorship. *Journal of Archaeological Science, 28*, 817–832.

Dolan, C. T., Brown, A. L., & Ritts, R. E. (1971). Microbiological examination of post mortem tissues. *Archeology and Pathology, 92*, 206–211.

Donnadieu, Y., Godderis, Y., & Pierrehumbert, R. et al. (2006). A GEOCLIM simulation of climatic and biogeochemical consequences of Pangea breakup. *Geochemistry Geophysics Geosystems, 7*, 1–21.

Downing, K. F., & Park, L. E. (1998). Geochemistry and early diagenesis of mammal-bearing concretions from the Sucker Creek Formation (Miocene) of southeastern Oregon. *Palaios, 13*, 14–27.

Dubiel, R. F., Parrish, J. T., Parrish, J. M., et al. (1991). The Pangaean megamonsoon: Evidence from the upper Triassic Chinle Formation, Colorado Plateau. *Palaios, 6*, 347–370.
Dynesius, M., & Jansson, R. (2000). Evolutionary consequences of changes in species' geographical distributions driven by Milankovitch climate oscillations. *Proceedings of the National Academy of Sciences of the United States of America, 97*, 9115–9120.
Einsele, G. (2000). *Sedimentary basins: Evolution, facies, and sediment budget*. New York: Springer.
Elder, R. L. (1985). Principles of aquatic taphonomy with examples from the fossil record. PhD dissertation. University of Michigan, Ann Arbor.
Elder, R. L., & Smith, G. R. (1988). Fish taphonomy and environmental inference in paleolimnology. *Palaeogeography, Palaeoclimatology, Palaeoecology, 62*, 577–592.
Engelmann, G. F., Chure, D. J., & Fiorillo, A. R. (2004). The implications of a dry climate for the paleoecology of the fauna of the Upper Jurassic Morrison Formation. *Sedimentary Geology, 167*, 297–308.
Enlow, D. H., & Brown, S. O. (1956). A comparative histological study of fossil and recent bone tissues, Part I. *Texas Journal of Science, 8*, 405–443.
Enlow, D. H., & Brown, S. O. (1957). A comparative histological study of fossil and recent bone tissues, Part II. *Texas Journal of Science, 9*, 186–214.
Enlow, D. H., & Brown, S. O. (1958). A comparative histological study of fossil and recent bone tissues, Part III. *Texas Journal of Science, 10*, 187–230.
Epstein, A. G., Epstein, J. P., & Harris, L. D. (1977). Conodont color alteration – An index to organic metamorphism. *US Geological Survey, Professional Papers 995*.
Evans, K. L., Warren, P. H., & Gaston, K. J. (2005). Species-energy relationships at the macroecological scale: A review of the mechanisms. *Biological Review, 80*, 1–25.
Fara, E. (2002). Sea-level variations and the quality of the continental fossil record. *Journal of the Geological Society 159*, 489–491.
Fastovsky, D. E. (1987). Paleoenvironments of vertebrate-bearing strata during the Cretaceous-Paleogene transition, eastern Montana and western North Dakota. *Palaios, 2*, 282–295.
Fastovsky, D. E. (1990). Rocks, resolution, and the record; a review of depositional constraints on fossil vertebrate assemblages at the terrestrial Cretaceous/Paleogene boundary, eastern Montana and western North Dakota. In V. L. Sharpton & P. D. Ward (Eds.), *Global catastrophes in Earth history; an interdisciplinary conference on impacts, volcanism, and mass mortality. Geological Society of America, Special Papers*. Snowbird, Utah.
Fastovsky, D. E., Badamgarav, D., Ishimoto, H., et al. (1997). The paleoenvironments of Tugrikin-Shireh (Gobi Desert, Mongolia) and aspects of the taphonomy and paleoecology of *Protoceratops* (Dinosauria: Ornithishichia). *Palaios, 12*, 59–70.
Fenner, J. (2001). Palaeoceanographic and climatic changes during the Albian, summary of the results from the Kirchrode boreholes. *Palaeogeography, Palaeoclimatology, Palaeoecology, 174*, 287–304.
Fernández-Jalvo, Y., Sánchez-Chillón, B., Andrews, P., et al. (2002). Morphological taphonomic transformations of fossil bones in continental environments, and repercussions on their chemical composition. *Archaeometry, 44*, 353–361.
Fiorillo, A. R. (1991). Prey bone utilization by predatory dinosaurs. *Palaeogeography, Palaeoclimatology, Palaeoecology, 88*, 157–166.
Fiorillo, A. R. (1999). Determining the relative roles of climate and tectonics in the formation of the fossil record of terrestrial vertebrates: A perspective from the Late Cretaceous of western North America. *Records of the Western Australian Museum Supplement, 57*, 219–228.
Flessa, K. W. (1975). Area, continental drift and mammalian diversity. *Paleobiology, 1*, 189–194.
Freeman, J. J., Wopenka, B., Silva, M. J., et al. (2001). Raman spectroscopic detection of changes in bioapatite in mouse femora as a function of age and in vitro fluoride treatment. *Calcified Tissue International, 68*, 156–162.
Gaines, R. R. (2008). Burgess Shale-type deposits worldwide share a common paleoenvironmental setting and origin. *Geological Society of America, Abstract with Program, 40*, 501.
Gaston, K. J. (2000). Global patterns in biodiversity. *Nature, 405*, 220–227.

Gibbs, S., Shackleton, N., & Young, J. (2004). Orbitally forced climate signals in mid-Pliocene nannofossil assemblages. *Marine Micropaleontology, 51*, 39–56.

Glimcher, M. J. (1984). Recent studies of the mineral phase in bone and its possible linkage to the organic matrix by protein-bound phosphate bonds. *Philosophical Transactions of the Royal Society of London. Series B: Biological Sciences, 304*, 479–508.

Goffredi, S. K., Orphan, V. J., Rouse, G. W., et al. (2005). Evolutionary innovation: A bone-eating marine symbiosis. *Environmental Microbiology, 7*, 1369–1378.

Goodwin, M. B., Grant, P. G., Bench, G., et al. (2007). Elemental composition and diagenetic alteration of dinosaur bone: Distinguishing micron-scale spatial and compositional heterogeneity using PIXE. *Palaeogeography, Palaeoclimatology, Palaeoecology, 253*, 458–476.

Graham, R. W., Lundelius, E. L., Graham, M. A., et al. (1996). Spatial response of mammals to Late Quaternary environmental fluctuations. *Science, 272*, 1601–1606.

Greenwald, D. N., & Brubaker, L. B. (2001). A 5000-year record of disturbance and vegetation change in riparian forests of the Queets River, Washington, USA. *Canadian Journal of Forest Research, 31*, 1375–1385.

Grynpas, M. D., & Omelon, S. (2007). Transient precursor strategy or very small biological apatite crystals? *Bone, 41*, 162–164.

Hackett, C. J. (1981). Microscopical focal destruction (tunnels) in exhumed human bones. *Medicine, Science and the Law, 21*, 243–265.

Haefner, J., Wallace, J., & Merritt, R. (2004). Pig decomposition in lotic aquatic systems: The potential use of algal growth in establishing a postmortem submersion interval (PMSI). *Journal of Forensic Sciences, 49*, 330–336.

Haglund, W. D., & Sorg, M. H. (2002). Human remains in water environments. In W. D. Haglund & M. H. Sorg (Eds.), *Advances in forensic taphonomy: Method, theory, and archaeological perspectives*. Boca Raton: CRC Press.

Hare, P. E. (1980). Organic geochemistry of bone and its relation to the survival of bone in the natural environment. In A. K. Behrensmeyer & A. P. Hill (Eds.), *Fossils in the making: Vertebrate taphonomy and paleoecology*. Chicago: University of Chicago Press.

Haskell, J. (2001). The latitudinal gradient of diversity through the Holocene as recorded by fossil pollen in Europe. *Evolutionary Ecology Research, 3*, 345–360.

Hawkins, B. A., Field, R., Cornell, H. V., et al. (2003). Energy, water, and broad-scale geographic patterns of species richness. *Ecology, 84*, 3105–3117.

Hedges, R. E. M. (2002). Bone diagenesis: An overview of processes. *Archaeometry, 44*, 319–328.

Hedges, R. E. M., & Millard, A. R. (1995). Bones and groundwater: Towards the modelling of diagenetic processes. *Journal of Archaeological Science, 22*, 155–164.

Hedges, R. E. M., Millard, A. R., & Pike, A. W. G. (1995). Measurements and relationships of diagenetic alteration of bone from three archaeological sites. *Journal of Archaeological Science, 22*, 201–209.

Henderson, J. (1987). Factors determining the preservation of human remains. In A. Boddington, A. N. Garland, & R. C. Janaway (Eds.), *Death, decay, and reconstruction: Approaches to archaeology and forensic science*. Manchester: Manchester University Press.

Hewadikaram, K. A., & Goff, M. L. (1991). Effect of carcass size on rate of decomposition and arthropod succession patterns. *The American Journal of Forensic Medicine and Pathology, 12*, 235–240.

Hobischak, N., & Anderson, G. (1999). Freshwater-related death investigations in British Columbia in 1995–1996. A review of coroners cases. *Canadian Society of Forensic Science Journal, 32*, 97–106.

Hobischak, N. R., & Anderson, G. S. (2002). Time of submergence using aquatic invertebrate succession and decompositional changes. *Journal of Forensic Sciences, 47*, 142–151.

Holz, M., & Schultz, C. L. (1998). Taphonomy of the south Brazilian Triassic herpetofauna: Fossilization mode and implications for morphological studies. *Lethaia, 31*, 335–345.

Horner, J. R. (1989). The Mesozoic terrestrial ecosystems of Montana. In D. E. French & R. F. Grabb (Eds.), *Geologic resources of Montana. Montana Geological Society field conference and symposium guidebook. Montana Geological Society, Billings.*

Horner, J. R., Varricchio, D. J., & Goodwin, M. B. (1992). Marine transgressions and the evolution of Cretaceous dinosaurs. *Nature, 358*, 59–61.

Horner, J. R., De Ricqles, A., & Padian, K. (2000). Long bone histology of the hadrosaurid dinosaur Maiasaura peeblesorum: Growth dynamics and physiology based on an ontogenetic series of skeletal elements. *Journal of Vertebrates Paleontology, 20*, 115–129.

Howe, J., & Francis, J. E. (2005). Metamorphosed palaeosols associated with Cretaceous fossil forests, Alexander Island, Antarctica. *Journal of Geological Society, 162*, 951–957.

Huey, R. B., & Ward, P. D. (2005). Hypoxia, global warming, and terrestrial Late Permian extinctions. *Science, 308*, 398–401.

Igamberdiev, A. U., & Lea, P. J. (2006). Land plants equilibrate O_2 and CO_2 concentrations in the atmosphere. *Photosynthesis Research, 87*, 177–194.

Jackson, J. B. C., & Erwin, D. H. (2006). What can we learn about ecology and evolution from the fossil record? *Trends in Ecology and Evolution, 21*, 322–328.

Jans, M. M. E., Nielsen-Marsh, C. M., Smith, C. I., et al. (2004). Characterisation of microbial attack on archaeological bone. *Journal of Archaeological Science, 31*, 87–95.

Johnsson, K. (1997). Chemical dating of bones based on diagenetic changes in bone apatite. *Journal of Archaeological Science, 24*, 431–437.

Kaiser, T. M. (2000). Proposed fossil insect modification to fossil mammalian bone from Plio-Pleistocene hominid-bearing deposits of Laetoli (Northern Tanzania). *Annals of the Entomological Society of America, 93*, 693–700.

Kellerman, G. D., Waterman, N. G., & Scharfenberger, L. F. (1976). Demonstration in vitro of post mortem bacterial migration. *American Journal of Clinical Pathology, 66*, 911–916.

Kelly, E. F., Chadwick, O. A., & Hilinski, T. E. (1998). The effect of plants on mineral weathering. *Biogeochemistry, 42*, 21–53.

Kerbis Peterhans, J. C., Wrangham, R. W., Carter, M. L., et al. (1993). A contribution to tropical rain forest taphonomy: Retrieval and documentation of chimpanzee remains from Kibale Forest, Uganda. *Journal of Human Evolution, 25*, 485–514.

Kidwell, S. M. (1986). Models for fossil concentrations: Paleobiologic implications. *Paleobiology, 12*, 6–24.

Kiessling, W. (2002). Radiolarian diversity patterns in the latest Jurassic-earliest Cretaceous. *Palaeogeography, Palaeoclimatology, Palaeoecology, 187*, 179–206.

Kreft, H., & Jetz, W. (2007). Global patterns and determinants of vascular plant diversity. *Proceedings of the National Academy of Sciences of the United States of America, 104*, 5925–5930.

Lambert, J. B., Simpson, S. V., Buikstra, J. E., et al. (1983). Electron microprobe analysis of elemental distribution in excavated human femurs. *American Journal of Physical Anthropology, 62*, 409–423.

Laskar, J., Robutel, P., Joutel, F., et al. (2004). A long-term numerical solution for the insolation quantities of the Earth. *Astronomy and Astrophysics, 428*, 261–285.

Lehman, T. M. (1997). Late Campanian dinosaur biogeography in the Western Interior of the United States. In D. L. Wolberg et al. (Eds.), *Dinofest international*. Philadelphia: Philadelphia Academy of Natural Sciences.

Leier, A. L., DeCelles, P. G., & Pelletier, J. D. (2005). Mountains, monsoons, and megafans. *Geology, 33*, 289–292.

Leng, Q., & Yang, H. (2003). Pyrite framboids associated with the Mesozoic Jehol Biota in northeastern China: Implications for microenvironment during early fossilization. *Progress in Natural Science, 13*, 206–212.

Lian, B., Hu, Q. N., Chen, J., et al. (2006). Carbonate biomineralization induced by soil bacterium Bacillus megaterium. *Geochimica et Cosmochimica Acta, 70*, 5522–5535.

Long, J. A., & Gordon, M. S. (2004). The greatest step in vertebrate history: A paleobiological review of the fish-tetrapod transition. *Physiological and Biochemical Zoology, 77*, 700–719.

Loope, D. B., Dingus, L., Swisher, C. C. I., et al. (1998). Life and death in a Late Cretaceous dune field, Nemegt Basin, Mongolia. *Geology, 26*, 27–30.

Lyman, R. L. (1994). *Vertebrate taphonomy*. Cambridge: Cambridge University Press.

Maas, M. C. (1985). Taphonomy of a Late Eocene microvertebrate locality, Wind River Basin, Wyoming (USA). *Palaeogeography, Palaeoclimatology, Palaeoecology, 52*, 123–142.

Magne, D., Pilet, P., Weiss, P., et al. (2001). Fourier transform infrared microspectroscopic investigation of the maturation of nonstoichiometric apatites in mineralized tissues: A horse dentin study. *Bone, 29*, 547–552.

Markwick, P. J. (1998). Crocodilian diversity in space and time: The role of climate in paleoecology and its implication for understanding K/T extinctions. *Paleobiology, 24*, 470–497.

Martill, D. M. (1988). Preservation of fish in the Cretaceous Santana Formation of Brazil. *Palaeontology, 31*, 1–18.

Martin, R. E. (1999). *Taphonomy: A process approach*. Cambridge: Cambridge University Press.

Martin, R. E. (2003). The fossil record of biodiversity: Nutrients, productivity, habitat area and differential preservation. *Lethaia, 36*, 179–193.

Martin, R. E., Goldstein, S. T., & Patterson, R. T. (1999). Taphonomy as an environmental science. *Palaeogeography, Palaeoclimatology and Palaeoecology, 149*, vii–viii.

Marui, Y., Chiba, S., Okuno, J., et al. (2004). Species-area curve for land snails on Kikai Island in geological time. *Paleobiology, 30*, 222–230.

McKean, A., Brooks, B., Nelson, S., et al. (2007). The relationship of geothermal alteration and relict organics to the color of fossil bone. *Journal of Veterinary Paleontology, 27*, 116A.

McNamara, M. E., Orr, P. J., Kearns, S. L., et al. (2006). High-fidelity organic preservation of bone marrow in ca. 10 Ma amphibians. *Geology, 34*, 641–644.

Mellen, P., Lowry, M., & Micozzi, M. (1993). Experimental observations on adipocere formation. *Journal of Forensic Sciences, 38*, 91–3.

Menczel, J., Posner, A. S., & Harper, R. A. (1965). Age changes in crystallinity of rat bone apatite. *Israel Journal of Medical Sciences, 1*, 251–252.

Middleton, G. V. (Ed.). (2003). *Encyclopedia of sediments and sedimentary rocks*. New York: Springer.

Miller, A. I. (2003). On the importance of global diversity trends and the viability of existing paleontological data. *Paleobiology, 29*, 15–18.

Miller, D. J., & Eriksson, K. A. (1999). Linked sequence development and global climate change: The Upper Mississippian record in the Appalachian basin. *Geology, 27*, 35–38.

Minshall, G. W., Hitchcock, E., & Barnes, J. R. (1991). Decomposition of Rainbow Trout (*Oncorhynchus mykiss*) carcasses in a forest stream ecosystem inhabited only by nonanadromous fish populations. *Canadian Journal of Fisheries and Aquatic Sciences, 48*, 191–195.

Montes, L., de Margerie, E., Castanet, J., et al. (2005). Relationship between bone growth rate and the thickness of calcified cartilage in the long bones of the Galloanserae (Aves). *Journal of Anatomy, 206*, 445–452.

Munoz-Duran, J., & Van Valkenburgh, B. (2006). The Rancholabrean record of Carnivora: Taphonomic effect of body size, habitat breadth, and the preservation potential of caves. *Palaios, 21*, 424–430.

Nicholson, R. A. (1996). Bone degradation, burial medium and species representation: Debunking the myths, an experiment-based approach. *Journal of Archaeological Science, 23*, 513–533.

Nielsen-Marsh, C. M., & Hedges, R. E. M. (1997). Dissolution experiments on modern and diagenetically altered bone and the effect on the infrared splitting factor. *Bulletin. Société Géologique de France, 168*, 485–490.

Nielsen-Marsh, C. M., Hedges, R. E. M., Mann, T., et al. (2000). A preliminary investigation of the application of differential scanning calorimetry to the study of collagen degradation in archaeological bone. *Thermochimica Acta, 365*, 129–139.

Nielsen-Marsh, C. M., Smith, C. I., Jans, M. M. E., et al. (2007). Bone diagenesis in the European Holocene II: Taphonomic and environmental considerations. *Journal of Archaeological Science, 34*, 1523–1531.

Norman, D. B. (1987). A Mass-accumulation of vertebrates from the Lower Cretaceous of Nehden (Sauerland), West Germany. *Proceedings of the Royal Society of London. Series B: Biological Sciences, 230*, 215–255.

Noth, S. (1998). Conodont color (CAI) versus microcrystalline and textural changes in Upper Triassic conodonts from Northwest Germany. *Facies, 38*, 165–173.

Noto, C. R. (2009). The influence of post-burial environment and plant-bone interactions on vertebrate preservation: an experimental taphonomic study. PhD dissertation. Stony Brook University, pp. 238.

O'Brien, T. G., & Kuehner, A. C. (2007). Waxing grave about adipocere: Soft tissue change in an aquatic context. *Journal of Forensic Sciences, 52*, 294–301.

Olsen, P. E., Kent, D. V., & Cornet, B. (1991). Thirty million year record of tropical orbitally-forced climate change from continental coring of the Newark early Mesozoic rift basin. *American Geophysics Union – Miner Society of America 1991 Spring Meeting, Baltimore 72*.

Paik, I. S. (2000). Bone chip-filled burrows associated with bored dinosaur bone in floodplain paleosols of the Cretaceous Hasandong Formation, Korea. *Palaeogeography, Palaeoclimatology, Palaeoecology, 157*, 213–225.

Pasteris, J. D., Wopenka, B., Freeman, J. J., et al. (2004). Lack of OH in nanocrystalline apatite as a function of degree of atomic order: Implications for bone and biomaterials. *Biomaterials, 25*, 229–238.

Pasteris, J. D., Wopenka, B., & Valsami-Jones, E. (2008). Bone and tooth mineralization: Why apatite? *Elements, 4*, 97–104.

Pawlicki, R., & Bolechala, P. (1987). X-ray microanalysis of fossil dinosaur bone: Age differences in the calcium and phosphorus content of *Gallimimus bullatus* bones. *Folia Histochemica et Cytobiologica, 25*, 241.

Pearson, P. N., & Palmer, M. R. (2000). Atmospheric carbon dioxide concentrations over the past 60 million years. *Nature, 406*, 695–699.

Person, A., Bocherens, H., Saliège, J.-F., et al. (1995). Early diagenetic evolution of bone phosphate: An X-ray diffractometry analysis. *Journal of Archaeological Science, 22*, 211–221.

Person, A., Bocherens, H., Mariotti, A., et al. (1996). Diagenetic evolution and experimental heating of bone phosphate. *Palaeogeography, Palaeoclimatology, Palaeoecology, 126*, 135–149.

Pfennig, D. W., Rice, A. M., & Martin, R. A. (2007). Field and experimental evidence for competition's role in phenotypic divergence. *Evolution, 61*, 257–271.

Pfretzschner, H. U. (2000). Pyrite formation in Pleistocene bones – A case of very early mineral formation during diagenesis. *Neues Jahrbuch für Geologie und Paläontologie–Abhandlungen, 217* 143–160.

Pfretzschner, H. U. (2001a). Iron oxides in fossil bone. *Neues Jahrbuch fiir Geologie und Paläontologie–Abhandlungen, 220*, 417–429.

Pfretzschner, H. U. (2001b). Pyrite in fossil bone. *Neues Jahrbuch fiir Geologie und Paläontologie–Abhandlungen, 220*, 1–23.

Pfretzschner, H. U. (2004). Fossilization of Haversion bone in aquatic environments. *Comptes Rendus Palevol, 3*, 605–616.

Pfretzschner, H. U. (2006). Collagen gelatinization: The key to understand early bone-diagenesis. *Paläontographica Abteilung A–Paläozoologie-Stratigraphie 278*, 135–148.

Piepenbrink, H. (1986). Two examples of biogenous dead bone decomposition and their consequences for taphonomic interpretation. *Journal of Archaeological Science, 13*, 417–430.

Pike, A. W. G., Nielsen-Marsh, C., & Hedges, R. E. M. (2001). Modelling bone dissolution under different hydrological regimes. In A. Millard (Ed.), *Archaeological Sciences 1997*. Durham: BAR International Series.

Plotnick, R. E., Baumiller, T., & Wetmore, K. L. (1988). Fossilization potential of the mud crab, *Panopeus* (Brachyura: Xanthidae) and temporal variability in crustacean taphonomy. *Palaeogeography, Palaeoclimatology, Palaeoecology, 63*, 27–43.

Post, E., & Forchhammer, M. C. (2002). Synchronization of animal population dynamics by large-scale climate. *Nature, 420*, 168–171.

Price, G. D., & Sellwood, B. W. (1997). "Warm" palaeotemperatures from high Late Jurassic palaeolatitudes (Falkland Plateau): Ecological, environmental or diagenetic controls? *Palaeogeography, Palaeoclimatology, Palaeoecology, 129*, 315–327.

Prothero, D. R., & Schwab, F. (1996). *Sedimentary geology: An introduction to sedimentary rocks and stratigraphy*. New York: WH Freeman and Company.

Randall, D., Burggren, W., & French, K. (1997). *Animal physiology: Mechanisms and adaptations*. New York: WH Freeman and Company.

Rees, P. M., Noto, C. R., Parrish, J. M., et al. (2004). Late Jurassic climates, vegetation, and dinosaur distributions. *Journal of Geology, 112*, 643–654.

Rees, P. M. & Noto, C. R. (2005). A new online database of dinosaur distributions. *Journal of Vertebrate Paleontology, 25*, 3 suppl.,103A.

Reich, P. B., Oleksyn, J., Modrzynski, J., et al. (2005). Linking litter calcium, earthworms and soil properties: A common garden test with 14 tree species. *Ecological Letters, 8*, 811–818.

Reisz, R. R., & Tsuji, L. A. (2006). An articulated skeleton of *Varanops* with bite marks: the oldest known evidence of scavenging among terrestrial vertebrates. *Journal of Vertebrate Paleontology, 26*, 1021–1023.

Retallack, G. J. (1990). *Soils of the past: An introduction to paleopedology*. Winchester: Unwin Hyman.

Retallack, G. J. (1997). Early forest soils and their role in Devonian global change. *Science, 276*, 583–585.

Retallack, G. J. (2005a). Earliest triassic claystone breccias and soil-erosion crisis. *Journal of Sedimentary Research, 75*, 679–695.

Retallack, G. J. (2005b). Were fossils exceptionally preserved in unusual times? *Geological Society of America, Abstract with Program, 37*, 117.

Retallack, G. J., Wynn, J. G., & Fremd, T. J. (2004). Glacial-interglacial-scale paleoclimatic change without large ice sheets in the Oligocene of central Oregon. *Geology, 32*, 297–300.

Rey, C., Renugopalakrishman, V., Collins, B., et al. (1991). Fourier transform infrared spectroscopic study of the carbonate ions in bone mineral during aging. *Calcified Tissue International, 49*, 251–258.

Rial, J. A. (2004). Abrupt climate change: Chaos and order at orbital and millennial scales. *Global and Planetary Change, 41*, 95–109.

Roberts, E. M., Rogers, R. R., & Foreman, B. Z. (2007). Continental insect borings in dinosaur bone: Examples from the Late Cretaceous of Madagascar and Utah. *Journal of Paleontology, 81*, 201–208.

Rogers, R. R. (1993). Systematic patterns of time-averaging in the terrestrial vertebrate record: A Cretaceous case study. In S. M. Kidwell & A. K. Behrensmeyer (Eds.), *Taphonomic approaches to time resolution in fossil assemblages*. Knoxville: The Paleontological Society.

Rogers, R. R. (2005). Fine-grained debris flows and extraordinary vertebrate burials in the Late Cretaceous of Madagascar. *Geology, 33*, 297–300.

Rogers, R. R., Arcucci, A. B., Abdala, F., et al. (2001). Paleoenvironment and taphonomy of the Chanares Formation tetrapod assemblage (Middle Triassic), northwestern Argentina: Spectacular preservation in volcanogenic concretions. *Palaios, 16*, 461–481.

Rogers, R. R., Krause, D. W., & Rogers, K. C. (2003). Cannibalism in the Madagascan dinosaur *Majungatholus atopus*. *Nature, 422*, 515–518.

Rohde, K. (1992). Latitudinal gradients in species-diversity – The search for the primary cause. *Oikos, 65*, 514–527.

Rosenzweig, M. L. (1995). Species diversity in space and time. Cambridge, Cambridge University Press, 436 p.

Rothman, D. H. (2001). Global biodiversity and the ancient carbon cycle. *Proceedings of the National Academy of Sciences of the United States of America, 98*, 4305–4310.

Rygel, M. C., Fielding, C. R., Frank, T. D., et al. (2008). The magnitude of Late Paleozoic glacioeustatic fluctuations: A synthesis. *Journal of Sedimentary Research, 78*, 500–511.

Sagemann, J., Bale, S. J., Briggs, D. E. G., et al. (1999). Controls on the formation of authigenic minerals in association with decaying organic matter: An experimental approach. *Geochimica et Cosmochimica Acta, 63*, 1083–1095.

Sander, P. M. (1987). Taphonomy of the lower Permian Geraldine Bonebed in Archer County, Texas. *Palaeogeography, Palaeoclimatology, Palaeoecology, 61*, 221–236.

Sander, P. M. (1989). Early Permian depositional environments and pond bonebeds in central Archer County, Texas. *Palaeogeography, Palaeoclimatology, Palaeoecology, 69*, 1–21.

Sansom, I. J., Smith, M. P., Armstrong, H. A., et al. (1992). Presence of the earliest vertebrate hard tissues in conodonts. *Science, 256*, 1308–1311.

Schaetzl, R. J., & Anderson, S. (2005). *Soils: Genesis and geomorphology*. New York: Cambridge University Press.

Scheckler, S. E., & Maynard, J. B. (2001). Effects of the middle to late Devonian spread of vascular land plants on weathering regimes, marine biotas, and global climate. In P. G. Gensel & D. Edwards (Eds.), *Plants invade the land: Evolutionary and environmental perspectives*. New York: Columbia University Press.

Schoeninger, M. J., Moore, K. M., Murray, M. L., et al. (1989). Detection of bone preservation in archaeological and fossil samples. *Applied Geochemistry, 4*, 281–292.

Schultze, H.-P., & Cloutier, R. (Eds.). (1996). *Devonian fishes and plants of Miguasha, Quebec, Canada*. Munich: Verlag Dr. Frederich Pfeil.

Schweitzer, M. H., Wittmeyer, J. L., & Horner, J. R. (2005). Gender-specific reproductive tissue in ratites and *Tyrannosaurus rex*. *Science, 308*, 1456–1460.

Schweitzer, M. H., Wittmeyer, J. L., & Horner, J. R. (2007). Soft tissue and cellular preservation in vertebrate skeletal elements from the Cretaceous to the present. *Proceedings of the Royal Society of London. Series B: Biological Sciences, 274*, 183–197.

Scotese, C. (2009). The Paleomap Project. www.scotese.com. Accessed 30 January 2009.

Sellwood, B. W., & Valdes, P. J. (2006). Mesozoic climates: General circulation models and the rock record. *Sedimentary Geology, 190*, 269–287.

Sephton, M. A., Looy, C. V., Brinkhuis, H., et al. (2005). Catastrophic soil erosion during the end-Permian biotic crisis. *Geology, 33*, 941–944.

Sepkoski, J. J. (1998). Rates of speciation in the fossil record. *Philosophical Transactions of the Royal Society of London. Series B: Biological Sciences, 353*, 315–326.

Sept, J. M. (1994). Bone distribution in a semi-arid riverine habitat in eastern Zaire: Implications for the interpretation of faunal assemblages at early archaeological sites. *Journal of Archaeological Science, 21*, 217–235.

Sereno, P. C. (1997). The origin and evolution of dinosaurs. *Annual Review of Earth and Planetary Sciences, 25*, 435–489.

Shalaby, O. A., deCarvalho, L. M. L., & Goff, M. L. (2000). Comparison of patterns of decomposition in a hanging carcass and a carcass in contact with soil in a xerophytic habitat on the island of Oahu, Hawaii. *Journal of Forensic Sciences, 45*, 1267–1273.

Shean, B. S., Messinger, L., & Papworth, M. (1993). Observations of differential decomposition on sun exposed vs shaded pig carrion in coastal Washington State. *Journal of Forensic Sciences, 38*, 938–949.

Sidor, C. A., O'Keefe, F. R., Damiani, R., et al. (2005). Permian tetrapods from the Sahara show climate-controlled endemism in Pangaea. *Nature, 434*, 886–889.

Silvertown, J. (1985). History of a latitudinal diversity gradient: Woody plants in Europe 13,000–1000 Years B.P. *Journal of Biogeography, 12*, 519–525.

Simon-Coinçon, R., Thiry, M., & Schmitt, J.-M. (1997). Variety and relationships of weathering features along the early Tertiary palaeosurface in the southwestern French Massif Central and the nearby Aquitaine Basin. *Palaeogeography, Palaeoclimatology, Palaeoecology, 129*, 51–79.

Sionkowska, A. (2005). Thermal denaturation of UV-irradiated wet rat tail tendon collagen. *International Journal of Biological Macromolecules, 35*, 145–149.

Smith, R. M. H. (1993). Vertebrate taphonomy of Late Permian floodplain deposits in the southwestern Karoo Basin of South Africa. *Palaios, 8*, 45–67.

Smith, R. M. H., & Botha, J. (2005). The recovery of terrestrial vertebrate diversity in the South African Karoo Basin after the end-Permian extinction. *Comptes Rendus Palevol, 4*, 623–636.

Smith, R. M. H., & Swart, R. (2002). Changing fluvial environments and vertebrate taphonomy in response to climatic drying in a mid-Triassic rift valley fill: The Omingonde Formation (Karoo Supergroup) of central Namibia. *Palaios, 17*, 249–267.

Smith, A. B., Gale, A. S., & Monks, N. E. A. (2001). Sea-level change and rock-record bias in the Cretaceous: A problem for extinction and biodiversity studies. *Paleobiology, 27*, 241–253.

Smoke, N. D., & Stahl, P. W. (2004). Post-burial fragmentation of microvertebrate skeletons. *Journal of Archaeological Science, 31*, 1093–1100.

Spencer, L. M., Van Valkenburgh, B., & Harris, J. M. (2003). Taphonomic analysis of large mammals recovered from the Pleistocene Rancho La Brea tar seeps. *Paleobiology, 29*, 561–575.

Stehli, F. G., Douglas, R. G., & Newell, N. D. (1969). Generation and maintenance of gradients in taxonomic diversity. *Science, 164*, 947–949.

Stone, L., Dayan, T., & Simberloff, D. (1996). Community-wide assembly patterns unmasked: The importance of species' differing geographical ranges. *The American Naturalist, 148*, 997–1015.

Storch, D., Evans, K. L., & Gaston, K. J. (2005). The species-area-energy relationship. *Ecological Letters, 8*, 487–492.

Symmons, R. (2005). New density data for unfused and fused sheep bones, and a preliminary discussion on the modelling of taphonomic bias in archaeofaunal age profiles. *Journal of Archaeological Science, 32*, 1691–1698.

Tappen, M. (1994). Bone weathering in the tropical rain forest. *Journal of Archaeological Science, 21*, 667–673.

Termine, J. D., Wuthier, R. E., & Posner, A. S. (1967). Amorphous-crystalline mineral changes during endochondral and periosteal bone formation. *Proceedings of the Society for Experimental Biology Medicine, 125*, 4–9.

Trueman, C. N., & Martill, D. M. (2002). The long-term survival of bone: The role of bioerosion. *Archaeometry, 44*, 371–382.

Trueman, C. N., & Tuross, N. (2002). Trace elements in recent and fossil bone apatite. In M. J. Kohn, J. Rakovan, & J. M. Hughes (Eds.), *Phosphates: Geochemical, geobiological and materials importance. Mineralogical Society of America, Reviews in Mineralogy and Geochemistry*, Washington, DC.

Trueman, C. N. G., Behrensmeyer, A. K., Tuross, N., et al. (2004). Mineralogical and compositional changes in bones exposed on soil surfaces in Amboseli National Park, Kenya: Diagenetic mechanisms and the role of sediment pore fluids. *Journal of Archaeological Science, 31*, 721–739.

Tucker, M. E. (1991). The diagenesis of fossils. In S. K. Donovan (Ed.), *The processes of fossilization*. New York: Columbia University Press.

Upchurch, G. R., Otto-Bliesner, B. L., & Scotese, C. R. (1999). Terrestrial vegetation and its effects on climate during the latest Cretaceous. *Geological Society of America Special Paper, 332*, 407–426.

Van der Zwan, C. J. (2002). The impact of Milankovitch-scale climatic forcing on sediment supply. *Sedimentary Geology, 147*, 271–294.

Van Valkenburgh, B., & Molnar, R. E. (2002). Dinosaurian and mammalian predators compared. *Paleobiology, 28*, 527–543.

Vandevivere, P., & Baveye, P. (1992). Effect of bacterial extracellular polymers on the saturated hydraulic conductivity of sand columns. *Applied and Environmental Microbiology, 58*, 1690–1698.

Von Endt, D. W., & Ortner, D. J. (1984). Experimental effects of bone size and temperature on bone diagenesis. *Journal of Archaeological Science, 11*, 247–253.

Voorhies, M. R. (1969). *Taphonomy and population dynamics of an Early Pliocene vertebrate fauna, Knox County*. Nebraska: University of Wyoming.

Walker, S. E., & Goldstein, S. T. (1999). Taphonomic tiering: Experimental field taphonomy of molluscs and foraminifera above and below the sediment-water interface. *Palaeogeography, Palaeoclimatology, Palaeoecology, 149*, 227–244.

Ward, P., Labandeira, C., Laurin, M., et al. (2006). Confirmation of Romer's Gap as a low oxygen interval constraining the timing of initial arthropod and vertebrate terrestrialization. *Proceedings of the National Academy of Sciences of the United States of America, 103*, 16818–16822.

Watson, J., & Alvin, K. L. (1996). An English Wealden floral list, with comments on possible environmental indicators. *Cretaceous Research, 17*, 5–26.

Weigelt, J. (1989). *Recent vertebrate carcasses and their paleobiological implications*. Chicago: University of Chicago Press.

Weiner, S., & Price, P. (1986). Disaggregation of bone into crystals. *Calcified Tissue International, 39*, 365–375.

Weiner, S., & Traub, W. (1992). Bone structure: From angstroms to microns. *The FASEB Journal, 6*, 879–885.

Weiner, S., & Wagner, H. D. (1998). The material bone: Structure–mechanical function relations. *Annual Review of Materials Science, 28*, 271–298.

Wendler, J., Graefe, K.-U., & Willems, H. (2002). Reconstruction of mid-Cenomanian orbitally forced palaeoenvironmental changes based on calcareous dinoflagellate cysts. *Palaeogeography, Palaeoclimatology, Palaeoecology, 179*, 19–41.

White, E. M., & Hannus, L. A. (1983). Chemical weathering of bone in archaeological soils. *American Antiquity, 48*, 316–322.

Wilborn, B. (2007). Deformation of dinosaur bones: Diagenetic and tectonic effects. *Journal of Veterinary Paleontology, 27*, 166A.

Williams, C. T., & Potts, P. J. (1988). Element distribution maps in fossil bones. *Archaeometry, 30*, 237–247.

Wings, O. (2004). Authigenic minerals in fossil bones from the Mesozoic of England: Poor correlation with depositional environments. *Palaeogeography, Palaeoclimatology, Palaeoecology, 204*, 15–32.

Winkler, D. A. (1983). Paleoecology of an Early Eocene Mammalian Fauna from Paleosols in the Clarks Fork Basin, Northwestern Wyoming (USA). *Palaeogeography, Palaeoclimatology, Palaeoecology, 43*, 261–298.

Wolfe, D. J., & Kirkland, J. I. (1998). *Zuniceratops chritopheri* nov. gen & nov. sp., a ceratopsian dinosaur from the Moreno Hill Formation (Cretaceous, Turonian) of west-Central New Mexico. *New Mexico Museum of Natural History and Science Bulletin, 14*, 303–317.

Wood, J. M., Thomas, R. G., & Visser, J. (1988). Fluvial processes and vertebrate taphonomy: The Upper Cretaceous Judith River Formation, south-central Dinosaur Provincial Park, Alberta, Canada. *Palaeogeography, Palaeoclimatology, Palaeoecology, 66*, 127–143.

Wopenka, B., & Pasteris, J. D. (2005). A mineralogical perspective on the apatite in bone. *Materials Science & Engineering C-Biomimetic and Supramolecular Systems, 25*, 131–143.

Wright, V. P., & Vanstone, S. D. (2001). Onset of Late Palaeozoic glacio-eustasy and the evolving climates of low latitude areas: A synthesis of current understanding. *Journal of Geological Society, 158*, 579–582.

Wright, S., Keeling, J., & Gillman, L. (2006). The road from Santa Rosalia: A faster tempo of evolution in tropical climates. *Proceedings of the National Academy of Sciences of the United States of America, 103*, 7718–7722.

Yang, W., Harmsen, F., & Kominz, M. A. (1996). Quantitative analysis of a cyclic peritidal carbonate sequence, the Middle and Upper Devonian Lost Burro Formation, Death Valley, California – A possible record of Milankovitch climatic cycles. *Journal of Sedimentary Research Section B: Stratigraphy and Global Studies, 65*, 306–322.

Yoshino, M., Kimijima, T., Miyasaka, S., et al. (1991). Microscopical study on estimation of time since death in skeletal remains. *Forensic Science International, 49*, 143–158.

Zazzo, A., Lécuyer, C., & Mariotti, A. (2004). Experimentally-controlled carbon and oxygen isotope exchange between bioapatites and water under inorganic and microbially-mediated conditions. *Geochimica et Cosmochimica Acta, 68*, 1–12.

Zhou, Z. G. (2006). Evolutionary radiation of the Jehol Biota: Chronological and ecological perspectives. *Geological Journal, 41*, 377–393.

Zhou, Z. H., Barrett, P. M., & Hilton, J. (2003). An exceptionally preserved Lower Cretaceous ecosystem. *Nature, 421*, 807–814.

Ziegler, A. M., Eshel, G., Rees, P. M., et al. (2003). Tracing the tropics across land and sea: Permian to present. *Lethaia, 36*, 227–254.

Chapter 9
Microtaphofacies: Exploring the Potential for Taphonomic Analysis in Carbonates

James H. Nebelsick, Davide Bassi, and Michael W. Rasser

Contents

Abstract The microtaphofacies of Paleogene carbonates from three time units (Middle Eocene, Late Eocene and Early Oligocene) from the circumalpine area are described and compared. These carbonates are dominated by various larger foraminiferal and coralline red algal facies with subordinate coral and bryozoan facies.

The taphonomy of different components and the taphonomic attributes for each facies type are detailed using a semi-quantified scheme describing four different taphonomic features (abrasion, fragmentation, encrustation and bioerosion). This allows the distribution and magnitude of taphonomic features to be determined along the shelf gradient and between different time units.

J.H. Nebelsick (✉)
Institute for Geosciences, University of Tübingen, Sigwartstrasse 10, 72076 Tübingen, Germany
nebelsiek@uni-tuebingen.de

D. Bassi
Dipartimento di Scienze della Terra, Università di Ferrara, Via Saragat 1, 44122 Ferrara, Italy

M.W. Rasser
Museum of Natural History Stuttgart, Rosenstein 1, 70191 Stuttgart, Germany

P.A. Allison and D.J. Bottjer (eds.), *Taphonomy: Process and Bias Through Time*,
Topics in Geobiology 32, DOI 10.1007/978-90-481-8643-3_9,

Taphonomic features vary between facies types and time units. Fragmentation, for example, is greatest in shallow water, inner shelf settings and is due to wave base water agitation. Abrasion and encrustation are variable throughout different facies whilst bioerosion varies through time. Middle Eocene facies are generally less taphonomically altered than the Late Eocene and Early Oligocene facies which seems to reflect the appearance of coralline algal dominated facies. The extinction events among larger foraminifera that dramatically influence the occurrence and distribution of facies have little effect on the distribution of taphonomic features.

1 Introduction

Taphonomic studies have mostly concentrated on the investigation and quantification of isolated macroscopic faunal and floral elements. These have determined how taphonomic pathways are influenced by: (1) the primary morphology and mineralogy of the faunal skeleton or floral supporting systems, (2) numerous environmental factors including water movement, salinity, temperature, oxygen levels etc., and (3) transport and time averaging which can mix environmental and temporal signals. A number of taphonomic processes and features including decay, abrasion, disarticulation, fragmentation, bioerosion, encrustation, micritization and corrosion which affect preservation have been identified and analysed for different organisms and in different environments. More encompassing studies using these taphonomic features have led to their inclusion in paleoecologic analysis and the establishment of taphonomic gradients and taphofacies as an alternative to bio- and sedimentary facies (e.g. Brett and Baird 1986; Cummins et al. 1986; Norris 1986; Staff et al. 1986; Speyer and Brett 1988; Allison and Briggs 1991; Donovan 1991a; Kidwell and Bosence 1991; Scoffin 1992; Flessa 1993; Kidwell and Flessa 1996; Wilson 1988; Martin 1999; Behrensmeyer et al. 2000; Yesares-García and Aguirre 2004). Actualistic studies have played a seminal role in elucidating these taphonomic processes and features.

Indurated carbonate rocks, in contrast to isolated macroscopic remains, do not readily offer material for taphonomic studies. There is, however, an enormous potential for analysing taphonomic features in carbonates by studying thin sections. The use of thin section analysis for very detailed ecological reconstruction of carbonate environments has been common since the inception of this technique (see numerous examples compiled in Flügel 1982, 2004). Microfacies analyses of limestones allow the identification and analysis of even highly damaged biogenic components at a wide range of magnifications. Various quantification techniques and subsequent statistical treatment of component distributions are in regular use. Temporal changes from initial deposition over early diagenetic effects to late diagenetic overprinting can also be determined. A further advantage of thin-section analysis of indurated rocks is that sampling can be spaces at regular intervals over long stratigraphic sections, a possibility not necessarily present when studying macroscopic fossils as a base for taphonomic investigations. These advantages of microfacies analysis have obvious potential for taphonomy.

2 Taphonomy in Carbonate Environments

2.1 Taphonomy as an Inherent Part of Microfacies Analysis

Microfacies studies and the subsequent interpretation of depositional environments inherently take taphonomic processes into account. The fact that the biotic components seen in thin section do not necessarily represent original skeletal architectures, mineralogies or distributions is not a fatal flaw. The component grains, their preservation and ultimately the limestone fabric is generally understood to result from a whole slew of taphonomic processes including in situ processes on the sea floor as well as early and late diagenetic effects.

Numerous reviews (e.g. Flügel 1982, 2004; Scoffin 1987, Tucker and Wright 1990) consider the role of mechanical destruction and biological breakdown of individual skeletons and bioclasts, the stability of carbonate build-ups, the production of micritic muds and the determination of calcium carbonate budgets. The destruction of biogenic hard substrates and skeletons occurs through diverse means:

1. Surface grazing by polyplacophores (Rasmussen and Frankenberg 1990; Barbosa et al. 2008), regular echinoids (Bak 1990, 1994; Steneck 1983) and parrot fish (Bellwood and Choat 1991). Surface grazing, although prevalent in Recent environments, is difficult if not impossible to recognize in the fossil record. In some cases rare traces, for example from echinoid teeth, can be identified, but generally they are not preserved. The differentiation of abrasion caused by grazing activity and that caused by sediment agitation is of course difficult.
2. Chemical and mechanical erosion results from the action of a variety of macroscopic borers (Neumann 1966; Bromley 1978; Perry 1996, 1998a, 2000; Perry and Bertling 2000; Scoffin and Bradshaw 2000; Wilson 2007) including clionid sponges (Goreau and Hartman 1963; Futterer 1974; Acker and Risk 1985) sipunculids (Perry 1998a) and boring bivalves (Kleemann 1994). Bioerosion not only weakens the skeletons so infested but also produces sediment (Hutchings 1986; Glynn 1997). Furthermore, since bioeroders are often filter feeding organisms, high infestation rates of macroborers are used as an indication of nutrient-rich carbonate environments. Boreholes can be recognized in thin section especially if the shells of the producers are still in the holes (as is often the case with lithophagid bivalves).
3. Infestation of bioclasts by bacterial and algal microborers (Golubic 1969, 1990; Golubic et al. 1975; Rooney and Perkins 1972; Budd and Perkins 1980; Tudhope and Risk 1985; Vogel 1993; Kiene et al 1995; Vogel et al. 1995, 2000; Perry 1998b; Glaub et al. 2007) is the factor that can be studied best in the rock record. This has led to exploration of the evolution of microboring taxa and boring strategies through time and in different reefal settings. Their use as ecological indicators

(such as light dependency related to depth and water clarity) is well known especially since relatively small samples are needed for analysis. The identification of these traces is problematic though since the taxonomy of micro-boreholes is based on three-dimensional reconstructions as recovered by various casting procedures.

4. Crushing of shells by predation and scavenging (e.g. Lipps 1988; Cate and Evans 1994) is commonplace in shallow water marine settings. Recognizing predation in thin section, as opposed to macroscopic evaluation of three dimensional wound morphologies and breakage patterns, is limited by the two-dimensionality of thin sections.
5. Grain destruction after ingestion by deposit feeders is often underestimated and is poorly studied (see discussion in Scoffin 1987). Numerous deposit feeders consume bioclastic components either whole or after mastication (depending on the consumer involved). The affects on the respective shells after passing through the digestive tracts is difficult to resolve, but must be important considering the high turnover rates in some carbonate environments by benthic animals (such as echinoids).

Perhaps the most important aspect of microboring for carbonate environments is micritization (Bathurst 1966; Alexandersson 1972; Kobluk and Risk 1977; Neugebauer 1978; Reid and Macintyre 2000) which can be either destructive or constructive (Flügel 2004). Pervasive micritization can make identification of carbonate grains impossible. Micritization, however, is of utmost importance for the recognition and identification of bioclastic grains originally consisting of aragonite such as dasycladalean algae, scleractinian corals and most gastropods and infaunal bivalves (see below). Since the outer surface of the grains becomes infested by miciritization, their morphology is conserved even after the aragonite has been completely dissolved and replaced by sparitic cements.

The differential response of various biogenic skeletons to transport has been elucidated from field observations and experiments (Chave 1964). A commonly quoted example of degradation is that of the aragonitic skeletons of *Halimeda* and other green calcareous algae which disaggregate into their constituent parts and the aragonite laths of which they are composed (Wefer 1980). The mechanical and biological degradation of green algae has been construed to be a major source of aragonite muds in tropical environments.

Diagenesis is obviously of prime importance for the transformation of loose skeletal and non-skeletal carbonate grains into indurated limestones (e.g. Flügel 1982, 2004; Dullo 1983; Schroeder and Purser 1986; Scoffin 1992; Tucker and Wright 1990). Of special importance from a taphonomic point of view is the preferential destruction of specific shell mineralogies at different stages of diagenesis.

Aragonitic components including dasycladalean algae, scleractinian corals, gastropods and bivalves, can dominate Recent carbonate environments. That the dissolution of aragonite can lead to the distortion of our perception of the original biofacies and carbonate budgets at hand has been a long standing issue (e.g. Bathurst 1964; Budd 1988; Palmer et al 1988; Canfield and Raiswell 1991; Budd and Hiatt 1993; Brachert and Dullo 2000). Aragonitic components can, however, be recognized in thin section if they are micritized thus preserving a recognizable shell outline or if they are encased in fine micritic muds before dissolution (see examples below).

Encrustation clearly plays an important role in carbonate environments, not only in reefal settings (e.g. Martindale 1992; Rasser and Riegl 2002), but also as a taphonomic agent affecting small substrates (Taylor 1979, 1992; McKinney 1995; Lescinsky et al. 2002; Taylor and Wilson 2003). Encrusting floral and faunal elements can produce complex multi-taxon overgrowth sequences which can be ideally resolved in thin section analysis (e.g. Reolid and Gaillard 2007; Reolid et al. 2007).

2.2 *Concepts and Definitions of Taphonomy in Thin Section Analysis*

There are few studies which have specifically approached the taphonomy of indurated carbonates using thin section analysis. The following terms have been used to describe the study of taphonomy in thin sections, in part derived from the different emphasis of these studies:

1. Brachert et al. (1998) coined the term "Microtaphofacies" and used taphonomic features recognized in thin section to augment standard microfacies analysis techniques in Miocene carbonates. This term will be used in this paper.
2. Reolid and Gaillard (2007) and Reolid et al. (2007) used the term "Microtaphonomy" in a similar way as Brachert et al. (1998). These authors use rigorous quantification techniques following various taphonomic indices of Olóriz et al. (2004) to assess specific taphonomic features in Jurassic carbonate facies from Spain. They use this term in the same sense as "microtaphofacies" – using taphonomic traits recognized in thin section to augment environmental information gained by component distributions.
3. Sanders and Krainer (2005) coined the term "Taphloss" when analysing Early Permian benthic assemblages from the Carnic Alps. As the term suggests, emphasis is placed on the loss of floral and faunal diversity and information in general due to taphonomic processes on the sea floor and later diagenesis (Sanders 1999). The difficulties induced by taphonomic loss in reconstructing carbonate budgets is emphasized (Sanders 2003).
4. "Microfacies taphonomy" (Wright and Burgess 2005) is used for information loss and stresses the problems involved in reconstructing paleoenvironments given the role and rate of taphonomic destruction.

The dichotomy of taphonomic "gain" and "loss" (Cummins et al. 1986; Thomas 1986; Wilson 1988) has also been carried into microfacies studies of carbonates. Taphonomic processes can either be seen as a source of information loss (often the case) or information gain as taphonomic processes can reveal ecologic patterns and developments not necessarily present in "normal" thin section analysis.

The present study provides an example of how taphonomic processes and features change across facies boundaries. It is based on the analyses shallow water carbonate

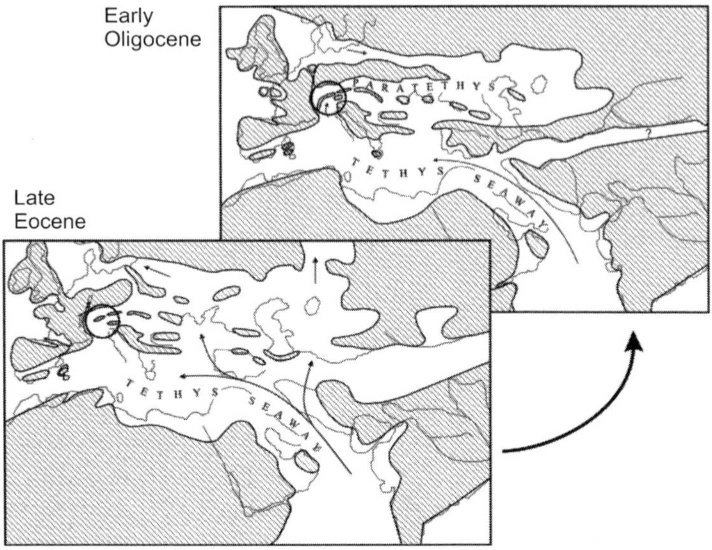

Fig. 1 Location of the study area within the changing Paleogene paleogeography of the Mediterranean and Paratethyan seaways. Study area denoted by a circle (maps after Rögl 1998)

facies during the Paleogene (specifically from the Middle Eocene to the Early Oligocene) of the circum alpine region (Fig. 1). This time period and geographic area are especially interesting due to dramatic developments in global climate, significant paleogeographic changes, extinction events and resulting shifts of major carbonate facies types. The studied carbonates are dominated by coralline algae and larger foraminifera with subordinate corals and bryozoans. This qualitative comparison will serve as a base for future, more detailed quantitative assessments of microtaphofacies.

3 Taphonomy of Paleogene Components in Thin Section

The taphonomic processes affecting coralline algae and larger foraminifera (Table 1 and Fig. 2) are diversely expressed (Figs. 3–11) although encrustation is especially important in the production of rhodoliths (which can reach diameters greater than 10 cm). Transport is particularly important to the accumulation of large foraminifera-dominated sediments, especially those containing *Nummulites* ("nummulithoclastic sediments" following Beavington-Penney 2004).

Corals have received intensive attention especially with respect to comparing faunal diversities of living corals to sub-fossil and fossil faunas (e.g. Scoffin 1992; Greenstein and Moffat 1996; Pandolfi and Minchin 1996; Pandolfi and Greenstein 1997; Greenstein and Pandolfi 2003; Meyer et al. 2003; Aronson 2007; Greenstein 2007; R. Wood, this

Table 1 Previous work on the taphonomy of Recent and fossil coralline red algae and larger benthic foraminifera

Main components	Taphonomic feature	Citations
Coralline algae	Reviews	Nebelsick and Bassi (2000)
	Disease and mortality	Littler and Littler (1995, 1997)
	Abrasion	Chave (1964); Bosence (1976); Testa (1997); Checconi et al. (2007)
	Fragmentation including maerl formation	Cabioch (1969); Adey and McKibbin (1970); Bosence (1976, 1980, 1983b); Freiwald et al. (1991); Freiwald (1993, 1995); Bordehore et al. (2003)
	Encrustation including rhodolith formation (Recent)	Bosellini and Ginsburg (1971); Adey and MacIntyre (1973); Bosence and Pedley (1982); Bosence (1976, 1983a–c, 1984, 1985b); Adey (1978); Scoffin et al. (1985); Sebens (1986); Reid and Macintyre (2000); Littler et al. (1991); Martindale (1992); Keats and Maneveldt (1994); Keats et al. (1994); Steller and Foster (1995); Piller and Rasser (1996); Foster et al. (1997); Rasser and Piller (1997); Basso (1998); Gherardi and Bosence (1999); Gischler and Pisera (1999); Marrak (1999); Foster (2001); Perry (2005); Piller and Rasser (2005); Hetzinger et al. (2006); Konar et al. (2006); Di Geronimo et al. (2002); Bassi et al. (2009)
	Encrustation including rhodolith formation (fossil)	Bosence and Pedley (1982); Braga and Martìn (1988); Iryu (1997); Bassi (1998, 2005); Hillis and Jones (2000); Braga and Aguirre (2001); Rasser (2000, 2001); Bassi et al. (2009)
	Bioerosion	Checconi et al. (2007)
	Predation (herbivory) and grazing	Adey and MacIntyre (1973); Lawrence (1975); van den Hoek et al. (1975); Wanders (1977); Brock (1979); Steneck (1983, 1987, 1997); Morse and Morse (1984); Figueiredo (1997); Johnson et al. (1997)
	Early diagenesis	Alexandersson (1972, 1974, 1977); Bosence (1985a, 1991); Martindale (1992)
Larger benthic foraminifera	Reviews	Beavington-Penney (2004); Beavington-Penney and Racey (2004)
	Abrasion	Peebles and Lewis (1988, 1991); Cottey and Hallock (1988); Yordanova and Hohenegger (2002); Beavington-Penney (2004)
	Fragmentation	Yordanova and Hohenegger (2002); Beavington-Penney (2004)
	Bioerosion	Kloos (1982); Serra-Kiel (1982); Serra-Kiel and Reguant (1984); Matteucci and Pignatti (1988)
	Dissolution	Cottey and Hallock (1988); Murray (1989); Peebles and Lewis (1988, 1991)

(continued)

Table 1 (continued)

Main components	Taphonomic feature	Citations
	Transport and sediment accumulation (Recent)	Davies (1970); Hohenegger and Yordanovea (2001); Hohenegger (2004); Severin and Lipps (1989); Yordanova and Hohenegger (2002)
	Transport and sediment accumulation (fossil)	Engel (1970); Aigner (1982, 1983, 1985); Serra-Kiel (1982); Serra-Kiel and Reguant (1984); Matteucci and Pignatti (1988); Eichenseer and Luterbacher (1992); Kondo (1995a, b); Racey (2001); Beavington-Penney (2004); Beavington-Penney and Racey (2004); Bassi (2005); Jorry et al. (2006)

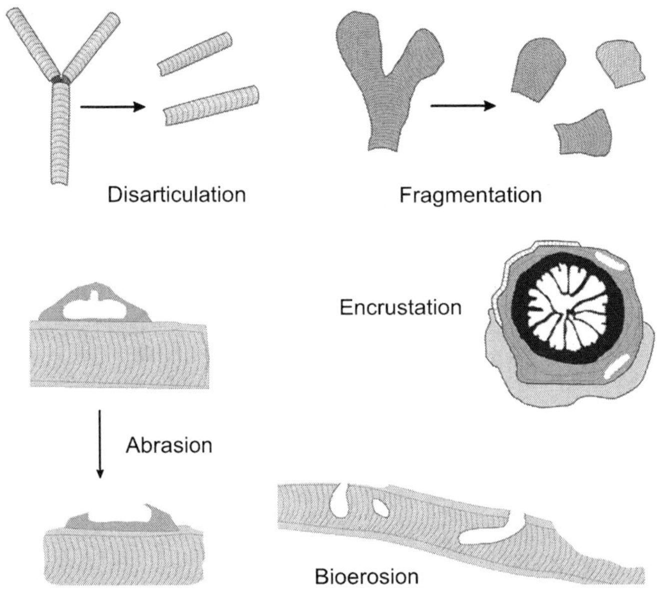

Fig. 2 Taphonomic features of coralline algae as seen in thin section. "Disarticulation" depicts a geniculate coralline alga, the rest depict non-geniculate coralline algae. "Fragmentation" is destroying a fructicose growth form. "Abrasion" shows the destruction of a conceptacle on the surface. "Encrustation" shows multi-taxonomic encrusting thalli on a coral. "Bioerosion" shows both surface removal by grazers as well as internal holes created by boring organisms (modified after Nebelsick and Bassi 2000)

volume). An increasing number of studies have also dealt with the taphonomy of corals in turbid, nutrient rich waters (Perry and Smithers 2006) and deep water environments (Freiwald and Wilson 1998). The role of taphonomy in reefs through time have been summarized with respect to the changing organisms involved in reef growth on the one hand, and the evolution of bioerosion and encrustation strategies on the other (e.g. Fagerstrom 1987, 1991; Vogel 1993; Wood 1998, 1999, this volume).

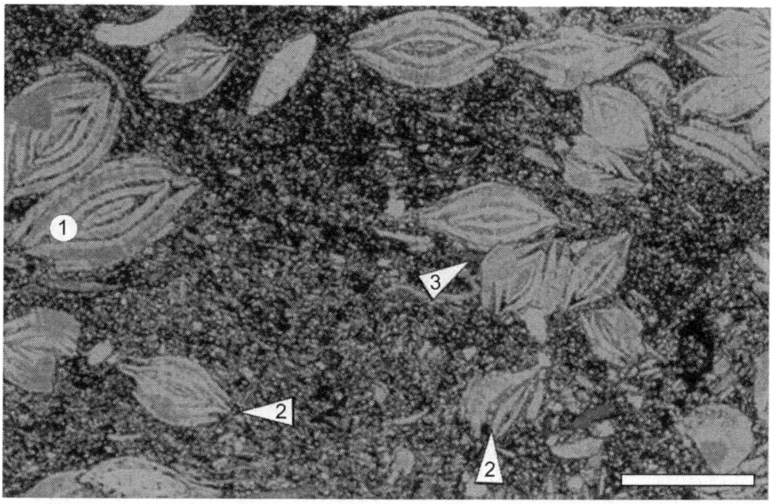

Fig. 3 Abrasion and Fragmentation; Small *Nummulites* Facies showing fragmented and abraded small *Nummulites* (1) in a terrigenous rich packstone matrix. Most *Nummulites* show abrasion and fragmentation to varying degrees (2). Pressure solution is also present (3). Late Eocene, Autochthonous Molasse, Upper Austria. Scale bar = 1 mm

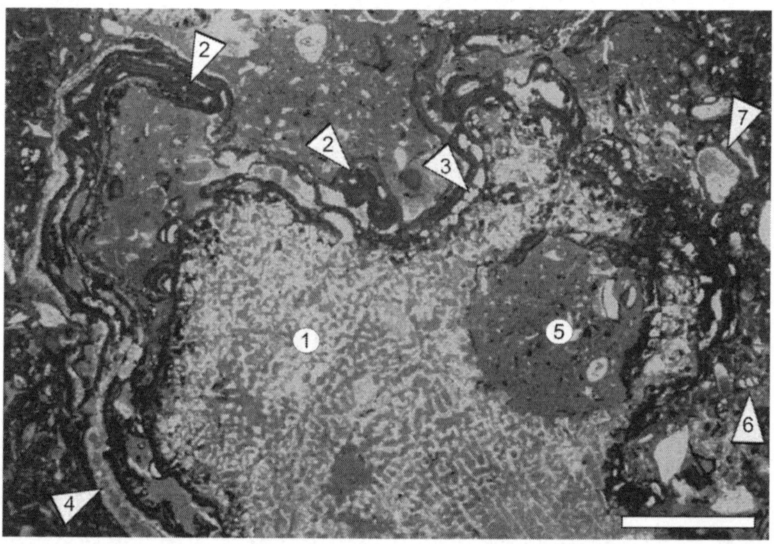

Fig. 4 Encrustation and Bioerosion: Detail of a coral dominated rudstone with a packstone matrix dominated by an encrusted coral. The components are very well preserved and show little abrasion and fragmentation. The coral colony (1) shows a complex multi-taxon encrustation sequence which includes coralline algae (2), encrusting foraminifera (3) and bryozoans (4). Bioerosion is present as a large rounded hole (5) probably representing a lithophagid borehole. Well preserved small benthic foraminifera are present. The aragonitic coral skeleton as well as an isolated gastropod (7) has been completely replaced by calcite. Early Oligocene, Gornji Grad formation, Slovenia. Scale bar = 2 mm

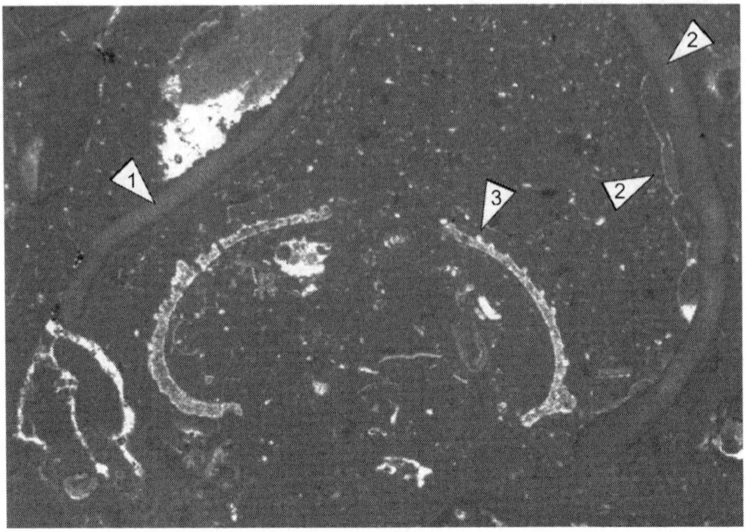

Fig. 5 Very well preserved components showing little or no post-depositional taphonomic features (precluding diagenesis). This section contains two thalli of very well preserved *Neogoniolithon* (1) and an echinoid from the Crustose Coralline Algal Facies. The coralline algal thallus on the right is encrusted by an encrusting acervulinid foraminifera (2). The complete regular echinoid test (3) shows distinct plates, tubercles and pores for the tube feet and has been eroded around the peristome and periproct. The *left hand side* with ambulacral pores is a section through an ambulacrum, the *right hand side* (without pores) is a section through the interambulacra. The high Mg-calcite of the echinoderm skeleton has been replaced by low Mg-calcite. Unidentified fragmented bioclastic material is present in the micritic matrix. Early Oligocene, Monti Berici, Northern Italy. Scale bar = 1 mm

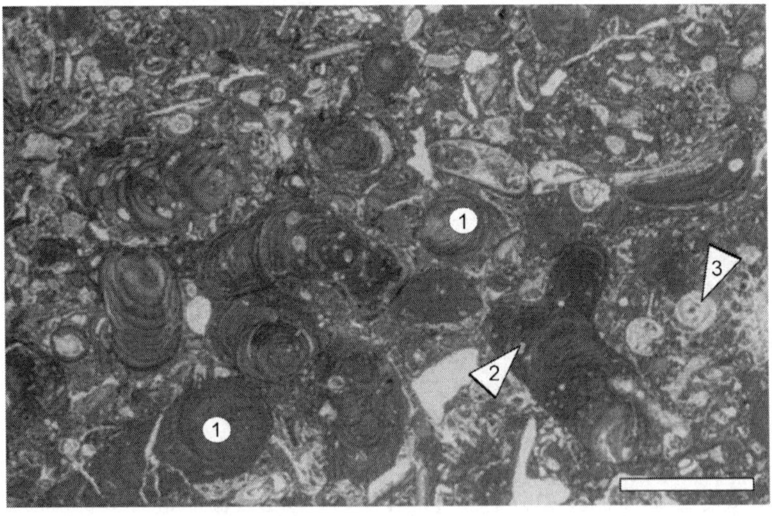

Fig. 6 Maerl Facies with coralline algal thalli (1) which are bioeroded (2). Protuberances are present with some branching. Some well preserved small benthic foraminifera (3) are also present. Late Eocene, Autochthonous Molasse, Upper Austria. Scale bar = 2 mm

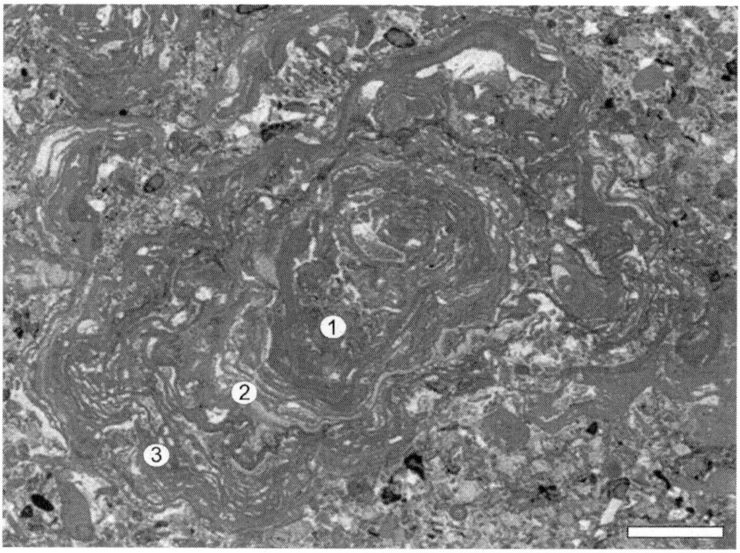

Fig. 7 Rhodolith Facies with a sub-ellipsoidal rhodolith showing complex multi-taxonomic coralline algal crusts. The rhodolith shows dense encrusting growth forms as well as protuberances. At least three growth generations are present (1, 2 and 3). Large spaces within the rhodolith are filled by skeletal matrix consisting of unidentified skeletal fragments. Late Eocene, Autochthonous Molasse, Upper Austria. Scale bar = 5 mm

The taphonomy of bryozoans has been especially studied in non-tropical environments (Nelson et al. 1988; Smith and Nelson 1994; Smith et al. 1996; Smith and Nelson 2003; Smith 1995) allowing specific taphonomic features such as abrasion, fragmentation and diagenesis to be identified and utilized in paleoenvironmental analysis. The differential mineralogy of bryozoans and the corresponding effect on diagenesis and dissolution is important (Smith et al. 1992; Steger and Smith 2005).

Subordinate components in Paleogene carbonates include smaller benthic foraminifera, molluscs and echinoderms (especially echinoids). Taphonomic studies on smaller benthic foraminifera are rare (e.g. Lipps 1988; Martin and Liddel 1991; Schroba 1993) and are mostly restricted to actualistic examples. The overwhelming majority of taphonomic studies concern Recent and fossil molluscs (e.g. Callender et al. 1994; Kowalewski et al. 1994; Best and Kidwell 2000; Zuschin and Stanton 2001, 2002; Lescinsky et al. 2002; Zuschin et al. 2003; Schneider-Storz et al. 2008). Molluscs, however, are relatively rare components in Paleogene carbonates and are either present as calcite shelled pectinid bivalves and oysters or as (dissolved) aragonite shelled gastropods and infaunal bivalves. Echinoderms are well studied as macrofossils (see reviews in Donovan 1991b; Brett et al. 1997; Ausich 2001; Nebelsick 2004), but comparatively little has been done on the taphonomy of echinoderms in thin sections.

Fig. 8 The Large *Nummulites* Facies dominated by prominent *Nummulites* (1) together with subordinate saddle-shaped orthophragminid foraminifera (2). Rare coralline algae (3) highly abraded and fragmented. Some components, such as a single isolated planktonic globigerinid foraminifer (4), are very well preserved. The components generally lie parallel to bedding, with, some "jamming" of components in a more inclined posture. There are no indication of encrustation and bioerosion. The foraminifera are generally well preserved, but can be slightly abraded (5) with some fragmentation. Post-depositional taphonomic features include in situ pressure solution (6) on components contacts. Pore spaces of the components (especially within the larger *Nummulites* specimens) are either filled by micritic mud, or by sparite. Middle Eocene, Monti Berici, Northern Italy. Scale bar = 2 mm

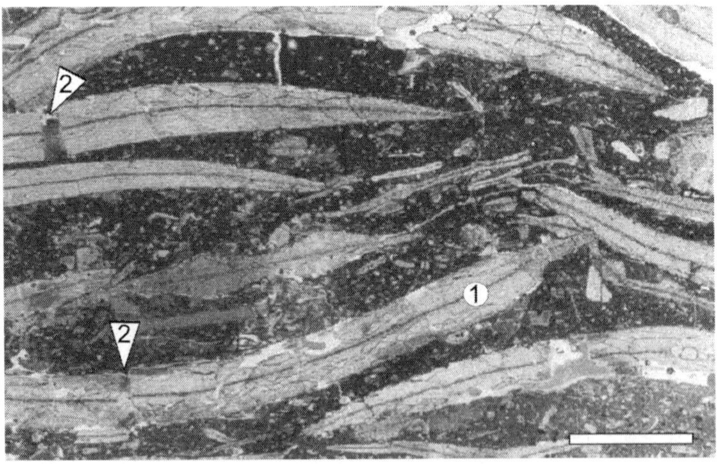

Fig. 9 Orthophragminid Facies dominated by saddle-shaped orthophragminid larger foraminifera (1) in a micritic matrix. The fabric is component-supported in matrix rich sediments. The larger foraminifera are generally well preserved. In situ fragmentation (2) is also present. The components lie more or less parallel to bedding. Late Eocene, Autochthonous Molasse, Upper Austria. Scale bar = 2 mm

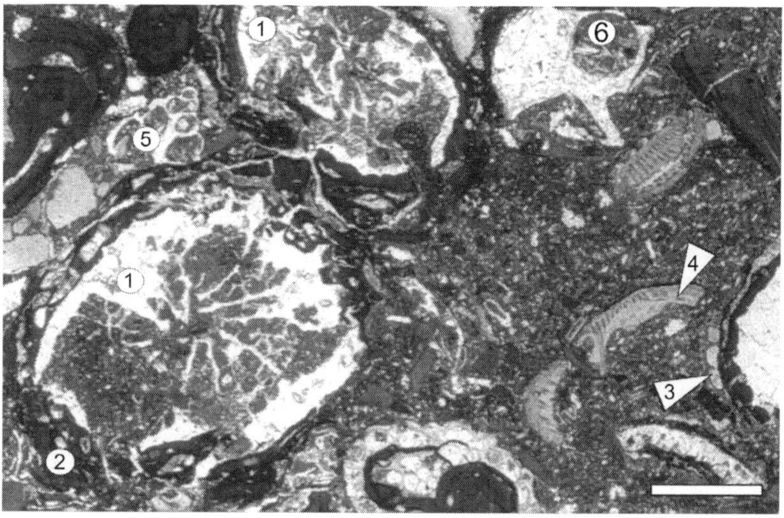

Fig. 10 Coral Facies showing corals (1) partially abraded and encrusted by coralline algae (2) and foraminfera (3). A broken fragment of a thecedine brachiopod (4) and a gastropod (5) are also present. Bioerosion (6) can be recognized. Late Eocene, Autochthonous Molasse, Upper Austria. Scale bar = 2 mm

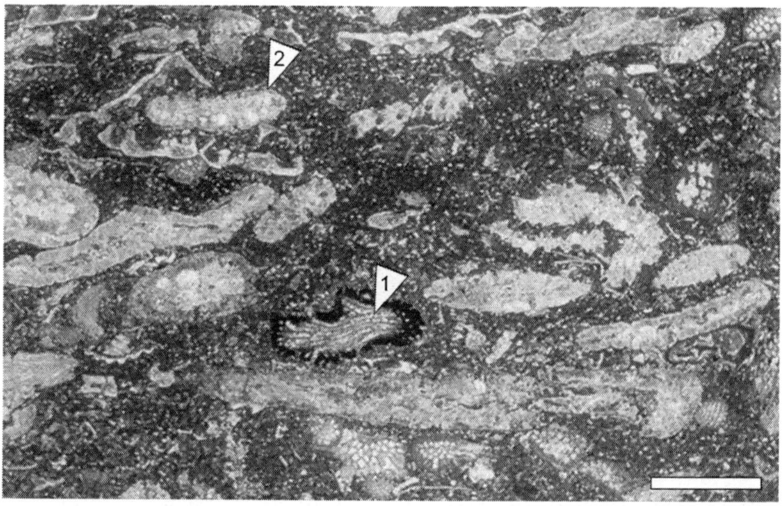

Fig. 11 Bryozoan Facies showing cyclostome (1) and cheilostome (2) bryozoans lying more or less parallel to bedding plane. Bilaminar upright growth-forms dominate. Some cylindrical branching forms are also present. The bryozoans are generally well preserved. Some fragmentation has occurred. The primary pore spaces are generally filled by calcite. Late Eocene, Autochthonous Molasse, Upper Austria. Scale bar = 2 mm

4 Taphonomic Attributes of Major Facies Types

4.1 Lateral and Temporal Facies Distribution

The Eocene/Oligocene boundary marks the transition from "greenhouse" to "icehouse" condition and is manifested by dramatic cooling at least in higher latitudes (e.g. Ivany et al. 2000; Zachos et al. 2001). It is, however, not clear how this cooling event affected the tropics (e.g. Pearson et al. 2007). This is relevant in the present study as the investigated carbonates lie at the northern edge of the "tropical" Tethys seaway. Relevant paleographic development during the studied time-frame include the establishment of Paratethys, a distinct geographic and paleogeographic unit north of the Alps reaching into central Asia (Rögl 1998; Harzhauser and Piller 2007). The connection from Tethys to the Atlantic remained open during the studied time frame (Fig. 1). Extinction events relevant to the development of carbonate facies include the disappearance of larger *Nummulites* species at the Middle-Late Eocene transition and the extinction of orthophragminid larger foraminifera at the Eocene/Oligocene boundary. It is not clear if and how these extinction events are related to climatic change.

The development of carbonate facies in the circum alpine area have been studied in detail by Nebelsick et al. (2003, 2005). A number of Major Facies Types (MFTs) were defined for the given time interval in the studied area. The definition of these MFTs is based on detailed microfacies analysis (Nebelsick et al. 2003, 2005) using quantitative techniques and statistical analysis of component relationships. The MFTs have been defined following dominating (name-giving) components, subordinate components and carbonate fabrics. The distributions of these facies along a shelf gradient were mapped in three time units (Middle Eocene, Late Eocene and Early Oligocene, see Figs. 11–19). Changes in the distribution of these facies along the shelf gradient were impacted by the extinction of major component types (i.e. larger foraminifera) and subsequent replacement of habitat space by other dominating components (and hence the MFT). Most of the MFTs are dominated by either coralline algae or larger foraminifera. Other MFTs are characterized by smaller benthic foraminifera, corals and bryozoans. In some cases, the MFT types are very distinct as far as composition and distribution are concerned. In other cases, they are less well defined. Examples for the former are the Peyssonneliacean MFT in the Late Eocene and the Acervulinid MFT in the Middle Eocene; both involve few taxa, occur in distinct shelf settings and are restricted to a single time slice. An example for the latter is the Rhodolith MFT which can involve various coralline algal taxa and growth-forms and occurs in a large range of depths and settings, and is present in all three time units. Although molluscs and other components such as echinoderms can be locally common, they do not constitute major facies components and are thus not listed as such.

TAPHONOMIC PROCESSES

Major Facies Types	PHYSICAL		BIOLOGICAL	
	ABRASION	FRAGMENTATION	ENCRUSTATION	BIOEROSION
1 Maerl	●●●	●●●●	●●	●●
2 Rhodolith	●●●	●●	●●●●	●●●
3 Crustose Coralline	●●●	●	●●●●	●●●
4 Coralline Algal Debris	●●●●	●●●●	●	●
5 Peyssonneliacean	●●	●	●●	●●
6 Larger *Nummulites*	●●	●●	●	●
7 Small *Nummulites*	●●	●●	●	●
8 Orthophragminid	●	●	●●	●
9 *Orbitolites*	●	●	●●	●
10 Smaller Miliolid	●●	●●	●	●
11 Alveolinid	●●	●	●	●
12 Acervulinid	●	●	●●●●	●●
13 Coral	●	●●	●●●●	●●●●
14 Bryozoan	●	●	●●	●

● low ●● moderate ●●● high ●●●● very high

Fig. 12 Compilation of taphonomic processes for 14 Major Facies Types of the Middle Eocene to Early Oligocene of the circumalpine region

4.2 *Facies Description and Distribution*

The following descriptions summarize the components of each MFT along with their texture and taphonomic traits. The stratigraphic distribution as well as the changes of distribution between the different time units is also noted (see Nebelsick et al. 2003, 2005 for detailed information).

4.2.1 Maerl Facies

The Maerl Facies is dominated by fragments of coralline algal branches often derived from rhodoliths. Nummulitid larger foraminifera, smaller miliolids and textulariid foraminifera are subordinate. It occurs as massively bedded rudstones with grain- and packstone matrix with little or no orientations or gradations. This facies represents a higher energy environment with grains being highly fragmented and abraded. Encrustation along with bioerosion is moderate. The taxonomic identification of highly fragmented algal remnants is difficult due to the lack of diagnostic characters.

The Maerl Facies first appears in the Late Eocene and continues to the Early Oligocene. It changes its distribution from the middle to inner outer shelf in the Middle Eocene to the inner to middle shelf in the Late Eocene and Early Oligocene.

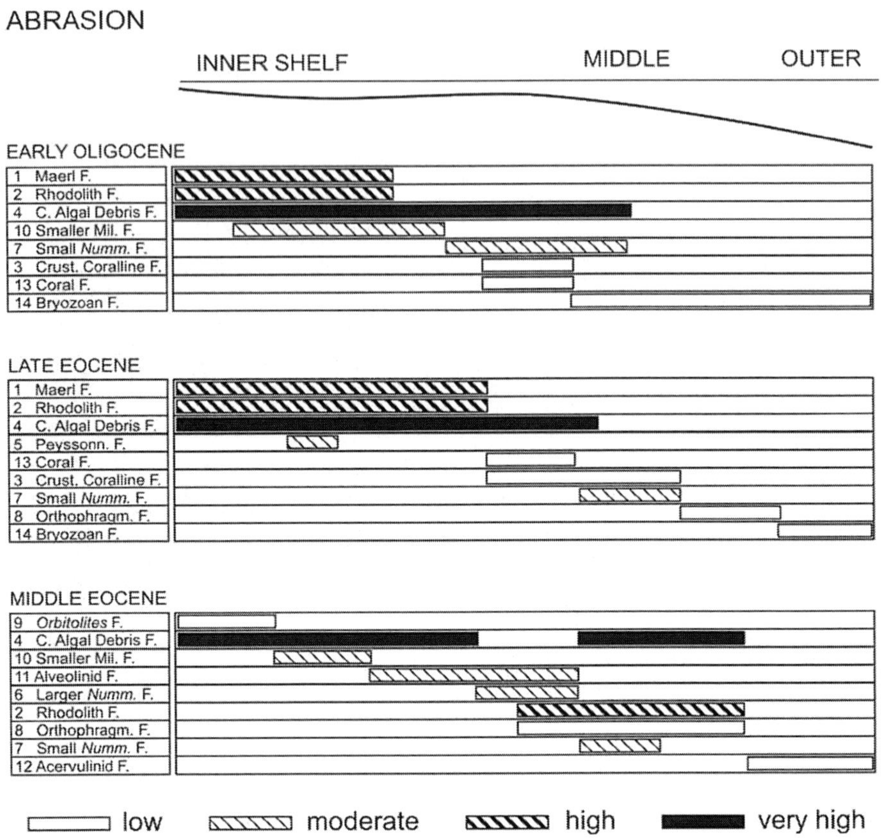

Fig. 13 Distribution of abrasion along a shelf gradient from the Early Eocene to the Early Oligocene

4.2.2 Rhodolith Facies

This facies is dominated by non-geniculate coralline algae, peyssonneliacean algae and encrusting acervulinid foraminifera. All three of these components contribute to rhodolith formation, though coralline algae dominate. Encrusting serpulids, agglutinated foraminifera (e.g. *Haddonia*) and unilaminate and multilaminate bryozoans also contribute to the rhodoliths which can reach sizes up to 10 cm. The rhodoliths occur as spherical, ellipsoidal, discoidal and boxwork shapes. The rhodoliths can be constrained to a single coralline algal taxon and growth-form, but often show complex encrusting sequences. The texture of these limestones is dominated by rudstones with grain- and packstone matrix. Abrasion rates are high and fragmentation rates moderate. Encrustation rates are, given the dominance of encrusting components, very high as is bioerosion which can be pervasive.

The Rhodolith Facies first appears in the Middle Eocene and continues into the Early Oligocene. It changes its distribution from the middle to inner outer shelf in the Middle Eocene to the inner to middle shelf in the Late Eocene and Early Oligocene.

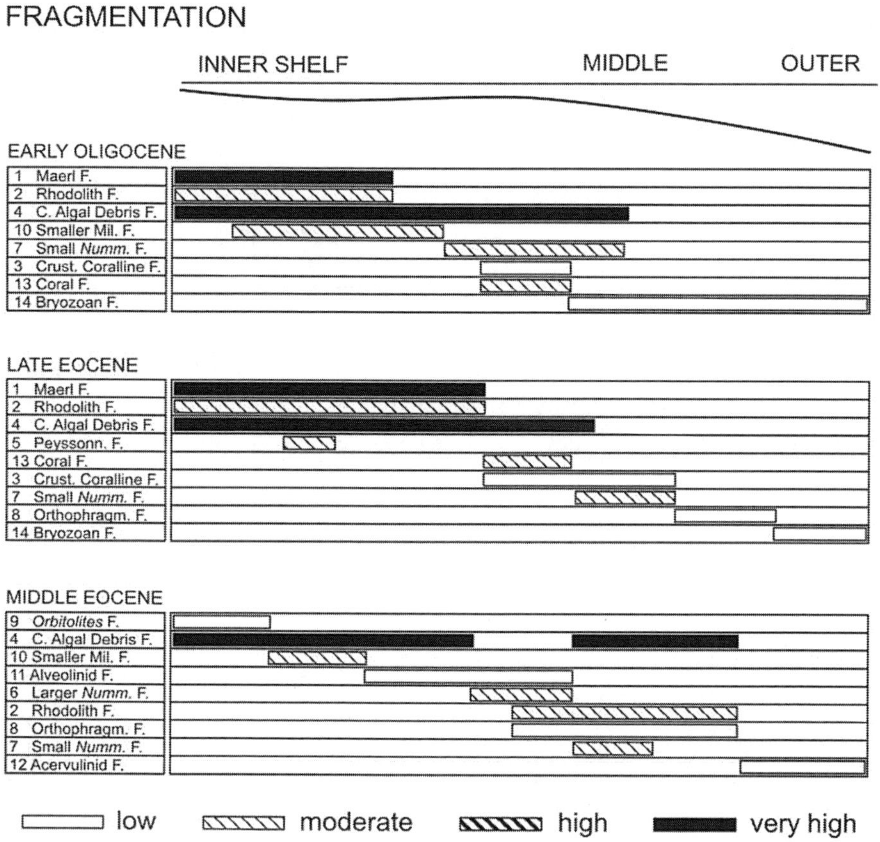

Fig. 14 Distribution of fragmentation along a shelf gradient from the Early Eocene to the Early Oligocene

4.2.3 Crustose Coralline Algal Facies

This facies is dominated by encrusting crustose coralline algae, along with rhodoliths, encrusting foraminifera and locally abundant nummulitids. Smaller miliolid and textulariid foraminifera, bryozoans, peyssonneliacean algae, echinoderms and serpulids are subordinate. Meter-thick bindstones composed of 0.5–1 mm thick algal crusts are characteristic of this facies. Due to their growth-forms, crustose coralline algae serve as binding agents and include sub-discoidal and sub-ellipsoidal rhodoliths with a loosely ordered inner arrangement. The algae themselves can be encrusted by encrusting foraminifera and bryozoans with unilaminate and multilaminate encrusting growth-forms. The algal crusts are predominantly horizontally oriented. Taphonomic features include very high rates of encrustation, high rates of bioerosion and low rates of abrasion and encrustation.

As in the previously described examples, this facies first appears in the Late Eocene and continues into the Early Oligocene. It reduces the distribution range from the middle to inner outer shelf in the Middle Eocene to the middle shelf in the Early Oligocene.

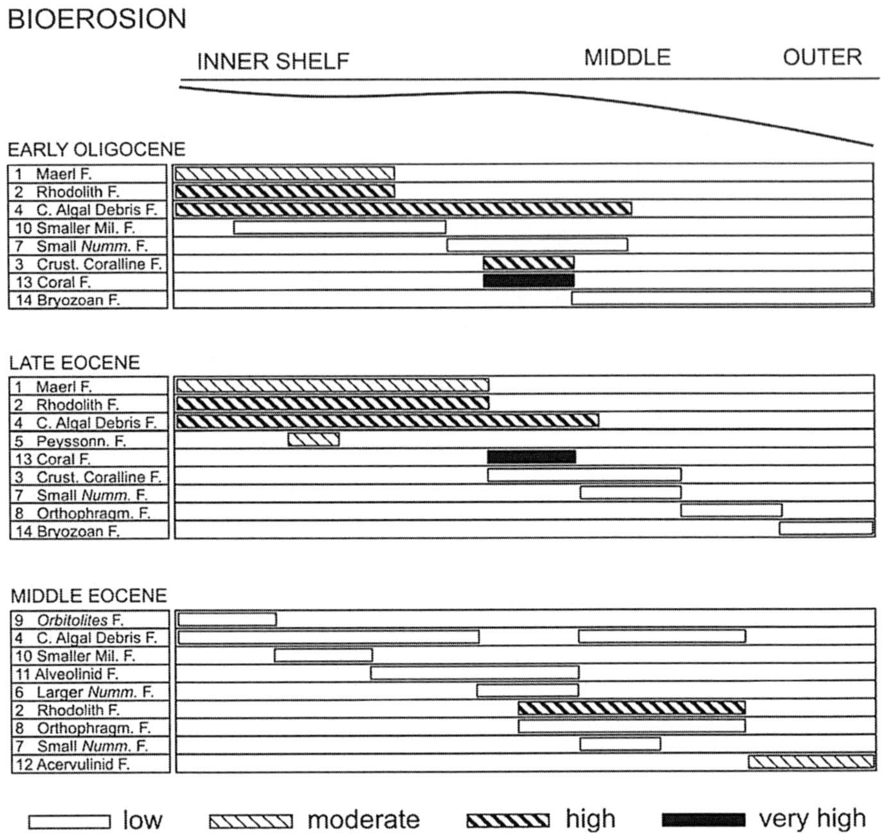

Fig. 15 Distribution of bioerosion a shelf gradient from the Early Eocene to the Early Oligocene

4.2.4 Coralline Algal Debris Facies

This widely distributed facies incorporates a wide range of sediments dominated by coralline algal debris, larger foraminifera, bryozoans, corals, peloids, and siliciclastics. Subordinate components include peyssonneliacean algae, molluscs and smaller benthic rotalid and textulariid foraminifera. The texture of the limestones of this facies is a rudstone with grainstone/packstone matrix. Taphonomic rates vary; abrasion and fragmentation rates can be very high as opposed to the low presence of encrustation and bioerosion. Highly fragmented and abraded components are difficult to identify.

This facies occurs in all three time units from the Middle Eocene to Early Oligocene. It shows a disjunct distribution in the Middle Eocene, and is found in the inner to middle shelf in the Middle Eocene to Late Oligocene.

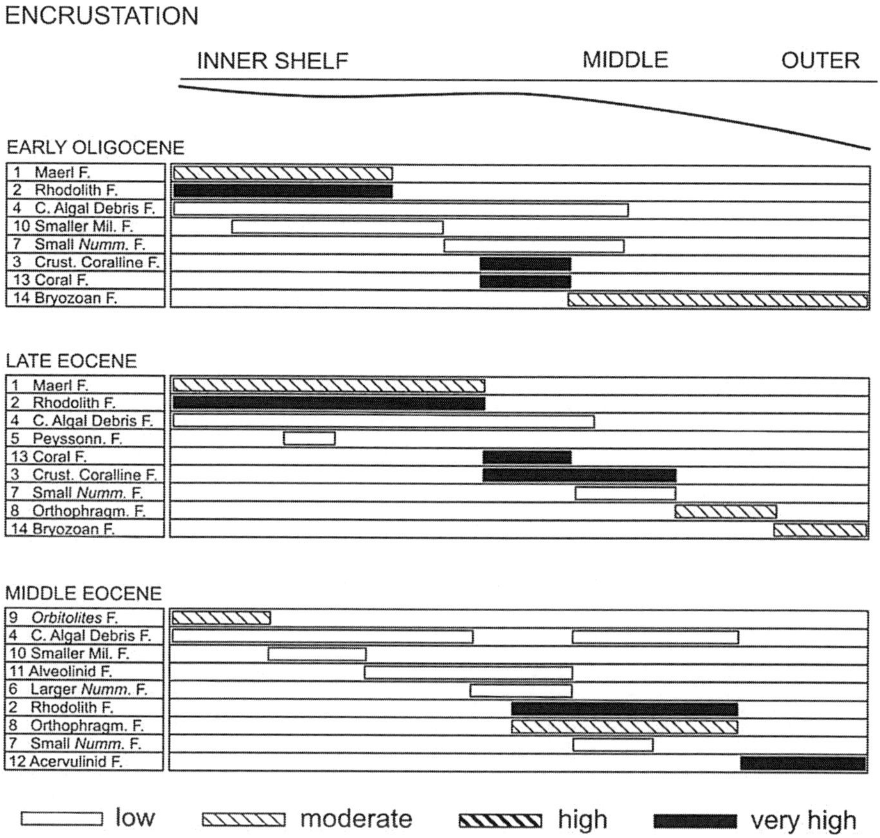

Fig. 16 Distribution of encrustation along a shelf gradient from the Early Eocene to the Early Oligocene

4.2.5 Peyssonneliacean Facies

In this facies, the peyssonneliacean species *Polystrata alba* forms sub-spheroidal rhodoliths with coralline algae as subordinate components. The fabric of these limestones consists of rudstones with a packstone matrix with no grading, sorting or preferred orientation. This facies is characteristic of low energy conditions and fragmentation rates are low and abrasion moderate. Encrustation is high due to the habitat of the dominating component, bioerosion is moderate. The Peyssonneliacean Facies is restricted to a few occurrences in the Late Eocene.

4.2.6 Larger *Nummulites* Facies

Various species of larger *Nummulite*s dominate this facies. Other larger and small benthic foraminifera, molluscs and echinoderms are subordinate. The components

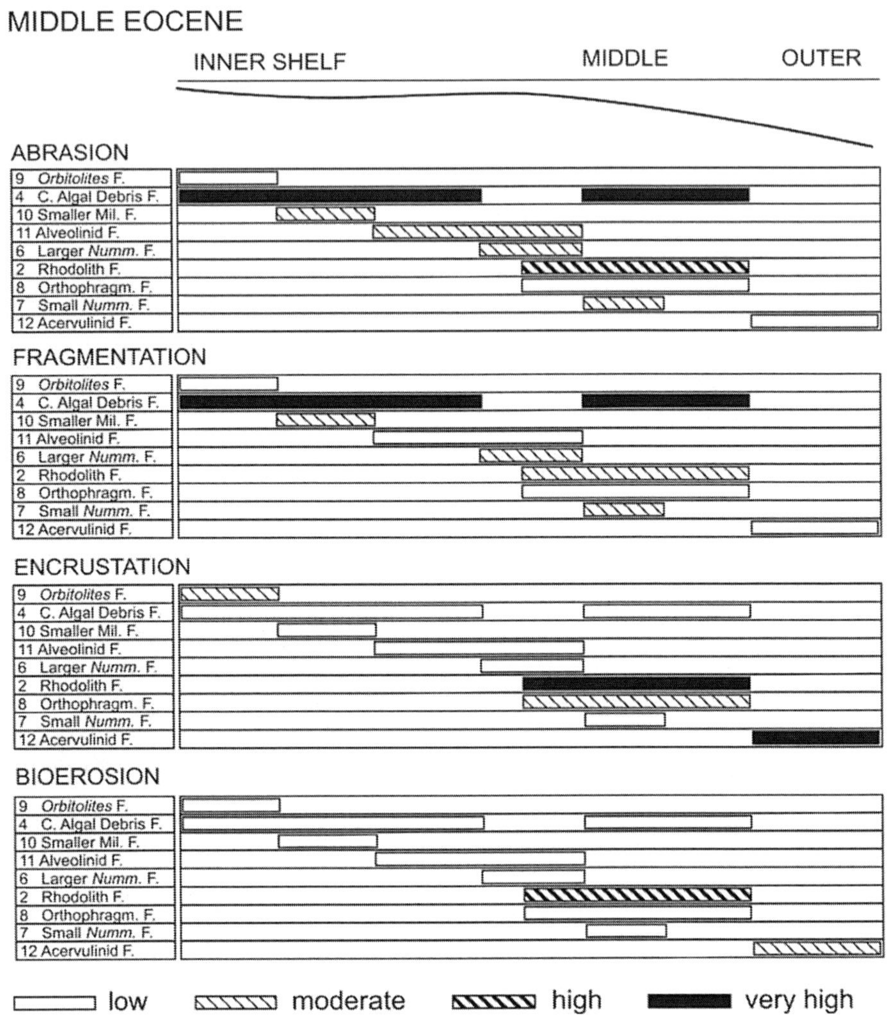

Fig. 17 Compilation of taphonomic processes for the Middle Eocene

can occur in very high densities and include numerous species of larger *Nummulites*. The largest *Nummulites* microspheric forms of *Nummulites gizehensis* can reach diameters of up to 10 cm. The texture is represented by rudstones with packstone matrix showing both orientated and chaotic fabrics.

Various reconstructions for this facies have been offered in the literature (e.g. Aigner 1985; Eichenseer and Luterbacher 1992; Racey 2001; Beavington-Penney and Racey 2004; Bassi 2005) for both autochthonous as well as allochthonous larger *Nummulites* dominating sediments. Edge abrasion and fragmentation leads to the production of abraded *Nummulites* fragments. Encrustation and bioerosion levels are low.

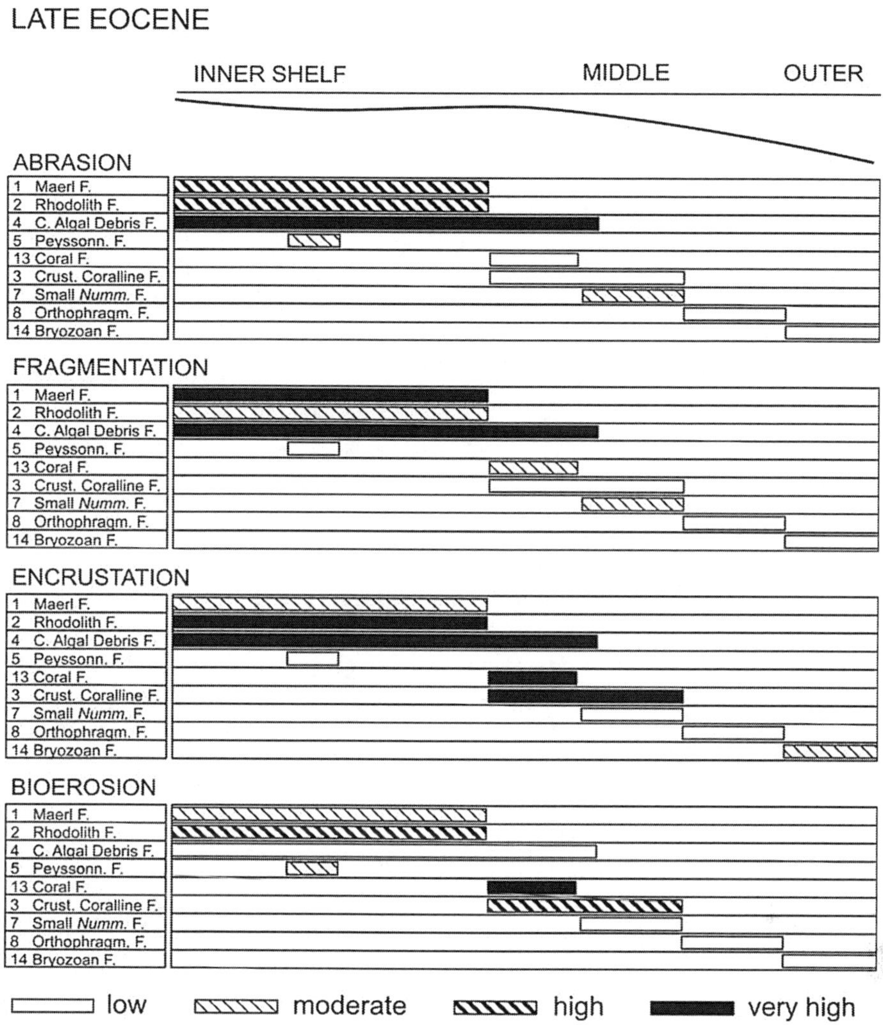

Fig. 18 Compilation of taphonomic processes for the Late Eocene

Larger *Nummulites* are common in the middle shelf of the Early to Middle Eocene and disappear at the Middle/Upper Eocene boundary.

4.2.7 Small *Nummulites* Facies

In this association, numerous different species of smaller *Nummulites* occur with subordinate coralline algal debris, other larger and smaller benthic foraminifera, bivalves, echinoids, brachiopods and corals. The texture consists of packstones and rudstones with packstone/grainstone matrix or siliciclastic matrix. Graded bedding

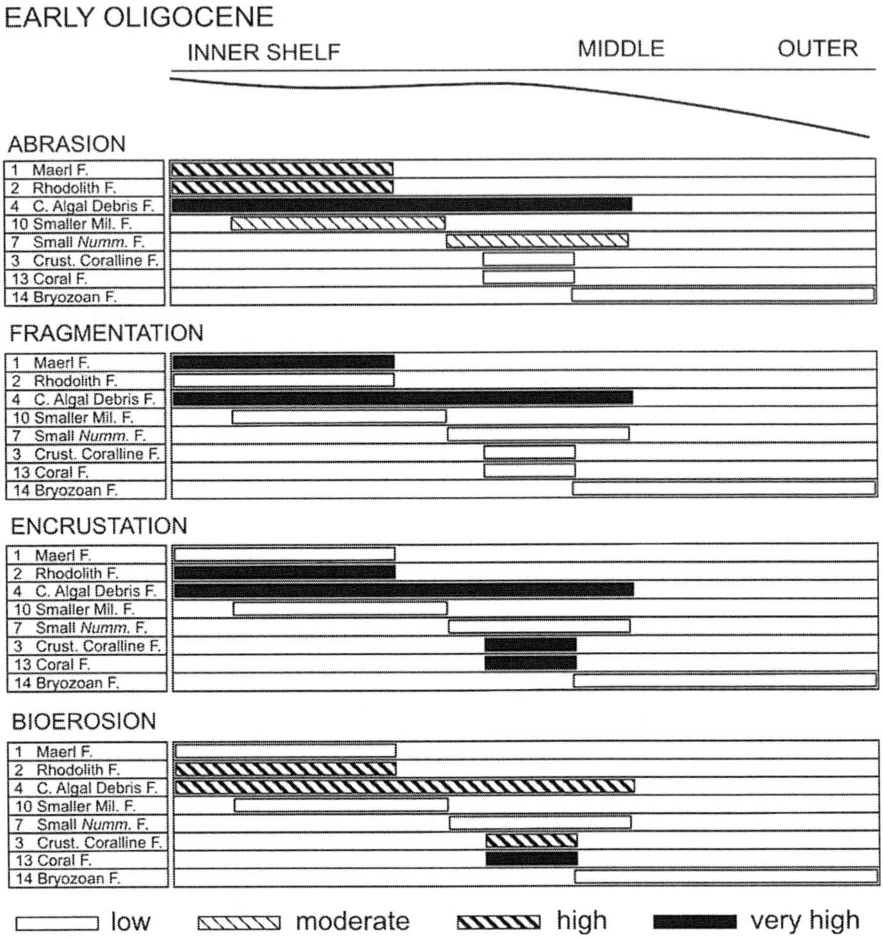

Fig. 19 Compilation of taphonomic processes for the Early Oligocene

can occur. Small *Nummulites* include moderate edge abrasion and fragmentation; encrustation and bioerosion are rare.

The Small *Nummulites* Facies occurs from the Middle Eocene to the Early Oligocene. It shows a dramatic shift in its distribution range from the outer middle shelf in the Middle Eocene to inner to middle shelf in the Late Eocene and Early Oligocene.

4.2.8 Orthophragminid Facies

This facies is dominated by various species of large, thin, disc- and saddle shaped orthophragminids along with coralline algal crusts. Larger and smaller benthic foraminifera (rotaliids and textulariids), bivalves and planktic foraminifera are subordinate. These components occur in rudstones with wacke- to packstone matrix. A horizontal orientation is usually present. The Orthophragminid Facies shows low values for all taphonomic features except for encrustation. The orthophragminids are often nested.

Orthophragminid occur from the Middle and Late Eocene after which these larger foraminifera disappear. This facies shifts its range from the middle and outer shelf in the Middle Eocene to the outer shelf in the Late Eocene.

4.2.9 *Orbitolites* Facies

This facies is dominated by the larger foraminifer *Orbitolites* along with small miliolid foraminifera. Peneroplid foraminifera, bivalves and gastropods are subordinated in rudstones with pack- to grainstone matrix. Graded bedding can be present. The *Orbitolites* Facies shows low values for all taphonomic features except for encrustation, despite the fact that they occur in shallow water settings. Encrustation can occur by coralline algae and bryozoans. This facies is restricted to the inner shelf of the Middle Eocene.

4.2.10 Smaller Miliolid Facies

Diverse small benthic miliolid foraminifera, peneroplids and alveolinid foraminifera dominate this facies along with subordinate textulariid foraminifera, *Sphaerogypsina*, bivalves, echinoderms, geniculate and non-geniculate coralline algae. The miliolid small benthic foraminifera are primarily quinqueloculine and triloculine forms. The fabric is represented by grainstones and packstones with moderately to well preserved components. No grading, sorting or orientation is present. Abrasion and fragmentation is moderate, with low values of encrustation and bioerosion. This facies expands its range from the Middle (inner shelf) to the Late Eocene (middle shelf).

4.2.11 Alveolinid Facies

This facies is dominated by alveolinids together with small miliolid benthic foraminifera, asterigerinid and nummulitid larger foraminifera. Small benthic rotaliid foraminifera, coralline algae, echinoderms are subordinated in pack- and grainstones. Both oriented and chaotic fabrics occur. Due to the rigid skeletons and high resistance of alveolinids, this facies shows low values for all taphonomic features except for abrasion. This facies shows a restricted range on the middle shelf of the Middle Eocene.

4.2.12 Acervulinid Facies

Acervulinid macroids formed by *Acervulina ogormani* and *A. linearis* and tubular aggregates of *A. multiformis* dominate in this association along with subordinate coralline algae, serpulids, homotrematid foraminifera as well as larger foraminifera and smaller miliolid benthic foraminifera. The acervulinids construct dense macroids which can show complex encrusting successions including various coralline algae, serpulids and other encrusting foraminifera within rudstones. The macroids can be up to 10 cm in diameter with laminar-encrusting growth-forms, and do not show grading, sorting or preferred orientations. Encrustation rates are correspondingly very high with moderate bioerosion. Fragmentation and abrasion is rare. The facies is restricted to the outer shelf of the Middle and Later Eocene.

4.2.13 Coral Facies

Corals and coralline algal crusts dominate this facies along with subordinate small *Nummulites*, bryozoans, thecideidean brachiopods and small benthic foraminifera. They occur in rudstones with wackestone to grainstone matrix and include both branching and encrusting corals as isolated colonies or in patches.

The aragonitic shelled corals are usually dissolved and often replaced by calcite. In many cases, corals are easily recognizable due to dense micritic muds which encase the specimens and fill in the space between the septa. In some cases dissolution is such that the corals are only recognizable as "ghost" structures. Corals are often at the core of complex encrustation successions being encrusted by coralline algae, foraminifera, unilaminate and multilaminate bryozoans and thecideidean brachiopods. These composite encrustation sequences are often heavily bioeroded. This facies is represented by low abrasion rates and moderate rates. It is most common on the middle shelf of the Late Eocene and Early Oligocene.

4.2.14 Bryozoan Facies

Both cheilostome and cyclostome bryozoans dominate in this facies along with smaller quantities of larger foraminifera, smaller benthic rotaliid foraminifera and coralline algae. The bryozoan growth forms are dominated by upright growing cylindrical and bilaminate forms. Bryozoans can be encrusted by coralline algae.

The bryozoans occur in rudstones with wackestone matrix as well as marly packstones. The components lie nearly parallel to bedding planes. The Bryozoan Facies represents a low energy system and includes well preserved components with moderate encrustation rates and sparse abrasion, fragmentation and bioerosion. This facies expands from the outer shelf in the Late Eocene to include the middle shelf in the Early Oligocene.

4.3 Taphonomic Processes in Paleogene Carbonates of the Study Area

The degree of abrasion, fragmentation, encrustation and bioerosion within each facies was qualitatively assessed (designated as low, moderate, high or very high;

Fig. 12) in the study area. These taphonomic features were then mapped with respect to facies distribution along a shelf gradient (Nebelsick et al. 2003, 2005).

The assessment of taphonomic features necessarily concentrates on the name-giving components. Taphonomic processes will obviously differentially affect the various components within the facies. Taphonomic gradients will also exist along shelf gradients, especially for those facies which show a wide distribution. The interpretation of taphonomic features is problematic as different processes can lead to similar features. Nonetheless, the first order comparison does show some general trends.

Abrasion represents destruction of surface characters and the rounding of particles. It can be caused by grain agitation during transport processes and/or biological activity. Abrasion is very high in the Coralline Algal Debris Facies and high in other coralline algal facies. It is low in the Orthophragminid, *Orbitolites*, Acervulinid, Coral and Byrozoan Facies. Both the Larger and Small *Nummulites* Facies show moderate values.

Fragmentation leads to diminution of components and is recognized by fragmented grains with sharp edges and abrupt termination of skeletal characters. As in abrasion, fragmentation can be caused by both grain agitation and biological activity. Fragmentation generally shows similar distributions to abrasion in its distribution among major facies types with the highest values in the Maerl and Coralline Algae Debris Facies, moderate values in the Larger and Small *Nummulites* Facies as well as the Rhodolith, Smaller Miliolid and Coral Facies. Fragmentation is least in some larger foraminiferal facies and the Bryozoans and the Crustose Coralline Algal Facies.

Encrustation is easily recognized in thin section by bio-immuration of components by encrusting organisms. Encrustation leads to an increase of (aggregate) component size and can stabilize components and sediment surfaces. Encrustation sequences can subsequently be affected by other taphonomic features especially bioerosion. Encrustation is especially high in the Rhodolith, Crustose Coralline Algal as well as Acervulinid and Coral Facies. It is rare in the Coralline Algal Debris, Small *Nummulites* and Acervulinid Facies.

Bioerosion can also be readily recognized in thin section if it extends into the biogenic substrates and can be caused by an array of micro- and macroborers. Bioerosion is very high in the Coral Facies and common in the Rhodolith and Crustose Coralline Algal Facies. The other facies show low or moderate values.

Other taphonomic processes which occur on or near the sediment–water surface include disarticulation which can be easily recognized in isolated genicula of geniculate coralline algae, disassociated echinoid spines, isolated elements of cellariform bryozoans and most obviously disjunct bivalve shells.

Post-depositional taphonomic features include dissolution of aragonitic components (see discussion above). In the study area, this affects dasycladalean and halimedacean algae, scleractinian corals, aragonitic shelled bivalves and gastropods. The presence of these components can often be easily recognized due the fact that they have been encased by a fine micritic matrix. High Mg-calcite skeletons are invariable transformed to low Mg-calcite in coralline algae and echinoderms. The later can be accompanied by syntaxial cements if enough pore space was present into which the cements could expand. Another post-depositional feature is contact breakage due to compaction. This is common in those facies in which the component packing is very dense including the larger *Nummulites* and the Orthophragminid Facies. Pressure solution is also a feature seen at contact points between components in these facies.

5 Discussion of the Distribution of Taphonomic Features Among and Between Time Units

Abrasion magnitude varies in time and space (Fig. 13). In the Middle Eocene, abrasion occurs in the middle shelf in the Larger *Nummulites* Facies where it may represent shoal settings. It is also very high in the Coralline Algal Debris Facies. In the Late Eocene and Early Oligocene, abrasion dominates in shallow water facies and is rare in the outer middle and outer shelf. The fact that abrasion does not follow a general depth gradient may reflect the fact that this taphonomic features can be caused not only by water movement and grain agitation, but also by surface grazing in deeper water.

Fragmentation shows the most clear cut spatial distribution with similar trends in all three time units (Fig. 14). Highest values are found in shallow water inner shelf settings. Moderate values are also present in shallow water and extend to middle shelf environments. Deeper water outer shelf facies show low rates of fragmentation. This may, in fact, reflect the simple correlation between water depth, wave base and water movement. The well preserved Middle Eocene *Orbitolites* Facies is the exception as it represents sheltered shallow water environment. The role of predation and bioturbation in the fragmentation of the constituent biogenic components is not well enough known in the studied facies to postulate on its influence.

Encrustation does not show as clear a distribution as the other taphonomic features (Fig. 15). Thus middle to outer shelf facies show the highest rates in the Middle Eocene. In the Late Eocene and Early Oligocene, higher values for encrustation are found in both inner and middle shelf setting. The inner shelf environments in the two younger time units, in fact, show three different levels of encrustation values. Encrustation seems to be least affected by different environmental parameters such as water energy levels or depth. The fact that components become encrusted seems to be more related to the size and encrusting potential of the specific components at hand. This is especially important for the encrusting foraminifera (acervulinids in the Acervulinid Facies) and calcareous algae which can encrust surfaces and construct self-encrusting macroids or rhodoliths (Rhodolith Facies and Peyssonneliacean Facies). Corals can also be encrusted during life as well as after death and offer various substrates for a number of different encrusters in both exposed and cryptic microhabitats. More detailed analysis of growth patterns and encrustation strategies are needed to interpret different encrustation types.

Bioerosion (Fig. 16) varies between Middle and Late Eocene-Early Oligocene. In the Middle Eocene bioerosion is highest in middle and outer shelf settings but in the Late Eocene and Early Oligocene it is high in the inner and middle shelf. Overall however, the highest values are found in the middle shelf (with the Coral Facies). This may reflect an ecological related to water depth. Of importance for the presence and preservation of bioerosion is the initial size and skeletal architecture of the potential biogenic substrates. Relatively large corals and rhodoliths are thus more conducive to the harbouring and preserving endolithic organisms than, for example, smaller foraminifers or branching bryozoans.

There is thus a general trend for those taphonomic features which can be related to physical processes such as transport and grain agitation to occur preferentially in shallower setting than in deeper setting. This is due to wave base and associated water movement. This is not always the case however as some shallow water facies (for example the *Orbitolites* Facies) represent quiet water settings and some abrasion and fragmentation may be biogenic in origin and may be affected by substate availability.

Of primary interest is the cause for differences in the distribution of taphonomic features between the time units. The main change in this respect occurs between the Middle Eocene on the one hand and the Late Eocene and Early Oligocene on the other. Middle Eocene facies (Fig. 17) generally show lower taphonomic values than the Late Eocene (Fig. 18) and Early Oligocene (Fig. 19). The extinction events that dramatically influence the occurrence and distribution of facies (extinction of larger *Nummulites* and most of the alveolinids at the Middle Eocene/Upper Eocene boundary and the extinction of orthophragminids at the Eocene/Oligocene boundary) have little effect on the distribution of taphonomic features. This is due to the fact that the Larger *Nummulites*, the Alveolinid and the Orthophragminid Facies show low to moderate values for all taphonomic features. More dramatic as far as taphonomy is concerned is the appearance in the study area of the Coral, Rhodolith and Maerl Facies in the Late Eocene. These facies show very high values of encrustation and bioerosion (Coral Facies) and abrasion and fragmentation (Rhodolith and Maerl Facies) and thus dominate the distribution of taphonomic features.

6 Conclusions

The long tradition of microfacies studies on carbonate rocks have resulted in vast collection of thin sections. These collections represent a largely untapped source of taphonomic data. The thin sections are typically all of the same size (depending on the specific tradition of the thin section laboratories – typically 5 × 5 cm in Central Europe for example), of uniform thickness, usually vertically orientated to the bedding plane, and taken at regular distances within stratigraphic sections. Furthermore, broad facies assignments are usually already known. Microtaphofacies analysis can potentially add new insight into paleoecological interpretations. More studies are not only needed on fossil carbonate successions, but also using an actualistic approach on modern components in different carbonate settings. The direct taphonomic analysis of particulate grains and the same material embedded in resin and cut into thin sections will allow two- and three dimensional taphonomic attributes to be compared.

This study represents a rather coarse, semi-quantitative treatment of taphonomic features at broad time scales (sub-stages) over a wide geographic range. Further studies could include more precise evaluation of taphonomic attributes and extend the analyses to further time units and a wider geographic area. Closer analyses can include detailed bed-by-bed quantification of specific taphonomic features (e.g. Reolid and Gaillard 2007) or document disparities among successive sequence stratigraphic units (e.g. Brachert et al. 1998). Inclusion of broader geographic areas would reveal taphonomic patterns across latitudinal and temperature gradients from more tropical (in this case further southwest in the tropical Tethys) to more temperate settings as well as longitudinal gradients between the study area and both the Eastern Tethyan and Caribbean provinces.

The quantification of taphonomic features such as encrustation and bioerosion can be used to follow not only the variations in the intensities of these features, but also the evolution of encrusting and bioeroding strategies (e.g. Fagerstrom 1987, 1991; Vogel 1993; Wood 1998, 1999, this volume). This can be extended to evaluate co-evolutionary scenarios between substrates and encrusters and bioeroders through time (e.g. Steneck 1983, 1986).

The carbonates which form the basis for this paper cross important stratigraphic boundaries marked by extinction events. Of great potential is the study of how taphonomic features change across other seminal boundaries including (1) major (and minor) extinction events, (2) the change in climatic (greenhouse/icehouse) regimes, and (3) marine geochemical turnovers (calcite/aragonite seas – see Palmer et al. 1988). As has been often appreciated (among taphonomists at least) taphonomy is an essential factor in determining the presence (and absence) of key biotic components on which biotic turnover is essentially measured. More detailed studies on the taphonomy of sedimentary sequences (including carbonate successions using the microtaphofacies approach presented here) can thus help to more completely understand key events in Earth history.

References

Acker, K. L., & Risk, M. J. (1985). Substrate destruction and sediment production by the boring sponge *Cliona caribbaea* on Grand Cayman Island. *Journal of Sedimentary Petrology, 55*, 705–711.

Adey, W. H. (1978). Algal ridges of the Caribbean Sea and West Indies. *Phycology, 17*, 361–367.

Adey, W. H., & Macintyre, I. G. (1973). Crustose coralline algae: A re-evaluation in the geological sciences. *Geological Society of America Bulletin, 84*, 883–904.

Adey, W. H., & McKibbin, D. L. (1970). Studies on the maerl species *Phymatolithon calcareum* (Pallas) nov. comb. and *Lithothamnium corallioides* Crouan in the Ria de Vigo. *Botanica Marina, 13*, 100–106.

Aigner, T. (1982). Event stratification in nummulite accumulations and in shell beds from the Eocene of Egypt. In G. Einsele & A. Seilacher (Eds.), *Cyclic and event stratification*. Berlin: Springer.

Aigner, T. (1983). Facies and origin of nummulitic buildups: An example from the Giza Pyramids Plateau (Middle Eocene, Egypt). *Neues Jahrbuch für Geologie und Paläontologie – Abhandlungen*, 347–368.

Aigner, T. (1985). Biofabrics as dynamic indicators in nummulite accumulations. *Journal of Sedimentary Research, 55*, 131–134.

Alexandersson, T. (1972). Micritization of carbonate particles: Process of precipitation and dissolution in modern shallow-water sediments. *Bulletin of Geological Institutions of the University of Uppsala N S, 3*, 201–236.

Alexandersson, T. (1974). Carbonate cementation in coralline algal nodules in the Skagerrak, North Sea: Biochemical precipitation in undersaturated waters. *Journal of Sedimentary Petrology, 44*, 7–26.

Alexandersson, T. (1977). Carbonate cementation in recent coralline algal constructions. In E. Flügel (Ed.), *Fossil algae, recent results and developments*. Berlin: Springer.

Allison, P. A., & Briggs, D. E. G. (Eds.). (1991). *Taphonomy: Releasing the data of the fossil record*. New York: Plenum.

Aronson, R. B. (Ed.). (2007). *Geological approaches to coral reef ecology*. New York: Springer.

Ausich, W. I. (2001). Echinoderm taphonomy. In M. Jangoux & J. M. Lawrence (Eds.), *Echinoderm Studies 6:171–227*. Lisse: Balkema.

Bak, R. P. M. (1990). Patterns of echinoid bioerosion in two Pacific coral reef lagoons. *Marine Ecology Progress Series, 66*, 267–272.

Bak, R. P. M. (1994). Sea urchin bioerosion in coral reefs: Place in the carbonate budget and relevant variables. *Coral Reefs, 13*, 99–103.

Barbosa, S. S., Byrne, M., & Kelaher, B. P. (2008). Bioerosion caused by foraging of the tropical chiton *Acanthopleura gemmata* at One Tree Reef, southern Great Barrier Reef. *Coral Reefs, 27*, 635–639.

Bassi, D. (1998). Coralline algal facies and their palaeoenvironments in the Late Eocene of northern Italy (Calcare di Nago, Trento). *Facies, 39*, 179–202.

Bassi, D. (2005). Larger foraminiferal and coralline algal facies in an Upper Eocene storm-influenced, shallow water carbonate platform (Colli Berici, north-eastern Italy). *Palaeogeography, Palaeoclimatology, Palaeoecology, 226*, 1–35.

Bassi, D., Nebelsick, J. H., Checconi, A., Hohenegger, J., & Iryu, Y. (2009). Present-day and fossil rhodolith pavements compared: their potential for analysing shallow-water carbonate deposits. *Sedimentary Geology, 214*, 74–84.

Basso, D. (1998). Deep rhodolith distribution in the Pontian Islands, Italy: a model for the paleoecology of a temperate sea. *Palaeogeography, Palaeoclimatology, and Palaeocology, 137*, 173–187.

Bathurst, R. G. C. (1964). The replacement of aragonite by calcite in the molluscan shell wall. In J. Imbrie & N. D. Newell (Eds.), *Approaches to paleoecology*. New York: Wiley.

Bathurst, R. G. (1966). Boring algae, micritic envelopes and lithification of molluscan biosparites. *Geological Journal, 5*, 15–32.

Beavington-Penney, S. J. (2004). Analysis of the effects of abrasion on the test of *Palaeonummulites venosus*: Implications for the origins of nummulithoclastic sediments. *Palaios, 19*, 143–155.

Beavington-Penney, S. J., & Racey, A. (2004). Ecology of extant nummulitids and other larger benthic foraminifera: Applications in palaeoenvironmental analysis. *Earth Science Reviews, 67*, 219–265.

Behrensmeyer, A. K., Kidwell, S. M., Gastaldo, & R. A. (2000). Taphonomy and paleobiology. In D. H. Erwin & S. W. Wing (Eds.), *Deep time. Paleobiology's perspective, Paleobiology Supplement 26* (pp. 103–147).

Bellwood, D. R., & Choat, J. H. (1991). A functional analysis of grazing parrot fishes (family Scaridae): the ecological implications. In P. F. Sale (Ed.), *The ecology of fishes on coral reefs*. San Diego: Academic.

Best, M. M. R., & Kidwell, S. (2000). Bivalve taphonomy in tropical mixed siliciclastic settings; I, Environmental variation in shell condition. *Paleobiology, 26*, 80–102.

Bordehore, C., Ramos-Esplá, A., & Riosmena-Rodríguez, R. (2003). Comparative study of two maerl beds with different otter trawling history, southeast Iberian Peninsula. *Acquatic Conservation Marine and Freshwater Ecosystems, 13*, 43–54.

Bosellini, A., & Ginsburg, R. N. (1971). Form and internal structure of Recent algal nodules (rhodolites) from Bermuda. *Journal of Geology, 79*, 669–682.

Bosence, D. W. J. (1976). Ecological studies on two unattached coralline algae from western Ireland. *Palaeontology, 19*, 365–395.

Bosence, D. W. J. (1980). Sedimentary facies, production rates and facies models for Recent coralline algal gravels, Co. Galway, Ireland. *Geological Journal, 15*, 91–111.

Bosence, D. W. J. (1983a). Coralline algal reef framework. *Journal of the Geological Society of London, 140*, 365–376.

Bosence, D. W. J. (1983b). The occurrence and ecology of recent rhodoliths – A review. In T. M. Peryt (Ed.), *Coated Grains*. Berlin: Springer.

Bosence, D. W. J. (1983c). Description and classification of rhodoliths (rhodoids, rhodolites). In T. M. Peryt (Ed.), *Coated grains*. Berlin: Springer.

Bosence, D. W. J. (1984). Construction and preservation of two modern coralline algal reefs, St Croix, Caribbean. *Paleontology, 27*, 549–574.

Bosence, D. W. J. (1985a). The "Coralligène" of the Mediterranean – A recent analogue for the Tertiary coralline algal limestone. In D. F. Toomey & M. H. Nitecki (Eds.), *Palaeoalgology: Contemporary research and applications*. Berlin: Springer.

Bosence, D. W. J. (1985b). The morphology and ecology of a mound-building coralline alga (*Neogoniolithon strictum*) from the Florida Keys. *Palaeontology, 28*, 189–206.

Bosence, D. W. J. (1991). Coralline algae: Mineralization, taxonomy and palaeoecology. In R. Riding (Ed.), *Calcareous algae and stromatolites*. Berlin: Springer.

Bosence, D. W. J., & Pedley, H. M. (1982). Sedimentology and palaeoecology of a Miocene coralline algal biostrome from the Maltese islands. *Palaeogeography, Palaeoclimatology, Palaeoecology, 38*, 9–43.

Brachert, T. C., & Dullo, W. C. (2000). Shallow burial diagenesis of skeletal carbonates: selective loss of aragonite shell material (Miocene to Recent, Queensland Plateau and Queensland Trough, NE Australia) – Implications for shallow cool-water carbonates. *Sedimentary Geology, 136*, 169–187.

Brachert, T. C., Betzler, C., Braga, J. C., & Martìn, J. M. (1998). Microtaphofacies of a warm-temperate carbonate ramp (uppermost Tortonian/lowermost Messinian, Southern Spain). *Palaios, 13*, 459–475.

Braga, J. C., & Aguirre, J. (2001). Coralline algal assemblages in upper Neogene reef and temperate carbonates in Southern Spain. *Palaeogeography, Palaeoclimatology, Palaeoecology, 175*, 27–41.

Braga, J. C., & Martìn, J. M. (1988). Neogene coralline-algal growth-forms and their palaeoenvironments in the Almanzora River Valley (Almeria, S.E. Spain). *Palaeogeography, Palaeoclimatology, Palaeoecology, 67*, 285–303.

Brett, C. E., & Baird, G. C. (1986). Comparative taphonomy: A key to paleoenvironmental interpretation based on fossil preservation. *Palaios, 1*, 207–227.

Brett, C. E., Moffat, H. A., & Taylor, W. L. (1997). Echinoderm taphonomy, taphofacies, and Lagerstätten. In J. A. Waters & C. G. Maples (Eds.), *Geobiology of Echinoderms. Paleontological Society Papers 3* (pp. 147–190).

Brock, R. E. (1979). An experimental study on the effects of grazing by parrotfishes and the role of refuges in the benthic community structure. *Marine Biology, 51*, 381–388.

Bromley, R. G. (1978). Bioerosion of Bermuda reefs. *Palaeogeography, Palaeoclimatology, Palaeocology, 23*, 169–197.

Budd, D. A. (1988). Aragonite to calcite transformation during freshwater diagenesis of carbonates: Insights from pore-water chemistry. *Geological Society of America Bulletin, 100*, 1260–1270.

Budd, D. A., & Hiatt, E. E. (1993). Mineralogical stabilization of high-Magnesium calcite: Geochemical evidence for intracrystal recrystallization within Holocene porcellaneous foraminifera. *Journal of Sedimentary Petrology, 63*, 261–274.

Budd, D. A., & Perkins, R. D. (1980). Bathymetric zonation and paleoecological significance of microborings in Puerto Rican shelf and slope sediments. *Journal of Sedimentary Petrology, 50*, 881–994.

Cabioch, J. (1969). Les fonds de maerl de la baie de Morlaix et leur peuplement végétal. *Cahiers de Biologie Marine, 10*, 131–161.

Callender, W. R., Powell, E. N., & Staff, G. M. (1994). Taphonomic rates of molluscan shells placed in autochthonous assemblages on the Louisiana continental slope. *Palaios, 9*, 60–73.

Canfield, D. E., & Raiswell, R. (1991). Carbonate precipitation and dissolution - its relevance to fossil preservation. In P. A. Allison & D. E. G. Briggs (Eds.), *Taphonomy – Releasing the data locked in the fossil record*. New York: Plenum.

Cate, A. S., & Evans, I. (1994). Taphonomic significance of the biomechanical fragmentation of live molluscan shell material by a bottom-feeding fish (*Pogonias cromis*) in Texas coastal bays. *Palaios, 9*, 254–274.

Chave, K. E. (1964). Skeletal durability and preservation. In J. Imbrie & N. D. Newell (Eds.), *Approaches to Paleoecology*. New York: Wiley.

Checconi, A., Bassi, D., Passeri, L., & Rettori, R. (2007). Coralline red algal assemblage from the Middle Pliocene shallow-water temperate carbonates of the Monte Cetona, Northern Appenines, Italy. *Facies, 53*, 57–66.

Cottey, T. L., & Hallock, P. (1988). Test surface degradation in *Archaias angulatus*. *Journal of Foraminiferal Research, 18*, 187–202.

Cummins, H., Powell, E. N., Stanton, R. J., Jr., & Staff, G. (1986). The rate of taphonomic loss in modern benthic habitats: How much of the potentially preservable community is preserved? *Palaeogeography, Palaeoclimatology, Palaeoecology, 52*, 291–320.

Davies, G. R. (1970). Carbonate bank sedimentation, eastern Shark Bay, Western Australia. *American Association of Petroleum Geologists, Memoirs, 13*, 85–168.

Di Geronimo, I., Di Geronimo, R., Rosso, A., & Sanfilippo, R. (2002). Structural and taphonomic analysis of a columnar coralline algal build-up from SE Sicily. *Geobios, 35*(Suppl 1), 86–95.

Donovan, S. K. (1991a). *The process of fossilization*. London: Belhaven.

Donovan, S. K. (1991b). The taphonomy of echinoderms; calcareous multi-element skeletons in the marine environment. In S. K. Donovan (Ed.), *The process of fossilization*. London: Belhaven.

Dullo, W. C. (1983). Fossildiagenese im miozänen Leithakalk der Paratethys von Österreich: Ein Beispiel für Faunenverschiebungen durch Diageneseunterschiede. *Facies, 8*, 1–112.

Eichenseer, H., & Luterbacher, H.-P. (1992). The marine Paleogene of the Tremp region (NE Spain): Depositional sequences, facies history, biostratigraphy and controlling factors. *Facies, 27*, 119–152.

Engel, W. (1970). Die Nummuliten-Breccien im Flyschbecken von Ajdovcina in Slowenien als Beispiel karbonatischer Turbidite. *Verhandlungen der Geologische Bundesanstalt, Wien, 4*, 570–582.

Fagerstrom, J. A. (1987). *The evolution of reef communities*. New York: Wiley Interscience.

Fagerstrom, J. A. (1991). Reef-building guilds and checklist for determining guild membership. *Coral Reefs, 10*, 47–52.

Figueiredo de, O. M. A. (1997). Colonization and growth of crustose coralline algae in Abrolhos, Brazil. In H. A. Lessios & I. G. Macintyre (Eds.), *Proceedings of the 8th International Coral Reef Symposium 1* (pp. 689–694).

Flessa, K. W. (1993). Time-averaging and temporal resolution in Recent marine shelly faunas. In S. M. Kidwell & A. K. Behrensmeyer (Eds.), *Taphonomic approaches to time resolution in fossil assemblages. Short Courses in Paleontology 6* (pp. 9–33). Paleontological Society, Knoxville.

Flügel, E. (1982). *Microfacies analysis of limestones*. Berlin: Springer.

Flügel, E. (2004). *Microfacies of carbonate rocks: Analysis interpretation and application*. Berlin: Springer.

Foster, M. S. (2001). Rhodoliths: Between rocks and soft places. *Journal of Phycology, 87*, 659–667.

Foster, M. S., Riosmena-Rodriguez, R., Steller, D. S., & Woelkerling, W. J. (1997). Living rhodolith beds in the Gulf of California and their implications for paleoenvironmental interpretation. In M. E. Johnson & J. Ledesma-Vásquez (Eds.), *Pliocene carbonates and related facies flaking the Gulf of California, Baja California. Geological Society of America, Special Papers, 318* (pp. 127–139).

Freiwald, A. (1993). Coralline algal maerl frameworks – Islands within the phaeophytic kelp belt. *Facies, 29*, 133–148.

Freiwald, A. (1995). Sedimentological and biological aspects in the formation of branched rhodoliths in northern Norway. *Beiträge Zur Paläontologie, 20*, 7–19.

Freiwald, A., & Wilson, J. B. (1998). Taphonomy of modern deep, cold-temperate water coral reefs. *Historical Biology, 13*, 37–52.

Freiwald, A., Henrich, R., Schäfer, P., & Willkomm, H. (1991). The significance of high Boreal to subartic maerl deposits in northern Norway to reconstruct Holocene climatic changes and sea level oscillations. *Facies, 25*, 315–340.

Futterer, D. K. (1974). Significance of the boring sponge *Cliona* for the origin of fine grained material of carbonate sediments. *Journal of Sedimentary Petrology, 44*, 79–84.

Gherardi, D. F. M., & Bosence, D. W. J. (1999). Modelling of the ecological succession of encrusting organisms in Recent coralline-algal frameworks from Atol das Rocas, Brazil. *Palaios, 14*, 145–158.

Gischler, E., & Pisera, A. (1999). Shallow water rhodoliths from Belize reefs. *Neues Jahrbuch für Geologie und Paläontologie – Abhandlungen, 214*, 71–93.

Glaub, I., Golubic, S., Gektidis, M., Radtke, G., & Vogel, K. (2007). Microborings and microbial endoliths: geological implications. In W. Miller III (Ed.), *Trace fossils: Concepts, problems, prospects*. Amsterdam: Elsevier.

Glynn, P. W. (1997). Bioerosion and coral-reef growth: A dynamic balance. In C. Birkeland (Ed.), *Life and death of coral reefs*. New York: Chapman & Hall.

Golubic, S. (1969). Distribution, taxonomy, and boring patterns of marine endolithic algae. *American Zoologist, 9*, 747–751.

Golubic, S. (1990). Shell boring microorganisms. In A. Boucot (Ed.), *The evolutionary paleobiology of behavior and coevolution*. Amsterdam: Elsevier.

Golubic, S., Perkins, R. D., & Lukas, K. J. (1975). Boring micro-organisms and microborings in carbonate substrates. In R. W. Frey (Ed.), *The study of trace fossils*. New York: Springer.

Goreau, T. F., & Hartman, W. D. (1963). Boring sponges as controlling factors in the formation and maintenance of coral reefs. In R.F. Sognnaes (Ed.), Mechanisms of hard tissue destruction. *American Association for the Advancement of Science, 75*, 25–54.

Greenstein, B. J. (2007). Taphonomy: Detecting critical events in fossil reef coral assemblages. In R. B. Aronson (Ed.), *Geological approaches to coral reef ecology*. New York: Springer.

Greenstein, B. J., & Moffat, H. A. (1996). Comparative taphonomy of modern and Pleistocene corals, San Salvador, Bahamas. *Palaios, 11*, 57–63.

Greenstein, B. J., & Pandolfi, J. M. (2003). Taphonomic alteration of reef corals: Effects of reef environment and coral growth form II: The Florida Keys. *Palaios, 18*, 495–509.

Harzhauser, M., & Piller, W. E. (2007). Benchmark data of a changing sea – Palaeogeography, Palaeobiogeography and events in the Central Paratethys during the Miocene. *Palaeogeography, Palaeoclimatology, Palaeoecology, 253*, 8–31.

Hetzinger, J., Halfar, J., Riegl, B., & Godinez-Orta, L. (2006). Sedimentology and acoustic mapping of modern rhodolith facies on a non-tropical carbonate shelf (Gulf of California, Mexico). *Journal of Sedimentary Research, 6*, 670–682.

Hillis, D. J., & Jones, B. (2000). Peyssonnelid rhodoliths from the Late Pleistocene Ironshore Formation, Grand Caymanm British West Indies. *Palaios, 15*, 212–224.

Hohenegger, H. (2004). Depth coenoclines and environmental considerations of Western Pacific larger foraminifera. *Journal of Sedimentary Research, 34*, 9–33.

Hohenegger, H., & Yordanovea, E. K. (2001). Depth-transport functions and erosions-deposition diagrams as indicator of slope inclination and time-averaged traction forces: applications in tropical reef environments. *Sedimentology, 48*, 1025–1046.

Hutchings, P. A. (1986). Biological destruction of coral reefs. *Coral Reefs, 4*, 239–252.

Ivany, L. C., Patterson, W. P., & Lohmann, K. C. (2000). Cooler winters as a possible cause of mass extinctions at the Eocene/Oligocene boundary. *Nature, 407*, 887–890.

Iryu, Y. (1997). Pleistocene fore-reef rhodoliths from the Ryukyu Islands, southeastern Japan. In H. A. Lessios & I. G. Macintyre (Eds.), *Proceedings of the 8th International Coral Reef Symposium 1*. Balboa: Smithsonian Tropical Research Institute.

Johnson, J. H., Knight, M. A., & Puesschel, C. M. (1997). Antifouling effects of epithallial shedding in three crustose coralline algae (Rhodophyta, Corallinales) on a coral reef. *Journal of Experimental Marine Biology and Ecology, 213*, 281–293.

Jorry, S. J., Hasler, C. A., & Davaud, E. (2006). Hydrodynamic behaviour of *Nummulites*: implications for depositional models. *Facies, 52*, 221–235.

Keats, D. W., & Maneveldt, G. (1994). *Leptophytum foveatum* Chamberlain & Keats (Rhodophyta, Corallinales) retaliates against competitive overgrowth by other encrusting algae. *Journal of Experimental Marine Biology and Ecology, 175*, 243–251.

Keats, D. W., Matthews, I., & Maneveldt, G. (1994). Competitive relationships in a guild of crustose algae in the eulittoral zone, Cape Province, South Africa. *South African Journal of Botany, 60*, 108–113.

Kidwell, S. M., & Bosence, D. W. J. (1991). Taphonomy and time-averaging of marine shelly faunas. In P. A. Allison & D. E. G. Briggs (Eds.), *Taphonomy: Releasing the data locked in the fossil record*. New York: Plenum.

Kidwell, S. M., & Flessa, K. W. (1996). The quality of the fossil record: Populations, species and communities. *Annual Review of Earth and Planetary Sciences, 24*, 433–464.

Kiene, W., Radtke, G., Gektidis, M., Golubic, S., & Vogel, K. (1995). Factors controlling the distribution of microborers in Bahamian reef environments. *Facies, 32*, 176–188.

Kleemann, K. H. (1994). Associations of coral and boring bivalves since the Late Cretaceous. *Facies, 31*, 131–140.

Kloos, D. P. (1982). Destruction of tests of the foraminifer *Sorites orbiculius* by endolithic microorganisms in a lagoon on Curacao (Netherlands Antilles). *Geologie en Mijnbouw, 61*, 201–205.

Kobluk, D. R., & Risk, M. J. (1977). Micritization and carbonate-grain binding by endolithic algae. *Bulletin of the American Association of Petroleum Geologists, 61*, 1069–1082.

Konar, B., Riosmena-Rodriguez, R., & Iken, K. (2006). Rhodolith bed: A newly discovered habitat in the Northern Pacific. *Botanica Marina, 2006*, 355–359.

Kondo, Y. (1995a). Density and relative abundance of microspheric and megalospheric forms of *Nummulites* as sedimentologic and taphonomic indicators. *Transactions and Proceedings of the Palaeontological Society of Japan, 177*, 59–64.

Kondo, Y. (1995b). Paleocurrent reconstruction, using imbricated tests of *Nummulites* in the Eocene of Hahajima, Ogasawara Islands. *Journal of the Geological Society of Japan, 101*, 228–234.

Kowalewski, M., Flessa, K. W., & Aggen, J. A. (1994). Taphofacies analysis of recent shelly cheniers (beach ridges), northeastern Baja California, Mexico. *Facies, 31*, 209–242.

Lawrence, J. M. (1975). On the relationships between marine plants and sea urchins. *Oceanography and Marine Biology: An Annual Review, 13*, 231–286.

Lescinsky, H. L., Edinger, E., & Risk, M. J. (2002). Mollusc shell encrustation and bioerosion rates in a modern epeiric sea: Taphonomy experiments in the Java Sea, Indonesia. *Palaios, 17*, 171–191.

Lipps, J. H. (1988). Predation on foraminifera by coral reef fish: Taphonomic and evolutionary implications. *Palaios, 3*, 315–326.

Littler, M. M., & Littler, D. S. (1995). Impact of CLOD pathogen on pacific coral reefs. *Science, 267*, 1356–1360.

Littler, M. M., & Littler, D. S. (1997). Disease-induced mass mortality of crustose coralline algae on coral reefs provides rationale for the conservation of herbivorous fish stocks. In H. A. Lessios & I. G. Macintyre (Eds.), *Proceedings of the 8th International Coral Reefs Symposium 1*. Balboa: Smithsonian Tropical Research Institute.

Littler, M. M., Littler, D. S., & Hanisak, M. (1991). Deep-water rhodolith distribution, productivity, and growth history at sites of formation and subsequent degradation. *Journal of Experimental Marine Biology and Ecology, 150*, 163–182.

Marrak, E. C. (1999). The relationship between water motion and living rhodolith beds in the southwestern Gulf of California, Mexico. *Palaios, 14*, 159–171.

Martin, R. E. (1999). *Taphonomy. A process approach*. Cambridge: Cambridge University Press.

Martin, R. E., & Liddel, W. D. (1991). Taphonomy of foraminifera in modern carbonate environments. In S. K. Donovan (Ed.), *The process of fossilization*. London: Belhaven Press.
Martindale, W. (1992). Calcified epibionts as palaeoecological tools: examples from the recent and Pleistocene reefs of Barbados. *Coral Reefs, 11*, 167–177.
Matteucci, R., & Pignatti, J. S. (1988). The taphonomy of *Nummulites*. In E. Robba (Ed.), *Proceedings of the 4th Symposium on ecology and paleoecology of Benthic Communities, Sorrento*. Turin: Museo Regionale di Scienze Naturali di Torino.
McKinney, F. K. (1995). Taphonomic effects and preserved overgrowth relationships among encrusting marine organisms. *Palaios, 10*, 279–282.
Meyer, D. M., Bries, J. M., Greenstein, B. J., & Debrot, A. O. (2003). Preservation of in situ reef framework in regions of low hurricane frequency: Pleistocene of Curaçao and Bonaire, southern Caribbean. *Lethaia, 36*, 273–286.
Morse, A. N. C., & Morse, D. E. (1984). Recruitment and metamorphosis of *Haliotis* larvae induced by molecules uniquely available at the surfaces of crustose red algae. *Journal of Experimental Marine Biology and Ecology, 75*, 191–215.
Murray, J. W. (1989). Syndepositional dissolution of calcareous foraminifera in modern shallow-water sediments. *Marine Micropaleontology, 15*, 117–121.
Nebelsick, J. H. (2004). Taphonomy of echinoderms: a review. In T. H. Heinzeller & J. H. Nebelsick (Eds.), *Echinoderms München. Proceedings of the 11th international echinoderm meeting*. Rotterdam: Taylor and Francis.
Nebelsick, J. H., & Bassi, D. (2000). Diversity, growth forms and taphonomy: key factors controlling the fabric of coralline algae dominated shelf carbonates. In E. Insalaco, P. W. Skelton, & T. J. Palmer (Eds.), *Carbonate platform systems: components and interactions. Geological Society London, Special Publications 178* (pp. 89–107).
Nebelsick, J. H., Rasser, M., & Bassi, D. (2003). The marine Eocene-Oligocene transition as recorded in shallow water, circumalpine carbonates. In D. R. Prothero, L. C. Ivany, & E. Nesbitt (Eds.), *Greenhouse to icehouse: The marine eocene-oligocene transition*. New York: Columbia University Press.
Nebelsick, J. H., Rasser, M. W., & Bassi, D. (2005). Facies dynamics in Eocene to Oligocene circumalpine carbonates. *Facies, 51*, 197–216.
Nelson, C. S., Hyden, F. M., Keane, S. L. M., Leask, W. A., & Gordon, D. P. (1988). Application of bryozoan zoarial growth-form studies in facies analysis of non-tropical carbonate deposits in New Zealand. *Sedimentary Geology, 60*, 301–322.
Neugebauer, J. (1978). Micritization of crinoids by diagenetic solution. *Sedimentology, 25*, 267–283.
Neumann, A. C. (1966). Observations on coastal erosion in Bermuda and measurement of the boring rate of the sponge, *Cliona lampa*. *Limnology and Oceanography, 11*, 92–108.
Norris, R. D. (1986). Taphonomic gradients in shelf fossil assemblages: Pliocene Purisima Formation. *Palaios, 1*, 256–270.
Olóriz, F., Reolid, M., & Rodríquez-Tovar, F. J. (2004). Microboring and taphonomy in Middle Oxfordian to lowermost Kimmeridgian (Upper Jurassic) from the Prebetic Zone (southern Iberia). *Palaeogeography, Palaeoclimatology, Palaeoecology, 212*, 181–197.
Palmer, T. J., Hudson, J. D., & Wilson, M. A. (1988). Palaeoecological evidence for early aragonite dissolution in ancient calcite seas. *Nature, 335*, 809–810.
Pandolfi, J. M., & Greenstein, B. J. (1997). Taphonomic alteration of reef corals: Effects of reef environment and coral growth form I: The Great Barrier Reef. *Palaios, 12*, 27–42.
Pandolfi, J. M., & Minchin, P. R. (1996). A comparison of taxonomic composition and diversity between reef coral life and death assemblages in Madang Lagoon, Papua New Guinea. *Palaeogeography, Palaeoclimatology, Palaeoecology, 119*, 321–341.
Pearson, P. N., van Dongen, B. E., Nicholas, C. J., Pancost, P. D., Schouten, S., Singano, J. M., et al. (2007). Stable warm tropical climate through the Eocene Epoch. *Geology, 35*, 211–214.
Peebles, M. W., & Lewis, R. D. (1988). Differential infestation of shallow-water benthic foraminifera by microboring organisms: Possible biases in preservation potential. *Palaios, 3*, 345–351.
Peebles, M. W., & Lewis, R. D. (1991). Surface textures of benthic foraminifera from San Salvador, Bahamas. *Journal of Foraminiferal Research, 21*, 285–292.

Perry, C. T. (1996). Distribution and abundance of macroborers in an Upper Miocene reef system, Mallorca, Spain: Implications for reef development and framework destruction. *Palaios, 11*, 40–56.

Perry, C. T. (1998a). Macroborers within coral framework at Discovery Bay, north Jamaica: species distribution and abundance, and effects on coral preservation. *Coral Reefs, 17*, 277–287.

Perry, C. T. (1998b). Grain susceptibility to the effects of microboring: implications for the preservation of skeletal carbonates. *Sedimentology, 45*, 39–51.

Perry, C. T. (2000). Macroboring of Pleistocene coral communities, Falmouth Formation, Jamaica. *Palaios, 15*, 483–491.

Perry, C. T. (2005). Morphology and occurrence of rhodoliths in siliciclastic, intertidal environments from a high latitude reef setting, southern Mozambique. *Coral Reefs, 24*, 201–207.

Perry, C. T., & Bertling, M. (2000). Spatial and temporal patterns of macroboring within Mesozoic and Cenozoic coral reef systems. In E. Insalaco, P. W. Skelton, T. J. Palmer (Eds.), *Carbonate platform systems: components and interactions. Geological Society of London, Special Publication* 178 (pp. 33–50).

Perry, C. T., & Smithers, S. G. (2006). Taphonomic signatures of turbid-zone reef development: Examples from Paluma Shoals and Lugger Shoal, inshore central Great Barrier Reef, Australia. *Palaeogeography, Palaeoclimatology, Palaeoecology, 242*, 1–20.

Piller, W. E., & Rasser, M. (1996). Rhodolith formation induced by reef erosion in the Red Sea, Egypt. *Coral Reefs, 15*, 191–198.

Racey, A. (2001). A review of Eocene nummulitic accumulations structure, formation and reservoir potential. *Journal of Petroleum Geology, 24*, 9–100.

Rasmussen, K. A., & Frankenberg, E. W. (1990). Intertidal bioerosion by the chiton *Acanthopleura granulata*; San Salvador, Bahamas. *Bulletin of Marine Science, 47*, 680–695.

Rasser, M. W. (2000). Coralline red algal limestones of the Late Eocene Alpine Foreland Basin in Upper Austria: component analysis, facies and paleoecology. *Facies, 42*, 59–92.

Rasser, M. W. (2001). Paleoecology and taphonomy of *Polystrata alba* (red alga) from the Late Eocene Alpine Foreland: a new tool for the reconstruction of sedimentary environments. *Palaios, 16*, 601–607.

Rasser, M., & Piller, W. E. (1997). Depth distribution of calcareous encrusting associations in the northern Red Sea (Safaga, Egypt) and their geological implications. In H. A. Lessios & I. G. Macintyre (Eds.), *Proceedings of the 8th International Coral Reef Symposium 1* (pp. 743–748).

Rasser, M. W., & Riegl, B. (2002). Holocene coral reef rubble and its binding agents. *Coral Reefs, 21*, 57–72.

Reid, R. P., & Macintyre, I. G. (2000). Microboring versus recrystallization: Further insight into micritization processes. *Journal of Sedimentary Research, A70*, 24–28.

Reolid, M., & Gaillard, C. (2007). Microtaphonomy of bioclasts and paleoecology of microencrusters from Upper Jurassic spongiolithic limestones (External Prebetic, Southern Spain). *Facies, 53*, 97–112.

Reolid, M., Gaillard, C., & Lathuilière, B. (2007). Microfacies, microtaphonomic traits and foraminiferal assemblages from Upper Jurassic oolitic-coral limestones: Stratigraphic fluctuations in a shallowing-upward sequence (French Jura, Middle Oxfordium). *Facies, 53*, 553–574.

Rögl, F. (1998). Palaeogeographic considerations for Mediterranean and Paratethys seaways (Oligocene to Miocene). *Annalen des Naturhistorischen Museums Wien, 99A*, 279–310.

Rooney, W. S., & Perkins, R. D. (1972). Distribution and geologic significance of microboring organisms within sediments of the Arlington reef complex, Australia. *Geological Society of America Bulletin, 83*, 1130–1150.

Sanders, D. (1999). Shell disintegration and taphonomic loss in rudist biostromes. *Lethaia, 32*, 101–112.

Sanders, D. (2003). Syndepositional dissolution of calcium carbonate in neritic carbonate environments: Geological recognition, processes, potential significance. *Journal of African Earth Sciences, 36*, 99–134.

Sanders, D., & Krainer, K. (2005). Taphonomy of Early Permian benthic assemblages (Carnic Alps, Austria). Carbonate dissolution versus biog carbonate precipitation. *Facies, 51*, 539–557.

Schneider-Storz, B., Nebelsick, J. H., Wehrmann, A., & Federolf, C. M. J. (2008). Comparative taphonomy of three bivalve species from a mass shell accumulation in the intertidal regime of North Sea tidal flats. *Facies, 54*, 461–478.

Schroeder, J. H., & Purser, B. H. (1986). Reef diagenesis: Introduction. In J. H. Schroeder & B. H. Purser (Eds.), *Reef diagenesis*. Berlin: Springer.

Scoffin, T. P. (1987). *Introduction to carbonate sediments and rocks*. Glasgow: Blackwell.

Scoffin, T. P. (1992). Taphonomy of coral reefs: a review. *Coral Reefs, 11*, 57–77.

Scoffin, T. P., & Bradshaw, C. (2000). The taphonomic significance of endoliths in dead- versus live-coral skeletons. *Palaios, 15*, 248–254.

Scoffin, T. P., Stoddart, D. R., Tudhope, A. W., & Woodroffe, C. (1985). Rhodoliths and coralliths of Muri Lagoon, Rarotonga, Cook Islands. *Coral Reefs, 4*, 71–80.

Sebens, K. P. (1986). Spatial relationships among encrusting marine organisms in the New England subtidal zone. *Ecological Monographs, 56*, 73–96.

Serra-Kiel, J. (1982). Contribució a la paleobiologia del *Nummulites*. *Butlletí de la Institució Catalana d'Història Natural. Secció de Geologia, 48*, 19–29.

Serra-Kiel, J., & Reguant, S. (1984). Paleoecological conditions and morphological variation in monospecific banks of *Nummulites*: An example. *Bulletin de Centres de Recherches Exploration-Production Elf Exploration, Mémoire, 6*, 557–563.

Severin, L. P., & Lipps, J. H. (1989). Weight-volume relationship of the test of *Alveolinella quoyi*: Implications for the taphonomy of large fusiform foraminifera. *Lethaia, 22*, 1–22.

Shroba, C. S. (1993). Taphonomic features of benthic foraminifera in a temperate setting: Experimental and field observations on the role of abrasion, solution and microboring in the destruction of foraminiferal tests. *Palaios, 8*, 250–266.

Smith, A. M. (1995). Paleoenvironmental interpretation using bryozoans: a review. In D. Bosence & P. Allison (Eds.), *Marine Palaeoenvironmental Analysis from Fossils. Geological Association of London, Special Publication 83* (pp. 231–243).

Smith, A. M., & Nelson, C. S. (1994). Selectivity in sea-floor processes: Taphonomy of bryozoans. In P. J. Hayward, J. S. Ryland, & P. D. Taylor (Eds.), *Biology and palaeobiology of bryozoans, proceedings of the 9th International Bryozoology Conference, 1992*. Fredensborg: Olsen & Olsen.

Smith, A. M., Nelson, C. S., & Grant-Mackie, J. A. (1996). Differential abrasion of bryozoan skeletons: Taphonomic implications of paleoenvironmental interpretation. In D. P. Gordon & A. M. Smith (Eds.), *Bryozoans in space and time. Proceedings of the 10th International Bryozoology Conference*. Wellington: Wilson & Horton.

Smith, A. M., & Nelson, C. S. (2003). Effects of early sea-floor processes on the taphonomy of temperate shelf skeletal carbonate deposits. *Earth Science Reviews, 63*, 1–31.

Smith, A. M., Nelson, C. S., & Danaher, P. J. (1992). Dissolution behaviour of bryozoan sediments: taphonomic implications for nontropical shelf carbonates. *Palaeogeography, Palaeoclimatology, Palaeoecology, 93*, 213–226.

Speyer, S. E., & Brett, C. E. (1988). Taphofacies models for epeiric sea environments: Middle Paleozoic examples. *Palaeogeography, Palaeoclimatology, Palaeoecology, 63*, 225–262.

Staff, G. M., Stanton, R. J., Jr., Powell, E. N., & Cummins, H. (1986). Time-averaging, taphonomy, and their impact on paleocommunity reconstruction: Death assemblages in Texas bays. *Geological Society of America Bulletin, 97*, 428–443.

Steger, K. K., & Smith, A. M. (2005). Carbonate mineralogy of free-living bryozoans (Bryozoa: Otionellidae), Otago shelf, southern New Zealand. *Palaeogeography, Palaeoclimatology, Palaeoecology, 218*, 195–203.

Steller, D. L., & Foster, M. S. (1995). Environmental factors influencing distribution and morphology of rhodoliths in Bahía Concepción, México. *Journal of Experimental Marine Biology and Ecology, 194*, 201–212.

Steneck, R. S. (1983). Escalating herbivory and resulting adaptive trends in calcareous algal crusts. *Paleobiology, 9*, 44–61.

Steneck, R. S. (1986). The ecology of coralline algal crusts: convergent patterns and adaptative strategies. *Annual Review of Ecological System, 17*, 273–303.

Taylor, P. D. (1979). Palaeoecology of the encrusting epifaunal of some British Jurassic bivalves. *Palaeogeography, Palaeoclimatology, Palaeoecology, 28*, 241–262.

Taylor, P. D. (1992). Encrusters. In D. E. G. Briggs & P. R. Crowther (Eds.), *Palaeobiology: A synthesis*. Oxford: Blackwell.

Taylor, P. D., & Wilson, M. A. (2003). Palaeoecology and evolution of marine hard substrate communities. *Earth Science Reviews, 62*, 1–103.

Testa, V. (1997). Calcareous algae and corals in the inner shelf of Rio Grande do Norte, NE Brazil. In H. A. Lessios, & I. G. Macintyre (Eds.), *Proceedings of the 8th International Coral Reefs Symposium 1*. Balboa: Smithsonian Tropical Research Institute.

Thomas, R. D. K. (1986). Taphonomy: Ecology's loss is sedimentology's gain. Theme-issue introduction. *Palaios, 1*, 206.

Tucker, M. E., & Wright, V. P. (1990). *Carbonate sedimentology*. Oxford: Blackwell.

Tudhope, A. W., & Risk, M. J. (1985). Rate of dissolution of carbonate sediments by microboring organisms, Davies Reef, Australia. *Journal of Sedimentary Petrology, 55*, 440–447.

van den Hoek, C., Cortel-Breeman, A. M., & Wanders, J. B. W. (1975). Algal zonation in the fringing coral reef of Curaçao, Netherlands Antilles, in relation of corals and gorgonians. *Aquatic Botany, 1*, 269–308.

Vogel, K. (1993). Bioeroders in fossil reefs. *Facies, 28*, 109–114.

Vogel, K., Bundschuh, M., Glaub, I., Hofmann, K., Radtke, G., & Schmidt, H. (1995). Hard substrate ichnocoenoses and their relations to light intensity and marine bathymetry. *Neues Jahrbuch für Geologie und Paläontologie – Abhandlungen, 195*, 49–61.

Vogel, K., Gektidis, M., Golubic, S., Kiene, W. E., & Radtke, G. (2000). Experimental studies on microbial bioerosion at Lee Stocking Island, Bahamas and One Tree Island, Great Barrier Reef, Australia: implications for paleoecological reconstructions. *Lethaia, 33*, 190–204.

Wanders, J. B. W. (1977). The role of benthic algae in the shallow water reef of Curaçao (Netherland Antilles), III: The significance of grazing. *Aquatic Botany, 3*, 357–390.

Wefer, G. (1980). Carbonate production by algae. *Halimeda*, *Penicillus* and *Padina*. *Nature, 285*, 323–324.

Wilson MVH (1988) Taphonomic processes: Information loss and information gain. Geoscience Canada 15:131-148

Wilson, M. A. (2007). Macroborings and the evolution of bioerosion. In W. Miller III (Ed.), *Trace fossils: Concepts, problems, prospects*. Amsterdam: Elsevier.

Wood, R. (1998). The ecological evolution of reefs. *Annual Review of Ecology and Systematics, 29*, 179–206.

Wood, R. (1999). *Reef evolution*. Oxford: Oxford University Press.

Wright, V. P., & Burgess, P. M. (2005). The carbonate factory continuum, facies mosaics and microfacies: An appraisal of some of the key concepts underpinning carbonate sedimentology. *Facies, 51*, 19–25.

Yesares-García, J., & Aguirre, J. (2004). Quantitative taphonomic analysis and taphofacies in lower Pliocene temperate carbonate-siliciclastic mixed platform deposits (Almería-Níjar basin, SE Spain). *Palaeogeography, Palaeoclimatology, Palaeoecology, 207*, 83–103.

Yordanova, E. K., & Hohenegger, J. (2002). Taphonomy of larger foraminifera: relationships between living individuals and empty tests on flat reef slopes (Sesoko Island, Japan). *Facies, 46*, 169–204.

Zachos, J., Pagani, M., Sloan, L., Thomas, E., & Billups, K. (2001). Trends, rhythms, and aberrations in global climate 65 Ma to present. *Science, 292*, 686–693.

Zuschin, M., & Stanton, R. J. (2001). Experimental measurements of shell strength and its taphonomic interpretation. *Palaios, 16*, 161–170.

Zuschin, M., & Stanton, R. J. (2002). Paleocommunity reconstruction from shell beds – A case study from the Main Glauconite Bed, Eocene, Texas. *Palaios, 17*, 602–614.

Zuschin, M., Stachowitsch, M., & Stanton, R. J. (2003). Patterns and processes of shell fragmentation in modern and ancient marine environments. *Earth Science Reviews, 63*, 33–82.

Chapter 10
Taphonomy of Reefs Through Time

Rachel Wood

Contents

Abstract Reefs are susceptible to multiple physical, chemical and biological taphonomic processes. Bioerosion, in particular has escalated through time and might be expected to have influenced the taphonomy of reefs. The following biases can be predicted: (1) In the absence of grain-reducing activities by reef biota (fish, echinoids, and clionid sponges) abrasion on Paleozoic reefs would have been dominated by physical processes and sediment grains may have been more coarse.

R. Wood (✉)
Grant Institute of Earth Sciences, School of Geosciences, University of Edinburgh, King's Buildings, West Mains Road, Edinburgh EH9 3JW, UK
e-mail: rwood4@staffmail.ed.ac.uk

P.A. Allison and D.J. Bottjer (eds.), *Taphonomy: Process and Bias Through Time*, Topics in Geobiology 32, DOI 10.1007/978-90-481-8643-3_10,

(2) Increased bioerosion since the Jurassic is such that modern reefs are quickly reduced to rubble and sand leaving only the resilient branching corals and thick coralline algae. By contrast, many pre-Jurassic reefs commonly preserve intact, in situ frameworks that include massive or laminar, often soft-sediment-dwelling, growth forms. (3) After the appearance of reef fish in the Eocene, sediment production and distribution within reef complexes is likely to have increased markedly but this has not yet been fully elucidated. (4) Escalation in rates of bioerosion from the Miocene onwards are such that it can be expected that substantial aprons of reef-slope sediment may not have been present on pre-Miocene reefs.

Evidence is persuasive that changing global seawater chemistry has exerted secular changes in the dominant carbonate mineralogy of reef organisms and early diagenetic cements but the subsequent effects upon reef taphonomy remain to be documented.

The current phase of climate change will exert a profound effect upon reef ecology and taphonomy. Reduction of reef herbivore populations will almost certainly lead to an increase in soft-bodied algal biomass, and a decrease in coral cover, particularly in areas of eutrophication or outbreaks of disease. Bleaching as a result of global warming may lead to significant or widespread coral mortality. Calcification rates are already between 6% and 20% lower than they were under pre-industrial conditions due to ocean acidification. These processes will reduce the structural integrity of reefs. Future death assemblages and the subsequent fossil record of reefs will be dominated by highly degraded coral fragments and grains with limited in situ reef frameworks, endolithic algal activity, and intense bioerosion.

1 Introduction

As records of in situ benthic communities, ancient reefs offer considerable paleoecological information, but the utility of this record is controlled by the fidelity of their geological expression. Reefs are complex environments of both hard (reef framework) and soft-substrate (reef sediment) communities that are exposed to a range of intense physical and biological exposure, transport, and burial processes. These processes have changed markedly over geological time due to extrinsic physicochemical controls and particularly an escalation in bioerosion.

The formation of a reef framework is dependent upon the maintenance of stability of an epibenthic marine community. But reef frameworks and their surrounding areas of soft-sediment support a huge variety of closely-interacting immobile and mobile organisms of varying skeletal durability and ability to withstand biological and physical attack. For example, reef sediment itself may be the result largely of abrasion and transport of skeletal debris derived from the reef framework. The varying rates of skeletal production, breakdown and subsequent transport, as well as overall sedimentary accumulation rates, control the concentration or dilution of any given component. Only through quantification of the degree of taphonomic information loss in reef death assemblages will comparisons between living, sub-fossil, and fossil assemblages be possible.

This review first summarizes the major controls on the post mortem history of modern reef communities, with an emphasis on the fate of coral skeletons. These include modes of life and death, abrasion, transport, burial and diagenetic processes. Reefs are defined here as *discrete carbonate structures that form by* in situ *or bound organic components that develop topographic relief upon the sea floor* (Wood 1999). As such, a reef (be it a deep-water mud mound or a shallow coral reef) is an elevated sessile benthic community that can resist the ambient hydrodynamic regime, but as reef formation involves both constructional and destructive processes these may present as any variant between an intact framework or a pile of skeletal debris in the geological record.

Physical disturbance, grazing pressure and spatial competition are important determinants of living coral reef community structure. Biological disturbances, such as predation, herbivory, and deep bioturbation, have evolved in tempo and strength over the Phanerozoic (Vermeij 1987). The second part of this review will explore the effects of this ecological escalation on taphonomic processes in reef communities.

Reefs worldwide are undergoing dramatic change, and many of these changes are historically recent (Hughes 1994; Jackson et al. 2001; Pandolfi et al. 2003; Hughes et al. 2003). Most notable is the decline of acroporoid corals in the Caribbean, and an increase in soft-bodied algal cover and biomass. The shift from coral to algal dominance has led to a marked reduction in coral biodiversity over whole regions and a notable decline in rates of reef calcification (Kleypas et al. 1999; Gardner et al. 2003; Kleypas 2007). In particular, the demise of *Acropora palmata* and *Acropora cervicornis* over the last few decades has removed zonation patterns now considered to have been characteristic of Caribbean reefs for at least the last 125 Kyr (Jackson 1991, 1992; Pandolfi and Jackson 2007).

The recent decline in coral reefs means that a geologic context is required to establish a baseline independent of any anthropogenic influence that also accounts for natural, often cyclical, factors (Bak and Nieuwland 1995). The final section reviews how the taphonomy of living coral reefs may change as a result of the current regime of marked human impact and climate change.

2 Spatial and Temporal Variation in Modern Coral Reef Communities

The modern reef primary framework is dominated by photosynthetic coralline algae or corals, with the more slow-growing suspension-feeding and filtering benthos being restricted to cryptic niches. Secondary, encrusting framebuilders flourish upon the primary framework above the level of accumulating sediment. Early lithification can aid frame integrity, but mechanical destruction and bioerosion reduces reef framework to rubble and sand (Fig. 1). Reefs however, generally have a high preservation potential, such that detailed ecological inter-relationships are often preserved in the ancient record (Fig. 2).

Modern coral reefs grow rapidly, with extension rates in branching corals exceeding 15 mm/year (Kleypas 1997), and so it has been supposed that some short-term processes may be preserved in the reef record (Jackson 1983). Coral growth decreases exponentially with depth and light. However, recently compiled data from cores show that reef accretion does not change significantly with either water depth or dominant coral species within the upper 20–30 m of the water column (Hubbard 2006). Bioerosion can progress at comparable rates to coral growth: reef accretion is therefore not constrained by rates of coral growth alone.

Physical disturbance, grazing pressure and spatial competition are all known to control the modern coral reef community structure (Wood 1999). Disturbance shows marked differences in distribution and intensity across a reef profile. Physical disturbance, predation (and herbivory) and bioturbation all decrease with depth,

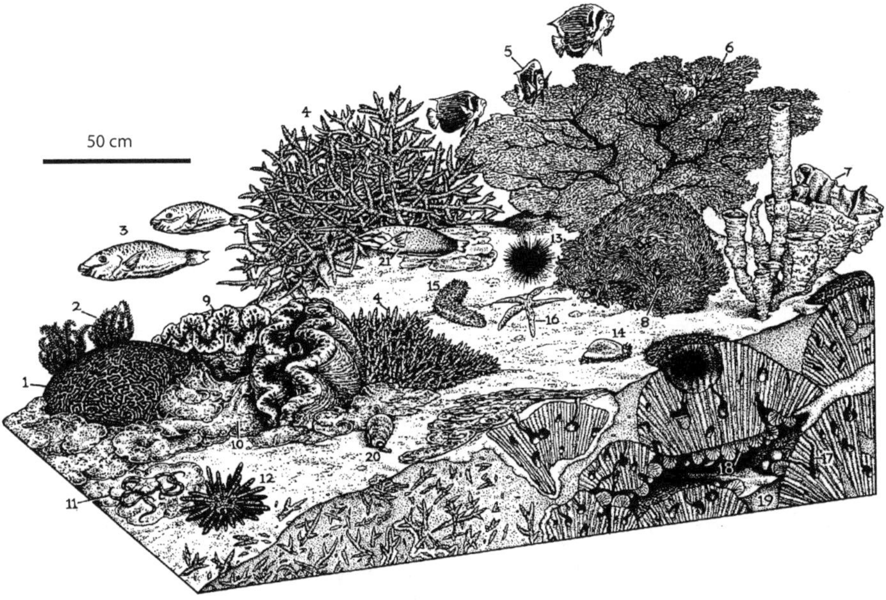

Fig. 1 Reconstruction of a modern Indo-Pacific coral reef and its sedimentological expression. 1. Brain coral (*Leptoria phrygia*); 2. Feather star (*Comanthus bennetti*); 3. Parrotfish (*Scarus* sp.); 4. Staghorn coral (*Acropora* sp.); 5. Emperor angelfish (*Pomacanthus imperator*); 6. Gorgonian; 7. Vase sponge (*Callyspongia* sp.); 8. Anemone with clown fish; 9. Giant clam (*Tridacna gigas*); 10. Encrusting corals (*Montipora* and *Hydnophora*); 11. Brittle star (*Ophiarachella gorgonia*); 12 and 13. Echinoids; 14. Cowrie gastropod; 15. Sea cucumber (*Thelenota ananus*); 16. Sea star; 17. Boring bivalve (*Lithophaga* sp.); 18. Cement botryoids; 19. Internal sediment; 20. Cone gastropod (*Conus textile*); 21. Wrasse (*Coris gaimard*) (From Wood 1999; copyright John Sibbick)

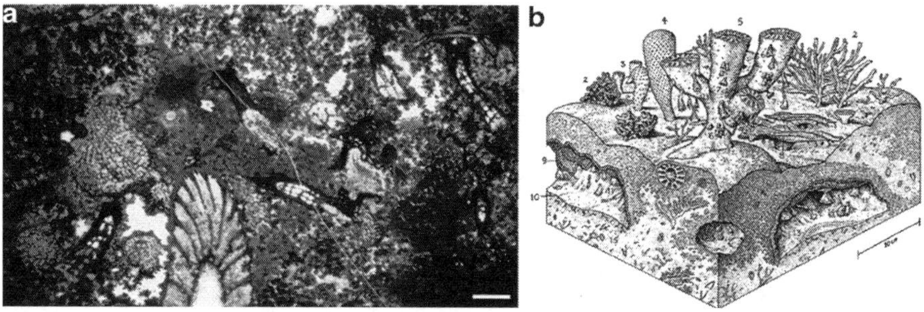

Fig. 2 (**a**) Lower Cambrian (Botomian) cryptic reef community showing a variety of pendent archaeocyath sponges and coralomorphs attached to the walls and ceiling of a crypt constructed by the calcimicrobes *Renalcis* (*upper left*) and *Ephiphyton* (*upper right*). Pockets of micrite within the crypt have been extensively microburrowed. Scale bar = 1 mm. (**b**) Reconstruction of a Lower Cambrian reef community (Atdabanian). 1. *Renalcis* (calcified cyanobacterium); 2. Branching archaeocyath sponges; 3. Solitary cup-shaped archaeocyath sponges; 4. Chancelloriid; 5. Radiocytahs; 6. Small archaeocyath sponges; 7. 'Coralomorphs' 8. *Okulitchicyathus* (archaeocyath sponge); 9. Fibrous cement; 10. Microburrows (traces of a deposit feeder); 11. Cryptic archaeocyath and coralomorphs. 12. Cribricyaths; 13. Trilobite trackway; 14. Botryoid cement; 15. Sediment with skeletal debris (From Wood 1999; copyright John Sibbick)

being usually greatest from the lower intertidal zone to about 20 m, particularly on reef slopes with substrates of high topographic complexity (Hay 1984). Herbivory is low above mean low water, often reaching a peak at 1–5 m depth on the forereef, and then declining rapidly with depth (Steneck 1988). Regardless of depth however, the effects of biological disturbance may be highly patchy and vary markedly according to local environmental differences.

Problems exist in extrapolating ecological processes to their manifestation in the geological record, in particular the results of experiments that operate over ecological timescales to observations in the fossil record. Inference of cause and effect require correlation between independent measures of environmental conditions and biological change, but reduced variability becomes apparent over broader temporal and spatial scales. Such issues impose an apparent uniformity on community structure that was, in fact, far more dynamic and labile.

Reef communities are often highly patchy by nature, such that differences in community structure apparent within a living reef, within core samples, or across restricted outcrop exposures may not reflect any significant changes in the community structure as a whole. In addition, methodological differences in data collection, e.g., quadrat vs. line transects vs chain transects, can produce significantly different results from the same modern reef (Hubbard 2006).

On a small scale, reef communities are clearly dynamic and to a large extent unpredictable, but on larger scales (over tens of kilometers and centuries to thousands of years) patterns that show considerable consistency become apparent (e.g. Pandolfi 1996, 2002). Variation at the smallest scales may be higher than even biogeographic differences. This suggests that 'order' in reef coral communities is lowest at smaller scales, highest at intermediate scales, and intermediate at the broadest spatial scales within the same biogeographic province (see Pandolfi and Jackson 2007). Similar trends in predictability are apparent over varying temporal scales (e.g. Tanner et al. 1994; Pandolfi 1996; Aronson and Precht 1997; Connell 1997).

3 Taphonomy of the Modern Coral Reef Environment

By definition, all reefs are autochthonous and produced by a local biota. Because they represent a record of a community that reflects ecological relationships modified by *pre* or post mortem disturbance and/or time averaging of generations, reef deposits can be termed an *association* (sensu Fürsich 1977).

The taphonomy of living reefs is controlled by the complex interaction and feedback of many factors (Scoffin 1992). These can be resolved simplistically into (a) the proportion of the community with preservable hard parts, (b) the source and rate of skeletal supply, (c) the resilience of both individual reef builders and reef framework to ambient physical and biological erosion, (d) the environment of accumulation, and (e) and time scale of accumulation (Fig. 3). Many feedbacks occur in this system. For example, the presence of skeletal hard parts provides substrates for further colonization, and the accumulation of skeletal material can influence pore water chemistry and hence subsequent diagenesis. Indeed, reef framework growth itself may be self-regulating as over-supply of framework-derived sediment will bury the framework, so terminating growth, arrest bioerosion, and reducing sediment production (Scoffin 1992).

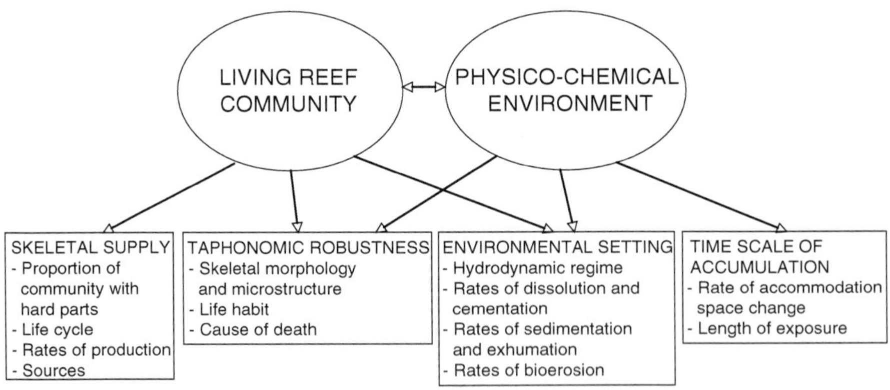

Fig. 3 The major factors affecting the preservation of reef communities (adapted from Kidwell and Bosence 1991)

3.1 Loss due to Non-Preservation

Many organisms on reefs have no preservable hard parts, and so this biota will only leave a record if their tissues have become bio-immured (Taylor and Todd 2001). Biota with skeletons composed of loose spicules (e.g. sponges and ascidians) will become dispersed upon death unless buried rapidly in fine-grained sediment.

One insurmountable problem is that the fossil record is virtually mute on many key ecological players and processes: for example, fleshy and filamentous algae leave at best a very poor fossil record, and the record of herbivorous reef fish and higher predators is highly incomplete.

Few studies have considered the proportion of skeletal taxa within reef communities. Open reef surfaces in Jamaica show an average of 70% skeletal taxa in shallow water (60 m), with deeper waters yielding progressively lower proportions decreasing to 1.8% skeletal taxa at 120 m depth (Liddell and Ohlhorst 1988). In a reef cave habitat, skeletal taxa represented less than 40% of total species richness and covered only 15% of the total surface area (Brett 1988).

3.2 Mode of Life, Skeletal Robustness and Rates of Skeletal Production

Reef environments offer substrate habitats ranging from hard substrates (rock; cemented substrates; other organisms), to rubble, gravel, sand or muddy soft sediments. The relative stability of these substrate types is broadly coincident with the energetics of the ambient hydrodynamic regime, with hard substrates dominating in the highest energy environments (the reef crest), and muddy sediments in the lowest (the lagoon). Hydrodynamic action can be provided by tidal currents, wave action, gravity flows, or intermittent storms.

The mode and timing of death relative to the life cycle will, in part, control abundance, size and state of preservation of reef material (Scoffin 1992). Skeletal organisms may be variously killed and crushed by predation, fragmented by storms, but left intact by pathogens, bleaching of photosymbionts, overgrowth by encrusters, or rapid burial by storm-generated sediment. All skeletal elements will suffer bioerosion and encrustation unless buried rapidly beyond the reach of bioturbators, bioeroders or physical reworking.

Many of the specific causes of mortality for either individuals or whole communities are either difficult or impossible to detect in fossil skeletal reef material. The range of bioerosive trace fossil morphologies is vast due to the diversity of organisms involved (Bromley 1992). Of these traces, however, very few are sufficiently characteristic to allow an unequivocal pairing of a particular predator with a given trace. For example, while Steneck (1983) noted considerable evidence of predatory damage in fossil solenoporacean and corallinacean algal thalli, he was unable to determine their origin.

Parrotfish (scarids) do produce distinctive stellate marks on the upper surfaces of scleractinian colonies or algal thalli. Likewise, camerodont echinoderms produce characteristic pentaradiate grazing traces (*Gnathichnus* and *Radulichnus*) due to the action of strengthened teeth in a stirodont lantern (Bromley 1975). Both these traces occur in living and dead modern coral material.

The rate and site of skeletal production, the organism longevity, as well as the aerial coverage, all determine the initial potential contribution of any given organism to the reef sedimentary record. Reef organisms vary greatly in their rates of skeletal production as well as their durability in the face of a multitude of destructive forces. While corals may occupy 90% of a reef framework and the green alga *Halimeda* only 10% (Scoffin 1992), the high rate of *Halimeda* growth and the robust nature of its skeleton results in 25% of all modern reef sediment being composed of *Halimeda* material. By contrast, the relatively fragile platy coral, *Agaricia*, while representing 54% of the living community on the shelf-edge of St. Croix, US Virgin Islands, is completely absent from cores taken though the underlying reef sediment (Hubbard et al. 1986).

Patterns of fidelity and time-averaging, that is the mixing of successive generations, are highly complex and there may be no general rules that can be applied consistently to all ancient reefs. Analyses show that the resolution provided by the fossil record will vary with different environments, habitat, and facies, such that each must be evaluated individually (Greenstein 2007). Inter-provincial differences are likely to be related to differences in live coral diversity, especially for branching species of *Acropora* which are difficult or impossible to distinguish when represented only as rubble.

Size-frequency distributions are controlled by recruitment, growth rate, and survivorship of a particular population or species (Scoffin 1992), but are also biased towards the larger size classes such that these data have most value in assessing post mortem transport history (Cummins et al. 1986). Staff et al. (1986) found that taxonomic composition, particularly of adults, and the biomass of the death assemblage most accurately reflect characteristics of the living community.

3.3 Bioerosion, Abrasion, Transport, and Burial

A considerable proportion of modern reefs are preserved in the geological record as rubble, sand, and voids as a result of physical and biological destruction (Hubbard et al. 1990). In addition, storms often remove reef sediment from its origin, and redistribute and reincorporate this material within the reef interior (Hubbard 1992).

Modern coral reefs are characterized by diverse active predators and herbivores, and non-predatory borers, which prey upon or otherwise attack sessile organisms and are capable of removing and ingesting calcareous skeletal material (Table 1). Epilithic predators feed directly upon sessile invertebrates or algae by etching, rasping or biting, so causing incidental skeletal or substrate damage. These include excavators that exert deep bites that result in the removal of large areas of substrate,

Table 1 Major groups of bioeroders and bioturbators on modern coral reefs and their first appearance in the fossil record

Group	Ecology	First appearance
Cyanobacteria*	Borers	?Neoproterozoic (Vermeij 1987)
Fungi*	Borers	?Cambrian (Vermeij 1987)
Chlorophyta	Borers	?Ordovician (Vermeij 1987)
Rhodophyta	Borers	?Ordovician (Vermeij 1987)
Porifera		
Clionidae*	Borers	?Jurassic (Vermeij 1987)
Annelida		
Spionidae (Polychaetes)	Deep burrowers	Triassic (Thayer 1983)
Mollusca		
Polyplacophora	Herbivores (scraping)	Late Cretaceous (van Belle 1977)
Gastropoda		
Patellacea*	Herbivores (scraping)	Late Cretaceous (Lindberg and Dwyer 1983)
Diverse	Deep burrowers	Late Triassic (Thayer 1983)
Bivalvia		
Lithophagidae*	Borers and live-borers	Boring: Jurassic (Vermeij 1987) Live boring: Eocene (Savazzi 1982)
Arthopoda		
Acrothoracica (Barnacles)	Borers	Boring: Live boring: Eocene (D.S. Jones, pers. comm.)
Decapoda	Deep burrowers	Early Jurassic (Thayer 1983)
Echinoderms		
Holothuroidea	Sediment disturbers	Devonian (Thayer 1983)
Echinodea		
Diadematoida*	Herbivores and corallivores	Late Triassic (Smith 1984)
Arbacioida	Herbivores (excavating)	
Echinoida	Herbivores (excavating)	
Spatangoida (Irregular echinoids)	Deep burrowers	Early Jurassic (Thayer 1983)
Pisces		
Chondrichthyes (Rays & Skates)	Sediment disturbers	Devonian (Vermeij 1987)
Scaridae*	Herbivores (excavating)	Miocene (Bellwood and Schulz 1991)
Mamallia		
Trichechidae (Manatees)	Sediment disturbers	Eocene (Thayer 1983)

*Indicates most important groups (After Vermeij 1987; Wood 1999)

and scrapers that have weaker jaw apparatuses that take smaller bite sizes with resultant limited substrate removal. The most important excavators and scrapers on modern coral reefs are limpets, chitons, some regular echinoids, and acanthuroids (surgeonfish) and scarids (parrotfish). Corallivores include crustaceans (hermit crabs), polychaetes (amphinomids), gastropods (prosobranchs and nudibranchs), echinoids (diadematoids), starfish, and numerous fish, which are arguably the most diverse of all reef predators.

Living and dead corals can carry massive and multiple simultaneous or successive infestations of endolithic organisms, particularly algae, fungi, sponges, and bivalves. Often only the final generation of borings is clearly preserved after fossilization (Scoffin 1992). The rise of the endolithic habit is thought to be a direct response to the rise of predation, as it provides protection from predators: many endoliths have reduced skeletal defences compared to open surface dwellers or their epifaunal ancestors (Harper and Skelton 1993). Endoliths severely weaken the skeleton, and may ultimately lead to the death of the coral, as well as the reduction of the skeleton to rubble or sediment. Some have estimated that the biomass of endoliths alone can equal, or exceed, that of the surface biota on coral reefs (Grassle 1973).

Sediment production by bioeroders in reef habitats varies from 0.2 to 16 kg/m^2/year (Scoffin 1987); some estimates suggest that up to 60% of all carbonate produced is reduced to sediment by bioerosion (Hubbard et al. 1990). Bioerosion by microboring is most prevalent in quieter water settings (see summary in Scoffin 1992), and rises substantially in areas of higher nutrient input. Highsmith (1980) has shown that the proportion of massive corals bored by bivalves increases proportionally with phytoplankton productivity.

Only a limited number of bioeroders produce diagnostic grains, e.g. clionid sponge chips. Although many grains may be the result of compounded bioerosion and physical abrasion, Scoffin (1987) found that in situ mechanical breakdown of skeletons produces a broad range of grain sizes (from 0.01 to 256 mm), whereas boring alone produces predominantly fine grains (0.016–4 mm).

The most important bioeroders on Indo-Pacific reefs are scarids (parrottfish), which are characteristic of reef crests and fronts. Scarids feed on living or dead convex surfaces, and pass substantial amounts of sediment through their guts which is then redistributed as fine grain sediments (0.063–1 mm; Scoffin 1987) at the base of the reef to form large sediment aprons. Estimates suggest that up to 5.6 kg m/year may be removed by excavating scarids at Lizard Island, on the Great Barrier Reef (Bellwood 1995). Here, a single male bumphead parrotfish can remove a staggering 5 t of reef per year (Bellwood 1996).

Scarids thus modify reefs by (a) direct erosion; (b) decrease in particle size due to erosion and sediment reworking, and (c) the net removal and transport of reef material directly from the area of most carbonate production (the reef crest and front) to deep reef sites. As such, they may significantly control the rate of reef progradation and removal of fine material from the reef system (Bellwood 1995).

Greenstein (2007) argues that any analysis of death or fossil reef assemblages must compensate for the facts that coral growth forms are differentially susceptible to degradation. In Papua New Guinea, Pandolfi and Minchin (1995) found that high-energy reef environments showed a greater loss in fidelity of coral composition between life and death assemblages than low energy reef environments. But while high energy environments produced the best-preserved corals, they also preserved the most biased assemblage. This was in contrast to that found in a comparable study of contemporary molluscan assemblages. In Florida Keys, deep-water death assemblages are more diverse than their living counterparts (Pandolfi and Greenstein 1997), perhaps due to either slower rates of coral growth and sedimentation.

The degradation of corals is determined by the residence time of dead coral material in the taphonomically active zone (TAZ) which extends a several centimetres below the sediment–water interface (Fig. 4). The majority of physical and biological destruction occurs to skeletal material post mortem. Massive, rather than branching or free-living, corals are both the preferred site for most borers (particularly worms, bivalves, and sponges) as well as showing higher rates of dissolution (Pandolfi and Greenstein 1997). In any reef environment, massive forms will survive longer in the TAZ than other forms, but in high energy settings they will be destroyed, transported or buried before extensive taphonomic alteration can occur (Greenstein 2007). In low–energy environments (leeward or deeper water sites), any colony growth form will survive longer in the TAZ than in higher energy environments. With the exception of encrusting foraminifera, however, epibiont encrustation was found to be higher in deep-reef (20–30 m) settings (Greenstein and Pandolfi 2003).

Encruster succession with reef frameworks or storm-generated coral debris can be very sensitive to decreasing light levels, so aiding interpretation of the history of reef framework burial (Scoffin and Hendry 1984).

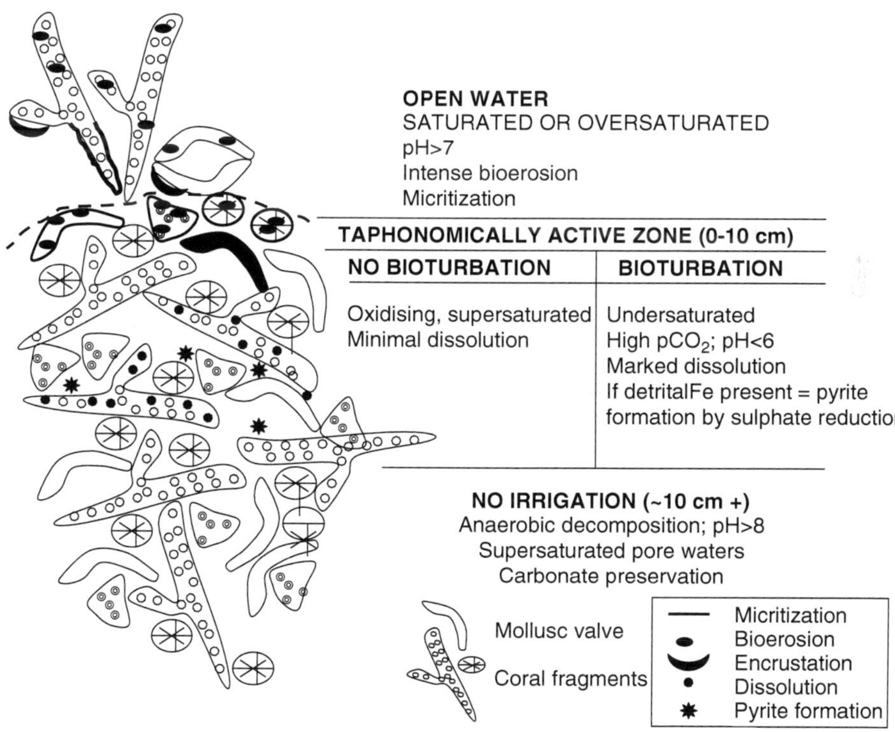

Fig. 4 The major taphonomic processes occurring within reef sediment

3.4 Early Diagenesis: Dissolution and Cementation

Aragonite is a metastable mineral that will tend to either neomorphose to calcite with loss of microstructural detail, or may dissolve completely to form a vug if exposed to an open system of undersaturated water. The solubility of high-Mg calcite can exceed that of aragonite; low-Mg calcite is relatively stable.

Rates of change are dependent upon sediment permeability, local water chemistry, and particularly climate (Fig 5). If reef material remains in contact with trapped interstitial sea water, mineralogical stabilization can take place over 1–3 million years, but exposure to freshwater may speed up this process to 100,000–200,000 years in the vadose zone, or 5,000–20,000 years in the phreatic zone (Humphrey et al. 1986; Matthews and Frohlich 1987).

Micritization occurs in warm, oversaturated waters (Alexandersson 1972), and chemical leaching or microbial attack can lead to chalky textures of both long-lived and dead skeletal material. Undersaturation of open or pore waters with respect to carbonate minerals can lead to etching, leaching, or dissolution of skeletal material. Cryptic biotas that inhabit caves may suffer preferentially the effects of corroding solutions which may be flushed through the reef framework (Scoffin 1972).

Smoothing and dissolution of skeletal material has been noted to be greatest in corals from reef-crest and patch reef environments; encrustation (except by foraminiferans) is highest in deep-reef settings (Greenstein 2007).

Upon shallow burial, reef sediment passes into the TAZ. Near-surface pore waters are generally oxidising and supersaturated due to good exchange with the overlying saturated water; this generally compensates for the acids produced by aerobic decomposition of organics (Fig. 4). By contrast, in areas of active bioturbation, undersaturated waters with high pCO_2 and low pH may develop which pro-

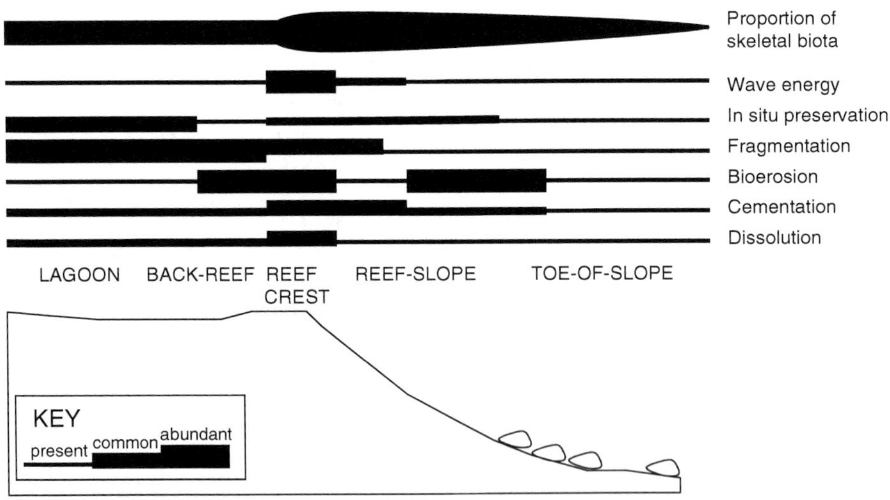

Fig. 5 Distribution of the key taphonomic determinants and processes across a generalized reef transect

motes carbonate dissolution. If reactive detrital Fe is present, diagenetic Fe-sulfides may form by local sulfate reduction. Below the zone of sediment irrigation (TAZ), anaerobic decomposition leads to high pH which promotes carbonate preservation, and early cementation (Fig. 4).

Peterson (1976) studied weight loss in buried shells over 7.5 months. After 50 cm-deep burial in sand or muddy sand, high-Mg calcite (echinoderm ossicles) lost 10–20% weight; aragonitic shells lost only 0–4%, and low-Mg calcite (scallop shells) lost 0.14%. In a similar study, Best et al. (2004) studied net weight change in mollusc and coral material from reef sites in Papua New Guinea. She found that the dominant control on taphonomic condition was the interaction of environmental energy with skeletal form and size (affecting exposure), with a secondary control of skeletal microstructure. Net weight change was positive for exposed bivalves and negative for buried ones, in both cases within 10% of the initial weight. By contrast, weight loss among corals was ubiquitous with the exception of a few *Acropora*, and often ranged between 10% and 20%. Both bivalves and corals showed lower surface alteration if originally buried; exposed specimen surfaces showed dull to chalky surface textures.

Cementation is pervasive in reef-fronts and reef-crests where water flux is high and de-gassing occurs as a result of the pumping action of waves (James et al. 1976). Walled reef complexes present a prominent steep surface to wave and current action so that the force of sea water flux is high; low-angle reef profiles undergo far less cementation (Kendall and Schlager 1981).

Modern reef cements range from aragonitic calcite crusts, fans, and botryoids, and high-Mg calcitic peloids, equant micrite, and acicular crusts or blades (Macintyre and Marshall 1988). Cements can grow remarkably rapidly in both shallow and deeper marginal parts (Grammer et al. 1999), but in high energy areas cementation occurs close to the framework surface, whereas in sheltered lagoonal patch reefs cementation takes place several centimeters below the sea floor (Scoffin 1992). Marine lithification presents further hard substrates for colonisation by reef biota, both encrusters and bioeroders. Pavement-like micrite crusts can form during a hiatus in reef growth, which can protect underlying reef deposits from diagenetic alteration (Macintyre 1985): rates of cementation are generally lower during rapid reef growth (Lightly 1985).

3.5 Changing Rates of Accumulation

Rates of terrigenous sediment supply will vary with the proximity of the reef to the land, with fringing reefs often being most affected. Increased clastic sediment will introduce nutrients into the system which will stimulate higher rates of bioerosion (Highsmith 1980).

Rates of reef sediment accumulation also affect the rates of taphonomic processes. The longer the period of accumulation, the more likely it is that the taxonomic and size composition of the assemblage will be modified by differential preservation (Kidwell and Bosence 1991). Geologically very short-term changes in reef community structure may be preserved only under sedimentation regimes that favour rapid burial of both living and dead corals, such as during periods of rapid sea-level rise and accommodation space increase that favours the growth of 'keep-up' reefs (Greenstein and Pandolfi 2003). During sea-level fall, reefs may be exposed to fresh water diagenesis and erosion.

3.6 Detection of Critical Events

In the past few decades, some modern reefs have been subject to ecologically critical events, such as the Caribbean-wide mass mortality in the early 1980s of the herbivorous sea urchin, *Diadema antillarum*, the outbreak of the crown-of-thorns starfish *Acanthaster planci* in the Indo-Pacific, and coral bleaching events and disease. All these highly significant and sometimes catastrophic occurrences have proven difficult or impossible to detect in the sedimentary record. For example, even though reef substrates were littered with *Diadema* spines and tests several weeks after the mass mortality, less than 1 year later, the impact of rapid sedimentation and bioturbation was such that evidence of this event was absent (Greenstein 1989).

Glynn (2000) outlines a variety of potential indicators of past mass bleaching events that might be applied to fossil material. These include isotopic and trace metal markers in coral cores indicative of ENSO events, alterations in skeletal banding, protuberant growths on massive corals, and accelerated bioerosion in reef sediments. All of these phenomena may, however, be caused by factors other than bleaching, so greatly limiting their utility. There is also evidence that some bleached corals may fail to secrete a growth band (see Halley and Hudson 2007). To date, no historical or fossil record of mass bleaching events at regional scales has been identified prior to 1982 (Glynn 1993).

Statistical methods, however, such as a probabilistic approach can help to place bounds on information loss in interpreted event preservation in sets of hierarchically sampled reef cores (Aronson and Ellner 2007). DeVantier and Done (2007) also offer a potential methodology to evaluate the frequency of feeding scars of starfish on living coral heads, so potentially enabling the detection of outbreaks in the geological record.

A signature for hurricane and storm events has been sought in coral death assemblages from San Salvador (Bishop and Greenstein 2001). All metrics of fidelity increased after Hurricane Floyd, suggesting that each reef setting received a pulse of storm-derived coral material. Such a signature would only be detectable where both the life and death assemblages were preserved, and could be distinguished, in the fossil record. In the Pleistocene of the Bahamas and the Dutch Antillies, reefs that grew in areas which today receive a lower frequency of hurricanes were found to have a greater proportion of in situ colonies (Meyer et al. 2003).

Using epibiont colonization sequences, Perry (2001) was able to distinguish between those *Acropora palmata*-dominated horizons that were derived from storm deposition, and those that had accumulated through normal reef accretion. Indeed, he noted repetition of the same reef succession following each storm horizon, each culminating in an *Acropora palmata* community.

The recent (~20 years) shift from coral-dominated (75–5%) to algal-dominated live (<5 to >65%) cover in Jamaica has also been detected in reef sediment. Prior to 1981, reef sediment was composed of >50% *Halimeda* and >35% coral, but post-1981, coral fragments have dominated due to widespread coral mortality and bioerosion (Precht and Aronson 1997). This suggests that for corals at least, their presence in ancient reef sediment may be indicative of widespread coral mortality.

Detection of critical events depends in part on the type of reef facies studied (Greenstein 2007). Due to the inverse relationship between wave energy and taphonomic alteration, high-energy reef facies could produce well-preserved fragile corals should rapid burial occur. However, in reefs with relatively low coral diversity the absence of a coral species from the death assemblage may be ecologically significant. Geologically very short-term changes in reef community structure may be preserved only under sedimentation regimes that favour rapid burial of both living and dead corals.

4 Taphonomic Bias in Ancient Reefs: Insight from the Pleistocene Record

Pleistocene and Holocene coral communities have been widely heralded as offering a record of pre-anthropogenic reef community ecology (Macintyre 1988; Jackson 1992; Greenstein et al. 1998; Greenstein 2007). While there is considerable ecological information preserved in Pleistocene reefs, numerous taphonomic processes have conspired to change, degrade or remove the evidence of events from future fossil communities that appear vital to understanding the functioning of present-day reefs (Greenstein and Moffatt 1996). Knowledge of which processes can be justifiably explored by analysis of the fossil record – and those that cannot – is therefore vital before any conclusions can be drawn.

Many authors have concluded that Pleistocene strata preserve a composite of both the living reef *and* the associated death assemblages (e.g. Goreau 1959; Ginsburg 1964; Edinger et al. 2001). Reef-coral death assemblages are therefore not reasonable proxies for fossil assemblages (Greenstein 2007), and it is possible that such composite assemblages where reef structure is integrated over ecological time may be the norm for all ancient reefs (Edinger et al. 2001).

Relative abundance data are available in fossil reefs and can be used to determine ecological patterns over broad temporal and spatial scales (Pandolfi and Jackson 2007), but other potential sources of data may be highly biased. For example, *Acropora cervicornis* growing in Pleistocene high-energy facies have been found to be significantly less degraded than these species from modern death assemblages; indeed branching growth forms are consistently over-represented in death assemblages due mainly to far higher rates of growth and fragmentation (Greenstein and Moffatt 1996).

It appears that patterns of fidelity and time-averaging are highly complex, and there may be no general rules that can be applied to all ancient reefs. What is clear is that the resolution provided by the fossil record will vary in different environments and within each habitat, and that facies must be evaluated individually.

5 Changes in Reef Taphonomy Through the Phanerozoic

The modern coral reef ecosystem is geologically very young. Scleractinian corals appeared in the mid-Triassic, and had almost certainly acquired photosymbionts by the late Triassic at the latest (Stanley and Swart 1995). Most modern coral genera appeared in the Eocene–Miocene (55–5.3 Ma), and many extant species extend back no further than the Pliocene (5.3–1.8 Ma) (Rosen 1984). Modern reef fish appeared in the Eocene (50 Ma), but the oldest record of parrotfish (scarid) remains are from Miocene sediments dated at 14 Ma.

During the Oligocene, the compression of climatic belts and the rise of the Isthmus of Panama created two distinct regions of reef growth to the Caribbean and Indo-Pacific. As a probable result of climatic cooling or habitat loss, a major episode of coral faunal turnover ensued between 4 and 1 Ma in the Caribbean (Budd et al. 1994). Extinction of genera in the Pocilloporidae and Agaricidae was most marked, but many of these genera continued to persist in the Indo-Pacific. A similar differential extinction coincident with corals removed all large excavating scarids, herbivorous siganids, and plantivorous caesionid fish from Atlantic reefs (Bellwood 1997).

Although acroporid corals appeared in the Eocene, pocilloporids appear to have dominated Caribbean reefs from 5 to 6 Ma, but following a 1 Myr transition period of mixed acroporid-pocilloporid asemblages, acroporids became dominant in reef communities in the early Pleistocene (approx. 1.6 Ma). Acroporids may not, however, have achieved levels of extreme abundance until the late Pleistocene (approx. 0.5 Ma) (Budd and Kievman 1994). With this as yet unexplained rise to dominance of branching *Acropora*, and a corresponding decline in massive, domal corals, coral reef communities with a completely modern aspect appeared about 0.5 Ma. Except for the extinction of *Pocillopora* in the Caribbean at about 60 ka, the patterns of community membership and dominance of coral species appears to have been highly predictable for at least the past 125 Kyr (Pandolfi and Jackson 2001).

The Phanerozoic witnessed major turnovers of reef biotas, mass and minor extinction events, and profound changes in the chemistry of sea water. This section explores the effects of biological innovations and extrinsic controls upon reef ecology and taphonony.

5.1 *Rise of Biological Disturbance*

Many researchers have emphasized the importance of herbivores and large marine vertebrates to the healthy functioning of coral reefs (see Wood 1999), and this is corroborated by analysis of the fossil record. A dramatic escalation of new organisms with innovative and destructive feeding methods occurred from the mid-Jurassic to Miocene (Table 1); indeed a taxon-independent morphological signal of herbivory is not recorded until the Eocene (Bellwood 2003). In particular, the arrival of piscine herbivores had the potential to fundamentally alter the dynamics of reef and other benthic marine communities.

In general, herbivorous grazers and carnivores throughout the Paleozoic and early Mesozoic were relatively small individuals with limited foraging ranges incapable of excavating calcareous substrates. A radiation during the Devonian of durophagous, mobile predators has been proposed by Signor and Brett (1984), but these forms probably relied upon manipulation only to crush or ingest (Harper and Skelton 1993). By the early Mesozoic, sessile organisms had to contend with an increasing battery of novel and more advanced feeding methods, as well as sediment disruption due to deep bioturbating activity (see summary in Vermeij 1987). Most notable was the rise of efficient excavation behaviours.

Bioerosion notably increased in intensity from the mid- to Late Jurassic. A radiation of endoliths occurred from the Triassic onwards, with deep borers (capable of penetration greater than 50 mm) appearing from the Jurassic. Clionid sponges – one of the major bioeroders on modern coral reefs – had become abundant by the latest Jurassic. The first live-borers are known from the Eocene (Krumm and Jones 1993), as are fishes similar to modern reef faunas (50 Ma) (Bellwood 1996). The ability for substantial excavation of hard substrata over large areas increased considerably from the latest Cretaceous-Early Tertiary when deep-grazing limpets, camerodont sea urchins, and especially the reef fishes appeared. The complex pharyngeal apparatus of labrids was present at this time, and major labrid clades were already differentiated (Bellwood 1997). Balistids first appeared in the Oligocene, and the oldest scarid fossil capable of deep excavation currently known is from the Miocene (14 Ma) (Bellwood and Schulz 1991). It seems likely that sometime during the Oligocene – Miocene, reef bioerosion gained a modern caste (Pleydell and Jones 1988).

The abundance of reef fishes is assumed to be of great importance on coral reefs, as evidenced by the dramatic increase of algal growth as a result of their decline on Jamaican reefs (Hughes 1994). Tropical marine hard substrata are usually sparsely vegetated, but a rich algal flora develops when herbivorous fish are excluded and/or nutrient input increases. Grazers not only promote the dominance of corals and coralline algae on coral reefs, they also contribute notably to carbonate sediment production and redistribution, algal ridge formation, and the maintenance of overall diversity. Like other predators, they can also ameliorate the effects of competition and may combine with physical controls to produce the characteristic zonation of modern coral reefs.

The major causes and indirect effects of predation, particularly herbivory, on coral reef communities (Table 2) are such that a series of effects on reef ecology and taphonomy can be predicted. In the sections following, these predictions are tested.

5.2 *Response to Increase in Disturbance*

Only skeletal anatomy and morphology, spatial distribution, and skeletal attack or breakage, and regeneration might be detected – or inferred – in the fossil record of reef organisms.

Table 2 Predicted changes in reef community ecology and taphonomy based on the rise to abundance of new predatory methods and endoliths as evidenced in the fossil record (After Wood 1999)

Event	Prediction	Timing
The rise of macroherbivores	A shift to more conspicuous, well-defended macroalgae (coralline algae) on reefs	Late Mesozoic-Eocene
The rise of specialized predators	Increase in diversity and retardation of dominance; reducing or preventing competition	
	Limiting of foraging ranges	Late Mesozoic-Eocene
	Zonation: Interaction of physical controls with differential effects of damselfish in the survival of different coral species	Eocene
The rise of excavatory grazers and predators	A shift to organisms with deterrent traits and those which tolerate partial mortality	Late Mesozoic to Miocene
	Increase in multiserial, branching corals	Cretaceous onwards
	Increase in the diversity of the cryptos and other spatial refugia	Jurassic onwards
	Algal ridge formation by coralline algae	Eocene
Rise of intense bioerosion and endoliths	Reduced reef framework preservation	Late Mesozoic to Miocene
	Sediment grain size reduction	Late Mesozoic to Miocene
	An increase in skeletal sediment production	Late Jurassic
	Increase in multiserial scleractinian corals	Throughout history of the group
The rise of parrotfish	Formation of sediment aprons	Miocene
	Thick coralline algal crusts	Miocene
	Reduction in rate of reef progradation	Miocene

Herein, the origin and diversification of such fossilizable traits are considered for Paleozoic reef-building cnidarians and skeletal sponges, coralline algae and scleractinian corals. The appearance of excavatory herbivores paralleled profound changes in reef ecology, including the rise of well-defended, highly tolerant coralline algae (Steneck 1985), a notable increase in branching corals since the Late Cretaceous (Jackson and McKinney 1991), and the loss of many functional organisms that prove to be intolerant to excavatory attack (Table 2). This suggests a cause-effect system where adaptation to predatory attack has been intimately bound to the origin and assembly of modern reefs.

5.2.1 Secure Attachment to a Hard Substrate

Organisms without secure attachment to a stable substrate are susceptible to the effects of disturbances. In modern shallow shelf seas, immobile epifauna are typically excluded from most soft-substrates, which are dominated by mobile deposit feeders. While immobile, but unattached corals are common today, they are restricted mainly to areas protected from high biological and physical disturbance. Most modern suspension-feeders require a hard-substrate, even if these are only isolated patches within areas of unstable, soft-substrate ('benthic islands', or in dense aggregations).

Possession of an edge zone in all but the most primitive scleractinian corals allows them to gain permanent attachment to a stable substrate. As a result, scleractinian corals dominate modern reef framework environments, especially those in high-energy settings where there is also an abundance of wave-swept, extensive hard substrata for colonization. Permanent attachment also allows the development of very large branching morphologies.

Cambrian archaeocyath sponges usually bore small holdfasts that enabled limited attachment to hard substrates (Fig. 2). But many mid- to late Paleozoic reefs were dominated by large, sheet-like invertebrates (stromatoporoid sponges, tabulate and rugose corals, and trepostome and cystoporate bryozoans) that were initially attached to small, ephemeral skeletal debris and then grew over the surrounding sediment (Fig. 6). Small, branching forms (some stromatoporoids and bryozoans) lacking extensive attachment sites were also common, and they were presumably partially rooted in soft-sediment. For most Paleozoic metazoan reef builders there is little evidence for any active recruitment onto extensive hard substrates; these forms were unspecialized and immobile.

The late Paleozoic decline of immobile epifauna coincides with the rise of major bulldozing taxa, which passed through the end-Permian extinction unscathed. This coincidence must remain conjectural until tested experimentally.

Since their inception 3.5 billion years ago reefs have developed zonation in response to environmental gradients (see summary in Wood 1999). Detecting the exact nature of the added affect of damselfish herbivory, known to be an important determinant of modern coral reef zonation, will therefore be highly problematic to assess. Likewise, metazoan reefs have been differentiated into open surface and cryptic reef communities from their inception (Wood et al. 2002), and this together with the taphonomic loss of soft-bodied and preferential dissolution of skeletal organisms from cryptic habitats makes any meaningful quantification of changing diversity of these two settings through geological time virtually impossible.

5.2.2 Resistance to Partial Mortality

Predation that actively excavates underlying skeleton often results in sub-lethal damage. In such cases, the capacity of the prey to heal or replace damaged areas of soft tissue becomes critical to survival. Strategies that rely upon herbivores/predators to remove competing algae therefore often entail the loss of the prey's own

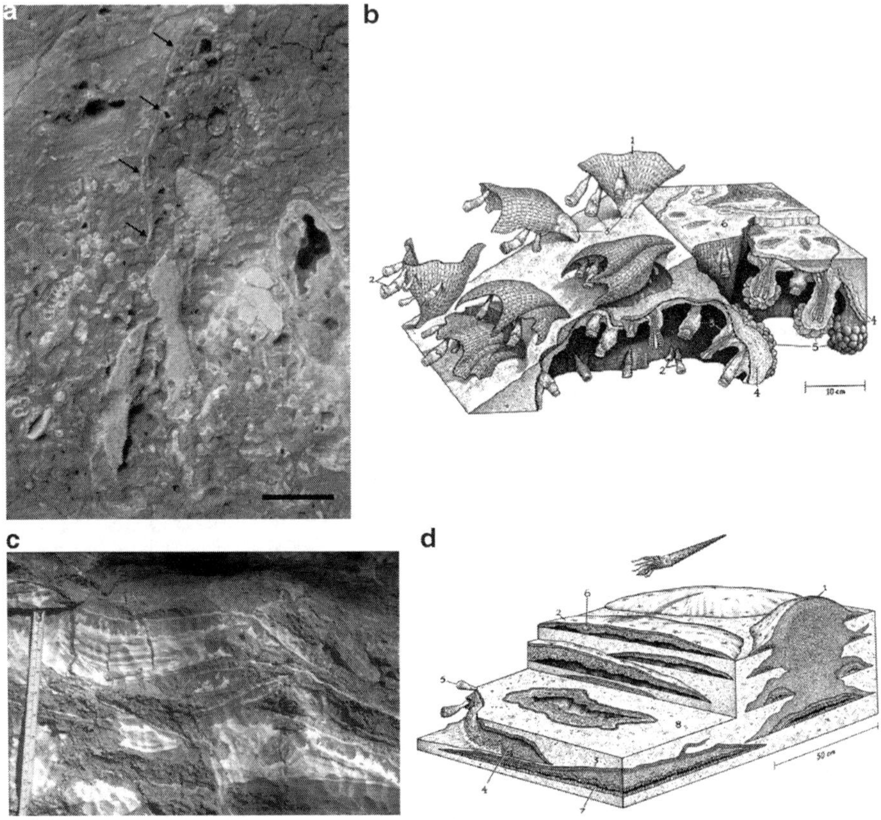

Fig. 6 (**a**) Permian Capitan reef (Middle Capitan) community of bryozoans (*arrowed*) with pendent sphinctozoan sponges. Remaining cavity space is filled with early botryoid cements, originally aragonitic now pseudomorphed to calcite; Upper Permian, Mckittrick Reef Trail, Texas, USA. Scale bar = 20 mm. (**b**) Reconstruction of Permian Capitan reef community 1. Frondose bryozoan (*Polypora* sp. and *Goniopora* sp.); 2. Solitary sphinctozoan sponges; 3. *Archaeolithoporella* (encrusting ?algae); 4. Microbialite; 5. Botryoidal cement; 6. Sediment (grainstone-packstone) (from Wood 1999; copyright John Sibbick), (**c**) Platy stromatoporoid sponge community, with cryptic *Shuguria*. Remaining cavity space is infilled with radiaxial calcite cement and sediment; Upper Devonian (Frasnian), Geikie Gorge, Western Australia. (**d**) Reconstruction of platy stromatoporoid sponge reef community 1. Domal stromatoporoid (*Actinostroma* sp.); 2. Laminar stromatoporoid (*Stachyodes australe*); 3. Tabular stromatoporoid; 4. *Shuguria* (calcified cyanobacterium); 3. Stalked lithistid sponge; 6. Spiny atrypid brachiopod; 7. Radiaxial fibrous calcite cement; 8. Sediment (From Wood 1999; copyright John Sibbick)

tissues. Algal turfs grow very rapidly and so can regenerate from basal portions that have escaped herbivory.

In coralline algae, a protective outer epithallus overlies the more delicate meristem, fusion cells allow the rapid translocation of photosynthates, and conceptacles that contain reproductive structures are enclosed within the perithallus. These structures have been demonstrated to protect the delicate reproductive anatomy from intensive grazing (Steneck 1982, 1983). Conceptacles are, however, no match for the deep excavation of parrotfishes, perhaps explaining why such structures are found only on non-tropical, thickened, crusts. Many coralline algae can also tolerate intense herbivory due to their ability to rapidly regenerate removed material (Steneck 1985, 1988). Thickened crusts are more tolerant to attack than thin encrusting or branching forms (Steneck 1985), but in modern reefs, the dominance of a particular growth form appears to be a trade-off between the cost of investment in increased defence, and the reduction in growth rate or competitive ability. As a result, thickened crusts dominate only in areas of high wave energy and biological disturbance. After the Eocene, herbivore-susceptible, delicately branched coralline algae reduced in abundance in the tropics, the proportion of thickened encrusting forms increased, and the first algal ridges appeared – all coincident with the rise of excavatory herbivorous fish (Steneck 1985, 1988).

Many sessile reef organisms possess a modular or colonial habit where partial predation and boring may remove either individual or a few modules, or large areas may be cleared of living tissue, sometimes together with the excavation of underlying skeleton. But the modular organization also reduces soft-tissue to a relatively thin veneer over a larger basal skeleton. This not only decreases accessibility and the ease of prey manipulation by predators, but also minimizes the tissue biomass while maximizing the cost of collection. For example, in a typical domal colony of *Porites*, only about 0.5% of the colony's radius is occupied by soft tissue (Rosen 1986). In branching and platy colony forms, the relative proportion of skeleton is even higher.

Cambrian archaeocyath sponges show a steady and marked increase in the proportion of complex modular forms during their history (Wood et al. 1992), as do scleractinian corals since the mid-Triassic, which appears to be uninterrupted by the end-Cretaceous extinction event (Coates and Jackson 1985).

5.2.3 Regeneration After Breakage

Some morphologies are more resistant to breakage than others. For example, colonies with closely spaced branches can make predator access difficult by forming hidden, protected areas. The flattening of branch terminations can also offer greater resistance to all forms of breakage and shearing, and this character is found in erect species of bryozoans, gorgonian corals and stylasterine corals.

A multi-serial modular organization, however, in addition to promoting architectural diversity and flexibility (Fig. 7), also allows compartmentalization of damage and enables some colonies to regenerate from fragments (Jackson and Hughes 1985).

Most significantly, branching corals also show tremendous powers of regeneration: *Acropora palmata* has one of the highest rates recorded (Bak 1983). Indeed, unlike massive, platy or encrusting forms, damage to branching corals often leads to an immediate increase in growth rate so causing an increase in size rather than simply repairing damaged tissue.

Populations of the staghorn coral (*Acropora cervicornis*) frequently form dense, monospecific stands on shallow Caribbean reefs, but there is little evidence of

Fig. 7 The variety of corals found on modern coral reefs showing flexibility of the modular habit, and the diversity of branching morphologies. 1. Cup-shaped soft-coral; 2. Columnar; 3. Free-living (solitary); 4. Digitate; 5. Encrusting; 6. Corymbose; 7. Caespitose; 8. Bottlebrush; 9. Massive; 10. Foliaceous (cup-shaped); 11. Foliaceous (whorl-forming); 12. Tables and plates; 13. Massive; 14. Arborescent (staghorn); 15. Arborescent (elkhorn) (From Wood 1999; copyright John Sibbick)

frequent sexual recruitment (Tunnicliffe 1981). The fragile organization of this species results in easy breakage due to high wave activity and bioerosion, especially by boring sponges that infest the colony bases. However, such corals are able to re-anchor fragments and rapidly regenerate and grow, often fusing with other colonies, at rates up to 150 mm/year (Tunnicliffe 1981). Such branching corals have turned adversity into considerable advantage, and appear to flourish because, and not in spite, of breakage.

The percentage of scleractinian erect species (mainly low integration phaceloid-dendroid growth forms) decreased until the Turonian, but increased markedly – particularly in multi-serial forms with inferred rates of rapid regeneration – after that time (Coates and Jackson 1985). This spectacular rise of various morphologies of branching forms (Fig. 7) was coincident with the appearance of new groups of predatory excavators.

All families of modern scleractinian corals that dominate reefs today spread throughout Tethys during the Eocene. The poritids, their relatives the actinids, and the favids (which had survived the Cretaceous extinction) dominate most coral reef communities throughout much of the Cenozoic (McCall et al. 1994). Although branching acroporoids appeared in the Eocene, they did not dominate reefs until early Pleistocene. The rise of this group – with its particularly remarkable powers of regeneration from fragmentation and rapid growth – would then seem to be independent of any known changes in predation style.

5.2.4 Patterns of Sediment Removal and Storage

Anecdotal evidence suggests that the proportion of reef framework preserved in situ before the Jurassic (Wood 1999) is greater than that occurring today. Many Paleozoic reefs commonly preserve intact reef frameworks, even of fragile biota such as frondose bryozoans (Fig. 6a, b) or platy stromatoporoid sponges (Fig. 6c, d). Such preservation was aided, in part, by abundant and probably rapid syn-sedimentary lithification, particularly cementation.

Almost nothing is known as to possible changes in the style of skeletal sediment production and distribution within reefs after the appearance of abundant bioerosion from the Late Jurassic, especially after the appearance of reef fish in the Eocene, and the rise of the scarids in the Miocene. We might predict that substantial aprons of sediment may not have been present on pre-Eocene reefs. Likewise in the absence of the grain size reduction activities of clionid sponges, echinoids and fish, mean sediment grain size may have been more coarse prior to the late Jurassic, perhaps resulting in a reduced net loss of carbonate to the system through the removal of fines. It is possible also that the modern style of coral reef lagoon may also not have appeared until the late Jurassic or later.

Also, barely explored are sedimentological consequences of differences in the geographical distribution of bioeroders – which is especially marked in fish populations due to differential extinction in the Atlantic during the mid–late Cenozoic (Bellwood 1997). This extinction resulted in the conspicuous loss of large excavating scarids from Caribbean reefs. Compared to the Atlantic, it may be predicted that modern Indo-Pacific reefs show the formation of larger slope sediment aprons, reduced rates of progradation of the reef crest, and a greater loss of carbonate in the form of fine grains in suspension from the system. These differences in sediment dynamics, however, require further quantification.

5.3 Response to Changing Seawater Chemistry: Secular Changes in Mineralogy

The dominant form of precipitated crystalline $CaCO_3$ has oscillated during the geological past, with both inorganic and organic production of aragonite and high-Mg calcite dominating carbonate formation during cool (icehouse) periods, and low-Mg calcite predominating during warm (greenhouse) periods (Sandberg 1983; Stanley and Hardie 1998). Such mineralogical shifts are interpreted as markers for major changes in seawater chemistry. Stanley and Hardie (1998) proposed that it was shifts in Mg:Ca that has controlled the predominance of calcite versus aragonite secretors, particularly reef builders, due to the inhibiting effect of high Mg^{2+} concentration on calcite secretion. Experimental work has subsequently confirmed the profound influence of Mg:Ca sea water ratios on modern reef builders, including scleractinian corals (Ries et al. 2004) and *Halimeda* algae (Ries 2006).

Scleractinian corals were dominant reef-builders in the Jurassic, but they did not build extensive reefs during the greenhouse period (calcite seas) of the Cretaceous. During this period, their species diversity remained high but with lower abundance on carbonate platforms compared to the Jurassic, and with a distribution shifted to outer platform settings and higher latitudes (~35–45°N; Rosen and Turnsek 1989). There are many hypotheses offered to explain these observations, including the high temperatures, restricted circulation, unstable sediment conditions of Cretaceous platforms, and the favouring of the calcite-producing rudist bivalves over aragonite corals (Wood 1999; Steuber 2002).

The role of changing seawater chemistry on the selective loss of aragonitic and high-Mg skeletal faunal is explored further in Cherns et al. (this volume). They argue that the fossil and skeletal grain record, particularly in siliciclastic and low-energy carbonate settings are markedly under-represented in these metastable carbonate minerals due to selective dissolution of during calcite seas. This is likely to hold true also to some extent in the reef record, but the loss would be predicted to be far less in in situ frameworks which became syn-sedimentarily encased by secondary framework and early marine cements.

5.3.1 Changing Styles of Early Diagenesis

Evidence is persuasive that changing global seawater chemistry has exerted secular changes in the dominant carbonate mineralogy of reef organisms (Stanley and Hardie 1998). It is likely, also, that seawater chemistry has also influenced the style of early diagenesis in carbonate regimes.

Hardgrounds, synsedimentary lithified seafloors, are found almost exclusively during periods of calcite seas (Wilson and Palmer 1992) due to the elevated abundance of calcium ions. Enhanced rates of calcite cementation during these times may have aided preservation of otherwise vulnerable biota to disturbance, particularly in crypts, and promoted rapid lithification of the reef framework, but this has yet to be documented.

The mineralogy of early marine reef cements also seems to follow the same secular changes (Wood 1999). For example, aragonitc botryoids are known exclusively from phases of aragonite seas (Early Cambrian, mid-Carboniferous to Early Jurassic, and mid-late Cenozoic), and while radiaxial calcite is unknown from the Quaternary, it is common in reefs that grew in calcite seas (particularly the Ordovician to Devonian).

There is also some limited evidence for enhanced sea-floor dissolution of aragonite during calcite seas (Palmer et al. 1988), but this requires further documentation. If present, such dissolution may have direct taphonomic consequences (explored further in Cherns et al. this volume). The exact nature of the control of sea water chemistry on all these diagenetic phenomena and the subsequent effects upon reef taphonomy remain to be quantified and tested experimentally.

6 Current Global Change and Taphonomy

How does the deep past, facilitate prediction of the taphonomic response of reefs to current global, anthropogenically-mediated, change, and to what extent might the processes that operated in the absence of anthropogenic change be at work today? This section concentrates on processes known to be important agents of current change and destruction in modern reefs.

6.1 Loss of Herbivores and Higher Predators

Many researchers have summarized the case for the importance of herbivores and large marine vertebrates to the healthy functioning of coral reefs. Jackson et al. (2001) present multiple historical data over a range of scales and biogeographic realms to show how overfishing of key marine vertebrates has been the major cause of the profound ecological changes seen on corals reefs (and other coastal ecosystems). These authors argue that overfishing may also be a necessary precondition for additional sources of degradation – such as eutrophication, and outbreaks of disease or gregarious species – to occur. The superimposition of multiple factors leads to feedbacks that cause increased vulnerability due to complex synergies, and these are far from understood.

Reduction of reef herbivore populations will almost certainly lead to an increase in soft-bodied algal biomass, and a decrease in coral cover. In turn, this may lead to enhanced rates of bioerosion, particularly in areas of eutrophication or outbreaks of disease. It is likely that such widespread predicted coral mortality will cause highly degraded coral fragments to dominate death assemblages and the subsequent fossil record due to widespread coral mortality, endolithic algal activity, and bioerosion (Precht and Aronson 1997).

6.2 Changing Storm Patterns

The behaviour of hurricanes and storms has been reviewed by Reigl (2007). The frequency of Atlantic hurricanes appears to follow 15–20 year cycles, and since the mid-1990s a period of more vigorous hurricane activity has begun. He suggests that the frequency of such storms is not predicted to increase under conditions of global warming, but peak intensities and their relative moisture content may increase, which will notably increase their powers of destruction. Tropical cyclone basins may also shift, so exposing more (or less) reef areas to their effects. This is likely to increase damage until acclimatization can take place.

Increasingly powerful tropical storms are predicted to reduce the proportion of in situ reef framework preserved and to increase all the metrics of coral death assemblage fidelity in the sedimentary record, but such a signature would only be detectable where both the life and death assemblages were preserved and distinguished, perhaps via changed in epibiont encrustation successions.

6.3 Rise in Sea Level

Sea level is expected to rise by about 0.5 m during this century (Houghton et al. 2001), two orders of magnitude less than the 120 m rise since the last glacial maximum. Reefs are not considered to be directly threatened by sea-level rise in terms of drowning (except where no suitable substrates for colonization are present) as the geological record of reefs shows extraordinary robustness in response to catastrophic sea-level change (Macintyre 2007). There may, however, be many other indirect effects of sea-level rise that could have an impact on some reefs: decreasing light-dependent calcification rates will severely restrict rates of reef growth potentially leading to drowning, and nutrients and sediments released from newly flooded coastlines could lead to degradation of water quality. Many of these scenarios will enhance bioerosion rates on reefs.

6.4 Rises in CO_2 and Global Temperature

According to the IPCC's Special Report on Emission Scenarios (Nakićenović and Swart 2000), atmospheric CO_2 concentrations within this century are predicted to reach between about 555 and 825 ppmv. Such a rise represents a doubling of the pre-industrial concentration by the middle of this century, and other greenhouse gases (CH_4, N_2O, H_2O) will increase as well (Houghton et al. 2001). The range of predicted temperature increase among models included in the Third IPCC Report is large (1.4–5.8°C for the period 1990–2100; Houghton et al. 2001), with most coupled models indicating greater warming at high latitudes than within the tropics (Kleypas 2007).

Carbonate-rich sediments at shallow ocean depths (<200 m) represent a major $CaCO_3$ reservoir that can rapidly react to decreasing saturation state of seawater with respect to carbonate minerals produced by rising atmospheric pCO_2. Kleypas (2007) suggest that the current rapid rate of increase in atmospheric CO_2 concentration is potentially catastrophic for regulation of Earth's climate and carbonate system, as the timescales of natural feedbacks required to return these systems to equilibrium are far greater than the timescale of fossil fuel burning. There is also the possibility that emergent diseases (pathogens) that thrive in warmer oceans will increase. This may lead to a synergistic effect in that such pathogens may preferentially attack an already vulnerable and weakened reef biota.

While reefs that formed during the Paleocene and Eocene may provide important clues in terms of certain physical reef characteristics such as calcification rates and distribution patterns, they are probably less useful as analogues for current ecological response as most of the dominant modern coral reef species, notably acroporids, had not appeared by that time.

The massive coral reef bleachings of the last two decades which have led to widspread coral mortality are probably unprecedented within this century and for several preceding centuries (Aronson et al. 2000). They are closely associated with abnormally warm sea surface temperature, and the clear inference is that global warming is their cause, exacerbated by other factors such as subaerial exposure, increased penetration of UV light, and decreased water circulation.

Reef-building corals and other symbiotic organisms can adapt to increasing temperatures through a range of mechanisms, including short-term acclimation, medium-term acclimatization, and even natural selection (Coles 2001). While corals will vary in their required acclimation periods, it is likely, however, that many corals will be unable to acclimatize sufficiently rapidly such that bleaching events with increased coral mortality will increase in frequency and strength over coming decades.

6.5 *Changes in Sea-Water Chemistry*

Although predicting seawater chemistry changes in the surface ocean over the short term (e.g. 100–200 years) is fairly straightforward, these predictions are complicated by biological response to increased pCO_2. Calcification of reef-building organisms decreases as pCO_2 increases, while organic carbon production may increase (Riebesell et al. 2001). Accurate prediction of surface seawater chemistry changes over the next one to two centuries will therefore depend on how well we predict both atmospheric CO_2 changes, and the biological responses and feedbacks (Kleypas 2007). Results from the Hamburg Model of the Ocean Carbon Cycle (HAMOCC) coupled with a carbonate sediment diagenesis model (Archer et al. 1997), predict that these processes will, however, require thousands of years to bring the carbonate system back to pre-industrial conditions.

All experimental work on natural reefs and in artificial conditions shows that calcification rates decrease and dissolution increases as the calcium carbonate saturation state declines (Gattuso et al. 1996; Suzuki and Kawahata 1999; Kayanne et al. 2003). Experimental evidence indicates that biogenic calcification rates are already 10–20% lower than they were under pre-industrial conditions. Kleypas et al. (1999) estimated that the average calcification rate on reefs may have already declined by 6–14% as atmospheric pCO_2 has increased from 280 ppmv to the present-day value of 370 ppmv. Halley and Yates (2000) estimated that on a reef in Hawaii, the dissolution rate will equal the calcification rate when atmospheric CO_2 concentrations reach double pre-industrial levels.

Atmospheric pCO_2 prior to the Miocene probably remained higher than today but that Mg:Ca ratio was probably lower than that of today (Wilson and Opdyke 1996), so that the ocean chemistry of the near future cannot be adequately compared to any past Tertiary time period (B.N. Opdyke, pers. comm. in Kleypas 2007). Ocean chemistry of the near future will be unique and extraordinary, mainly because the rapidity of the increase in atmospheric CO_2 will drive the system out of equilibrium.

Rates of reef cement precipitation are also likely to change in response to seawater chemistry changes. As outlined in the review of Kleypas et al. (2001) these factors together with increased rates of dissolution will almost certainly lower the net carbonate deposition on reefs, and so reduce reef-building potential globally.

7 Summary

Reefs have been subject to markedly shifting changes in taphonomic processes through the Phanerozoic (Fig. 8). Biological disturbance has clearly escalated since the Mesozoic. Reef biotas have responded with the proliferation of traits with proven anti-predatory benefits, particularly rapid regeneration after partial mortality. Indeed some modern dominant reef taxa, such as branching corals and coralline

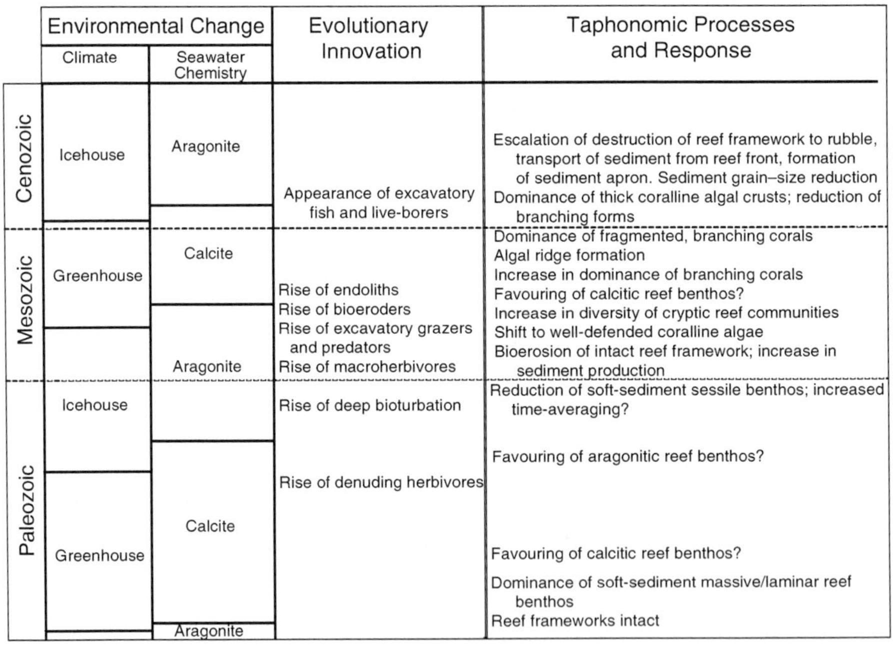

Fig. 8 Summary of major environmental changes and evolutionary innovations through the Phanerozoic that created taphonomic processes and responses on reefs. Approximate position of global climate states (Icehouse; Greenhouse) from Fischer (1983), and seawater chemistry states (aragonite or calcite seas) from Sandberg (1983)

algae, appear not only to thrive, but actually require conditions of considerable disturbance for their survival in shallow tropical seas.

Many modern reefs are largely reduced to rubble and sand via physical abrasion and particularly bioerosion. These reefs, their fossil assemblages, and final geological expression, are all dominated by branching corals due to their high diversity and abundance in living communities, propensity to proliferate via fragmentation, and resilience to taphonomic destruction. By contrast, many pre-Jurassic reefs, prior to the escalation in bioerosion, show the common preservation of intact, in situ frameworks, and a dominance of massive or laminar growth forms. In addition, most of the unattached, soft-sediment dwelling organisms typical of Paleozoic reefs appear to have become largely absent from shallow marine tropical reef biotas during the late Paleozoic to early Mesozoic, perhaps due to intolerance of deep burrowing taxa and excavatory attack.

Scleractinian corals, in particular branching taxa, show a marked increase in the proportion of forms with complex modularity from the Eocene onwards, even though corals displayed the full range of morphological forms and corallite size by the Late Triassic. Highly defended, thick crusts in coralline algae become more dominant, and branching forms also become noticeably less conspicuous, on reefs from the Eocene onwards. This major reorganization of the coral reef ecosystem coincides with the rapid appearance and radiation of herbivorous and corallivorous reef fish, but this remains to be tested experimentally.

We know little as to possible changes in the style of skeletal sediment production and distribution within reefs after the appearance of abundant bioerosion in the Late Jurassic, especially after the appearance of reef fish in the Eocene, and particularly with the rise of the scarids in the Miocene. In addition to the loss of common intact reef frameworks, we might predict that substantial aprons of reef-slope sediment may not have been present on pre-Miocene reefs, and likewise in the absence of the grain-reduction activities of reef fish, echinoids, and clionid sponges, the size distribution of sediment grains may have been more coarse due to the dominance of physical abrasion on Palaeozoic reefs. We also need to quantify the taphonomic and sedimentological consequences of differences in the geographical distribution of bioeroders, particularly a comparison between the Cenozoic history of Caribbean reefs that lack large excavating scarids after the Miocene, and Indo-Pacific reefs where they remained and flourish to this day (Bellwood 1997). It is also possible that the modern style of coral reef lagoon may also not have appeared until the late Mesozoic-Cenozoic, and that the carbonate budget of reef systems has shifted to greater net loss due to removal of fines since the late Jurassic.

Evidence is persuasive that changing global seawater chemistry has exerted secular changes in the dominant carbonate mineralogy of reef organisms (Stanley and Hardie 1998), and it is likely, also, that seawater chemistry has influenced the style of early diagenesis in carbonate regimes. Enhanced calcite cementation during phases of elevated Ca^{2+} availability (calcite seas) may have aided preservation of otherwise vulnerable reef biota to disturbance, particularly within cavities, and promoted rapid lithification of the reef framework, but in turn may have promoted selective dissolution of aragonitic and high-Mg skeletal biota. Aragonitic reef framework biota preserved by secondary encrusters and early (aragonitic) cements that grew in aragonitic seas may not suffer such loss. The nature of the control of sea water chemistry on all these diagenetic phenomena and their subsequent effects upon reef taphonomy remain to be documented.

Climate change and extinction are a persistent motif of the geological record and mass-extinctions are likely to impact upon taphonomic processes (Fraiser et al. this volume). The current phase of climate change and extinction will exert a profound effect upon reef ecology and taphonomy of reefs. Reduction of reef herbivore populations will almost certainly lead to an increase in soft-bodied algal biomass, and a decrease in coral cover, particularly in areas of eutrophication or outbreaks of disease. Bleaching as a result of global warming may lead to significant or widespread coral mortality. The indirect effects of sea-level rise and increased storm intensity may severely restrict rates of reef growth potentially leading to drowning; nutrients and sediments released from newly flooded coastlines could lead to eutrophication. Evidence indicates that biogenic calcification rates are already between 6% and 20% lower than they were under pre-industrial conditions due to sea water chemistry changes. Continuing accelerating rates of dissolution and a reduction in reef calcification will almost certainly lower the net carbonate deposition on reefs, and reduce reef-building potential globally. All these processes will cause loss of the structural integrity of reefs. Future death assemblages and the subsequent fossil record of reefs will be dominated by highly degraded coral fragments and grains with limited in situ reef frameworks, widespread coral mortality, endolithic algal activity, and intense bioerosion.

References

Alexandersson, T. (1972). Micritization of carbonate particles: Processes of precipitation and dissolution in modern shallow-marine sediments. *Bulletin Geological Institute University Uppsala NS, 3*, 201–236.

Archer, D., Kheshgi, H., & Maier-Reimer, E. (1997). Multiple timescales for neutralization of fossil fuel CO_2. *Geophysical Research Letters, 24*, 405–408.

Aronson, R. B., & Ellner, S. P. (2007). Biotic turnover events on coral reefs: A probabilistic approach. In R. B. Aronson (Ed.), *Geological approaches to coral reef ecology, ecological studies*. New York: Springer.

Aronson, R. B., & Precht, W. F. (1997). Stasis, biological disturbance, and community structure of a Holocene coral reef. *Paleobiology, 23*, 326–346.

Aronson, R. B., Precht, W. F., Macintyre, I. G., & Murdoch, T. J. T. (2000). Coral bleach-out in Belize. *Nature, 405*, 36–38.

Bak, R. P. M. (1983). Neoplasia, regeneration and growth in the reef-building coral *Acropora palmata*. *Marine Biology, 77*, 221–227.

Bak, R. P. M., & Nieuwland, G. (1995). Long-term change in coral communities along depth gradients over leeward reefs in the Netherlands Antilles. *Bulletin of Marine Science, 56*, 609–619.

Bellwood, D. R. (1995). Carbonate transport and within-reef patterns of bioerosion and sediment release by parrotfishes (Scaridae) on the GBR. *Marine Ecology Press Series, 117*, 127–136.

Bellwood, D. R. (1996). Coral reef crunchers. *Nature Australia, 25*, 48–55.

Bellwood, D. R. (1997). Reef fish biogeography; habitat associations, fossils and phylogenies. Proceedings of the 8th international coral reef symposium. *Panama, 2*, 1295–1300.

Bellwood, D. R. (2003). Origins and escalation of herbivory in fishes: A functional perspective. *Paleobiology, 29*, 71–83.

Bellwood, D. R., & Schulz, O. (1991). A review of the fossil record of the parrotfishes (family Scaridae) with a description of a new *Calatomus* species from the middle Miocene (Badenian) of Austria. *Annalen Naturhistorisches Museum Wien, 92*, 55–71.

Best, M. M. R., Burniaux, P., & Pandolfi, J. M. (2004). Experimental bivalve taphonomy in reefs of Madang Lagoon, Papua New Guinea. In M. M. R. Best & J. B. Caron (Eds.), *Canadian paleontology conference proceedings, no. 2. Geological Association of Canada Special Publication*, Calgary.

Bishop, D., & Greenstein, B. (2001). The effects of hurricane Floyd on the fidelity of coral life and death assemblages in San Salvador, Bahamas: Does a hurricane leave a signature in the fossil record? *Geological Society of America, Abstracts with Programs, 34*(2), A7.

Brett, C. E. (1988). Paleoecology and evolution of marine hard communities: An overview. *Palaios, 3*, 374–378.

Bromley, R. G. (1975). Comparative analysis of fossil and recent echinoid bioerosion. *Palaeontology, 18*, 725–739.

Bromley, R. G. (1992). Bioerosion: Eating rocks for fun and profit. In C. G. Maples & R. R. West (Eds.), *Trace fossils: Short courses in paleontology 5* (pp. 121—129). University of Tennessee.

Budd, A. F., & Kievman, C. M. (1994). Coral assemblages and reef environments in the Bahamas Drilling Project cores. In Final draft report of the Bahamas drilling project, 3. Coral Gables, Florida. Rosensteil School of Marine and Atmospheric Science, University of Miami.

Budd, A. F., Stemann, T. A., & Johnson, K. G. (1994). Stratigraphic distribution of genera and species of Neogene to Recent Caribbean reef corals. *Journal of Paleontology, 68*, 951–977.

Coates, A. G., & Jackson, J. B. C. (1985). Morphological themes in the evolution of clonal and aclonal marine invertebrates. In J. B. C. Jackson, L. W. Buss, & R. E. Cook (Eds.), *Population biology and evolution of clonal organisms*. New Haven: Yale University Press.

Coles, S. L. (2001). Coral bleaching: What do we know and what can we do? In *Proceedings of the workshop on mitigating coral bleaching impact through MPA design*, Honolulu, HI.

Connell, J. H. (1997). Disturbance and recovery of coral assemblages. *Coral Reefs, 16*, 101–113.

Cummins, H., Powell, E. N., Stanton, R. J., Jr., & Staff, G. (1986). The size-frequency distribution in palaeoecology: Effects of taphonomic processes during formation of molluscan death assemblages in Texas bays. *Palaeontology, 29*, 495–518.

DeVantier, L. M., & Done, T. J. (2007). Inferring past outbreaks of the crown-of-thorns seastar from scar patterns on coral heads. In R. B. Aronson (Ed.), *Geological approaches to coral reef ecology, ecological studies 192* (pp. 85–125). New York: Springer.

Edinger, E. N., Pandolfi, J. M., & Kelley, R. A. (2001). Community structure of quaternary coral reefs compared with recent life and death assemblages. *Paleobiology, 27*, 669–694.

Fischer, A. G. (1983). Long-term climatic oscillations recorded in stratigraphy. In W. Berger (Ed.), *Climate in earth history*. Washington, DC: National Academy of Sciences.

Fürsich, F. T. (1977). Corallian (Upper Jurassic) marine benthic associations from England and Normandy. *Palaeontology, 20*, 337–385.

Gardner, T. A., Côté, I. M., Gill, J. A., Gran, A., & Watkinson, A. R. (2003). Long-term region-wide declines in Caribbean corals. *Science, 301*, 958–960.

Gattuso, J.-P., Pichon, M., Delesalle, B., Canon, C., & Frankignoulle, M. (1996). Carbon fluxes in coral reefs. I. Lagrangian measurement of community metabolism and resulting air-sea CO_2 disequilibrium. *Marine Ecology Progress Series, 145*, 109–121.

Ginsburg, R. N. (1964). *South Florida carbonate sediments. GSA annual meeting, Guidebook for field trip #1*. GSA, Boulder.

Glynn, P. W. (1993). Coral reef bleaching: Ecological perspectives. *Coral Reefs, 12*, 1–7.

Glynn, P. W. (2000). El Niño-Southern Oscillation mass mortalities of reef corals: A model of high temperature marine extinctions? In E. Insalaco, P. W. Skelton, & T. J. Palmer (Eds.),

Carbonate platform systems: Components and interactions, Geological Society of London, Special Publication 178 (pp. 117—133).

Goreau, T. F. (1959). The ecology of Jamaican coral reefs 1. Species composition and zonation. *Ecology, 40*, 67–90.

Grammer, G. M., Crescini, C. M., McNeill, D. F., & Taylor, L. H. (1999). Quantifying rates of syndepositional marine cementation in deeper platform environments-new insight into a fundamental process. *Journal of Sedimentary Research, 69*, 202–207.

Grassle, J. F. (1973). Variety in coral reef communities. In O. A. Jones & R. Endean (Eds.), *Biology and geology of coral reefs II, biology 1* (pp. 247–270). New York: Academic.

Greenstein, B. J. (1989). Mass mortality of the west Indian echinoid *Diadema antillarum* (Echinodermata: Echinoidea): A natural experiment in taphonomy. *Palaios, 4*, 487–492.

Greenstein, B. J. (2007). Taphonomy: Detecting critical events in fossil reef-coral assemblages. In R. G. Aronson (Ed.), *Geological approaches to coral reef ecology, ecological studies 192* (pp. 31–60). Springer: New York.

Greenstein, B. J., & Moffatt, H. A. (1996). Comparative taphonomy of modern and Pleistocene corals. *San Salvador, Bahamas, Palaios, 11*, 57–63.

Greenstein, B. J., & Pandolfi, J. M. (2003). Taphonomic alteration of reef corals: Effects of reef environment and coral growth form II: The Florida keys. *Palaios, 18*, 495–509.

Greenstein, B. J., Curran, H. A., & Pandolfi, J. M. (1998). Shifting ecological baselines and the demise of *Acropora cervicornis* in the Western Atlantic and Caribbean Province: A Pleistocene perspective. *Coral Reefs, 17*, 249–261.

Halley, R. B., & Hudson, J. H. (2007). Fidelity of annual growth in Montastrea faveolata and the recentness of coral bleaching in Florida. In R. G. Aronson (Ed.), *Geological approaches to coral reef ecology, ecological studies 192* (pp. 126–160). New York: Springer.

Halley, R. B., & Yates, K. K. (2000). Will reef sediments buffer corals from increased global CO_2? *9th international coral reef symposium*, Bali, Abstract.

Harper, E. M., & Skelton, P. W. (1993). The Mesozoic marine revolution and epifaunal bivalves. *Scripta Geologica (Special Issue), 2*, 127–153.

Hay, M. E. (1984). Patterns of fish and urchin grazing on Caribbean coral reefs: are previous results typical? *Ecology, 65*, 446–454.

Highsmith, R. C. (1980). Geographic patterns in coral bioerosion: A productivity hypothesis. *Journal of Experimental Marine Biology and Ecology, 46*, 177–196.

Houghton, J. T., Ding, Y., Griggs, D. J., Noguer, M., van der Linden, P. J., & Xiaosu, D. (Eds.). (2001). *IPCC third assessment report: Climate change 2001: The scientific basis*. UK: Cambridge University Press. 944 pp.

Hubbard, D. K. (1992). Hurricane-induced sediment transport in open-shelf tropical systems; an example from St. Croix, U.S. Virgin Islands. *Journal of Sedimentary Research, 62*, 946–960.

Hubbard, D. K. (2006). Coral growth versus reef accretion: Problems of scale, taphonomy and perception. *Geological Society of America Abstracts with Programs, 38*(4), 16.

Hubbard, D. K., Burke, R. B., & Gill, I. P. (1986). Styles of reef accretion along a steep, shelf-edge reef, St. Croix, U.S. Virgin Islands. *Journal of Sedimentary Research, 56*, 848–861.

Hubbard, D. K., Miller, A. I., & Scaturo, D. (1990). Production and cycling of calcium carbonate in a shelf-edge reef system (St. Croix, US Virgin Islands): Applications to the nature of reef systems in the fossil record. *Journal of Sedimentary Petroleum, 60*, 335–360.

Hughes, T. P. (1994). Catastrophes, phase shifts and large-scale degradation of a Caribbean coral reef. *Science, 265*, 1547–1551.

Hughes, T. P., Baird, A. H., Bellwood, D. R., Card, M., Connelly, S. R., Folke, C., et al. (2003). Climate change, human impacts, and the resilience of coral reefs. *Science, 301*, 929–933.

Humphrey, J. D., Ransom, K. L., & Matthews, R. K. (1986). Early meteoric diagenetic control of upper Smackover production, Oaks Field, Louisiana. *American Association of Petroleum Geologists Bulletin, 70*, 70–85.

Jackson, J. B. C. (1983). Biological determinants of present and past sessile animal distributions. In M. Tevesz & P. W. McCall (Eds.), *Biotic interactions in recent and fossil benthic communities* (pp. 39–120). New York: Plenum.

Jackson, J. B. C. (1991). Adaptation and diversity of reef corals. *BioScience, 41*, 475–482.

Jackson, J. B. C. (1992). Pleistocene perspectives of coral reef community structure. *American Zoology, 32*, 719–731.

Jackson, J. B. C., & Hughes, T. P. (1985). Adaptive strategies of coral-reef invertebrates. *American Science, 75*, 265–274.

Jackson, J. B. C., & McKinney, F. K. (1991). Ecological processes and progressive macroevolution of marine clonal benthos. In R. M. Ross & W. D. Allmon (Eds.), *Causes of evolution* (pp. 173–209). Chicago: University of Chicago Press.

Jackson, J. B. C., Kirby, M. X., Berger, W. H., Bjorndal, K. A., Botsford, L. W., Bourque, B. J., et al. (2001). Historical overfishing and the recent collapse of coastal ecosystems. *Science, 293*, 629–638.

James, N. P., Ginsburg, R. N., Marszalek, D. S., & Choquette, P. W. (1976). Facies and fabric specificity of early subsea cements in shallow Belize (British Honduras) reefs. *Journal of Sedimentary Petroleum, 46*, 523–544.

Kayanne, H., Kudo, S., Hata, H., Yamano, H., Nozaki, K., Kato, K., Negishi, A., Saito, H., Akimoto, F., & Kimoto, H. (2003). Integrated monitoring system for coral reef water pCO_2, carbonate system and physical parameters. *Proceedings 9th international coral reef symposium,* Bali.

Kendall, G. S. C., & Schlager, W. (1981). Carbonates and relative changes in sea level. In M. B. Cita & W. B. F. Ryan (Eds.), *Carbonate platforms of the passive-type continental margins, present and past. Marine Geology 44*, 181–212).

Kidwell, S. M., & Bosence, D. W. J. (1991). Taphonomy and time-averaging of marine shelly faunas. In P. A. Allison & D. E. G. Briggs (Eds.), *Taphonomy: Releasing the data locked in the fossil record* (Vol. 560, pp. 115–209). New York: Plenum.

Kleypas, J. A. (1997). Modeled estimates of global reef habitat and carbonate production since the last glacial maximum. *Paleoceanography, 12*, 533–545.

Kleypas, J. A. (2007). Constraints on predicting coral reef response to climate change. In R. G. Aronson (Ed.), *Geological approaches to coral reef ecology, ecological studies 192* (pp. 386–424). New York: Springer.

Kleypas, J. A., Buddemeier, R. W., Archer, D., Gattuso, J.-P., Langdon, D., & Opdyke, B. N. (1999). Geochemical consequences of increased atmospheric CO_2 on coral reefs. *Science, 284*, 118–120.

Kleypas, J. A., Buddemeier, R. W., & Gattuso, J.-P. (2001). The future of coral reefs in an age of global change. *International Journal of Earth Sciences, 90*, 426–437.

Krumm, D. K., & Jones, D. S. (1993). A new coral-bivalve association (*Actinastrea-Lithophaga*) from the Eocene of Florida. *Journal of Paleontology, 67*, 945–951.

Liddell, W. D., & Ohlhorst, S. L. (1988). Hard substrata community patterns, 1–120 m, North Jamaica. *Palaios, 3*, 413–423.

Lightly, R. G. (1985). Preservation of internal reef porosity and diagenetic sealing of submerged early Holocene barrier reef, southeast Florida shelf. In C. Schniedermann & P. M. Harris (Eds.), *Carbonate cements, SEPM 36* (pp. 123–151).

Lindberg, D. R., & Dwyer, K. R. (1983). The topography, formation and mode of home depression of *Collisella scabra* (Gould) (Gastropoda: Acmaeidae). *Veliger, 25*, 229–234.

Macintyre, I. G. (1985). Pre-burial and shallow-subsurface alteration of modem scleractinian corals. *Palaeontography of America, 54*, 229–244.

Macintyre, I. G. (1988). Modern coral reefs of western Atlantic: New geological perspective. *Bulletin of the American Association of Petroleum Geologists, 72*, 1360–1369.

Macintyre, I. G. (2007). Demise, regeneration, and survival of some western Atlantic reefs during the Holocene transgression. In R. G. Aronson (Ed.), *Geological approaches to coral reef ecology, ecological studies 192* (pp. 181–200). New York: Springer.

Macintyre, I. G., & Marshall, J. F. (1988). Submarine lithification in coral reefs: Some facts and misconceptions, *Sixth international coral reef congress, Townsville, Australia 1* (pp. 263–272). Townsville, Australia: Australian Institute of Marine Science.

Matthews, R. K., & Frohlich, C. (1987). Forward modeling of bank-margin carbonate diagenesis. *Geology, 15*, 673–676.

McCall, J., Rosen, B. R., & Darrell, J. (1994). Carbonate deposition in accretionary prism settings: Early Miocene coral limestones and corals of the Makhran Mountain Range in southern Iran. *Facies, 31*, 141–178.

Meyer, D. M., Bries, J. M., Greenstein, B. J., & Debrot, A. O. (2003). Preservation of an in situ reef framework in regions of low hurricane frequency: Pleistocene of Curacao and Bonaire, southern Caribbean. *Lethaia, 36*, 273–286.

Nakićenović, N., & Swart, R. (Eds.). (2000). *Emission scenarios 2000, Special Report of the IPCC*. Cambridge: Cambridge University Press. 570 pp.

Palmer, T. J., Hudson, J. D., & Wilson, M. A. (1988). Palaeoecological evidence for early aragonite dissolution in ancient calcite seas. *Nature, 335*, 809–810.

Pandolfi, J. M. (1996). Limited membership in Pleistocene reef coral assemblages from the Huon Peninsula, Papua New Guinea: Constancy during global change. *Paleobiology, 22*, 152–176.

Pandolfi, J. M. (2002). Coral community dynamics at multiple scales. *Coral Reefs, 21*, 13–23.

Pandolfi, J. M., & Greenstein, B. J. (1997). Preservation of community structure in death assemblages of deep water corals. *Limnology and Oceanography, 42*, 1505–1516.

Pandolfi, J. M., & Jackson, J. B. C. (2001). Community structure in Pleistocene coral reefs of Curacao reefs, Netherlands Antilles. *Ecological monographs, 72*, 49–67.

Pandolfi, J. M., & Jackson, J. B. C. (2007). Broad-scale patterns in Pleistocene coral reef communities from the Caribbean: Implications for ecology and management. In R. G. Aronson (Ed.), *Geological approaches to coral reef ecology, ecological studies 192* (pp. 201–236). New York: Springer.

Pandolfi, J. M., & Minchin, P. R. (1995). A comparison of taxonomic composition and diversity between reef coral life and death assemblages in Magang Lagoon, Papua New Guinea. *Palaeoceanography, Palaeoclimatology, Palaeoecology, 119*, 321–341.

Pandolfi, J. M., Bradbury, R. H., Sala, E., Hughes, T. P., Bjorndal, K. A., Cooke, R. J., et al. (2003). Global trajectories of the long-term decline of coral reef ecosystems. *Science, 301*, 955–958.

Perry, C. (2001). Storm-induced coral rubble deposition: Pleistocene records of natural reef disturbance and community response. *Coral Reefs, 20*, 171–183.

Peterson, C. H. (1976). Relative abundance of living and dead molluscs in two Californian lagoons. *Lethaia, 9*, 137–148.

Pleydell, S. M., & Jones, B. (1988). Boring of various faunal elements in the Oligocene-Miocene Bluff Formation of Grand Cayman. *Journal of Paleontology, 62*, 348–367.

Precht, W. B., & Aronson, R. B. (1997). Compositional changes in reef sediments related to changes in coral reef community structure. *AAPB Bulletin, 81*, 1561.

Reigl, B. (2007). Extreme climatic events and coral reefs: How much short-term threat from global change? In R. G. Aronson (Ed.), *Geological approaches to coral reef ecology, ecological studies 192* (pp. 315–341). New York: Springer.

Riebesell, U., Zondervan, I., Rost, B., Tortell, P. D., Zeebe, R. E., & Morel, F. M. M. (2001). Reduced calcification of marine plankton in response to increased atmospheric CO_2. *Nature, 407*, 364–368.

Ries, J. G. (2006). Aragonite algae in calcite seas: Effect of seawater Mg/Ca on codiacean biomineralization. *Journal of Sedimentary Research, 76*, 515–523.

Ries, J. G., Stanley, S. M., & Hardie, L. A. (2004). Scleractinian corals produce calcite, and grow more slowly, in artificial Cretaceous seawater. *Geology, 37*, 525–528.

Rosen, B. R. (1984). Reef coral biogeography and climate through the Late Cainozoic: just islands in the sun or a critical pattern of islands? *Geology Journal Special Issue, 11*, 201–262.

Rosen, B. R. (1986). Modular growth and form of corals: A matter of metamers? *Philosophical Transactions of the Royal Society, London, B313*, 115–142.

Rosen, B. R., & Turnsek, D. (1989). Extinction patterns and biogeography of scleractinian corals across the Cretaceous/Tertiary boundary. *Memoir of the Association of Australasian Paleontologists, 8*, 355–370.

Sandberg, P. A. (1983). An oscillating trend in Phanerozoic non-skeletal carbonate mineralogy. *Nature, 305*, 19–22.

Savazzi, E. (1982). Commensalism between boring mytilid bivalves and a soft bottom coral in the upper Eocene of Northern Italy. *Palaontologisches Zeitschrift, 56*, 165–175.

Scoffin, T. P. (1972). Fossilization of Bermuda patch reefs. *Science, 178*, 1280–1282.

Scoffin, T. P. (1987). *Introduction to carbonate sediments and rocks*. New York: Chapman & Hall/ Methuen. 272 pp.

Scoffin, T. P. (1992). Taphonomy of reefs: A review. *Coral Reefs, 11*, 57–77.

Scoffin, T. P., & Hendry, M. D. (1984). Shallow-water sclerosponges on Jamaican reefs and a criterion for recognition of hurricane deposits. *Nature, 307*, 728–729.

Signor, P. W., III, & Brett, C. E. (1984). The mid-Paleozoic precursor to the Mesozoic marine revolution. *Paleobiology, 10*, 229–245.

Smith, A. B. (1984). *Echinoid palaeobiology*. London: Allen & Unwin.

Staff, G. M., Stanton, R. J., Jr., Powell, R. N., & Cummins, H. (1986). Time-averaging, taphonomy and their impact on paleocommunity reconstruction: Death assemblages in Texas bays. *Geological Society of American Bulletin, 97*, 428–443.

Stanley, S. M., & Hardie, L. A. (1998). Secular oscillations in the carbonate mineralogy of reef-building and sediment-producing organisms driven by tectonically forced shifts in seawater chemistry. *Palaeogeography, Palaeoclimatology Palaeoecology, 144*, 3–19.

Stanley, G. D., & Swart, P. W. (1995). Evolution of the coral-zooxanthellae symbiosis during the Triassic: A geochemical approach. *Paleobiology, 21*, 179–199.

Steneck, R. S. (1982). Adaptive trends in the ecology and evolution of crustose coralline algae (Rhodophyta, Corallinaceae). Ph.D. Dissertation, The John Hopkins University, Baltimore, MD, USA.

Steneck, R. S. (1983). Escalating herbivory and resulting adaptive trends in calcareous algal crusts. *Paleobiology, 9*, 44–61.

Steneck, R. S. (1985). Adaptations of crustose coralline algae to herbivory: Patterns in space and time. In D. F. Toomey & M. H. Nitecki (Eds.), *Paleobiology: Contemporary research and applications* (pp. 352–366). Berlin: Springer.

Steneck, R. S. (1988). Herbivory on coral reefs: A synthesis, *Proceedings of the 6th international coral reef symposium 1* (pp. 37–49), Townsville, Queensland.

Steuber, T. (2002). Plate tectonic control on the evolution of Cretaceous platform-carbonate production. *Geology, 30*, 259–262.

Suzuki, A., & Kawahata, H. (1999). Partial pressure of carbon dioxide in coral reef lagoon waters: Comparative study of atolls and barrier reefs in the Indo-Pacific Oceans. *Journal of Oceanography, 55*, 731–745.

Tanner, J. E., Hughes, T. P., & Connell, J. H. (1994). Species co-existence, keystone species, and succession: A sensitivity analysis. *Ecology, 75*, 2204–2219.

Taylor, P. D., & Todd, J. A. (2001). Bioimmuration. In D. E. G. Briggs & P. R. Crowther (Eds.), *Paleobiology II* (pp. 285–289). Oxford: Blackwell.

Thayer, C. W. (1983). Sediment-mediated biological disturbance and the evolution of marine benthos. In M. Tevesz & P. W. McCall (Eds.), *Biotic interactions in recent and fossil Benthic communities* (pp. 480–625). New York: Plenum.

Tunnicliffe, V. (1981). Breakage and propagation of the stony coral *Acropora cervicornis*. *Proceedings of the National Academic Science USA, 78*, 2427–2431.

Van Belle, R. A. (1977). Sur las classification des Polyplacophora. III. Classification systematic des Subterenochitonidae et des Ischnochitinonidae (Neoloricata: Chitonina). *Info Societe Belgique Malacologique, 5*, 15–40.

Vermeij, G. J. (1987). *Evolution and escalation: An ecological history of life*. Princeton: Princeton University Press. 527 pp.

Wilson, P. A., & Opdyke, B. N. (1996). Equatorial sea surface temperatures for the Maastrichtian revealed through remarkable preservation of metastable carbonate. *Geology, 24*, 555–558.

Wilson, M. A., & Palmer, T. J. (1992). Hardgrounds and hardground faunas. *University of Wales, Aberystwyth, Institute of Earth Studies Publications, 9*, 1–131.

Wood, R. (1999). *Reef evolution*. Oxford: Oxford University Press. 414 p.

Wood, R., Yu Zhuravlev, A., & Debrenne, F. (1992). Functional biology and ecology of Archaeocyatha. *Palaios, 7*, 131–156.

Wood, R. A., Grotzinger, J. P., & Dickson, J. A. D. (2002). Proterozoic modular biomineralized metazoan from the Nama Group. *Science, 296*, 2383–2386.

Chapter 11
Silicification Through Time

Susan H. Butts and Derek E.G. Briggs

Contents

Abstract Silicification, which requires dissolution of original shell material and precipitation of silica, is mediated by numerous biological and environmental factors. The processes and controls on silicification, the environments and conditions under which it occurs, and the temporal and spatial distribution of silicified deposits through the Phanerozoic are reviewed. Selective dissolution of original shell material in certain taxa or lithological settings results in a taphonomic bias which impacts our understanding of paleoecology and patterns of diversity over time. The temporal pattern of silicification is mediated mainly by global ocean chemistry and climate,

S.H. Butts (✉)
Division of Invertebrate Paleontology, Peabody Museum of Natural History, Yale University, P.O. Box 208118, New Haven, CT 06520-8118, USA
e-mail: Susan.Bulls@yale.edu

D.E.G. Briggs
Department of Geology and Geophysics, Yale University, P. O. Box 208109, New Haven, CT 06520-8109, USA and Peabody Museum of Natural History, Yale University, P.O. Box 208118, New Haven, CT 06520-8118, USA

P.A. Allison and D.J. Bottjer (eds.), *Taphonomy: Process and Bias Through Time*, Topics in Geobiology 32, DOI 10.1007/978-90-481-8643-3_11,

particularly changes in carbonate solubility (the contrast in calcite and aragonite seas). On a finer scale, silicification of shelly fossils is influenced by taxonomic factors (shell mineralogy, organic material in the soft tissues and within the shell, shell ultrastructure) and by depositional factors (from porewater geochemistry, lithology, porosity, permeability, position within a stratigraphic sequence and basin characteristics, to global ocean chemistry). Silicification is more prevalent in the Paleozoic than in younger strata. The relationship between the abundance of silicified faunas, the greenhouse/icehouse signal, and fluctuations in carbonate rock volume are complex; the observed pattern may reflect sampling of exceptionally preserved silicified faunas rather than a global signal in silicification. The influence of shifts in biodiversity, in carbonate skeletons susceptible to silicification, and siliceous skeletons that provide a source of silica, remains to be determined.

1 Introduction

Silicification of shells normally requires the concurrent dissolution of calcium carbonate and precipitation of silica, and is mediated by numerous taxonomic and environmental factors in natural systems. Here we review the processes and controls on silicification, the environments and conditions under which it occurs, and the temporal and spatial distribution of silicified deposits through the Phanerozoic. The impact of selective silicification on biotas is evaluated as an important taphonomic bias.

Major advances have been made in our understanding of the taphonomy of marine fossils, particularly the preservation of non-biomineralized tissues, scales of resolution, taphonomic filters and their impact on compositional fidelity, and megabiases (large-scale variations through time) (Behrensmeyer and Kidwell 1985; Schubert et al. 1997; Behrensmeyer et al. 2000). The dissolution and replacement of shelly invertebrates, however, particularly in silica, is an important filter that remains poorly understood. It is clear that silicification of skeletal material has a significant effect on the representation of some groups, particularly bivalved molluscs and brachiopods, and that it occurs far more frequently in Paleozoic than in post-Paleozoic units, introducing a potential megabias (Schubert et al. 1997).

Some aspects of silica replacement, such as the chemical constraints on dissolution of carbonate mineral phases and the transformation of opal-A to quartz, are well understood. Fundamental issues, however, such as the conditions under which silicification occurs in the natural environment, are poorly constrained even though they determine, on a local scale, whether or not dissolution and silicification occur. Replacement of calcareous skeletal material by silica results in fossils that can be extracted and investigated in three dimensions, with consequent advantages for systematic description. When silicification is concurrent with dissolution of shell material the original community may be preserved with exceptional fidelity (Wright et al. 2003). On the other hand, when silicification post-dates the degradation of skeletal material or selects particular shell mineralogies or textures, or when silica neomorphism destroys morphological detail, the resulting fossil record is skewed.

Selective dissolution and silicification may impact seriously the data available for paleoecological analysis of fossil communities. Brachiopods, for example, are generally less susceptible to dissolution than molluscs, although there is some variability within these taxa. A comparison of coeval Lower Jurassic units in South Wales revealed a 65% loss of bivalve genera as a result of early aragonite dissolution relative to their diversity in assemblages preserved by early silicification (Wright et al. 2003). A similar loss of mainly molluscan faunas has been documented in the Silurian of Gotland, Sweden (Cherns and Wright 2000). Despite the striking loss of diversity in these two specific cases, tests comparing silicified faunas with nonsilicified faunas are rare.

Controls on dissolution/silicification may have important consequences for patterns of diversity as revealed by the fossil record. Analyses of the occurrence of silicification through time reveal a notable decline in its prevalence following four out of five of the major extinction events (Kidder and Erwin 2001), but such patterns are poorly known at present. Bias in silicification may be mediated by changes in ocean water chemistry, in the skeletal mineralogy of important organisms at any given period, and in the source of silica (including the diversification and abundance of siliceous organisms over geologic time).

2 Processes and Controls

Silicification of shelly fossils results in the partial or complete replacement of the original mineralized skeleton (Fig. 1). Examples of high fidelity morphological replacement show an intimate relationship between the dissolution of calcium carbonate and the precipitation of silica (Newell et al. 1953; Cooper and Grant 1972; Boyd and Newell 1972; Jacka 1974; Laufeld and Jeppsson 1976; Schmitt and Boyd 1981; Holdaway and Clayton 1982; Misík 1995; Daley and Boyd 1996; Erwin and Kidder 2000; Butts 2007). The fidelity of replacement (e.g., complete or partial, pervasive or non-pervasive, as a fine-scale fabric or beekite rings) is influenced by factors such as the degree to which carbonate phases have been transformed to more stable forms (e.g., aragonite to calcite), the differential solubility of these phases, the geochemical conditions generated by their alteration, and the conditions that give rise to the dissolution and precipitation of silica. Silica availability is often a limiting factor. Beekite rings form by liesegang diffusion (Hodges 1932) in conditions of limited or episodic silica supply (Holdaway and Clayton 1982) or fluctuating carbonate dissolution. The formation of chert in sediments is probably subject to environmental controls similar to those affecting silicification of shells, but it is influenced less by the unique nature of skeletal material and is not treated here.

The timing and rate of silicification also influence the fidelity of replacement. More than one silicification texture may be evident in a specimen. In thin section, silicification of bioclasts is best observed under cross nicols with insertion of the

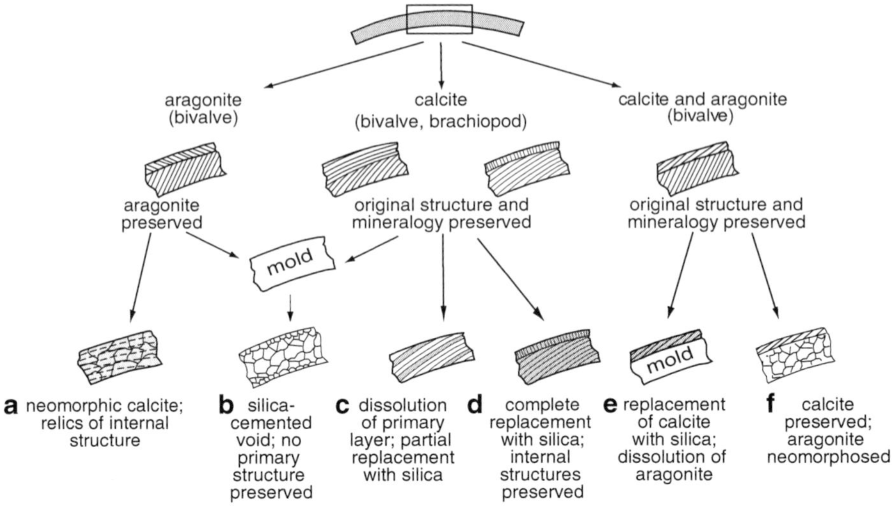

Fig. 1 Diagenetic pathways of bioclastic materials (After Scholle and Ulmer-Scholle 2003)

first order red (gypsum) plate (Scholle and Ulmer-Scholle 2003). As the stage is rotated, carbonate changes between first order white and purple, while quartz (generally as cryptocrystalline chalcedony which has replaced the skeletal fabric) alternates between first order red, yellow, and blue (Daley 1987). Complete silicification (Fig. 2a, m) involves replacement of the entire shell. Fine-scale textural replacement (fabric replacive) preserves the shell ultrastructure (barring subsequent silica transformation) including punctae in spiriferinide brachiopods (Fig. 2b).

Fig. 2 Silicification types and textures in specimens from the Permian of west Texas (**a, c, d, g, i, j, k**), the Carboniferous of east-central Idaho (**b, e**), and the Devonian of New York (**f, h**) and in thin section the Carboniferous of east-central Idaho (**l–n**). (**a, b**) Complete pervasive silicification of brachiopod shells. (**a**) Fine scale replacement of the entire shell of an athyridide; a portion of the shell has been removed to reveal the silicified spiralia (×3.42). (**b**) Fine scale replacement preserving ultrastructure, including punctae (×4.7). (**c, d**) Incomplete silicification of brachiopod shells leaving parts of the shell in calcite and subject to dissolution (×3.2). (**e**) Non-pervasive silicification of a brachiopod shell affects only the external layers and leaves the inner shell as calcite (×1.8). (**f**) Partial silicification of a crinoid ossicle, in which the inner organic canal is also preserved in silica (×3.8). In this case silicification forms a surface-coating "crust" or "rim" on the bioclast, rather than replacing skeletal material. (**g**) Granular silicification of a bivalve shell (×3.2). (**h**) Spongy silicification of a gastropod shell (×6.8). (**i, j**) Beekite rings. (**i**) A lyttoniidine brachiopod (×3.7). (**j**) A pectinid bivalve in which the ribs and inter-rib areas are differentially replaced even though the shell thickness was uniform (×2.7). (**k**) Spherulitic chalcedony, which commonly occurs as a surface-coating on bioclasts but is not evident at the scale of the other textures (×15). (**l–n**) Brachiopod shells in thin section. (**l**) *Composita*, an athyridide, showing the tertiary prismatic shell layer (p) in original calcite and secondary layer replaced with spherulitic chalcedony (sc) (scale bar = 1 mm). (**m**) Pervasive fine scale replacement in silica (s) (scale bar = 0.5 mm). (**n**) Non-pervasive fine scale replacement showing a calcite interior (c) and silicified exterior (s) (scale bar = 0.5mm)

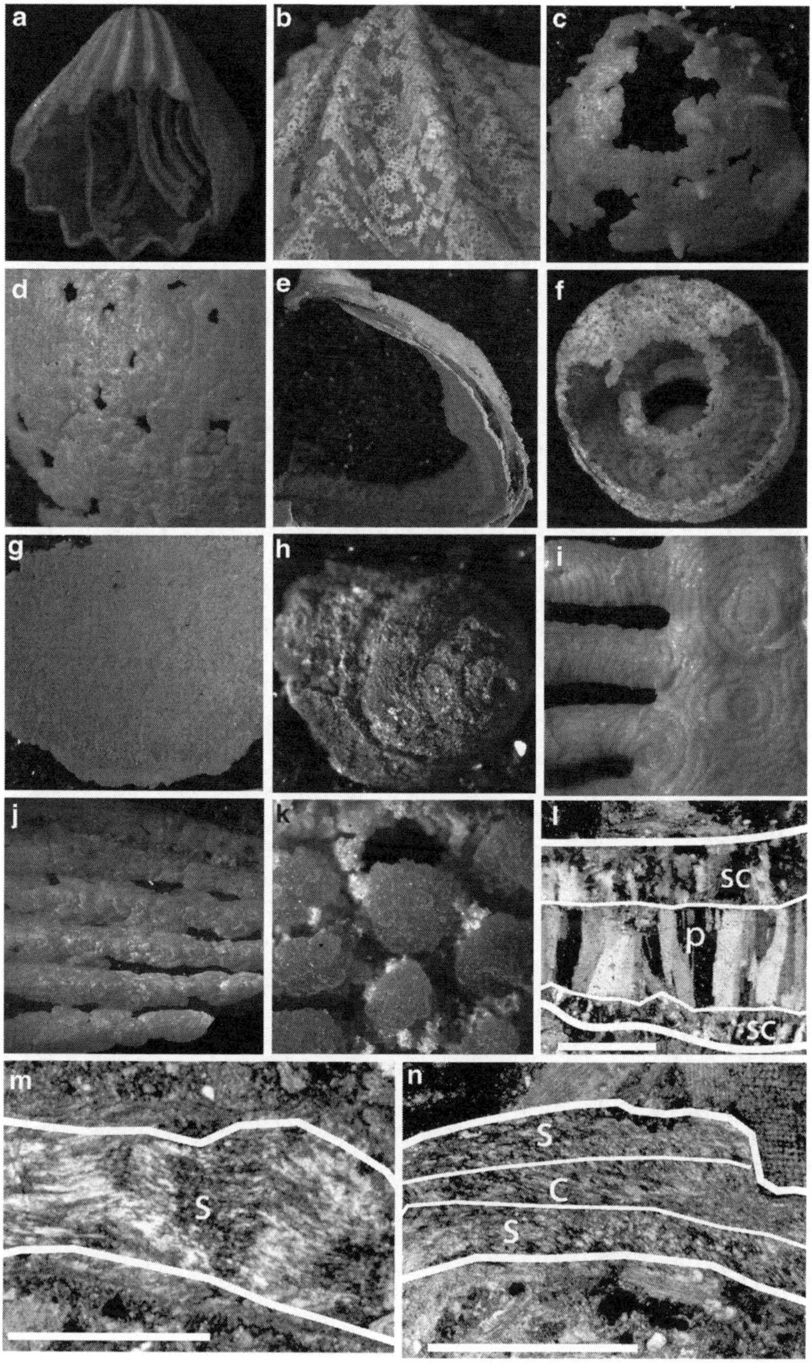
a
b
c
d
e
f
g
h
i
j
k
l
sc
p
sc
m
S
n
S
C
S

Partial silicification is characterized by a surface irregularly replaced with silica. Typically, variably developed beekite rings fail to replace the entire shell resulting in distinctly "scalloped" edges surrounding voids (Fig. 2c, d) or "lacy" silicified areas.

Pervasiveness refers to the degree to which the shell thickness is silicified. The shell surface, which was in contact with the matrix, may be preserved, while inner shell material, still preserved in calcite, is etched away during acidification (Fig. 2e, l, n). Silicification of the outer shell layers enhances taphonomic stability by creating a relatively non-reactive barrier; however, it may prevent pervasive silicification. Non-pervasive replacement is evident in thin section where silicified outer shell margins enclose a calcitic shell interior (Fig. 2n); in etched hand samples void spaces occur most notably where the shell is thickest, as in the umbonal region of many brachiopods. Silicification also may be inhibited by the formation of a siliceous crust on the surface of a bioclast (Fig. 2f).

Where silicification is not fine-scale fabric replacive a granular (Fig. 2g) or spongy (Fig. 2h) texture, or beekite rings, may result (Fig. 2i, j). Finely crystalline granular texture may retain surface microornament but ultrastructure is no longer evident in thin section. Granular texture may reflect the gradual enlargement of silica crystals during silica diagenesis. Spongy texture often occurs in originally aragonitic skeletons, but is common in all kinds of shelly fossils in fine-grained siliciclastic rocks. Spongy silicification varies in completeness and pervasiveness but in some cases preserves ultrastructural features. Beekite rings are concentric rings of silica each delineated by more or less well developed grooves (Fig. 2i). Where replacement as beekite rings is complete and pervasive it may retain fine details of the morphology, allowing taxonomic determination, but ultrastructure is destroyed. Where replacement is partial, on the other hand, identification may be more difficult (Fig. 2j). Beekite rings may be relatively flat to nearly spheroidal and in some cases have the appearance of "blebs" rather than well-formed concentric rings. Chalcedony spherules are much smaller than other common silicification textures (Fig. 2k, l); they typically form a "crust" or rim on the external surface or skeletal/shell layer. At higher magnification, the botryoidal nature of the silica is readily apparent (Fig. 2k).

Early silicification, i.e., concurrent with the dissolution of the original carbonate shell, shows complete or partial ultrastructural replacement (Fig. 2a,b) (Holdaway and Clayton 1982; Maliva and Siever 1988; Cherns and Wright 2000; Butts 2005, 2007). Beekite textures (Fig. 2i, j) and spherulitic chalcedony rims (Fig. 2f, l) may form during early or late silicification. Late silicification, when the mineral precipitates as a cast (Fig. 1b), is characterized by mega-quartz grains (which grow perpendicular to the shell boundary and may increase in size inwards), euhedral crystals and concentric quartz laminae (Boyd and Newell 1972; Bathurst 1975; Schmitt and Boyd 1981). Silica permineralization is a different process which involves filling voids within organic tissues such as wood with silica-rich fluids (Briggs 2003). Here our primary concern is with early silicification as a major control on the completeness of the marine shelly fossil record.

2.1 *Experiments*

Attempts to determine the controls on silicification using laboratory experiments have met with only limited success. Bacteria, algae, fungi (Francis et al. 1978) and wood (Leo and Barghoorn 1976) have been permineralized successfully with silica in experiments, using tetraethyl orthosilicate to deliver silicic acid, but this only involves precipitation in association with organic matter. Klein and Walter (1995) showed that some amorphous silica will precipitate on calcite surfaces in a matter of hours at 25–50°C but they could not replicate the replacement of shells with silica in the laboratory. They suggested that organic material in shells may have an effect that was not considered in their study. Bivalve shells, however, had been replaced by silica in the laboratory some 20 years earlier (Paraguasso 1976). In these experiments shells were suspended by threads in a solution of sodium metasilicate at a pH adjusted to allow the external shell form, shell ultrastructure, and organic matrix of the shell to remain stable during dissolution of carbonate (trials showed that a solution of 2 g/l of SiO_2 at pH 2 resulted in successful replacement of a 0.5 g shell). After 3 months, silica encrusted the shells and after 8 months shells removed from the tank and dehydrated in ambient air were composed of 75% silica with some organic matter and minor constituents (Paraguasso 1976). The original shell was replaced with a silica gel, which shrunk slightly during dehydration, but it was only ~10% of the mass of the original calcium carbonate. Replacement in natural systems incorporates much more silica and presumably involves much slower rates.

2.2 *Skeletal Factors*

The mineralogy and distribution of organic material within individual shells influences their susceptibility to silicification. The original mineralogy and abundance and location of organic material (within the shell or in the associated soft tissues) vary by taxon, which may result in a taphonomic bias. In some cases dissolution destroys skeletal material (e.g., readily soluble aragonite) prior to silicification, resulting in taphonomic loss.

2.2.1 Original Mineralogy

Three biogenic carbonate phases are common in marine organisms: high magnesium calcite (HMC), which contains >4 mol% $MgCO_3$ (14 mol% on average: Bathurst 1975); low magnesium calcite (LMC), which contains <4 mol% $MgCO_3$ (Stanley and Hardie 1998); and aragonite, an orthorhombic calcite polymorph (Tucker and Wright 1990). Although aragonite is unstable, original aragonite is known in shells as old as Devonian and aragonite preservation is ubiquitous even in some Paleozoic units (e.g., the Upper Carboniferous Breathitt Formation of Kentucky: Brand 1983). Skeletal aragonite neomorphoses to coarsely crystalline calcite, but traces of the original aragonite may be retained as crystallites within the neomorphic calcite or as relict textures (Fig. 1a, f). Due to the high solubility of aragonite, dissolution of shell material frequently occurs even prior to burial.

Skeletons may consist of a single carbonate phase or include different mineral phases (as in the shell layers of pectinid bivalves). Differences in the stability of aragonite, HMC and LMC are reflected in differences in solubility and susceptibility to neomorphosis, all of which affect the replacement of fossils in silica. The concentration of magnesium in skeletal material varies with the latitude at which organisms live (Chave 1954). To complicate matters, laboratory studies have shown that an organism with a particular (fixed) skeletal mineralogy may secrete a carbonate skeleton with a different Mg/Ca ratio depending on the ambient water geochemistry (Ries and Blaustein 2003) (see Section 3.2), so skeletal composition in a single taxon may vary through geologic time.

2.2.2 Distribution of Organic Material

The decay of organic material creates conditions conducive to both carbonate dissolution and silica precipitation. Shell ultrastructure, i.e., the size of carbonate crystallites or grains that make up the shell, influences the overall dissolution rate (Henrich and Wefer 1986; Glover and Kidwell 1993; Harper 2000). Shell structure and the location and content of organic matter vary between different taxa (e.g., brachiopods and mollusks) and within taxonomic groups (e.g., between brachiopod orders). Organic material forms sheaths surrounding the calcium carbonate crystallites, and also occurs within them. Specialized morphological features, such as mantle outgrowths within caeca and spines, are present in some brachiopod shells (Williams 1997). Organic material may also cover the shell surface (e.g., the periostracum of mollusks and brachiopods).

The organic content of a brachiopod (soft-tissue and shell) is 2.5–4.1% of its total mass compared to 15–30% in bivalve molluscs (Peck 1993). Forty to fifty percent of the organic material in all brachiopods is normally in the shell (Curry and Ansell 1986; Curry et al. 1989), largely in the caeca, which are outgrowths of the mantle that line the punctae (an organic content of up to 75% has been reported in the shell of one example: Peck et al. 1987). But to what extent do organic materials within shells influence the likelihood of silicification? As organic material decays, CO_2 and acidity increase and promote the dissolution of carbonate shell material and the precipitation of silica (Froelich et al. 1979; Holdaway and Clayton 1982). Observations of degradation of modern punctate (terebratulid) and impunctate (rhynchonellid) brachiopods in marine settings show a higher rate of shell degradation in the former due in part to the greater concentration of organic material in the shell (Tomasovych and Rothfus 2005). Unfortunately, most investigations of silicification have involved petrographic examination of the shells of extinct organisms, without direct information on the distribution and composition of organic material, so the influence of shell degradation on silicificiation of organic matter-rich shells (for example, punctate brachiopods) compared to shells that are relatively poor in organic matter (impunctate brachiopods) is unknown.

2.2.3 Shell Ultrastructure

Studies of silicified faunas reveal a correlation between shell microstructure and degree of silicification, resulting in a taphonomic bias. Silicification is often initiated in the secondary shell layer of brachiopods (Holdaway and Clayton 1982); the primary layer frequently is lost through abrasion, erosion (Williams 1997) or dissolution. The secondary shell layer is fibrous in the rhynchonellates and cross-bladed laminar in the strophomenates, the two most abundant classes of brachiopods (Buening 2001). A tertiary prismatic shell layer is also present in a few brachiopods (Williams 1990). Spiriferids and athyridids (rhynchonellates) show a greater tendency to silicify than strophomenates (Schmitt and Boyd 1981; Loope and Watkins 1989; Tucker 1991; Daley and Boyd 1996; Butts 2005) and punctate brachiopods are more prone to silicification than impunctate (Newell et al. 1953; Dott 1958). Among bivalves, differences in shell structure and mineralogy within shells also may affect silicification. The calcitic outer shell layer of pectinids, for example, may be silicified while the inner aragonitic material is lost through dissolution (Fig. 1e; Schmitt and Boyd 1981). Similar patterns occur in Permian pseudomonotid bivalves from the Glass Mountains (Newell and Boyd 1970). Laufeld and Jeppsson (1976) noted that bryozoans, brachiopods, and corals in the Silurian of Gotland are generally more susceptible to silicification than other taxa. This 'hierarchy of silicification' is also found in the Ordovician Tanner Creek Formation of Indiana (Fox 1962) and in the Permian Glass Mountains of West Texas (Newell et al. 1953).

2.3 *Diagenesis: Coupled Dissolution/Precipitation*

Dissolution of calcium carbonate and precipitation of silica occur in conditions where pH declines from an initial state of pH > 9 (Correns 1969) in environments supersaturated in silica and undersaturated in calcite (Knauth 1979). The chemical conditions required for the dissolution and precipitation of calcium carbonate are well documented (Canfield and Raiswell 1991). Dissolution is affected by the thermodynamic stability of the calcite phase, the saturation state of the pore water solution, and the reactive surface area of the shell or other bioclastic particle (Walter 1985). HMC with over 8.5 mol% $MgCO_3$ is slightly more soluble than aragonite, which is more soluble than LMC (<4 mol% $MgCO_3$) in normal (Mg/Ca ratio) marine seawater (Berner 1975). In field and experimental studies, however, HMC has been found to dissolve at the same rate as aragonite (Berner et al. 1976; Walter and Morse 1985). Due to the relative instability of aragonite, it has the potential to invert to calcite (Land 1967) under favorable conditions at any point in its diagenetic history.

The reactive surface area of skeletal material varies with microstructure (Walter 1983; Walter and Morse 1984) which therefore affects its solubility. Laboratory studies on crushed bioclastic material (results on larger fragments or whole shells might differ) revealed a rank order of degree of dissolution that could not be predicted simply on the basis of original mineralogy, and varied with the calcite/aragonite saturation of the seawater used (Walter 1985; Walter and Morse 1985). Aragonite in the green alga *Halimeda*, for example, dissolved more rapidly than less stable magnesian calcites in red algae and forams. The familiar rank order of susceptibility to dissolution and diagenetic alteration (HMC → aragonite → LMC) applied only in solutions near or above saturation for aragonite. The influence of microstructure on dissolution is greatest where calcite is undersaturated and most aragonite dissolves. Composition becomes a more important influence than microstructure as calcite and finally aragonite reach saturation; in the last case only HMC dissolves (Walter 1985).

The precipitation of silica involves the polymerization of silicic acid to a silica gel which dehydrates to form opal (unstable), chalcedony, and quartz (Fairbridge 1983). In carbonate sediments, which are alkaline and typically Mg-enriched, a compound with a ratio of Mg/Ca ~ 1:2 acts as a nucleation site for opal CT lepispheres by attracting silanol groups (Williams and Crerar 1985). The concentration of dissolved magnesium is increased by the conversion of HMC to LMC. This change may be reflected in the formation of early diagenetic dolomite rhombs adjacent to mineralogically stabilized clasts as in the Middle Permian Gateway Limestone Member of the Cherry Canyon Formation of Texas and New Mexico (Jacka 1974). Ca^{2+} released during the solubilization of calcite favors silica precipitation because it decreases silica solubility (Paraguasso 1976). Nucleation and polymerization of silica is slower at lower pH values (Hinman 1987). Electrical charges distributed through chains of proteins and amino acids in organic matter aid in complexing silica molecules (Paraguasso 1976; Iler 1979). Organic matter decomposition by sulfate-reducing bacteria creates low pH conditions (Schieber 1996), favoring carbonate dissolution and silica precipitation (Loope and Watkins 1989) and pyrite has been observed in association with silicified skeletal material (e.g., framboidal pyrite in the caeca of brachiopods: Gaspard and Roux 1974).

Following precipitation, silica goes through a well understood series of diagenetic transformations: opal-A (amorphous) → opal A′ (secondary) → opal CT →opal CT (reordered phase) → cryptocrystalline quartz or chalcedony → microcrystalline quartz (Williams and Crerar 1985; Hesse 1989, 1990). The presence of organic matter (Hinman 1990) and certain ions derived from carbonate and clay minerals (Lancelot 1973; Hinman 1998) may influence the rate of silica phase changes. Silicification does not always involve such a transformation; some of the mineral variation may be primary (Holdaway and Clayton 1982). The increase in crystal size associated with neomorphism may obliterate structures required for taxonomic determination (Fig. 1a,f).

2.4 *Influence of Depositional Environment*

Controls on silicification operate hierarchically. The agents of carbonate dissolution and silicification are a function of the composition of the sediment and pore water chemistry, and hence the sedimentary environment. The chemical environment is also affected by the sediment: composition, grain size, porosity and organic content.

2.4.1 Sequence Stratigraphic Framework

Preliminary observations indicate that the distribution of silicified faunas is reflected in sequence and parasequence boundaries. In the Ordovician of Tennessee (Holland and Patzkowsky 1997, 1998), for example, chertification occurs in laminated mudstones and bioturbated packstones and wackestones below sequence boundaries that are overlain by transgressive surfaces. Intervals of subaerial exposure are evidenced in these same units by zones of aragonite dissolution (Railsback et al. 2003). However, this pattern may be obscured by the nature of the lithologies within stratigraphic sequences. In the Arco Hills Formation of east-central Idaho (S.H. Butts, unpublished), for example, silicified units are most common in the lower part of parasequences, probably as a result of lithology and porewater geochemistry in the depositional environment. The Arco Hills Formation parasequences are capped by either calcareous quartz arenite, pelmatazoan grainstone (encrinite) with syntaxial cements, or fine-grained siliciclastics, all of which are unfavorable for silicification. The high porosity quartz arenite was presumably originally undersaturated in calcite and aragonite and later cemented with calcite. The early syntaxial cements in pelmatazoan grainstone eliminated porosity and precluded silicification. Low-porosity lithologies may act as aquitards to protect fossil-bearing units from diagenetic waters (as in the Breathitt Formation: Brand 1983). Silicification is reduced where low porosity clay-rich lithologies cap parasequences. In contrast, where carbonate-rich lithologies cap sequences or parasequences, silicification may be common. Subaerial exposure may also provide access for silica-rich meteoric waters (Meyers 1977). Thus an understanding of how sequence boundaries affect the geochemistry of the depositional environment and the early diagenetic history of sediments may allow them to be used as predictors of the occurrence of silicified fossils within a succession.

2.4.2 Silica Source

Extensive silicification requires sufficient available silica to replace dissolved carbonate shell material. In marine environments, dissolved silica is mainly from three sources: detrital, volcanic/hydrothermal, and biogenic (Tréguer et al. 1995). Input of dissolved silica (generally as monomeric silicic acid, $Si(OH)_4$) to the world ocean is primarily riverine, comprising 80% of the total input (Tréguer et al. 1995). To a lesser extent silica is derived from eolian, submarine weathering, and hydrothermal sources, and from the dissolution of skeletons (Tréguer et al. 1995). These inputs are balanced to some extent by coastal and abyssal deposition, but mainly by the generation of siliceous skeletons by organisms using dissolved silica (Tréguer et al. 1995). The concentration of dissolved silica in marine waters is approximately 3 ppm, but it reaches 10–50 ppm in pore waters, depending on the local availability of siliceous shells in the sediment (Lancelot 1973).

The skeletons of modern siliceous organisms – diatoms, radiolarians, and silicoflagellates – are composed of Opal-A (Tucker 1991). These organisms, as well as ebridians, sponges (Maliva et al. 2005), alcyonarians and holothurians (Fairbridge 1983), are the common sources of biogenic silica today. During the Paleozoic, the primary source of biogenic silica was likely sponge spicules and radiolarians (Maliva et al. 2005). Simultaneous or penecontemporaneous replacement of siliceous bioclasts with carbonate, and carbonate clasts with silica, may generate an abundant source of silica, as in the Permian of the Glass Mountains (Newell et al. 1953) and the Lake Valley Formation (Mississippian) of New Mexico, where siliceous sponge spicules are replaced with carbonate (Meyers 1977). Devitrified volcanic ash may also be an important source of silica: silicification is prevalent in association with bentonite beds (Laufeld and Jeppsson 1976; Cherns and Wright 2000; Butts 2004).

2.4.3 Other Factors

In certain circumstances organisms may be protected from silicification in spite of favorable conditions for its occurrence. Early isopachous cements, typically calcite or aragonite, may shield a shell from interaction with pore waters. Such syntaxial cements commonly occur on crinoid ossicles, for example. The crystals of crinoid ossicles are large, giving them "high microstructural stability and thus low susceptibility to replacement" (Maliva 1992, p. 169). Crinoid grainstone with pervasive syntaxial cements has very limited porosity (Butts 2005). Other syngenetic and early diagenetic coarse cements likewise limit porosity and permeability, and preclude silicification.

Diagenetic "crusts" of low-magnesium calcite or silica, which form on the surfaces of skeletal material in contact with the matrix, may impede or prevent dissolution or may prevent pervasive silicification. Specimens from the mid-Carboniferous Arco Hills Formation of southeast Idaho and the Pennsylvanian/Permian faunas from the Guadalupe Mountains region of Texas preserve shells that are silicified only on the external and internal surfaces of the valves. This suggests that silicification was initiated by contact with the surrounding matrix/pore water (Newell et al. 1953), but ceased with depletion of silica or buffering of the system beyond conditions allowing silicification.

Ecology and post mortem transport of an organism may influence the chances of the skeleton becoming silicified. Infaunal invertebrates are favored over epifaunal, because the chemical environment above the sediment–water interface is more likely to promote rapid carbonate dissolution (Davies et al. 1989). Bioclasts in storm deposits are "buried" before organic matter breaks down within the shell; its subsequent degradation promotes carbonate dissolution and silica precipitation (Erwin and Kidder 2000). Biological reworking of sediments also influences the dissolution of carbonate bioclasts. Dissolution of carbonate bioclasts directly below the sediment–water interface is enhanced when the burrowing activity of organisms oxygenates the sediment locally, promoting aerobic respiration and degradation of organic material, thus increasing acidity and skeletal dissolution (Aller 1982). Burrowing may also rework organic-rich anoxic sediments to the sediment–water interface where sulfide oxidation occurs and sulfuric acid is produced (Aller 1982).

2.5 *Models of Silicification*

The silicification of shelly fossils and the formation of chert have been explained in a number of ways. The mixing zone model of Knauth (1979) was based on the presumption that diagenetic fluids are undersaturated with calcium carbonate and supersaturated with silica. Such conditions may prevail at coastal locations due to the mixing of meteoric and marine waters. In many examples of silicification, however, there is no link to terrestrial emergence, and the conditions that favor silicification are generated by the decay of organic matter.

Early silicification commonly takes place in the upper layers of the sedimentary column where active decay occurs. Organic matter may act as a nucleation site for silica precipitation (Maliva and Siever 1988) and decay-induced production of bicarbonate ions (HCO_3^-) reduces pH and promotes the dissolution of carbonate (Cherns and Wright 2000). Silicification is associated with decaying organic matter in a wide range of contexts, as evidenced by examples of silica precipitation restricted to bioclasts and absent or very limited in non-bioclastic sediments (Newell et al. 1953; Schmitt and Boyd 1981; Holdaway and Clayton 1982; Misík 1995; Daley and Boyd 1996).

Late stage silicification may occur where shells are in contact with siliciclastic grains, as reported in the Devonian Oriskany Sandstone of New York (Maliva 1992), perhaps as a result of pressure solution in pore fluids. The force of crystallization during the growth of syntaxial quartz cements may be responsible for the dissolution of calcite minerals (Maliva 1992). Dissolution/silicification involving cavity filling does not usually retain shell structure, and is characterized by megaquartz.

3 Silicified Faunas Through Time

The primary control on the distribution of silicified faunas over time is the presence and abundance of organisms with siliceous tests, which provide a source of biogenic silica. Also important is the relative abundance of different carbonate skeletons at any time and their relative stability within the global ocean. Controls on dissolution/silicification may operate on a global scale resulting in megabiases in the fossil record (Kidder and Erwin 2001). Prior to the Paleozoic, early diagenetic silica was deposited readily as chert in peritidal deposits. During the Cambrian radiation, silica was incorporated in skeletons, particularly of sponges (Maliva et al. 1989). During the radiation of radiolarians in the Ordovician, there was increased incorporation of silica into skeletons (Maliva et al. 1989). Silicified bioclasts are common throughout the Paleozoic, but far less common in post-Paleozoic strata. However, patterns in the temporal occurrence of silicified faunas are not well established, even on the scale of geologic period, and have not been tied directly to any large-scale geologic events (e.g., widespread volcanism or ocean anoxia). Only broad associations between the occurrence of silicification and other geologic processes have been noted, such as the dissolution of aragonite in calcite saturated seas (Palmer et al. 1988; Cherns and Wright 2000; Brennan and Lowenstein 2002). Fluctuations in global ocean chemistry and abiogenic sources of silica, and variation in the relative abundance of organisms with different skeletal mineralogies, were important controls on the silicification of shells and resulted in predictable megabiases in the fossil record. Unfortunately, however, the complexity of factors that control the silicification of fossils complicates the interpretation of the pattern.

3.1 *Temporal Patterns*

Silicification is not distributed randomly through the Phanerozoic. Based on data from 1,863 published papers, Schubert et al. (1997) determined that 21% of Paleozoic faunas are silicified, compared to 4% of post-Paleozoic faunas. They attributed this pattern to variation in the abundance of siliceous sponges as a silica source and the increase in aragonitic faunas in the Mesozoic and Cenozoic. Within the Paleozoic, the Cambrian, Ordovician, Devonian, and Permian show notably high levels of silicification whereas silicified faunas are rarer in the Silurian and Carboniferous (Schubert et al. 1997). The abundance of silicified faunas does not correlate with marine outcrop area, carbonate rock volume, changes in sea level or long-term climate, volcanism, or the amount of nearshore chert (Schubert et al. 1997). Analyses of the occurrence of silicification within series revealed a notable decline following four out of five of the major extinction events, including a precipitous drop at the end of the Permian (Kidder and Erwin 2001). Peaks in the Ordovician and Devonian are associated with increases in sponge abundance and diversity (Schubert et al. 1997).

Throughout the Ordovician cherts occur in increasing water depths, from peritidal settings to deep marine basins (Maliva et al. 1989), indicating a transition in silica source and possibly reflecting declining levels of dissolved silica in shallow water (Kidder and Mumma 2003). At the Permian-Triassic boundary there appears to be a similar shift in silica sequestration as chert deposits transitioned from shallow environments toward basinal environments (Schubert et al. 1997). The decline in silicification in the Mesozoic may correlate with the change in dominant shell chemistry from calcite to aragonite (Wilkinson 1979; Railsback and Anderson 1987; but see Stanley and Hardie 1998), and a reduction in the diversity and abundance of siliceous sponges (Kidder and Erwin 2001).

3.2 Global Ocean Chemistry

The magnesium/calcium (Mg/Ca) ratio and saturation state of calcium carbonate are susceptible to the CO_2 content of the atmosphere and oceans, which is different in a greenhouse world (when seas are calcite saturated and undersaturated in aragonite) to that in an icehouse world (when seas are aragonite saturated) (Fischer 1982). Throughout the Phanerozoic, there is a shift between calcite and aragonite seas (Fig. 3). Mg/Ca ratios in seawater are known to affect the phase of inorganic carbonates in cements and ooids (Brennan and Lowenstein 2002). They also may facilitate the secretion of skeletons of a particular mineralogy (Stanley and Hardie 1998) and thus influence the predominance of particular organisms over time; however, modern organisms that normally secrete HMC have been shown to secrete LMC skeletons in low Mg/Ca environments (Ries and Blaustein 2003). No correlation had been detected between the abundance of biogenic silica (as bedded cherts and silicified fossils) and the occurrence of greenhouse versus icehouse conditions (Kidder and Erwin 2001). Higher production of siliceous tests, however, may occur when calcareous organisms are in crisis, such as during the Frasnian-Famennian and end-Cretaceous mass extinction events (Racki 1999). Global changes in ocean chemistry may result in biases in preservation potential, particularly in the case of aragonitic faunas (Smith 2003). Extensive loss of aragonite bioclasts and cements occurs in the Upper Ordovician and Jurassic, for example, when calcite seas ($Mg^{2+}/Ca^{2+} < 2$: Brennan and Lowenstein 2002) were undersaturated with

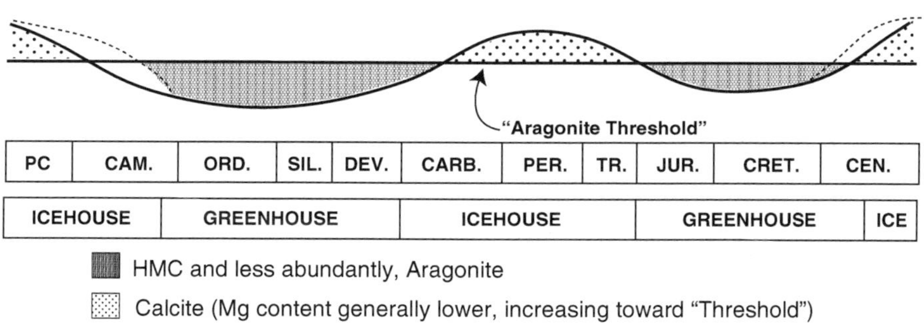

Fig. 3 Secular variation in non-skeletal carbonate mineralogy in seawater and associated climatic episodes (After Sandberg 1983)

aragonite (Palmer et al. 1988). On a more local scale, a period of major calcite dissolution occurred in the Norwegian-Greenland Sea during the most recent deglaciation, but the effects were not global (Henrich 1985).

3.3 Spatial Patterns

Tectonic setting and type of sedimentary basin play a role in controlling the prevalence of silicification. Basins may be carbonate or siliciclastic dominated, or mixed carbonate-siliciclastic; they may be active or passive tectonically; they may occur in greenhouse or icehouse regimes, and through all ranges of latitude. These factors affect a range of phenomena including rates of weathering, sediment accumulation and accommodation, facies distribution (vertical and horizontal), faunal composition, water temperature, and intra-basinal water depths. Weathering rates, the input of detrital silica, and volcanism also vary depending on the tectonic regime of a basin (active or passive). Basin characteristics therefore can influence the distribution of silicified faunas in space and time.

Stratigraphic "traps", such as reefs (Newell et al. 1953) and Waulsortian mounds (Meyers 1977), create sedimentological barriers that control local environmental conditions. Silicification has been reported on the units flanking such traps (e.g., Tierra Blanca, Dona Ana, and Alamogordo Members of the Carboniferous Lake Valley Formation of the Sacramento Mountains of New Mexico: Meyers 1977) but there is no widely accepted explanation for this localization. Silicified shells of gastropods, bivalves, chitons and scaphopods are associated with sponge bioherms in the Permian limestones of the Glass and Guadalupe mountains of West Texas (Cooper and Grant 1972). The distribution of silicification in association with such stratigraphic traps may reflect differences in porosity and permeability. Silicification occurs on the outer margin of reef talus deposits within the Permian of West Texas (Newell et al. 1953). Pore fluid migration between the reef, with early diagenetic cements, and the more permeable flanking reef debris (or between the outer flank deposits and adjacent sediments) was probably inhibited. Within Waulsortian mounds, a similar barrier may develop between mud-supported and grain-supported lithologies.

4 Taphonomic Bias of Selective Silicification

Silicification of shelly fossils is influenced by the nature of organisms (shell mineralogy, organic material in the soft tissues and within the shell, shell ultrastructure) and by the depositional environment (porewater geochemistry, lithology, porosity and permeability, position within a stratigraphic sequence, basin characteristics, and global ocean chemistry). The non-uniform nature of silicification reflects taphonomic biases that impact interpretations of paleoecology and diversity, particularly in Paleozoic sequences.

4.1 Diversity Through Time

Selective silicification of taxa may skew the evidence of faunal diversity in a rock unit and bias the interpretation of diversity through time. Very early silicification (preceding aragonite dissolution) provides evidence of total diversity, which is lacking in similar assemblages where silicification occurs after aragonite dissolution. In the Jurassic of South Wales, for example, there is a 65% decrease in the diversity of bivalve genera where dissolution of aragonite preceded silicification (Wright et al. 2003). Shallow infaunal bivalves and small epifaunal gastropods were more susceptible to dissolution than large burrowing bivalves and large ammonites.

The transition from predominantly calcite skeletons in the Paleozoic to a greater proportion of aragonite skeletons thereafter may explain the paucity of post-Paleozoic silicified faunas. The faunal turnover at the end Permian extinction resulted in the decline of calcitic stalked echinoderms and trilobites (Schubert et al. 1997) and the replacement of calcitic brachiopods and rugose and tabulate corals with faunas dominated by aragonitic bivalves and scleractinian corals (Railsback and Anderson 1987). Changes in conditions during the Triassic favored the precipitation of aragonitic skeletons through an increased concentration of seawater sulfate, higher temperatures and possibly high Mg/Ca, all of which affect the rate of dissolution and precipitation of carbonates (Railsback and Anderson 1987). Aragonitic faunas are less susceptible to silicification, because dissolution and diagenesis of aragonite normally outpaces silica precipitation.

4.2 Paleoecology

The potential for silicification varies from taxon to taxon (Section 2.2) and this clearly has implications for paleoecological inferences based on fossil assemblages. A good example is provided by a comparison of the Arco Hills Formation (Butts 2007) of east-central Idaho and the Bluestone Formation (particularly the Bramwell Member) of West Virginia and Virginia (Henry and Gordon 1992), both Mississippian in age. These formations are similar in lithologies and depositional environments; they consist of interbedded or heterolithic carbonate-siliciclastic units and yield a diversity of fossils. The silicified brachiopod-rich assemblages in the Arco Hills Formation are nearly identical to those in the Bluestone Formation except that they lack molluscs. The mollusc component of the fauna has presumably been lost as a result of early dissolution. Preservation in the Bluestone Formation, in contrast, is primarily calcareous, resulting in a mollusc-dominated assemblage with a significant brachiopod component, more closely reflecting the original community structure, as exemplified by the Bramwell Member. A comparison of silicified and nonsilicified faunas from the Lower Jurassic of Wales suggests that the shells of shallow infaunal marine taxa were more impacted by dissolution, due to high rates of microbial degradation and consequent acidity, than those of deeper burrowers (Wright et al. 2003).

Silicification is more prevalent in carbonates than in fine grained siliciclastics; conditions in carbonate and siliciclastic environments differ in terms of the pH of pore waters, concentration of ions, saturation state of carbonate, ρCO_2, permeability and porosity, and there is considerable variation locally. The Devonian Kalkberg Formation (Helderberg Group) of New York, for example, is a mixed carbonate and siliciclastic unit which includes the Bald Hills volcanics with K-bentonite ash layers (Butts 2004). Carbonate bioclasts are preserved in the fine-grained siliciclastic beds (Fig. 4a, b) but they are normally silicified where they occur in carbonate-rich beds (Fig. 4c, d). It is clear that dissolution and loss of the carbonate clasts at any stage in the diagenetic history of the unit would have resulted in a significant loss of preserved diversity and consequent distortion of paleoecological interpretations of the fauna.

Lithology may also affect the degree and fidelity of silicification. High fidelity replacement tends to be more common in open marine, siliciclastic-poor sediments (Erwin and Kidder 2000; Butts 2007). Spongy silicification is most prevalent in shallow-water calcareous fine-grained siliciclastic-rich sediments (Butts 2007). Shells in offshore deeper deposits are more prone to dissolution than those in

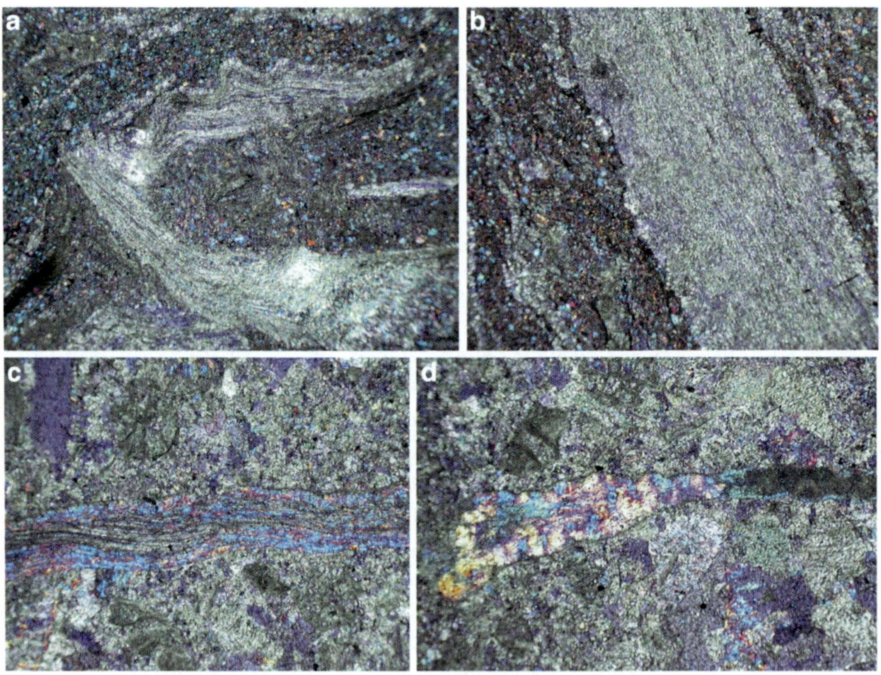

Fig. 4 Silicification of brachiopod shells in the Kalkberg Formation, Helderberg Group, Devonian, New York. (**a, b**) In fine-grained siliciclastic lithologies shells retain their original carbonate composition. (**c, d**) In carbonate lithologies shells are silicified. (**c**) Fine-scale, non-pervasive replacement with length-slow chalcedony. (**d**) Spherulitic length-slow chalcedony on outer margins of the shell and fine-scale textural replacement in the interior

nearshore shallow settings, because the clean, high energy, oxidized carbonate sands of the latter lack organic material for microbial degradation and the resulting increase in acidity (Wright et al. 2003).

5 Conclusion

Fine scale textural silicification of organisms provides high quality data for taxonomic analysis, preserving shell ultrastructure and morphological features in high resolution. Likewise early silicification can preserve paleocommunities with extremely high fidelity. However, silicification is influenced by a range of biological and lithological controls, resulting in taphonomic biases in the record of biodiversity through time, particularly in the Paleozoic. The nature of such biases is poorly constrained with our present knowledge of silicification patterns and processes. Large scale trends are influenced by global climate and ocean chemistry as well as shifts in the relative abundance and composition of invertebrate skeletons through time. Most notable are the transition from dominantly calcite faunas during the Paleozoic to post-Paleozoic aragonite faunas, and the diversification and extinction of groups with siliceous skeletons, such as sponges and radiolarians. Silicification is biased toward preservation of lower solubility LMC organisms; aragonite and, to a lesser extent, HMC skeletons are susceptible to dissolution prior to silicification. Silicification is also biased by depositional conditions, including lithology and porewater geochemistry. Degradation of organic matter can generate conditions for the dissolution and silicification of skeletons, a role that has yet to be investigated experimentally.

The different dissolution rates of carbonate phases in global and local conditions determine the abundance and selectivity of silicified faunas through time. In global greenhouse climates deposition is characterized by thick sequences of carbonate lithologies with little variation laterally and vertically. Silicification is favored in carbonate lithologies, and therefore may be more prevalent in greenhouse climates. However, calcite saturated greenhouse seas promote the dissolution of aragonite thus biasing the composition of silicified faunas (Palmer et al. 1988). In icehouse climates, in contrast, seas saturated with aragonite favor the dissolution of HMC and LMC shells while increased deposition of fine grained siliciclastics may reduce the abundance of silicified faunas. High amplitude, high frequency sea level changes repeatedly expose sediment to meteoric diagenesis thereby promoting bioclast dissolution and replacement in silica. Icehouse climates may also create porosity barriers that likewise favor silicification locally but do not generate a global signal. In the absence of a correlation between the abundance of silicified faunas and a greenhouse/icehouse signal (Kidder and Erwin 2001) or fluctuations in carbonate rock volume (Schubert et al. 1997), the pattern of silicification through time may reflect sampling of exceptionally preserved, compartmentalized, silicified faunas rather than a global signal. There remains the task of unravelling the influence of shifts in biodiversity, specifically the abundance and diversity of carbonate skeletons that are susceptible to silicification, and of the siliceous skeletons that provide a source of silica.

References

Aller, R. C. (1982). Carbonate dissolution in nearshore terrigenous muds: The role of physical and biological reworking. *Journal of Geology, 90*, 79–95.

Bathurst, R. G. C. (1975). *Carbonate sediments and their diagenesis* (2nd ed., 620 p.). *Developments in Sedimentology 12*. Amsterdam: Elsevier.

Behrensmeyer, A. K., & Kidwell, S. M. (1985). Taphonomy's contributions to paleobiology. *Paleobiology, 11*, 105–119.

Behrensmeyer, A. K., Kidwell, S. M., & Gastaldo, R. A. (2000). Taphonomy and paleobiology. *Paleobiology, 26*(Supplement to part 4), 103–147.

Berner, R. A. (1975). The role of magnesium in the crystal growth of calcite and aragonite from sea water. *Geochimica et Cosmochimica Acta, 39*, 489–504.

Berner, R. A., Berner, E. K., & Keir, R. S. (1976). Aragonite dissolution on the Bermuda Pedestal: Its depth and geochemical significance. *Earth and Planetary Science Letters, 30*, 169–178.

Boyd, D. W., & Newell, N. D. (1972). Taphonomy and diagenesis of a Permian fossil assemblage from Wyoming. *Journal of Paleontology, 46*, 1–14.

Brand, U. (1983). Mineralogy and chemistry of the Lower Pennsylvanian Kenrick Fauna, eastern Kentucky, U.S.A., 3. Diagenetic and paleoenvironmental analysis. *Chemical Geology, 40*, 167–181.

Brennan, S. T., & Lowenstein, T. K. (2002). The major-ion composition of Silurian seawater. *Geochimica et Cosmochimica Acta, 66*, 2683–2700.

Briggs, D. E. G. (2003). The role of decay and mineralization in the preservation of soft-bodied fossils. *Annual Review of Earth and Planetary Sciences, 31*, 275–301.

Buening, N. (2001). Brachiopod shells: Recorders of the present and keys to the past. In S. J. Carlson & M. R. Sandy (Eds.), *Brachiopods ancient and modern, paleontological society special papers 7* (pp. 117–143). Paleontological Society.

Butts, S. H. (2004). Silica diagenesis in the Lower Devonian Helderberg Group of New York. *Geological Society of America, Abstracts with Programs, 36*, 383–384.

Butts, S. H. (2005). Latest Chesterian (Carboniferous) initiation of Gondwanan glaciation recorded in facies stacking patterns and brachiopod communities of the Antler foreland basin. *Idaho, Palaeogeography, Palaeoclimatology, Palaeoecology, 223*, 275–289.

Butts, S. H. (2007). Silicified carboniferous (Chesterian) brachiopoda of the Arco Hills Formation. *Idaho, Journal of Paleontology, 81*, 48–63.

Canfield, D. E., & Raiswell, R. (1991). Carbonate precipitation and dissolution: Its relevance to fossil preservation. In P. A. Allison & D. E. G. Briggs (Eds.), *Taphonomy: Releasing the data locked in the fossil record* (pp. 411–453). New York: Plenum.

Chave, K. E. (1954). Aspects of the biogeochemistry of magnesium 1. *Calcareous Marine Organisms, Journal of Geology, 62*, 266–283.

Cherns, L., & Wright, V. P. (2000). Missing molluscs and evidence of large-scale, early skeletal aragonite dissolution in a Silurian sea. *Geology, 28*, 791–794.

Cooper, G. A., & Grant, R. E. (1972). *Permian Brachiopods of West Texas, I, Smithsonian Contributions to Paleobiology 14*, 183 p.

Correns, C. W. (1969). *Introduction to mineralogy*. New York: Springer. 484 p.

Curry, G. B., & Ansell, A. D. (1986). Tissue mass in living brachiopods. *Biostratigraphie du Paléozoique, 4*, 231–241.

Curry, G. B., Peck, L. S., Ansell, A. D., & James, M. (1989). Physiological constraints in fossil and recent brachiopods. *Transactions of the Royal society of Edinburgh: Earth Sciences, 80*, 255–262.

Daley, R. L. (1987). Patterns and controls of skeletal silicification in a Mississippian fauna, northwestern Wyoming. Unpublished masters thesis, University of Wyoming, 140 p.

Daley, R. L., & Boyd, D. W. (1996). The role of skeletal microstructure during selective silicification of brachiopods. *Journal of Sedimentary Research, 66*, 155–162.

Davies, D. J., Powell, E. N., & Stanton, R. J., Jr. (1989). Relative rates of shell dissolution and net sediment accumulation; a commentary, can shell beds form by the gradual accumulation of biogenic debris on the sea floor? *Lethaia, 22*, 207–212.

Dott, D. H., Jr. (1958). Cyclical patterns in mechanically deposited Pennsylvanian limestones of northeastern Nevada. *Journal of Sedimentary Petrology, 28*, 3–15.

Erwin, D. H., & Kidder, D. L. (2000). Depositional controls on selective silicification of Permian fossils, southwestern United States. In B. R. Wardlaw, R. E. Grant, & D. M. Rohr (Eds.), *Guadalupian symposium, Smithsonian contributions to earth science 32* (pp. 407–415). Washington, DC: Smithsonian Institution Press.

Fairbridge, R. W. (1983). Syndiagenesis-anadiagenesis-epidiagenesis; phases in lithogenesis. In G. Larsen & G. V. Chilingar (Eds.), *Diagenesis in sediments and sedimentary rocks 2* (pp. 17–113). Amsterdam: Elsevier.

Fischer, A. G. (1982). Long-term climatic oscillations recorded in stratigraphy. In W. Berger (Ed.), *Climate in earth history, National Research Council, studies in geophysics* (pp. 97–104). Washington, DC: National Academy Press.

Fox, W. T. (1962). Stratigraphy and paleoecology of the Richmond Group in Southeastern Indiana. *Geological Society of America Bulletin, 73*, 621–642.

Francis, S., Barghoorn, E. S., & Margulis, L. (1978). On the experimental silicification of microorganisms; III. Implications of the preservation of the green prokaryotic alga Prochloron and other coccoids for interpretation of the microbial fossil record. *Precambrian Research, 7*, 377–383.

Froelich, P. N., Klinkhamemer, G. P., Bender, M. L., Luedtke, N. A., Heath, G. R., Cullen, D., et al. (1979). Early oxidation of organic matter in pelagic sediments of the eastern equatorial Atlantic: Suboxic diagenesis. *Geochimica et Cosmochimica Acta, 43*, 1075–1090.

Gaspard, D., & Roux, M. (1974). Quelques aspects de la fossilisation des tests chez les Brachiopodes et les Crinoides; relation entre la presence de matiere organique et le developpement d'agregats ferrifères. *Geobios, 7*, 81–89.

Glover, C. P., & Kidwell, S. M. (1993). Influence of organic matrix on the postmortem destruction of molluscan shells. *Journal of Geology, 101*, 729–747.

Harper, E. M. (2000). Are calcitic layers an effective adaptation against shell dissolution in the Bivalvia? *Journal of Zoology, 251*, 179–186.

Henrich, R. (1985). A calcite dissolution pulse in the Norwegian-Greenland Sea during the last deglaciation. *Geologische Rundschau, 75*, 805–827.

Henrich, R., & Wefer, G. (1986). Dissolution of biogenic carbonates; effects of skeletal structure. *Marine Geology, 71*, 341–362.

Henry, T. W., & Gordon, M. Jr. (1992). Middle and Upper Chesterian brachiopod biostratigraphy, eastern Appalachians, Virginia and West Virginia. In P. K. Sutherland & W. L. Manger (Eds.), *Recent advances in Middle Carboniferous biostratigraphy – A symposium, Oklahoma Geological Survey Circular 94* (pp. 1–21).

Hesse, R. (1989). Silica diagenesis; origin of inorganic and replacement cherts. *Earth Science Reviews, 26*, 253–284.

Hesse, R. (1990). Origin of cherts: Diagenesis of biogenic siliceous sediments. In I. A. McIlreath & D. W. Morrow (Eds.), *Diagenesis: Geological Association of Canada, Geoscience Canada Reprint series 4.* (pp. 227–252).

Hinman, N. W. (1987). *Organic and inorganic chemical controls on the rates of silica diagenesis; a comparison of a natural system with experimental results.* Unpublished Ph.D. dissertation. University of California, San Diego, 402 p.

Hinman, N. W. (1990). Chemical factors influencing the rates and sequences of silica phase transitions: Effects of organic constituents. *Geochimica et Cosmochimica Acta, 54*, 1563–1574.

Hinman, N. W. (1998). Sequences of silica phase transitions: Effects of Na, Mg, K, Al, and Fe ions. *Marine Geology, 147*, 13–24.

Hodges, E. S. (1932). *Liesegang rings and other periodic structures*. London: Chapman & Hall. 122 p.

Holdaway, H. K., & Clayton, C. J. (1982). Preservation of shell microstructure in silicified brachiopods from the Upper Cretaceous Wilmington Sands of Devon. *Geological Magazine, 119*, 371–382.

Holland, S. M., & Patzkowsky, M. E. (1997). Distal orogenic effects on peripheral bulge sedimentation: Middle and Upper Ordovician of the Nashville Dome. *Journal of Sedimentary Research, 67*, 250–263.

Holland, S. M., & Patzkowsky, M. E. (1998). Sequence stratigraphy and relative sea-level history of the Middle and Upper Ordovician of the Nashville Dome, Tennessee. *Journal of Sedimentary Research, 68*, 684–699.

Iler, R. K. (1979). *The chemistry of silica: Solubility, polymerization, colloid and surface properties, and biochemistry*. New York: Wiley. 866 p.

Jacka, A. D. (1974). Replacement of fossils by length-slow chalcedony and associated dolomitization. *Journal of Sedimentary Petrology, 44*, 421–427.

Kidder, D. L., & Erwin, D. H. (2001). Secular distribution of biogenic silica through the Phanerozoic: Comparison of silica-replaced fossils and bedded cherts at the series level. *Journal of Geology, 109*, 509–522.

Kidder, D. L., & Mumma, S. A. (2003). Silica-replaced oolites, bedded shelf cherts and Paleozoic changes in the silica cycle. *Sedimentary Geology, 162*, 159–166.

Klein, R. T., & Walter, L. M. (1995). Interactions between dissolved silica and carbonate minerals: An experimental study at 25–50°C. *Chemical Geology, 125*, 29–43.

Knauth, L. P. (1979). A model for the origin of chert in limestone. *Geology, 7*, 274–277.

Lancelot, Y. (1973). Chert and silica diagenesis in sediments from the central Pacific. *Initial Reports of the Deep Sea Drilling Project, 17*, 377–405.

Land, L. S. (1967). Diagenesis of skeletal carbonates. *Journal of Sedimentary Petrology, 37*, 914–920.

Laufeld, S., & Jeppsson, L. (1976). Silicification and bentonites in the Silurian of Gotland. *Geologiska Föreningens i Stockholm Föhandlingar, 98*, 31–44.

Leo, R. F., & Barghoorn, E. S. (1976). Silicification of wood. *Botanical Museum Leaflets, Harvard University, 25*, 1–47.

Loope, D. B., & Watkins, D. K. (1989). Pennsylvanian fossils replaced by red chert: Early oxidation of pyritic precursors. *Journal of Sedimentary Petrology, 59*, 375–386.

Maliva, R. G. (1992). Selective silicification of fossils by syntaxial overgrowths on quartz sand, Oriskany Sandstone (Lower Devonian), New York. *Sedimentary Geology, 78*, 165–170.

Maliva, R. G., & Siever, R. (1988). Mechanisms and controls of silicification of fossils in limestones. *Journal of Geology, 96*, 387–398.

Maliva, R. G., Knoll, A. H., & Siever, R. (1989). Secular change in chert distribution: A reflection of evolving biological participation in the silica cycle. *Palaios, 4*, 519–532.

Maliva, R. G., Knoll, A. H., & Simonson, B. M. (2005). Secular change in the Precambrian silica cycle; insights from chert petrology. *Geological Society of America Bulletin, 117*, 835–845.

Meyers, W. J. (1977). Chertification in the Mississippian lake valley formation, Sacramento Mountains, New Mexico. *Sedimentology, 24*, 75–105.

Misík, M. (1995). Selective silicification of calcitic fossils and bioclasts in the West-Carpathian limestones. *Geologica Carpathica, 46*, 151–159.

Newell, N. D., & Boyd, D. W. (1970). Oyster-like Permian bivalvia. *Bulletin of the American Museum of Natural History, 143*, 217–282.

Newell, N. D., Rigby, J. K., Fischer, A. G., Whiteman, A. J., Hickox, J. E., & Bradley, J. S. (1953). *The Permian reef complex of the Guadalupe mountains region, Texas and New Mexico; A study in paleoecology*. San Francisco: W.H. Freeman. 236 p.

Palmer, T. J., Hudson, J. D., & Wilson, M. A. (1988). Palaecological evidence for early aragonite dissolution in ancient calcites seas. *Nature, 335*, 809–810.

Paraguasso, A. B. (1976). Experimental replacement of silica. *Revista Brasiliera de Geociências, 6*, 89–94.

Peck, L. S. (1993). The tissues of articulate brachiopods and their value to predators. *Philosophical Transactions of the Royal Society of London, 339*, 17–32.

Peck, L. S., Clarke, A., & Holmes, L. J. (1987). Size, shape, and the distribution of organic matter in the Recent Antarctic brachiopod Liothyrella uva. *Lethaia, 20*, 33–40.

Racki, G. (1999). Silica-secreting biota and mass extinctions: Survival patterns and processes. *Palaeogeography, Palaeoclimatology, Palaeoecology, 154*, 107–132.

Railsback, B. L., & Anderson, T. F. (1987). Control of Triassic seawater chemistry and temperature on the evolution of post-Paleozoic aragonite-secreting faunas. *Geology, 15*, 1002–1005.

Railsback, L. B., Holland, S. M., Hunter, D. M., Jordon, E. M., Díaz, J. R., & Crowe, D. E. (2003). Controls on geochemical expression of subaerial exposure in Ordovician limestones from the Nashville Dome, Tennesee, USA. *Journal of Sedimentary Research, 73*, 790–805.

Ries, J. B., & Blaustein, M. K. (2003). Modern high-magnesium calcite organisms produce low-magnesium calcite in experimental Mid-Cretaceous sea water. *Geological Society of America, Abstracts with Programs, 35*, 204–205.

Sandberg, P. A. (1983). An oscillating trend in Phanerozoic non-skeletal carbonate mineralogy. *Nature, 305*, 19–22.

Schieber, J. (1996). Early diagenetic silica deposition in algal cysts and spores; a source of sand in black shales? *Journal of Sedimentary Research, 66*, 175–183.

Schmitt, J. G., & Boyd, D. W. (1981). Patterns of silicification in Permian pelecypods and brachiopods from Wyoming. *Journal of Sedimentary Petrology, 51*, 1297–1308.

Scholle, P. A., & Ulmer-Scholle, D. S. (2003). *A color guide to the petrography of carbonate rocks: grains, textures, porosity, diagenesis.* American Association of Petroleum Geologists Memoir 77, Tulsa, OK, 474 p.

Schubert, J. K., Kidder, D. L., & Erwin, D. H. (1997). Silica-replaced fossils through the Phanerozoic. *Geology, 25*, 1031–1034.

Smith, A. B. (2003). Getting the measure of diversity. *Paleobiology, 29*, 34–36.

Stanley, S. M., & Hardie, L. A. (1998). Secular oscillations in the carbonate mineralogy of reef-building and sediment-producing organisms driven by tectonically forced shifts in seawater chemistry. *Palaeogeography, Palaeoclimatology, Palaeoecology, 144*, 3–19.

Tomasovych, A., & Rothfus, T. A. (2005). Differential taphonomy of modern brachiopods (San Juan Islands, Washington State): Effect of intrinsic factors on damage and community-level abundance. *Lethaia, 38*, 271–292.

Tréguer, P., Nelson, D. M., Van Bennkeom, A. J., DeMaster, D. J., Leynaert, A., & Quéguiner. (1995). The silica balance in the world's ocean: A reestimate. *Science, 268*, 375–379.

Tucker, M. E. (1991). The diagenesis of fossils. In S. K. Donovan (Ed.), *The processes of fossilization* (pp. 84–104). New York: Columbia University Press.

Tucker, M. E., & Wright, V. P. (1990). *Carbonate sedimentology*. Oxford/England: Blackwell. 482 p.

Walter, L. M. (1983). *The dissolution kinetics of shallow water carbonate grain types; effects of mineralogy, microstructure, and solution chemistry* (347 p). Unpublished Ph.D. dissertation, University of Miami.

Walter, L. M. (1985). Relative reactivity of skeletal components during dissolution. In N. Schneidermann & P. M. Harris (Eds.), *Carbonate cements, SEPM Special Publication 36* (pp. 3–16).

Walter, L. M., & Morse, J. W. (1984). Reactive surface area of skeletal carbonates during dissolution; effect of grain size. *Journal of Sedimentary Petrology, 54*, 1081–1090.

Walter, L. M., & Morse, J. W. (1985). The dissolution kinetics of shallow marine carbonates in seawater: A laboratory study. *Geochimica et Cosmochimica Acta, 49*, 1503–1513.

Wilkinson, B. H. (1979). Biomineralization, paleoceanography, and the evolution of calcareous marine organisms. *Geology, 7*, 524–527.

Williams, A. (1990). Biomineralization in the lophophorates. In J. G. Carter (Ed.), *Skeletal biomineralization: patterns, trends, and evolutionary processes* (Vol. 1, pp. 67–82). New York: Van Nostrand Reinhold.

Williams, A. (1997). Shell Structure. In R. Kaesler (Ed.), *Treatise on invertebrate paleontology, Part H, revised, Brachiopoda* (Vol. 1, pp. 267–320). Lawrence: Geological Society of America and University of Kansas Press.

Williams, L. A., & Crerar, D. A. (1985). Silica diagenesis, II. General Mechanisms. *Journal of Sedimentary Petrology, 55*, 312–321.

Wright, V. P., Cherns, L., & Hodges, P. (2003). Missing molluscs: Field testing taphonomic loss in the Mesozoic through early large-scale aragonite dissolution. *Geology, 31*, 211–214.

Chapter 12
Phosphatization Through the Phanerozoic

Stephen Q. Dornbos

Contents

Abstract Phosphatization of soft tissues and skeletal remains has varied temporally and taxonomically through the Phanerozoic. During the Cambrian through early Ordovician, microscopic arthropods and animal embryos were preferentially preserved. Phosphatization was uncommon during the rest of the Paleozoic, as recalcitrant tissues of a few taxa were preserved in hospitable microenvironments. The Cretaceous through Eocene saw another expansion of phosphatization, with a strong bias toward fish remains already enriched in apatite. Throughout its Phanerozoic history, phosphatization exhibited a taphonomic bias toward taxa with recalcitrant tissues that could resist the early stages of organic decay, taxa with organic structures already enriched in calcium phosphate, and, in many cases, taxa with small body sizes. The pulse of phosphatization during the Cambrian through Early Ordovician may have been facilitated by the generally lower levels of mixed layer development in the upper few centimeters of seafloor sediments during that time period, whereas the Cretaceous through Eocene increase in phosphatization was possibly related to the enlargement of euxinic epicontinental seaways.

S.Q. Dornbos (✉)
Department of Geosciences, University of Wisconsin-Milwaukee,
Milwaukee, WI 53201-0413, USA
e-mail: sdornbos@uwm.edu

P.A. Allison and D.J. Bottjer (eds.), *Taphonomy: Process and Bias Through Time*, Topics in Geobiology 32, DOI 10.1007/978-90-481-8643-3_12,

1 Introduction

Phosphatization is a rare yet astounding mode of fossil preservation that has the capability of preserving soft tissues to cellular and even subcellular levels. While it is high fidelity in nature, phosphatization is also a highly selective and biased taphonomic pathway. It favors the preservation of recalcitrant tissues that are resistant to decay and those that are already enriched with phosphate. There is also a size bias in many deposits because phosphatization has either occurred at a limited depth within seafloor sediments or in a minute microenvironment within/around an organism.

The goal of this chapter is to examine the Phanerozoic history of phosphatization in search of large-scale patterns over geologic time. Are there times in Earth history when phosphatization is more prevalent? What are possible explanations for this temporal distribution? Are there biases toward the preservation of certain taxa, and how has that changed over time? These are some of the questions that this chapter will explore through a review of the available literature.

2 Phosphatization Processes and Biases

2.1 *Phosphatization Processes*

Dissolved phosphate in the oceans is derived from continental weathering and is a limiting nutrient for marine phytoplankton. Most phosphate is remobilized or remineralized in the photic zone, but in upwelling zones a larger amount of organic material makes it to the seafloor and phosphate-enriched deposits can result. Such phosphatic deposits account for only a small percentage of phosphate buried in marine sediments, making phosphatic fossil preservation exceedingly rare. Phosphatization of soft tissues and replacement of calcareous skeletons by phosphate, most commonly in a form of apatite called francolite (carbonate flourapatite), can take place in either these broad phosphogenic depositional environments or in a local phosphatizing microenvironment created by the decay of soft tissues in the organism itself. When phosphatization does take place in a phosphogenic depositional environment, there are some general paleoenvironmental, biological, and chemical parameters that play important rolls in the process. Recent phosphorite deposits are typically forming in environments characterized by low sedimentation rates, strong seafloor currents, and a large influx of organic material due to high primary productivity in the water column above (e.g. Föllmi 1996; Baturin 1999). The low sedimentation rates and large amount of organic input in these environments combine to create sediment enriched in organic material (e.g. Föllmi 1996; Baturin 1999). The organic material in the sediment then begins to decay and releases phosphate into the pore waters. Although many details about the process are not yet understood, the dissolved phosphate content of the pore waters reaches supersaturation, allowing phosphogenesis and the phosphatization of organisms to begin (e.g. Baturin 1999). The concentration of dissolved phosphate within sediment pore waters is probably strongly assisted by the sealing of microbial mats on the seafloor (e.g. Briggs and Kear 1993; Wilby et al. 1996; Soudry 2000).

Although sediment pore water is typically saturated with respect to phosphate, under normal seawater pH conditions (~8) calcium carbonate is more stable than phosphate and thereby inhibits phosphate precipitation. One way in which phosphate can precipitate is if the pH of the pore waters is reduced to around 7, which destabilizes calcium carbonate and allows phosphate to begin precipitating (Lucas and Prevot 1991; Briggs and Wilby 1996; Trappe 1998). It is thought that bacterial metabolic processes serve to lower the pH of the pore waters during this process, which also results in the release of phosphate that is bound to iron hydroxides derived from continental weathering (Froelich et al. 1988; Baturin 1999; Sagemann et al. 1999; Briggs 2003). Once precipitation of phosphate begins, it is estimated that the process can phosphatize soft tissues in rapid timescales ranging anywhere from days to weeks (e.g. Föllmi 1996). Bacteria coating the surfaces of the tissue often appear to serve as the nucleation points for phosphatization, and are typically phosphatized themselves (Baturin 1999).

After phosphogenesis and phosphatization have taken place currents remobilize the sediment, transporting and redepositing it elsewhere. A new round of phosphogenesis then begins in the newly deposited sediment and the cycle continues (Föllmi 1996). This process of remobilization and redeposition of the sediment results in different generations of phosphate being present in the same sediment (Baturin 1999). Resulting phosphogenic sediments often include a mixture of phosphatic lithologies including primary phosphatic mud, hardgrounds, pellets, and concretions, all of which can contain fossils (Glenn et al. 1994). Phosphatic concretions containing phosphatized fossils can also form in non-phosphogenic settings.

Other models, including strictly geochemical and hydrothermal explanations, have been proposed to explain phosphogenesis (e.g. Kholodov and Paul 1995), but the biogenic model currently has the greatest amount of support because it explains the most about both ancient and Recent phosphogenic settings (Baturin 1999). In fact, recent work has indicated that giant sulfur-oxidizing bacteria, such as *Thiomargarita, Beggiatoa,* and *Thioploca*, which are abundant in modern phosphogenic settings may play an important role in phosphogenesis by releasing large amounts of dissolved phosphate into their surrounding environment in anoxic conditions (Reimers et al. 1990; Krajewski et al. 1994; Schulz and Schulz 2005). This modern correlation between phosphorites, upwelling zones, and sulfur-oxidizing bacteria capable of mediating pore water phosphate concentrations have led various authors to suggest that certain ancient phosphatized microfossils represent sulfur-oxidizing bacteria (e.g. Reimers et al. 1990; Bailey et al. 2007).

When phosphatization takes place in non-phosphogenic depositional environments, it occurs when an organism creates its own phosphatization microenvironment through organic decay (Briggs and Kear 1993). Such fossils are often preserved in phosphatic nodules within broader non-phosphatic facies, usually carbonates (e.g. Walossek et al. 1993). Preservational completeness varies tremendously within single organisms, with more labile soft tissues often absent or heavily decayed and more recalcitrant tissues such as carapaces or connective tissues extremely well preserved (e.g. Klug et al. 2005).

The bacterial decay of the labile soft tissues evidently produced enough dissolved phosphate and lowered the pH sufficiently to phosphatize the remaining undecayed tissues (Klug et al. 2005). This process would have taken place after the organism had been buried in the sediment, likely below the oxic–anoxic interface.

2.2 *Phosphatization Biases*

Phosphatization is a highly biased mode of preservation. These biases primarily involve the original composition of organic material, the decay resistance of organic material, and the size of the organism being preserved. There does appear to be a phosphatization bias toward organic material that is already enriched in phosphate. This is true of fish coprolites in the Cretaceous through Eocene phosphorites of the southern Mediterranean, as well as the phosphatized fish and whale bones found in Pleistocene and Recent phosphorites offshore of Namibia and Peru (Lamboy et al. 1994; Resig and Glenn 1997; Baturin and Dubinchuk 2003). The beginnings of bacterial decay in these organic remains releases large amounts of dissolved phosphate, rapidly creating a phosphatizing microenvironment that preserves the undecayed portions of the remains (e.g. Lamboy et al. 1994). Organic remains that are not enriched in phosphate are certainly also known to become phosphatized, but this bias toward previously enriched remains is prevalent in many deposits.

Another persistent bias is that toward the preservation of more recalcitrant organic tissues. The more labile, decay-prone, portions of organisms are often decayed to the point of absence, while the resistant organic structures are phosphatized (e.g. Klug et al. 2005). This bias is particularly strong in the phosphatized animal embryos and microarthropods of the Cambrian to Early Ordovician. Both of these types of fossils are nearly hollow on the inside while the outer surface is phosphatized (Maas et al. 2003; Dong et al. 2005). The labile inner organic remains clearly decayed heavily while the outer cuticle remained relatively intact and became phosphatized as the decay process created the necessary environmental conditions for phosphatization. Most of the preserved animal embryos were likely those of ecdysozoans, so initial cuticular development probably contributed greatly to their preservation (e.g. Donoghue et al. 2006).

Size is also an important bias in phosphatization. In phosphogenic environments, the thin zone of phosophogenesis in the upper few millimetres of the sediment, as well as the constant reworking of the sediments, places strong size limitations on any organic material that is going to become phosphatized or survive the reworking to end up in the final deposit (e.g. Dornbos et al. 2006). In non-phosphogenic settings, small body size increases the chances of preservation because the phosphatizing microenvironment is often quite small itself. These biases are likewise evident in the animal embryos and microarthropods found in the Cambrian through Early Ordovician (e.g. Walossek et al. 1993; Maas et al. 2003; Donoghue et al. 2006). The preserved embryos and microarthropods are both microscopic. Later developmental stages are not preserved in the same rocks as the embryos and larger animals are not phosphatized with the meiofaunal arthropods (e.g. Maas et al. 2003; Donoghue et al. 2006). Phosphatization gives paleontologists an extremely powerful yet painfully narrow view of ancient life.

3 Temporal Distribution with Examples

The following subsections describe important instances of phosphatized fossils through the Phanerozoic fossil record. It includes examples of soft tissue replication as well as replacement of skeletons by apatite. This section is not intended to be an exhaustive listing of all known occurrences of Phanerozoic phosphatization, but a description of certain occurrences that are important for understanding long-term trends. The examples and relevant references are summarized in Table 1.

3.1 Paleozoic Phosphatization

3.1.1 Cambrian Phosphatization

The Cambrian contains a wealth of phosphatized fossils relative to the remainder of the Phanerozoic. The lowermost Cambrian rocks of the Dengying Formation, Shaanxi area, China, for example, contain an ontogenetic series of putative cnidarian embryos, known as *Olivooides* (Yue and Bengtson 1999). These embryos and early larval stages exhibit exceptional preservation of their exterior surfaces, which are both coated and impregnated with apatite. Similarly preserved cnidarian-like embryos are also described from the Lower Cambrian Manykay Formation of Siberia (Kouchinsky et al. 1999). In contrast to the putative animal embryos of the Ediacaran Doushantuo Formation, southwest China, these Cambrian embryos are not internally impregnated with phosphate (e.g. Xiao and Knoll 1999; Dornbos et al. 2005). Instead their interiors are mostly hollow with thin strands of phosphate running between the outer phosphatized walls (Yue and Bengtson 1999). These are likely the phosphatized remnants of the decayed interior of these embryos.

Similarly preserved animal embryos are also known from the Lower Cambrian Kuanchuanpu Formation of southern Shaanxi, China (Steiner et al. 2004). The external features of these embryos are magnificently preserved while the interiors are mostly hollow. Several possible species of embryo are preserved in these rocks, but categorizing them taxonomically is difficult. Likely developmental sequences have been reconstructed for two of these groups of embryos, one of which includes larval stages with the beginnings of limb development. Affinities of these embryos with protostomes, specifically ancestral arthropods, are proposed (Steiner et al. 2004).

Phosphatized microscopic arthropods preserved in three dimensions are also known from the lower Cambrian. They are found preserved in the Petrolenus

Table 1 Phanerozoic phosphatization examples discussed in the text

Age	Formation/unit/stage	Location	Fossil content	Representative reference(s)
Early Cambrian	Dengying formation	China	Animal embryos	Yue and Bengtson (1999)
Early Cambrian	Manykay formation	Russia	Animal embryos	Kouchinsky et al. (1999)
Early Cambrian	Kuanchuanpu formation	China	Animal embryos	Steiner et al. (2004)
Early Cambrian	Petrolenus limestone	UK	Microarthropods	Siveter et al. (2001)
Middle Cambrian	Monastery creek phosphorite fm.	Australia	Animal embryos, microarthropods	Walossek et al. (1993); Donoghue et al. (2006)
Middle Cambrian	Gowers formation	Australia	Microarthropods	Walossek et al. (1993)
Middle Cambrian	Inca Shale formation	Australia	Microarthropods	Walossek et al. (1993)
Middle Cambrian	Devoncourt limestone fm.	Australia	Microarthropods	Walossek et al. (1993)
Middle Cambrian	Kuonamka formation	Russia	Microarthropods	Müller et al. (1995)
Middle Cambrian	Wagcun and Wa'ergang sections	China	Animal embryos	Dong et al. (2004)
Late Cambrian	Orsten	Sweden	Microarthropods	Maas et al. (2003)
Late Cambrian	Orsten	Poland	Microarthropods	Walossek and Szaniawski (1991)
Late Cambrian	Bitiao formation	China	Animal embryos, microarthropods	Dong et al. (2005)
Late Cambrian	Wa'ernang & Paibi sections	China	Animal embryos	Dong et al. (2005)
Early Ordovician	Orsten	Sweden	Microarthropods	Andres (1989)
Early Ordovician	Middle-arm point formation	Canada	Microarthropods	Roy and Fåhraeus (1989)
Early Ordovician	Green point formation	Canada	Microarthropods	Walossek et al. (1994)
Early Ordovician	Fairview formation	USA	Animal embryos	Donoghue et al. (2006)
Early Silurian	Bardo beds	Poland	Acritarchs	Kremer (2005)
Late Silurian	Panuara group	Australia	Anthozoans, echinoderms	Bischoff (1978); Bischoff and Hall (1980)
Late Devonian	Gogo formation	Australia	Fish	Trinajistic et al. (2007)
Late Mississippian	Bear Gulch limestone	USA	Fish, invertebrates	Lund et al. (1993); Grogan and Lund (1997)
Late Pennsylvanian	Coffeyville formation	USA	Cephalopods	Tanabe et al. (2001)

Permian	Arcturus formation	USA	Cephalopods	Mapes et al. (2002, 2003)
Early Triassic	Olenekian stage	Svalbard	Ostracods, ciliates	Weitschat (1995)
Middle Triassic	Muschelkalk	Germany	Bivalves	Klug et al. (2005)
Middle Triassic	Grenzbitumenzone & Meride limestone	Italy	Reptiles, nothosaurs, fish	Etter (2002); Renesto and Avanzini (2002)
Ealry Jurassic	Lombardische Kieselkalk formation	Italy	Crustaceans, fish, coleoids, polychaetes, nematodes	Pinna (1985)
Middle-Late Jurassic	Oxford clay	UK	Coleoids	Allison (1988)
Late Jurassic	Portland limestone formation	UK	Bivalves	Wilby and Whyte (1995)
Early Cretaceous	Santana formation	Brazil	Fish	Martill (1988)
Middle Cretaceous	Marnes Bleues formation	France	Bacteria, fungi	Bréhéret (1991)
Middle Cretaceous	DSDP site 369	East Atlantic	Bacteria, fungi	Bréhéret (1991)
Late Cretaceous	Austin group	Mexico	Fish, crustaceans, ammonites, pterosuars	Stinnesbeck et al. (2005)
Late Cretaceous	Colon formation	Venezuela	Foraminifera	Ghosh (1984)
Late Cretaceous	Various formations	Tunisia	Fish coprolites	Lamboy et al. (1994)
Late Cretaceous	Various formations	Morocco	Fish coprolites	Lamboy et al. (1994)
Late Cretaceous	Various formations	Mauritania	Fish coprolites	Lamboy et al. (1994)
Late Cretaceous	Various formations	Egypt	Fish coprolites	Lamboy et al. (1994)
Late Cretaceous	Duwi formation	Egypt	Diatoms	Ahmed and Kurzweil (2002)
Paleogene (Paleocene)	Various formations	Egypt	Fish coprolites	Lamboy et al. (1994)
Paleogene (Paleocene-Eocene)	Various formations	Tunisia	Fish coprolites	Lamboy et al. (1994)
Paleogene (Paleocene-Eocene)	Various formations	Morocco	Fish coprolites	Lamboy et al. (1994)

(continued)

Table 1 (continued)

Age	Formation/unit/stage	Location	Fossil content	Representative reference(s)
Paleogene (Paleocene-Eocene)	Various formations	Mauritania	Fish coprolites	Lamboy et al. (1994)
Paleogene (Eocene)	Various formations	Senegal	Fish coprolites	Lamboy et al. (1994)
Paleogene (Oligocene-Miocene)	Riversleigh limetones	Australia	Insects, fungi, bacteria	Duncan et al. (1998)
Neogene (Miocene)	Monterey formation	USA	Fish, foraminifera, bacteria	Garrison et al. (1990)
Neogene (Miocene)	Libros basin	Spain	Amphibians, birds, snakes, leaves	McNamara et al. (2004)
Pleistocene	Continental margin	India	Stromatolites, bacteria	Purnachandra Rao et al. (2000), 2002)
Recent	Outer shelf	Namibia	Fish, whales	Baturin and Dubinchuk (2003)
Recent	Outer shelf/slope	Peru	Fish, whales	Resig and Glenn (1997)

Limestone of Great Britain (Siveter et al. 2001). Much like the *Olivooides* embryos of China, these microscopic arthropod fossils have largely hollow interiors within a thin zone of phosphatized soft tissue. These fossils are preserved in phosphatic nodules within a carbonate facies. This is the earliest known example of "Orsten" style preservation of microscopic arthropods, a preservational mode within nodules that is particularly common in the upper Cambrian of Sweden (Siveter et al. 2001).

Similarly preserved arthropods are also found in the middle Cambrian of Australia. Their preservational style is classically Orsten, with phosphatized external features and nearly hollow interiors (Walossek et al. 1993). These fossils are found in the Monastery Creek Phosphorite, the Gowers, the Inca Shale, and the Devoncourt Limestone Formations, all of western Queensland (e.g. Walossek et al. 1993). Small phosphatized shields of microscopic arthropods, without soft part preservation, are also preserved in the middle Cambrian of Kuonamka Formation of western Siberia (Müller et al. 1995).

Phosphatized animal embryos are also known from the middle Cambrian of Australia and China (Dong et al. 2004; Donoghue et al. 2006). As with those of the lower Cambrian, their exteriors are phosphatized and their interiors are mainly hollow with some phosphatized decayed organic remnants. The Australian embryos are preserved in the Monastery Creek Phosphorite Formation of western Queensland. These embryos, named *Markuelia*, have a distinctive enrolled morphology and have been interpreted as the late-stage embryos of cycloneuralian nemathelminths (Donoghue et al. 2006). The Chinese embryos are also *Markuelia*, and are found in the Wangcun and Wa'ergang Sections of western Hunan Province (Dong et al. 2004).

Just as in the middle Cambrian, the upper Cambrian is known to contain phosphatized micrscopic arthropods as well as the cycloneuralian nemathelminth embryo *Markuelia*. Most of these arthropods are found in the classic Orsten localities in Västergötland, Sweden and have the typical Orsten style preservation (Maas et al. 2003). These arthropods include larval stages of the trilobite *Agnostus*, a chelicerate larvae, crustaceans, and phosphatocopines. Similar fossils are known from the Isle of Öland in Sweden (Maas et al. 2003). Upper Cambrian rocks of the Hel Peninsula of Poland also contain Orsten arthropods, including a stem group crustacean and phosphatocopines (Walossek and Szaniawski 1991). One phosphatocopine arthropod is found in the Bitiao Formation, Wangcun Section, western Hunan Province (Dong et al. 2005).

Upper Cambrian rocks in China also contain the embryo *Markuelia* (Dong et al. 2005). These embryos exhibit the same preservational style as those of the middle Cambrian. They are preserved in the Bitiao Formation, Wangcun Section, western Hunan Province, and the Wa'ernang and Paibi Sections, also of western Hunan Province (Dong et al. 2005).

3.1.2 Ordovician Phosphatization

Although phosphatization becomes a scarcer phenomenon after the Cambrian, there are several known occurrences in the remainder of the Paleozoic. In fact, typical Orsten-style preservation and the embryo *Markuelia* are both known from the Lower Ordovician. The Lower Ordovician of the Isle of Öland in Sweden contains abundant microscopic arthropods (Andres 1989). Other microscopic phosphatized arthropods are known from the Lower Ordovician of Newfoundland, Canada. These fossils are preserved in the Middle-Arm Point Formation and the Green Point Formation (Roy and Fåhraeus 1989; Walossek et al. 1994). Lower Ordovician *Markuelia* are preserved in the Fairview Formation of Nevada, western USA (Donoghue et al. 2006).

3.1.3 Silurian Phosphatization

Phosphatized acritarchs, known as Mazuelloids, are known from the Lower Silurian Bardo Beds of Poland (Kremer 2005). The acritarchs are thought to have sunk to the seafloor as part of macroaggregates that were later overgrown by cyanobacterial mats. Phosphatization, which took place on the outer and inner surfaces of the organic acritarch walls, may have been facilitated by microbial mats – the presence of which is suggested by the co-occurrence of coccoidal microfossils that resemble modern cyanobacteria (Kremer 2005). They are preserved in black radiolarian cherts deposited in a bathyal setting (Kremer 2005).

Silurian phosphatization is also known from Upper Silurian strata in central New South Wales, Australia (Bischoff 1978; Bischoff and Hall 1980). These fossils consist of phosphatized anthozoan soft parts and echinoderm skeletal elements found in the Boree Creek Formation and Borenore Limestone of the Panuara Group. The echinoderm skeletal elements include crinoid and asteroid fragments that exhibit exceptional preservation of the original stereom structure, providing critical insight into the postlarval skeletal growth of echinoderms. All of these fossils are preserved within carbonates (Bischoff 1978; Bischoff and Hall 1980).

3.1.4 Devonian Phosphatization

The muscle fibers, circulatory tissues, and nerve tissues of placoderm fish are preserved in the Late Devonian (Frasnian) Gogo Formation of the Canning Basin, Western Australia (Trinajistic et al. 2007). These structures are preserved in three dimensional phosphate likely mineralized under the influence of bacteria in the small infillings under the headshield of the fish. Currently the oldest soft tissues described from such fish, the fossilized material supports the hypothesis that the placoderms are the sister group of the other gnathostomes (Trinajistic et al. 2007).

3.1.5 Carboniferous Phosphatization

Similarly preserved fish soft parts, including phosphatized muscles, are known from the upper Mississippian Bear Gulch Limestone of the Heath Formation of central Montana, USA (Lund et al. 1993). Skin coloration, eye pigments, circulatory system structures, and internal organs are sometimes preserved (Grogan and Lund 1997). One of the most diverse fish fossil deposits in the world, with over 100 fish species described, the Bear Gulch also contains a diverse array of invertebrates (Hagadorn 2002). Frequent turbidity currents across the shallow marine basin likely facilitated such exceptional preservation, as fish and other organisms appear to have been simultaneously asphyxiated and buried (Grogan and Lund 1997). Phosphatized cephalopod mouthparts and a radula are preserved in the Upper Pennsylvanian Coffeyville Formation of Oklahoma, central USA (Tanabe et al. 2001). The fossil contains an almost complete jaw apparatus and a well-preserved radula, both in life orientation. They are preserved in a phosphate nodule found within black shales likely deposited during low-oxygen conditions associated with a marine transgression. Based on the morphology of the mouthparts, this fossil is likely that of an ammonoid (Tanabe et al. 2001).

3.1.6 Permian Phosphatization

Partially phosphatized cephalopod fossils are also known from carbonate nodules in the Permian Arcturus Formation of Nevada, USA (Mapes et al. 2002, 2003). The specimens have phosphatized intracameral membranes and siphuncular tissues. These are more labile tissues than the mouthparts and radula preserved in the Carboniferous example. These fossils, however, are also preserved in nodules within shales representing deposition in low-oxygen settings (Mapes et al. 2002, 2003).

Although the Permian Phosphoria Formation of the western USA contains phosphoritic shales with abundant phosphatic crusts and ooids, no evidence for the phosphatization of soft tissues or original shell material has been found (Martindale 1986). Considering that apatite in the Phosphoria often encrusts the insides of gastropod and brachiopod shells (Martindale 1986), it would appear that phosphogenesis took place after the soft tissues of these organisms had already fully decayed.

3.2 Mesozoic Phosphatization

3.2.1 Triassic Phosphatization

Phosphatized ostracods and associated ciliates are preserved in the Lower Triassic Olenekian Stage of Svalbard (Weitschat 1995). The ostracods are preserved within the living chamber of the large ammonoid *Keyserlingites*. Their valves and appendages are both preserved through phosphatization. The phosphatized ciliates are preserved within the ostracod valves, usually on the epipodial appendages of the ostracod. It is thought that the ostracods were scavenging on the decaying ammonoid tissue when the ammonoids were buried and phosphatizing microenvironments formed within the living chamber of the ammonoid and within he ostracod valves (Weitschat 1995).

Middle Triassic bivalves with some soft parts phosphatized are found in the Muschelkalk carbonates of Germany (Klug et al. 2005). Phosphatized tissues include the mantle, adductor muscles, inhalant and exhalant siphons, gills and gill supports, and the labial palps (Klug et al. 2005). These preserved features were found within internal molds (steinkerns) of articulated trigoniid bivalves. The infaunal life mode of these bivalves likely contributed to the phosphatization of their tissues because they were already buried upon death. The gill supports show the finest level of preservation, with microscopic ultrastructures visible in SEM, whereas the other structures are preserved by more massive, undifferentiated phosphate mineralization (Klug et al. 2005). This is likely because the gill supports were chitinous, while other more labile tissues had begun decaying by the time of phosphatization.

The Middle Triassic Monte San Giorgio Lagerstätte of southern Switzerland and northern Italy contains rare phosphatized soft tissues. These include the skin of the reptile *Macrocnemus bassanii* (Renesto and Avanzini 2002; Etter 2002), skin of nothosaurs, and the digestive tracts of fish. These fossils are preserved within the bituminous dolomites and bituminous black shales of the Grenzbitumenzone and the Meride Limestone, interpreted as being deposited in a stagnant basin surrounded by carbonate reefs. The anoxic-dysoxic boundary is thought to have been at the seafloor and prokaryotic microbial mats covered the seafloor. These environmental conditions created the right conditions for exceptional preservation, including some phosphatization (Etter 2002).

3.2.2 Jurassic Phosphatization

Phosphatized soft tissues are preserved in the Lower Jurassic Lombardische Kieselkalk Formation of northern Italy, known as the Osteno Lagerstätte (Pinna 1985). Soft part preservation through phosphatization in this deposit includes the muscles and branchia of crustaceans, fish tissues, and the digestive tracts of coleoids, polychaetes, and nematodes. These fossils are interpreted as having been preserved in a stagnant, restricted basing with anoxic conditions likely within the sediment pore waters (Pinna 1985).

The Middle to Upper Jurassic Oxford Clay of England contains phosphatized soft-bodied squids (Allison 1988). They are preserved as films of calcium phosphate in the Peterborough Member, interpreted as deposited in a low-oxygen setting. The bodies of these squids are almost completely phosphatized, with some decay evident on their tentacles (Allison 1988).

Bivalves exhibiting a similar style of preservation as those preserved in the Triassic Muschelkalk of Germany are found in the Upper Jurassic Portland Limestone Formation of Great Britain (Wilby and Whyte 1995). As in the Triassic example, these are articulated trigoniid bivalves preserved as internal molds, in which the phosphatized soft parts are visible (Wilby and Whyte 1995). The preserved features include muscle fibers from the mantle and the gill axis. Preservation took place through phosphatization of microbes on the surface of the soft tissues, resulting in external molds with no preservation of internal features (Wilby and Whyte 1995). The source of the phosphate for the preservation of these structures was likely the decay of the more labile tissues of the bivalve itself. This decay elevated the dissolved phosphate levels within the articulated valves, facilitating the phosphatization of more refractory soft tissues (Wilby and Whyte 1995).

3.2.3 Cretaceous Phosphatization

Phosphatization becomes more common again during the Cretaceous and early Cenozoic. One example is the Lower Cretaceous Santana Formation of Brazil, well known for its fossil fish. These fish are preserved in three dimensions within calcareous concretions in bituminous shale and limestone facies (Martill 1988). Phosphate replaces muscle fibers in the myomeres and encrusts the bones of many of these fish. This phosphatization happened extremely early, perhaps even preburial. It seems likely that the earliest phases of decay within the fish bodies created the microenvironment necessary for rapid phosphatization of these tissues (Martill 1988).

The Mid-Cretaceous (Aptian-Albian) Marnes Bleues Formation in the Vocontian Basin of southeast France and the black shales (Albian) of site DSDP 369 on the African continental slope in the eastern central Atlantic both contain abundant evidence for phosphatized bacteria and possibly fungi (Bréhéret 1991). These fossils are preserved in phosphatic nodules within black shales that were deposited in bathyal, low-oxygen settings. Occasional phosphatized fish bones are found in the nodules as well. The original source of the phosphate in these nodules is unclear, but it may be that the bacterial or fungal mats themselves created phosphogenic microenvironments in sporadic localities (Bréhéret 1991). The phosphatized bacteria include bacilli and cocci, sometimes arranged in rosettes. The possible fungi are preserved as monolayered mat fragments (Bréhéret 1991).

The Upper Cretaceous Austin Group at El Rosario, northeastern Mexico also contains phosphatized fish soft parts and intestinal contents. Such preserved structures include myotomes, gill filaments, scales, and fins (Stinnesbeck et al. 2005). Other phosphatized soft tissues found include decapod crustacean carapaces, ammonite siphos, and parts of pterosaurs. These fossils are preserved in laminated, platy limestones and marls, interpreted as representing deposition in a basin with dysoxic to anoxic bottom water conditions near the southernmost portion of the Western Interior Seaway (Stinnesbeck et al. 2005).

Another example of Upper Cretaceous phosphatization is found in the Tres Esquinas Member of the Colon Formation, Venezuela (Ghosh 1984). Only a few meters thick, this phosphorite unit was deposited as a condensed sequence during a trangression and the beginning of a sea level highstand (Ghosh 1984). Although no soft tissues are preserved, most calcareous shell material has been replaced by apatite. Such fossils are mostly foraminiferal.

Upper Cretaceous phosphatization is also known from a series of phosphorites in Tunisia, Morocco, Mauritania, and Egypt (Lamboy et al. 1994). These phosphorites span well into the Cenozoic, reaching as young as Eocene in age. A few of them will be discussed in the following section. The phosphatized fossils in these deposits are elongate coprolites, which make up a large percentage of the grains in these lithologies. Bacteria likely controlled their preservation by causing the initial decay necessary for the rest of the coprolites to be phosphatized (Lamboy et al. 1994). This scenario is supported by the microstructure of the coprolites which, when examined with an SEM, are morphologically consistent with abundant phosphatized bacteria (Lamboy et al. 1994). Some of these coprolites contain evidence of fish remains, so it is thought that predatory fish produced them. Excrement containing fish remains would be highly concentrated with phosphate, making only a small amount of bacterial decay necessary before the proper microenvironment is produced for phosphatization (Lamboy et al. 1994). This may explain why no other phosphatized fossils are known from these deposits despite them representing a broadly phosphogenic environment.

The Upper Cretaceous phosphorites of Egypt, particularly the Campanian-Maastrichtian Duwi Formation, also contain phosphatized diatoms (Ahmed and Kurzweil 2002). The Duwi Formation was deposited during the second of four phosphatization events preserved in the Upper Cretaceous of Egypt, each of which is linked to a trangression (Ahmed and Kurzweil 2002). These phosphatized diatoms include both concentric and pinnate forms. Their frustules are either replaced or molded on a fine scale by phosphatic minerals. It is likely that the diatoms themselves were an important source of the phosphate necessary for the deposition of these phosphoritic deposits, as in seen in other phosphorites around the world (Ahmed and Kurzweil 2002).

3.3 *Cenozoic and Recent Phosphatization*

3.3.1 Paleogene Phosphatization

Eocene phosphorites in Senegal contain abundant phosphatized fish coprolites (Lamboy et al. 1994). These are found in strata equivalent to part of the previously discussed Cretaceous to Eocene Mediterranean phosphorite sequence. As such, these Senegalese phosphorites contain elongate fish coprolites that were likely formed through initial bacterial decay of the phosphate-rich fecal pellets themselves. Eocene fossils of this type are also found in Tunisia, Morocco, and Mauritania (Lamboy et al. 1994).

A fauna of phosphatized insects is found in latest Oligocene/earliest Miocene limestones of the Riversleigh area of Australia (Duncan et al. 1998). These fossils are preserved in three dimensions and include coleopterans, trichopterans, myriapodans, and isopods. The internal soft tissues of these arthropods are not preserved, only portions of the cuticle and eyes are phosphatized (Duncan et al. 1998).

The eyes are extremely well preserved, allowing for detailed examination of their morphologies. Fungi and bacteria are also phosphatized within the specimens, indicating that substantial decay had begun when the cuticle and eyes were phosphatized (Duncan et al. 1998). The environment of deposition is interpreted as a shallow lime-rich pool in a rain forest (Duncan et al. 1998).

3.3.2 Neogene Phosphatization

Evidence for Miocene phosphogenesis is quite abundant and a common example is the Monterey Formation of California, USA (Garrison et al. 1990). While predominantly a siliceous diatomite, the Monterey Formation contains richly phosphatic units in which carbonate skeletal material has been replaced by phosphate, most commonly with foraminiferal tests (Garrison et al. 1990). Phosphatic fish bones are also typically preserved. Evidence for soft tissue phosphatization exists in the form of abundant phosphatized bacterial material. These rocks were deposited in a biologically productive upwelling zone where bacterial mats likely played a crucial role in creating phosphogenic settings in the midst of a normally siliceous deposit (Reimers et al. 1990).

The Libros Basin of northeast Spain contains a Late Miocene assemblage of amphibians, birds, snakes, and leaves, with phosphatization as a common mode of preservation (McNamara et al. 2004). The fossils are found in laminated mudstones within lacustrine deposits. Frog fossils are particularly abundant in these deposits. They exhibit a high degree of articulation and their outer skin is often phosphatized. Carbonaceous bacteria are preserved within the skin, evidence that decay of the inner layers of the skin had begun when the outer portion of the skin was phosphatized (McNamara et al. 2004).

3.3.3 Pleistocene and Recent Phosphatization

Pleistocene phosphatization of bacteria within phosphatic stromatolites is known from phosphorites on the continental margin of southeast India (Purnachandra Rao et al. 2000, 2002). Found at depths between 186 and 293 m, these stromatolites contain phosphatized microfilaments and coccoid objects consistent with the cells of cyanobacteria. These stromatolites formed when sea level was lower during the Pleistocene and are analogous with Cambrian phosphatic stromatolites (Purnachandra Rao et al. 2000, 2002). The mats sealed off the pore waters from the ocean waters above, creating conditions favourable to phosphatization when buried microbial mats decayed, producing dissolved phosphate from the organic material as well as buried continental detritus (Purnachandra Rao et al. 2000, 2002). Although the fossils are not phosphatized, the phosphorites in these deposits frequently encrust and encapsulate calcareous fossils of bivalves, gastropods, foraminifera, scleractinians, barnacles, serpulids, and algal rhodoliths.

Recent phosphogenic environments exist in multiple localities around the world including the continental margins of southeast Africa, Peru, Chile, and Mexico (Burnett 1977; Jahnke et al. 1983; Baturin et al. 1998). This process is particularly active on the outer shelf off the coast of Namibia, where phosphates are forming in a high-productivity shelf zone created by upwelling (Baturin and Dubinchuk 2003). Periodic blooms of poisonous zooplankton species in this area causes episodic mass mortality events. Because of these events, the seafloor in this region is covered with bones of marine vertebrates (Baturin and Dubinchuk 2003). These bones become phosphatized, and detailed mineralogical studies of some of these bones recovered from benthic trawls have been performed. These analyses of fish and whale bones indicate that the primary fibrous structure of the original bone apatite was altered to a colloform substance that formed globular clots, which were then crystallized. So, although these bones are being phosphatized, their microstructures are not being precisely replicated (Baturin and Dubinchuk 2003).

Similarly phosphatized fish and whale bones are common in the Recent active phosphogenic zone on the slope and outer shelf off the coast of Peru (Resig and Glenn 1997). These phosphatized bones, as well as the phosphatic seafloor crusts common in this region, are inhabited by encrusting foraminifera, serpulid worms, bryozoans, cnidarians, boring organisms, and grazing chitons. Fossil examples of such phosphatic hardground communities are known from the Jurassic-Lower Cretaceous of southern Spain and the Upper Cretaceous of the Czech Republic (Zítt and Nekvásilova 1993; Martín-Algarra and Sánchez-Navas 1995). The Upper Cretaceous biota is most analogous to that seen in the modern Peruvian phosphogenic setting, containing a diverse suite of foraminifera, brachiopods, sponges, scleractinians, octocorals, bryozoans, bivalves, and worm tubes (Zítt and Nekvásilova 1993).

4 Temporal Distribution Hypotheses

Although the examples described here do not encompass every known instance of phosphatization in the Phanerozoic fossil record, they do broadly reflect the temporal distribution and biases of phosphatization during the past 543 million years. There do appear to be two important pulses in phosphatization during the Phanerozoic: the Cambrian through Early Ordovician and the Cretaceous through Eocene (Table 1). In most other times in Earth history, phosphatization is an exceedingly rare taphonomic window that usually depends on the formation of amenable microenvironments.

This temporal pattern of phosphatic fossil preservation is similar, but not identical, to the Phanerozoic distribution of marine phosphogenic events, which additionally shows increased phosphorite occurrence during the Permian and early Mesozoic (e.g. Föllmi 1996). This discrepancy further emphasizes the rare nature of phosphatization because it does not universally become more common during intervals of increased marine phosphogenesis. The reasons for this remain unclear, but perhaps, as may have been possible with the Permian Phosphoria Formation, phosphogenesis was occurring at a diagenetically later time when soft tissues had already decayed away.

The Cambrian through Early Ordovician record consists of phosphatized animal embryos and microarthropods (Table 1). These fossils are preserved both within phosphoritic facies and within nodules in non-phosphoritic facies across a broad spectrum of marine shelf depositional environments. Similarly preserved animal embryos and meiofaunal arthropods are not yet known from younger strata.

Perhaps the explanation for the closure of this taphonomic window lies in the evolution of animal life itself. Animal sediment mixing in marine settings did not reach typical Paleozoic levels until the Ordovician radiation (e.g. Droser 1987). The creation of a persistent mixed layer in the upper few centimeters of seafloor sediments may have permanently narrowed the possible range of phosphatizing microenvironments in the oceans. This mixed layer would have lowered the oxic–anoxic interface within sediments, making it more difficult for organic remains to become buried in low-oxygen settings where mineralization of soft tissues is more feasible. Similar hypotheses have been proposed for the decline of Burgess Shale-type soft bodied biotas during the Cambrian (Allison and Briggs 1993), but the phosphatization window apparently persisted until Ordovician bioturbation levels made these taphonomic processes exceedingly difficult. Thereafter, phosphatization was restricted to low-oxygen and/or phosphogenic environmental settings, as well as microenvironments.

The pulse of phosphatization during the Cretaceous to Eocene may have to do with broad paleoceanographic patterns. It corresponds well with a general increase in phosphogenic environments, which may be related to the development of large euxinic epicontinental seas during this greenhouse time. Cretaceous to Eocene phosphorites throughout northern Africa, for example, are strongly related to a series of transgressions by the Tethys Sea (Ahmed and Kurzweil 2002). These transgressions would have brought zones of high-productivity and low-oxygen bottom waters onto the continent, where phosphogenesis could take place and phosphorite deposits could accumulate.

5 Biases Through Time

Phosphatization during the earliest Phanerozoic is highly biased toward ecdysozoans and their embryos (Table 1). Both the Orsten-style microarthropods and the animal embryos found in the Cambrian through Early Ordovician have well-preserved exterior layers with almost no internal preservation (e.g. Maas et al. 2003; Dong et al. 2005). In both cases, their recalcitrant cuticles survived initial decay to become phosphatized, likely as a direct result of the organic decay of their interiors. In both cases, there is also a strong size bias. Later developmental stages of the embryos, including adult forms, were not preserved and the arthropod assemblages only contain the meiofauna (e.g. Maas et al. 2003; Dong et al. 2005).

Phosphatization is much rarer in the remainder of the Paleozoic, and shows a bias toward taxa with recalcitrant tissues (Table 1). Arcritarch organic walls, echinoderm plates, and cephalopod mouthparts all would remain intact for phosphatization well after the rest of the organism began decaying (Bischoff and Hall 1980; Tanabe et al. 2001; Mapes et al. 2002, 2003; Kremer 2005). All of these remaining Paleozoic examples rely on phosphatizing microenvironments. None of them are found in phosphoritic rocks.

This same bias toward the preservation of recalcitrant tissues continues into the Mesozoic, but a wider array of organisms are preserved and the strong preservation bias toward fish remains begins to be seen (Table 1). During the Triassic and Jurassic, phosphatization takes place in either phosphatizing microenvironments, such as those within the valves of infaunal bivalves, or within larger low-oxygen environmental settings that preserve a broad range of taxa (Pinna 1985; Allison 1988; Weitschat 1995; Wilby and Whyte 1995; Etter 2002; Klug et al. 2005). The Cretaceous does contain phosphoritic facies with phosphatized organic remains, mostly fish skeletal elements and fish coprolites (e.g. Lamboy et al. 1994). The enrichment of fish bones in calcium phosphate creates a strong bias toward their secondary mineralization with apatite that begins in the Mesozoic, particularly the Cretaceous, and continues into the Recent (Table 1).

Cenozoic and Recent phosphatization exhibits this bias toward fish, with whales also becoming common in Recent settings (Table 1). Phosphatization does seem to follow phosphogenesis during the Cenozoic, becoming rare as phosphogenesis decreases in the Neogene. Fish remains dominate Paleocene through Miocene examples, with a diverse group of taxa being preserved more infrequently (Garrison et al. 1990; Lamboy et al. 1994; Duncan et al. 1998; Purnachandra Rao et al. 2000, 2002; McNamara et al. 2004). Recent examples are only known from phosphogenic settings and contain fish and whale bones (Resig and Glenn 1997; Baturin and Dubinchuk 2003). This also exemplifies the bias toward recalcitrant remains, since the only part of the fish and whales that are preserved are typically their secondarily mineralized bones. Bacteria are the one group of fossils that transcends any broad Phanerozoic bias patterns. Probably owing to their importance during the phosphatization process, putative phosphatized bacteria are common throughout the Phanerozoic. Putative bacterial fossils have been listed in Table 1 only when they are one of the primary fossil forms found in the deposit, but they are described as associated with nearly every one of the fossil occurrences discussed in this chapter (Table 1).

6 Summary

Phosphatization is a high resolution yet strongly biased taphonomic process. Review of the available literature indicates pulses of phosphatization during the Cambrian through Early Ordovician and the Cretaceous through Eocene. The Phanerozoic record of phosphatization is biased toward taxa with recalcitrant tissues, those with body parts enriched in phosphate, and those with small body sizes. Phosphatization can take place in phosphogenic environmental settings, but does not always do so. It is also likely to occur in local phosphatizing microenvironments created by a decaying organism. The soft tissues of a wide range of Phanerozoic taxa, from vertebrates down to bacteria, are preserved through phosphatization, making this taphonomic process a powerful tool in understanding the details of the history of life on Earth.

References

Ahmed, E. A., & Kurzweil, J. (2002). Sedimentological, mineralogical and geochemical characteristics of Upper Cretaceous Egyptian phosphorites with special reference to the microbial role in phosphogenesis. In M. Wagreich (Ed.), *Aspects of Cretaceous stratigraphy and palaeobiogeography*.

Allison, P. A. (1988). Phosphatized soft-bodied squids from the Jurassic Oxford Clay. *Lethaia, 21*, 403–410.

Allison, P. A., & Briggs, D. E. G. (1993). Exceptional fossil record: Distribution of soft-tissue preservation through the Phanerozoic. *Geology, 21*, 605–608.

Andres, D. (1989). Phosphatisierte Fossilien aus dem unteren Ordoviz von Südschweden. *Berliner geowissenschaftliche Ahandlungen (A), 106*, 9–19.

Bailey, J. V., Joye, S. B., Kalanetra, K. M., Flood, B. E., & Corsetti, F. A. (2007). Evidence of giant sulphur bacteria in Neoproterozoic phosphorites. *Nature, 445*, 198–201.

Baturin, G. N. (1999). Hypothoses of phosphogenesis and ocean environment. *Lithology and Mineral Resources, 34*, 411–430.

Baturin, G. N., & Dubinchuk, V. G. (2003). The composition of phosphatized bones in Recent sediments. *Lithology and Mineral Resources, 38*, 265–274.

Baturin, G. N., Zhegallo, E. A., & Isaeva, A. B. (1998). The formation of phosphate grains in sediments of the Namibian Shelf. *Okeanologiya, 38*, 260–269.

Bischoff, G. C. O. (1978). Internal structures of conulariid tests and Circonulariina n. suborder (Cnidaria, Scyphozoa). *Senckenbergiana Lethaea, 59*, 275–327.

Bischoff, G. C. O., & Hall, S. J. (1980). Growth of the post-larval echinoderm endoskeleton documented by phosphatized Silurian remains. *Senckenbergiana Lethaea, 61*, 145–171.

Bréhéret, J. G. (1991). Phosphatic concretions in black facies of the Aptian-Albian Marnes bleues Formation of the Vocontian basin (SE France), and at site DSDP 369: Evidence of benthic microbial activity. *Cretaceous Research, 12*, 411–435.

Briggs, D. E. G. (2003). The role of decay and mineralization in the preservation of soft-bodied animals. *Annual Review of Earth and Planetary Sciences, 31*, 275–301.

Briggs, D. E. G., & Kear, A. J. (1993). Fossilization of soft tissue in the laboratory. *Science, 259*, 1439–1442.

Briggs, D. E. G., & Wilby, P. R. (1996). The role of the calcium carbonate-calcium phosphate switch in the mineralization of soft-bodied fossils. *Journal of the Geological Society of London, 153*, 665–668.

Burnett, W. C. (1977). Geochemistry and origin of phosphorite deposits from off Peru and Chile. *Bulletin of the Geological Society of America, 88*, 813–823.

Dong, X., Donoghue, P. C. J., Cheng, H., & Liu, J. B. (2004). Fossil embryos from the Middle and Late Cambrian Period of Hunan, south China. *Nature, 427*, 237–240.

Dong, X., Donoghue, P. C. J., Liu, Z., Liu, J., & Peng, F. (2005). The fossils of Orsten-type preservation from Middle and Upper Cambrian in Hunan, China – Three-dimensionally preserved soft-bodied fossils. *Chinese Science Bulletin, 50*, 1352–1357.

Donoghue, P. C. J., Kouchinsky, A., Waloszek, D., Bengtson, S., Dong, X., Val'kov, A. K., et al. (2006). Fossilized embryos are widespread but the record is temporally and taxonomically biased. *Evolution and Development, 8*, 232–238.

Dornbos, S. Q., Bottjer, D. J., Chen, J.-Y., Oliveri, P., Gao, F., & Li, C. W. (2005). Precambrian animal life: Taphonomy of phosphatized metazoan embryos from southwest China. *Lethaia, 38*, 101–109.

Dornbos, S. Q., Bottjer, D. J., Chen, J.-Y., Gao, F., Oliveri, P., & Li, C. W. (2006). Environmental controls on the taphonomy of phosphatized animals and animal embryos from the Neoproterozoic Doushantuo Formation, southwest China. *Palaios, 21*, 3–14.

Droser, M. L. (1987). *Trends in depth and extent of bioturbation in Great Basin Precambrian-Ordovician strata*, California, Nevada, and Utah. Unpublished Ph.D. Thesis, University of Southern California, 365 pp.

Duncan, I. J., Briggs, D. E. G., & Archer, M. (1998). Three-dimensionally mineralized insects and millipedes from the Tertiary of Riversleigh, Queensland, Australia. *Palaeontology, 41*, 835–851.

Etter, W. (2002). Monte San Giorgio: Remarkable Triassic marine vertebrates. In D. J. Bottjer, W. Etter, J. W. Hagadorn, & C. M. Tang (Eds.), *Exceptional fossil preservation: A unique view on the evolution of marine life*. New York: Columbia University Press.

Föllmi, K. B. (1996). The phosphorus cycle, phosphogenesis and marine phosphate-rich deposits. *Earth Science Reviews, 40*, 55–124.

Froelich, P. N., Arthur, M. A., Burnett, W. C., Deakin, M., Hensley, V., Jahnke, R., et al. (1988). Early diagenesis of organic material in Peru continental margin sediments: Phosphorite precipitation. *Marine Geology, 80*, 309–343.

Garrison, R. E., Kastner, M., & Reimers, C. E. (1990). Miocene phosphogenesis in California. In W. C. Burnett & S. R. Riggs (Eds.), *Phosphate deposits of the world: volume 3, Neogene to Modern phosphorites*. Cambridge: Cambridge University Press.

Ghosh, S. K. (1984). Late Cretaceous condensed sequence, Venezuelan Andes. In W. E. Bonini, R. B. Hargraves, & R. Shagam (Eds.), *The Caribbean-South American plate boundary and regional tectonics. Geological Society of America Memoirs 162*

Glenn, C. R., Föllmi, K. B., Riggs, S. R., Baturin, G. N., Grimm, K. A., Trappe, J., et al. (1994). Phosphorus and phosphorites: Sedimentology and environments of formation. *Eclogae Geologicae Helvetiae, 87*, 747–788.

Grogan, E. D., & Lund, R. (1997). Soft tissue pigments of the upper Mississippian chondrenchelyid Harpagofututor volsellorhinus (Chondrichthyes, Holocephalie) from the Bear Gulch Limestone, Montana, USA. *Journal of Paleontology, 71*, 337–342.

Hagadorn, J. W. (2002). Bear Gulch: An exceptional Upper Carboniferous Plattenkalk. In D. J. Bottjer, W. Etter, J. W. Hagadorn, & C. M. Tang (Eds.), *Exceptional fossil preservation: A unique view on the evolution of marine life*. New York: Columbia University Press.

Jahnke, R. A., Emerson, A. R., Roe, K. K., & Burnett, W. C. (1983). The present formation of apatite in Mexican continental margin sediments. *Geochimica et Cosmochimica Acta, 47*, 259–266.

Kholodov, V. N., & Paul, R. K. (1995). The Black Sea: A geochemical model of phosphate deposition. *Litol Polezn Iskop, 6*, 563–581.

Klug, C., Hagdorn, H., & Montenari, M. (2005). Phosphatized soft-tissue in Triassic bivalves. *Palaeontology, 48*, 833–852.

Kouchinsky, A., Bengtson, S., & Gershwin, L. A. (1999). Cnidarian-like embryos associated with the first shelly fossils in Siberia. *Geology, 27*, 609–612.

Krajewski, K. P., Van Cappellen, P., Trichet, J., Kuhn, O., Lucas, J., Martin-Algarra, A., et al. (1994). Biological processes and apatite formation in sedimentary environments. *Eclogae Geologicae Helvetiae, 87*, 701–745.

Kremer, B. (2005). Mazuelloids: Product of post-mortem phosphatization of acanthomorphic acritarchs. *Palaios, 20*, 27–36.

Lamboy, M., Purnachandra Rao, V., Ahmed, E., & Azzouzi, N. (1994). Nannostructure and significance of fish coprolites in phosphorites. *Marine Geology, 120*, 373–383.

Lucas, J., & Prevot, L. E. (1991). Phosphates and fossil preservation. In P. A. Allison & D. E. G. Briggs (Eds.), *Taphonomy: Releasing the data locked in the fossil record*. New York: Plenum.

Lund, R., Feldman, H., Lund, W. L., & Maples, C. G. (1993). The depositional environment of the Bear Gulch Limestone, Fergus County, Montana. In L. D. V. Hunter (Ed.), *Energy and mineral resources of central Montana: 1993 Field Conference Guidebook*. Billings: Montana Geological Society.

Maas, A., Waloszek, D., & Müller, K. J. (2003). Morphology, ontogeny, and phylogeny of the Phosphatocopina (Crustacea) from the Upper Cambrian 'Orsten' of Sweden. *Fossils and Strata, 49*, 1–238.

Mapes, R. H., Landman, N. H., Tanabe, K., & Maeda, H. (2002). Intracameral membranes in Permian ammonoids from the Buck Mountain, Nevada Lagerstätte. *Geological Society of America, Annual Meeting Abstracts with Programs 34*.

Mapes, R. H., Maeda, H., Piercey, P., & Landman, N. (2003). An unusual taphonomic scenario for the Buck Mountain (Permian) cephalopod fauna in Nevada. *Geological Society of America, Annual Meeting Abstracts with Programs 35* (496 p.).

Martill, D. M. (1988). Preservation of fish in the Cretaceous Santana Formation of Brazil. *Palaeontology, 31*, 1–18.

Martín-Algarra, A., & Sánchez-Navas, A. (1995). Phosphate stromatolites from condensed cephalopod limestones, Upper Jurassic, southern Spain. *Sedimentology, 42*, 893–919.

Martindale, S. G. (1986) Depositional environments and phosphatization of the Meade Peak Phosphatic Shale Tongue of the Phosphoria Formation, Leach Mountains, Nevada. Contributions to Geology, University of Wyoming *24*, 143–156.

McNamara, M. E., Orr, P. J., Alcala, L., Anadon, P., Penalver Molla, E. (2004). Exceptionally preserved frogs from the Miocene of Libros, NE Spain. *Geological Society of America, Annual Meeting Abstracts with Programs 36* (384 p.).

Müller, K. J., Walossek, D., & Zakharov, A. (1995). Orsten type phosphatized soft-integument preservation and a new record from the Middle Cambrian Kuonamka Formation in Siberia. *Neues Jahrbuch für Geologie und Paläontologie, Abhandlungen, 197*, 101–118.

Pinna, G. (1985). Exceptional preservation in the Jurassic of Osteno. *Philosophical Transactions of the Royal Society of London, B, 311*, 171–180.

Purnachandra Rao, V., Mohan Rao, K., & Raju, D. S. N. (2000). Quaternary phosphorites from the continental margin off Chennai, southeast India: Analogs of ancient phosphate stromatolites. *Journal of Sedimentary Research, 70*, 1197–1209.

Purnachandra Rao, V., Michard, A., Naqvi, S. W. A., Böttcher, M. E., Krishnaswamy, R., Thamban, M., et al. (2002). Quaternary phosphorites off the southeast coast of India. *Chemical Geology, 182*, 483–502.

Reimers, C. E., Kastner, M., & Garrison, R. E. (1990). The role of bacterial mats in phosphate mineralization with particular reference to the Monterey Formation. In W. C. Burnett & S. R. Riggs (Eds.), *Phosphate deposits of the world: volume 3, Neogene to Modern phosphorites*. New York: Cambridge University Press.

Renesto, S., & Avanzini, M. (2002). Skin remains in a juvenile *Macrocnemus bassanii* Nopsca (Reptilia, Prolacertiformes) from the Middle Triassic of Italy. *Neues Jahrbuch für Geologie und Paläontologie, Abhandlungen, 224*, 31–48.

Resig, J. M., & Glenn, C. R. (1997). Foraminifera encrusting phosphoritic hardgrounds of the Peruvian upwelling zone: Taxonomy, geochemistry, and distribution. *Journal of Foraminiferal Research, 27*, 133–150.

Roy, K., & Fåhraeus, L. E. (1989). Tremadocian (Early Ordovician) nauplius-like larvae from the Middle Arm Point Formation, Bay of Islands, western Newfoundland. *Canadian Journal of Earth Sciences, 26*, 1802–1806.

Sagemann, J., Bale, S. J., Briggs, D. E. G., & Parkes, R. J. (1999). Controls on the formation of authigenic minerals in association with decaying organic matter: An experimental approach. *Geochimica et Cosmochimica Acta, 63*, 1083–1095.

Schulz, H. N., & Schulz, H. D. (2005). Large sulfur bacteria and the formation of phosphate. *Science, 307*, 416–418.

Siveter, D. J., Wiliams, M., & Waloszek, D. (2001). A phosphatocopid crustacean with appendages from the Lower Cambrian. *Science, 293*, 479–481.

Soudry, D. (2000). Microbial phosphate sediment. In R. E. Riding & S. M. Awramik (Eds.), *Microbial sediments*. Berlin: Springer.

Steiner, M., Zhu, M. Y., Li, G. X., & Erdtmann, B. D. (2004). New Early Cambrian bilaterian embryos and larvae from China. *Geology, 32*, 833–836.

Stinnesbeck, W., Ifrim, C., Schmidt, H., Rindfleisch, A., Buchy, M. C., Frey, E., et al. (2005). A new lithographic limestone deposit in the Upper Cretaceous Austin Group at El Rosario, county of Múzquiz, Coahuila, northeastern Mexico. *Revista Mexicana de Ceincias Geológicas, 22*, 401–418.

Tanabe, K., Napes, R. H., & Kidder, D. L. (2001). A phosphatized cephalopod mouthpart from the Upper Pennsylvanian of Oklahoma, USA. *Paleontological Research, 5*, 311–318.

Trappe J (1998) Phanerozoic phosphorite depositional systems: A dynamic model for a sedimentary resource system. Lecture Notes in Earth Science 76:1–316.

Trinajistic, K., Marshall, C., Long, J., & Bifield, K. (2007). Exceptional preservation of nerve and muscle tissues in Late Devonian placoderm fish and their evolutionary implications. *Biology Letters, 3*, 197–200.

Walossek, D., & Szaniawski, H. (1991). *Cambrocaris baltica* n. gen. n. sp., a possible stem-lineage crustacean from the Upper Cambrian of Poland. *Lethaia, 24*, 363–378.

Walossek, D., Hinz-Schallreuter, I., Shergold, J. H., & Müller, K. J. (1993). Three-dimensional preservation of arthropod integument from the Middle Cambrian of Australia. *Lethaia, 26*, 7–15.

Walossek, D., Repetski, J. E., & Müller, K. J. (1994). An exceptionally preserved parasitic arthropod, Heymonsicambria taylori n. sp. (Arthropoda incertae sedis: Pentastomida), from Cambrian-Ordovician beds of Newfoundland, Canada. *Canadian Journal of Earth Sciences, 31*, 1664–1671.

Weitschat, W. (1995). First evidence of fossil ciliates – Preserved in phosphatized ostracods from the Lower Triassic of Spitsbergen. *Ostracoda and Biostratigraphy, 426*.

Wilby, P. R., & Whyte, M. A. (1995). Phosphatized soft tissues in bivalves from the Portland Roach of Dorset (Upper Jurassic). *Geological Magazine, 132*, 117–120.

Wilby, P. R., Briggs, D. E. G., Bernier, P., & Gaillard, C. (1996). Role of microbial mats in the fossilization of soft tissues. *Geology, 24*, 787–790.

Xiao, S., & Knoll, A. H. (1999). Fossil preservation in the Neoproterozoic Doushantuo phosphorite Lagerstätte, South China. *Lethaia, 32*, 219–240.

Yue, Z., & Bengtson, S. (1999). Embryonic and post-embryonic development of the Early Cambrian cnidarian Olivooides. *Lethaia, 32*, 181–195.

Zítt, J., & Nekvásilova, O. (1993). Octocoral encrusters of rock substrates in the Upper Cretaceous of Bohemia. *Journal of Czech Geological Society, 38*, 71–78.

Chapter 13
Three-Dimensional Morphological (CLSM) and Chemical (Raman) Imagery of Cellularly Mineralized Fossils

J. William Schopf, Anatoliy B. Kudryavtsev, Abhishek B. Tripathi, and Andrew D. Czaja

Contents

J.W. Schopf (✉)
Department of Earth and Space Sciences, Institute of Geophysics and Planetary Physics (Center for the Study of Evolution and the Origin of Life), Molecular Biology Institute, and NASA Astrobiology Institute, University of California, Los Angeles, CA 90095, USA
e-mail: schopf@ess.ucla.edu

A.B. Kudryavtsev
Institute of Geophysics and Planetary Physics (Center for the Study of Evolution and the Origin of Life) and NASA Astrobiology Institute, University of California, Los Angeles, CA 90095, USA

A.B. Tripathi
Advanced Projects Office, Constellation Program, NASA Johnson Spacecraft Center, 77058, Houston, TX, USA

A.D. Czaja
Department of Earth and Space Sciences and Institute of Geophysics and Planetary Physics (Center for the Study of Evolution and the Origin of Life), University of California, 90095, Los Angeles, CA, USA

P.A. Allison and D.J. Bottjer (eds.), *Taphonomy: Process and Bias Through Time*, Topics in Geobiology 32, DOI 10.1007/978-90-481-8643-3_13,

Abstract Of all modes of fossilization, cellular mineralization, whether by the non-biologic process of permineralization ("petrifaction") or by microbially mediated mineral precipitation ("authigenic mineralization"), is the most faithful to the preservation of life-like cells and tissues that is known, yielding fossils that are among the biologically and taphonomically most informative available from the geological record. Such preservation spans all forms of life, ranging from vascular plants, such as those permineralized in calcitic coal balls; to organic-walled algae, fungi and bacterial prokaryotes, permineralized most commonly in fine-grained quartz; to metazoans that exhibit preserved soft tissues, such as those mineralized in apatite. Though such fossils can be preserved in exquisite cellular detail, two deficiencies have long hampered their study: (1) an inability to document fully their three-dimensional morphology at micron-scale spatial resolution; and (2) the lack of a means to analyze in situ and at such resolution the chemistry of the carbonaceous matter (kerogen) that comprises their structurally preserved anatomy. These needs have now been met by two techniques newly introduced to paleobiology, three-dimensional confocal laser scanning microscopy (CLSM) and two- and three-dimensional Raman imagery.

We here document the use of these techniques to elucidate the fine-scale structure and kerogenous composition of representative fossils of each of the major biologic groups (animals, plants, fungi, algal protists, and microbes) preserved in phosphorites, cherts, and carbonates, the three principal rock types in which cellular mineralization occurs. The examples presented include an apatite-mineralized ctenophore embryo preserved in a Cambrian phosphorite; quartz-permineralized Eocene fern rhizomes and a fungal-infested Devonian plant axis preserved in carbonaceous cherts; a calcite-permineralized plant stem preserved in a calcareous Carboniferous coal ball; and quartz-permineralized acritarchs (phytoplanktonic algae), cyanobacteria, and especially ancient fossil microbes permineralized in Precambrian cherts.

Use of CLSM and Raman imagery can provide new information about the morphology, cellular anatomy, taphonomy, carbonaceous composition and geochemical maturity of organic-walled mineralized fossils, whereas Raman imagery used alone can document the mineralogy of the enclosing matrix and the spatial relations between such fossils and their embedding minerals. Not only can the use of these techniques elucidate the sequence of events and taphonomic processes involved in the cellular mineralization of organic-walled fossils, but the use of Raman to document the geochemical maturity of their kerogenous constituents can provide new evidence of the biases of such preservation over time. Because both techniques are non-intrusive and non-destructive, both can be applied to specimens archived in museum collections. Taken together, the two techniques represent a major advance in the study of ancient fossils.

1 Introduction

Taphonomic studies of fossils depend critically on the fidelity of the preservation of the specimens studied and on an ability to investigate and to document at appropriate spatial resolution the structure of the biologic remnants preserved.

Among all fossils, cellularly mineralized organisms, studied typically in petrographic thin sections or cellulose acetate peels, are among the best preserved and, thus, the biologically and taphonomically most informative. Nevertheless, until recently it has not been possible to document accurately the three-dimensional morphology and cellular anatomy of such fossils at high spatial resolution, a deficiency particularly troublesome to studies of the fine-structural features of megascopic fossils and the morphology and cellular make-up of minute fossil microorganisms. Moreover, there has been no means available by which to analyze directly the molecular-structural composition and geochemical maturity of the coal-like carbonaceous organic matter (kerogen) that comprises permineralized fossils, factors crucial to an assessment of the fidelity of their preservation. This chapter documents the use of two techniques newly introduced to paleobiology that meet these needs: three-dimensional confocal laser scanning microscopy, "CLSM" (Schopf et al. 2006a) and two- (Schopf et al. 2002, 2005) and three-dimensional Raman imagery (Schopf and Kudryavtsev 2005; Schopf et al. 2007).

Shown here to be applicable to cellularly mineralized fossils whether studied in thin sections or in acetate peels, CLSM and Raman imagery, together, can demonstrate, in three dimensions and at micron-scale spatial resolution, a one-to-one match of cellular form and kerogenous composition. Particularly useful for studies of coccoidal and sinuous filamentous Precambrian microbes and comparably minute Phanerozoic organic-walled microfossils, both techniques are applicable also to studies of higher plants – where they can provide evidence of the fine structure and molecular-structural composition of fossilized cells unavailable by any other means – and of metazoans, where they can yield insight into the nature and three-dimensional preservation of mineralized soft tissues. Used in tandem, the two techniques can provide valuable taphonomic information about the biological degradation of diverse organically preserved specimens (e.g., the fungal infestation of plant axes, the enzymatic breakdown of the middle lamellae of vascular plant cell walls, and the preferential decay of specific cell wall components in fossil cyanobacteria). Moreover, their use can also elucidate the sequence of events that led to the not uncommonly exquisite preservation of such fossils (e.g., apatite-mineralization of the organic matter of the soft tissues of metazoans followed by calcitic infilling of interstices and fluid-filled cavities) and their fidelity of preservation (measured by use of the Raman index of preservation, "RIP," a metric that documents the geochemical maturity of the kerogenous components of such fossils; Schopf et al. 2005). Properly executed, both techniques are non-intrusive and non-destructive – factors that permit their application to specimens archived in museum collections – and unlike standard two-dimensional optical photomicrographs, the three-dimensional digitized images provided by CLSM and Raman imagery can be rotated and examined from multiple perspectives (e.g., in video presentations), a telling advance over standard optical microscopy of particular relevance to investigations of taphonomy.

1.1 Cellularly Mineralized Fossils

Of all modes of fossil preservation, cellular mineralization, whether by the non-biologic process of permineralization ("petrifaction") or by microbially mediated mineral precipitation ("authigenic mineralization"), provides the most faithful life-like representation of biologic anatomy and cellular morphology known in the geological record.

Of the two processes, permineralization (known commonly, but incorrectly, as "petrification") was the first to be described, more than a century ago (White 1893). The misnomer "petrifaction" (from petrify, "to convert into stone") dates from this early study when the process was misinterpreted as reflecting "molecule-by-molecule…replacement" of cellular organic matter by minerals (White 1893) – an error repeated in numerous popular books on ancient life and even in some modern textbooks. In contrast with this stoichiometrically implausible (if historically interesting) account of the process, permineralization is now known to result from the pervasion of mineral-charged solutions into biologic tissues during the early stages of diagenesis, prior to complete decay and cellular disintegration (Schopf 1975). The organic structures of such fossils, in many specimens preserved in microscopic detail, are not mineral-replaced. Rather, the permeating fluids infill micellar, intercellular, and intracellular spaces – replacing the watery milieu of the biomolecular components, not the organic structures themselves – to produce a mineral-infused inorganic-organic mix that serves to preserve such physically robust organic-rich structures as the cell walls of plants, algae, fungi and microbes, and, much less commonly, the soft tissues of animals (which lack robust cell walls).

Despite such understanding of the products of permineralization, the chemistry of the process has yet to be defined. It may well be that permineralization merely reflects the occurrence of infiltration and fine-scale embedding like that performed routinely by the use of organic resins and/or waxes in the preparation of living specimens for sectioning and study by optical or transmission electron microscopy. However, if chemical bonding plays a role in permineralization, the most likely such chelation would be that involving hydrogen-bonding between the peripheral hydrogen atoms of the interlinked polycyclic aromatic hydrocarbons ("PAHs") that comprise the kerogen of permineralized fossils and oxygen atoms of their embedding minerals (Schopf et al. 2005). The sequence of events implied by such a process – from the mineralic infilling of (and chemical bonding with?) the biomolecules of decaying tissues to the conversion of such material to a chemically bonded mineral-PAH mix – has yet to be demonstrated. In short, although it is well established that permineralization results in the fine-scale preservation of mineral-infused, physically robust cellular organic structures, the chemistry of such preservation has yet to be elucidated.

In most cases, the fine-scale structural preservation of the soft tissues of animals, for example in apatite, occurs by a different process: authigenic mineralization. As an over-simplification, permineralization is a non-biologic chemical process whereas authigenic mineralization is microbially mediated (for recent detailed discussions of this process, see Briggs 2003; Briggs et al. 2005). Although most soft-tissue preservation in apatite is evidently due to bacterially driven authigenic mineralization, it is possible that some phosphorites represent settings in which concentrations of phosphate were so high that mineralization of soft tissues in apatite was more akin to permineralization than to authigenic mineralization (D.E.G. Briggs, pers. comm. 2007). Similarly, preservation in pyrite, whether of the cellular structure of fossil plants (e.g., Scheckler 1986; Grimes et al. 2002) or of the soft tissues of metazoans (e.g., Briggs et al. 1996), is typically due to authigenic mineralization, sulfate reducing bacteria using the hydrogen of decaying organic matter to produce hydrogen sulfide that unites with ferrous iron to produce mineralizing fine-grained pyrite.

Cellularly mineralized fossils are known from all of the major groups of life – plants, animals, protists, fungi, and bacteria. Among these, plant fossils have been the most intensively studied, with in-depth knowledge of the Carboniferous flora, for example, having been derived almost entirely from studies of plant parts permineralized in calcitic coal balls (e.g., Taylor and Taylor 1993). Similarly, current understanding of the Precambrian fossil record is dependent largely on studies of organic-walled acritarchs (planktonic algal protists) and cellular microbes permineralized in cherts (Schopf 1999). Permineralized fungi (Taylor and Remy 1992), animals (e.g., Chen et al. 2007) and protozoal protists (Bloeser et al. 1977; Schopf 1992, p. 592; Mus and Moczydlodłowska 2000) have also been reported. Although relatively rare occurrences of plant axes (e.g., Scheckler and Banks 1971; Scheckler 1986; Grimes et al. 2002) and metazoan soft tissues (e.g., Stuermer 1970; Stuermer and Bergström 1973; Cisne 1974; Briggs et al. 1996, 2005; Briggs 2003) authigenically mineralized in pyrite have been recorded, cellularly mineralized fossils are typically infused by three principal minerals: apatite (as in Cambrian metazoans preserved in phosphatic limestones); calcite (as in plants permineralized in Carboniferous coal balls); and quartz (as in Phanerozoic plants and fungi, and Precambrian microorganisms permineralized in bedded cherts and the cherty parts of stromatolitic carbonates). The use of CLSM and Raman to analyze fossil animals, plants, fungi, algae and bacteria from each of these matrices is documented below.

2 Techniques

2.1 Confocal Laser Scanning Microscopy (CLSM)

The confocal laser scanning microscope was developed in the mid-1980s for use in biology to image in three dimensions the structural components of cells (Amos and White 2003). In such microscopes, the aperture-constrained light path serves to restrict the focus of the system to a discrete focal plane. By thus suppressing the image-blurring input of out-of-focus planes above and below this focal plane,

CLSM provides a crisp image of a thin in-focus plane that cannot be provided by standard optical microcopy. The laser beam of such systems excites fluorescence in the material analyzed, for modern cells derived typically from biomolecules having conjugated ring systems or from introduced fluorescent dyes, but for kerogenous fossils, such as those analyzed here, emitted from the interlinked PAHs of which they are primarily composed (Schopf et al. 2005). Such kerogen-derived fluorescence, maximum in the red to near-infrared region of the spectrum (Schopf et al. 2005, 2006a), is then collected by the detector of the system. By rapidly rastering the laser beam of the system across a kerogenous fossil at precisely defined increasing depths and then processing the digitized sequential series of images acquired – by use first of the 3-D rendering software of the CLSM system and, if desired, by use of the more advanced VolView 3-D rendering computer program (Kitware, Inc., Clifton Park, NY) – CLSM can produce a three-dimensional image of a specimen at sub-micron spatial resolution (Schopf et al. 2006a).

CLSM has only recently been introduced to paleobiology. Most such work has focused on Phanerozoic organic-walled palynomorphs freed from their encompassing matrices by acid maceration (Scott and Hemsley 1990; Foster et al. 1990; Talyzina 1997; Feist-Burkhardt and Pröss 1999; Mus and Moczydlodłowska 2000; Feist-Burkhardt and Monteil 2001; Nix and Feist-Burkhardt 2003; Hochuli and Feist-Burkhardt 2004). Few studies have applied this technique to fossils in petrographic thin sections (Mus and Moczydlodłowska 2000; Nix and Feist-Burkhardt 2003; Schopf et al. 2006a; Chi et al. 2006) and its use to document the anatomy of mineralized Phanerozoic metazoans (Chen et al. 2007) and the cellular structure of Precambrian protozoans (Mus and Moczydlodłowska 2000), acritarchs and microbes (Schopf et al. 2006a) has been barely tapped. The present work records the first application of CLSM to Phanerozoic plant fossils and infesting fungi, and its first use for studies of specimens prepared in cellulose acetate peels.

2.2 Raman Spectroscopy

Raman spectroscopy is an analytical technique used widely in geochemistry for the identification and molecular-structural characterization of minerals (e.g., McMillan and Hofmeister 1988; Williams et al. 1997), including graphite and graphite-like mineraloids, studies that have focused on the use of such carbonaceous materials as indices of high-grade metamorphic alteration (e.g., Pasteris and Wopenka 1991; Jehlička and Beny 1992; Wopenka and Pasteris 1993; Yui et al. 1996; Spötl et al. 1998; Kelemen and Fung 2001; Jehlička et al. 2003). In contrast, our use of Raman has centered on the carbonaceous components of less altered and, thus, potentially fossiliferous, unmetamorphosed to low-grade (greenschist) metamorphic units, geochemically relatively immature organic matter that has been little investigated by Raman. Similarly, unlike previous Raman studies – virtually all of which have been devoted to analysis of carbonaceous matter preserved by compression (in shales, slates, coals, gneisses and the like) – our work has centered on organic materials preserved by cellular mineralization, three-dimensionally embedded in their enclosing mineral matrices (Kudryavtsev et al. 2001; Schopf et al. 2002, 2005, 2007; Schopf 2004a, b; Schopf and Kudryavtsev 2005; Chen et al. 2007).

In analyses of cellularly mineralized carbonaceous matter, CLSM and Raman studies are complementary, both being used to measure signals derived from properties of the kerogenous materials analyzed: for CLSM, laser-induced fluorescence derived chiefly from the interlinked PAHs that predominate in kerogen (Schopf et al. 2006a); for Raman, laser-induced vibrational transitions in the bonds of such PAHs and their associated functional groups (Schopf et al. 2005). Only recently introduced to paleobiology (Roberts et al. 1995; Arouri et al. 2000; Kudryavtsev et al. 2001; Schopf et al. 2002), Raman can be used to characterize both the molecular structure and the geochemical maturity of the kerogenous cell walls of organically preserved fossils – whether megascopic (e.g., plant axes: Dietrich et al. 2001; Kudryavtsev et al. 2001; Nestler et al. 2003) or microscopic (e.g., metazoan embryos: Chen et al. 2007; acritarchs and cyanobacteria: Arouri et al. 2000; Kudryavtsev et al. 2001; Schopf et al. 2002, 2005, 2007; Kaufman and Xiao 2003; Schopf 2004a, b; Schopf and Kudryavtsev 2005); and whether such fossils have been preserved by compression (Roberts et al. 1995; Arouri et al. 2000; Kaufman and Xiao 2003; Marshall et al. 2005) or by cellular mineralization (Kudryavtsev et al. 2001; Schopf et al. 2002, 2005, 2006a, 2007; Nestler et al. 2003; Schopf 2004a, b; Schopf and Kudryavtsev 2005; Chen et al. 2007).

Two-dimensional Raman imagery, an advanced application of Raman spectroscopy recently introduced to paleobiologic studies (Kudryavtsev et al. 2001; Schopf et al. 2002, 2005, 2007; Schopf 2004a, b; Schopf and Kudryavtsev 2005; Chen et al. 2007), has proven especially useful for analyses of permineralized kerogenous fossils and associated carbonaceous matter present in petrographic thin sections. In this technique, documented by Schopf et al. (2005), a large number of point spectra of the type measured in standard Raman spectroscopy are acquired over a defined area of a fossil specimen to provide a molecular-structural map in two dimensions and at micron-scale resolution. Such maps show the distribution of the carbonaceous and mineralic matter comprising such fossils in which varying intensities of the detected Raman signal correspond to the relative concentrations of the molecular structures present, maps that permit direct spatial correlation of the carbonaceous material with optically discernable morphology (Kudryavtsev et al. 2001; Schopf et al. 2002, 2005, 2007).

Three-dimensional Raman imagery of permineralized fossils (Schopf and Kudryavtsev 2005; Schopf et al. 2007) and micron-scale mineral assemblages (McKeegan et al. 2007), even more recent extensions of Raman imagery, can be achieved by the computerized "stacking" and VolView-processing of two-dimensional images acquired at sequentially increasing depths throughout a thin section-embedded specimen. The 3-D image thus obtained maps the chemistry – the molecular-structural characteristics – of the specimen analyzed. Such images can be compared readily with CLSM images or with photomicrographs obtained by standard optical techniques. But unlike optical photomicrographs, the digitized images produced can be rotated or otherwise manipulated in three dimensions to reveal detailed features of morphology, including those of the interiors of such specimens that are not discernable by standard optical microscopy. Moreover, because the Raman data used to prepare such images record the distribution not only of carbonaceous matter but also of associated minerals, this technique can be used to document the three-dimensional spatial relations between cellularly mineralized carbonaceous fossils and their encompassing rock matrices.

3 Applications

Our goal here is to document the paleobiologic application of CSLM and Raman spectroscopy to diverse cellularly mineralized organisms, with particular emphasis on deciphering their taphonomy. The relevant limitations of the two techniques are discussed elsewhere (Schopf and Kudryavtsev 2005; Schopf et al. 2005, 2006a; Tripathi 2007). Below, we discuss results obtained from studies of a fossil metazoan and fossil plants, fungi, acritarchs and microbes, including examples of fossils that are among the oldest known in the geological record.

4 Mineralized Soft Tissues of Metazoans

Because carbonaceous animal fossils are typically preserved by compression rather than cellular mineralization, intact preservation of the soft tissues of fossil animals, such as that reported for the musculature of mineralized Devonian sharks (Dean 1902) and Jurassic horseshoe crabs (Briggs et al. 2005), is notably rare. Nevertheless, exquisitely well-preserved metazoan embryos have recently been described from latest Precambrian (Xiao et al. 1998; Xiao and Knoll 2000; Chen et al. 2000, 2006) and Early Cambrian (Bengtson and Zhao 1997; Chen 2004; Chen et al. 2004, 2007; Steiner et al. 2004a, b) phosphorites and phosphatic limestones of China, including specimens preserved in such detail that they exhibit convincing evidence of the early stages of cellular cleavage. Late-stage embryos from these deposits have been relatively little studied, but the one such example discussed below provides important insight into the taphonomy of such preservation.

4.1 Apatite-Mineralized Ctenophore Embryo

Shown in Fig. 1 are optical, CLSM, and Raman images of a thin section-embedded ctenophore embryo from the Lower Cambrian (~540-Ma-old) Kuanchuanpu Formation of Ningqiang County, Shaanxi Province, China. Described by Chen et al. (2007), this is the oldest comb jelly and the only embryonic ctenophore

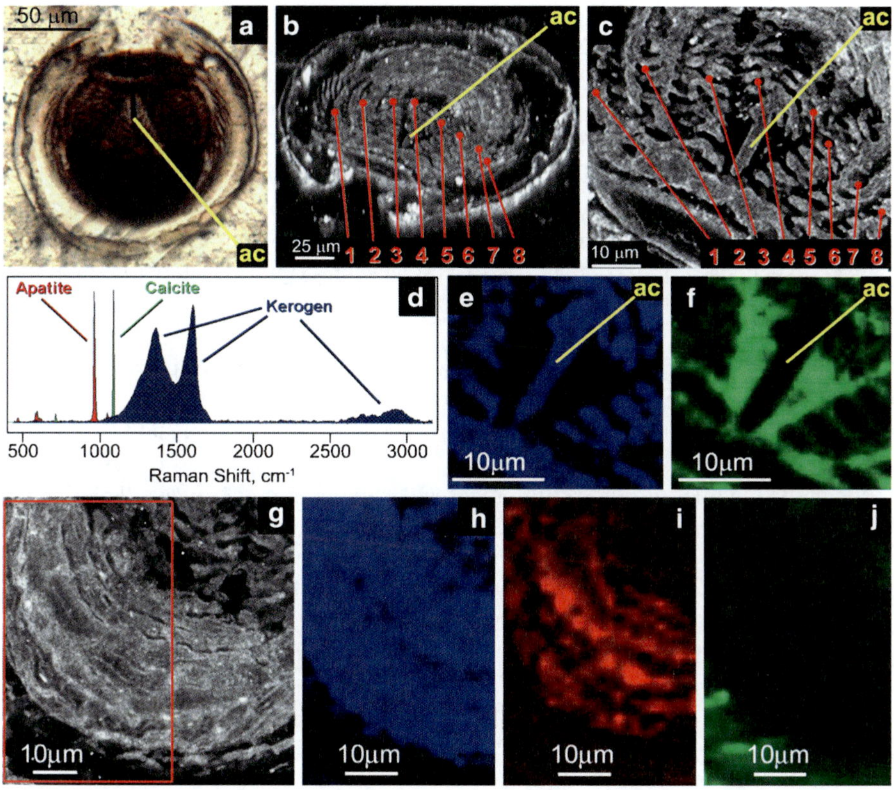

Fig. 1 Optical image (**a**), CLSM images (**b**, **c**, **g**), Raman spectra (**d**), and Raman images (**e**, **f**, **h**–**j**) of a thin section-embedded apatite-mineralized ctenophore embryo from the Lower Cambrian Kuanchuanpu Formation of Shaanxi Province, China (Chen et al. 2007). (**a**) Optical image of the complete specimen; ac = aboral canal (denoted also in **b**, **c**, **e**, **f**). (**b**) Rotated (VolView-processed) CLSM image of the embryo showing the comb rows, numbered one through eight. (**c**) CLSM image of the aboral region showing the numbered comb rows. (**d**) Overlapping Raman spectra showing the major bands of the apatite (~965 cm^{-1}), calcite (~1,087 cm^{-1}), and kerogen (~1,350 cm^{-1}, ~1,600 cm^{-1}, ~2,800 cm^{-1}) that comprise the fossil (baseline subtracted). (**e**, **f**) Raman images of the aboral canal and surroundings showing the distribution of kerogen (**e**, blue; acquired in a spectral window centered at ~1,600 cm^{-1}) and calcite (**f**, green; acquired at ~1,087 cm^{-1}). (**g**) CLSM image of the distal portion of the embryo in which the red rectangle denotes the region in **h**–**j**. (**h**–**j**) Raman images acquired in spectral windows centered at the major Raman bands of kerogen (**h**), apatite (**i**), and calcite (**j**)

known from the geological record. The ctenophoran affinities of this egg-enclosed embryo in late development, prior to hatching, are well established by its superb anatomical preservation that includes such features as the egg membrane, the apical (aboral) organ, meridional canals, the aboral canal (Fig. 1a–c, e, f) and, perhaps most notably, the eight sets of ctene-composed comb rows that are diagnostic of comb jellies (Fig. 1b, c). Such characteristics are especially evident in the CLSM images that, as is shown in Fig. 1b,c, can be rotated to reveal the anatomy of the embryo in three dimensions. Moreover, because such images are derived from the laser-excited fluorescence emitted by the kerogenous components of the specimen, they provide a firm indication of its carbonaceous composition.

Figure 1d shows the Raman spectra of the apatite, calcite, and kerogen of which this embryo is composed. Two-dimensional Raman images (Fig. 1e,f,h–j) show that the soft tissues of the specimen are composed of carbonaceous kerogen (Fig. 1e,h) infused by mineralizing apatite (Fig. 1i), and that the interstices between such structures (Fig. 1f, j), as well as such originally fluid-filled cavities as meridional canals (Chen et al. 2007), have been secondarily infilled by calcite.

Backed by optical microscopy (Fig. 1a), studies of this specimen by the combined use of CLSM and Raman imagery (1) establish its ctenophoran affinities, (2) demonstrate its anatomy and molecular-structural composition, and (3) elucidate the sequence of mineral emplacement that resulted in its preservation. Use of these techniques, together, provides biological and taphonomic data unavailable by any other means. Such data, combined with knowledge of the biology of modern comb jellies, of the anatomy of their embryos, and of the morphology and paleoecology of adult members of the group preserved by compression in other Cambrian units have provided important new insights into the early evolutionary history of the Ctenophora (Chen et al. 2007).

5 Permineralized Plants

Unlike the cells of animals, those of plants – and of algae (including phytoplanktonic acritarchs), fungi, and microbes – are enclosed by robust organic cell walls. In living organisms the chemistry of such walls varies substantially, for plants and algae being composed primarily of cellulose with the tracheary xylem of many plants having been infused secondarily by lignin; the walls of fungi being composed largely of nitrogen-containing "fungal chitin"; and those of bacterial microbes, such as cyanobacteria, being composed chiefly of mucopeptides such as peptidoglycans. During the diagenesis and resulting geochemical maturation of such biomolecules, all are driven toward the same end-product, kerogen, carbonaceous matter composed mainly of stacked, interlinked, polycyclic aromatic hydrocarbons (Schopf et al. 2005; Vandenbroucke and Largeau 2007), a precursor of metamorphically produced graphite. Because of its composition, such kerogen is amenable to investigation by CLSM and Raman, in cellularly mineralized fossils providing the basis for three-dimensional micron-scale imaging. To illustrate the paleobiologic usefulness of these techniques in studies of such fossils, the following section of this chapter deals with their application to quartz- and calcite-permineralized Phanerozoic plants. Subsequent sections demonstrate their use in the study of Precambrian fossil microorganisms.

5.1 Quartz-Permineralized Plant Axes

In Fig. 2 are shown optical (Fig. 2a, b, d, f) and CLSM images (Fig. 2c, e, g) of a transverse section of a quartz-permineralized rhizome (underground stem) of an Eocene (~45-Ma-old) fossil fern preserved in chert from the Clarno Formation of Oregon, USA. The CLSM image of the preserved aerenchyma (Fig. 2c), the oxygen-supplying tissue of such underground stems, provides appreciably more information than that of a standard optical image (Fig. 2b). This superiority is exhibited also by CLSM images of the thick-walled cortical cells of this specimen, whether imaged in a thin section (compare Fig. 2e, a CLSM image, with Fig. 2d, an optical photomicrograph) or in a cellulose acetate peel (compare Fig. 2g with Fig. 2f). In the CLSM image of the cortical cells in thin section (Fig. 2e), intracellular middle lamellae – initially composed of calcium pectate rather than cellulose, the principal component of the adjacent cell walls – and the spaces resulting from the preferential degradation of these lamellae, are well displayed. The CLSM image of the same cells of this specimen in a cellulose acetate peel provides additional information, showing imprints of the permineralizing cryptocrystalline grains of quartz (Fig. 2g). Such information, relevant both to the preferential degradation of intercellular organic cement and the spatial distribution of permineralizing quartz crystals, are crucial to an understanding of the taphonomy of such specimens.

Figure 3 shows a chert-permineralized Clarno rhizome imaged in a peel of a longitudinal section that similarly illustrates the three-dimensional detail provided by CLSM imagery (Fig. 3c) in comparison with that of standard optical microscopy (Fig. 3b).

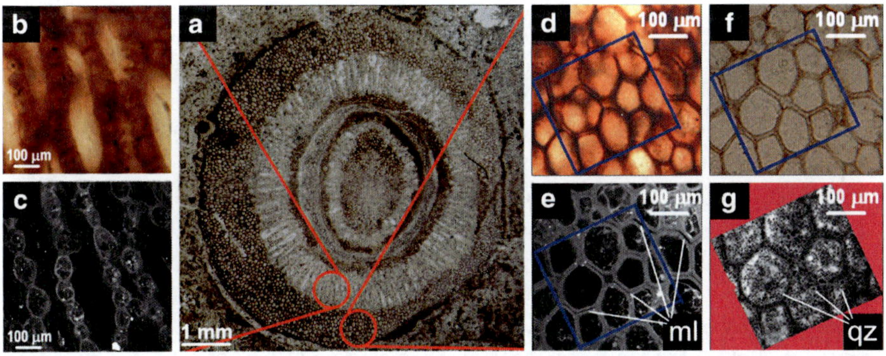

Fig. 2 Optical (**a**, **b**, **d**, **f**) and CLSM images (**c**, **e**, **g**) of a transverse section of a quartz-permineralized rhizome of a fossil fern (*Dennstaedtiopsis aerenchymata*) in carbonaceous chert from the Eocene Clarno Formation of Oregon, USA (Arnold and Daugherty 1964), shown in a petrographic thin section (**b**–**e**) and a cellulose acetate peel (**a**, **f**, **g**), the area shown in (**g**) being denoted by the blue rectangles in (**d**–**f**). (**a**) Optical image of the rhizome denoting the areas illustrated in (**b**–**g**). (**b**, **c**) Optical (**b**) and CLSM (**c**) images of aerenchymatous strands and associated interstrand spaces. (**d**–**g**) Optical (**d**, **f**) and CLSM images (**e**, **g**) of cortical cells in a thin section (**d**, **e**) and a peel (**f**, **g**); intercellular middle lamellae (ml) are evident in the CLSM image of the thin section-embedded cells (**e**) as are imprints of permineralizing quartz grains (qz) in the CLSM image of the cells in a peel (**g**)

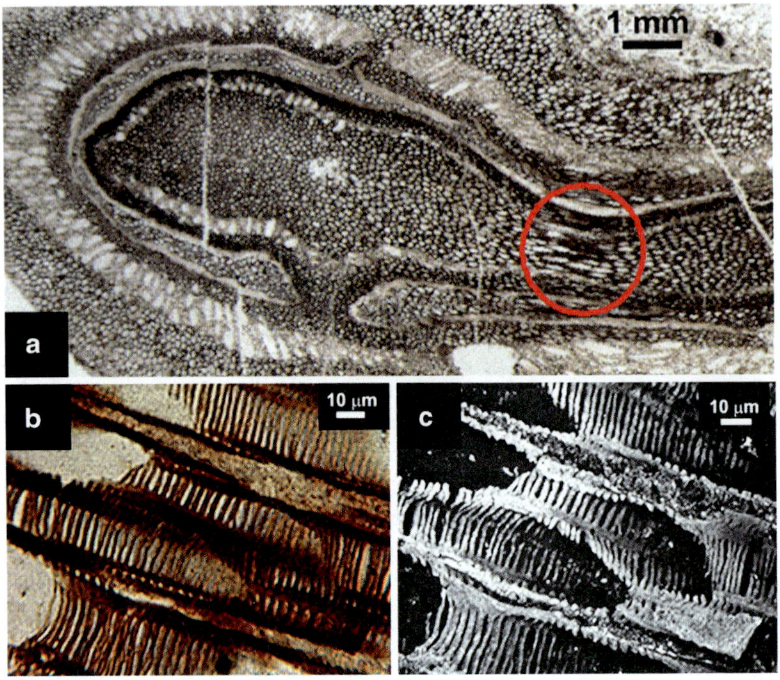

Fig. 3 Optical (**a**, **b**) and CLSM images (**c**) of a longitudinal section of a quartz-permineralized rhizome of *Dennstaedtiopsis aerenchymata* (cf. Fig. 2) in a cellulose acetate peel. (**a**) Optical image of the rhizome in which the red circle denotes the area illustrated in (**b**, **c**). (**b**, **c**) Optical (**b**) and CLSM (**c**) images of tracheids illustrating their prominent annular thickenings

In Fig. 4 are shown optical and CLSM images of a quartz-permineral-ized axis of *Aglaophyton*, a prototypical early-evolved land plant, from the Lower Devonian Rhynie Chert of Aberdeenshire, Scotland. Unlike the other permineralized plants illustrated here, this specimen was studied in a rock slice several millimeters thick. Despite the fact that the specimen could therefore not be studied by transmitted light optical microscopy, CLSM is shown here to be capable of "seeing into" this optically opaque specimen, analyzing its uppermost surface and underlying cellular structure, by detection of backscattered kerogen-emitted fluorescence that provides a high quality image of the xylem cells that compose its haplostelic core (Fig. 4b). Moreover, as is shown in Fig. 4c, CLSM studies of cellulose acetate peels lifted from the surface of this fossil provide excellent images of the sporangia of the chitrid fungi that infest the cortical tissue of this specimen, evidencing the use of this technique to investigate the biological degradation of such tissues, even in optically opaque rock slices.

5.2 Calcite-Permineralized Plant Axes

The most anatomically informative diverse assemblage of plant fossils known from the Paleozoic geological record is that preserved by calcite-permineralization in

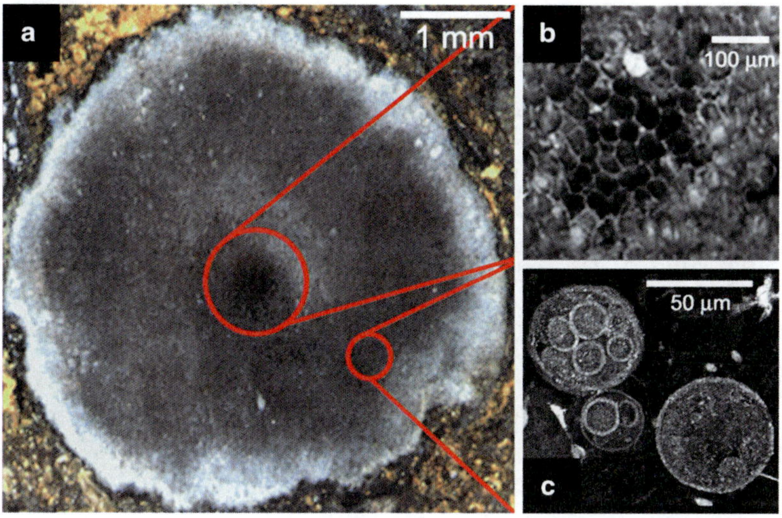

Fig. 4 Optical (**a**) and CLSM images (**b**, **c**) of a transverse section of a quartz-permineralized axis of a rhyniophyte (*Aglaophyton*) from the Devonian Rhynie Chert (Kidston and Lang 1917), exposed at the surface of a ~3-mm-thick rock slice (**a**, **b**) and shown in a peel (**c**). (**a**) Optical image of the axis denoting the areas shown in (**b**, **c**). (**b**) CLSM image of xylem elements of the central protostele. (**c**) CLSM image of fungal (chitrid) sporangia, containing spheroidal spores, permineralized in the cortical tissue of the specimen

Carboniferous coal balls. Though such fossils are not uncommonly excellently preserved, the processes involved in their permineralization, events documented to have occurred prior to lithification of the coalified peat beds in which they occur, are incompletely defined (Schopf 1975; see also Scott et al. 1996). Shown here in Fig. 5 are optical and CLSM images of one such fossil, the stem of an herbaceous sphenophyte related to modern *Equisetum* (the "horsetail" or "scouring-brush" plant). Importantly, not only do the CLSM images provide crisp images of the cell walls of the secondary xylem of this specimen, both in a thin section (Fig. 5b) and in a peel (Fig. 5c), but they also show the morphology of the sparry calcite crystals in which the cells of this axis have been permineralized. CLSM studies of coal ball-preserved plant fossils (cf. Schopf et al. 2006b; Tripathi 2007) provide a promising new approach to investigations of the taphonomy of calcite-permineralization.

6 Permineralized Organic-Walled Microorganisms

Because of the micron-scale spatial resolution provided both by CLSM and by two- and three-dimensional Raman imagery, these techniques are especially useful for investigation of permineralized microscopic organisms such as those that comprise the bulk of the 3,000 million years of Precambrian biologic history. Indeed, before the recent introduction of these techniques to the study of such ancient minute fossils (Schopf and Kudryavtsev 2005; Schopf et al. 2005, 2006a), no methods had

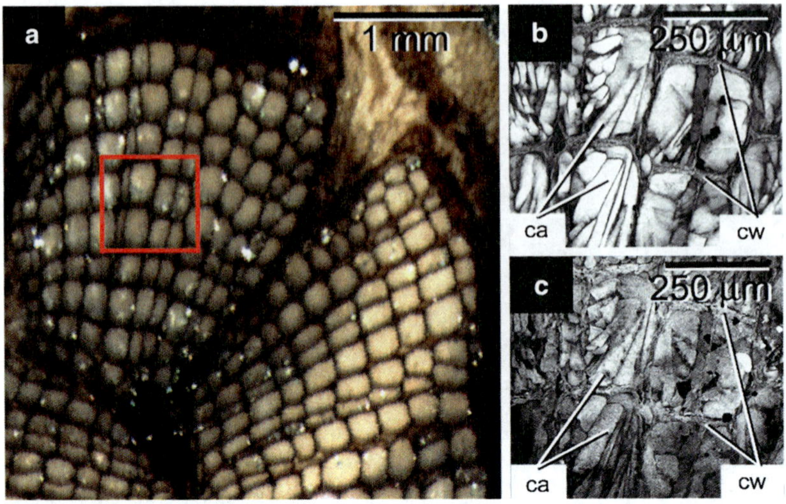

Fig. 5 Optical (**a**) and CLSM images (**b**, **c**) of a transverse section of a calcite-permineralized sphenophyte (*Sphenophyllum*) in a calcareous coal ball from the Carboniferous of Illinois, USA (cf. Schopf 1941), shown in a petrographic thin section (**a**, **b**) and a cellulose acetate peel (**c**). (**a**) Optical image of the axis in which the red rectangle denotes the area shown in (**b**, **c**). (**b**, **c**) CLSM images of the secondary xylem showing the sparry calcite crystals (ca) that have infilled cell lumina (enclosed by well defined cell walls, cw), particularly evident in the image of the petrified cells in thin section (**b**)

been available by which to objectively depict and firmly establish on the basis of direct analyses in situ either the three-dimensional cellular morphology or the molecular-structural composition of such specimens. The following sections of this chapter illustrate the applicability of these techniques to studies of Precambrian sphaeromorph acritarchs and coccoidal and filamentous microbes; demonstrate their use to elucidate the taphonomy of such microfossils, including the indirect evidence they can provide of original biochemistry; and document the cellular preservation and molecular-structural composition of certain of the oldest (~3,465-Ma-old) fossils now known.

6.1 Quartz-Permineralized Acritarchs

In Fig. 6 are shown optical and CLSM images of a sphaeromorph acritarch quartz-permineralized in a thin section of a Neoproterozoic (~650-Ma-old) cherty stromatolite. The optical photomicrographs in Fig. 6a–c show, respectively, the uppermost surface ("north pole"), equatorial plane, and lowermost "south pole" surface of the specimen, three views at sequentially increasing focal depths of the type that might typically be presented in a scientific publication. For comparison, Fig. 6d shows a CLSM image of the entire acritarch, demonstrating the fine structural detail

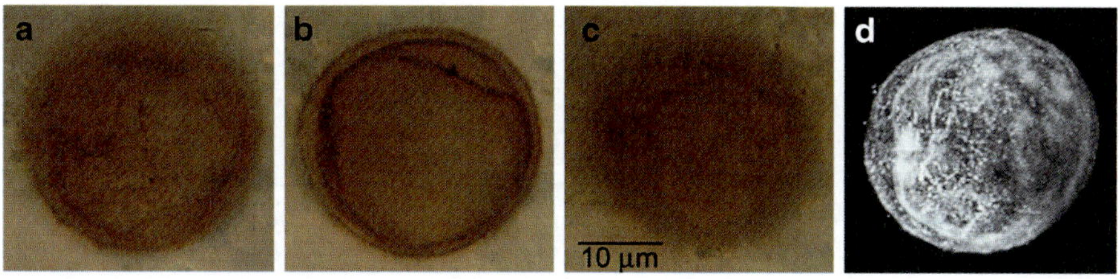

Fig. 6 Optical (**a**–**c**) and CLSM images (**d**) of an ensheathed sphaeromorph acritarch in a petrographic thin section of a conical cherty stromatolite from the ~650-Ma-old Chichkan Formation of southern Kazakhstan (Schopf and Sovietov 1976); scale in (**c**) applies also to (**a**, **b**, **d**). (**a**–**c**) Optical photomicrographs showing the uppermost surface (**a**), equatorial plane (**b**), and lowermost surface (**c**) of the quartz-permineralized unicell for comparison with (**d**), a CLSM image of the specimen

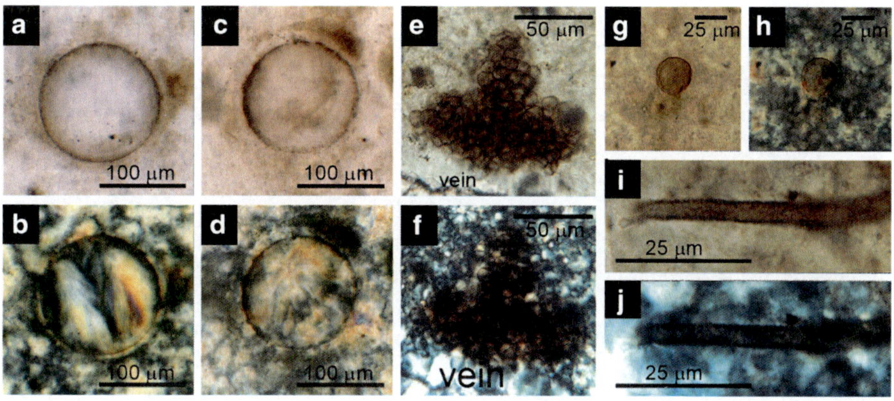

Fig. 7 Optical photomicrographs of quartz-permineralized microfossils in petrographic thin sections of stromatolitic cherts of the ~650-Ma-old Chichkan Fm. of southern Kazakhstan (**a**–**d**; Schopf and Sovietov 1976) and the ~750-Ma-old Bitter Springs Formation of central Australia (**e**–**j**; Schopf 1968; Schopf and Blacic 1971). (**a**–**d**) Transmitted (**a**, **c**) and plane-polarized light (**b**, **d**) images of large-celled sphaeromorph acritarchs. (**e**, **f**) Transmitted (**e**) and plane-polarized light (**f**) images of a small-celled colonial cyanobacterium (*Myxococcoides minor*), truncated at the lower left by a quartz-filled vein. (**g**, **h**) Transmitted (**g**) and plane-polarized light (**h**) images of a ~25-mm-diameter unicell (*Myxococcoides* sp.). (**i**, **j**) Transmitted (**i**) and plane-polarized light (**j**) images of a ~4-mm-diameter tubular cyanobacterial sheath (*Eomycetopsis* sp.)

provided by only a single such image, the information-rich content of which is even better depicted in rotating three-dimensional video views of the specimen.

Although typically encompassed by an interlocking three-dimensional mosaic of irregularly shaped grains of cryptocrystalline quartz, large-sized quartz-permineralized acritarchs (>40 μm in diameter), such as the sphaeromorphs shown in Figs. 6 and 7a–d, are virtually always filled by swaths of fibrous quartz, known commonly as "flame chalcedony" (Fig. 7b,d), the robust cell walls of such specimens defining the boundary between the two forms of quartz. In such fossils, irregularities on the inner surfaces of the cell walls have served as points of nucleation for the formation of cell lumina-infilling chalcedony. In contrast, quartz-permineralized small-celled spheroidal fossils, whether colonial (Fig. 7e, f) or unicellular (Fig. 7g, h), as well as small-diameter tubular microbial filaments (Fig. 7i, j), are almost always embedded within and thoroughly infilled by cryptocrystalline quartz, the grain boundaries of which transect but do not disrupt their kerogenous walls. Evidently, the particular mineral phase involved in such quartz-permineralization is dependent primarily on the dimensions and cell wall thickness of the fossil preserved, small thin-walled cells being infused by and embedded within small-sized quartz grains, whereas larger cells, outlined by their relatively thick cell walls, are preserved by an infilling of their cell lumina by chalcedonic quartz. As is shown above, such permineralization by large cell-filling crystals is exhibited also by the sparry calcite that infills the cell lumina of permineralized higher plants (Fig. 5b, c), a similarity suggesting that detailed taphonomic analyses by CLSM can be expected to yield useful insight into the processes and products both of quartz- and calcite-permineralization.

In Fig. 8 is shown an additional quartz-permineralized Precambrian sphaeromorph, included here to illustrate the method used to obtain three-dimensional chemical (Raman) images of such acritarchs (Schopf and Kudryavtsev 2005). Figure 8a–e shows representative optical photomicrographs from a sequential "through-focus" series that extends from the uppermost to the lowermost surface of the specimen, whereas Fig. 8f–j shows an equivalent series of two-dimensional Raman images. Computer-aided stacking and processing of such Raman images yielded a micron-resolution three-dimensional "chemical map" of the acritarch as viewed from immediately above its uppermost surface (Fig. 8k) and below its lowermost surface (Fig. 8l). The digitized data used to create such images – like many of the other images shown here acquired in situ, measured on a specimen entirely embedded within a petrographic thin section – can be rotated in three dimensions or otherwise manipulated to provide additional useful information. Thus, for example, Fig. 8m shows a view of this specimen from which its bottom hemisphere has been removed, documenting the texture of the inner surface of the spheroid and the prominent grooves in its upper surface, diagenetically produced tears in the specimen that are all but indiscernible in optical photomicrographs (Fig. 8a, b).

6.2 *Quartz-Permineralized Filamentous Microbes*

Illustrated here by their use for the study of Precambrian microscopic organisms, CLSM and two- and three-dimensional Raman imagery are particularly applicable to investigations of minute, sinuous, filamentous microbes.

6.2.1 Precambrian Cyanobacteria

In Fig. 9 are shown an optical image, an interpretive drawing, CLSM images, and a three-dimensional Raman image of a quartz-permineralized kerogenous cellular cyanobacterial filament imaged within a petrographic thin section of the ~750-Ma-old

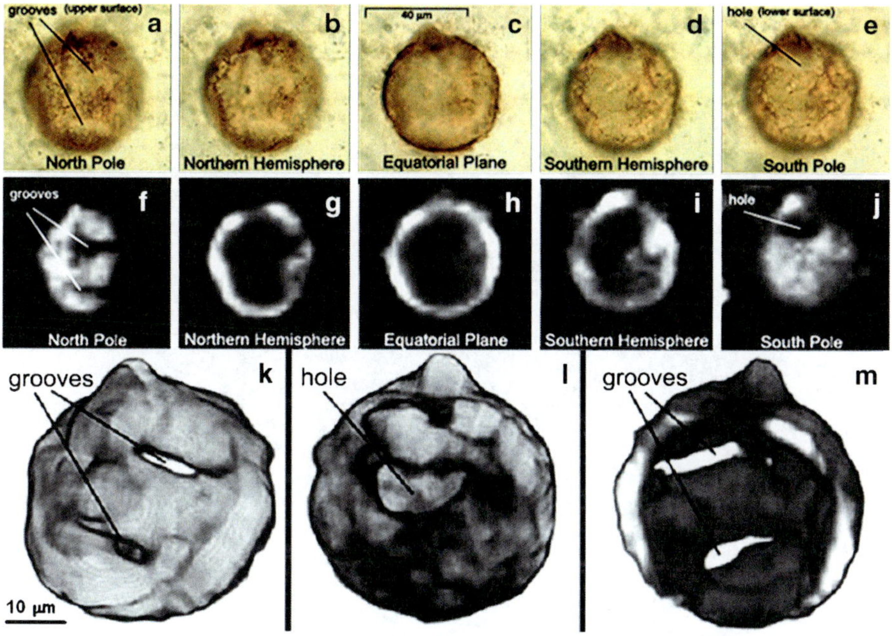

Fig. 8 Optical (**a**–**e**) and Raman images (**f**–**m**) of a sphaeromorph acritarch in a petrographic thin section of a conical cherty stromatolite from the ~650-Ma-old Chichkan Formation of southern Kazakhstan (Schopf and Sovietov 1976); Raman images were acquired in a spectral window centered at the ~1,600 cm^{-1} band of kerogen; scale in (**c**) applies also to (**a**, **b**, **d**–**j**); scale in (**k**) applies also to (**l**, **m**). (**a**–**e**) Optical photomicrographs showing representative images from a sequential "through-focus" series from the uppermost surface (*left*) to the lowermost surface (*right*) of the quartz-permineralized specimen. (**f**–**j**) Two-dimensional Raman images acquired at the same focal planes as the corresponding optical images in (**a**–**e**). (**k**–**m**) Three-dimensional Raman images as viewed from (**k**) above the "north pole" of the specimen, showing the two grooves in its uppermost surface; (**l**) beneath its "south pole," showing the hole in its lowermost surface; and (**m**) its interior, looking outward toward its "north pole," showing the inner surface of the spheroid and the open grooves in its uppermost surface

Bitter Springs Formation of central Australia. The optical image (Fig. 9a) is a photomontage composed of ten photomicrographs of the medial plane of the fossil, a presentation necessitated by the minute size of the specimen (and the resultant need for its optical documentation by use of a high-magnification, but narrow focal-plane, microscope objective) and its sinuosity, plunging from the upper surface of the thin section (at the right end of the filament in Fig. 9a–c, and e) to ~20 μm beneath this surface (at its left end, Fig. 9a–c). The interpretive drawing (Fig. 9b), a stippled tracing of this photomontage that presents a somewhat more life-like rendering of the specimen, shares the same deficiencies as the photomontage: both are based on subjective interpretations of the specimen (resulting from the pasting together of photomicrographs acquired at differing focal depths). Neither can be regarded as depicting accurately and objectively the exact morphology of the fossil. In contrast, the CLSM images of this filament (Fig. 9c,d) faithfully show the sinuosity and cellularity of the

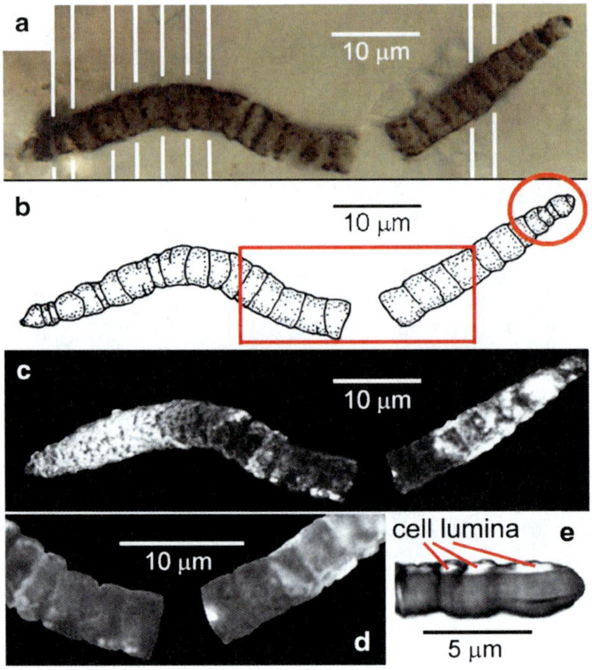

Fig. 9 Optical image (**a**), interpretive drawing (**b**), CLSM images (**c**, **d**), and a three-dimensional Raman image (**e**) of a quartz-permineralized tapering cyanobacterial trichome (*Cephalophytarion laticellulosum*) in a petrographic thin section of a flat-laminated cherty stromatolite from the ~750-Ma-old Bitter Springs Formation of central Australia (Schopf and Blacic 1971; holotype specimen, Harvard University Paleobotanical Collections No. 58571). (**a**, **b**) Traditional renderings of a sinuous specimen such as this, in which the area shown at higher magnification in the CLSM image in (**d**) is outlined by the red rectangle and the area imaged in three dimensions by Raman spectroscopy (**e**) is denoted by the red circle, with (**a**) showing a photomontage composed of 10 optical photomicrographs (demarcated by the *white lines*) and (**b**) illustrating the specimen by an interpretive drawing. (**c**, **d**) CLSM images of the specimen, the right end of which transects the thin section surface and the left end being situated at a depth of 20 mm within the section. (**e**) Three-dimensional Raman image (acquired in a spectral window centered at the ~1,600 cm^{-1} band of kerogen) of the terminal several cells of the specimen, VolView-processed and rotated to show the flat uppermost surface of the cells (where they transect the surface of the thin section), that demonstrates the kerogenous composition of its lateral and transverse cell walls (*grey*) and shows the quartz-filled cell lumina (*white*) that they enclose

specimen (even better depicted in rotating three-dimensional video views of the specimen), whereas the three-dimensional Raman image (Fig. 9e) shows not only its cellularity but provides data that establish that its kerogen-defined cell lumina are infilled by quartz. Optical and CLSM images of an additional filamentous cyanobacterium from the same geologic unit are shown in Fig. 10.

As is evident from a comparison of the optical and CLSM images of both of these examples (viz., Fig. 9a vs. c, and Fig. 10a vs. b), use of confocal laser scanning microscopy can provide appreciably more information about the fine

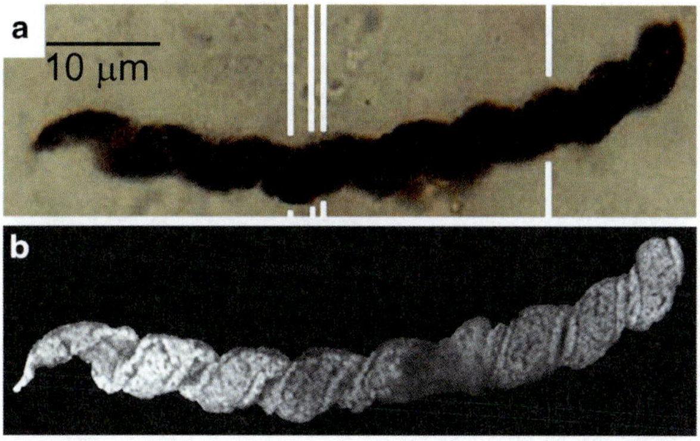

Fig. 10 Optical (**a**) and CLSM images (**b**) of a quartz-permineralized helical cyanobacterial trichome (*Heliconema funiculum*) in a petrographic thin section of a flat-laminated cherty stromatolite from the ~750-Ma-old Bitter Springs Formation of central Australia (Schopf and Blacic 1971; holotype specimen, Harvard University Paleobotanical Collections No. 58595); scale in (**a**) applies also to (**b**). (**a**) Photomontage composed of five optical photomicrographs (demarcated by the *white lines*). (**b**) CLSM image, showing the fine structural detail that can be depicted by use of CLSM

structural morphology and, thus, the biological affinities and taphonomy of such specimens, than can standard optical microscopy alone.

Taphonomic Evidence of Original Biochemistry For fossil microbes, CLSM and Raman imagery can also provide evidence, albeit indirect, of original biochemistry. Figure 11 shows a many-celled portion of a broken (and at this break, partially offset) originally ensheathed cyanobacterial trichome compared by Schopf and Sovietov (1976) to the living oscillatoriacean *Lyngbya majuscula*. Shown also are three-dimensional CLSM (Fig. 11b) and Raman images (Fig. 11c, d) of a seven-celled segment of the specimen. The image in Fig. 11c shows the contours of the two-dimensional Raman images (oriented parallel to the thin section surface) that have been combined to produce the three-dimensional image of the portion of the specimen shown in Fig. 11d. Each of these three-dimensional images has been rotated to an orientation that permits examination of the central core of the specimen. Notably, all show the core of the trichome to be "hollow", a quartz-filled cavity that in its central region is devoid of the carbonaceous matter that would evidence the presence of transverse cell walls. The absence or only partial presence of the central region of such cross walls – on the basis of optical microscopy, cross walls universally assumed to be preserved in such specimens – is typical of many such Precambrian cyanobacterial trichomes (Schopf et al. 2006a). In Fig. 12, this same specimen is shown in an optical photomicrograph (Fig. 12a) and in CLSM images (Fig. 12b–d) that by illustrating the differing degrees of image quality obtainable by use of various excitation laser wavelengths and filter arrays (cf. Tripathi 2007) further show the fine structural detail that can be acquired by use of CLSM (e.g., Fig. 12d), data that for this fossil confirm the near-absence of transverse walls and the "hollow" (i.e., quartz-filled) nature of its trichomic cavity.

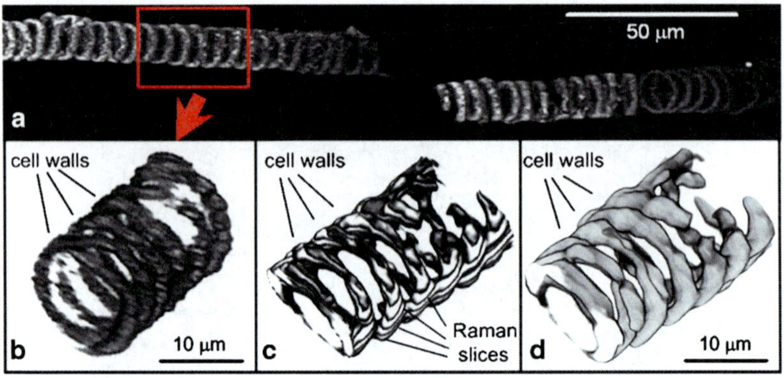

Fig. 11 CLSM images (**a**, **b**) and Raman images (**c**, **d**) of a quartz-permineralized Lyngbya-like cyanobacterial trichome composed of disc-shaped cells, in a petrographic thin section of a conical cherty stromatolite from the ~650-Ma-old Chichkan Formation of southern Kazakhstan (Schopf and Sovietov 1976); scale in (**d**) applies also to (**c**). (**a**, **b**) Rotated CLSM images of the specimen in which the red rectangle in (**a**) denotes the portion of the trichome shown in the VolView-processed CLSM image in (**b**). (**c**, **d**) Rotated, VolView-processed, three-dimensional Raman images (acquired in a window centered at the ~1,600 cm^{-1} kerogen band) of the same part of the specimen shown in (**b**), illustrating in (**c**) the spatial relations between the preserved cell walls and the two-dimensional Raman slices used to prepare the three-dimensional image in (**d**), a more accurate representation of the distribution of the kerogenous components of the specimen

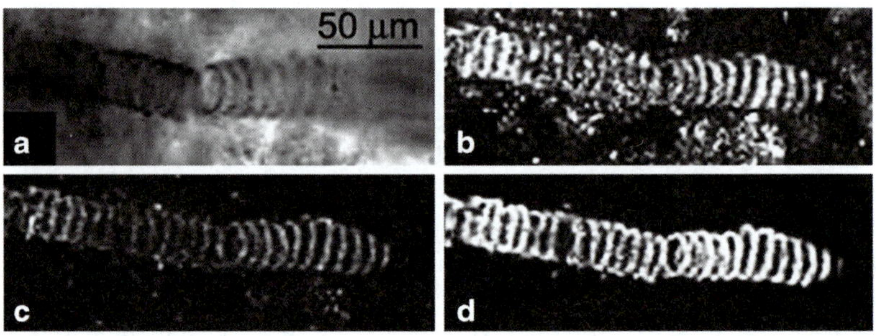

Fig. 12 Optical photomicrograph (**a**) and CLSM images (**b–d**) of the thin section-embedded quartz-permineralized specimen shown in Fig. 11 (but unlike those in Fig. 11, shown here in non-rotated images); scale in (**a**) applies also to (**b**, **c**). These images illustrate the increased depth of focus provided by CLSM in comparison with that of optical microscopy (**a**) and differences in the quality of CLSM images acquired by use of excitation wavelengths of 488 nm (**b**, filtered detection window = 520–560 nm), 543 nm (**c**, window = 560–600 nm), and 633 nm (**d**, window = >660 nm), the last providing the sharpest image of the specimen

This example of the use of CLSM and Raman imagery provides insight into the taphonomic history of such filamentous fossil microbes that reflects their original biochemistry. Cell division in oscillatoriacean cyanobacteria occurs by invagination of septations that grow centripetally from the periphery of the cells to ultimately divide them into new daughter cells. At their inception termed partial septations, these inward-growing transverse cell walls are thinner than and differ in biochemical composition from the exterior (lateral) walls of such microbes. As discussed by Drews (1973), the lateral cell walls of oscillatoriaceans are composed of four layers surrounded on the outside in ensheathed taxa by a sheath or slime layer and enclosing within them, toward the interior of such cells, the cytoplasmic membrane. About half of the total thickness of such lateral walls is made up of the outermost two layers, the other half by the inner pair of layers (Drews 1973). Notably, the transverse walls that define the cellular segmentation of such oscillatoriaceans are composed only of the two innermost layers (Pankratz and Bowen 1963; Lamont 1969; Halfen and Castenholz 1971; Drews 1973) and, thus, are typically only about half as thick as the lateral walls that define the organismal form of such microbes. Moreover, biosynthesis of a principal constituent of one of the two innermost layers, peptidoglycan (known also as murein or mucopeptide and the rigidifying component of oscillatoriacean lateral cell walls) ceases in transverse walls after their initial stages of growth so that the partial septations from which they are derived are peptidoglycan-rich only near the periphery of such cells (Frank et al. 1962; Halfen and Castenholz 1971). Relatively thin and peptidoglycan-deficient, such transverse walls are less robust and relatively more susceptible to diagenetic degradation than the lateral, organismal-form defining cell walls of such microorganisms (Van Baalen and Brown 1969). Thus, the absence or only partial preservation of such transverse cell walls in the fossil oscilliatoriacean illustrated in Figs. 11 and 12 meshes well with expectations based on the biochemical and fine-structural morphology of comparable microorganisms living today.

6.2.2 Raman Index of Preservation (RIP)

Among the paleobiologically useful attributes of Raman imagery is its ability to characterize the molecular-structural composition of the materials analyzed, being applicable, as shown above, both to minerals and to mineralized fossils. Moreover, Raman spectra resulting from studies of the kerogenous materials that comprise such fossils can themselves be analyzed to yield their Raman Index of Preservation (Schopf et al. 2005), RIP values that provide a firm basis for the assessment of their diagenetic alteration, their geochemical maturity (Schopf et al. 2005).

In Fig. 13 are shown seven Raman spectra, obtained from organic-walled Precambrian microfossils preserved at various stages of geochemical maturation (Schopf et al. 2005). As is there illustrated, the two major Raman bands of kerogen change markedly as a function of increasing geochemical alteration: the left-most "D" band becomes increasingly more peaked (and, correspondingly, less broad and "bumpy") as the right-most "G" band becomes increasingly narrow and ultimately bifurcated. Such data, obtainable from organic-walled fossils permineralized in rocks subjected even to greenschist facies metamorphism (Schopf et al. 2002, 2005), can provide definitive insight into the geochemical maturity (degree of alteration) of fossilized organic matter that is unavailable by any other means.

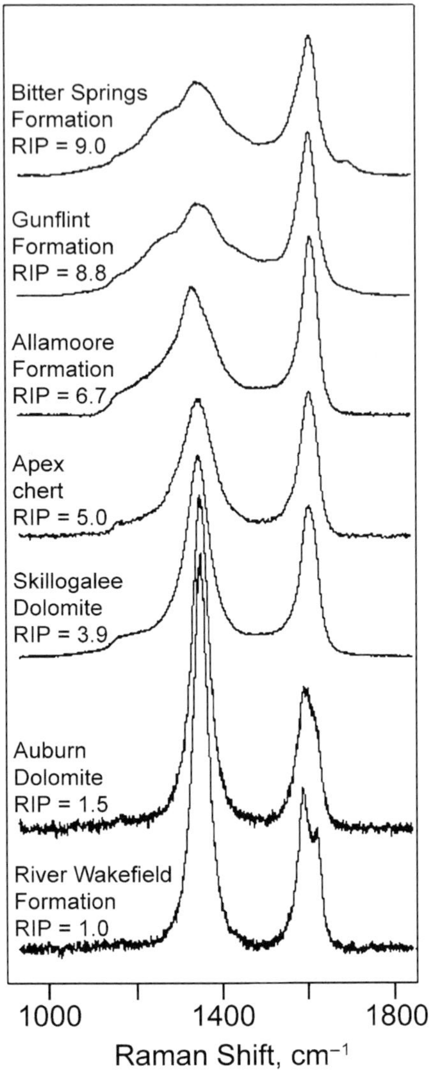

Fig. 13 Raman spectra of kerogenous microfossils permineralized in cherts of the ~750-Ma-old Bitter Springs Fm., central Australia (cf. Figs. 8 and 9); the ~1,900-Ma-old Gunflint Fm., Ontario, Canada; the ~1,050-Ma-old Allamoore Fm., Texas, USA; the ~3,465-Ma-old Apex chert, Western Australia (cf. Fig. 14); and of the ~760-Ma-old Skillogalee Dolomite, ~720-Ma-old Auburn Dolomite, and ~775-Ma-old River Wakefield Fm. of South Australia (Schopf et al. 2005). The spectra are ordered by their RIP values (Schopf et al. 2005) from less (*top*) to more (*bottom*) geochemically mature

6.2.3 Archean Bacteria

In recent years, questions have been raised about the biogenicity of certain of the oldest putative records of life now known (Brasier et al. 2002, 2005), reported from especially ancient, Archean (>2,500-Ma-old), geological units. Indeed, it has even been suggested that "true consensus for life's existence" dates only from "the bacterial fossils of 1.9-billion-year-old Gunflint Formation of Ontario" (Moorbath 2005). According to this view, all supposed evidence of earlier life, "the many claims of life in the first 2.0–2.5 billion years of Earth's history," is in doubt (Moorbath 2005).

Notwithstanding such skepticism, the evidence for Archean life seems compelling (Schopf 2004a, b, 2006a, b; Altermann 2005; Altermann et al. 2006; Schopf et al. 2007). Though markedly less abundant and almost always less well-preserved than biologic remnants of the younger, Proterozoic, Precambrian – a result, primarily, of the paucity of Archean rocks that have survived to the present and their pervasive metamorphic alteration (Schopf 2006a; Schopf et al. 2007) – diverse microbially produced stromatolites are known from 48 Archean deposits; 14 such units contain some 40 morphotypes of described microfossils; and hundreds of carbon isotopic measurements consistent with the presence of biologic activity have been reported from Archean rock units dating to 3,500 Ma ago (Strauss and Moore 1992; Schopf 2006a, b). Even more significantly, units 3,200 to 3,500 Ma in age contain abundant evidence of life: 10 such units are known to be stromatolitic; 11 contain organic-walled microfossils; and carbon isotopic data consistent with biologic CO_2-fixation are available for nine such deposits (Schopf 2006a, b). In addition, the oldest metasediments now known, >3,830-Ma-old units of southwestern Greenland, have recently been shown by 3-D Raman imagery to contain apatite-enclosed graphitic carbonaceous matter determined by secondary ion mass spectrometry to have an isotopic value similarly consistent with biological CO_2-fixation – a strong hint of microbial activity arguably suggesting that "the record of life on Earth is as old as the oldest sedimentary rocks now known" (McKeegan et al. 2007).

Studies of the taphonomy of ancient fossils by CLSM and Raman imagery have played a pivotal role in resolving the uncertainty about life's early existence. Shown in Fig. 14a–c are specimens from the most contentious of the known Archean microfossil assemblages (Brasier et al. 2002, 2005), minute filamentous structures reported from the ~3,465-Ma-old Apex chert of northwestern Australia that have been interpreted to be composed of carbonaceous, kerogenous, cells (Schopf 1993). As recently documented (Schopf 2004a, 2006a; Schopf et al. 2007), these fossil-like filaments meet ten separate tests of biogenicity: all exhibit (1) *biological morphology*, including (2) *structurally distinct carbonaceous cell walls* that define (3) *cell lumina*. All occur in (4) a *multi-member population* that includes (5) *numerous taxa*, members of which exhibit (6) *variable preservation*. All are (7) *preserved three-dimensionally by permineralization* in fine-grained quartz, shown above (Figs. 9–12) to be a common mode of preservation of such fossils. And all have (8) *biological size ranges*, as measured for several hundred specimens, and exhibit a (9) *Raman signal of biogenic kerogen*, carbonaceous matter that has an (10) *isotopic composition typical of biologically produced organic matter*.

Perhaps primary among these criteria for establishment of the biogenicity of these fossil-like filaments are their organic (kerogenous) and cellular composition. That they are composed of carbonaceous organic matter that is indistinguishable from the kerogen of bona fide fossils, an interpretation supported by numerous lines of evidence (e.g., De Gregorio and Sharp 2003, 2006; De Gregorio et al. 2005),

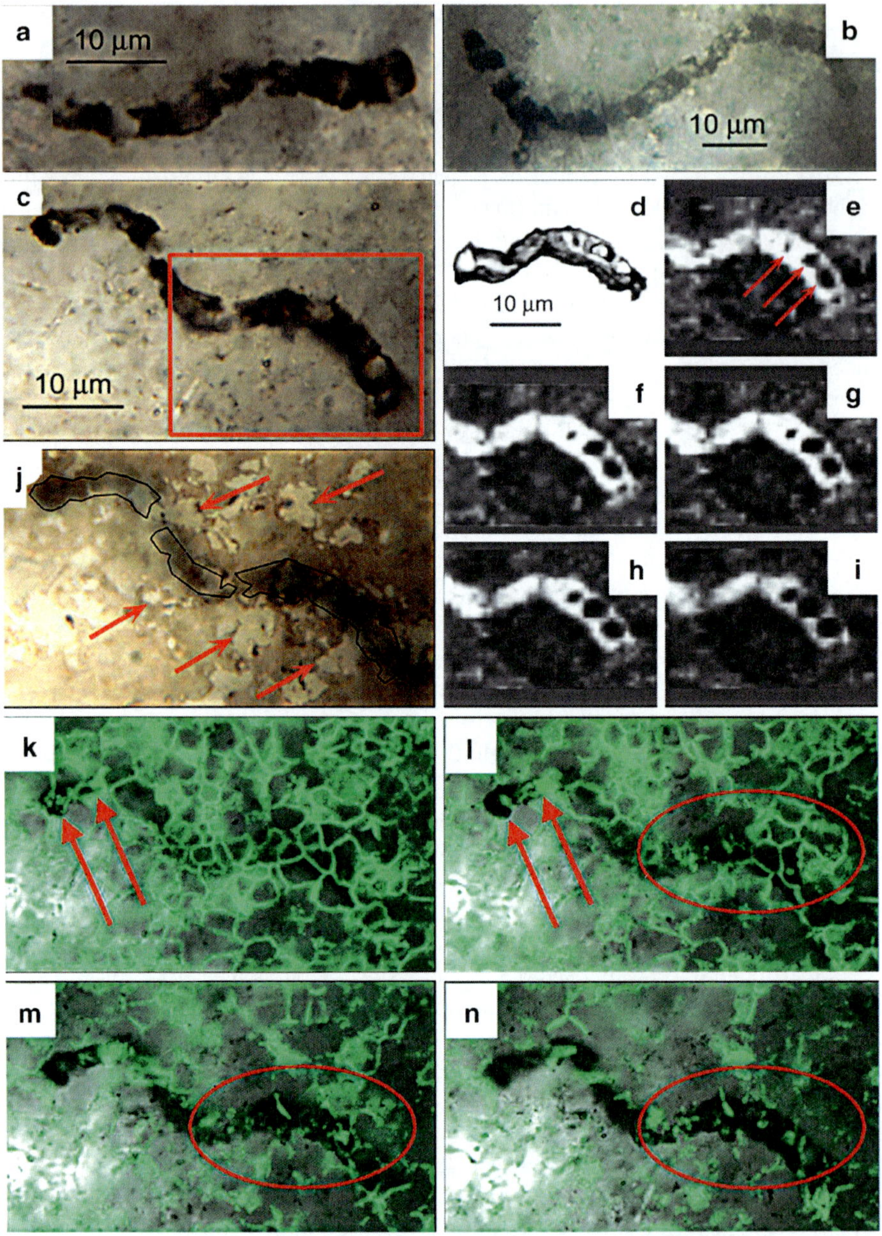

Fig. 14 Optical (**a–c**, **j**), Raman (**d–i**), and CLSM images (**k–n**) of permineralized kerogenous filaments (*Primaevifilum amoenum*) in petrographic thin sections of the ~3,465-Ma-old Apex chert of Western Australia; (**a**) Natural History Museum, London V.63164[5]; (**b**) V.63166[1]; (**c–n**) V.63164[6] (Schopf 1993); Raman images were acquired in a spectral window centered at the ~1,600 cm^{-1} band of kerogen; scale in (**c**) applies also to (**j–n**); scale in (**d**) applies also to (**e–i**); (**a–c**) show photomontages. (**a–c**) Photomicrographs of three specimens of *P. amoenum*, that in (**c**) ranging from 3 µm (*left end*) to 9 µm (*right end*) below the section surface with the red

is well shown by the Raman spectra presented in Fig. 13: the Apex filaments, having an RIP value of 5.0, exhibit an intermediate grade of organic preservation, being neither as well preserved as the Bitter Springs (Figs. 7, 9, and 10) and other relatively little-altered Precambrian microbes, but exhibiting an appreciably greater fidelity of preservation than fossils preserved in more metamorphosed Precambrian geological units. And the cellularity of the Apex filaments is firmly established by three-dimensional Raman imagery (cf. Schopf et al. 2007). Shown in Fig. 14d is a three-dimensional Raman image of a portion of the Apex filament illustrated in Fig. 14c. As is shown by the two-dimensional Raman images in Fig. 14e–i, this specimen (like numerous others from the deposit; Schopf et al. 2007) is composed of box-like cells defined by carbonaceous (kerogenous) walls. Such walls are not a result of petroleum-like carbonaceous fluids having enveloped quartz grains during recrystallization (Brasier et al. 2005). As is shown by the CLSM images in Fig. 14k–n, permeation of organic fluids into the Apex chert results in formation of a three-dimensional chicken wire-like mosaic, not in the formation of discrete, cylindrical, microbe-like sinuous filaments composed of regularly aligned uniseriate strands of cell-like segments.

Backed by additional factors and subfactors that seem similarly indicative of biogenicity – including a firm fit with all other reported evidence of comparably ancient life (Schopf 2004a, b, 2006a, b; Schopf et al. 2007) – demonstration of organic-walled cellularity in putative filamentous microfossils such as these is a strong indicator of their biological origin. Such organic-walled cellular structure is a defining characteristic of bona fide microbial filaments, both extant and fossil. Indeed, particulate carbonaceous matter like that comprising the Apex filaments is not known to be produced by any non-biologic means, and pseudofossils that exhibit such carbonaceous uniseriate cell-like structure are unknown from the entire geological record, reported not even from petroleum- or anthraxolite-rich

Fig. 14 (continued) rectangle outlining the part shown in (**d–i**). (**d**) Three-dimensional Raman image showing the cylindrical structure of the kerogenous filament (*gray*) infilled by permineralizing quartz (*white*). (**e–i**) Two-dimensional Raman images of the part of the filament shown in (**d**) acquired at sequential depths below the filament surface (**e**, at 0.75 μm; **f**, 1.5 μm; **g**, 2.25 μm; **h**, 3.0 μm; **i**, 3.75 μm) demonstrating that it is composed of quartz-filled cell lumina (black "voids" denoted by the arrows in e, evident also in **f–i**) defined by kerogenous cell walls (*white*). (**j**) Photomicrograph of the upper surface of the thin section showing that the specimen (*black outline*) is embedded in a chert matrix composed of irregularly shaped quartz grains (*arrows*). (**k–n**) CLSM images of the filament at sequential depths below the thin section surface (**k**, at 4 μm; **l**, 5 μm; **m**, 6 μm; **n**, 7 μm). Heating of the specimen-containing ~150-μm-thick section during its remounting at the Natural History Museum, London (P. Hayes, pers. comm. to J.W.S. 2005) separated quartz grains at its upper surface that permitted microscopy immersion oil to permeate at grain boundaries to a depth of ~7 μm within the section. This separation enabled imaging of the outlines of quartz grains at the section surface without the use of polarized optics (**j**) and the fluorescence emission of the permeating oil permitted CLSM imaging of grain margins within the upper few microns of the section. Arrows in (**k, l**) point to oil-filled grain boundaries that transect the uppermost (4- to 5-μm-deep) part of the filament; ellipses in (**l–n**) denote deeper parts of the filament (cf. **g–i**) to which fluorescent oil permeated only partially deposits where they might be expected to be abundant. As this example shows, CLSM and Raman imagery, together, can provide important insight into both the taphonomy and the biogenicity of ancient microscopic fossils.

7 Summary

Cellularly mineralized fossils are among the biologically and taphonomically most informative known from the geological record. Spanning all of biology, from metazoans and vascular plants to algae, fungi and bacteria, such fossils can be preserved in exquisite detail. Two techniques newly introduced to paleobiology and documented here, three-dimensional confocal laser scanning microscopy (CLSM) and two- and three-dimensional Raman imagery, provide a means to establish the three-dimensional morphology as well as the molecular-structural composition and geochemical maturity of the carbonaceous kerogen that comprises such fossils.

Illustrated here is the use of these techniques to elucidate the preserved anatomy and cellular structure of examples of all of the major biologic groups (animals, plants, fungi, algal protists, and microbes), preserved in the three principal cellularly mineralizing rock types (phosphorite, chert, and carbonate). As is shown, CLSM and Raman imagery, together, can provide new information about the morphology, cellular anatomy, taphonomy, carbonaceous composition, and geochemical maturity of organic-walled mineralized fossils, and Raman imagery can be used as well to document the mineralogy of the fossil-enclosing matrix and the spatial relations between such fossils and their embedding minerals. Together, the two techniques can provide definitive evidence of the sequence of taphonomic events involved in such preservation (exemplified here by the study of a phosphate-mineralized ctenophore embryo) and the biological degradation of diverse organically preserved specimens (shown by the fungal infestation of plant axes, the enzymatic breakdown of the middle lamellae of vascular plant cell walls, and the preferential decay of specific cell wall components in fossil cyanobacteria). Similarly, the data presented that permit comparison of chert-permineralized Phanerozoic plants and Precambrian microbes, and of large-celled and small-celled organic-walled mineralized microfossils – coupled with the in situ measurements of the geochemical maturity of their kerogenous constituents afforded by Raman spectroscopy – provide new means for assessment of the biases of such preservation over time. Taken together, these non-intrusive and non-destructive techniques can provide important new knowledge of ancient fossils and the history of life.

Acknowledgments We thank D.E.G. Briggs, J. Shen-Miller, and the editors of this volume for helpful comments on the manuscript. The participation of A.B.K. in this work was supported by CSEOL, the IGPP Center for Study of the Origin and Evolution of Life at UCLA, and by the UCLA administration in support of UCLA's membership in the NASA Astrobiology Institute. Both A.D.C. (supported in part during these studies by a pre-doctoral NSF Fellowship) and A.B.T. are recent recipients of Ph.D. degrees from UCLA, supported during their graduate studies by CSEOL Fellowships and by the principal source of funding for this work, CSEOL and NASA Exobiology Grant NAG5-12357 (to J.W.S).

References

Altermann, W. (2005). The 3.5 Ga apex fossil assemblage? Consequences of an enduring discussion. ISSOL'05, International Society of Study Origin Life Triennial Mtg., Beijing, Program Abstracts, pp. 136–137 [abstract].

Altermann, W., Kazmierczak, J., Oren, A., & Wright, D. T. (2006). Cyanobacterial calcification and its rock-building potential during 3.5 billion years of Earth history. *Geobiology, 4*, 147–166.

Amos, W. B., & White, J. G. (2003). How the confocal laser scanning microscope entered biological research. *Biology of the Cell, 95*, 335–342.

Arnold, C. A., & Daugherty, L. H. (1964). A fossil dennstaedtioid fern from the Eocene Clarno Formation of Oregon. *Contributions from the Museum of Paleontology, University of Michigan, 19*, 65–88.

Arouri, K. R., Greenwood, P. F., & Walter, M. R. (2000). Biological affinities of Neoproterozoic acritarchs from Australia: Microscopic and chemical characterization. *Organic Geochemistry, 31*, 75–89.

Bengtson, S., & Zhao, Y. (1997). Fossilized metazoan embryos from the earliest Cambrian. *Science, 277*, 1645–1648.

Bloeser, B., Schopf, J. W., Horodyski, H. J., & Breed, W. J. (1977). Chitinozoans from the Late Precambrian Chuar Group, Arizona. *Science, 195*, 676–679.

Brasier, M. D., Green, O. R., Jephcoat, A. P., Kleppe, A. K., Van Kranendonk, M. J., Lindsay, J. F., et al. (2002). Questioning the evidence of Earth's oldest fossils. *Nature, 416*, 76–81.

Brasier, M. D., Green, O. R., Lindsay, J. F., McLoughlin, N., Steele, A., & Stoakes, C. (2005). Critical testing of Earth's putative fossil assemblage from the 3.5 Ga Apex chert, Chinaman Creek, Western Australia. *Precam Res, 140*, 55–102.

Briggs, D. E. G. (2003). The role of decay and mineralization in the preservation of soft-bodied fossils. *Annual Review of Earth and Planetary Sciences, 31*, 275–301.

Briggs, D. E. G., Raiswell, R., Bottrell, S. H., Hatfield, D., & Bartels, C. (1996). Controls on the pyritization of exceptionally preserved fossils: An analysis of the Lower Devonian Hunsrück Slate of Germany. *American Journal of Science, 296*, 633–663.

Briggs, D. E. G., Moore, R. A., Shultz, J. W., & Schweigert, G. (2005). Mineralization of soft-part anatomy and invading microbes in the horseshoe crab *Mesolimulus* from the Upper Jurassic Lagerstätte of Nusplingen, Germany. *Proceedings of Royal Society of London, B272*, 727–632.

Chen, J.-Y. (2004). *The dawn of animal world*. Nanjing, China: Jiangsu Science and Technology.

Chen, J.-Y., Oliveri, P., Li, C.-W., Zhou, G.-Q., Gao, F., Hagadorn, J. W., et al. (2000). Precambrian animal diversity: Putative phosphatized embryos from the Doushantuo Formation of China. *Proceedings of the National Academy of Sciences of the United States of America, 97*, 4457–4462.

Chen, J.-Y., Braun, A. Waloszek, D., Peng, Q.-Q., & Maas, A. (2004). Lower yolk-pyramid embryos from Southern Shaanxi, China. In J. Zhu & M. Steiner (Eds.), *Biological and geological processes of the Cambrian explosion, progress in natural science*, Special Issue, 2004.

Chen, J.-Y., Bottjer, D. J., Davidson, E. H., Dornbos, S. Q., Gao, X., Yang, J.-H., et al. (2006). Phosphatized polar lobe-forming embryos from the Precambrian of southwest China. *Science, 312*, 1644–1646.

Chen, J.-Y., Schopf, J. W., Bottjer, D. J., Zhang, C.-Y., Kudryavtsev, A. B., Tripathi, A. B., et al. (2007). Raman spectra of a Lower Cambrian ctenophore embryo from southwestern Shaanxi, China. *Proceedings of the National Academy of Sciences of the United States of America, 104*, 6289–6292.

Chi, H. M., Xiao, Z. D., Fuk, D. G., & Lu, Z. H. (2006). Analysis of fluorescence from algae fossils of the Neoproteozoic Doushantuo Formation of China by confocal laser scanning microscopy. *Microscopic Research Technique, 69*, 253–259.

Cisne, J. L. (1974). Trilobites and the origin of arthropods. *Science, 186*, 13–18.

De Gregorio, B. T., & Sharp T. G. (2003). Determining the biogenicity of microfossils in the Apex chert, Western Australia, using transmission electron microscopy. *Lunar and Planetary Science, XXXIV*, 1267 [abstract].

De Gregorio, B. T., & Sharp, T. G. (2006). The structure and distribution of carbon in 3.5 Ga Apex chert: Implications for the biogenicity of Earth's oldest putative microfossils. *American Mineralogist, 91*, 784–789.

De Gregorio, B. T., Sharp, T. G., & Flynn, G. F. (2005). A comparison of the structure and bonding of carbon in Apex chert kerogenous material and Fischer-Tropsch-Type carbons. *Lunar and Planetary Science, XXXVI*, 1866.

Dean, B. (1902). The preservation of muscle fibers in sharks of the Cleveland shale. *The American Geologist, 30*, 273–278.

Dietrich, D., Witke, K., Röbler, R., & Marx, G. (2001). Raman spectroscopy of *Psaronius* sp.: A contribution to the understanding of the permineralization process. *Applied Surface Science, 179*, 230–233.

Drews, G. (1973). Fine structure and chemical composition of the cell envelopes. In N. G. Carr & B. A. Whitton (Eds.), *The biology of blue-green algae, Botanical Monographs* (Vol. 9). CA, Berkeley: University of California Press.

Feist-Burkhardt, S., & Monteil, E. (2001). Gonyaulacacean dinoflagellate cysts with multi-plate precingular archaeopyle. *Neues Jahrbuch für Geologie und Paläontologie, 219*, 33–81.

Feist-Burkhardt, S., & Pröss, J. (1999). Morphological analysis and description of middle Jurassic dinoflagellate cyst marker species using confocal laser scanning microscopy, digital optical microscopy and conventional light microscopy. *Bull du Centre de Recherche Elf Exploring, 22*, 103–145.

Foster, B., Williams, V. E., Witmer, R. J., & Piel, K. M. (1990). Confocal microscopy: A new technique for imaging micro-organisms and morphology in three-dimensions. *Palynology, 14*, 212 [abstract].

Frank, H., Lefort, M., & Martin, H. H. (1962). Elektronenoptische und chemische Untersuchungen an Zellwäden der Blaualgen, *Phormidium unicinatum. Zeitschrift für Naturforschung B, 17*, 262–268.

Grimes, S. T., Davies, K. L., Butler, I. B., Brock, F., Edwards, D., Rickard, D., et al. (2002). Fossil plants from the Eocene London Clay: The use of pyrite textures to determine the mechanism of pyritisation. *Journal of Geological Society, 159*, 493–501.

Halfen, L. N., & Castenholz, R. W. (1971). Gliding motility in the blue-green alga *Oscillatoria princeps. J Phycol, 7*, 133–145.

Hochuli, P., & Feist-Burkhardt, S. (2004). An early boreal cradle of Angiosperms? Angiosperm-like pollen from the Middle Triassic of the Barents Sea (Norway). *Journal of Micropalaeontology, 23*, 97–104.

Jehlička, J., & Beny, C. (1992). Application of Raman microspectrometry in the study of structural changes in Precambrian kerogens during regional metamorphism. *Organic Geochemistry, 18*, 211–213.

Jehlička, J., Urban, A., & Pokorny, J. (2003). Raman spectroscopy of carbon and solid bitumens in sedimentary and metamorphic rocks. *Spectrochimica Acta, A59*, 2341–2352.

Kaufman, A. J., & Xiao, S. (2003). High CO_2 levels in the Proterozoic atmosphere estimated from analyses of individual microfossils. *Nature, 425*, 279–282.

Kelemen, S. R., & Fung, H. L. (2001). Maturity trends in Raman spectra from kerogen and coal. *Energy and Fuels, 15*, 653–658.

Kidston, R., & Lang, W. H. (1917). On Old Red Sandstone plants showing structure from the Rhynie Chert Bed, Aberdeenshire, part 1: Rhynie *Gwynne-vaughani,* K. and L. *Transactions on Royal Society of Edinburgh, 52*, 761–784.

Kudryavtsev, A. B., Schopf, J. W., Agresti, D. G., & Wdowiak, T. J. (2001). *In situ* laser-Raman imagery of Precambrian microscopic fossils. *Proceedings of the National Academy of Sciences of the United States of America, 98*, 823–826.

Lamont, H. C. (1969). Sacrificial cell death and trichome breakage in an oscillatoriacean blue-green alga – the role of murein. *Archiv für Mikrobiologie, 69*, 237–259.

Marshall, C. P., Javaux, E. J., Knoll, A. H., & Walter, M. R. (2005). Combined micro-Fourier transform infrared (FTIR) spectroscopy and micro-Raman spectroscopy of Proterozoic acritarchs: A new approach to paleobiology. *Precambrian Research, 138*, 208–224.

McKeegan, K. D., Kudryavtsev, A. B., & Schopf, J. W. (2007). Raman and ion microscopic imagery of graphitic inclusions in apatite from >3830 Ma Akilia supracrustals, West Greenland. *Geology, 35*, 591–594.

McMillan, P. F., & Hofmeister, A. M. (1988). Infrared and Raman spectroscopy. *Reviews in Mineralogy, 18*, 99–159.

Moorbath, S. (2005). Dating earliest life. *Nature, 434*, 155.

Mus, M. M., & Moczydlodłowska, M. (2000). Internal morphology and taphonomic history of the Neoproterozoic vase-shaped microfossils from Visingsö Group, Sweden. *Norsk Geologisk Tidsskrift, 80*, 213–228.

Nestler, K., Dietrich, D., Witke, K., Rößler, R., & Marx, G. (2003). Thermogravimetric and Raman spectroscopic investigations on different coals in comparison to dispersed anthracite found in permineralized tree fern *Psaronius* sp. *Journal of Molecular Structure, 661–662*, 357–362.

Nix, T., & Feist-Burkhardt, S. (2003). New methods applied to the microstructure analysis of Messel shale: Confocal laser scanning microscopy (CLSM) and environmental scanning electron microscopy (ESEM). *Geological Magazine, 140*, 469–478.

Pankratz, H. S., & Bowen, C. C. (1963). Cytology of blue-green algae I. The cells of *Symploca muscorum*. *American Journal of Botany, 50*, 387–399.

Pasteris, J. D., & Wopenka, B. (1991). Raman spectra of graphite as indicators of degree of metamorphism. *The Canadian Mineralogist, 29*, 1–9.

Roberts, S., Tricker, P. M., & Marshall, J. E. A. (1995). Raman spectrometry of chitinozoans as a maturation indicator. *Organic Geochemistry, 23*, 223–238.

Scheckler, S. E. (1986). Geology, floristics and paleoecology of Late Devonian coal swamps from Appalachian Laurentia. *Annales Société Géologique de Belgique, 109*, 209–222.

Scheckler, S. E., & Banks, H. P. (1971). Anatomy and relations of some Devonian progymnosperms from New York. *American Journal of Botany, 58*, 737–751.

Schopf, J. M. (1941). Contributions to Pennsylvanian paleobotany. *Mazocarpon oedipternum* sp. nov. and sigillarian relationships. *Illinois State Geological Survey Report of Investigations, 75*, 1–53.

Schopf, J. W. (1968). Microflora of the Bitter Springs Formation, Late Precambrian, central Australia. *Journal of Paleontology, 42*, 651–688.

Schopf, J. M. (1975). Modes of fossil preservation. *Review of Palaeobotany and Palynology, 20*, 27–53.

Schopf, J. W. (1992). Evolution of the Proterozoic biosphere: Benchmarks, tempo, and mode. In J. W. Schopf & C. Klein (Eds.), *The Proterozoic biosphere, a multidisciplinary study*. New York: Cambridge University Press.

Schopf, J. W. (1993). Microfossils of the Early Archean Apex chert: New evidence of the antiquity of life. *Science, 260*, 640–646.

Schopf, J. W. (1999). *Cradle of life: The discovery of Earth's earliest fossils*. Princeton, NJ: Princeton University Press.

Schopf, J. W. (2004a). Earth's earliest biosphere: Status of the hunt. In P. G. Eriksson, W. Altermann, D. R. Nelson, W. U. Mueller, O. Cateneanu (Eds.), *The Precambrian earth: Tempos and events, Developments in Precambrian Geology 12*, Amsterdam: Elsevier.

Schopf, J. W. (2004b). Geochemical and submicron-scale morphologic analyses of individual Precambrian microorganisms. In R. J. Hill, Z. Aizenshtat, M. J. Baedecker, G. Claypool, R. Eanhouse, R. Goldhaber, M. Goldhaber, J. Lenventhal, & K. Peters (Eds.), *Geochemical investigation in earth and space science, Publication No. 6*. St. Louis, MO: The Geochemical Society.

Schopf, J. W. (2006a). Fossil evidence of Archaean life. *Philosophical Transactions of Royal Society of London B, 361*, 869–885.

Schopf, J. W. (2006b). The first billion years: When did life emerge? *Elements, 2*, 299–233.

Schopf, J. W., & Blacic, J. M. (1971). New microorganisms from the Bitter Springs Formation (late Precambrian) of the north-central Amadeus Basin, Australia. *Journal of Paleontology, 45*, 925–960.

Schopf, J. W., & Kudryavtsev, A. B. (2005). Three-dimensional imagery of Precambrian microscopic organisms. *Geobiology, 3*, 1–12.

Schopf, J. W., & Sovietov, Y. K. (1976). Microfossils in *Conophyton* from the Soviet Union and their bearing on Precambrian biostratigraphy. *Science, 193*, 143–146.

Schopf, J. W., Kudryavtsev, A. B., Agresti, D. G., Wdowiak, T. J., & Czaja, A. D. (2002). Laser-Raman imagery of Earth's earliest fossils. *Nature, 416*, 73–76.

Schopf, J. W., Kudryavtsev, A. B., Agresti, D. G., Czaja, A. D., & Wdowiak, T. J. (2005). Raman imagery: A new approach to assess the geochemical maturity and biogenicity of permineralized Precambrian fossils. *Astrobiology, 5*, 333–371.

Schopf, J. W., Tripathi, A. B., & Kudryavtsev, A. B. (2006). Three-dimensional confocal optical imagery of Precambrian microscopic organisms. *Astrobiology, 6*, 1–16.

Schopf, J. W., Kudryavtsev, A. B., Czaja, A. D., & Tripathi, A. B. (2006b). Three-dimensional morphological (CLSM) and chemical (Raman) imagery of permineralized plants and organic-walled microorganisms. *Prog Ann Mtg Bot Soc Amer, Chico, California* (p. 171) [abstract].

Schopf, J. W., Kudryavtsev, A. B., Czaja, A. D., & Tripathi, A. B. (2007). Evidence of Archean life: Stromatolites and microfossils. *Precambrian Research, 158*, 141–155.

Scott, A. C., & Hemsley, A. R. (1990). A comparison of new microscopical techniques for the study of fossil spore wall ultrastructure. *Review of Palaeobotany and Palynology, 67*, 133–139.

Scott, A. C., Mattey, D. P., & Howard, R. (1996). New data on the formation of Carboniferous coal balls. *Review of Palaeobotany and Palynology, 93*, 317–31.

Spötl, C., Houseknecht, D. W., & Jaques, R. C. (1998). Kerogen maturation and incipient graphitization of hydrocarbon source rocks in the Arkoma Basin, Oklahoma and Arkansas: A combined petrographic and Raman study. *Organic Geochemistry, 28*, 535–542.

Steiner, M., Zhu, M., Li, G., Quian, Y., & Erdtmann, B.-D. (2004). New Early Cambrian bilaterian embryos and larvae from China. *Geology, 32*, 833–836.

Steiner, M., Li, G., Quian, Y., & Erdtmann, B.-D. (2004). Lower Cambrian small shelly faunas from Zhejiang (China), and their biostratigraphic importance. *Geobios, 37*, 59–275.

Strauss, H., & Moore, T. B. (1992). Abundances and isotopic compositions of carbon and sulfur species in whole rock and kerogen samples. In J. W. Schopf & C. Klein (Eds.), *The Proterozoic biosphere, a multidisciplinary study*. New York: Cambridge University Press.

Stuermer, W. (1970). Soft parts of cephalopods and trilobites: Some surprising results of X-ray examination of Devonian slates. *Science, 170*, 1300–1302.

Stuermer, W., & Bergström, J. (1973). New discoveries on trilobites by X-rays. *Paläontologishe Zeitscrift, 47*, 104–141.

Talyzina, N. M. (1997). Fluorescence intensity in early Cambrian acritarchs from Estonia. *Review of Palaeobotany and Palynology, 100*, 99–108.

Taylor, T. N., & Remy, W. H. H. (1992). Fungi from the Lower Devonian Rhynie Chert – Chytridiomycetes. *American Journal of Botany, 79*, 1233–1241.

Taylor, T. N., & Taylor, E. L. (1993). *The biology and evolution of fossil plants*. New York: Prentice Hall.

Tripathi, A. B. (2007). *Three-dimensional confocal imagery and spectral analysis of ancient cellularly preserved fossils*. Ph.D. dissertation, Department of Earth and Space Sciences, University of California, Los Angeles.

Van Baalen, C., & Brown, R. M., Jr. (1969). The ultrastructure of the marine blue-green alga *Trichodesmium erythraeum*, with special reference to the cell wall, gas vacuoles, and cylindrical bodies. *Archiv für Mikrobiologie, 69*, 79–91.

Vandenbroucke, M., & Largeau, C. (2007). Kerogen origin, evolution and structure. *Organic Geochemistry, 38*, 719–833.

White, C. A. (1893). The character and origin of fossil remains. *Smithsonian Institution, Annual Report for the year ending June 3, 1982, Report of the US National Museum 245(368)*, 251–267.

Williams, K. P. J., Nelson, J., & Dyer, S. (1997). *The Renishaw Raman database of gemological and mineralogical materials*. Gloucestershire, England: Renishaw Tranducers Systems Division.

Wopenka, B., & Pasteris, J. D. (1993). Structural characterization of kerogens to granulite-facies graphite: Applicability of Raman microprobe spectroscopy. *The American Mineralogist, 78*, 533–557.

Xiao, S.-H., & Knoll, A. H. (2000). Phosphatized animal embryos from the Neoproterozoic Doushantuo Formation at Weng'an, Guizhou, South China. *Journal of Paleontology, 74*, 767–788.

Xiao, S.-H., Zhang, Y., & Knoll, A. H. (1998). Three-dimensional preservation of algae and animal embryos in a Neoproterozoic phosphorite. *Nature, 391*, 553–558.

Yui, T.-F., Huang, E., & Xu, J. (1996). Raman spectrum of carbonaceous material: A possible metamorphic grade indicator for low-grade metamorphic rocks. *Journal of Metamorphic Geology, 14*, 115–124.

Chapter 14
Taphonomy in Temporally Unique Settings: An Environmental Traverse in Search of the Earliest Life on Earth

Martin D. Brasier, David Wacey, and Nicola McLoughlin

Contents

Abstract There is an apparent preservational paradox in the early rock record. Cellularly preserved and ensheathed microfossils which are remarkably preserved from the late Archaean (*c*.2700 Ma) onward, have rarely been found in the earlier rock record and when they are their biogenicity is debated. Likewise, the abundance and morphological complexity of stromatolites appears much reduced in the early Archaean and even these lack compelling associations with organic remains of microbial mats. This 'preservational dark age' may have arisen because microfossils and

M.D. Brasier (✉)
Department of Earth Sciences, Oxford University, Parks Road, Oxford OX1 3PR, UK
e-mail: martinbrasier@yahoo.co.uk

D. Wacey
Centre for Microscopy, Characterisation and Analysis + School of Earth and Environment, The University of Western Australia, 35 Stirling Highway, Crawley, WA 6009, Perth, Australia

N. McLoughlin
Department of Earth Sciences and centre of Excellence in Geobiology, University of Bergen, 5020 Bergen, Norway

P.A. Allison and D.J. Bottjer (eds.), *Taphonomy: Process and Bias Through Time*, Topics in Geobiology 32, DOI 10.1007/978-90-481-8643-3_14,

microbial mats were absent, because conditions for their preservation were rare or, as we suggest here, because scientists have largely been looking in the wrong places.

To illustrate the potential of looking far beyond 'chertified Bahamian lagoons', we make a traverse across the key potential habitats for early life on Earth and identify some exciting and new taphonomic windows, in the search for Earth's earliest microfossils, trace fossils and stromatolites. Such habitats include hitherto little explored pillow lavas, hydrothermal vents and beach sandstones. These new windows are already starting to provide surprising insights into the nature of the earliest vital processes.

1 Introduction: A Preservational Dark Age?

The fossil record of the Archaean, the interval of time before 2500 Ma BP, is a preservational paradox. Promising rocks such as isotopically light carbonaceous cherts are widespread but signals of life are enigmatic and hard to decipher, creating a so-called 'preservational dark age' within the fossil record. This is surprising given the high fidelity of the younger, Proterozoic (c. 2500–542 Ma) microfossil record in cherts and carbonates (e.g., Knoll 2003; Brasier and Armstrong 2005) and the ostensible ease with which microbes can be silicified in modern settings (e.g., Konhauser et al. 2003). A conventional explanation for this paradox has been the relatively low abundance of ancient rocks, most of which have been consumed or greatly modified by erosion, subduction or metamorphism over the last 3 billion years (e.g., Schopf 1999). But against this, one may argue that remarkable cellular preservation is not an unreasonable expectation of the early Archaean rock record. This is because many of the conditions necessary for preservation would seem to have been prevalent at this time. For example, low levels of atmospheric oxygen, abundant carbonaceous matter and high levels of silica supersaturation all seem to have been the norm in the early Archaean. Together, these should have helped to deliver a plethora of respectable morphological remains and chemical signatures for life in rocks of this age. So what, exactly, has been the problem?

It has been argued (Brasier et al. 2005, 2006) that most reports of early microfossils (e.g., Schopf 1999) and stromatolites (e.g., Hoffman et al. 1999; Allwood et al. 2006) are not readily distinguishable from self-organizing structures (SOS) and have yet to pass the null hypothesis of Brasier et al. (2002, 2004). This hypothesis states that microfossils and stromatolite-like structures of early Archaean age should not be accepted as being of biological origin until appropriate hypotheses for their abiogenic origin have been tested and falsified (see also Grotzinger and Rothman 1996; Brasier et al. 2005 and references therein). Although there have been many reports (e.g., reviews in Schopf 2006; Brasier et al. 2006), it emerges that rigorously tested examples of cellular preservation from the early Archaean Dark Age are scarce and still widely debated. In addition, abiotic scenarios capable of replicating many of the candidate geochemical signatures for life in these earliest rocks have not been entirely excluded (e.g., Van Zuilen et al. 2002). Thus, there is as yet, no consensus as to the oldest verifiable evidence of life on Earth and many of the existing claims need further analysis and testing.

One explanation for this poverty of the early cellular fossil record is that, until recently, we may have been applying search images that are too restrictive. Armed with misleading questions, it has become easy to overlook more favourable habitats and taphonomic windows. The traditional focus on Archaean cherts and silicified sediments has, for instance, meant that informative lithological windows such as volcanic glasses, siliciclastics and pyritic deposits have been relatively neglected. We argue that these taphonomic windows may yet help us to fill the many gaps in our knowledge about the origins and history of life on Earth.

2 Early Eden or Distant Planet?

Our understanding of Early Archaean Earth environments greatly shapes the strategies adopted for seeking the earliest evidence of life on Earth. A conventional model for Archaean surface environments is one that can be termed the Early Eden Hypothesis (Brasier et al. 2004). This hypothesis, which has dominated thinking for several decades, takes familiar and habitable environments in which primitive microbes abound today, such as Bahamian tropical lagoons or Shark Bay in Western Australia, and then uses these to make predictions about the surface of the Early Earth. This is, of course, a tried and tested method – the so-called Uniformitarian Principle – advanced by Sir Charles Lyell (1830). This uniformitarian method can be argued to work reasonably well when applied to the rock record from the Quaternary back into the Proterozoic (2500–1600 Ma BP) and even as early as the late Archaean (3000–2500 Ma BP). However, the Principle of Uniformity can be pushed beyond its limits when extended back into the earliest Archaean. In its most extreme expression, the Early Eden Hypothesis predicts the presence, on the early Earth, of continents, subduction zones, carbonate platforms, an oxygenated atmosphere and oxygenic photosynthesis. Examination of the earliest sedimentary rocks, however, coupled with an ever-increasing understanding of the nature of the solar system, suggests that Lyell's much vaunted Principle of Uniformity may lead towards mistaken conclusions (see Rose et al. 2006). It is useful to remember the warning of Sir Francis Bacon here: "The subtlety of nature is greater many times over than the subtlety of the senses and understanding; so that all those specious meditations, speculations, and glosses in which men indulge are quite from the purpose, only there is no one by to observe it" (Bacon 1620). In other words, we need to remain aware of the huge gaps in our understanding at this time. To encourage this caution we recommend that all scientists view the young Earth as though it were a distant planet.

Once we take this rather unwelcome monster on board, we can see that the early Earth may have been stranger than we imagine and, perhaps, stranger than we *can* imagine. Consider, for example, the following list of conditions is now thought by many to pertain at the surface of the Earth in the early Archaean: solar luminosity some 20% lower than now (e.g., Sagan and Mullen 1972); an atmosphere of reducing gases that largely lacked oxygen (e.g., Kasting and Catling

2003; Lowe and Tice 2004); no ozone layer to protect life from ultraviolet light (e.g., Konhauser et al. 2001); much higher rates of solar and cosmic rays (e.g., Delsemme 1998); high rates of meteoritic bombardment, with many over 10km in diameter (e.g., Byerly et al. 2002; Moorbath 2005); a lack of large continental landmasses (e.g., Lindsay and Brasier 2002); a hot young crust, with higher rates of heat flux and hotter oceans (e.g., Knauth and Lowe 2003; Knoll 2003); the predominance of oceanic crust over granitic crust (Lowe 1994b) and a lack of extensive, modern style subduction zones and crustal recycling (McCall 2003; Van Kranendonk et al. 2004).

Given these radically different boundary conditions acting upon the early Earth, it appears that the planet's endogenic energy was potentially a much greater source for the early biosphere than was the solar energy of our star, the sun. A first consequence is that the highly metaliferous crust of the early Earth, when combined with enormous outflows of energy emanating from hydrothermal and volcanic systems, is likely to have played a significant role in both the genesis and sustenance of the earliest forms of life. This message is also delivered to us by the discovery of thriving life forms around black smokers and modern deep-sea vents (e.g. Jannasch and Mottl 1985; Teske et al. 2002). In addition, theoretical and chemical studies have certainly confirmed that a 'hydrothermal cradle for life' is indeed plausible (see Nisbet and Fowler 1996). A second consequence of this view of the early Earth as a distant planet is that oxygenic photosynthesis need not have been the foundation for all other forms of life as it might seem to us today. We will return to many of these concepts below.

3 New Taphonomic Windows for Old

For a generation, conventional wisdom has encouraged us to search for the earliest cells within bedded siliceous sediments such as Banded Iron Formations (BIF) and related lithologies (e.g., Schopf and Klein 1992). Such cherts do, indeed, have an excellent track record, that ranges from the exquisitely preserved cells of microbes and early land floras in the Lower Devonian Rhynie Chert of Scotland (Trewin and Rice 2004) to the microbial assemblages of the 1900 Ma Gunflint Chert of Canada (Barghoorn and Tyler 1965). In both those settings, silica supersaturation appears to have been achieved as the consequence of high levels of dissolved silica coupled to low levels of biological silica extraction (Maliva et al. 2005). Preservation of cellular fossils has then been achieved by their immuration within glassy silica derived from the surrounding environment, either during life or soon after death (cf. Konhauser et al. 2003).

The poor cellular fossil record of the early to middle Archaean (3500–3000 Ma) therefore appears puzzling, given that silica supersaturation was common within the water column (cf. Maliva et al. 2005). One possible explanation for this (Brasier et al. 2005, 2006), is that the post-depositional history of these sedimentary cherts is less simple than was at first believed (cf. Schopf 1992a, b, 1993).

This becomes clear when the lithogenesis of these cherts is mapped out on scales that range from microns to kilometres (e.g. Brasier et al. 2002, 2005). Most importantly here, we find that the original sedimentary protolith which might be expected to contain the indigenous cellular fossils has typically been modified, rather drastically, in one or more of the following ways after burial: by remobilization of the silica (especially of carbonaceous cherts); recrystallization of the cryptocrystalline silica components; replacement of one silica phase by another; dilation, displacement and intrusion of the protolith by many subsequent siliceous phases, some of which may be quite young in age; metamorphic modification of the silica, carbon and other phases; and finally, modification of silica and carbon phases during the prolonged episodes of weathering and exposure in near surface environments (see Brasier et al. 2005).

Unfortunately, such a convoluted diagenetic history now appears to have been typical for nearly all banded sedimentary cherts of Archaean age. A pre-requisite for finding remains of the earliest life in such rocks is, therefore, to attempt to map out, date and distinguish each of the silica and other mineral phases within the host rock. This requires time-consuming macro- and micro- scale mapping and stratigraphy and such a program of work is only just beginning (see for example: Brasier et al. 2002, 2005; Tice and Lowe 2006). When adopted, this approach has revealed that some putative microfossil like structures, once widely accepted (e.g. Schopf 1993) are not actually located within the primary protolith at all, but reside in later, probably much younger, post-depositional phases.

There are therefore, a great many concerns regarding the veracity of the earliest fossil record. Even so, there is a clear way forward – but only if we are prepared to search for new taphonomic windows onto the early Earth. In subsequent sections, we describe several rock types in which the post mortem histories are potentially much less complicated and much better preserved than more conventional materials, so that there is a reasonable hope of discovering, and of constraining, some of the earliest signs of life on Earth. Three lithologies or taphonomic windows now appear especially promising in this respect: the formerly glassy margins of early Archaean pillow basalts (Furnes et al. 2004; Banerjee et al. 2006); the pyritic layers within hydrothermal black smoker deposits (Rasmussen 2000); and the clasts and matrix of the earliest beach sediments, comprising quartzose and pyritic sandstones (Brasier et al. 2006; Wacey et al. 2006, 2008). It is from within these newly explored habitats, as we explain below, that the *nature* of the earliest life now seems likely to emerge.

4 Cellular Lagerstätten

The cell is the fundamental unit of life. The eminent naturalist Jean Baptiste Lamarck (1809) discovered this major truth, some 150 years after Robert Hooke (1665) had first described both living and fossil cells. Arguably, many of the most fundamental steps in evolution have taken place at the cellular level (e.g. Cavalier-Smith et al. 2006).

We here focus on the most primitive type of cellular organisms known as the prokaryotes, which predate unequivocal eukaryotic cells in the fossil record by perhaps 2000 Ma or more. Prokaryotes are distinguished from the more advanced eukaryotes (e.g. algae) by their lack of cellular organelles, including the nucleus.

There must have been many crucial pre-cellular steps leading towards the origins of life and the first prokaryotes. These steps are likely to have included the development of an information transfer mechanism (e.g. Cairns-Smith 1985) and the appearance of a cell wall to hold and concentrate the prebiotic chemicals (e.g. Hanczyc et al. 2003). Locating these prebiotic precursors in the rock record is difficult and has not yet been attempted. This means that we are currently required to focus entirely upon the emergence of cells themselves. The cell performs three vital roles that help to sustain life. The cell wall provides a compartment in which chemical reactions can be concentrated and controlled and biological products can be stored. The intra-cellular chromosomes are made of DNA, which acts as the information store for living cells and reproduction. As a whole, the cell participates in metabolic processes – chemical reactions – that sustain the cell. These three actions – compartmentalization, reproduction and metabolism – may have evolved separately, but they are together responsible for the enormous success of the cellular unit.

The preservation potential of each of these three features of the cell is rather different. The products of metabolic processes arguably have the highest chance of preservation. Although these processes may have little morphological expression, they must inevitably modify the chemistry in and around the site of life. It is these chemical signatures that can be preserved. Typical examples of this include metabolic fractionation of isotopes such as $^{13}C/^{12}C$ (Schidlowski 2001) and/or $^{34}S/^{32}S$ (Shen et al. 2001). To verify such biosignatures in the rock record, however, it is necessary to discount the role of fractionations arising from purely abiogenic processes. Plausible abiogenic processes may involve so-called Fischer-Tropsch type reactions for the fractionation of carbon isotopes (Sherwood Lollar et al. 2002; Horita and Berndt 1999; McCollum and Seewald 2006), or hydrothermal and photochemical fractionations of sulfur isotopes (cf. Grassineau et al. 2001). Other key indicators of cellular metabolic processes may involve, for example, the highly localized storage of biologically-significant, or even biolimiting elements. Enrichments in nitrogen and phosphorus, as well as Fe, Co, V, Mo and other trace elements are now being identified, within cellular bodies using high-resolution techniques such as nanoSIMS (e.g. Robert et al. 2005; Oehler et al. 2006; Wacey et al. 2008).

The characteristic of prokaryote cells with the lowest chances of preservation is that of reproduction and its associated reproductive apparatus. This may be because RNA and DNA molecules are intrinsically unstable and are readily degraded over geological timescales by heat and pressure. And while there are examples of nuclear preservation in eukaryotes, the nucleus is absent from the prokaryotes under discussion here.

The cell membrane in the early fossil record seems to have only a low to intermediate chance of preservation. The cell membrane of bacteria is largely composed of a mureine which, although tougher than the phospholipids of higher plants, is weaker than cellulose and can be readily degraded, albeit less rapidly than the cell contents themselves. A morphological record of cell contents and membranes therefore appears unlikely, but a chemical expression of these cellular components may nonetheless survive in the rock record. A promising avenue for research involves the use of 'molecular fossils' – cell membrane lipids which can be preserved as soluble hydrocarbons in sediments. Where sediments are sufficiently well preserved, these hydrocarbons may yet indicate the former presence of specific groups of organisms, such as cyanobacteria (e.g., Summons et al. 1999).

For bacterial cells to preserve, there is therefore a requirement for rapid immuration of the cell wall within the preservational medium. This medium can include glassy silica gel (e.g., the 1900 Ma old Gunflint Chert), iron sulfide (e.g., the 3200 Ma Sulfur Springs deposit), iron oxide (e.g., *Galionella* in modern hot-springs) and calcium phosphate (e.g., Doushantuo Formation; see Brasier et al. this volume). This immuration may be a consequence of the metabolic processes within the cell itself. For example, encrustation with a mineral precipitate may act as a UV shield for the organism (cf. Phoenix et al. 2006); or serve to increase the proton motive force across the cell membrane, as with some iron oxidising bacteria (e.g. Chan et al. 2004). Conversely, the precipitate may be deleterious to the cell, restricting the diffusion of reactants and waste products to and from the cell (e.g., Fortin 2004).

Aggregates of prokaryotic cells are often surrounded by communal extracellular polymeric substances (EPS) that can have a relatively high chance of preservation. A good example is the extracellular cytoplasmic sheaths or envelopes found around the cells of cyanobacteria. The sheath is often preserved when the cells themselves have collapsed and decomposed, for example in the Bitter Springs Formation (Oehler et al. 2006). The glutinous substances which comprise EPS have adhesive qualities, which trap and bind sediment particles onto biofilms and bioaggregates, leading to the formation of wrinkle structures and stromatolites (e.g., Noffke et al. 2003). These organo-sedimentary structures have a much higher preservation potential than the constructing organisms themselves.

5 The Challenge of Pseudofossils

The 'burden of proof' needed for the demonstration of the earliest cellular life is very great indeed. Any proposal of this kind requires the demonstration of multiple, in situ and mutually supporting lines of evidence, including: a well-constrained age and geological context, a morphology unique to biology, and more than a single line of geochemical evidence for metabolic cycling. In addition, there must be falsification of all plausible abiogenic scenarios (see Brasier et al. 2002, 2004, 2005; Altermann and Kazmierczak 2003; Cady et al. 2003; Westall 2005; Rose et al. 2006). Evidence for age and context comes from geological mapping at scales from kilometres to metres, supported by mapping of petrographic thin sections in order to show that candidate structures are truly syngenetic and ancient (e.g., Cady et al. 2003;

Brasier et al. 2005 and references therein). Additional evidence for syngenicity can come from laser Raman spectra (Pasteris and Wopenka 2002, 2003) or atomic force microscopy (AFM; Altermann and Kazmierczak 2003), though equivocal results are commonplace here. Evidence for a uniquely biogenic morphology can be obtained, by in situ imaging and mapping, to distinguish the fields of biotic and abiotic morphology and by comparing these with self-organising structures (see below). Geochemical evidence for life requires high-resolution, sub-micron scale, in situ three-dimensional mapping and analysis, using more than a single line of contaminant-free evidence. Examples include the in situ study of C and S isotopes and oxidation states (e.g., House et al. 2000; Ueno et al. 2001; Wacey et al. 2008), major and trace element mapping (Kamber and Webb 2001) and biomarker analysis (Summons et al. 1999) from putative microfossils and host rocks.

A significant but widely ignored challenge in early life studies, however, concerns our reliance upon inductive lines of reasoning. More specifically, there has tended to be too much reliance upon evidence that is 'consistent with' microbes, without falsifying or rejecting (sensu Popper 1959) other possible non-biological scenarios that may likewise be consistent. In particular, the criterion of 'morphological complexity' is widely used as a keystone characteristic for testing the earliest fossils (e.g., Buick et al. 1981; Buick 1990; Schopf 1999). However, an appreciation of both self-organizing structures (SOS) and complexity theory suggests that complex structures do not require complex causes (d'Arcy Thompson 1917). Complexity can arise naturally in physico-chemical systems through 'chaotic' behaviour and it is possible for a spectrum of 'life-like' signals to be generated completely without biology (Brasier et al. 2006, Fig. 2). In other words, a range of physio-chemical gradients can alone lead to macroscopic stromatoloids and, of course, ripples, as well as to macrofossil and microfossil-like structures generated by the growth of dendrites, 'coffee-ring' effects, polygonal crystal rims and spherulites. In each case, these arise from a 'symmetry-breaking cascade', which is a particularly conspicuous phenomenon during the growth and re-crystallization of spherulites, leading to a natural assemblages of structures that can range from spheroidal (broadly rotational symmetry), to dendritic (reflectional to slide symmetry), to arcuate (no clear symmetry). The resulting SOS include spheroids, filaments, septate filaments, wisps and fluffs (Brasier et al. 2006).

6 An Early Earth Taphonomic Traverse

Now we shall take a tour, like a time traveller, across a spectrum of those early Archaean habitats in which life should be sought. We will start in deeper waters around hydrothermal vents and in associated pillow lavas, then work towards the earliest known shoreline and beach deposits (Fig. 1). In each section, we will assess the quality of cellular fossil preservation that may be found in that setting, and show how true fossils may be usefully distinguished from the bewildering plethora of pseudofossils.

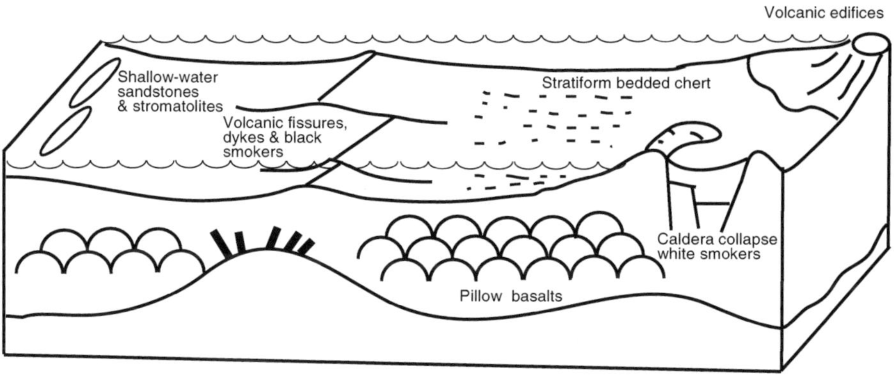

Fig. 1 Types of environments in which to search for the earliest signs of life

Two Archaean geological domains provide the basis for this traverse: the rocks of the Barberton Greenstone Belt, South Africa and those of the Pilbara Craton in Western Australia (see Wacey et al. 2008 for an overview). Early Archaean Barberton rocks are placed in the Swaziland Supergroup, which comprises the Onverwacht, Fig Tree and Moodies Groups (Anhaeusser 1973; Lowe and Byerley 1999). The Onverwacht Group, being the oldest, is of most interest and spans the time interval ~3500–3200 Ma (Armstrong et al. 1990). It is composed of komatiitic and tholeiitic basaltic rocks interbedded with thin sedimentary units of silicified ash and black chert, together with rare felsic volcaniclastic and intrusive rock.

The Pilbara craton of Western Australia comprises the three ancient granite greenstone terranes of East Pilbara, West Pilbara and Kurrana. The East Pilbara terrane houses the oldest rocks, as ancient as 3515 Ma. The 3515–3420 Ma Warrawoona Group consists mostly of mafic volcanic rocks interspersed with thin chert horizons and felsic volcanics. The Kelly Group lies unconformably above these and it, in turn, is unconformably overlain by the ~3240 Ma Sulfur Springs Group (for detailed stratigraphy see Van Kranendonk 2006). Together, these rock units are home to some of the Earth's oldest purported microfossils, trace fossils and stromatolites.

6.1 Pillow Basalts

We begin our search for life within rock substrates themselves, especially from volcanic pillow lavas on the ancient seafloor. We seek micron-sized cavities created by the metabolic activities of microorganisms (e.g., Bromley 2004). These trace fossils can preserve evidence for microbial behaviour, ecology and metabolism in their selection and modification of rock substrates. Endolithic microborings have long been known from silicified carbonate sediments younger than *c.* 1600 Ma (e.g., Zhang and Golubic 1987) but have more recently been reported from the glassy margins of pillow basalts from modern to ancient volcanic rocks (Thorseth et al. 1992; Fisk et al. 1998; Furnes et al. 2001; Staudigel et al. 2008).

A rock-dwelling 'endolithic' mode of life in the Archaean oceanic crust may indeed have offered many attractions to early life including: proximity to geothermal heat; a source of reductants, principally Fe and Mn which are abundant in basalts; and access to both oxidants and carbon sources carried by circulating fluids. In the early Archaean, especially, an endolithic mode of life would also have offered protection from the elevated UV radiation, meteoritic and cometary impacts. The latter may have severely hampered the emergence of life in surface environments. In addition, given that volcanic pillow lavas constitute an estimated 99% of greenstone successions from the Barberton and Pilbara cratons, they represent perhaps the largest potential habitat for early life. We first review of what is known about these organisms and their trace fossil record in modern volcanic rocks (see also McLoughlin et al. 2009). Then we will compare these with mineralized, tubular structures from the Archaean to assess their biogenicity and possible taphonomic pathways.

Traces of euendolithic microbes have been documented over the last 10 years or more from both in situ oceanic crust world-wide and from Phanerozoic ophiolites (for a recent review see Furnes et al. 2007). They are preserved as microtubular and granular structures at the interface of fresh and altered glass, along fractures in the rims of pillow basalts and around the margins of volcanic glass fragments in hyaloclastites (Fig. 2). Importantly, they are both texturally and chemically distinct from abiotic, palagonite alteration textures found in basalts (cf. Thorseth et al. 2001) so that, in many samples, evidence for episodes of both biotic and abiotic alteration can be found along fracture planes. Studies of recent material have found nucleic-acids, bacterial and archeal RNA concentrated within these bioalteration textures

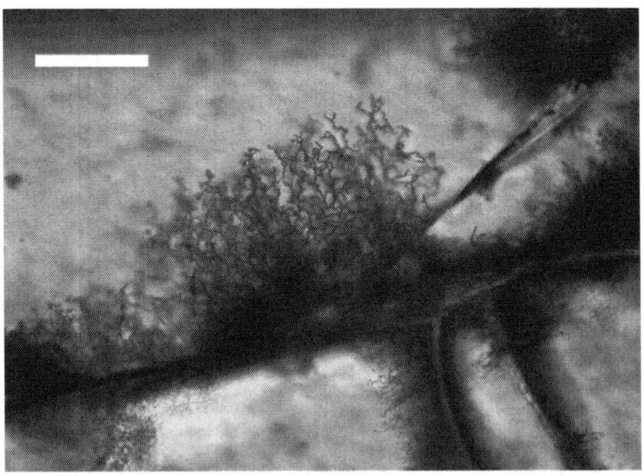

Fig. 2 Photomicrograph of endolithic microborings in ~10 Ma volcanic glass from ODP Hole 396B in the Mid-Atlantic. This branched ichnotaxon is termed *Tubulohyalichnus stipes* (McLoughlin et al. 2009). Such evidence for modern microborings provides an exciting new search image for signs of early life in early Archaean basaltic glass. Scale bar is 10 μm

(e.g., Torsvik et al. 1998; Santelli et al. 2008). These alteration zones may later be mineralized by zeolites and clays that can typically preserve localized enrichments in C, N, and P along the margins of the bioalteration textures themselves. These concentrations are therefore interpreted to represent the chemical effects of decayed cellular remains (e.g., Furnes and Muehlenbachs 2003).

Quantitative studies of the distribution and abundance of alteration textures with depth in the modern oceanic crust have found that, in the upper ~350 m of the crust, a 'granular' type of alteration is dominant. This component decreases steadily down through the drill core to become subordinate at temperatures of about 115°C (e.g., Staudigel et al. 2006). The microtubular alteration textures, meanwhile, constitute only a small fraction of the total zone of alteration and show a clear maximum at ~120–130m depth, corresponding to temperatures of about 70°C. Abiotic alteration is seen to dominate at progressively greater depths. Comparisons of seafloor and drill core samples of different age now suggest that bioalteration commences early and may take place largely during the first ~6 Ma years after crystallization of the basalt flows (Furnes et al. 2001).

In the Archaean, pillow basalts may well have been more widespread than today and microtubular bioalteration textures were first reported from the formerly glassy rims of pillow basalts and inter-pillow hyaloclastites from the Barberton Greenstone Belt of South Africa (Furnes et al. 2004). These titanite ($CaTiO_3$) mineralized microtubes are now preserved in greenschist facies meta-volcanic glasses that have been described from various units but some of the best preserved examples have come from the upper Hooggenoeg Formation, dated to about ~3472–3456 Ma (Banerjee et al. 2006). These structures are typically 1–10 μm in width and up to 200 μm in length (Fig. 3a). They extend away from "root zones" of fine grained titanite associated with fractures within the basaltic glass that were later annealed. These Archaean microtubes can have a segmented appearance brought about by overgrowths of metamorphic chlorite. Morphologically comparable microtubular structures have also been reported from inter-pillow hyaloclastite layers within the 3350 Ma Eurobasalt Formation of Western Australia (Fig. 3b; Furnes et al. 2006). The latter are also infilled with titanite that has now been dated directly using U-Pb systematics. Such dates confirm that the microtubes formed prior to a late Archaean (c. 2700 Ma) phase of metamorphism (Banerjee et al. 2007). In other words, these microtubes are likely to have formed during, or shortly after, seafloor colonization of the basaltic lava flows and are therefore unlikely to be younger contaminants.

Studies thus far have found that microtubular bioalteration textures tend to predominate in the Archaean Era, and that granular textures are much less common at this early date. This may, in part, be due to the enhancement and masking of titanite grains. The early precipitation of titanite within the larger microtubular textures is suggested to enhance microtube preservation by means of limiting those morphological changes that would otherwise be caused by re-crystallization of the host rock (see Fig 7 in McLoughlin et al. (2010a)). It is also possible, of course, that the smaller granular textures have been obscured by recrystallization of the glass.

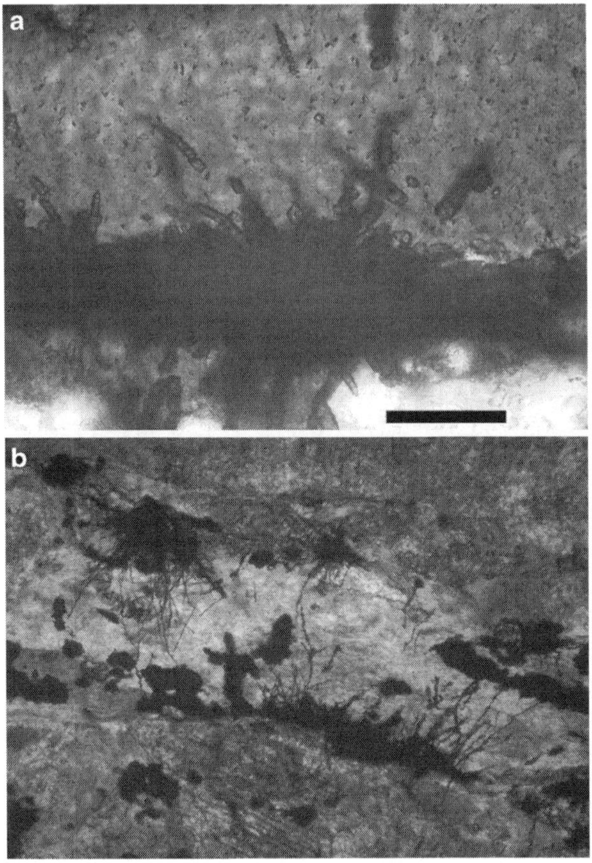

Fig. 3 Photomicrographs of microtubular structures in the glassy margins and inter pillow breccias of early Archaean basalts. These microtubes are infilled with titanite and emanate from early fractures in a way that closely resembles modern microborings of biological origin. (**a**) From the ~3472–3456 Ma Hooggenoeg Formation, South Africa (Furnes et al. (2004)); (**b**) from the ~3350 Ma Eurobasalt Formation, Western Australia (Banerjee et al. (2007)). Scale bar is 50 μm for **a**, and 250 μm for **b**

6.2 Black Smokers

As we continue along our Archaean environmental traverse, we come across hydrothermal vents with chimney shaped deposits of iron sulfide, much like those from modern mid-oceanic ridges (cf. Corliss et al. 1979; Rona et al. 1986; Von Damm et al. 1995) and back arc settings (Fouquet et al. 1991). While sulfide-rich black cherts are well known from hydrothermal rocks some 3500–3400 Ma old in Australia (e.g., Brasier et al. 2002, 2005; Orberger et al. 2006), it is not until much later, in the c. 3240 Ma Sulfur Springs Group on the Pilbara craton of Western Australia, that we can see such hydrothermal black smoker deposits convincingly preserved in the rock record (Vearncombe et al. 1995). This Sulfur Springs deposit is associated with a sequence of komatiites, basalts, dacites and rhyolites that erupted on the seafloor (Van Kranendonk 2006). Thin sections through the well-preserved drill core materials from the Sulfur Springs region show a wide range of volcanigenic and hydrothermal fabrics, including laminated pyrite nodules, chalcedonic silica, vein quartz and hydrocarbon globules (Vearncombe et al. 1995; Rasmussen and Buick 2000).

Pyritic filaments from within this massive sulfide deposit were first reported by Rasmussen (2000) and interpreted by him as the fossilized remains of thread like thermophilic, chemotrophic prokaryotes. These filaments are 0.5–2.0 μm in width and up to 300 μm long, can be straight, curved or sinuous and exhibit putative biological behaviour including preferred orientations, clustering and intertwining (Fig. 4). They only occur in phases of paragenetically early chert plus (interestingly) coarse-grained quartz that are clearly cross cut by later fractures. The null hypothesis that needs to be rejected here is that the Sulfur Springs filaments are abiogenic mineral growths that grew within the hydrothermal setting, and that were later replaced by pyrite. Abiogenic fibrous mineral growths are a well known feature of hydrothermal ore deposits, and many of these have been questionably interpreted as of microbial origin (e.g., Little et al. 2004). We have since recollected and re-examined this material. Preliminary analyses indicate that these filaments differ from abiogenic ones in being unbranched, of constant diameter, and distinctively entangled. There is as yet, however, no evidence for cellular organization nor for metabolic processing. Even so, this is an intriguing discovery that is at least consistent with the hypothesis of a thermophilic habitat for primitive life forms, in the vicinity of sub-marine hydrothermal vents (cf. Nisbet 2000; Shock 1990; Stetter 1996).

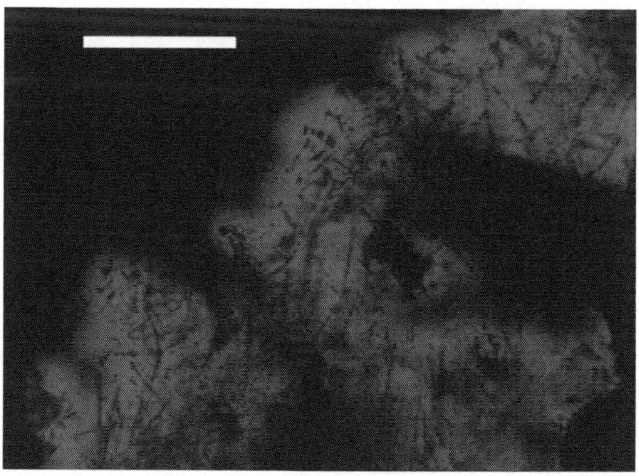

Fig. 4 Photomicrograph of pyrite filaments from the ~3200 Ma Sulfur Springs Group, Western Australia. The dark areas are stromatoloidal pyrite laminae. The pale areas are of macrocrystalline quartz containing pyrite filaments. These filaments have a morphology and context consistent with their formation by hyperthermophile archaea living in a black smoker setting in the Archaean. Scale bar is 100 μm

6.3 White Smokers

Continuing our traverse along hydrothermal vent systems, we encounter chimneys of barium sulfate (barite), silica and subordinate sulfide minerals known today as white smoker deposits (Mills and Elderfield 1995). Comparable, barite-rich chert veins and sediments are widely preserved in ~3500–3400 Ma old rocks from the Pilbara Craton of West Australia. For example, the Dresser Formation (*c.*3490 Ma, Nijman et al. 1998; Van Kranendonk 2006) and the Apex Basalt (c. 3465 Ma, Brasier et al. 2002, 2005), chert-barite veins are both associated with sequences of tholeiitic basalts and felsic tuffs that erupted on the seafloor, seemingly at times of granitic intrusion and caldera collapse (Nijman et al. 1998; Van Kranendonk et al. 2001). These veins extend to a depth of up to a kilometre or more down growth faults and elemental analyses provide evidence for the upward advection of Ba, Pb, Ni and As along with silica through these structures (Brasier et al. 2002; Orberger et al. 2006). Taking these observations together, several authors have advanced a hydrothermal, white smoker type model for these units (e.g. Nijman et al. 1998; Brasier et al. 2002, 2005; Orberger et al. 2006). Such white smokers with sulfates tend to form at lower temperatures than sulfide-containing black smokers, and they thereby increase the spectrum of temperature and venting conditions that were available to primitive forms of life and proto-life. This conjecture is supported by the close association observed between white smoker deposits and black cherts with ^{13}C depleted carbonaceous matter (e.g. Ueno et al. 2004). Thin sections through these deposits show a wide range of volcanigenic and hydrothermal fabrics, including hydrobreccias, laminated chalcedonic and carbonaceous silica, carbonaceous clots and clasts, as well as barite domes and veins, and vein quartz (Brasier et al. 2005; Orberger et al. 2006). In each case, the cherts are found to record a complex history in which the protolith (typically basalt, felsic tuff and black 'shale') has been extensively injected by, and replaced by, fine grained hydrothermal silica. Such displacive-replacive rocks have often been mistaken for the seafloor sediments themselves (e.g., Schopf 1993; Orberger et al. 2006), making the interpretation of putative biological signals in ancient white smoker type environments a difficult task.

6.4 Seafloor Banded Cherts

An unusual lithology across large areas of the Archaean seafloor is that of black, grey and white silica deposits. Such deposits make up less than about 1% of the thickness of greenstone belts in the Barberton and Pilbara cratons. It seems that these cherts were deposited as seafloor or stratiform deposits during the final parts of volcanic cycles through intrusion induced doming and fracturing of seafloor crust (Van Kranendonk et al. 2001).

Such banded cherts have, until recently, provided the primary search image for the earliest cellular preservation in the Archaean. That is, perhaps, because the silicification of microfloras is familiar to us from within much younger banded cherts, such as the 1900 Ma Gunflint Chert (e.g., Barghoorn and Tyler 1965). In the Precambrian world, without silica-secreting organisms such as sponges and radiolarians, much of the ocean was supersaturated with respect to silica (Maliva et al. 2005). Hence, it may have been relatively easy to precipitate silica in a wide range of settings in which precipitation could not happen today. Banded cherts of Archaean to Proterozoic age have indeed been found to range from shallow water and lagoonal environments, through photic zone depths with current rippled sands (e.g., the Buck Reef chert of South Africa, Tice and Lowe 2004), down to deeper water, more-distal environments in the vicinity of hydrothermal vents, such as the Apex chert (Brasier et al. 2005), and ferruginous laminites like the Banded Iron Formations (BIFs, e.g., Klein 2005).

A wide range of potentially biological signals has been reported from carbonaceous material in such banded cherts, the morphologies of which include 'wisps', 'fluffs', 'filaments', 'spheroids' and 'spindles'. Each of these morphologies has been described in detail by Brasier et al. (2006), accompanied by an explanation of the plausible abiotic scenarios that need to be excluded in each case. Here, we will briefly review these signals, with the exception of septate filaments from the 3465 Ma Apex chert, which we will discuss in more detail later in the chapter.

'Wisps' are microscopic carbonaceous wrinkled laminae (Fig. 5a). When found in laminated modern to late Archaean deposits, they are widely interpreted as biological features derived from microbial biofilms (e.g. Noffke 2000; Noffke et al. 2001, 2003). Wisp-like structures are found in bedded cherts both from the Pilbara (e.g., Brasier et al. 2005) and the Barberton (e.g., Westall et al. 2001). Using morphological comparisons with modern day examples, as well as their depth-restricted distribution, and the presence of roll up structures (Fig. 5b), they have often been interpreted as the remains of anaerobic, photosynthetic mats (Walsh and Lowe 1999; Tice and Lowe 2004). In these earliest rocks however, an origin for wispy and finely laminated textures from colloidal sediments, volcaniclastic sediments and prebiotic, abiogenic films will always need to be falsified. This problem has been highlighted by recent experimental studies that show how laminated micro-stromatolites and wrinkle structures can be generated by the diffusion-limited aggregation of synthetic colloids (McLoughlin et al. 2008).

The role that biology has to play in the generation of 'fluff' textures is even more equivocal. Modern carbonaceous 'fluff', sometimes termed marine snow, forms as a result of decaying planktonic matter settling through the water column, forming discrete layers within deep-sea sediments. In the Archaean, 'fluffy' carbonaceous grains are common in bedded cherts (Fig. 5b; Walsh and Lowe 1999), but they are also common in subsurface dyke cherts (Lindsay et al. 2005) where they can form layers of bush-like shrubs within hydrothermal cavern systems. These bushes arise from the growth of self-organising dendrites, meaning that similar abiogenic scenarios cannot yet be excluded for comparable carbonaceous 'fluff' textures found in seafloor cherts.

Carbonaceous filaments (Fig. 6a) have been at the centre of much controversy in the search for earliest life. The problem here is that while filaments can be easily compared with younger examples of prokaryotic microfossils (e.g., Schopf 2006),

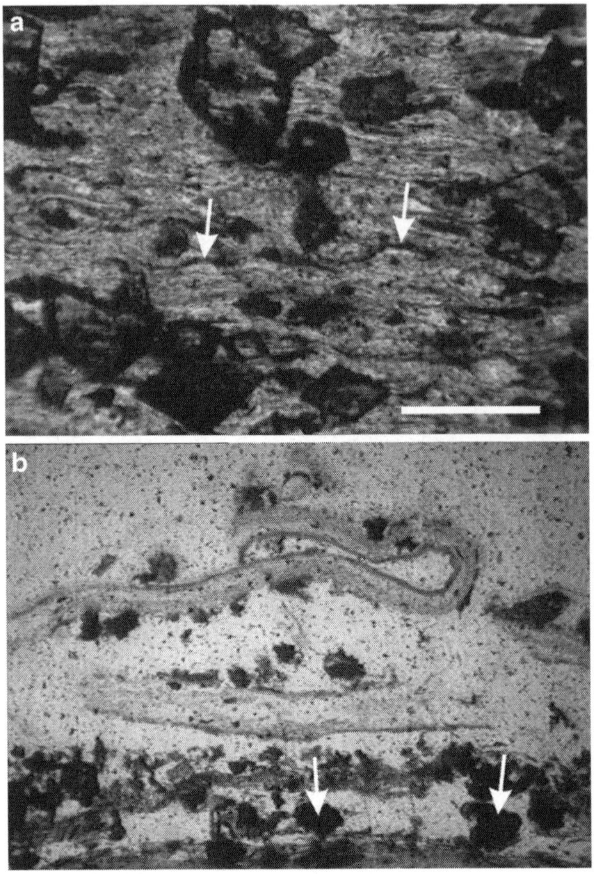

Fig. 5 Carbonaceous structures from banded cherts of early Archaean age. Such structures have been used to argue for the presence of cohesive microbial 'mats' on the seafloor at this time. (**a**) Carbonaceous wisps (*arrowed*) from the 3465 Ma Apex chert, Western Australia; (**b**) fluffy composite carbonaceous grains (*arrowed*) and a 'roll up structure' from the 3416 Ma Buck Reef Chert, South Africa. Scale bar is 50 μm for both **a** and **b**

they are also one of the most easily formed self organising structures (Brasier et al. 2005, 2006). Filaments can result from the breaking of polygonal, spheroidal or circular symmetry during crystal growth (see also Buick 1984, 1988; Deegan 2000). In addition, complex filaments that resemble the earliest Archaean microfossils can be generated in simple experiments by the precipitation of metallic salts in silica gels (Fig. 6b) and by subsequent nucleation of carbonaceous material (Garcia-Ruiz et al. 2003). Furthermore, hollow bacteria-like filaments can be generated by spark-discharge or FTT-like synthesis of organic polymers in prebiotic experiments (Folsome 1977; Baker and Harris 1978). This matters because Fischer-Tropsch-like processes may well have operated in Archaean hydrothermal systems, while spark discharges are likely to have accompanied all major volcanic eruptions (Lindsay et al. 2005).

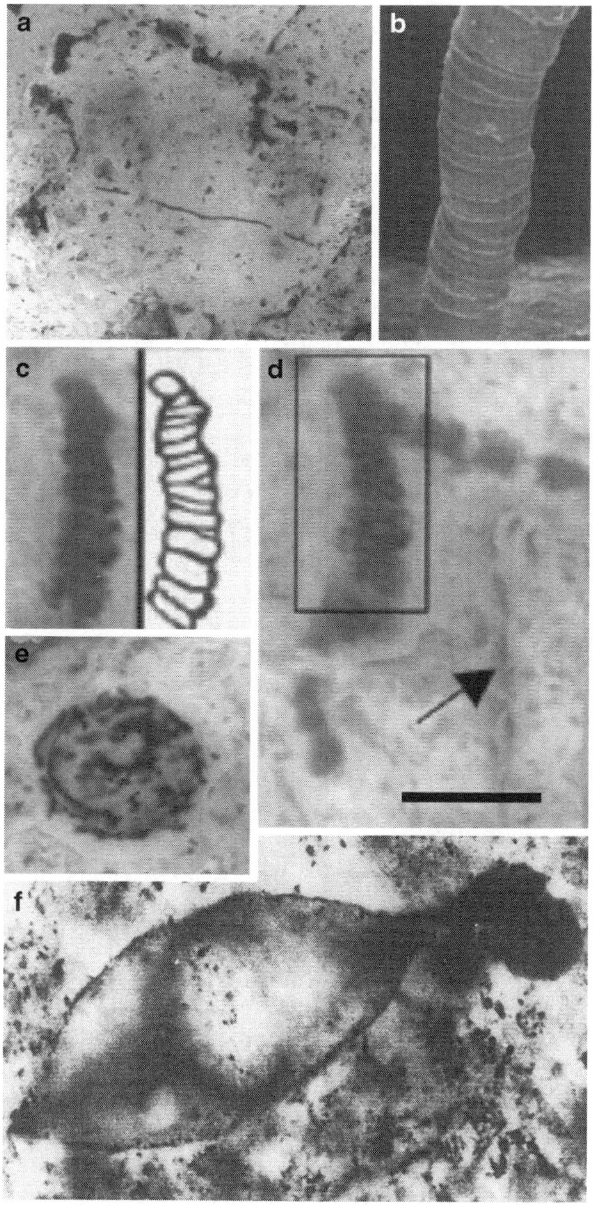

Fig. 6 Carbonaceous filaments, spheres and spindles from banded cherts of claimed but questionable biological origin (**a**, **c–f**) and of certain abiological origin (**b**) The biogenicity of such structures is proving difficult to demonstrate because they can also arise from complex abiological self organising structures (see Fig. 7). (**a**) carbonaceous filament from the 3465 Ma Apex chert, Western Australia; (**b**) twisted filamentous pseudofossil made experimentally by precipitating barium-carbonate crystals in sodium silicate gel (image courtesy of A Cannerup); (**c**) septate filament from the Apex chert interpreted as putative cyanobacterium *Archaeoscillatoriopsis disconformis* (Schopf 1993) now explained as **d**, an abiogenic self-organising structure (*boxed area* equates to structure shown in **c**) formed around a rhombic crystal (*arrowed*); (**e**) solitary sphere from the Apex chert formerly interpreted as a coccoid cell; (**f**) spindle structure from the 3400 Ma lower Kromberg Formation, South Africa (Walsh 1992). Scale bar is 100μm for **a**; 15 μm for **b** and **e**; 25 μm for **c**; 30 μm for **d** and **f**

Carbonaceous *septate* filaments have been seen as a 'Holy Grail' in searches for the earliest life because they can most closely approach the appearance of younger authentic prokaryotic microfossils, owing to the presence of cell-like subdivisions (Fig. 6c). Such filaments have understandably been interpreted as the remains of bacteria, and at times compared with photosynthetic cyanobacteria because of their size range (Awramik et al. 1983; Schopf and Packer 1987; Awramik 1992; Schopf 1992a, 1993, 1999; Ueno et al. 2001). On cross-examination, however, many of these claims falter. For example, it has been shown that the early Archaean, Apex chert 'microfossils' (Schopf 1992a, 1993, 1999) are in truth a population of artefacts (e.g., Fig. 6d) that occur within the complex boundary zones of re-crystallized silica spherulites and crystal rhombs, as well as within jaspilitic and carbonaceous cherts, volcanic glass and rhyolites. The most parsimonious explanation for these structures involves their formation during the recrystallization of amorphous glassy silica to spherulitic chalcedony and other hydrothermal fabrics, as part of a symmetry-breaking cascade from spheroidal – to dendritic – to arcuate artefacts (see Brasier et al. 2002, 2004, 2005 for details). A spectrum of artefacts is thereby produced which depends upon the size of the spherulites, and the purity (carbonaceous content) of the chert, as illustrated by Fig. 7. Further inaccuracies in the original reports of the Apex microfossils, in particular the nature of the depositional setting, their occurrence in late stage fabrics, and the nature of branching, have also been found by our detailed mapping at a range of scales (e.g. Brasier et al. 2005). The combined evidence must therefore lead to the rejection of the biological nature of these putative Apex chert fossils. It also casts doubt on the veracity of other reported occurrences of early Archaean septate 'microfossils'.

Carbonaceous spheroids (e.g., Fig. 6e) are also commonplace within Archaean cherts and some have been regarded as microfossils based upon comparisons with modern coccoid and baccilate bacteria. The problem with spheroids, however, is their relatively simple morphology which can be generated by purely physicochemical mechanisms in the form of fluid inclusions, vesicles (bubbles), globules, rings, and spheroidal crystallites (see Folsome 1977; Deegan 2000; Brasier et al. 2006). This makes it difficult to demonstrate the biogenicity of either solitary (e.g., Knoll and Barghoorn 1977; Walsh 1992) or clustered spheroids (e.g., Schopf and Packer 1987; Sugitani et al. 1998, 2006; Westall et al. 2001). The same can be said for structures which have been regarded as 'cells in the process of division' (e.g., Schopf 1993, 2006); these likewise can form naturally within complex self-organizing systems, such as mineral growths (Brasier et al. 2005, 2006).

A further structure of note within banded cherts of the Barberton are ~40 μm diameter 'spindles' (Fig. 6f; Walsh 1992; Westall et al. 2001). These intriguing morphologies have been interpreted as being either the outer sheaths of colonies of bacterial cells or as the abiogenic, carbonaceous coatings of ghosted gypsum crystals (Walsh 1992). A further explanation, advanced by Westall et al. (2001) is that they are similar to the fenestrae of stromatolites and are thus created by bacterially-produced gas. These scenarios certainly merit further investigation, especially in light of the recent discovery of similar structures in Western Australia (Sugitani et al. 2006).

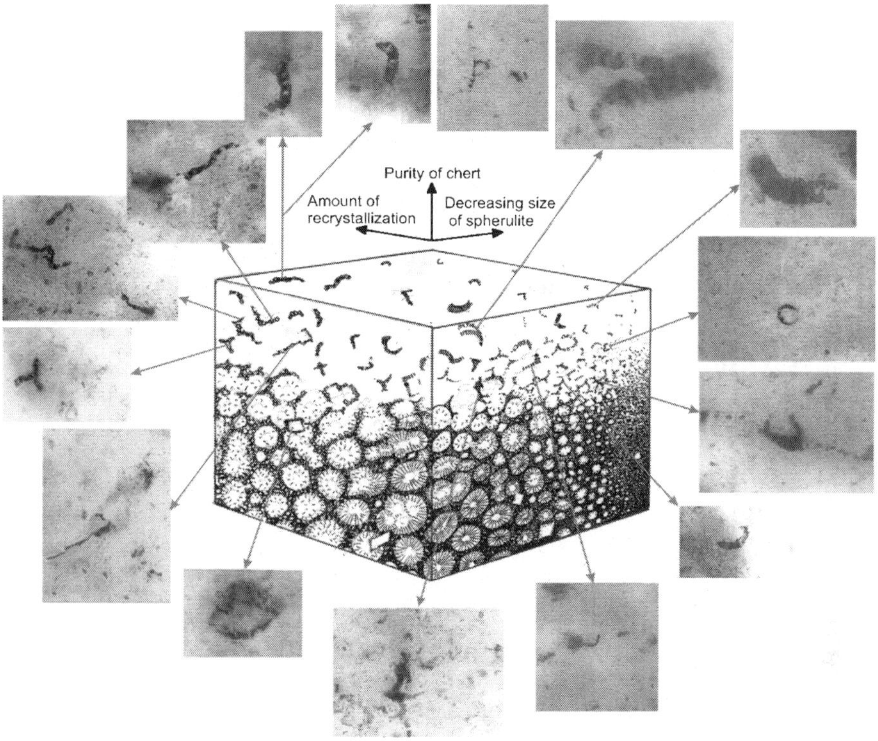

Fig. 7 This three dimensional morphospace model (*centre block*) of the famous Apex Chert 'microfossils'(*outer images*) shows how this spectrum of microfossil-like structures was most likely created entirely by physicochemical controls during recrystallization of the chert and the redistribution of carbonaceous material around spherulite and crystal margins. The key controls here were the relative purity of the chert (*vertical axis*), the degree of recrystallization of the fibrous chalcedony to equigranular microcrystalline chert (*left horizontal axis*), and the diameter of the spherulites (*right horizontal axis*). Arrows link theoretical with observed and reported microfossil-like artefacts sharing similar morphologies

6.5 Stromatolites

Moving into Archaean shallow water environments, our classic expectation is to find stromatolites. Stromatolites have provided a key search image for the emergence of life on Earth because they are assumed by many workers to be organosedimentary structures that require a microbial component in order to grow (e.g., Walter 1976). This view is largely based upon analogous reasoning from studies of modern examples in Shark Bay, Western Australia and from the Bahamas, both of which accrete largely as a result of microbial processes of trapping, binding and cementation. In many ancient examples, however, and most especially in the early Archaean (where the diagenetic destruction of microbial microfabrics and chemical biomarkers is pervasive), demonstration of such a biological component to growth is notoriously difficult to demonstrate (e.g., Buick et al. 1981; Lowe 1994a).

It is now becoming increasingly apparent that abiotic, chemical precipitation is an important component of stromatolite accretion and such processes may well have been prevalent during the earliest periods of Earth history (Grotzinger and Knoll 1999). These so called "chemical stromatolites" (sensu Pope et al. 2000) appear to have been common during periods of high levels of seawater carbonate saturation. Such forms can display a wide-range of morphologies and are characterized by isopachous laminae (i.e. of uniform thickness) with extreme lateral continuity and a high degree of vertical inheritance of topography from one layer to the next. In many instances, therefore, demonstrating an active role for microbes in the growth of such stromatolites has proved extremely difficult (e.g., Pope and Grotzinger 2000). For that reason, a non-genetic definition of a stromatolite is adopted here: i.e., an attached, laminated, lithified sedimentary growth structure that accretes away from a point or limited surface of initiation (Semikhatov et al. 1979).

Interestingly, both the abundance and diversity of Archaean stromatolites is much lower than seen in the succeeding Proterozoic interval (e.g., Hofmann 2000) and their morphologies tend to be less complex over a range of scales. It is also notable that microfossils of the kind usually inferred to have built these and related structures have never been found in association with Archaean stromatolites. This absence of evidence may, of course, be attributed to the low preservation potential of microfossils in stromatolites generally.

Some of the oldest putative stromatolites have been reported from the ~3490 Ma Ga Dresser Formation of the Warrawoona Group (Fig. 8a). These occur at several localities in the North Pole Dome, both in syn-depositional barite mounds and dykes from a hydrothermal complex (Van Kranendonk et al. 2001; Nijman et al. 1998) as well as within intercalated and silicified, ferruginous carbonates (Walter et al. 1980). The stromatolites originally described by Walter et al. (1980) were reviewed by Buick et al. (1981) who concluded that they were only "probable or possible" biogenic stromatolites. More recent studies have also described domal and stratiform stromatolites from around the 'vents' of barite dykes at the North Pole and some authors have argued that these mounds were constructed by hyperthermophilic microbes (Van Kranendonk 2006). The macro-morphology of these stromatolites is largely controlled by the thickness of the precipitated barite crusts and draping chert layers, however, and their distribution more likely reflects the supply of supersaturated solutions from which they were precipitated. Robust micro-textural and isotopic evidence for the involvement of microbial mats in the growth of these baritic stromatolites has not yet been reported, casting some doubt upon their biogenicity.

Fuel for this debate about Archaean stromatolite biogenicity has been provided by the discovery of a second Pilbara stromatolite locality, in the ~3430 Ma Ga Strelley Pool Formation, a marker horizon between the Warrawoona and Kelly Groups (Hoffman et al. 1999). Conical stromatolites (Fig. 8b) are a characteristic

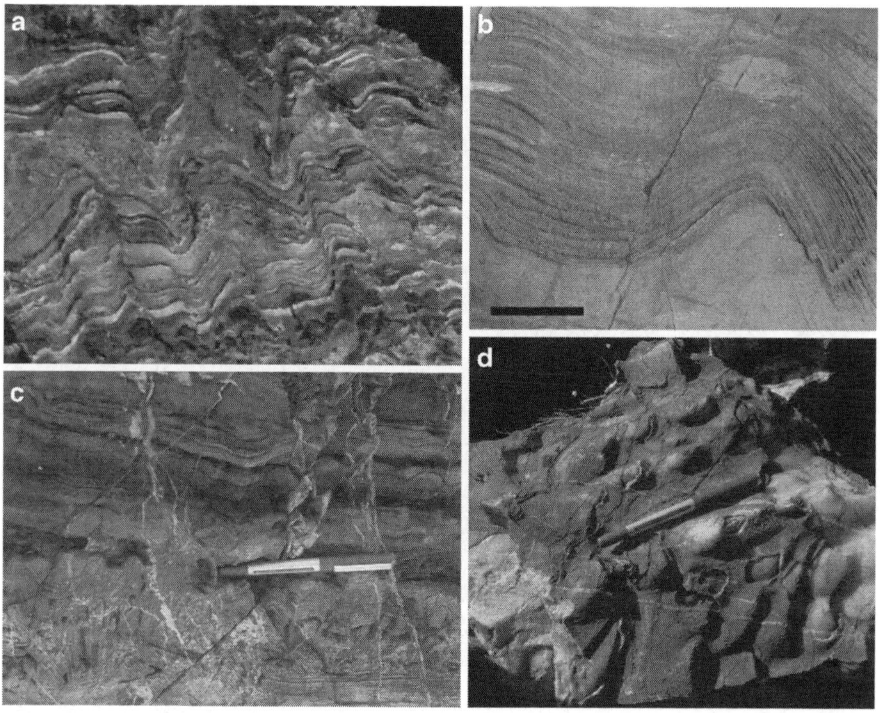

Fig. 8 Stromatolites of uncertain origin from the Pilbara of Western Australia. (**a**) Small domal stromatolite from the ~3490 Ma Dresser Formation interpreted as abiogenic by many authors; **b–d** are from the ~3430 Strelley Pool Chert and are of the kind that have been recently claimed to have a biological origin, but here we show abiological features that include isopachous laminae and reversible symmetry (**b**), accretion above crystal fans (**c**), and intergradation with asymmetrical linguoid ripples (**d**). Scale bar is 2 cm for **a**; 5 cm for **b**; pen is 15 cm long in **c** and **d** (see also Wacey et al. 2010)

feature of this unit and these were originally considered to be of biogenic origin (Lowe 1980), a claim that was then rescinded in favour of an abiogenic origin by means of evaporitic sedimentation (Lowe 1994a). The 'Trendall locality' (Hoffman et al. 1999) is notable for possessing an unusually diverse range of 'conical' and 'columnar' morphologies, plus one example of so-called 'branching'. Morphological arguments together with rare Earth element studies have then been used to argue for their shallow marine setting and their biological origin (Hoffman et al. 1999; Van Kranendonk et al. 2003; Allwood et al. 2006; Allwood et al. 2009).

Tellingly, the model put forward by Allwood et al. (2006) for stromatolites in the Strelley Pool Formation at the 'Trendall locality' fails to apply to the same unit in other areas. In the East Strelley greenstone belt, studied in detail by us (McLoughlin 2006; Wacey (2010a)), small unbranched 'coniform' stromatolites are typical and these do not show any changes in morphology or distribution with varying depth across the region. Like the Dresser Formation examples discussed above, they show a close interrelationship with crystal fan arrays (Fig. 8c). Sadly, this points clearly towards a strong chemical component for their growth. We also find that the 'cones' intergrade with linguoid and linear current ripples, highlighting a major role for physical processes during their accretion (Fig. 8d). In the absence of supporting microtextural and geochemical evidence, the biogenicity of the early Archaean stromatolites from much of the Strelley Pool Formation remains to be demonstrated (but see especially Wacey 2010).

The case for an entirely abiotic origin for at least some Precambrian stromatolites was advanced by Grotzinger and Rothman (1996), who used the Kadar Paris Zhang (KPZ) equation of interface growth (Kadar et al. 1986), to argue that the morphologies of some stromatolites can be modelled by abiotic processes alone. Although some authors dispute their interpretation of the KPZ equation (see Jogi and Runnegar 2003), this study has reinvigorated the debate surrounding biogenicity of the earliest stromatolites. More recently, McLoughlin et al. (2008) have shown that synthetic stromatolites, 'grown' abiogenically in colloidal media by diffusion-limited aggregation, can display features at one time believed to reflect some level of biological participation (Buick et al. 1981), i.e., convex upwards laminae; laminae that vary in thickness across stromatolite columns (non-isopachous); and laminae with several orders of curvature. We have found that columnar, branched and digitate stromatolites can all be generated abiologically in our laboratory experiments (Fig. 9).

It is curious that the capability of gelatinous or colloidal sediments to produce stromatolites and wrinkle mat-like fabrics has been largely overlooked, given their role in laminar to dendritic agate synthesis (e.g., Hopkinson et al. 1998). In the Precambrian oceans, with a benthic boundary layer that was supersaturated with silica, diffusion-limited deposition of colloidal sediments such as silica gel must

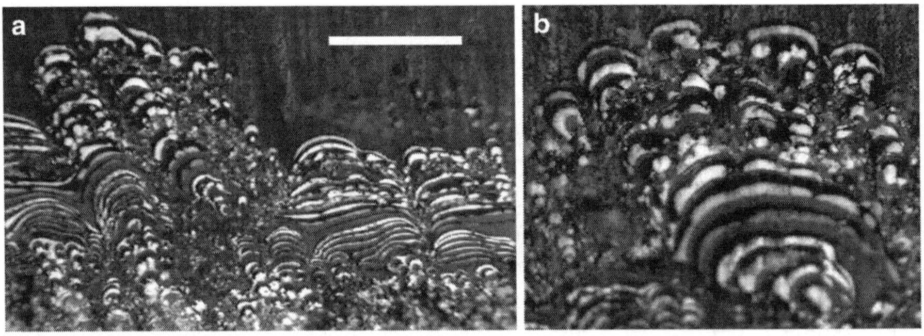

Fig. 9 Inclined digitate stromatolite structures generated abiologically in the laboratory by means of diffusion-limited aggregation of three alternating coloured colloids (paints). Here we show that features such as anisopachous laminae, wrinkled laminae and inclined columns, which have hitherto been regarded as biological features, can be generated abiologically. (**a**) Cross section of columnar digitate paint stromatolite inclined towards the sediment source on the *left hand side* with bridging laminae between the columns; (**b**) cross section of the bulbous head of a paint stromatolite with multiple branches. Scale bar is 1 mm for both **a** and **b** from McLoughlin et al. (2008)

have been capable of generating both laminar wrinkle mat and stromatolite textures in the absence of microbes (cf. McLoughlin et al. 2008). Given the lack of compelling microbial mat or microfossil remains in many early Archaean stromatolites, and their close association with non-equilibrium hydrothermal systems supersaturated with silica, questions must therefore remain as to whether, alone, stromatolites have anything useful to tell us about microbes or early biology. We would tend to agree, rather pessimistically with the statement that "it is perhaps impossible, 'to prove beyond question' that the vast majority of reported stromatolites…are assuredly biogenic" (Schopf 2006).

6.6 Siliclastics

Moving further towards the Archaean shoreline we encounter quartz arenites. These have proved to be rather rare because the area of exposed land that could provide the source material for these sediments was still very small at this early stage in Earth history (Buick et al. 1995). Nonetheless, quartz arenites are turning out to provide promising windows into the earliest biosphere, not least because of the relative ease with which the complex depositional and diagenetic history of sandstones can be untangled compared with rock types such as cherts and basalts.

A silicified sandstone unit at the base of the ~3430 Ma Strelley Pool Formation in Western Australia (Brasier et al. 2006; Wacey et al. 2006, 2008, 2010b) is currently revealing multiple, supporting lines of evidence consistent with a variety of biological activities at this time. The presence of low angle cross bedding and channels (e.g., Lowe 1983) shows, together with evidence for relatively high textural and compositional maturity, that deposition took place during the course of a shallow marine transgression, arguably the oldest such deposit in the rock record. The sandstones contain well rounded detrital grains of pyrite that, together with associated rounded grains of chromite, rutile, and zircon, indicate the formation of heavy mineral placer deposits within the beach setting (Wacey et al. 2010b). The pyrite grains are associated with carbonaceous biofilms and pits and channels that are interpreted as microbial trace fossils (Wacey et al. 2010b). A number of mineral precipitates, including iron oxides and sulfates, formed in close proximity to the pyrite surfaces and biofilms. These have been interpreted by Wacey et al. (2010b) as biomineral products of microbial pyrite oxidation.

A second kind of micro-structure present in this sandstone horizon (and others in the Pilbara) is that of 'ambient inclusion trails' (AIT) (Fig. 10). These are enigmatic structures have, in the past, been confused with both microfossils and endolithic microborings. However, they can be distinguished by the following features: (1) presence of a mineral crystal (e.g., a metal sulfide or oxide) at one end of an AIT, of equivalent diameter to the tube, which may be pseudomorphed by later minerals (e.g. silica, metallic oxide or phosphate); (2) longitudinal striations on the AIT created by facets of the propelled mineral

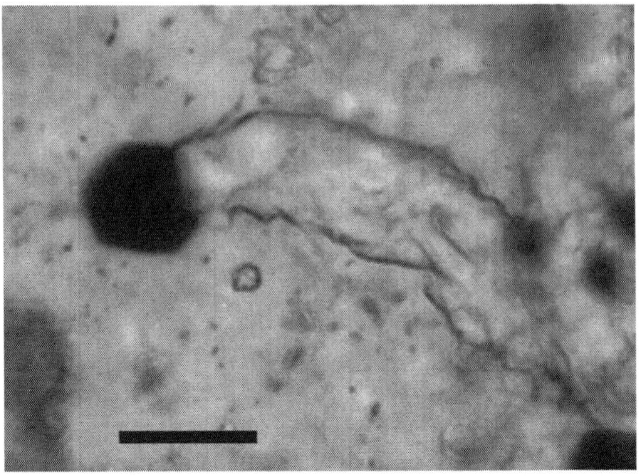

Fig. 10 Photomicrograph of silica-filled ambient inclusion trail (AIT) in a cryptocrystalline silica matrix from the ~3200 Ma Kangaroo Caves Formation, Western Australia. Such AIT are microtubes which typically have striated margins and a pyrite crystal at one end. They have often been mistakenly interpreted for microbial borings, though they may originate through biological decomposition processes. Scale bar is 15μm

crystal (which may, however, be obscured by later mineral infill); (3) curved or twisted paths, particularly towards their ends as impedance of the host lithology affects movement; (4) tendency of AITs to crosscut or form branches of a different diameter (i.e., where the propelled mineral becomes fragmented or a second crystal is intercepted), and to make sharp turns; (5) the AIT will likely have a polygonal cross sectional profile that matches the geometry of the propelled crystal.

Initially, AIT were thought to be a completely inorganic phenomenon (Tyler and Barghoorn 1963) but a conjecture was later advanced for their formation from the degassing of decomposing biological material during burial and/or metamorphism (Knoll and Barghoorn 1974). This hypothesis has now been confirmed by us using high-resolution mass spectrometry (NanoSIMS) coupled to detailed field and petrographic mapping (Wacey et al. 2008). Further discussion of these AIT formation mechanisms and a summary of criteria to distinguish them from microtunnels in a range of rock substrates including sediments and volcanic glass can be found in McLoughlin et al. (2010b).

Siliclastic deposits of the ~3.2 Ga Moodies Group of S Africa contain hollow spheroidal organic-walled structures comparable with many younger 'acritarchs' (Javaux et al. 2010). These structures pass syngenicity and endogenicity tests and appear to be the oldest acritarch-like microfossils yet reported. The null hypothesis here is for an origin from benthic prokaryotic cysts, contemporaneous with benthic microbial 'wrinkle structures' reported from the same rocks (e.g. Noffke et al. 2006). More speculative is their interpretation as bacterial plankton, or even eukaryotic cells (e.g. Buick 2010). These discoveries will help to define the search images needed for life in very ancient siliciclastic sediments (see Fig. 11).

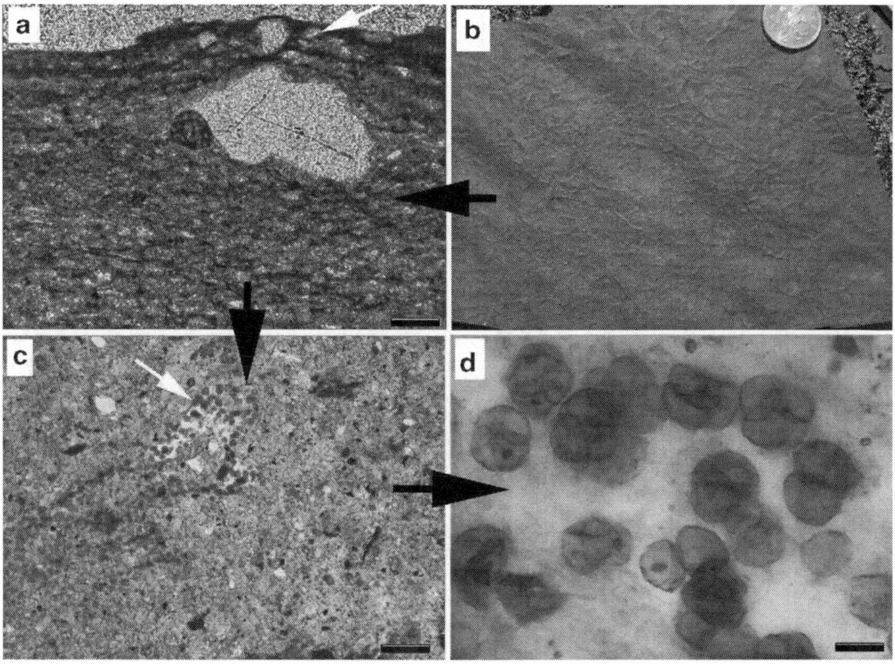

Fig. 11 Examples of features sought from early life within siliciclastic sediments, as here found within some of the earliest known terrestrial ecosystems. (**a**) petrographic evidence in transverse section for trapping and binding of sediment grains within organic polymers (arrow); (**b**) bedding plane evidence for microbially-induced sedimentary structures in the form of wrinkles or domes; (**c**) evidence in horizontal section for organization of cell-like bodies into sheets or mats; (**d**) detail of (**c**) showing evidence for cell walls, cell contents, and growth strategies including binary fission. All images are from ~1000 Ma siliciclastic lake beds, Torridonian of Scotland. Scale bar (**a**) and (**c**) = 100 micromillimetres; (**d**) = 10 micromillimetres

7 Summary

In this chapter we have advocated the view that the early Archaean Earth should be considered as a distant planet. We have reviewed the traditional taphonomic windows, especially carbonaceous cherts, through which the Archaean biosphere has long been studied. The importance of understanding self-organising structures has been stressed, along with ways scientists can refute such scenarios when working to establish the veracity of candidate Archaean fossils. A traverse across early Archaean environments has highlighted the importance of promising new taphonomic windows into earliest life. These include pillow lavas, pyritic deposits and siliciclastic sediments, suggesting that life may have been widely distributed at this time. Further research involving detailed mapping, petrography and geochemistry is now needed to pin down the specific life processes operating on the early Earth.

References

Allwood, A. C., Walter, M. R., Kamber, B. S., Marshall, C. P., & Burch, I. W. (2006). Stromatolite reef from the Early Archaean era of Australia. *Nature, 441*, 714–718.

Allwood, A. C., Grotzinger, J. P., Knoll, A. H., Burch, I. W., Anderson, M. S., Coleman, M. L., et al. (2009). Controls on development and diversity of Early Archean stromatolites. *PNAS, 106*, 9548–9555.

Altermann, W., & Kazmierczak, J. (2003). Archaean microfossils: A reappraisal of early life on Earth. *Research in Microbiology, 154*, 611–617.

Anhaeusser, C. R. (1973). The evolution of the early Precambrian crust of South Africa. *Philosophical Transactions of the Royal Society, London, A273*, 359–388.

Armstrong, R. A., Compston, W., de Wit, M. J., & Williams, L. S. (1990). The stratigraphy of the 3.5–3.2 Ga Barberton Greenstone Belt revisited: A single zircon ion microprobe study. *Earth and Planetary Science Letters, 101*, 90–106.

Awramik, S. M. (1992). The oldest records of photosynthesis. *Photosynthesis Research, 33*, 75–89.

Awramik, S. M., Schopf, J. W., & Walter, M. R. (1983). Filamentous Fossil Bacteria from the Archaean of Western Australia. *Precambrian Research, 20*, 357–374.

Bacon, F. (1620). *Novo Organum*. London.

Baker, R. T. K., & Harris, P. S. (1978). The formation of filamentous carbon. In P. L. Walker & P. A. Thrower (Eds.), *Chemistry and physics of carbon* (pp. 2–165). New York: Dekker.

Banerjee, N. R., Furnes, H., Muehlenbachs, K., Staudigel, H., & de Wit, M. (2006). Preservation of 3.4–3.5 Ga microbial biomarkers in pillow lavas and hyaloclastites from the Barberton Greenstone Belt, South Africa. *Earth and Planetary Science Letters, 241*, 707–722.

Banerjee, N. R., Simonetti, A. S., Furnes, H., Muehlenbachs, K., Staudigel, H., Heaman, L., et al. (2007). Direct dating of Archean microbial ichnofossils. *Geology, 35*, 487–490.

Barghoorn, E. S., & Tyler, S. A. (1965). Microorganisms from the Gunflint Chert. *Science, 147*, 563–577.

Brasier, M. D., & Armstrong, H. (2005). *Microfossils*. Science: Blackwell. 304pp.

Brasier, M. D., Green, O. R., Jephcoat, A. P., Kleppe, A. K., Van Kranendonk, M. J., Lindsay, J. F., et al. (2002). Questioning the evidence for Earth's oldest fossils. *Nature, 416*, 76–81.

Brasier, M. D., Green, O. R., & Mcloughlin, N. (2004). Characterization and critical testing of potential microfossils from the early Earth: The Apex 'microfossil debate' and its lessons for Mars sample return. *International Journal of Astrobiology, 3*, 1–12.

Brasier, M. D., Green, O. R., Lindsay, J. F., McLoughlin, N., Steele, A., & Stoakes, C. (2005). Critical testing of Earth's oldest putative fossil assemblage from the 3.5 Ga Apex Chert, Chinaman Creek Western Australia. *Precambrian Research, 140*, 55–102.

Brasier, M. D., McLoughlin, N., & Wacey, D. (2006). A fresh look at the fossil evidence for early Archaean cellular life. *Philosophical Transactions of the Royal Society B, 361*, 887–902.

Bromley, R. G. (2004). A stratigraphy of marine bioerosion. *Geological Society, Special Publication, 228*, 455–479.

Buick, R. (1984). Carbonaceous filaments from North Pole, Western Australia: Are they fossil bacteria in Archaean stromatolites? *Precambrian Research, 24*, 157–172.

Buick, R. (1988). Carbonaceous filaments from North Pole, Western Australia: Are they fossil bacteria in Archaean stromatolites? A reply. *Precambrian Research, 39*, 311–317.

Buick, R. (1990). Microfossil recognition in Archaean rocks: An appraisal of spheroids and filaments from 3500 M.Y old chert-barite at North Pole, Western Australia. *Palaios, 5*, 441–459.

Buick, R., Dunlop, J. S. R., & Groves, D. I. (1981). Stromatolite recognition in ancient rocks: An appraisal of irregularly laminated structures in an early Archaean chert-barite unit from North Pole, Western Australia. *Alcheringa, 5*, 161–181.

Buick, R., Thornett, J. R., McNaughton, N. J., Smith, J. B., Barley, M. E., & Savage, M. (1995). Record of emergent continental crust 3.5 billion years ago in the Pilbara craton of Australia. *Nature, 375*, 574–577.

Buick, R. (2010). Ancient acritarchs. *Nature, 463*, 885–886.

Byerly, G. R., Lowe, D. R., Wooden, J. L., & Xiaogang, X. (2002). An Archean impact layer from the Pilbara and Kaapvaal Cratons. *Science, 297*, 1325–1327.

Cady, S. L., Farmer, J. D., Grotzinger, J. P., Schopf, J. W., & Steele, A. (2003). Morphological biosignatures and the search for life on Mars. *Astrobiology, 3*, 351–368.

Cairns-Smith, A. G. (1985). *Seven clues to the origin of life*. Cambridge: Cambridge University Press. 154pp.

Cavalier-Smith, T., Brasier, M. D., & Embley, T. M. (2006). How and when did microbes change the world? *Philosophical Transactions of the Royal Society B, 361*, 845–850.

Chan, C. S., De Stasio, G., Welch, S. A., Girasole, M., Frazer, B. H., Nesterova, M. V., et al. (2004). Microbial polysaccharides template assembly of nancrystal fibres. *Science, 303*, 1656–1658.

Corliss, J. B., Dymond, J., Gordon, L. I., Edmond, J. M., Herzen, R. P. V., Ballard, R. D., et al. (1979). Submarine thermal springs on the Galapagos Rift. *Science, 203*, 1073–1083.

Deegan, R. D. (2000). Pattern formation in drying drops. *Physical Review, E61*, 475–485.

Delsemme, A. H. (1998). Cosmic origin of the biosphere. In A. Brock (Ed.), *The molecular origins of life: Assembling the pieces of the puzzle* (pp. 100–118). Cambridge: Cambridge University Press.

Fisk, M. R., Giovannoni, S. J., & Thorseth, I. H. (1998). Alteration of oceanic volcanic glass: Textural evidence of microbial activity. *Science, 281*, 978–980.

Folsome, C. (1977) Synthetic organic microstructures as model systems for early protobionts. In C. Ponnamperuma (Ed.), *Chemical evolution of the early Precambrian* (pp. 171–179).

Fortin, D. (2004). What biogenic minerals tell us. *Science, 303*, 1618–1619.

Fouquet, Y., Vonstackelberg, U., Charlou, J. L., Donval, J. P., Foucher, J. P., Muhe, R., et al. (1991). Hydrothermal activity in the Lau Back-Arc Basin – Sulfides and water chemistry. *Geology, 19*, 303–306.

Furnes, H., & Muehlenbachs, K. (2003). Bioalteration recorded in ophiolitic pillow lavas. In Y. Dilek & P. T. Robinson (Eds.), *Ophiolites in earth's history. Geological Society of London, Special Publication 218* (pp. 415–426).

Furnes, H., Staudigel, H., Thorseth, I. H., Torsvik, T., Muehlenbachs, K., & Tumyr, O. (2001). Bioalteration of basaltic glass in the oceanic crust. G^3 2, 2000GC000150.

Furnes, H., Banerjee, N. R., Muehlenbachs, K., Staudigel, H., & de Wit, M. (2004). Early life recorded in Archaean pillow lavas. *Science, 304*, 578–581.

Furnes, H., Banerjee, N. R., Staudigel, H., Muehlenbachs, K., Simonetti, A., de Wit, M., et al. (2006). Earth's oldest microbial biomarkers in pillow lavas: A new geological setting in the search for early life. *Geophysical Research Abstracts, 8*, 11078.

Furnes, H., Banerjee, N. R., Staudigel, H., Muehlenbachs, K., McLoughlin, N., de Wit, M., & Van Kranendonk, M. (2007). Comparing petrographic signatures of bioalteration in recent to Mesoarchean pillow lavas: Tracing subsurface life in oceanic igneous rocks. Precambrian Research. *Precambrian Research, 158*, 156–176.

Garcia-Ruiz, J. M., Hyde, S. T., Carnerup, A. M., Christy, A. G., Van Kranendonk, M. J., & Welham, N. J. (2003). Self-assembled silica-carbonate structures and detection of ancient microfossils. *Science, 302*, 1194–1197.

Grassineau, N. V., Nisbet, E. G., Bickle, M. J., Fowler, C. M. R., Lowry, D., Mattey, D. P., et al. (2001). Antiquity of the biological sulphur cycle: Evidence from sulphur and carbon isotopes in 2700 million-year old rock of the Belingwe Belt. *Zimbabwe Proceedings of the Royal Society of London B, 268*, 113–119.

Grotzinger, J. P., & Knoll, A. H. (1999). Stromatolites in Precambrian carbonates; evolutionary mileposts or environmental dipsticks? *Annual Review of Earth and Planetary Sciences Letters, 27*, 313–358.

Grotzinger, J. P., & Rothman, D. H. (1996). An abiotic model for stomatolite morphogenesis. *Nature, 383*, 423–425.

Hanczyc, M. M., Fujikawa, S. M., & Szostak, J. W. (2003). Experimental models of primitive cellular compartments: Encapsulation, growth, and division. *Science, 302*, 618–622.

Hofmann, H. J. (2000). Archean stromatolites as microbial archives. In R. E. Riding & S. M. Awramik (Eds.), *Microbial sediments* (pp. 315–327). Berlin: Springer.

Hoffman, H. J., Grey, K., Hickman, A. H., & Thorpe, R. (1999). Origin of 3.45Ga coniform stromatolites in Warawoona Group, Western Australia. *Geological Society of America Bulletin, 111*, 1256–1262.

Hopkinson, L., Roberts, S., Herrington, R., & Wilkinson, J. (1998). Self-organisation of submarine hydrothermal siliceous deposits: Evidence form the TAG hydrothermal mound, 26°N Mid-Atlantic Ridge. *Geology, 26*, 347–350.

Horita, J., & Berndt, M. E. (1999). Abiogenic methane formation and isotopic fractionation under hydrothermal conditions. *Science, 285*, 1055–1057.

House, C. H., Schopf, J. W., McKeegan, K. D., Coath, C. D., Harrison, T. M., & Stetter, K. O. (2000). Carbon isotopic compositions of individual Precambrian microfossils. *Geology, 28*, 70–710.

Hooke, R. (1665) *Micrographia*. or, Some physiological descriptions of minute bodies made by magnifying glasses. London. J. Martyn and J. Allestry.

Jannasch, H. W., & Mottl, M. J. (1985). Geomicrobiology of deep-sea hydrothermal vents. *Science, 229*, 717–725.

Javaux, E. J., Marshall, C. P., Bekker, A. (2010) Organic-walled microfossils in 3.2-billion-year-old shallow-marine siliclastic deposits. *Nature, 463*, 934–938.

Jogi, P., & Runnegar, B. (2003). Theoretical support for biological activity in ancient stromatolites. Abstract, NASA Astrobiology Institute.

Kadar, M., Parisi, G., & Zhang, Y. (1986). Dynamic scaling of growing interfaces. *Physical Review Letters, 56*, 889–892.

Kamber, B. S., & Webb, G. E. (2001). The geochemistry of late Archaean microbial carbonate: Implications for ocean chemistry and continental erosion history. *Geochimica et Cosmochimica Acta, 65*, 2509–2525.

Kasting, J. F., & Catling, D. C. (2003). Evolution of a habitable planet. *Annual Review of Astronomy and Astrophysics, 41*, 429–463.

Klein, C. (2005). Some Precambrian banded iron formations (BIFs) from around the world: Their age, geologic setting, mineralogy, metamorphism, geochemistry, and origins. *American Mineralogist, 90*, 1473–1499.

Knauth, L. P., & Lowe, D. R. (2003). High Archean climatic temperature inferred from oxygen isotope geochemistry of cherts in the 3.5 Ga Swaziland Supergroup, South Africa. *Bulletin of Geological Society of America, 115*, 566–580.

Knoll, A. H. (2003). *Life on a young planet: The first three billion years of evolution on Earth.* Princeton/Chichester: Princeton University Press. 277pp.

Knoll, A. H., & Barghoorn, E. S. (1974). Ambient pyrite in Precambrian chert: New evidence and a theory. *Proceedings of National Academy of Sciences of the United States of America, 71*, 2329–2331.

Knoll, A. H., & Barghoorn, E. S. (1977). Archean microfossils showing cell division from the Swaziland system of South Africa. *Science, 198*, 396–398.

Konhauser, K. O., Phoenix, V. R., Bottrell, S. H., Adams, D. G., & Head, I. M. (2001). Microbial-silica interactions in Icelandic hot spring sinter: Possible analogues for some Precambrian siliceous stromatolites. *Sedimentology, 48*, 415–433.

Konhauser, K. O., Jones, B., Reysenbach, A. L., & Renault, R. W. (2003). Hot spring sinters: Key to understanding Earth's earliest life forms. *Canadian Journal of Earth Sciences, 40*, 1713–1724.

Lamarck, J. B. (1809). *Philosophie zoologique ou exposition des considérations relatives à l'histoire naturelle des animaux*. Paris. 467pp.

Lindsay, J. F., & Brasier, M. D. (2002). Did global tectonics drive early biosphere evolution? Carbon isotope record from 2.6 to 1.9 Ga carbonates of Western Australian basins. *Precambrian Research, 114*, 1–34.

Lindsay, J. F., Brasier, M. D., McLoughlin, N., Green, O. R., Fogel, M., Steele, A., et al. (2005). The problem of deep carbon – an Archaean Paradox. *Precambrian Research, 143*, 1–22.

Little, C. T. S., Glynn, S. E., & Mills, R. A. (2004). Four hundred and ninety million year record of bacteriogenic iron oxide precipitation at sea-floor hydrothermal vents. *Geomicrobiology Journal, 21*, 415–429.

Lowe, D. R. (1980). Stromatolites 3,400-Myr old from the Archean of Western Australia. *Nature, 284*, 441–443.

Lowe, D. R. (1983). Restricted shallow-water sedimentation of early Archaean stromatolitic and evaporitic strata of the Strelley Pool chert, Pilbara block, Western Australia. *Precambrian Research, 19*, 239–283.

Lowe, D. R. (1994a). Abiological origin of described stromatolites older than 3.2 Ga. *Geology, 22*, 387–390.

Lowe, D. R. (1994b). Early environments: Constraints and opportunities for early evolution. In S. Bengston (Ed.), *Early life on earth, Nobel symposium 84* (pp. 24–35).

Lowe, D. R., & Byerley, G. R. (1999). Geologic evolution of the barberton greenstone belt, South Africa. *Geological Society of American Special Papers 329*, Boulder, Colorado.

Lowe, D. R., & Tice, M. M. (2004). Geologic evidence for Archean atmospheric and climatic evolution: Fluctuating levels of CO_2, CH_4, and O_2 with an overriding tectonic control. *Geology, 32*, 493–496.

Lyell, C. (1830). *Principles of Geology*. 1st edn. vol 3. John Murray, London. 168pp.

Maliva, R. G., Knoll, A. H., & Simonson, B. M. (2005). Secular change in the Precambrian silica cycle: Insights from chert petrology. *G.S.A. Bulletin, 117*, 835–845.

McCall, G. J. H. (2003). A critique of the analogy between Archaean and Phanerozoic tectonics based on regional mapping of the Mesozoic-Cenozoic plate convergent zone in the Makran. *Iran Precambrian Research, 127*, 5–17.

McCollum, T. M., & Seewald, J. S. (2006). Carbon isotope composition of organic compounds produced by abiotic synthesis under hydrothermal conditions. *Earth and Planetary Science Letters, 243*, 74–84.

McLoughlin, N. (2006). Earth's earliest biosphere: Western Australia. PhD Thesis, University of Oxford.

McLoughlin, N., Wilson, L., & Brasier, M. D. (2008). Growth of synthetic stromatolites and wrinkle structures in the absence of microbes – implications for the early fossil record. *Geobiology, 6*, 95–105.

McLoughlin, N., Furnes, H., Banerjee, N. R., Muehlenbachs, K. and Staudigel, H. (2009). Ichnotaxonomy of Microbial Trace Fossils in Volcanic Glass. J. Geol. Soc. London *166*, 159–170.

McLoughlin, N., Fliegel, D. J., Furnes, H., Staudigel, H., Simonetti, A., Zhao, G. C., and Robinson, P. T. (2010a). Assessing the Biogenicity and Syngenicity of Candidate Bioalteration Textures in Pillow Lavas of the ~2.52 Ga Wutai Greenstone Terrane of China. *Chinese Science Bulletin, 55*, 188–199.

McLoughlin, N., Staudigel, H., Furnes, H., Eickmann, B. and Ivarsson, M. (2010b in press). Mechanisms of microtunneling in Rock Substrates – Distinguishing Endolithic Biosignatures from Abiotic Microtunnels. Geobiology.

Mills, R. A., & Elderfield, H. (1995). Rare Earth element geochemistry of hydrothermal deposits from the active TAG mound, 26°N Mid-Atlantic ridge. *Geochimica et Cosmochimica Acta, 59*, 3511–3524.

Moorbath, S. (2005). Oldest rocks, earliest life, heaviest impacts, and the Hadean–Archaean transition. *Applied Geochemistry, 20*, 819–824.

Nijman, W., De Bruin, K., & Valkering, M. (1998). Growth fault control of early Archaean cherts, barite mounds, and chert-barite veins, North Pole Dome, Eastern Pilbara, Western Australia. *Precambrian Research, 88*, 25–52.

Nisbet, E. G. (2000). The realms of Archaean life. *Nature, 405*, 625–626.

Nisbet, E. G., & Fowler, M. R. (1996). The hydrothermal imprint on life: Did heat-shock proteins, metalloproteins and photosynthesis begin around hydrothermal vents? In C. J. MacLeod, P. A. Tyler, & C. L. Walker (Eds.), *Tectonic, magmatic, hydrothermal and biological segmentation of mid-ocean ridges, Geological Society, Special Publications 118* (pp. 239–251).

Noffke, N. (2000). Extensive microbial mats and their influences on the erosional and depositional dynamics of a siliciclastic cold water environment (Lower Arenigian, Montagne Noire, France). *Sedimentary Geology, 136*, 207–215.

Noffke, N., Gerdes, G., Klenke, T., & Krumbein, W. E. (2001). Microbially induced sedimentary structures – a new category within the classification of primary sedimentary structures. *Journal of Sedimentary Research, 71*, 649–656.

Noffke, N., Hazen, R., & Nhleko, N. (2003). Earth's earliest microbial mats in a siliciclastic marine environment (2.9 Ga Mozaan Group, South Africa). *Geology, 31*, 673–676.

Noffke, N, Eriksson, K. A., Hazen, R. M. & Simpson, E. L. (2006). A new window into Early Archean life: Microbial mats in Earth's oldest siliciclastic tidal deposits (3.2 Ga Moodies Group, South Africa). *Geology, 34*, 253–256.

Oehler, D. Z., Robert, F., Mostefaoui, S., Meibom, A., Selo, M., & McKay, D. S. (2006). Chemical mapping of Proterozoic organic matter at submicron spatial resolution. *Astrobiology, 6*, 838–850.

Orberger, B., Rouchon, V., Westall, F., de Vrise, S. T., Pinti, D. L., Wagner, C., et al. (2006). Microfacies and origin of some Archean cherts (Pilbara, Australia). *G.S.A. Special Paper, 405*, 133–156.

Pasteris, J. D., & Wopenka, B. (2002). Images of the Earth's oldest fossils? (discussion and reply). *Nature, 420*, 476–477.

Pasteris, J. D., & Wopenka, B. (2003). Necessary, but not sufficient: Raman identification of disordered carbon as a signature of ancient life. *Astrobiology, 3*, 727–738.

Phoenix, V. R., Bennett, P. C., Engel, A. S., Tyler, S. W., & Ferris, F. G. (2006). Chilean high-altitude hot-spring sinters: A model system for UV screening mechanisms by early Precambrian cyanobacteria. *Geobiology, 4*, 15–28.

Pope, M. C., & Grotzinger, J. P. (2000). Controls on fabric development and morphology of tufas and stromatolites, uppermost Pethei group 1.8 Ga, Great Slave Lake, NW Canada. In *Carbonate sedimentation and diagenesis in the evolving Precambrian world. S.E.P.M. Special Publications 67* (pp. 103–121).

Pope, M. C., Grotzinger, J. P., & Schreiber, B. C. (2000). Evaporitic subtidal stroms produced by in situ precipitation: Textures, facies associations, and temporal significance. *Journal of Sedimentary Research, 70*, 1139–1151.

Popper, K. (1959). *The logic of scientific discovery*. London: Hutchinson Press.

Rasmussen, B. (2000). Filamentous microfossils in a 3, 235-million-year-old volcanogenic massive sulphide deposit. *Nature, 405*, 676–679.

Rasmussen, B., & Buick, R. (2000). Oily old ores: Evidence for hydrothermal petroleum generation in an Arcgean massive sulfide deposit. *Geology, 28*, 731–734.

Robert, F., Selo, M., Hillion, F., & Skrzypczak, A. (2005). NanoSIMS images of Precambrian fossil cells. *Lunar and Planetary Science, XXXVI*, abstract 1314.

Rona, P. A., Klinkhammer, G., Nelson, T. A., Trefry, J. H., & Elderfield, H. (1986). Black smokers, massive sulphides and vent biota at the Mid-Atlantic Ridge. *Nature, 321*, 33–37.

Rose, E. C., Mcloughlin, N., & Brasier, M. D. (2006). Ground truth: The epistemology of searching for the earliest life on Earth. In J. Seckbach (Ed.), *Life as we know it: Cellular origin, life in extreme habitats and astrobiology 10*. New York: Springer. 650 pp.

Sagan, C., & Mullen, G. (1972). Earth and Mars: Evolution of atmospheres and surface temperatures. *Science, 177*, 52–56.

Santelli C. M., Orcutt B. N., Banning E., et al. (2008). Abundance and Diversity of microbial Life in ocean crust. *Nature, 453*, 653–657.

Schidlowski, M. (2001). Carbon isotopes as biogeochemical recorders of life over 3.8 Ga of Earth history: Evolution of a concept. *Precambrian Research, 106*, 117–134.

Schopf, J. W. (1992a). The oldest fossils and what they mean. In J. W. Schopf (Ed.), *Major events in the history of life* (pp. 29–63). Boston: John & Bartlett.

Schopf, J. W. (1992b). Paleobiology of the Archaean. In J. W. Schopf & C. Klein (Eds.), *The Proterozoic biosphere: A multidisciplinary study* (pp. 25–39). New York: Cambridge University Press.

Schopf, J. W. (1993). Microfossils of the Early Archaean Apex Chert: New evidence for the antiquity of life. *Science, 260*, 640–646.

Schopf, J. W. (1999). *The cradle of life* (p. 367). New York: Princeton University Press.

Schopf, J. W. (2006). Fossil evidence of Archean life. *Philosophical Transactions of the Royal Society B, 361*, 869–886.

Schopf, J. W., & Klein, C. (1992). *The Proterozoic biosphere: A multidisciplinary study*. New York: Cambridge University Press.

Schopf, J. W., & Packer, B. M. (1987). Early Archaean (3.3 Billion to 3.5 Billion-year-old) microfossils from Warawoona Group, Australia. *Science, 237*, 70–73.

Semikhatov, M. A., Gebelein, C. D., Cloud, P., Awramik, S. M., & Benmore, W. C. (1979). Stromatolite morphogenesis: Progress and problems. *Canadian Journal of Earth Sciences, 16*, 992–1015.

Shen, Y., Buick, R., & Canfield, D. E. (2001). Isotopic evidence for microbial sulphate reduction in the early Archean era. *Nature, 410*, 77–81.

Sherwood Lollar, B., Westgate, T. D., Ward, J. A., Slater, G. F., & Lacrampe-Couloume, G. (2002). Abiogenic formation of alkanes in the Earth's crust as a minor source for global hydrocarbon reservoirs. *Nature, 416*, 522–524.

Shock, E. L. (1990). Geochemical constraints on the origin of organic compounds in hydrothermal systems. *Origins of Life and Evolution of the Biosphere, 20*, 331–367.

Staudigel, H., Furnes, H., McLoughlin, N., Banerjee, N. R., Connell, L. B., Templeton, A. (2008) 3.5 Billion years of glass bioalteration: Volcanic rocks as a basis for microbial life? *Earth Science Reviews, 89*, 156–176.

Staudigel, H., Furnes, H., Banerjee, N. R., Dilek, Y., & Muehlenbachs, K. (2006). Microbes and volcanoes: A tale from the oceans, ophiolites and greenstone belts. *GSA Today, 16*(10), 4–11.

Stetter, K. O. (1996). Hyperthermophiles in the history of life. In G. R. Bock & J. A. Goode (Eds.), *Evolution of hydrothermal ecosystems on Earth (and Mars?)* (pp. 1–18). Chichester: Wiley.

Sugitani, K., Yamamoto, K., Adachi, M., Kawabe, I., & Sugisaki, R. (1998). Archaean cherts derived from chemical, biogenic and clastic sedimentation in a shallow restricted basin; examples from the Gorge Creek Group in the Pilbara Block. *Sedimentology, 45*, 1045–1062.

Sugitani, K., Nagaoka, T., Mimura, K., Grey, K., Van Kranendonk, M., Minami, M., et al. (2006). Discovery of possible microfossils from c. 3.4 Ga Strelley Pool Chert, Kelly Group, Pilbara Craton: Evidence for antiquity of life and biotic diversity? *Geophysical Research Abstracts, 8*, 02562.

Summons, R. E., Jahnke, L. L., Hope, M., & Logan, G. A. (1999). 2-methylhopanoids as biomarkers for cyanobacterial oxygenic photosynthesis. *Nature, 400*, 554–557.

Teske, A., Hinrichs, K.-U., Edgcomb, V., de Vera Gomez, A., Kysela, D., Sylva, S. P., et al. (2002). Microbial diversity of hydrothermal sediments in the Guaymas Basin: Evidence for anaerobic methanotrophic communities. *Applied and Environmental Microbiology, 68*, 1994–2007.

Thompson, D. A. W. (1917). *On growth and form*. Cambridge: Cambridge University Press.

Thorseth, I. H., Furnes, H., & Heldal, M. (1992). The importance of microbiological activity in the alteration of natural basaltic glass. *Geochimica et Cosmochimica Acta, 56*, 845–850.

Thorseth, I. H., Torsvik, V., Daae, F. L., Pederson, R. B., & Keldysh-98 Scientific Party. (2001). Diversity of life in ocean floor basalts. *Earth and Planetary Science Letters, 194*, 31–37.

Tice, M. M., & Lowe, D. R. (2004). Photosynthetic microbial mats in the 3, 416-Myr-old ocean. *Nature, 431*, 549–552.

Tice, M. M., & Lowe, D. R. (2006). The origin of carbonaceous matter in pre-3.0 Ga greenstone terrains: A review and new evidence from the 3.42 Ga Buck Reef Chert. *Earth Science Reviews, 76*, 259–300.

Torsvik, T., Furnes, H., Muehlenbachs, K., Thorseth, I. H., & Tumyr, O. (1998). Evidence for microbial activity at the glass-alteration interface in oceanic basalts. *Earth and Planetary Science Letters, 162*, 165–176.

Trewin, N. H., & Rice, C. M. (2004). The Rhynie hot-spring system: Geology, biota and mineralization. *Transactions of the Royal Society of Edinburgh: Earth Sciences 94* (246 pp.).

Tyler, S. A., & Barghoorn, E. S. (1963). Ambient pyrite grains in Precambrian cherts. *American Journal of Science, 261*, 424–432.

Ueno, Y., Isozaki, Y., Yurimoto, H., & Maruyama, S. (2001). Carbon isotopic signatures of individual Archean microfossils (?) from Western Australia. *International Geology Review, 43*, 196–212.

Ueno, Y., Yoshioka, H., Maruyama, S., & Isozaki, Y. (2004). Carbon isotopes and petrography of kerogens in 3.5 Ga hydrothermal silica dykes in the North Pole area, Western Australia. *Geochimica Cosmochimica Acta, 68*, 573–589.

Van Kranendonk, M. J. (2006). Volcanic degassing, hydrothermal circulation and the flourishing of early life on Earth: A review of the evidence from c. 3490–3240 Ma rocks of the Pilbara Supergroup, Pilbara Craton, Western Australia. *Earth Science Reviews, 74*, 197–240.

Van Kranendonk, M. J., Hickman, A. H., Williams, I. R., & Nijman, W. (2001). *Archean geology of the east Pilbara granite-greenstone Terrane Western Australia – A field guide.* Western Australia Geologic Survey, Record 2001/9, 134 pp.

Van Kranendonk, M. J., Webb, G. E., & Kamber, B. S. (2003). Geological and trace element evidence for a marine sedimentary environment of deposition and biogenicity of 3.45Ga stromatolitic carbonates in the Pilbara Craton, and support for a reducing Archaean ocean. *Geobiology, 1*, 91–108.

Van Kranendonk, M. J., Collins, W. J., Hickman, A., & Pawley, M. J. (2004). Critical tests of vertical vs. horizontal tectonic models for the Archaean East Pilbara Granite-Greenstone Terrane, Pilbara Craton, Western Australia. *Precambrian Research, 131*, 173–211.

Van Zuilen, M. A., Lepland, A., & Arrhenius, G. (2002). Reassessing the evidence for the earliest traces of life. *Nature, 418*, 627–630.

Vearncombe, S., Barley, M. E., Groves, D. I., McNaughton, N. J., Mikuchi, E. J., & Vearncombe, J. R. (1995). 3.26 Ga black-smoker type mineralization in the Strelley Belt, Pilbara Craton, Western Australia. *Journal of Geological Society London, 152*, 587–590.

Von Damm, K. L., Oosting, S. E., Kozlowski, R., Buttermore, L. G., Colodner, D. C., Edmond, J. M., et al. (1995). Evolution of east Pacific rise hydrothermal vent fluids following a volcanic eruption. *Nature, 375*, 47–50.

Wacey, D., McLoughlin, N., Green, O. R., Stoakes, C. A., & Brasier, M. D. (2006). The 3.4 billion-year-old Strelley Pool Sandstone: A new window into early life on Earth. *International Journal of Astrobiology, 5*, 333–342.

Wacey, D., Kilburn, M. R., McLoughlin, N., Parnell, J., Stoakes, C. A., & Brasier, M. D. (2008). Use of NanoSIMS to investigate early life on Earth: Ambient inclusion trails in a c.3400 Ma sandstone. *Journal of the Geological Society London, 165*, 43–53.

Wacey, D., McLoughlin, N., & Brasier, M. D. (2008). Looking through windows onto the earliest history of life on Earth and Mars. In J. Seckbach & M. Walsh (Eds.), *From fossils to astrobiology* (pp. 39–68). Springer: Dordrecht, Netherlands.

Wacey, D. (2010). Stromatolites in the ~3400 Strelley Pool Formation, Western Australia: examining biogenicity from the macro- to the nano-scale. *Astrobiology, 10*, 381–395.

Wacey, D (2010a). Stromatolites in the ~3400 Ma Strelley Pool Formation, Western Australia: examining Biogenicity from the Macro- to Nano- Scale. *Astrobiology, 10*, 381–395.

Wacey, D., Saunders, M., & Kilburn, M. R. (2010b in review) Microbially-mediated pyrite oxidation in a 3.4 billion-year-old sedimentary environment. A new pyrite-based microbial metabolism on the early Earth.

Walsh, M. M. (1992). Microfossils and possible microfossils from the early Archaean Onverwacht Group, Barberton Mountain Land, South Africa. *Precambrian Research, 54*, 271–292.

Walsh, M. M., & Lowe, D. R. (1999). Modes of accumulation of carbonaceous matter in the Early Archaean: A petrographic and geochemical study of the carbonaceous cherts of the Swaziland Supergroup. In D. R. Lowe & G. R. Byerley (Eds.), *Geologic evolution of the Barberton greenstone belt, South Africa, Geological Society of America, Special Papers, 329* (pp. 115–132). Colorado: Boulder.

Walter, M. R. (1976). *Stromatolites*. Amsterdam: Elsevier. 790 pp.

Walter, M. R., Buick, R., & Dunlop, J. S. R. (1980). Stromatolites, 3, 400–3, 500 Myr old from the North Pole area, Western Australia. *Nature, 284*, 443–445.

Westall, F. (2005). Life on the early Earth: A sedimentary view. *Science, 308*, 366–367.

Westall, F., de Wit, M. J., van der Dann, J., de Gaast, S., Ronde, C. E. J., & Gerneke, D. (2001). Early Archaean fossil bacteria and biofilms in hydrothermally-influenced sediments from the Barberton greenstone belt, South Africa. *Precambrian Research, 106*, 93–116.

Zhang, Y., & Golubic, S. (1987). Endolithic microfossils (Cyanophyta) from early Proterozoic stromatolites, Hebei, China. *Acta Micropalaeontologica Sinica, 4*, 1–12.

Chapter 15
Evolutionary Trends in Remarkable Fossil Preservation Across the Ediacaran–Cambrian Transition and the Impact of Metazoan Mixing

Martin D. Brasier, Jonathan B. Antcliffe, and Richard H.T. Callow

Contents

Abstract A unifying model is presented that explains most of the major changes seen in fossil preservation and redox conditions across the Precambrian–Cambrian transition. It is proposed that the quality of cellular and tissue preservation in Proterozoic and Cambrian sediments is much higher than it is in more recent marine deposits. Remarkable preservation of cells and soft tissues occurs in Neoproterozoic to Cambrian cherts, phosphates, black shales, siliciclastic sediments and carbonates across a wide range of environmental conditions. The conditions for remarkable preservation were progressively restricted to more marginal environments through time, such as those now found in stagnant lakes or beneath upwelling zones. These paradoxes can no longer be adequately explained by recourse to a series of ad hoc explanations, such as those involving unusually tough organic matter in the Ediacaran, or unusual seawater chemistry, or even the role of microbial biofilms alone. That is because the exceptions to these are now too many. Instead, we suggest that elevated pore water ion concentrations, coupled with the almost complete lack of infaunal bioturbation, and hence the lack of a sediment Mixed-layer, provided an ideal environment for microbially-mediated ionic concentrations at or near the sediment–water interface. These strong ionic gradients encouraged early

M.D. Brasier (✉) J.B. Antcliffe, and R.H.T. Callow
Department of Earth Sciences, University of Oxford, Parks Road, Oxford OX1 3PR, UK
e-mail: martinbrasier@yahoo.co.uk

P.A. Allison and D.J. Bottjer (eds.), *Taphonomy: Process and Bias Through Time*, Topics in Geobiology 32, DOI 10.1007/978-90-481-8643-3_15,

cementation and lithification of sediments, often prior to complete decomposition of delicate organic structures. Seen in this way, not only did the biosphere evolve across the Precambrian–Cambrian transition. Fossilization itself has evolved through time, and never more dramatically so than across this interval.

1 Introduction

"If my theory be true, it is indisputable that before the lowest [Cambrian] stratum was deposited, long periods elapsed, as long, or possibly far longer than the whole interval from the [Cambrian] age to the present day: and that during these vast yet quite unknowable, periods of time the world swarmed with living creatures." (Darwin 1859).

It took a hundred years of research for Darwin's words of 1859 to be seen for what they were: a remarkable prediction about the nature of the Precambrian fossil record. For most of the time since Darwin, there was for example, no concept of the vast expanse of Precambrian time, nor was there any evidence for a distinct biota. But we now realize that the Precambrian world was indeed 'teeming with life'. Furthermore, it can now be argued that the fossil record is qualitatively better than anyone of Darwin's time could ever have dared to imagine (e.g. Brasier 2009).

Analysis of taphonomy in the latest Precambrian (the Ediacaran Period) is intricately linked to one of the most exciting questions in paleobiology: just how real was the Cambrian explosion? Was it an explosion of animals or merely an explosion of fossils? To answer this, we need to understand not only the nature of fossil preservation in the Cambrian but also in the preceding Ediacaran Period (c. 635–542 Ma). Herein we review the concept of a bias in the fossil record *towards* remarkable preservation in the Ediacaran interval.

Good preservation can, of course, take place in a variety of ways. Understandably, the various changes in the quality of preservation, particularly of unmineralized tissues, across the Ediacaran–Cambrian boundary have received a range of distinct explanations. Most of these have tended to focus upon oceanic phenomena such as sea water chemistry, or upon superficial features such as surface mats. Hence, the decline away from high-resolution siliceous, phosphatic and tufa-like calcareous preservation of cellular materials have been explained by chemical causes, such as a decline in sea water silica (Maliva et al. 1989, 2005), phosphate (Brasier 1992a, b), or $p\mathrm{CO_2}$ and alkalinity (Arp et al. 2001; Riding 2006a), whereas the reduction in siliciclastic preservation has been attributed to a physical cause, namely the loss of benthic microbial mats (Gehling 1999). Each explanation has its merits but each shares a common problem too – a lack of universal explanatory power. Put another way, why should each of these different factors have coincided in time? Could there have been a single ultimate cause or trigger? In the following review, we consider these ideas and place them alongside the explanatory potential of the hypothesis illustrated in Fig. 1 (see Callow and Brasier 2009b).

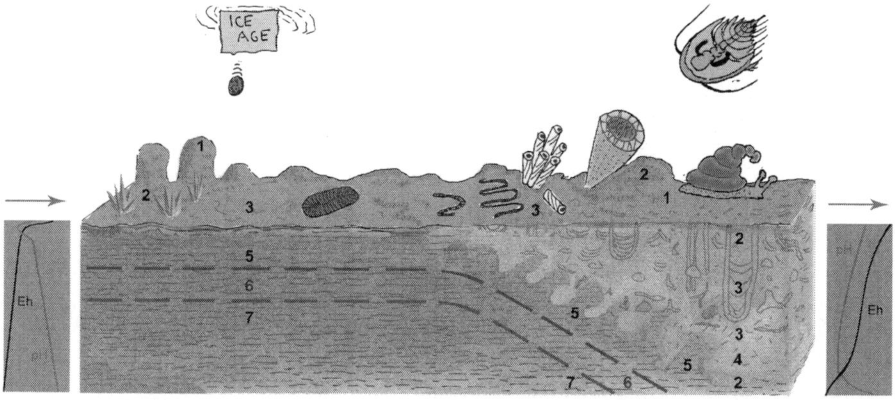

Fig. 1 Model showing how the biosphere revolution from Ediacaran (*at left*) to Cambrian times and later (*at right*) shifted the position of the biogeochemical cycles and hence the quality of seafloor preservation. The evolution of burrowing, grazing and scavenging across the Ediacaran–Cambrian boundary introduced an actively maintained mixed layer (see Seilacher and Pflüger 1994; McIlroy and Logan 1999; Seilacher 1999; Bottjer et al. 2000; Droser et al. 2002, 2004; Bailey et al. 2006 and references in text). This not only brought about the disruption of formerly pervasive microbial mats (Seilacher and Pflüger, 1994), but it also brought about seminal changes in the position of important redox boundaries. Each of these five taphonomic windows discussed in the text was extremely sensitive to Eh and pH. In the Ediacaran, the redox boundary was rather sharp and typically lay high in the sediment profile so that high levels of mineral saturation could build up near the sediment-water interface. Early lithogenesis could often entomb fossil remains before their decay. During and after the Cambrian, expansion in both the extent and depth of bioturbation pushed down the redox boundary and made it more diffuse. This increased the recycling of organic matter before it could become fossilized, and lowered the pH. The associated explosion of biomineralized shells helped to buffer the falling pH sediments. Zones of ionic saturation and early lithogenesis lay further down within the sediment profile. The numbered metabolic processes are broadly as follows: (1) Oxygenic photosynthesis, including cyanobacteria: $CO_2 + H_2O \rightarrow CH_2O + O_2$. (2) Calcium carbonate precipitation: $Ca^{2+} + 2HCO_3^- \rightarrow CaCO_3 + CO_2 + H_2O$. This requires raised pH. (3) Aerobic respiration, including metazoans: $CH_2O + O_2 \rightarrow CO_2 + H_2O$. This tends to reduce pH and Eh. (4) Calcium carbonate dissolution: $CaCO_3 + CO_2 + H_2O \rightarrow Ca^{2+} + 2HCO_3^-$. This tends to raise pH. (5) Calcium phosphate precipitation. This requires Ca availability and some alkalinity. (6) Anaerobic respiration by sulfate-reducing bacteria: $2CH_2O + SO_4^{2+} \rightarrow 2HCO_3^- + HS + H^-$. This tends to increase pH and reduce Eh. (7) Anaerobic respiration by methanogenic bacteria: $2CH_2O + H_2O \rightarrow CH_4 + HCO_3^- + H^-$. This tends to increase pH and reduce Eh. Adapted from Callow & Brasier (2009)

The model shown in Fig. 1 is focussed upon the role of the bioturbated surface layer – the so-called 'mixed layer' (sensu Bromley and Ekdale 1984) – and its associated subsurface chemistry. In the Cambrian to modern ocean (shown at right), this top 10 cm or so of sediment was, and still is, typically mixed and processed by aerobic activities including metazoan burrowing and grazing (McIlroy and Logan 1999) plus metazoan to microbial oxidation of organic matter (cf. Martin and Sayles 2003). Processes including vertical and lateral particle mixing and bioirrigation within this zone (e.g. Aller 1978, 1982, 1984, 1994; Martin and Sayles 2003; Burdige 2006) have a considerable effect upon the distribution of reactive solids such as organic matter, iron and sulfur minerals, as well as solutes such as O_2, CO_2, CH_4, HCO_3^-, and PO_4^{3-} (see Burdige 2006). Exposure to oxidation in this mixed layer can be quite long, with a typical residence time for C_{org} of ~10^4 years (Martin and Sayles 2003). This gives ample time for aerobic metabolism to decompose much (about 80%) of the reactive organic materials. Build-up of metabolic CO_2 also takes place in this zone, leading to a downward decrease in pH that is typically buffered by dissolution of $CaCO_3$ shells (e.g. Morse 2003). Added to this is the significant 'weathering' effect of particle digestion within the digestive tracts of metazoans (McIlroy et al. 2003). Anaerobic processes, such as denitrification, sulfate reduction and methanogenesis together oxidize 30% or less of the remaining organic matter, mainly in microbial zones beneath the mixed layer. Since mixing encourages upward diffusion of products, some of these (especially iron and sulfur compounds) are able to participate repeatedly as electron donors and acceptors (Martin and Sayles 2003; Burdige 2006). In this way, organic materials will typically be consumed before all available electron acceptors have been used up. This means that relatively little C_{org} is left (c. 10%; e.g. Martin and Sayles 2003) to enter the rock record.

Conditions on the seafloor in the Ediacaran and earlier periods (Fig. 1, shown at left) were significantly different from those of today (Fig. 1; Seilacher 1956; Brasier 1979, 1992b; Seilacher and Pflüger 1994; Droser et al. 1999, 2002, 2004; Hagadorn and Bottjer 1999; McIlroy and Logan 1999; Bottjer et al. 2000; Jensen 2003; Jensen et al. 2005). Before the evolution of metazoan burrowers and in the presence of benthic microbial mats (Seilacher and Pflüger 1994), the mixed layer must have been confined to the effects of solute diffusion, perhaps compressed within the top ~1 cm below the sediment–water interface. That being so, the redox boundary will have lain much closer to the surface, with sulfate-reduction and methanogenesis playing much more significant roles, as can be seen in some modern estuaries and lacustrine systems (cf. Martin and Sayles 2003). The contribution of alkaline solutes, arising from both sulfate-reduction and methanogenesis could then have been much more important than now. Being released closer to the sediment surface, they will have increased pore-water alkalinity, encouraging the precipitation of both calcium phosphate and calcium carbonate (cf. Morse 2003). Microbial mats and biofilms at or near the surface would have further limited diffusion at the sediment – water interface (e.g. Gehling 1999) and would have provided ideal sites for crystal nucleation.

In brief, this model predicts that conditions in the Ediacaran to earliest Cambrian were well-suited to both rapid and high quality impregnation and cementation of organic materials by a variety of taphonomic mechanisms. This was because the important zones of fossil lithogenesis lay at, or near, the sediment–water interface. High quality cellular preservation of this kind may also be connected to higher levels of oceanic stagnation (e.g. Briggs and Crowther 2001), as indicated by studies of Cryogenian to Cambrian carbon and sulfur isotopes (Brasier 1992a, b; Shields et al. 1997; Kimura and Watanabe 2001; Fike et al. 2006; Schröder and Grotzinger 2007) and iron contents (Canfield et al. 2007).

The biological revolution at the base of the Cambrian is defined (see Brasier et al. 1994) by metazoan recycling of carbonaceous matter through the activities of bioturbation as well, of course, as by grazing (including zooplankton), scavenging and biomineralization. The appearance of these new Phanerozoic strategies led to the development of an actively controlled sediment mixed layer for the first time in Earth history (Seilacher and Pflüger 1994). Metazoan burrowing led to downward stretching of the aerobic zone from ~1 to 10 cm or more. A significant change in the pH of sediments followed from metazoan inputs of respiratory CO_2 that was buffered by raised rate of dissolution of carbonate grains, not least by the 'newly invented' $CaCO_3$ shells of metazoans. Together, these new process are predicted to have led to an increase in the average depth at which lithogenesis was taking place in the sediment. Furthermore, these changes are likely to have brought about long-term decreases in the quality of cellular preservation, as discussed below.

The following review examines the evidence for taphonomic changes, especially of unmineralized tissues, within five different modes of preservation: siliceous; phosphatic; siliciclastic; carbonaceous; and carbonate. The review then goes on to consider various competing models and explanations for these phenomena, including the role played by the evolution of the mixed layer itself.

2 Siliceous (Gunflint-type) Preservation

Precambrian silica deposits are truly non-uniformitarian (Perry and Lefticariu 2003). From about 2700 to 1900 Ma, occasionally fossiliferous, siliceous banded iron formations (BIFs; Fig. 2a) dominated deep-sea silica sedimentation in a world generally believed to have significantly lower levels of atmospheric oxygen (Han and Runnegar 1992; Bjerrum and Canfield 2002). Although the genesis of these unusual sediments is far from understood, their demise and disappearance after 1800 Ma may be related in some way to evolution of the atmosphere (see Holland 2006) and/or to ocean pH and temperature (see Perry and Lefticariu 2003). These laterally extensive sediments are significant in the Precambrian because they preserve organic-walled microfossils, including coccoid cells and filaments, as for example in the Gunflint Chert (Barghoorn and Tyler 1965). BIF-like sediments reappear during a brief interval in the Neoproterozoic (c. 720–580 Ma), coincident with the so-called 'snowball earth' intervals (Hoffman and Schrag 2002), although younger BIF-like deposits are only known from settings of intense hydrothermal activity (e.g. the Red Sea in the Cenozoic; see Butuzova et al. 1990). Both BIFs and other kinds of widespread seafloor chert precipitation (e.g. seafloor bedded cherts) largely disappeared after c. 1800 Ma, to be replaced by nodular or lenticular chert within carbonate sediments, often formed within evaporative and peritidal environments, which sometimes bear exceptionally preserved microfossil assemblages (Maliva et al. 1989, 2005).

Fossiliferous cherts of Mesoproterozoic to Cambrian age are found across a range of depositional environments from deep marine to peritidal settings (Figs. 2, 3, 5) where the silicification of organic cellular materials and microbial sheaths can be remarkably common (Fig. 2b; Table 1). The petrifaction of microfloras within peritidal cherts is common in these Meso- to Neoproterozoic cherts, such as the Boorthanna Chert of Western Australia (Fig. 2b) or the Bitter Springs Chert of central Australia (e.g. Barghoorn and Tyler 1965; Schopf 1968; Schopf and

Fig. 2 (**a**) before the Cambrian, siliceous sediments, like these banded iron formations (BIFs), can be found across a wide range of settings from the deep aphotic zone to the supratidal zone and may be occasionally fossiliferous (e.g. Gunflint Chert). The millimetre scale laminations (*arrow*) are rich in iron and maybe seasonal in origin. From the Hammersley Iron Formation Dale Gorge, Karijini National Park, Western Australia (c. 2400 Ma). Lens cover c. 5cm in diameter. (**b**) the early silicification of cells, such as those of *Eoentophysalis* sp. from the c. 925 Ma Boorthanna chert of Western Australia, is common in many Proterozoic sediments (see also Table 1). Arrow shows probable photosynthetic cyanobacterial coccoid cells undergoing binary fission. Scale bar 1mm for (**b**)

Fairchild 1973; Schopf and Klein 1992). Such cherts seemingly acted as 'traps' that show a taphonomic bias towards small organic structures including cellulose cell walls, mucilaginous sheaths, and possibly even subcellular structures (Oehler 1977) or molecular biomarkers (Hod et al. 1999). Nucleation sites for silica formation near the sediment surface also appear to have been provided by decaying organic matter (Knoll 1985). It appears that it was relatively easy, therefore, for silica and/or silicates to precipitate directly or diagenetically within a range of Proterozoic–Cambrian marine settings.

Table 1 The changing pattern of soft tissue preservation in siliceous deposits through Earth history, and in particular, across the Ediacaran–Cambrian transition. This shows how the preservation of coccoid cells and microbial filaments by silicification was abundant and common throughout the Meso- to Neoproterozoic. The silicification of soft tissues, including cells, is known through the Cambrian, but appears to decrease in frequency and in quality throughout the remainder of the Phanerozoic, where silicified delicate cellular of subcellular materials and less common. Despite occasional reports of exceptional preservation in normal marine conditions, silicification in the Phanerozoic tends to be confined either to unusual environments (sinters, alkaline lakes) or to unusually recalcitrant organic materials (e.g. lignin or biominerals). From sources cited in the text and in references

	Coccoid benthic cells	Filament sheaths	In situ mats	Acritarch (phytoplonkton)	Eggs (embryos)	Micro-arthropods	Micro-faecal pellets
Cainozoic							
Cretac		+					
Jurassic							
Triassic							
Permian							
Carbonif							
Devonian				+		+	
Silurian				+		+	
Ordovician					+	+	
U. Camb		+			+	+	+
M. Camb		+			+	+	+
L. Camb		+			+	+	+
U. Ediac		+			+		
M. Ediac	+	+		+	+		
Proteroz	+	+		+			

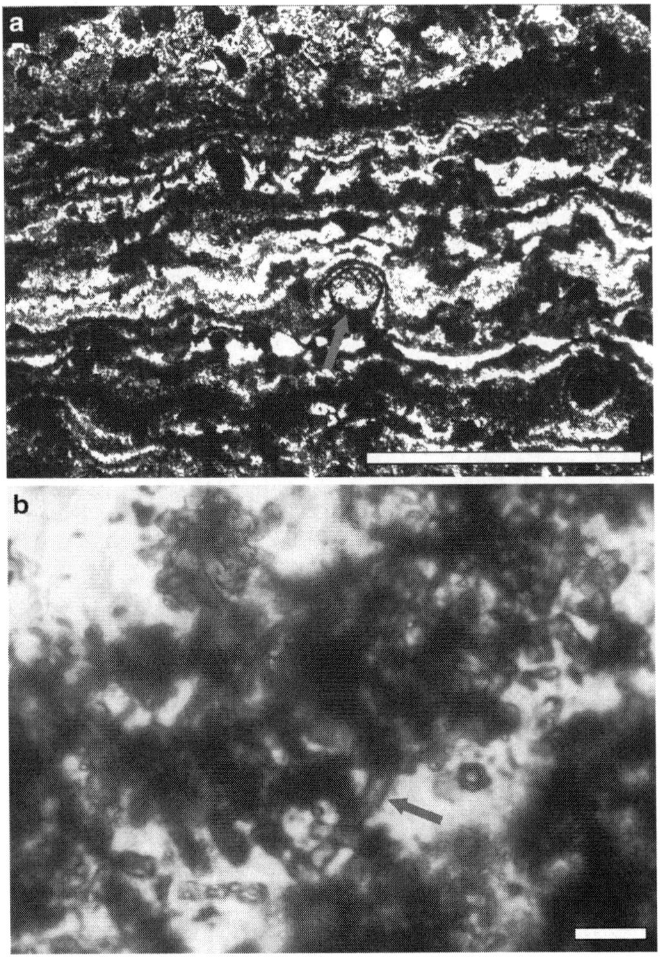

Fig. 3 (**a**) Relatively deep-water siliceous sediments are widely encountered across the Ediacaran–Cambrian transition, like these finely laminated cherts from the Tal Formation in the Lesser Himalaya of India (c. 543–530 Ma). The white layers consist of purer chert while the darker layers are rich in calcium phosphate and organic matter. Note the cross section through a sphaeroidal, embryo-like structure (*arrow*). From Brasier & Callow (2007). (**b**) Close up image of the alternating dark and white layers shows the presence of abundant small filaments and sheaths (*arrow*) of probable benthic microbial origin, from the latest Ediacaran to basal Cambrian Tal Formation. Scale bar 1 mm for (**a**) and 10 mm for (**b**)

Cellular preservation of acritarchs and other organic-walled microfossils in silica continued through the Ediacaran interval (e.g. Table 1; Tiwari and Knoll 1994; Xiao 2004). Some of these cherts, such as those of the Doushantuo Formation of China, were deposited close to storm wave base where they preserve multicellular algae (Xiao 2004), giant spiny acritarchs (Zhou et al. 2006) and putative sponge spicules (Li et al. 1998; Yin et al. 2001). But it is across the Precambrian–Cambrian

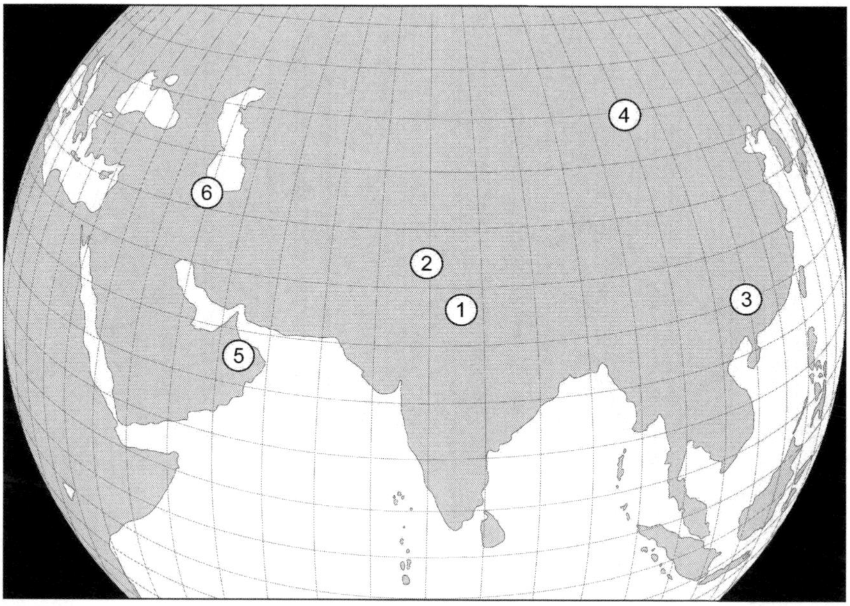

Fig. 4 Modern geographic map showing localities studied herein for examples of silicification, phosphatization or black shale (carbonaceous) preservation during the Ediacaran to early Cambrian. 1, Tal Formation chert, phosphorite and black shale, Uttar Pradesh, Lesser Himalaya, India (c. 545–530 Ma). B, Hazara chert and phosphorite of Pakistan (c. 545–530 Ma). 3, Doushantuo Formation chert, phosphorite and black shale of South China Platform (c. 630–580 Ma); Meishucun Formation phosphorite (c. 535 Ma), Badaowan Formation black shale and chert (c. 530 Ma), and Chengjiang 'black shale' biota (c. 525 Ma), all from the South China Platform. 4, Khubsugul chert and phosphorite of NW Mongolia and Tsagaan Oloom phosphate and chert beds of SW Mongolia (both c. 550–545 Ma). 5, Fara Formation chert and phosphorite of north Oman, and Ara Group black shale and chert ('Athel silicilyte') of south Oman (c. 545–540 Ma). 6, Soltanieh Formation black shale and phosphorite of the Elburz Mountains of NW Iran (c. 545–530 Ma). For further details and sources, see the text

boundary that widespread silicification of organic materials in subtidal settings again becomes prominent, often in association with phosphates and black shales (Fig. 3; see Mazumdar and Banerjee 1998; Shen and Schidlowski 2000; Amthor et al. 2005). These lithologies can be used as indicators of high productivity and eutrophic conditions (e.g. Brasier 1995) and this seems to be the first interval of Earth history in which these distinctive lithologies can be found as a 'nutrient trinity'. It is also during this interval that the first volumetrically significant silica skeletons emerge, including those of hexactinellid sponges and radiolarians (e.g. Brasier et al. 1997, but see also Porter et al. 2003). Siliceous preservation of soft tissues in the Ediacaran–Cambrian boundary interval (Fig. 3) seems to be limited to relatively resistant microbial sheaths, and lacks the delicate coccoidal cellular clusters known from earlier Proterozoic times (Fig. 2b). This is in spite of abundant

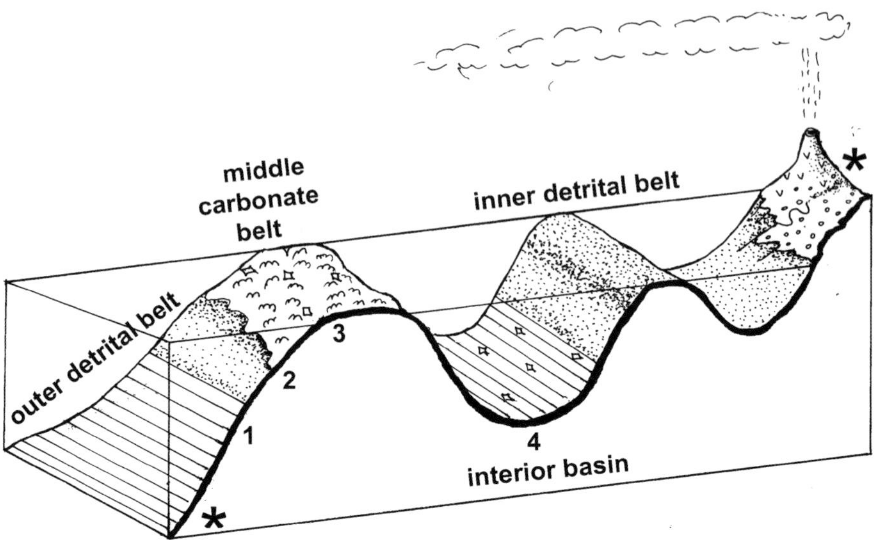

Fig. 5 Model contrasting the wide range of Ediacaran to early Cambrian environments where cellular preservation occurs by silicification (numbers 1 to 4, outlined below) with the very limited range of Cretaceous to modern settings, including terrestrial hot springs (*asterisk at right*) and biogenic diatomites, radiolarites and flint nodules (*asterisk at left*). This enormous contraction in the zone of silica deposition and preservation shows the extraordinary effects of the Cambrian explosion upon the silica cycle. The Ediacaran to Cambrian examples studied are as follows: 1–2, Doushantuo Formation of China, Tsagaan Oloom Formation of southwest Mongolia, and Tal Formation of India. 3, Khufai, Buah and Ara Formations of Oman. 4, Athel silicilyte of southern Oman. From sources in text and references

silica-rich deposits of this age, such as those from Arabia (Gorin et al. 1982; Amthor et al. 2005; Schröder et al. 2005). The preservation of possible fungal filaments alongside discoidal microfossils within aluminosilicate minerals (Callow and Brasier 2009a; Brasier et al. 2009b) may provide further evidence for the preservation of unmineralized tissues by silica/silicate minerals during this period.

Silicified cells from marine environments after the Precambrian–Cambrian boundary become increasingly rare (Table 1). Examples include silicified *Michrystridium*-like acritarchs from the Lower Cambrian Yurtus Formation of South China (Yao et al. 2005), silicified embryo-like structures in Middle Cambrian cherts in China (Lin et al. 2006) and poorly preserved cells of cyanobacteria in the Upper Cambrian to Lower Ordovician Durness Formation of Scotland (Brasier 1977 and unpublished data).

Chert-rich sediments remain common throughout the Phanerozoic. Common examples include flint nodules within the Cretaceous chalk of southern England, or the radiolarites and diatomites often associated with upwelling zones (e.g. Schubert et al. 1997; Kidder and Erwin 2001). Silica-rich sediments such as diatomites are

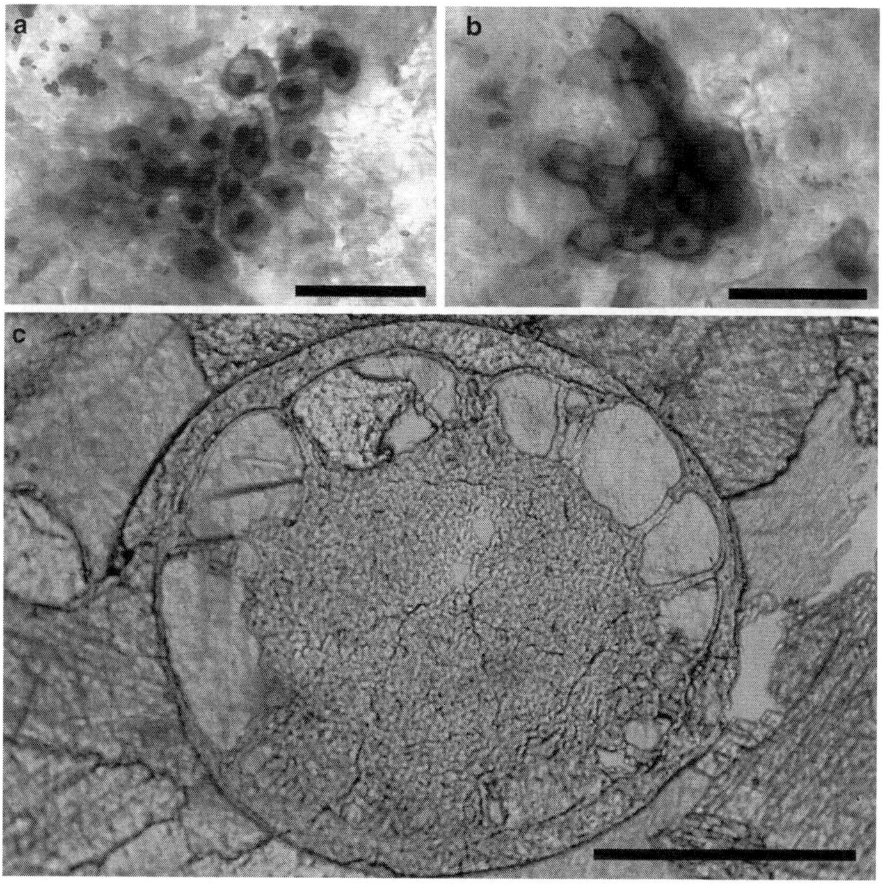

Fig. 6 Examples of the remarkably high-quality of phosphatic preservation that can be seen in Neoproterozoic phosphorites. (**a**) and (**b**) Colonies of presumed photosynthetic cells, from the Torridon Group of northwest Scotland (c. 1000 Ma). Note the presence of dark structures within the cells, which may represent contracted cell contents. (**c**) Cross section through one of the clusters of cells for which the Doushantuo phosphorite is rightly renowned (c. 630–580 Ma). Such forms have been regarded as cnidarian polyps or stalks. Photo courtesy of Zhou Chuanming. Scale bar Scale bar 5 mm for (**a**) and (**b**) and 200 mm for (**c**) From Brasier & Callow (2007)

often used as indicators of high levels of nutrient supply or upwelling (Brasier 1995). However, although chert nodules and silica-rich sediments can be common, the quality of organic and cellular preservation within younger cherts remains poor in the vast majority of marine examples. Typically, only the most resistant organic materials (e.g. wood) or relatively resistant cyanobacterial sheaths are preserved (Table 1). In other cases, chert can be often be seen replacing biomineral skeletons (e.g. Mu and Riding 1983; Schubert et al. 1997; Kidder and Erwin 2001), although this is often a late stage diagenetic process and the chert preserves little of the original biomineral or its ultrastructure. Examples of high-quality soft tissue and cellular preservation in Phanerozoic cherts (see Tobin 2004) include reports of preserved coccoid cells or filaments from evaporative peritidal or marginal marine settings (Wardlaw and Collinson 1978) and the exceptional preservation of arthropods in the alkaline lakes of the Miocene Barstow Formation (Park 1995; Park and Downing 2001). Terrestrial sinter deposits around hot springs, such as the Lower Devonian Rhynie Chert of Scotland (Trewin and Rice 2004) and silicified algae and bacteria forming in situ around modern hot spring systems, represent further examples of exceptional preservation of organic materials by silica (Konhauser et al., 2001; Jones et al. 2007).

Modern laboratory experiments (e.g. Toporski et al. 2002) and geological observations around modern hot springs (Konhauser et al. 2001; Jones et al. 2007) demonstrate that petrifaction of cells is aided by raised concentrations and rapid precipitation of silica. In modern oceans, silica deposition is typically restricted to areas of high nutrient flux, as seen for example in regions of equatorial upwelling, where large volumes of biological opaline silica formed by radiolarians, diatoms and sponges are deposited on the seafloor. Only about 10% of this silica enters the geological record, owing to the high surface area:volume ratio of opaline silica skeletons and their ready dissolution within the undersaturated water masses and pore waters typical of the modern ocean (e.g. Martin and Sayles 2003). Cementation within Cenozoic diatomites and radiolarites also seems to be rather slow and late, taking place around subsurface concentrations of organic matter to form nodules around faecal pellets, sponges and burrow systems. Cellular materials therefore degrade before they can be encased in silica, allowing only the more-resistant organic materials such as spores, cysts and wood, to enter the fossil record. All this clearly suggests that biological innovations near the start of the Cambrian could have directly influenced both the time and place of silica authigenesis.

Conditions for siliceous preservation on the Ediacaran to early Cambrian seafloor were markedly different from those found today. Extraction of silica by diatoms was lacking and that by radiolarians and sponges was limited (Maliva et al. 2005, but see Porter et al. 2003). This is thought to have favoured significantly higher silica saturation states in the water column, with potential for direct silica precipitation and chert formation in deeper neritic to peritidal settings. Added to this is the likelihood of greater alkalinity in waters near the sediment–water interface, from the greater activities of sulfate-reduction and methanogenesis (see Fig. 1). This would have raised the dissolved concentrations of silica in both pore waters and the local water column. There was, it seems, still considerable potential at this time for the rapid precipitation and preservation of delicate organic materials during very early diagenetic lithification by silica. It can therefore be argued that high-quality cellular and sub-cellular silicification appears to have been more common in many marine settings during the Meso- to Neoproterozoic (Fig. 5) in comparison with the Phanerozoic. The zone of exceptional silicification appears to have moved away from the shallow marine environments and into areas where unusual geochemical conditions are present, such as evaporative basins, lacustrine settings and terrestrial hot spring environments.

3 Phosphatic (Doushantuo-type) Preservation

The preservation of fossil organisms by means of diagenetic phosphate arguably provides a litmus test for the quality of fossil preservation during the emergence of Metazoa (Cook and Shergold 1984; Xiao and Knoll 1999; Hagadorn et al. 2006; Brasier and Callow 2007; Brasier 2009; Dornbos 2009 this volume). As a key biolimiting nutrient, phosphate ions (PO_4^{2-}) are rapidly utilized by photoautotrophs and are typically undersaturated in the surface layer of the modern oceans. Phosphate ions increase in concentration beneath the photic zone in the Oxygen Minimum Zone (OMZ), where microbial processes remineralize organic materials, thereby releasing phosphate ions (Föllmi 1996; Compton et al. 2000; Martin and Sayles 2003; Ruttenberg 2003; Burdige 2006). Similar processes of microbially mediated phosphate ion-release also operate within the sediment profile at the redox boundary (Fig. 1). Phosphate ions are highly sensitive to the redox state. In oxidizing conditions such as those found within the modern sediment mixed layer and the upper well-mixed layer of the oceans, phosphate ions tend to complex with ferric oxides, which removes bioavailable phosphate from both pore waters and the water column (Föllmi 1996; Compton et al. 2000; Ruttenberg 2003; Burdige 2006). This process can, however, be reversed under anaerobic and acidic conditions, as for example during burial beneath the sediment mixed layer or within the oxygen minimum zone (e.g. Van Cappellen and Ingall 1994; Föllmi 1996; Ruttenberg 2003). Precipitation of phosphate on the modern seafloor therefore occurs in reducing conditions, such as those beneath upwelling zones, typically at water depths of c. 200–400m on the continental slope (Piper and Link 2002). In these settings, phosphatization typically occurs in moderately alkaline environments where abundant phosphate ions are supplied by the remineralization of organic matter. In modern settings such as this, it is typically materials such as faecal pellets which become phosphatized.

There is good evidence that phosphate concentrations and distributions are strongly controlled by microbial processes (Krajewski et al. 1994; Wilby et al. 1996), although in most cases there is no evidence that organisms or microbes act as preferential sites for phosphate nucleation. Preservation of fossils by diagenetic phosphate minerals can occur in a number of ways. These include the formation of phosphatic internal moulds within shells, the replacement of calcium carbonate biominerals, or the replacement or encrustation of organic tissues (Brasier 1990; Xiao and Knoll 1999). Phosphatic internal moulds or casts provide little or no information about the soft parts of an organism, or about the details of unstable or ephemeral biomineral phases. The phosphatic replacement of primarily calcareous skeletons (e.g. Lamboy 1993) is common, as for example in early

Cambrian shelly fossils such as *Anabarites* (see Kouchinsky and Bengtson 2002; Feng and Sun 2003) and other small shelly fossils (see Porter 2004). This style of preservation can replicate the original microstructure of primary biomineral phases (e.g. unstable aragonite) and therefore provides information which is typically lost during fossil preservation. The encrustation or replacement of unmineralized tissues (e.g. Briggs et al. 2005) also provides valuable paleobiological information about soft tissues and examples from the Ediacaran–Cambrian interval include the mucilaginous sheaths of putative fossil cyanobacterium *Spirellus* (Fig. 7a) and the putative fossil eggs, embryos and hatchlings of cnidarians (Bengtson and Zhao 1997; Koushinsky et al. 1999; Donoghue et al. 2006b; Hagadorn et al. 2006).

Some of the earliest described examples of phosphatic preservation are dated to c. 1000 Ma from the Torridonian of Scotland (Peach et al. 1907; Peat and Diver 1982; Turnbull et al. 1996; Brasier 2009). These reveal a remarkable quality of preservation in both cells and cell contents (Figs. 6a, b). Phosphate

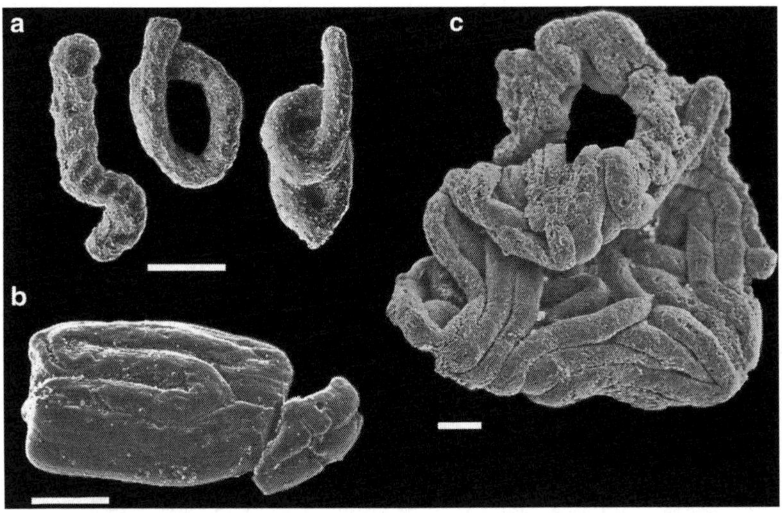

Fig. 7 A dramatic transformation took place in the phosphatic preservation of organic matter between the late Ediacaran and the late Cambrian that is interpreted to be related to a downward shift of the oxygen minimum zone and the associated zone of phosphogenesis. (**a**) Phosphatized cells and extracellular sheaths of spirally twisted cyanobacterium *Spirellus*, from the Tal Formation phosphorite of India of Ediacaran–Cambrian boundary age (c. 545–530 Ma). This style of preservation, from within the photic zone, is widely known from the phosphorite localities labelled in Fig. 4. (**b**) and (**c**) Similar examples of late Cambrian age typically show compacted filaments, which are here interpreted as the gut contents of zooplankton and animal grazers living below the photic zone, from the Orsten Biota, *Agnostus pisiformis* Zone of Kinnekulle, Sweden (c. 480 Ma). After the Cambrian, preservation of phytodetritus became increasingly rare because the phosphogenic zone began to fall even further below the photic zone. For the majority of the Phanerozoic it is processed materials and faecal matter that constitutes most phosphate deposits. Scale bar 100 mm for (**a**)–(**c**). From Brasier & Callow 2007

does not appear interbedded with other members of the 'nutrient trinity' (i.e. chert and black shale) until much later, however, as for example in the Doushantuo Formation of China (see Fig. 4, >580 Ma, Condon et al. 2005). These Ediacaran phosphorites preserve what many now regard as the oldest animal embryos (Fig. 6c; see Xiao et al. 1998; Xiao and Knoll 1999; Hagadorn et al. 2006), although they have been interpreted as vesicles of giant sulfur oxidizing bacteria (Bailey et al. 2007). It is also at this time that the primary calcium carbonate tubes of *Cloudina* can be replaced by diagenetic phosphate, preserving details of what is thought to be the original biomineral ultrastructure (Feng et al. 2003; Hua et al. 2005).

The zenith of phosphatic preservation at any time in Earth history was reached during the earliest stages of the Cambrian, in particular along the fabled 'Silk Route' (Fig. 4; see Shergold and Brasier 1986; Brasier 1989, 1992a, b). This zone once lay along the northern margins of a vast ocean (McKerrow et al. 1992) whose anoxic water masses upwelled into shallow water carbonate lagoons. These early Cambrian phosphorites typically reveal two modes of phosphatic replacement (Brasier 1990). The first involves replacement of organic tissues (cf. Briggs et al. 2005), such as can be seen in the mucilaginous sheaths of putative fossil cyanobacterium *Spirellus* (Fig. 7a) as well as in putative fossil eggs, embryos and hatchlings of cnidarians (Bengtson and Zhao 1997; Koushinsky et al., 1999), some of which may represent aphotic fungal microbes (Brasier et al., in press). The second mode of phosphatization typically involves replacement of primarily calcareous skeletons (cf. Lamboy 1993), as seen in early Cambrian *Anabarites* (see Kouchinsky and Bengtson 2002; Feng and Sun 2003) and other small shelly fossils (Porter 2004).

Patterns of phosphatization through time shows several interesting trends (Tables 2 and 3; Brasier and Callow 2007; Dornbos 2009 this volume). Examples from the c. 1000 Ma Torridonian, for example, as well as those from the >580 Ma Doushantuo Formation, include clear evidence for preservation of cell walls, and potentially for sub-cellular architecture (see Fig. 6a, b). By the start of the Cambrian, however, such remarkable preservation becomes much harder to detect. This is especially curious given the vast abundance of phosphatic deposits at this time (e.g. Brasier 1992b). Nor is there evidence for high-quality cellular to sub-cellular preservation in any marine, post-Ordovician phosphates known to us (other than of resistant acritarch vesicles, see below).

Of relevance here may be a trend that also can be discerned in the kinds of organisms that are phosphatized (Table 1). Both the Torridonian and Doushantuo phosphatic assemblages consist largely of algal thalli, acritarch vesicles and embryo-like cell clusters. The presence of well preserved algal thalli suggests rapid phosphatization of the shallow seafloor *within the photic zone* (e.g. Xiao and Knoll 1999). Preservation of large masses of coccoid benthic algae, however, becomes rare from the base of the Cambrian. Here, the remains of primary producers seem to be confined to bundles of cyanobacteria-like sheaths and filaments, like those of *Spirellus* (Fig. 7a; see also Zhegallo et al. 2000; Brasier and Callow 2007). By middle and late Cambrian times, phosphatic preservation of photoautotrophs in the

Table 2 Tables 2 and 3 show the changing pattern of soft bodied preservation in phosphatic deposits through Earth history, and in particular, from the Ediacaran Period and through the Phanerozoic. The tables show how well-preserved embryo-like structures, coccoid benthic algae and cyanobacteria and microbial filaments are typical of Ediacaran to early Cambrian deposits. The shift towards the preservation of recalcitrant materials, faecal pellets and putative zooplankton later in the Cambrian is suggested to be related to a deepening of the redox boundary within the water column, coeval with a similar lowering in the sediment profile. This was followed by continued downwards and offshore migration of the phosphogenic zone through the Phanerozoic. From sources cited in the text and in references

	Subcellular structures	Cell walls	Eggs/Embroyos	Microbial sheaths
Recent	x	x	Laboratory simulations of Artemia egg and larvae decay Gostling et al. 2009	Cyanobacteria sheaths from atolls. Trichet and Fikn 1997
Neogene	x	x	x	Cyanobacterial sheaths from NW Arabian Sea. Rao et al. 2008
Paleogene	x	x	x	x
Cretaceous	x	Coccoids cells from Voronezh, Maleokina 2003	x	Cyanobacterial sheaths from Voronezh Maleokina 2003
Jurassic	x	Coccoid cells from Nusplingen lagerstatle, Briggs et al. 2005	x	Filaments from Nusplingen lagerstatle. Briggs et al. 2005
Triassic	x	x	x	x
Permian	x	x	x	x
Carboniferous	x	x	x	x
Devonian	x	Mazuelloids (acritarchs), Kremer 2005	x	x
Silurian	x	Mazuelloids from Holy Cross mils of Poland Kremer 2005	x	x
Ordovician	x	Mazuelloids (acritarchs), Kremer 2005	Markuelia from Vinni Fm. Donoghue et al. 2006a	x

Upper Cambrian	x	x	Arthropod embryos from Duyun s. China Zhang and Pratt. 1994	x
Middle Cambrian	x	x	Markuelia from shallow marine Georgina Basin of Queensland Donoghue et al. 2006a	
Lower Cambrian	Possible subcellular structures within embroyos from the L Cambrian Kuanchuanpu Fm. Donoghue et al. 2006a	x	Olivooides and Markuelia from L. Cam shallow water carbonates. Bengtson and Zhao 1997	Filaments of the sheath Spirellus from L. Cam Tal Fm, Brasier and Callow 2007
Neoproterozoic	Possible organelles within embryos from 580 Ma shallow marine Doushantuo Fm. Hagadorn et al, 2006	Coccoi cyanobacteria and algal thalli from 580 Ma shallow marine Doushantuo Fm. Hagadorn et al. 2006	Possible animal embryos from 580 Ma shallow marine Doushantuo Fm. Hagadorn et al. 2006	Cyanobacterial sheaths and filaments from 580 Ma shallow marine Doushantuo Fm Hagadorn et al. 2006
Mesoproterozoic	Possible algal nuclei or | plasmolysed cyanobacterial cell contents from lacustrine Torridon Group of Scotland Brasier and Callow 2007	Coccoid and filamentous cyanobacteria and algae from lacustrine Torridon Group of Scotland. Brasier and Callow 2007	x	Cyanobacterial sheaths from lacustrine Torridon Group of Scotland. Brasier and Callow 2007
Paleoproterozoic	x	x	x	x
Archean	x	x	x	x

Table 3 See Table 2 caption

	Faecal pellets	Biominerals	Unmineralized arthropods	Other Metazoan tissues
Recent	Lab experiments on lobster faecal pellets (McIlroy pers Comm.)	√	Various arthopods in 16th century cesspits of York. UK McCobb et al. 2004	x
Neogene	√	√	Insects from Riversteigh. Queenstand, Arena 2008	x
Paleogene	√	√	Insects from Oligocene of Ronheim. Germany Hellmund and Hellmund 1996	x
Cretaceous	√	√	Ostracods and copepods from Sanatana Fm of Brazil, Wilkinson et al. 2007	Fish tissues from Sanatana Fm of Brazil Martill 1988
Jurassic	√	√	x	Squids from Oxford clay. Allison 1988
Triassic	√	√	Ostracods from Spitsbergen Weitschat 1983	Bivalve soft tissues from Muschelkalk Klug 2005
Permian	√	√	x	Goniatite cameral membranes. Polizotto et al. 2007
Carboniferous	√	√	x	Goniatite cameral membranes. Polizotto et al. 2007
Devonian	√	√	x	x
Silurian			x	x
Ordovician	√	√	Zooplankton crustaceans from deep water Orsten biota Maas et al. 2006	x
Upper Cambrian	Bundles of packaged cyanobacteria from deep water Orsten biota of Sweden, Brasier and Callow 2007	√	Zooplankton crustaceans from deep water Orsten biota Maas et al. 2006	x

Middle Cambrian	Faecal strings from Mt Cap Fm, Butterfield 2001	Molluscs. SSFs and problematica common Porter 2004	Arthropod integument from shallow marine Georgina Basin. Walossek et al 1993	Burgess Shale Gut contents from Burgess Shale Butterfield 2002
Lower Cambrian	x	Molluscs. SSFs problematica in shallow water limestones from around the world Bengtson et al. 1990	Arthropods from shallow water Comley Limestone of Shropshire. Siveter et al. 2001	x
Neoproterozoic	x	Cloudina and Sinotubulites from shallow marine latest Ediacaran Dengying Fm. Feng et al. 2003	x	x
Mesoproterozoic	x	x	x	x
Paleoproterozoic	x	x	x	x
Archean	x	x	x	x

so-called 'Orsten biotas' is largely restricted to rare clots of putative cyanobacterial material (Fig. 7b, c). Here the filaments can be compressed and distorted in a way consistent with metazoan faecal processing (Butterfield 2001). Benthic algal or microbial remains preserved in phosphate are seldom seen or reported after this date, with the exception of a few filaments (e.g. Trichet and Fikri 1997; Rao et al. 2000; Maleokina 2003).

Trends in the preservation of giant spiny acritarchs are equally curious. Acritarchs like those from the Doushantuo Formation appear almost globally in the mid Ediacaran (Vidal 1990; Tiwari and Knoll 1994) but are barely known through the Cambrian to Ordovician. They then 'reappear' in some Silurian and Devonian phosphate deposits, where they are known as mazuelloids or muellerisphaerids (Zhou et al. 2001; Kremer 2005). Their morphology compares with that of resting-cysts like *Baltisphaeridium*, a Paleozoic acritarch of widely assumed pyrrhophyte affinity and phytoplanktonic mode of life (Kremer 2005). The much larger size of these phosphatized acritarchs (c. 300 μm) has accordingly been attributed to high levels of nutrients in the water column (Zhou et al. 2001; Kremer 2005), though this planktonic interpretation is open to question (Butterfield 2007).

We also draw attention to the limited time span over which embryo- or egg capsule-like structures are preserved in phosphate through time (Table 2 and 3). They appear in the middle Ediacaran of South China (Hagadorn et al. 2006; Yin et al. 2007) and remain common in the earliest Cambrian phosphatic deposits (Koushinsky et al. 1999; Donoghue et al. 2006b; Pyle et al. 2006) but then dwindle to a few records in the middle and late Cambrian and the early Ordovician (Cheng and Liu 2004; Donoghue et al. 2006a), with later examples largely restricted to large, priapulid-like *Markuelia*. This decline is interesting because laboratory experiments show that real animal embryos can be fairly resistant to decay (e.g. Martin et al. 2000; Raff et al. 2006).

Phosphatic preservation of small arthropods, including putative zooplankton, likewise shows distinctive patterns (Tables 2 and 3). They first appear in the lower Cambrian of England (Siveter et al. 2001) and become widespread within middle to late Cambrian phosphates from Sweden (Maas et al. 2006), Newfoundland (Walossek et al. 1994), Siberia (Müller et al. 1995) and China (Dong et al. 2005). Younger marine deposits typically lack remarkably preserved small arthropods, however, despite major phosphatic deposits in the Permian (e.g. Piper and Link 2002), the Cretaceous (e.g. Maleokina 2003) and to a lesser extent the Jurassic (e.g. Allison 1988; Wilby et al. 1996). Phosphatized small arthropods are found, however, within lacustrine sediments of Cretaceous to Miocene age (e.g. Bate 1972; Müller 1985; Park and Downing 2001). The rather poor quality of preservation in most Phanerozoic phosphatic sediments (see below) is not inconsistent with phosphogenesis taking place fairly slowly at the sediment–water interface, or at greater depth in the sediment, so that organic materials have degraded before being encased by phosphate.

Prior to widespread disruption by animal activity across the Ediacaran–Cambrian transition, the locations and mechanisms of phosphatic preservation appear to have been significantly different from those observed today. Firstly, it is predicted that the phosphorus-rich redox boundary layer of the early oceans is likely to have been much shallower and much sharper (Figs. 1 and 8), making very early phosphatization possible within the shallow photic zone (Fig. 8). This, and the associated scarcity of benthic grazers and burrowers, can together explain the preservation of benthic algae and other photic-zone flora and fauna (e.g. Xiao and Knoll 1999). From near the start of the Cambrian, however, increasing oxygenation of the upper water column by both nekton and zooplankton (Signor and Vermeij 1994; Logan et al. 1995; Butterfield 2007) and of the sediment surface by bioturbation (McIlroy and Logan 1999) is likely to have forced the phosphogenic zone downward, not only through the water column but also deeper into the sediment (Fig. 1). Shallow bioturbation will also have lowered the pH within the upper mixed layer, encouraging $CaCO_3$ dissolution, raising Ca^{2+} levels and increasing Ca-phosphate saturation states yet further (Dr. G. Shields, pers. comm. 2007).

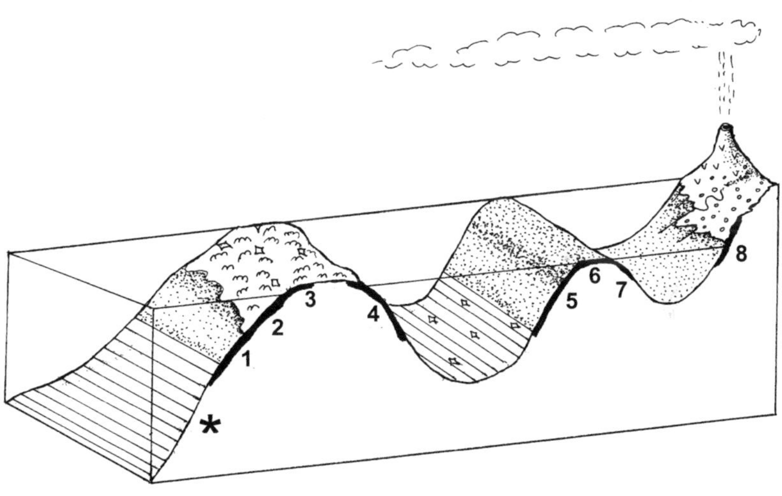

Fig. 8 Model contrasting Ediacaran to early Cambrian settings where cellular preservation occurs in phosphate (numbers 1 to 8, outlined below) with the limited range of Cretaceous to modern settings, including deep slope phosphorites (*asterisk at left*). This demonstrates the effects of the Cambrian explosion upon the phosphorus cycle. Ediacaran to Cambrian examples studied are as follows: 1–2, Doushantuo Formation of China, Khybsugul phosphorite of northwest Mongolia, Tal Formation of India. 3, Soltanieh Formation of Iran, Dengying Formation of China. 4, Fara Formation of Oman. 5–7, Chapel Island Formation of Newfoundland. 8, Torridon Group of Scotland, St John's Group of Newfoundland. From sources in text and references

4 Siliciclastic (Ediacara-type) Preservation

'Siliciclastic' preservation is here used to refer to the style of preservation best-known from the macrofossils that are commonly known as 'the Ediacara biota', including, *Dickinsonia* (Figs. 9a and 10a; Brasier and Antcliffe 2008), *Charnia* (Fig. 9b; Gehling et al. 2005; Antcliffe and Brasier 2007), *Rangea* (Fig. 11a) and a diverse array of related forms analysed in detail by Brasier and Antcliffe (2009). Such fossils are widely reported from around the world and are typically regarded as impressions made by macroscopic, flexible, soft-bodied organisms that came to be preserved beneath event beds, such as storm sands and volcanic ashes (Narbonne 2005; Droser et al. 2006; Fedonkin et al. 2007).

The earliest example of this type of three-dimensional preservation of soft-bodied organisms within siliciclastic sediments include the unusual 'string of beads' markings on bedding-planes, *Horodyskia*, of possible protistan-grade (Dong et al. 2008), which are reported from the 1500 Ma Belt Supergroup (Fedonkin and Yochelson 2002). The Ediacaran Period witnessed the greatest abundance of this style of preservation (see below), whereas similar environments from later Phanerozoic successions typically lack comparable soft-bodied impressions. Whatever the nature of the taphonomic window, it appears to have narrowed during the Cambrian. Late Ediacaran examples are scarcer but include the discoidal *Nimbia* structures of Crimes and McIlroy (1999) from Norway, *Beltanelliformis* markings from England (McIlroy et al. 2005), while rare Cambrian examples include petalonamaean-type fossils described by Hagadorn et al. (2000) from the Lower Cambrian of Nevada, and medusae impressions from the Upper Cambrian of Wisconsin (Hagadorn et al. 2002). Post-Cambrian examples are even more rare and are of highly limited diversity and include the possible cnidarians from the Ordovician of Morocco (Samuelsson and Butterfield 2001; Alessandrello and Bracchi 2003), the enigmatic worm-like organisms from the Devonian of New York (Conway-Morris and Grazhdankin 2005) and possible medusoid cnidarians from the Cretaceous (Bell et al. 2001).

Ediacaran fossils preserved within siliciclastic sediments are distributed across a remarkably wide range of facies that, unfortunately, show rather limited temporal and geographic overlap (see Grazhdankin 2004). Each taphofacies also tends to preserve its own distinctive assemblage of fossils, further compounding the problem. Consequently, it can appear difficult to state whether changes in the biota seen from one region to another are the result of taphonomic and facies difference alone, or due to the evolution of the creatures themselves.

Three main types of Ediacaran preservation–lower surface, upper surface and within-bed (Figs. 9, 10a and 11a) can be distinguished. Rapid cohesion or cementation of the *lower bed* of sediments is the mode of preservation typically found across the Avalon terrane (e.g. England and Newfoundland). Most examples of this kind are dated to between 575 and 555 Ma (e.g. Brasier and Antcliffe 2004; Narbonne 2005). Such preservation (Fig. 9b) is accompanied in most cases by Pompeii-like smothering of frondose fossils beneath layers of waterlain volcanic ash. With fossils like

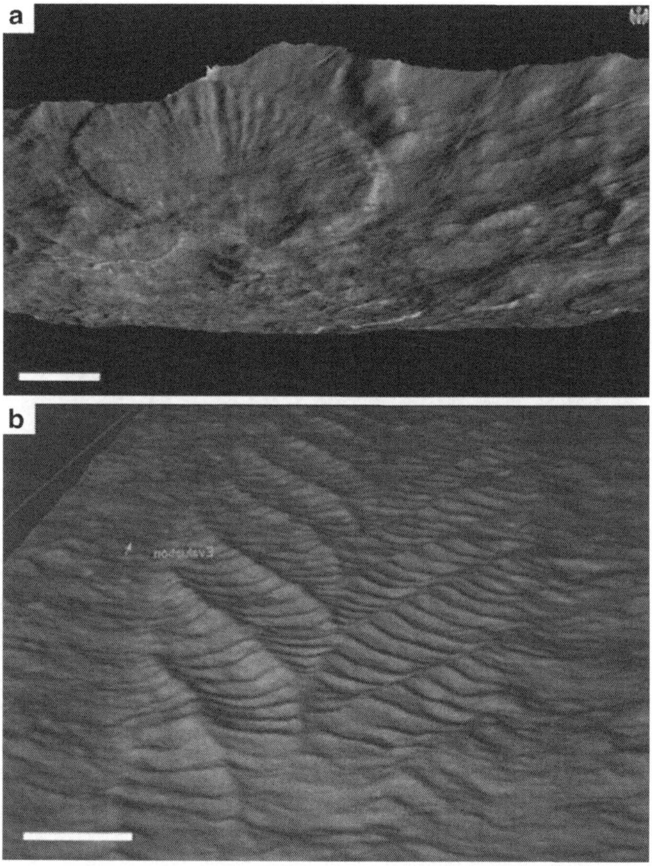

Fig. 9 Laser scanned images of Ediacara biota preservation of soft bodied macrofossils, showing the difference between lower- and upper surface preservation of wrinkle-marked sediment layers. (**a**) Lower-surface type preservation of *Dickinsonia costata* here preserves the form of its top surface topography on the lower bedding surface of a slab from the Rawnsley Quartzite (c. 555 Ma), Ediacara sheep station, Flinders Ranges, South Australia. (**b**) Upper-surface type preservation of the holotype of *Charnia masoni* here preserves the form of its bottom surface topography, on the upper bedding surface of fine grain volcanic tuffs from the Maplewell Series (c. 560 Ma), Charnwood Golf Course, Leicestershire, England. Scale bar 1cm for (**a**) and (**b**)

Charnia, *Charniodiscus* and *Bradgatia*, the bottom surface was, then, mainly preserved as negative impressions made by the organism as it lay against the substrate (see Brasier and Antcliffe 2009). Such fossils often show some degree of transport by bottom currents. Interestingly, however, these frondose organisms clearly lived well below the photic zone, as shown by the evidence for deposition on volcanoclastic talus slopes well below storm wave base. This means that the wrinkle-marked or 'elephant-skin' top surfaces with which they are often associated (e.g. Bailey 2002) are likely to have been made by microbes of a heterotrophic or chemoautotrophic nature rather than by photoautotrophs like cyanobacteria (Brasier et al., in press).

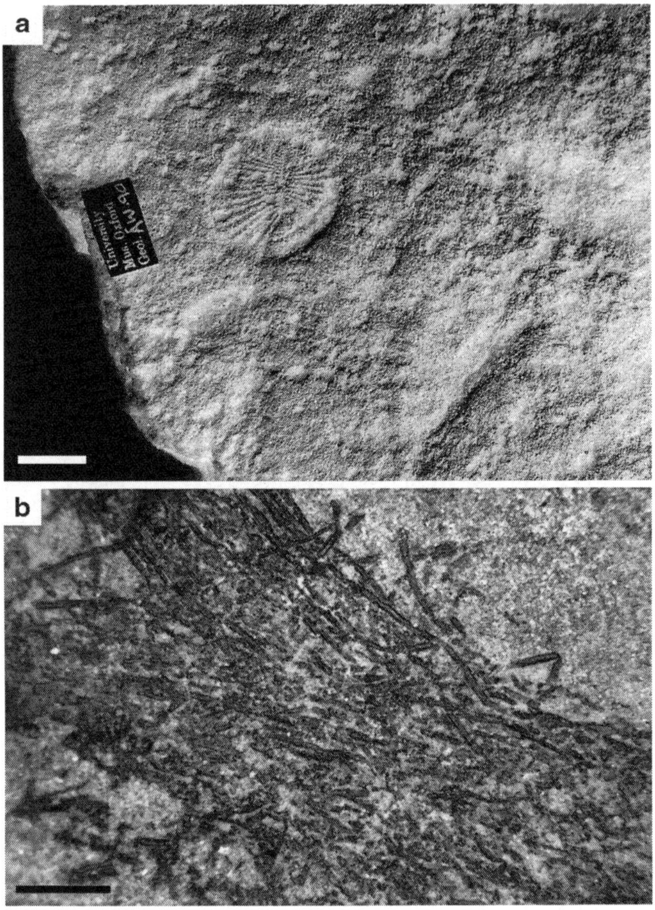

Fig. 10 Details showing the nature of preservation of the Ediacara biota in association with ancient microbial mats, here from lower-surface type preservation. (**a**) Preservation of *Dickinsonia costata* on the under surface of a rock slab, here shown as a optically inverted digital image to indicate how the fossil and the surrounding seafloor may have looked before its burial by sand. Note the undulose and pustular structures of inferred microbial origin that not only surround but also underlie the structur, showing that the fossil was extremely thin. From lower bedding surface of a slab from the Rawnsley Quartzite (c. 555 Ma), Ediacara sheep station, Flinders Ranges, South Australia. From Callow & Brasier (2009). (**b**) Close up of such a surface directly adjacent to a mould of *Dickinsonia costata*, showing both parallel and entwined microbial filaments replaced by pyrite (*arrow*). From lower bedding surface of a slab from the White Sea area of Russia (c. 555Ma). From Callow & Brasier (2009). Scale bar 1cm for A and 1mm for (**b**)

Upper surface preservation takes the form of negative moulds *on the base of the overlying bed*, and is best known from South Australia (Gehling 1999) and the White Sea region of Russia (Grazhdankin 2004). Both assemblages are dated to about 558–550 Ma (Martin et al. 2000). In these rocks, it is usually the top surface

Fig. 11 Standard photographic images of further kinds of preservation of Ediacaran soft tissue. (**a**) Within-bed type preservation of *Rangea schneidehorni*, here preserves its three dimensional surface topography (*arrow*) within a slab of the Kuibis Quartzite from Namibia (c. 550 Ma), from the Hans Pflug Collection, Geological Survey, Windhoek, Namibia. (**b**) Bottom surface of a slab of quartz sandstone bearing the intertwined impressions (*arrow*) of possible filamentous microbial or algal impressions known as 'Arumberia', from the Masirah Bay Formation (c. 600 Ma), Kufai Dome, Huqf mountains, central Oman. Scale bar 2cm for (**a**) and (**b**)

of an organism like *Dickinsonia*, that is preserved (Fig. 9a and 10a; Gehling et al. 2005; Brasier and Antcliffe 2008), though positive casts of lower surfaces of less resistant organisms or structures are also known (Narbonne 2005). Such upper surface preservation has at times been attributed to the presence of tissues of great durability (Wade 1968; Seilacher 1992) perhaps like that of modern lichens (Retallack 1994, but see also Waggoner 1995). A more favoured suggestion, explored below, is that the fossils were preserved by microbial mats that formed a kind of 'death mask', maintaining selective aspects of external shape (Gehling 1999; Gehling et al. 2005; Narbonne 2005; Droser et al. 2006; Mapstone and McIlroy 2006).

In South Australia, the sedimentary layers beneath these Ediacaran fossils were at one time interpreted as a muddy seafloor laid down during relatively quiet conditions at mainly subtidal depths (e.g. Goldring and Curnow 1967). Further studies have shown, however, that surfaces preserving Ediacara biota often lack mud. Instead, they typically display a wrinkled texture (Fig. 10a; often called 'elephant skin texture') like that seen on modern microbial mats (Gehling 1987, 1991; Seilacher 1999; Noffke et al. 2001). Associated features typically include pustules, over-steepened ridges, current-induced folding, contortion and tearing, suspended quartz grains, and concentrations of authigenic minerals such as mica in the upper layers and pyrite beneath (see Hagadorn and Bottjer 1997, 1999; Noffke et al. 2001, 2002). These mats may also have aided preservation by trapping and binding of sediment (Narbonne 1998; Gehling 1991, 1999; Noffke et al. 2001).

Well-preserved material from the White Sea region of Russia includes surfaces that were once covered, and locally surrounded, by a mat of filamentous, pyrititized microbes (Fig. 10b; see also Fedonkin and Waggoner 1997; Gehling 1999; Steiner and Reitner 2001; Dzik 2005; Grazhdankin 2004; Gehling et al. 2005). In South Australia, Newfoundland and the Ukraine, comparable surfaces are usually iron-stained, presumably owing to the oxidation of this pyrite to haematite (e.g. Gehling 1999). Rapid preservation of soft-bodied fossils from these regions have therefore been attributed to early formation of a death mask of pyrite (Fig. 10b; see Dzik 2003; Gehling et al. 2005) and/or to the early growth of authigenic clay minerals and mica (Hagadorn and Bottjer 1997; Mapstone and McIlroy 2006) soon after burial. Preservation of Ediacaran fossils within the Khatyspyt Formation of Siberia shows some parallels (Dzik 2005) but here, preservation is due to early lithification by calcium carbonate. A popular hypothesis for upper layer preservation, therefore involves this cohesive mat of filamentous microbes (perhaps sulfur-oxidizing, beggiatoan bacteria) upon the seafloor, typically with sulfate-reducing bacteria thriving *just beneath the surface.* That sulfate-reduction took place on a massive scale from the Ediacaran to late Cambrian is clearly confirmed by the sulfur isotope record (e.g. Shields et al. 1997; Hurtgen et al. 2005). Sulfate-reducers were then able to produce a thin, post mortem layer of pyrite, especially after the organism was smothered by an influx of sand. Sand from the underlying beds could then be mobilized upwards to cast the fossil from below (e.g. *Dickinsonia* in Gehling et al. 2005, fig. 2).

Cohesive microbial mats of this kind survived into the Cambrian in places (e.g. Bailey et al. 2006). Their progressive disruption by new metazoan activities has been used to explain the scarcity of similar preservation at later times in the Phanerozoic (e.g. Allison and Briggs 1991; Bottjer et al. 2000). Microbial mat preservation cannot, however, explain the increasing number of observations in which such a death mask was not involved. Examples of such *within-bed* preservation of macrofossils are well seen in sandstones from the Nama Group of Namibia (Fig. 11a), dated to about 549–542 Ma. Here, soft-bodied fossils such as *Pteridinium*, *Rangea* and *Ernietta* are typically preserved as three-dimensional moulds and casts within the sandstone layer itself. Within-bed preservation of soft-bodied organisms is also known from various taxa in Australia (Glaessner and

Wade 1966), the White Sea (Grazhdankin 2004; Dzik 2005), south-western USA (Hagadorn and Waggoner 2000) and Newfoundland (Narbonne 2004). This has led to the controversial suggestion that such organisms may have lived infaunally (Grazhdankin and Seilacher 2002; but see also Narbonne 2005). It is important to emphasize here that the disappearance of within-bed preservation during the Cambrian cannot be explained by the effects of Phanerozoic bioturbation alone. This is because suitable substrates (well sorted and micaceous quartz arenites without bioturbation) remained common from this time onward; examples of this are legion from lower Cambrian quartzites of Avalonia and Baltica. Excellent within-bed preservation of soft-bodied organisms within siliciclastic sediment beds is also a puzzle because such shallow-water sandstones from near-shore, oxidising, siliciclastic settings are generally found to have very low preservation potential for organic materials at later times. One possibility worth exploring, therefore, is that silica levels in the ocean were still high because of the negligible influence sponges and the absence of radiolarians at this time. Low pH and Eh in surface layers (see Fig. 1) then allowed early silicate (including phyllosilicate) cementation before the body walls had any chance to decay.

Recent discoveries of preserved microbes from bedding planes in argillaceous rocks of Ediacaran age (Callow and Brasier 2009a; Callow and Brasier, 2009b) have highlighted the potential for the preservation of a variety of microbes in a style similar to that of macrofossils during the Ediacaran. Detailed impressions and moulds of filaments and discoids can occur in high densities on siliciclastic bedding-planes and may constitute an important and hitherto unrecognized style of microbial preservation in ancient siliciclastic rocks (Callow and Brasier 2009a, Callow and Brasier 2009b).

Unusual cohesiveness of sediments may also be used to explain enigmatic structures called *Arumberia* (Glaessner and Walter 1975) and *Aspidella* (see Gehling et al. 2000), both largely confined to the Ediacaran Period. *Arumberia* has been reported from numerous sections around the world at this time, including Australia, France, England and Newfoundland (Bland 1984; McIlroy et al. 2005) and Oman (herein). This fossil comprises gently curved or linear subparallel markings, typically preserved as epichnial grooves or hypichnial ridges. Such markings can cover bedding-planes for hundreds of square kilometres in Oman (Fig. 11b). In Australia, they were first interpreted as the remains of a bag-shaped organism (Glaessner and Walter 1975) but later reinterpreted as abiogenic hydraulic structures caused by turbulent flow (Brasier 1979). *Arumberia* is most typically seen on the bottom surfaces of storm event beds (Mapstone and McIlroy 2006) and seems to have been enhanced by the presence of a cohesive substrate stabilized by microbial mats (McIlroy and Walter 1997; McIlroy et al. 2005). New material from the Masirah Bay Formation of Oman shows that, while the markings clearly reflect the flow of bottom currents, they can overlie each other or be intertwined in different directions (Fig. 11b). This suggests that some or all of these lines represent the remains and impressions of long bundles of organic filaments. At one locality in the Longmyndian of England, interwoven carbonaceous filaments some 50μm diameter are preserved in mudrocks from about the same stratigraphic level as

Arumberia (Fig. 13b; see also Peat 1984, Callow and Brasier 2009a). This suggests that *Arumberia* structures arose from microbially stabilized substrates somewhat akin to those indicated by pyritized filaments around *Dickinsonia* from the White Sea region (Fig. 10b) but without the formation of extensive pyrite.

Arumberia markings can be associated with circular impressions called *Aspidella*, both in Australia (Mapstone and McIlroy 2006) and in Avalonia (McIlroy et al. 2005; Narbonne 2005). *Aspidella* has recently been upgraded from a fossil of dubious biogenic origin to an all-encompassing name for discoid impressions (Gehling et al. 2000). Some *Aspidella* may indeed represent the attachment sites of Ediacaran fronds but others seem likely to be microbial (e.g. Grazhdankin and Gerdes 2007) and algal impressions or even abiogenic sedimentary structures (Jensen et al. 2002). Whatever the cause of these circular markings, their sharp three-dimensional preservation on successive stacks of sedimentary laminae seems to be largely absent from Phanerozoic marine sandstones and mudrocks.

In summary, the Ediacara-type biota is preserved across a remarkably wide variety of habitats (Fig. 12) in ways that are barely seen since then (Callow & Brasier 2009b). This pattern of preservation can best be explained by early cohesion and lithification of sedimentary laminae on or just beneath the seafloor, before compaction could erase all topographic expression.

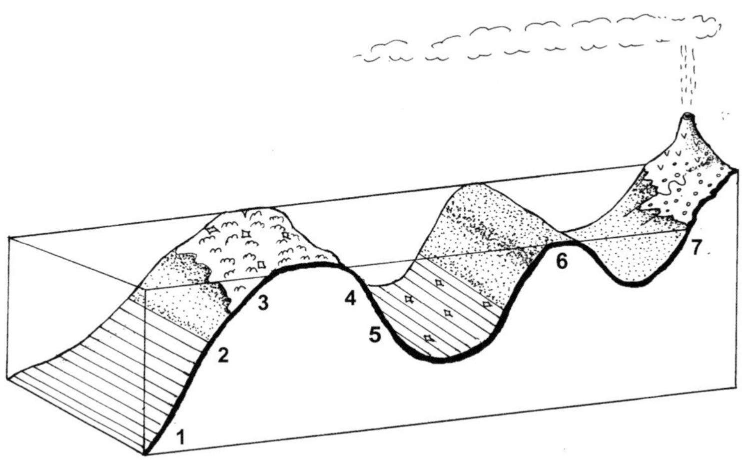

Fig. 12 Model contrasting the wide range of Ediacaran environments where soft-bodied organisms can become preserved in situ within siliciclastic or calcareous sediments (numbers 1 to 7, outlined below) with the general lack of such preservation throughout the Phanerozoic. The disappearance of this kind of preservation is here attributed to slower sediment lithification and rising levels of oxygenation on the seafloor after the Cambrian explosion. The examples are as follows: 1, Charnian Supergroup volcanoclastics of England, and similar rocks of the Conception Group in Newfoundland (c. 580–555 Ma). 2, Longmyndian Supergroup of England and Drook Formation of Newfoundland showing microbial preservation. 3, Khatyspyt Formation of Siberia, and Dengying Formation of South China with calcareous preservation (c. 560–545 Ma). 4, Shuram Formation of Oman. 5, Masirah Bay Formation of Oman. 6, Rawnsley Quartzite of Flinders Ranges, South Australia, White Sea biota of Russia (c. 560–550 Ma); Nama Group of Namibia (c. 550–542 Ma). 7, St John's Group of Newfoundland. From sources in text and references

5 Carbonaceous Film (Miaohe-type) Preservation

Preservation of macrofossils as compressions and carbonaceous films is well known from Proterozoic mudrocks (see Hofmann 1994; Steiner 1994; Zhu et al. 2000). The ~1.9 Ga, possible alga or bacterium *Grypania* is perhaps the first known example (Han and Runnegar 1992), and Proterozoic fossils such as *Chuaria circularis* and *Tawuia* are also typical of this style of preservation (see Hofmann 1994; Steiner 1994; Zhu et al. 2000; Dutta et al. 2006). By Ediacaran times, such assemblages commonly contain disc-shaped macroscopic fossils such as *Beltanelloides* (Fig. 13a).

Fig. 13 (**a**) Black shale bedding-plane showing clusters of macroscopic, carbonaceous discs of *Beltanelloides sorichaevi*, which show concentric wrinkles and folds. From the latest Ediacaran Pusa Shale (c. 545 Ma), Montes de Toledo, central Spain. The field of view is 15cm. (**b**) Petrographic thin-section of shales bearing darker layers packed with abundant entwined carbonaceous filaments. From the Lightspout Formation, Longmyndian Supergroup (c. 556 Ma), Shropshire, England (see Callow and Brasier 2009a). Scale bar 400 mm

Once regarded as eukaryote cells or cell colonies, many of these rounded fossils have been reinterpreted as the compressed envelopes of prokaryote colonies, perhaps like those of living cyanobacterium *Nostoc* (Steiner 1994; Steiner and Reitner 2001; Xiao et al. 2002, but see also Hofmann 1994). Bedding-planes many tens of square metres across can be packed with such carbonaceous compressions during the Ediacaran period, as for example from the Miaohe Formation of China (Xiao et al. 2002), the Pusa Shale of Spain (Fig. 13a; Brasier et al. 1979), the Chapoghlu Shale within the Soltanieh Formation of Iran (Ford and Breed 1973). It is suggested that in some settings, similar vesicles can be preserved in three dimensions and infilled with sediment (*Nemiana*; see Fedonkin 1990; Hofmann 1994). Interestingly, such giant vesicles tend to disappear from levels above the Precambrian–Cambrian boundary.

Petalonamaean organisms such as *Charnia* can also be preserved as carbonaceous films, as for example in the carbonate hosted assemblages of arctic Siberia (Grazhdankin et al. 2008). Elongate carbonaceous ribbons and filaments are also common in the Ediacaran Period (Fig. 13b; e.g. Hofmann 1994). Best known of these is the Miaohe assemblage from the Doushantuo Formation of China, with over twenty taxa of putative algal remains (Xiao et al. 2002). Ribbon-like compressions of *Vendotaenia* and *Tyrasotaenia* are found from the Precambrian–Cambrian boundary interval in both Europe and Newfoundland (Urbanek and Rozanov 1983; Peat 1984; Landing et al. 1988; Vidal and Moczydlowska 1992; Callow and Brasier 2009a). Indeed, carbonaceous preservation reaches a peak during a global anoxic event at this time (Brasier 1992a; b; Kimura and Watanabe 2001; Schröder and Grotzinger 2007). Higher in the Cambrian, simple algal fossils continue to appear alongside carbonaceous compression fossils of the Chengjiang biota (such as arthropod cuticles, Gabbott et al. 2004; Hou et al. 2004) and they can range well into the middle Cambrian (e.g. Briggs et al. 1993; Yang and Zhao 2000).

Carbonaceous ribbons with transverse markings have also been found in several Ediacaran assemblages (Peat 1984; Hofmann 1994; Sun 1994; Fedonkin 2003). These have sometimes been interpreted as the remains of invertebrate fossils, perhaps even of bilaterians (but see Steiner 1994; Xiao et al. 2002). The first carbonaceous remains of likely animal and possible bilaterian origin are the organic-walled tubes of *Sabellidites* from the latest Precambrian and basal Cambrian of Newfoundland and the east European Platform (Urbanek and Rozanov 1983; Gnilovskaya 1996). Simple carbonaceous ribbons known as *Vendotaenia* are known from around the world during the Ediacaran and have been suggested to be among of the most abundant of organisms from this interval (Cohen et al. 2009).

Several factors appear to have allowed the frequent preservation of carbonaceous compression fossils during Ediacaran times. In a world before burrowers and grazers, microbial mats were able to colonize the shallow seafloor during intervals of relatively clay-rich input, directly leading to carbonaceous preservation (Schieber 1986). By the Early Cambrian, when bioturbation and scavenging were becoming more widespread, such preservation begins to disappear. Real carbonaceous mats are not seen, for example, in either the lower Cambrian Chengjiang biota of south China (e.g. Babcock et al. 2001; Gabbott et al. 2004; Hou et al. 2004) nor in the middle Cambrian Burgess Shale-type biotas of North America, though strands of algal material occur in both. These famous invertebrate lagerstätte were favoured by rapid accumulation of clays and silts beneath poorly oxygenated water masses, somewhat below wave base (Fig. 14). Such stagnant conditions were then able to help retard the rates of microbially-induced decay (see Allison and Brett 1995; Butterfield and Nicholas 1996; Hagadorn 2002; Butterfield 2003; Gaines et al. 2005). This 'Burgess Shale type' of preservation is rarely observed after the Cambrian and, even then, is typically limited to a few isolated specimens (Hagadorn 2002; Butterfield 2003).

A complicating factor in studies of carbonaceous preservation is that in many cases, the original organic materials can be transformed by diagenetic reactions into secondary phases such as clay minerals or pyrite (Fig. 10b; Schieber 2002; Gabbott et al. 2004; Page et al. 2008). The high fidelity pyritization of carbonaceous films, via the activities of sulfate-reducing bacteria (Grimes et al. 2001), is known to be associated with Ediacaran fronds and discs and can be recognized by the presence of ancient pyritic laminae (e.g. Mapstone and McIlroy 2006) or by the remains of pyritized filaments themselves (Fig. 10b). Pyritized microbial mats and stromatolites are common across the Precambrian–Cambrian boundary level (e.g. the Tal Formation of India) but thereafter largely disappear from the fossil record. The high-quality pyritization of carbonaceous materials is known sporadically from the

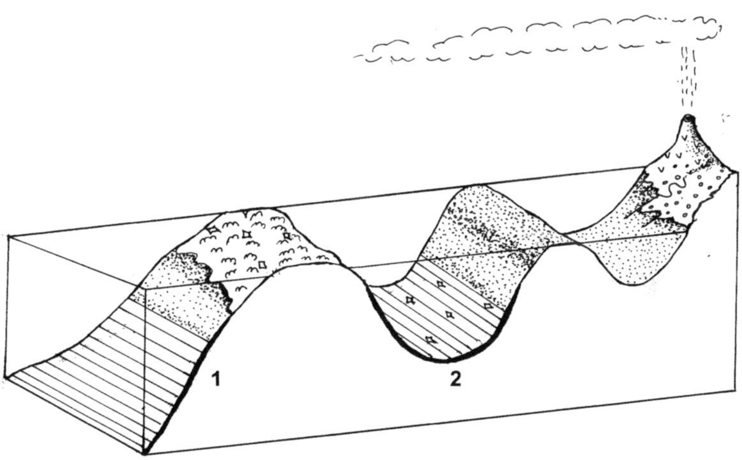

Fig. 14 Model contrasting the range of Ediacaran to early Cambrian showing carbonaceous preservation in marine 'black shales' (numbers 1 and 2, outlined below) with the near absence of such preservation in the marine realm from the Cretaceous onwards. The disappearance of this kind of preservation is here attributed to more efficient recycling and remineralization of organic materials and rising levels of oxygenation on the seafloor since the Cambrian. The examples studied include the following: 1, Miaohe biota of South China (c. 550 Ma); *Vendotaenia*, *Tyrasotaenia* and *Sabellidites* biota of Baltica and Avalonia (c. 550–540 Ma); *Beltanelloides* biota of the Pusa Shales in central Spain, and of Chapoghlu Shale, Soltanieh Formation, Iran (c. 545 Ma); Chengjiang biota of South China (c. 525 Ma); Burgess Shale of British Columbia (c. 500 Ma). 2, Athel silicilyte of South Oman. From sources in text and references

Phanerozoic, as for example in sulfide-rich, reducing settings of the Maotianshan Shales at Chengjiang, China (e.g. Gabbott et al. 2004) or Beecher's trilobite bed in New York State (Briggs et al. 1991). In the majority of cases however, Phanerozoic pyritization is largely confined to the steinkern infills within isolated reducing microenvironments such as shelly fossils or burrows, as seen in the Lower Cambrian of southeast Newfoundland and the East European Platform (e.g. Urbanek and Rozanov 1983; Landing et al. 1988, 1989), which preserve no details of primary soft-tissue morphology or cellular structures. In other Phanerozoic examples, organic materials can be replaced during volatilization reactions by diagenetic aluminosilicate phyllosilicates, as for example in the Burgess Shale and Paleozoic graptolites (Page et al. 2008).

There is a clear contrast between the Proterozoic and Ediacaran intervals, where the carbonaceous-pyritic preservation of microbial and/or algal materials abounds, and Cambrian examples such as the Burgess Shale, which teem with animal fossils but where algal or microbial remains appear to be more rare. A number of factors can be identified which appear to help explain these observations. It can be argued that in a world without burrowers and grazers and with a shallow redox boundary, buried carbonaceous materials were not effectively scavenged by metazoans and more rapidly reached potential zones of preservation within the sediment. This resulted in a greater potential for unmineralized and carbonaceous materials to enter the rock record (Fig. 1). By the early Cambrian, when bioturbation and scavenging were becoming more widespread, such preservation begins to disappear, because all available organic materials are rapidly remineralized by metazoan and microbial processes. Settings where comparable preservation could occur were favoured by rapid smothering of sediment and the accumulation of clays and silts beneath poorly oxygenated water masses below wave base, as for example in the lower Cambrian Chengjiang biota (e.g. Babcock et al. 2001; Gabbott et al. 2004; Hou et al. 2004) and the Middle Cambrian Burgess Shale (Fig. 14). Such poorly mixed and stagnant conditions may have mimicked those of the Ediacaran. In these cases it is commonly pyritic veneers and aluminosilicate layers that are preserved, rather than carbonaceous films themselves, which are rarely seen after the Middle Cambrian (see Hagadorn 2002).

6 Carbonate (Tufa-like) Preservation

Biological activities such as photosynthesis have major influences upon the aqueous carbonate cycle. For instance, as a consequence of the uptake of CO_2 by cyanobacteria or algae during photosynthesis, the saturation state of carbonate in surrounding fluids is increased and this can lead to carbonate precipitation on microbes and their sheaths as well as on and within algal thalli (cf. Lowenstam 1981; Pentecost and Spiro 1990). While cyanobacterially-induced microbial mats may have been present from as early as 2.9 Ga (Noffke et al. 2008), the earliest widely accepted evidence for calcification of microbes is known from the c. 2.5 Ga Campbelrand

Supergroup of South Africa (Kazmierczak and Altermann 2002; Altermann et al. 2006), where putative cyanobacterial filaments can be found within euhedral dolomite crystals, possibly in association with minute aragonite needles. Photosynthetic microbes might have been involved in the construction of earlier Archean stromatolites, although no body fossils are yet known (e.g. Sakurai et al. 2005). Oxygenic cyanobacteria were arguably among the major players in facilitating the so-called 'Great Oxidation Event' (c. 2.5 Ga), which further suggess the existence of cyanobacteria by this time (see also Konhauser et al. 2009).

A curious feature of the early fossil record is the poverty of evidence for calcified microbes and cyanobacterial sheaths before ~800 Ma (Riding 2006b), in spite of their abundance in diagenetic silica deposits. By the late Ediacaran to earliest Cambrian, however, there was a bloom in the abundance of marine microbial and algal carbonate fossils (e.g. *Angulocellularia*, *Renalcis*, *Epiphyton*, *Girvanella*; Grant et al. 1991; Wood 1998; Riding 2006a, b), which may relate to the evolution of carbon dioxide concentrating mechanisms (CCMs) within cyanobacteria (Riding 2006b). Some of these calcified cyanobacteria are known to occur alongside the first putative metazoan carbonate skeletons, including tubular *Cloudina* (Fig. 15c; Germs 1972), goblet-shaped *Namacalathus* (Fig. 15b; Grotzinger et al. 2000), and the large, modular, possibly colonial fossil *Namapoikia* (Wood et al. 2002), not long before the Ediacaran-Cambrian boundary (c. 549 Ma, Grotzinger et al. 1995). Such microbial carbonates reached an acme during the early to middle Cambrian and declined thereafter (Riding 2006b). The late Ediacaran-Cambrian also saw rapid seafloor carbonate cementation in the form of thrombolites and stromatolites with isopachous laminae, giant oolite grains, carbonate flat pebble breccias, edgewise conglomerates, 'molar tooth' carbonate, carbonate crystal fans, tidal flat 'tufas', botryoids and fabric-retentive early diagenetic dolostones (e.g. McCarron 1999; Pratt 1998; Grotzinger et al. 2000; Shields 2002; Sumner and Grotzinger 2004). This range of features can be seen across a wide region, from Namibia, Oman, Siberia, Mongolia and China to North and South America (e.g. Mattes and Conway-Morris 1990; Turner et al. 1993). This suite of features supports high levels of carbonate ion saturation and raises the possibility that abiogenic carbonate precipitation was also able to take place widely onto abiogenic, as well as microbial and/or metazoan templates (i.e. 'tufa-style' precipitation).

Some of these indicators of supersaturation (e.g. calcified microbes) are still found today in settings of unusual geochemical conditions (Fig. 15a). Around many springs for example, carbonate-saturated waters reach the surface and lead to rapid deposition of calcium carbonate crusts (tufa) around microbial filaments and even around chitinous larval skeletons (Fig. 15a), bryophytes, tree stumps, or even around man-made objects (e.g., Brasier et al. 2009). It is generally accepted (e.g. Arp et al. 2001; but see also Riding 2006a, b) that the role of organisms in this style of precipitation is limited to the provision of a suitable substrate plus the involuntary promotion of crystal nucleation by extracellular polymeric substances (Turner and Jones 2005). The subsequent oxidation of the carbonaceous substrates leaves behind only their external moulds in carbonate minerals.

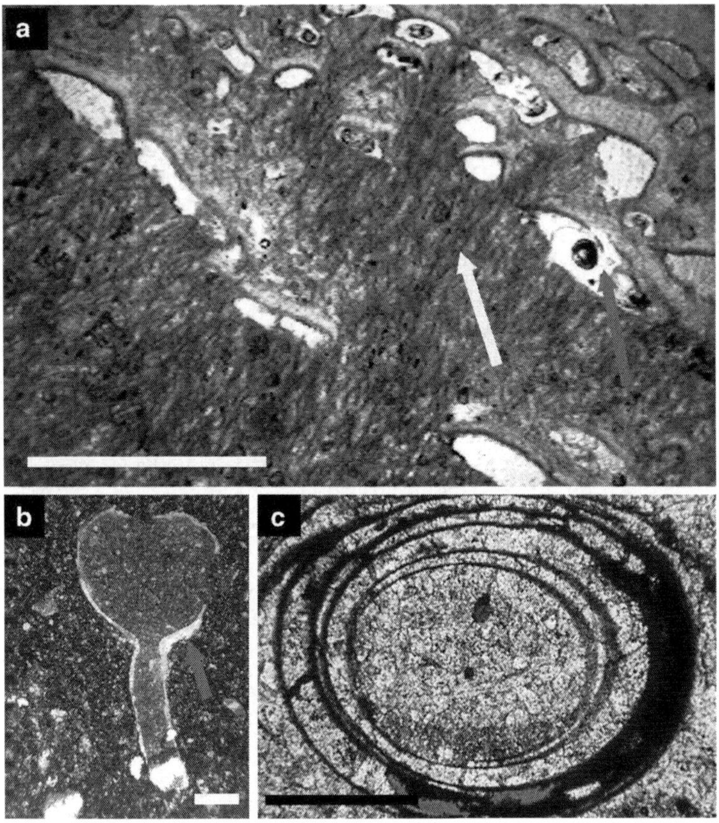

Fig. 15 Examples demonstrating biologically controlled biomineralization, biologically induced mineralization and secondary tufa-like calcification in the Ediacaran and Quaternary. (**a**) Petrographic thin-section through calcified cyanobacterial filaments (cf. *Rivularia* sp., *white arrow*) which have induced the precipitation of calcium carbonate by the photosynthetic absorption of carbon dioxide. These dark filaments alternating with abiogenically calcified organic cases of chironomid midge larvae which appear as open vesicles (*red arrow*). From Quaternary tufa at Zemeno, Greece, image courtesy of Dr A.T. Brasier. (**b**) Longitudinal section through the calcified fossil *Namacalathus* showing the irregular thickness of the wall (*red arrow*) and stalk, from the late Ediacaran of Namibia (c. 549 Ma). This is here interpreted as having formed in a similar way to the external calcification of the vesicular midge larvae in (**a**): by 'tufa-like', abiogenic calcification. (**c**) Transverse sections through thin calcareous shell layers of *Cloudina* from the latest Ediacaran Ara Group of Oman, representing real biologically controlled mineralization. Scale bar is 1mm for (**a**)–(**c**)

Such 'accidental' calcification relies largely upon pH shifts and hence is favoured by raised alkalinity (due to degassing or removal of CO_2), often from the effects of turbulence rather than photosynthesis (e.g. Pentecost and Spiro 1990). Interestingly, our studies of calcification from Namibia show features in *Namacalathus* (Fig. 15b; but not in *Cloudina,* Fig. 15c) consistent with the tufa-like encrustation of an otherwise unmineralized organism. A comparable tufa-like phenomenon may, we suggest, also explain the curious calcification of tiny canal-like spaces between soft tissues of

Ediacaran fronds in South China (Xiao et al. 2005). These anomalous carbonates could therefore be the product of a world in which *p*CO2 was falling from previously very high levels (Riding 2006a), though other factors may have been involved, such as high surface temperature, the absence of crystal inhibitors, an abundance of calciphilic molecules such as aspartic acid on the seafloor (see Morse 2003) and, of course, widespread sediment stagnation (Fig. 1; Shields et al. 1997).

Biologically controlled (enzymatically mediated) calcium carbonate biomineralization seemingly began with *Cloudina* in the latest Ediacaran and expanded dramatically at the base of the Cambrian, coincident with the appearance of the major modern animal phyla, changing the nature of the marine carbonate cycle and the fossil record forever (Fig. 1, 16; Brasier et al. 1996; Bengtson 2004). One of the consequences of this evolutionary event was that these biominerals acted to greatly reduce the overall saturation state of carbonate in the oceans (Shields 2002). These first biominerals were often extremely thin and delicate (e.g. Brasier 1990), allowing their ready dissolution and thereby raising the local pH of pore waters within the mixed layer. Together, these processes resulted in a new kind of carbonate, that of pink nodular bioclastic 'griotte' limestones, which first appear not far above the base of the Cambrian in Avalonia and Siberia (e.g. Brasier et al. 1992). Interestingly, we have observed that such limestones became progressively more offshore in their distribution (e.g. Devonian 'griotte' and 'cephalopodenkalk', Jurassic 'ammonitico rosso') and then largely disappeared after the evolution of coccolithic-foraminiferid carbonate oozes in the Cretaceous.

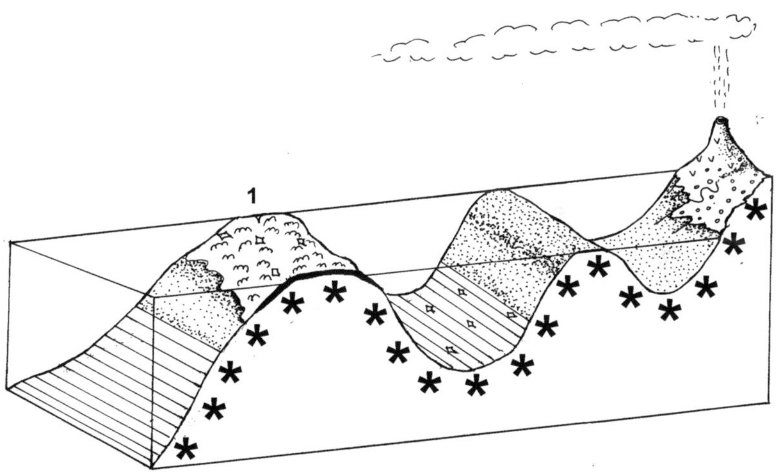

Fig. 16 Model contrasting the restricted range of Ediacaran environments where carbonate biominerals are preserved in marine settings (number 1) with the almost ubiquitous presence of biomineral preservation in modern times (*asterisks*). The first appearance of carbonate biominerals on carbonate platforms is here attributed to the relatively high levels of carbonate saturation states in such settings in the Ediacaran. Examples of early carbonate biominerals studied by us are as follows: 1, Nama Group carbonates of Namibia; Ara Group carbonates of Oman; Dengying carbonates of South China; olistostrome carbonates of central Spain; Reed Dolomite of California (c. 550–540 Ma). From sources in text and references

7 Conclusion

The need for taphonomic studies of the fossil record during the Ediacaran and Cambrian periods is of fundamental evolutionary importance and can hardly be overstated. We argue here for an unexpected bias in the fossil record *towards* remarkable preservation of organic remains on the Ediacaran and Cambrian seafloor. Conditions appear to have been especially favourable to the rapid lithogenesis of surface sediments at this time. This is especially well seen in the changing nature, and decreasing quality, of phosphatic preservation of soft tissues through time. Soft-bodied preservation in sandstones also began to decline after the onset of the 'Cambrian explosion'. Comparable trends can be traced, such as the reducing incidence and quality of silicification, calcification and carbonaceous-pyritic preservation of organic matter within marine sediments, especially after the Cambrian. Suitable conditions seem to have become more and more restricted in the marine world, though they continued to occur in a few non-marine settings.

Many of these taphonomic changes can be accounted for by a progressive depression in the depth of the redox boundary and changes in alkalinity, both within the water column and within the sediment, forcing the zones of lithification both deeper and later, effectively closing up several important taphonomic windows. The fact that comparable modes of preservation (especially of cellular features) are seldom seen again within the marine realm is suggestive of a trigger related to the 'Cambrian explosion'. In other words, this inferred 'fall' in redox, pH and the zone of mineral lithogenesis was a likely consequence of major evolutionary innovations taking place, notably in metazoan respiratory recycling of carbonaceous matter through the activities of bioturbation, grazing and zooplankton. Of prime importance here is the directly visible and potentially testable impact of increasingly deep and complex metazoan bioturbation upon both seafloor porosity and biogeochemistry (see Brasier 1992b; McIlroy and Logan 1999; Jensen et al. 2005). Carbonate, phosphate and silica were also being removed at an increasing rate from the water column by new skeleton builders such as molluscs, brachiopods and sponges. Conceivably, these organisms chose their biominerals in response to their ready availability as solutes within the still 'primitive water masses' across the Precambrian–Cambrian boundary interval (see Brasier 1986). Such removal also helped to prevent carbonate-, phosphate- and silica-saturated fluids from rapidly building up their concentrations to levels approaching the saturated conditions found so widely in earlier surface sediments. Extreme oscillations in carbon isotopic signatures of Neoproterozoic carbonates and their falling amplitudes during the Cambrian (e.g. Lindsay et al. 2005) could likewise reflect the growing influence of bioturbation upon the carbon cycle, reducing the impact of methane and its oxidized products.

We argue, therefore, that the nature of fossil preservation was progressively transformed by the impact of a biological revolution across the Precambrian–Cambrian transition. Given the rather remarkable quality of the Ediacaran fossil record, we conclude that the 'Cambrian explosion' is likely to have been a real biological revolution of very great magnitude.

References

Alessandrello, A., & Bracchi, G. (2003). *Eldonia berbera* n. sp., a new species of the enigmatic genus Eldonia Walcott, 1911 from the Rawtheyan (Upper Ordovician) of Anti-Atlas (Erfoud, Tafilalt, Morocco). *Atti della Società italiana di scienze naturali e del museo civico di storia naturale di Milano, 144*(2), 337–358.

Aller, R. C. (1978). Experimental studies of changes produced by deposit feeders on pore water, sediment and overlying water chemistry. *American Journal of Science, 278*, 1185–1234.

Aller, R. C. (1982). The effects of macrobenthos on chemical properties of marine sediments and overlying water. In P. L. McCall & M. J. S. Tevesz (Eds.), *Animal-sediment relations* (pp. 53–102). New York: Plenum.

Aller, R. C. (1984). The importance of relict burrow structure and burrow irrigation in controlling sedimentary solute distributions. *Geochimica et Cosmochimica Acta, 48*, 1929–1934.

Aller, R. C. (1994). Bioturbation and remineralization of sedimentary organic matter: Effects of redox oscillations. *Chemical Geology, 114*, 331–345.

Allison, P. A. (1988). Phosphatized soft-bodied squids from the Jurassic Oxford Clay. *Lethaia, 21*, 403–410.

Allison, P. A., & Brett, C. E. (1995). In situ benthos and paleo-oxygenation in the Middle Cambrian Burgess Shale. *Geology, 23*(12), 1079–1082.

Allison, P. A., & Briggs, D. E. G. (1991). Taphonomy of nonmineralized tissues. In P. A. Allison & D. E. G. Briggs (Eds.), *Taphonomy: Releasing the data locked in the fossil record* (pp. 25–70). New York: Plenum.

Altermann, W., Kazmierczak, J., Oren, A., & Wright, D. T. (2006). Cyanobacterial calcification and its rock-building potential during 3.5 billion years of earth history. *Geobiology, 4*, 147–166.

Amthor, J. E., Ramseyer, K., Faulkner, T., & Lucas, P. (2005). Stratigraphy and sedimentology of a chert reservoir at the Precambrian-Cambrian boundary: the Al Shomou Silicilyte, South Oman Salt Basin. *GeoArabia, 10*, 89–122.

Antcliffe, J. B., & Brasier, M. D. (2007). Charnia at fifty: new evolutionary models for Ediacaran fronds. *Palaeontology, 51*(1), 11–26.

Arena, D. A. (2008). Exceptional preservation of plants and invertebrates by phosphatization, Riversleigh, Australia. *Palaios, 23*(7), 495–502.

Arnold, C. A. (1931). *On Callixylon newberryi* (Dawson) Elkins et Wieland. Contributions from the Museum of Paleontology of the University of Michigan, *3*(12), 207–232.

Arp, G., Reimer, A., & Reitner, J. (2001). Photosynthesis-induced biofilm calcification and calcium concentrations in Phanerozoic oceans. *Science, 292*, 1701–1704.

Babcock, L. E., Zhang, W., & Leslie, S. A. (2001). The Chengjiang biota: record of the Early Cambrian diversification of life and clues to exceptional preservation of fossils. *GSA Today, 11*(2), 4–9.

Bailey, R. H. (2002). Microbially induced sedimentary structures and preservation of Ediacaran-like fossils in the Boston Bay Group, Massachusetts. Abstract, Geological Society of America, Denver Meeting, Paper No. 187–133.

Bailey, J. V., Corsetti, F. A., Bottjer, D. J., & Marenco, K. N. (2006). Microbially-mediated environmental influences on metazoan colonization of matground ecosystems: evidence from the Lower Cambrian Harkless Formation. *Palaios, 21*, 215–226.

Bailey, J. V., Joye, S. B., Kalanetra, K. M., Flood, B. E., & Corsetti, F. A. (2007). Evidence of giant sulphur bacteria in Neoproterozoic phosphorites. *Nature, 445*, 198–201.

Barghoorn, E., & Tyler, S. (1965). Microfossils from the Gunflint chert. *Science, 147*, 563–577.

Basinger, J. F., & Rothwell, G. W. (1977). Anatomically preserved plants from the Middle Eocene (Allenby Formation) of British Columbia. *Canadian Journal of Botany, 55*, 1984–1990.

Bate, R. H. (1972). Phosphatised ostracods with appendages from the Lower Cretaceous of Brasil. *Palaeontology, 15*, 379–393.

Bell, C. M., Angseesing, J. P. A., & Townsend, M. J. (2001). A Chondrophorine (Medusoid Hydrozoan) from the Lower Cretaceous of Chile. *Palaeontology, 44*(5), 1011–1023.

Bengtson, S. (2004). Early skeletal fossils. In: J. H. Lipps and B. M. Waggoner (Eds.), *Neoproterozoic-Cambrian biological revolutions* (pp. 67–77). The Palaeontological Society Papers 10.

Bengtson, S., Conway Morris, S., Cooper, B. J., Jell, P. A., & Runnegar, B. N. (1990). Early Cambrian fossils from South Australia. *Memoirs of the Australasian Association of Palaeontologists, 9*, 1–364.

Bengtson, S., & Zhao, Y. (1997). Fossilized metazoan embryos from the earliest Cambrian. *Science, 277*, 1645–1648.

Bjerrum, C. J., & Canfield, D. E. (2002). Ocean productivity before about 1.9Ga limited by phosphorous adsorption onto iron oxides. *Nature, 417*, 159–162.

Bland, B. H. (1984). Arumberia Glaessner & Walter, a review of its potential for correlation in the region of the Precambrian–Cambrian boundary. *Geological Magazine, 121*(6), 625–633.

Bottjer, D. J., Hagadorn, J. W., & Dornbos, S. Q. (2000). The Cambrian substrate revolution. *GSA Today, 10*(9), 1–7.

Brasier, A. T., Andrews, J. E., Marca-Bell, A. D. & Dennis, P. F. (2009). Depositional continuity of seasonally laminated tufas. Implications for $\delta^{18}O$ based seasonal temperatures. *Global and Planetary Change, 71*, 160–167.

Brasier, M. D. (1977). An early Cambrian chert biota and its implications. *Nature, 268*, 719–720.

Brasier, M. D. (1979). The Cambrian radiation event. In: M. R. House (Ed.), *The origin of major invertebrate groups* (pp. 103–159). The Systematics Association Special Volume No. 12. London: Academic.

Brasier, M. D. (1986). Why do lower plants and animals biomineralize? *Palaeobiology, 12*(3), 241–250.

Brasier, M. D. (1989). China and the palaeotethyan belt (India, Pakistan, Iran, Kazakhstan, and Mongolia). In: J. W. Cowie and M. D. Brasier (Eds.), *The Precambrian–Cambrian boundary* (pp. 40–74). Oxford Monographs on Geology and Geophysics No. 12. Oxford: Oxford University Press.

Brasier, M. D. (1990). Phosphogenic events and skeletal preservation across the Precambrian–Cambrian boundary interval. In: A. G. Norholt and I. Jarvis (Eds.), *Phosphorite research and development* (289–303). Special Publication of the Geological Society of London 52(V).

Brasier, M. D. (1992a). Nutrient-enriched waters and the early skeletal fossil record. *Journal of the Geological Society of London, 149*(4), 621–629.

Brasier, M. D. (1992b). Palaeoceanography and changes in the biological cycling of phosphorus across the Precambrian–Cambrian boundary. In J. Lipps & P. Signor (Eds.), *Origins of the Metazoa* (pp. 483–523). New York: Plenum.

Brasier, M. D. (1995). Fossil indicators of nutrient levels. 1: Eutrophication and climate change. In: D. J. W. Bosence and P. A. Allison (Eds.), *Marine micropalaeontological analysis from fossils* (pp. 113–132). Geol. Soc. London Spec. Publ., 83.

Brasier, M. D. (2009). *Dawrin's Lost World. The hidden history of animal life*. Oxford: Oxford University Press. 304 pp.

Brasier, M. D., Anderson, M. M., & Corfield, R. M. (1992). Oxygen- and carbon-isotope stratigraphy of early Cambrian carbonates in southeastern Newfoundland and England. *Geological Magazine, 129*, 265–279.

Brasier, M. D., & Antcliffe, J. B. (2004). Decoding the Ediacaran Enigma. *Science, 305*, 1115–1117.

Brasier, M. D., & Antcliffe, J. B. (2008). Dickinsonia from Ediacara: A new look at morphology and body construction. *Palaeogeography, Palaeoclimatology, Palaeoecology, 270*(3–4), 311–323.

Brasier, M. D., & Antcliffe, J. B. (2009). Evolutionary relationships within the Avalonian Ediacara biota: new insights from laser analysis. *Journal of the Geological Society, London, 166*, 363–384.

Brasier, M. D., & Callow, R. H. T. (2007). Changes in the patterns of phosphatic preservation across the Proterozoic–Cambrian transition. *Memoirs of the Association of Australasian Palaeontologists, 34*, 377–389.

Brasier, M. D., Callow, R. H. T., Menon, L. R. and Liu, A. G. (2009b). Osmotrophic biofilms: from modern to ancient. In: J. Seckbach (Ed.), *Microbial mats*. Berlin: Springer-Verlag.

Brasier, M. D., Cowie, J. W., & Taylor, M. E. (1994). Decision on the Precambrian–Cambrian boundary stratotype. *Episodes, 17*, 3–8.

Brasier, M. D., Green, O. R., & Shields, G. (1997). Ediacaran sponge spicule clusters fom southwestern Mongolia and the origins of the Cambrian fauna. *Geology, 25*(4), 303–306.

Brasier, M. D., Perejon, A., & De San Jose, M. A. (1979). Discovery of an important fossiliferous Precambrian–Cambrian sequence in Spain. *Estudios Geológicos, 35*, 379–383.

Brasier, M. D., Shields, G., Kuleshov, V., & Zhegallo, E. A. (1996). Integrated chemo- and biostratigraphic correlation of early animal evolution: Neoproterozoic–Early Cambrian of southwest Mongolia. *Geological Magazine, 133*(4), 445–485.

Briggs, D. E. G., Bottrell, S. H., & Raiswell, R. (1991). Pyritization of soft-bodied fossils: Beecher's Trilobite Bed, Upper Ordovician, New York State. *Geology, 19*, 1221–1224.

Briggs, D. E. G., & Crowther, P. (2001). *Palaeobiology II*. Blackwell Science: Oxford. 583 pp.

Briggs, D. E. G., Erwin, D., & Collier, F. J. (1993). *Fossils of the Burgess Shale of British Columbia*. Washington, DC: Smithsonian Institution. 238 pp.

Briggs, D. E. G., Moore, R. A., Shultz, J. W., & Schweigert, G. (2005). Mineralization of soft-part anatomy and invading microbes in the horseshoe crab Mesolimulus from the Upper Jurassic lagerstätte of Nusplingen Germany. *Proceedings of the Royal Soceity B, 272*, 627–632.

Bromley, R., & Ekdale, A. A. (1984). Chondrites: a trace fossil indicator of anoxia in sediments. *Science, 224*, 872–874.

Burdige, D. J. (2006). *Geochemistry of marine sediments*. Princeton: Princeton University Press. 630 p.

Butterfield, N. J. (2001). Cambrian food webs. In: D. E. G. Briggs & P. R. Crowther (Eds.), *Palaeobiology II* (pp. 40–43). Oxford: Blackwell Science.

Butterfield, N. (2002). Leanchoilia guts and the interpretation of three-dimensional structures in Burgess Shale-type fossils. *Paleobiology, 28*(1), 155–171.

Butterfield, N. J. (2003). Exceptional fossil preservation and the Cambrian explosion. *Integrative and Comparative Biology, 43*, 166–177.

Butterfield, N. J. (2007). Macroevolution and macroecology through deep time. *Palaeontology, 50*(1), 41–55.

Butterfield, N. J., & Nicholas, C. J. (1996). Burgess Shale-type preservation of both non-mineralizing and 'shelly' Cambrian organisms from the Mackenzie Mountains, northwestern Canada. *Journal of Paleontology, 70*(6), 893–899.

Butuzova, G. Yu., Drits, V. A., Morozov, A. A., & Gorschov, A. I. (1990). Processes of formation of iron-manganese oxyhydroxides in the Atlantic-II and Thetis Deeps of the Red Sea. In: J. Parnell, Y. Lianjun, & C. Changming (Eds.), *Sediment-hosted mineral deposits* (pp. 57–72). International Association of Sedimentolgists Special Publication, 11.

Callow, R. H. T., & Brasier, M. D. (2009a). A solution to Darwin's dilemma of 1859: exceptional preservation in Salter's material from the late Ediacaran Longmyndian Supergroup, England. *Journal of Geological Society, London, 166*(1), 1–4.

Callow, R. H. T., & Brasier, M. D. (2009b). Remarkable preservation of microbial mats in Neoproterozoic siliciclastic settings: implications for Ediacaran taphonomic models. *Earth-Science Reviews, 96*, 207–219.

Cameron, B. (1985). Discovery of silicified lacustrine microfossils and stromatolites: Triassic-Jurassic Fundy Group, Nova Scotia. *Geological Society of America*. Abstracts with Programs, *17*, 98.

Campbell, J. A. (1970). Stratigraphy of the Chaffee Group (Upper Devonian) west-central Colorado. *American Association of Petroleum Geologists Bulletin, 54*, 313–325.

Canfield, D. E., Poulton, S. W., & Narbonne, G. M. (2007). Late-Neoproterozoic deep-ocean oxidation and the rise of animal life. *Science, 315*, 92–95.

Carson, G. A. (1991). Silicification of fossils. In P. A. Allison & D. E. G. Briggs (Eds.), *Taphonomy: releasing the data locked in the fossil record* (pp. 455–499). New York: Plenum Press.

Cheng, H., & Liu, J. (2004). Fossil embryos from the Middle and Late Cambrian period of Hunan, south China. *Nature, 427*, 237–240.

Cohen, P. A., Bradley, A., Knoll, A. H., Grotzinger, J. P., Jensen, S., Abelson, J., et al. (2009). Tubular compression fossils from the Ediacaran Nama Group, Namibia. *Journal of Paleontology, 83*(1), 110–122.

Compton, J., Mallinson, D., Glenn, C. R., Filipelli, G., Föllmi, K., Shields, G., et al. (2000). Variations in the global phosphorus cycle. *SEPM. Special Publication, 46*, 21–33.

Condon, D., Zhu, M., Bowring, S., Wang, W., Yang, A., & Jin, Y. (2005). U-Pb ages for the Neoproterozoic Doushantuo Formation, China. *Science, 308*, 95–98.

Conway-Morris, S., & Grazhdankin, D. (2005). Enigmatic worm-like organisms from the Upper Devonian of New York: an apparent example of Ediacaran-like preservation. *Palaeontology, 48*(2), 395–410.

Cook, P. J., & Shergold, J. H. (1984). Phosphorus, phosphorites and skeletal evolution at the Precambrian–Cambrian boundary. *Nature, 308*, 231–236.

Crimes, T. P., & McIlroy, D. (1999). A biota of Ediacaran aspect from Lower Cambrian strata on the Digermul Peninsular, Arctic Norway. *Geological Magazine, 136*(6), 633–642.

Dong, X., Donoghue, P. C. J., Liu, Z., Liu, J., & Peng, F. (2005). The fossils of Orsten-type preservation from Middle and Upper Cambrian in Hunan, China – Three dimensionally preserved soft-bodied fossils (arthropods). *Chinese Science Bulletin, 50*(13), 1352–1357.

Dong, L., Xiao, S., Shen, B., & Zhou, C. (2008). Silicified Horodyskia and Palaeopascichnus from upper Ediacaran cherts in South China: tentative phylogenetic interpretation and implications for evolutionary stasis. *Journal of the Geological Society, London, 165*(1), 367–378.

Donoghue, P. C. J., Bengtson, S., Dong, X., Gostling, N. J., Huldtgren, T., Cunningham, J. A., et al. (2006). Synchrotron X-ray tomographic microscopy of fossil embryos. *Nature, 442*, 680–683.

Donoghue, P. C. J., Kouchinsky, A., Waloszek, D., Bengtson, S., Dong, X., Val'kov, A. K., et al. (2006). Fossilized embryos are widespread but the record is temporally and taxonomically biased. *Evolution & Development, 8*(2), 232–238.

Dornbos, S. Q., Droser, M. L., Gehling, J. G., & Jensen, S. (1999). When the worm turned: concordance of Early Cambrian ichnofabric and trace-fossil record in siliciclastic rocks of South Australia. *Geology, 27*(7), 625–628.

Droser, M. J., Gehling, J. G., & Jensen, S. R. (2006). Assemblage palaeoecology of the Ediacara Biota: the unabridged edition. *Palaeogeography Palaeoclimatology Palaeocology, 232*, 131–147.

Droser, M. J., Jensen, S., & Gehling, J. G. (2002). Trace fossils and substrates of the terminal Proterozoic-Cambrian transition: implications for the record of early bilaterians and sediment mixing. *Proceedings of the National Academy of Sciences of the United States of America, 99*, 12572–12576.

Droser, M. J., Jensen, S., & Gehling, J. G. (2004). Development of early Cambrian ichnofabrics: evidence from shallow marine siliciclastics. In: D. McIlroy (Ed.), *The application of ichnology to palaeoenvironmental and stratigraphic analysis* (pp. 383–396). *Geological Society of London, Special Publication* 228.

Dutta, S., Steiner, M., Banerjee, S., Erdtmann, B.-D., Jeevanjumar, S., & Mann, U. (2006). Chuaria circularis from the early Mesoproterozoic Suket Shale, Vindhyan Supergroup, India: Insights from light and electron microscopy and pyrolysis-gas chromatography. *Journal of Earth System Science, 115*(1), 99–112.

Dzik, J. (2003). Anatomical information content in the Ediacaran fossils and their possible zoological affinities. *Integrative and Comparative Biology, 43*(1), 114–126.

Dzik, J. (2005). Behavioral and anatomical unity of the earliest burrowing animals and the cause of the "Cambrian explosion". *Paleobiology, 31*(3), 503–521.

Fedonkin, M. A. (1990). Systematic description of the Vendian Metazoa. In: B. S. Sokolov & A. B. Iwanowski (Eds.), *The Vendian System. Palaeontology* (Vol. 1, pp. 71–120). Berlin: Springer-Verlag.

Fedonkin, M. A. (2003). The origin of the Metazoa in the light of the Proterozoic fossil record. *Paleontology Research, 7*(1), 9–41.

Fedonkin, M. A., Gehling, J. G., Grey, K., Narbonne, G. M., & Vickers-Rich, P. (2007). *The rise of animals*. Baltimore: The John Hopkins University Press, 326pp.

Fedonkin, M. A., & Waggoner, B. M. (1997). The late Precambrian fossil Kimberella is a mollusc like bilaterian organism. *Nature, 388*, 868–871.

Fedonkin, M. A., & Yochelson, E. L. (2002). Middle Proterozoic (1.5 Ga) Horodyskia moniliformis Yochelson and Fedonkin, the oldest known tissue-grade colonial Eukaryote. *Smithosonian Contributions to Paleobiology, 94*, 29 p.

Feng, W., Chen, Z., & Sun, W. (2003). Diversification of skeletal microstructures of organisms from the interval from the latest Precambrian to the early Cambrian. *Science in China (Series D), 46*(10), 977–985.

Feng, W., & Sun, W. (2003). Phosphate replicated and replaced microstructure of molluscan shells from the earliest Cambrian of China. *Acta Palaeontologica Polonica, 48*(1), 21–30.

Fike, D. A., Grotzinger, J. P., Pratt, L. M., & Summons, R. E. (2006). Oxidation of the Ediacaran Ocean. *Nature, 444*, 744–747.

Föllmi, K. (1996). The phosphorus cycle, phosphogenesis and marine phosphate-rich deposits. *Earth-Science Reviews, 40*, 55–124.

Ford, T. D., & Breed, W. J. (1973). Late Precambrian Chuar Group, Grand Canyon, Arizona. *Geological Society of America Bulletin, 84*, 1273–1260.

Gabbott, S., Xian-guang, H., Norry, M., & Siveter, D. (2004). Preservation of Early Cambrian animals of the Chengjiang biota. *Geology, 32*(10), 901–904.

Gaines, R. R., Kennedy, M. J., & Droser, M. L. (2005). A new hypothesis for organic preservation of Burgess Shale taxa in the middle Cambrian Wheeler Formation, House Range, Utah. *Palaeogeogr. Palaeocl., 220*, 193–205.

Gehling, J. G. (1987). Earliest-known echinoderm – a new fossil from the Pound Subgroup of South Australia. *Alcheringa, 11*, 337–345.

Gehling, J. G. (1991). The case for the Ediacaran fossil roots to the Metazoan Tree. *Geological Society of India Memoir, 20*, 181–224.

Gehling, J. G. (1999). Microbial mats in terminal Proterozoic siliciclastics: Ediacaran death masks. *Palaios, 14*, 40–57.

Gehling, J. G., Droser, M. L., Jensen, S. R., & Runnegar, B. N. (2005). Ediacara organisms: relating form to function. In: D. E. G. Briggs (Ed.), *Evolving form and function: fossils and development*, Peabody Museum of Natural History, New Haven, 43–66.

Gehling, J. G., Narbonne, G. M., & Anderson, M. M. (2000). The first named Ediacaran body fossil, *Aspidella terranovica*. *Palaeontology, 43*, 427–456.

Germs, G. J. B. (1972). New shelly fossils from Nama Group, South West Africa. *American Journal of Science, 272*, 752–761.

Glaessner, M. F., & Wade, M. (1966). The late Precambrian fossils from Ediacara, South Australia. *Palaeontology, 9*(4), 599–628.

Glaessner, M. F., & Walter, M. R. (1975). New Precambrian fossils from the Arumbera Sandstone, Northern Territory, Australia. *Alcheringa, 1*, 59–69.

Gnilovskaya, M. B. (1996). New Vendian Saarinids from the Russian platform. *Doklady Akademii Nauk, 348*, 89–93.

Goldring, R., & Curnow, C. N. (1967). The stratigraphy and facies of the late Precambrian at Ediacara, South Australia. *Journal of the Geological Society of Australia, 14*, 195–214.

Gorin, G. E., Racz, L. G., & Walter, M. R. (1982). Late Precambrian-Cambrian sediments of Huqf Group, Sultanate of Oman. *AAPG Bulletin, 66*, 2609–2627.

Gostling, N. J., Dong, X., & Donoghue, P. C. J. (2009). Ontogeny and taphonomy: an experimental taphonomy study of the development of the brine shrimp Artemia salina. *Palaeontology, 52*(1), 169–186.

Grant, S. W. F., Knoll, A. H., & Germs, G. J. B. (1991). Probable calcified metaphytes in the latest Proterozoic Nama Group, Namibia: origin, diagenesis and implications. *Journal of Palaeontology, 65*(1), 1–18.

Grazhdankin, D. (2004). Patterns of distributions in the Ediacaran biotas: facies versus biogeography and evolution. *Paleobiology, 30*, 203–221.

Grazhdankin, D. V., Balthasar, U., Nagovitsin, K. E., & Kochnev, B. B. (2008). Carbonate-hosted Avalon-type fossils in arctic Siberia. *Geology, 36*(10), 801–806.

Grazhdankin, D., & Gerdes, G. (2007). Ediacaran microbial colonies. *Lethaia, 40*, 201–210.

Grazhdankin, D., & Seilacher, A. (2002). Underground Vendobionta from Namibia. *Palaeontology, 45*(1), 57–78.

Grimes, S. T., Brock, F., Rickard, D., Davies, K. L., Edwards, D., Briggs, D. E. G., et al. (2001). Understanding fossilization: experimental pyritization of plants. *Geology, 29*(2), 123–126.

Grotzinger, J. P., Bowring, S. A., Saylor, B. Z., & Kaufman, A. J. (1995). Biostratigraphic and geochronologic constraints on early animal evolution. *Science, 270*, 598–604.

Grotzinger, J. P., & Knoll, A. H. (1999). Stromatolites in Precambrian carbonates; evolutionary mileposts or environmental dipsticks? *Annual Review of Earth and Planetary Sciences, 27*, 313–358.

Grotzinger, J. P., Watters, W. A., & Knoll, A. H. (2000). Calcified metazoans in thrombolite-stromatolite reefs of the terminal Proterozoic Nama Group, Namibia. *Paleobiology, 26*(3), 334–359.

Hagadorn, J. W. (2002). Bear Gulch: an exceptional Upper Carboniferous plattenkalk. In D. J. Botter et al. (Eds.), *Exceptional Fossil Preservation: a unique view on the evolution of marine life* (pp. 167–183). Columbia University Press: New York.

Hagadorn, J. W., & Bottjer, D. J. (1997). Wrinkle structures: microbially mediated sedimentary structures common in subtidal siliciclastic settings at the Proterozoic–Phanerozoic transition. *Geology, 25*, 1047–1050.

Hagadorn, J. W., & Bottjer, D. J. (1999). Restriction of a late Neoproterozoic biotope: suspect-microbial structures and trace fossils at the Vendian–Cambrian transition. *Palaios, 14*, 73–85.

Hagadorn, J. W., Dott, R. H., Jr., & Damrow, D. (2002). Stranded on a Late Cambrian shoreline: medusae from central Wisconsin. *Geology, 30*, 147–150.

Hagadorn, J. W., Fedo, C. M., & Waggoner, B. M. (2000). Early Cambrian Ediacaran-type fossils from California. *Journal of Paleontology, 74*, 731–740.

Hagadorn, J. W., & Waggoner, B. (2000). Ediacaran fossils from the southwestern Great Basin, United States. *Journal of Paleontology, 74*, 349–359.

Hagadorn, J. W., Xiao, S., Donoghue, P. C. J., Bengtson, S., Gostling, N. J., Pawlowska, M., et al. (2006). Cellular and subcellular structure of Neoproterozoic animal embryos. *Science, 314*, 291–294.

Han, T. M., & Runnegar, B. (1992). Megascopic eukaryotic algae from the 2.1-billion-year-old negaunee iron-formation, Michigan. *Science, 257*, 232–235.

Hellmund, M., & Hellmund, W. (1996). Zum Fortpflanzungsmodus fossiler Kleinlibellen (Insecta, Odonata, Zygoptera). *Paläontologishe Zeitscrift, 70*(1–2), 153–170.

Hod, I. M., Schouten, S., Jelleman, J., & Sinninghe Damste, J. S. (1999). Origin of free and bound mid-chain methyl alkanes in oils, bitumens and kerogens of the marine, Infracambrian Huqf formation (Oman). *Organic Geochemistry, 30*, 1411–1428.

Hoffman, P. F., & Schrag, D. P. (2002). The snowball Earth hypothesis: testing the limits of global change. *Terra Nova, 14*(3), 129–155.

Hofmann, H. J. (1994). Proterozoic carbonaceous compressions ("metaphytes" and "worms"). In: S. Bengtson (Ed.), *Early life on Earth* (pp. 342–357). Nobel Symposium No. 84. New York: Columbia University Press.

Holland, H. D. (2006). The oxygenation of the atmosphere and oceans. *Philosophical Transactions of the Royal Society B, 361*, 903–916.
Horodyski, R. J., & Donaldson, J. A. (1983). Distribution and significance of microfossils in cherts of the Middle Proterozoic Dismal lakes Group, District of Mackenzie, Northwest Territories, Canada. *Journal of Paleontology, 57*(2), 271–288.
Hou, X. -G., Aldridge, R. J., Bergström, J., Sieveter, D. J., Siveter, D. J., & Feng, X. H. (2004). *Cambrian fossils of Chengjiang, China*. Oxford: Blackwell Scientific. 233 pp.
Hua, H., Chen, Z., Yuan, X., Zhang, L., & Xiao, S. (2005). Skeletogenesis and asexual reproduction in the earliest biomineralizing animal Cloudina. *Geology, 33*(4), 277–280.
Hurtgen, M. T., Arthur, M. A., & Halverson, G. P. (2005). Neoproterozoic sulfur isotopes, the evolution of microbial sulfur species, and the burial efficiency of sulfide as sedimentary pyrite. *Geology, 33*, 41–44.
Jensen, S. (2003). The Proterozoic and earliest Cambrian trace fossil record; patterns, problems and perspectives. *Integrative and Comparative Biology, 43*(1), 219–228.
Jensen, S. R., Droser, M. L., & Gehling, J. G. (2005). Trace fossil preservation and the early evolution of animals. *Palaeogeography, Palaeoclimatology, Palaeoecology, 220*, 19–29.
Jensen, S. R., Gehling, J. G., Droser, M. L., & Grant, S. F. (2002). A scratch circle origin for the medusoid fossil Kullingia. *Lethaia, 35*(4), 291–299.
Jones, B., de Ronde, C. E. J., Renaut, R. W., & Owen, R. B. (2007). Siliceous sublacustrine spring deposits around hydrothermal vents in Lake Taupo, New Zealand. *Journal of the Geological Society, London, 164*(1), 227–242.
Kazmierczak, J., & Altermann, W. (2002). Neoarchean biomineralization by benthic cyanobacteria. *Science, 298*, 2351.
Kidder, D. L., & Erwin, D. H. (2001). Secular distribution of biogenic silica through the Phanerozoic: comparison of silica-replaced fossils and bedded cherts and the series level. *Journal of Geology, 109*, 509–522.
Kimura, H., & Watanabe, Y. (2001). Oceanic anoxia at the Precambrian–Cambrian boundary. *Geology, 29*(11), 995–998.
Klug, C., Hagdorn, H., & Montenari, M. (2005). Phosphatized soft-tissue in Triassic bivalves. *Palaeontology, 48*(4), 833–852.
Knoll, A. H. (1985). Exceptional preservation of photosynthetic organisms in silicified carbonates and silicified peats. *Philosophical Transactions of the Royal Society B, 311*, 111–122.
Konhauser, K. O., Pecoits, E., Lalonde, S. V., Papineau, D., Nisbet, E. G., Barley, M. E., et al. (2009). Oceanic nickel depletion and a methanogen famine before the Great Oxidation Event. *Nature, 458*, 750–753.
Konhauser, K. O., Phoenix, V. R., Bottrell, S. H., Adams, D. G., & Head, I. M. (2001). Microbial-silica interactions in Icelandic hot spring sinter: possible analogues for some Precambrian sinterous stromtolites. *Sedimentology, 48*, 415–433.
Kouchinsky, A., & Bengtson, S. (2002). The tube wall of Cambrian Anabaritids. *Acta Palaeontologica Polonica, 47*(3), 431–444.
Koushinsky, A., Bengtson, S., & Gershwin, L.-A. (1999). Cnidarian-like embryos associated with the first small shelly fossils in Siberia. *Geology, 27*(7), 609–612.
Krajewski, K. P., Van Cappellen, P., Trichet, J., Kuhn, O., Lucas, J., Martín-Algarra, A., Prévot, L., Tewari, V. C., Gaspar, L., Knight, R.I., & Lamboy, M. (1994). Biological processes and apatite formation in sedimentary environments. In: K. B. Föllmi (Ed.), *Concepts and controversies in phosphogenesis. Eclogae Geologicae Helvetiae, 87*, 701–745.
Kremer, B. (2005). Mazuelloids: product of post-mortem phosphatization of acanthomorphic acritarchs. *Palaios, 20*(1), 27–36.
Lamboy, M. (1993). Phosphatization of calcium carbonate in phosphorites: microstructure and importance. *Sedimentology, 40*(1), 53–62.
Landing, E., Myrow, P. M., Benus, A. P., & Narbonne, G. M. (1989). The Placentian Series: appearance of the oldest skeletalized faunas in southeastern Newfoundland. *Journal of Paleontology, 63*(6), 739–769.

Landing, E., Narbonne, G. M., & Myrow, P. (1988). Trace fossils, small shelly fossils and the Precambrian–Cambrian boundary. Proceedings. New York State Museum Bulletin 463, 81 pp.

Larson, E. R., & Langenheim, R. L. Jr. (1979). The Mississippian and Pennsylvanian (Carboniferous) systems in the United States, Nevada. *United States Geological Survey Professional Paper*, 1110–BB.

Laurie, J. R. (1997). Silicified Late Cambrian brachiopods from the Georgina Basin, western Queensland. *Alcheringa, 21*(3), 179–189.

Li, C.-W., Chen, J.-Y., & Hua, T.-E. (1998). Precambrian sponges with cellular structures. *Science, 279*, 879–882.

Lin, J.-P., Scott, A. C., Li, C.-W., Wu, H.-J., Ausich, W. I., Zhao, Y.-L., & Hwu, Y.-K. (2006). Silicified egg clusters from a Middle-Cambrian Burgess Shale-type deposit, Guizhou, south China. *Geology, 34*(12), 1037–1040.

Lindsay, J. F., Kruse, P. D., Green, O. R., Hawkins, E., Brasier, M. D., Cartlidge, J., & Corfield, R. M. (2005). The Neoproterozoic-Cambrian record in Australia: a stable isotope study. *Precambrian Research, 143*, 113–133.

Logan, G. A., Hayes, J. M., Hieshima, G. B., & Summons, R. E. (1995). Terminal Proterozoic reorganisation of biogeochemical cycles. *Nature, 376*, 53–56.

Lowenstam, H. A. (1981). Minerals formed by organisms. *Science, 211*, 1126–1131.

Maas, A., Braun, A., Dong, X.-P., Donoghue, P. C. J., Müller, K. J., Olempska, E., et al. (2006). The 'Orsten' – more than a Cambrian konservat-lagerstätte yielding exceptional preservation. *Palaeoworld, 15*, 266–282.

Maleokina, S. Y. (2003). Phosphatized algal-bacterial assemblages in Late Cretaceous phosphorites of the Voronezh Anteclise. In: R. B. Hoover, A. Yu. Rozanov, & J. H. Lipps (Eds.), *Instruments, methods and missions for astrobiology VI.*

Maliva, R. G., Knoll, A. H., & Siever, R. (1989). Secular change in chert distribution: a reflection of evolving biological participation in the silica cycle. *Palaios, 4*, 519–532.

Maliva, R. G., Knoll, A. H., & Simonson, B. M. (2005). Secular change in the Precambrian silica cycle: insights from chert petrology. *Geological Society of America Bulletin, 117*, 835–845.

Mapstone, N. B., & McIlroy, D. (2006). Ediacaran fossil preservation: taphonomy and diagenesis of a discoid biota from the Amadeus Basin, Central Australia. *Precambrian Research, 149*, 126–148.

Martill, D. M. (1988). Preservation of fish in the Cretaceous Santana Formation of Brazil. *Palaeontology, 31*(1), 1–18.

Martin, D., Briggs, D. E. G., & Parkes, R. J. (2005). Decay and mineralization of invertebrate eggs. *Palaios, 20*(6), 562–572.

Martin, M. W., Grazhdankin, D. V., Bowring, S. A., Evans, D. A. D., Fedonkin, M. A., & Kirschvink, J. L. (2000). Age of Neoproterozoic bilaterian body and trace fossil, White Sea, Russia. Implications for metazoan evolution. *Science, 288*, 841–845.

Martin, W. R., & Sayles, F. L. (2003). The recycling of biogenic material at the seafloor. *Treatise on Geochemistry, 7*, 27–65.

Mattes, B. W., & Conway-Morris, S. (1990). Carbonate/evaporite deposition in the Late Precambrian–Early Cambrian Ara Formation of southern Oman. In: A. H. F. Robertson, M. P. Searle, & A. C. Ries (Eds.), *The geology and tectonics of the Oman Region* (pp. 617–636). Geological Society Special Publication 49.

Mazumdar, A., & Banerjee, D. M. (1998). Siliceous sponge spicules in the early Cambrian chert-phosphorite member of the lower Tal Formation, Krol Belt, Lesser Himalaya. *Geology, 26*(10), 899–902.

McCarron, M. E. G. (1999). The Sedimentology and Chemostratigraphy of the Nafun Group, Huqf Supergroup, Oman. Unpublished D.Phil Thesis, University of Oxford.

McCobb, L. M. E., Briggs, D. E. G., Hall, A. R., & Kenward, H. K. (2004). The preservation of invertebrates in 16th century cesspits at St Saviourgate, York. *Archaeometry, 46*(1), 157–169.

McIlroy, D., Crimes, T. P., & Pauley, J. C. (2005). Fossils and matgrounds from the Neoproterozoic Longmyndian Supergroup, Shropshire, UK. *Geological Magazine, 142*(4), 441–455.

McIlroy, D., & Logan, G. A. (1999). The impact of bioturbation on infaunal ecology and evolution during the Proterozoic–Cambrian transition. *Palaios, 14*(1), 58–72.
McIlroy, D., & Walter, M. R. (1997). A reconsideration of the biogenicity of Arumberia banksi Glaessner and Walter. *Alcheringa, 21*, 79–80.
McIlroy, D., Worden, R. H., & Needham, S. J. (2003). Faces, clay minerals and reservoir potential. *Journal of the Geological Society, London, 160*, 489–493.
McKerrow, W. S., Scotese, C. R., & Brasier, M. D. (1992). Early Cambrian continental reconstructions. *Journal of the Geological Society, London, 149*, 599–606.
McMenamin, M. A. S., & Ryan, T. E. (2002). Cambrian echinoderms, brachiopods and silicified microfossils from the Peerless Formation, Colorado. *Geological Soceity of America, Annual Meeting, Denver*, 31–10.
Morse, J. W. (2003). Formation and diagenesis of carbonate sediments. *Treatise on Geochemistry, 7*, 67–98.
Mu, X., & Riding, R. (1983). Silicified Gymnocodiacean algae from the Permian of Nanjing, China. *Palaeontology, 26*(2), 261–276.
Müller, K. J. (1985). Exceptional preservation in calcareous nodules. *Philosophical Transactions of the Royal Society B, 311*, 67–73.
Müller, K. J., Walossek, D., & Zakharov, A. (1995). 'Orsten' type phosphatizaed soft-integument preservation and a new record from the Middle Cambrian Kuonamka Formation in Siberia, *Neues Jahrbuch für Geologie und Paläontologie – Abhandlungen, 197*, 101–118.
Narbonne, G. M. (1998). The Ediacara biota; a terminal Neoproterozoic experiment in the evolution of life. *GSA Today, 8*(2), 1–6.
Narbonne, G. M. (2004). Modular construction in the Ediacara biota. *Science, 315*, 1141–1144.
Narbonne, G. M. (2005). The Ediacara biota: Neoproterozoic origin of animals and their ecosystems. *Annual Review of Earth and Planetary Sciences, 33*, 421–442.
Noffke, N., Beukes, N., Bower, D., Hazen, R. M., & Swift, D. J. P. (2008). An actualistic perspective into Archean worlds – (cyano-)bacterially induced sedimentary structures in the siliciclastic Nhlazatse Section, 2.9 Ga Pongola Supergroup, South Africa. *Geobiology, 6*, 5–20.
Noffke, N., Gerdes, G., Klenke, T., & Krumbein, W. E. (2001). Microbially induced sedimentary structures – a new category within the classification of primary sedimentary structures. *Journal of Sedimentary Research, 71*, 649–656.
Noffke, N., Knoll, A. H., & Grotzinger, J. P. (2002). Sedimentary controls on the preservation of microbial mats in siliciclastic deposits: a case study from the Upper Neoproterozoic Nama Group, Namibia. *Palaios, 17*(6), 533–544.
Oehler, D. Z. (1977). Pyrenoid-like structures in Late Precambrian algae from the Bitter Springs Formation of Australia. *Journal of Paleontology, 51*(5), 885–901.
Page, A., Gabbott, S. E., Wilby, P. R., & Zalasiewicz, J. A. (2008). Ubiquitous Burgess Shale-style 'clay templates' in low grade metamorphic mudrocks. *Geology, 36*(11), 855–858.
Park, L. E. (1995). Geochemical and paleoenvironmental analysis of lacustrine arthropod-bearing concretions of the Barstow Formation, southern California. *Palaios, 10*(1), 44–57.
Park, L. E., & Downing, K. F. (2001). Paleoecology of an exceptionally preserved arthropod fauna from lake deposits of the Miocene Barstow Formation, Southern California, USA. *Palaios, 16*(2), 175–184.
Peach, B. N., Horne, J., Gunn, W., Clough, C. T., Hinxman, L. W., & Teall, J. J. H. (1907). *The geological structure of the North-West Highlands of Scotland.* Memoirs of the Geological Survey of Great Britain.
Peat, C. (1984). Precambrian microfossils from the Longmyndian of Shropshire. *Proceedings of the Geologists' Association, 5*, 17–22.
Peat, C., & Diver, W. (1982). First signs of life on Earth. *New Scientist, 95*, 776–778.
Pentecost, A., & Spiro, B. (1990). Stable carbon and oxygen isotope composition of calcites associated with modern freshwater cyanobacteria and algae. *Geomicrobiology Journal, 8*, 17–26.
Perry, E. C., & Lefticariu, L. (2003). Formation and geochemistry of Precambrian cherts. *Treatise on Geochemistry, 7*, 99–113.

Philippe, M., Szakmany, Gy, Gulyas-Kis, Cs, & Jozsa, S. (2000). An Upper Carboniferous-Lower Permian silicified wood in the Miocene conglomerate from the western Mecsek Mts. (southern Hungary). *Neues Jahrbuch für Geologie und Paläontologie, Monatshefte, 4*, 193–204.

Piper, D. Z., & Link, P. K. (2002). An upwelling model for the Phosphoria Sea: A Permian, ocean-margin sea in the northwest United States. *American Association of Petroleum Geologists Bulletin, 86*(7), 1217–1235.

Pires, E. F., & Sommer, M. G. (2008). Plant-arthropod interaction in the Early Cretaceous (Berriasian) of the Araripe Basin, Brazil. *Journal of South American Earth Sciences, 27*(1), 50–59.

Polizzotto, K., Landman, N. H., & Mapes, R. H. (2007). Cameral Membranes in Carboniferous and Permian goniatites: description and relation to pseudosutures. In N. H. Landman, R. A. Davis, & R. H. Mapes (Eds.), *Cephalopods present and past: new insights and fresh perspectives* (pp. 181–204). Berlin: Springer.

Porter, S. M. (2004). Closing the phosphatization window: Testing for the influence of taphonomic megabias on the pattern of small shelly fossil decline. *Palaios, 19*(2), 178–183.

Porter, S. M., Meisterfield, R., & Knoll, A. H. (2003). Vase-shaped microfossils from the Neoproterozoic Chuar Group, Grand Canyon: a classification guided by modern testate amoebae. *Journal of Paleontology, 77*(3), 409–429.

Powell, W. (2007). Silicification as a common mode of preservation in North American Cambrian lagerstätten. *Geological Society of America*. Abstracts with Programs, 39(6), 333.

Pratt, B. R. (1998). Molar-tooth structure in Proterozoic carbonate rocks: Origin from synsedimentary earthquakes, and implications for the nature and evolution of basins and marine sediment. *GSA Bulletin, 110*(8), 1028–1045.

Pyle, L. J., Narbonne, G., Nowlan, S., Xiao, S., & James, N. P. (2006). Early Cambrian metazoan eggs, embryos and phosphatic microfossils from northwestern Canada. *Journal of Paleontology, 80*(5), 811–825.

Raff, E. C., Villinski, J. T., Turner, F. R., Donoghue, P. C. J., & Raff, R. A. (2006). Experimental taphonomy shows the feasibility of fossil embryos. *Proceedings of the National Academy of Science of the United States America, 103*, 5846–5851.

Rao, V. P., Hegner, E., Naqvi, S. W. A., Kessarkar, P. M., Ahmad, S. M., & Raju, D. S. (2008). Miocene phosphorites from the Murray Ridge, northwestern Arabian Sea. *Palaeogeography, Palaeoclimatology, Palaeoecology, 260*(3–4), 347–358.

Rao, V. P., Rao, K. M., & Raju, D. S. N. (2000). Quaternary phosphorites from the continental margin off Chennai, southeast India: analogs of ancient phosphate stromatolites. *Journal of Sedimentary Research, 70*(5), 1197–1209.

Retallack, G. J. (1994). Were the Ediacaran fossils lichens? *Palaeobiology, 20*(4), 523–544.

Riding, R. (2006a). Cyanobacterial calcification, carbon dioxide concentrating mechanisms, and Proterozoic–Cambrian changes in atmospheric composition. *Geobiology, 4*(4), 299–316.

Riding, R. (2006b). Microbial carbonate abundance compared with fluctuations in metazoan diversity over geological time. *Sedimentary Geology, 185*, 229–238.

Ruttenberg, K. C. (2003). The global phosphorous cycle. *Treatise on Geochemistry, 8*, 585–643.

Sakurai, R., Ito, M., Ueno, Y., Kitajima, K., & Maruyama, S. (2005). Facies architecture and sequence-stratigraphic features of the Tumbiana Formation in the Pilbara Craton, northwestern Australia: Implications for depositional environments of oxygenic stromatolites during the Late Archean. *Precambrian Research, 138*, 255–273.

Samuelsson, J., & Butterfield, N. J. (2001). Neoproterozoic fossils from the Franklin Mountains, northwestern Canada: stratigraphic and palaeobiological implications. *Precambrian Research, 107*(3), 235–251.

Schieber, J. (1986). The possible role of benthic microbial mats during the formation of carbonaceous shales of shallow Mid-Proterozoic basins. *Sedimentology, 33*(4), 521–536.

Schieber, J. (2002). Sedimentary pyrite: a window into the microbial past. *Geology, 30*(6), 531–534.

Schopf, J. W. (1968). Microflora of the bitter springs formation, Late Precambrian Central Australia. *Journal of Paleontology, 42*, 651–688.

Schopf, J. W., & Fairchild, T. R. (1973). Late Precambrian microfossils: a new stromatolitic biota from Boorthanna, South Australia. *Nature, 242*, 537–538.

Schopf, J. W., & Klein, C. (1992). *The Proterozoic biosphere: a multidisciplinary study*. Cambridge University Press: New York. 1374 pp.

Schröder, S., & Grotzinger, J. P. (2007). Evidence for anoxia at the Ediacaran–Cambrian boundary: the record of redox-sensitive trace elements and rare earth elements in Oman. *Journal of the Geological Society, London, 164*(1), 175–187.

Schröder, S., Grotzinger, J. P., Amthor, J. E., & Matter, A. (2005). Carbonate deposition and hydrocarbon reservoir development at the Precambrian–Cambrian boundary: the Ara Group in South Oman. *Sedimentary Geology, 180*, 1–28.

Schubert, J. K., Kidder, D. L., & Erwin, D. H. (1997). Silica-replaced fossils through the Phanerozoic. *Geology, 25*(11), 1031–1034.

Seilacher, A. (1956). Der Beginn des Kambriums als biologische Wende. *Neues Jahrbuch für Geologie und Paläontologie – Abhandlungen, 103*, 155–180.

Seilacher, A. (1992). Vendobionta and Psammocorallia: lost constructions of Precambrian evolution. *Journal of the Geological Society, London, 149*, 607–613.

Seilacher, A. (1999). Biomat-related lifestyles in the Precambrian. *Palaios, 14*(1), 86–93.

Seilacher, A., & Pflüger, F. (1994). From biomats to benthic agriculture: a biohistoric revolution. In W. E. Krumbein, D. M. Paterson, & L. Stal (Eds.), *Biostabilization of Sediments* (pp. 97–105). Oldenburg, Germany: Bibliotheks und Informations system der Universität Oldenburg.

Shen, Y., & Schidlowski, M. (2000). New C isotope stratigraphy from southwest China: implications for the placement of the Precambrian–Cambrian boundary on the Yangtze Platform and global correlations. *Geology, 28*(7), 623–626.

Shergold, J. H., & Brasier, M. D. (1986). Proterozoic and Cambrian phosphorites – specialist studies: biochronology of Proterozoic and Cambrian phosphorites. In P. J. Cook & J. H. Shergold (Eds.), *Phosphate deposits of the world* (Vol. 1, pp. 295–345). Cambridge: Cambridge University Press.

Shields, G. A. (2002). 'Molar-tooth microspar': a chemical explanation for its disappearance 750 Ma. *Terra Nova, 14*, 108–113.

Shields, G., Stille, P., Brasier, M. D., & Atudorei, N.-V. (1997). Stratified oceans and oxygenation of the Late Precambrian environment: a post glacial geochemical record from the Neoproterozoic of W Mongolia. *Terra Nova, 9*, 218–222.

Signor, P. W., & Vermeij, G. J. (1994). The plankton and benthos: origins and early history of an evolving relationship. *Paleobiology, 20*(3), 297–319.

Siveter, D. J., Williams, M., & Waloszek, D. (2001). A phosphatocopid crustacean with appendages from the Lower Cambrian. *Science, 293*, 479–481.

Steiner, M. (1994). Die neoproterozoischen Megaalgen Südchinas. *Berliner Geowissenschaftliche Abhandlungen, E, 15*, 1–146.

Steiner, M., & Reitner, J. (2001). Evidence of organic structures in Ediacara-type fossils and associated microbial mats. *Geology, 29*(12), 1119–1122.

Sumner, D. Y., & Grotzinger, J. P. (2004). Implications for Neoarchaean ocean chemistry from primary carbonate mineralogy of the Cambellrand-Malmani Platform, South Africa. *Sedimentology, 51*, 1273–1299.

Sun, W. (1994). Early multicellular fossils. In: S. Bengtson (Ed.), *Early life on Earth* (pp. 358–369). Nobel Symposium No. 84. New York: Columbia University Press.

Swett, K. (1964). Petrology and paragenesis of the Ordovician Manitou Formation along the Front Range of Colorado. *Journal of Sedimentary Research, 34*, 615–624.

Taylor, T. N., Hass, H., Kerp, H., Krings, M., & Hanlin, R. T. (2005). Perithecial ascomycetes from the 400 million year old Rhynie chert: an example of ancestral polymorphism. *Mycologia, 97*(1), 269–285.

Tiwari, M., & Knoll, A. H. (1994). Large acanthomorphic acritarchs from the Infrakrol Formation of the Lesser Himalaya and their stratigraphic significance. *Journal of Himalayan Geology, 5*, 193–201.

Tobin, K. J. (2004). A survey of Paleozoic microbial fossils in chert. *Sedimentary Geology, 168*, 97–107.

Toporski, J. K., Steele, A., Westall, F., Thomas-Keprta, K. L., & McKay, D. S. (2002). The simulated silicification of bacteria – new clues to the modes and timing of bacterial preservation and implications for the search for extraterrestrial microfossils. *Astrobiology, 2*(1), 1–26.

Trewin, N. H., Fayers, S. R., & Kelman, R. (2003). Subaqueous silicification of the contents of small ponds in an Early Devonian hot-spring complex, Rhynie, Scotland. *Canadian Journal of Earth Sciences, 40*(11), 1697–1712.

Trewin, N. H., & Rice, C. M. (2004). The Rhynie hot-spring system: geology, biota and mineralization. *Transactions of the Royal Society of Edinburgh, Earth Sciences, 94*, 1–246.

Trichet, J., & Fikri, A. (1997). Organic matter in the genesis of high-island atoll peloidal phosphorites; the lagoonal link. *Journal of Sedimentary Research, 67*(5), 891–897.

Turnbull, M. J. M., Whitehouse, M. J., & Moorbath, S. (1996). New isotopic age determinations for the Torridonian, NW Scotland. *Journal of the Geological Society, London, 153*(6), 955–964.

Turner, E. C., & Jones, B. (2005). Microscopic calcite dendrites in cold-water tufa: implications for nucleation of micrite and cement. *Sedimentology, 52*, 1043–1066.

Turner, E. C., Narbonne, G. M., & James, N. P. (1993). Neoproterozoic reef microstructures from the Little Dal Group, northwestern Canada. *Geology, 21*, 259–262.

Urbanek, A., & Rozanov, A. Yu. (1983). *Upper Precambrian and Cambrian Palaeontology of the East-European Platform.* Wydawnictwa Geologiczne, Warszawa: Publishing House, 158 pp, 94pls.

Van Cappellen, P., & Ingall, E. D. (1994). Benthic phosphorus regeneration, net-primary production, and ocean anoxia: a model of the coupled marine biogeochemical cycles of carbon and phosphorus. *Paleoceanography, 9*(5), 677–692.

Vidal, G. (1990). Giant acanthomorphic acritarchs from the upper Proterozoic in southern Norway. *Palaeontology, 33*, 287–298.

Vidal, G., & Moczydlowska, M. (1992). Patterns of phytoplankton radiation across the Precambrian–Cambrian boundary. *Journal of the Geological Society, London, 149*(4), 647–654.

Wade, M. (1968). Preservation of soft-bodied animals in pre-Cambrian sandstones at Ediacara, South Australia. *Lethaia, 1*, 238–267.

Waggoner, B. M. (1995). Ediacaran lichens: a critique. *Paleobiology, 21*, 393–397.

Walossek, D., Hinz-Schallreuter, I., Shergold, J. H., & Müller, K. (1993). Three-dimensional preservation of arthropod integument from the Middle Cambrian of Australia. *Lethaia, 26*(1), 7–15.

Walossek, D., Repetski, J. E., & Müller, K. J. (1994). An exceptionally preserved parasitic arthropod, Heymonsicambria taylori n. sp. (Arthropoda incertae sedis: Pentastomida), from Cambrian-Ordovician boundary beds of Newfoundland, Canada. *Canadian Journal of Earth Science, 31*, 1664–1671.

Wardlaw, B. R., & Collinson, J. W. (1978). Biostratigraphic zonation of the Park City Group. United States Geological Survey Professional Paper 1163-D.

Weitschat, W. (1983). Myodocopid ostracodes with preserved appendages from the Lower Triassic of Spitzbergen. *Paläontologishe Zeitscrift, 57*, 309–323.

Westall, F. (1994). Silicified bacteria and associated biofilm from the deep-sea sedimentary environment. *Darmstädter Beiträge zur Naturgeschichte, 4*, 29–43.

Westall, F., Boni, L., & Guerzoni, E. (1995). The experimental silicification of microorganisms. *Palaentology, 38*(3), 495–528.

Wilby, P. R., Briggs, D. E. G., Bernier, P., & Gaillard, C. (1996). The role of microbial mats in the fossilization of soft-tissues. *Geology, 24*(9), 787–790.

Wilkinson, I., Wilby, P. R., Williams, P., Siveter, D. J., & Vannier, J. (2007). Ostracod carnivory through time. In A. M. T. Elewa (Ed.), *Predation in organisms: a distinct phenomenon* (pp. 39–57). Heidelberg: Springer-Verlag.

Wood, R. A. (1998). The ecological evolution of reefs. *Annual Review of Ecology and Systematics, 29*, 179–206.
Wood, R. A., Grotzinger, J. P., & Dickson, J. A. D. (2002). Proterozoic modular biomineralized Metazoan from the Nama Group, Namibia. *Science, 296*, 2383–2386.
Xiao, S. (2004). New multicellular algal fossils and acritarchs in Doushantuo chert nodules. *Journal of Paleontology, 78*, 393–401.
Xiao, S., & Knoll, A. H. (1999). Fossil preservation in the Neoproterozoic Doushantuo phosphorite lagerstatte, South China. *Lethaia, 32*(3), 219–240.
Xiao, S., Shen, B., Zhou, C., Xie, G., & Yuan, X. (2005). A uniquely preserved Ediacaran fossil with direct evidence for a quilted body plan. *Proceedings of the National Academy of Sciences of the United States of America, 102*, 10227–10232.
Xiao, S., Yuan, X., Steiner, M., & Knoll, A. H. (2002). Macroscopic carbonaceous compressions in a terminal Proterozoic shale: A systematic reassessment of the Miaohe Biota, South China. *Journal of Paleontology, 76*(2), 347–376.
Xiao, S., Zhang, Y., & Knoll, H. (1998). Three-dimensional preservation of algae and animal embryos in a Neoproterozoic phosphorite. *Nature, 391*, 553–558.
Yang, R., & Zhao, Y. (2000). Discovery of Corallina fossil from the Middle Cambrian of Taijiang County, Guizhou Province, China. *Chinese Science Bulletin, 45*, 544–547.
Yao, J., Xiao, S., Yin, L., Li, G., & Yuan, X. (2005). Basal Cambrian microfossils from the Yurtus and Xisanblaq Formations (Tarim, North West China): systematic revision and biostratigraphic correlation of Michrystridium-like acritarchs. *Palaeontology, 48*(4), 687–708.
Yin, C., Bengtson, S., & Zhao, Y. (2004). Silicified and phosphatized Tianzhushania, spheroidal microfossils of possible animal origin from the Neoproterozoic of South China. *Acta Palaeontologica Polonica, 49*(1), 1–12.
Yin, L., Xiao, S., & Yuan, X. (2001). New observations on spicule-like structures from Doushantuo phosphorites at Weng'an. *Guizhou Province. Chinese Science Bulletin, 46*(21), 1828–1832.
Yin, L., Zhu, M., Knoll, A. H., Yuan, X., & Hu, J. (2007). Doushantuo embryos preserved inside diapause egg cysts. *Nature, 446*, 661–663.
Zhang, Z. (1997). A new Palaeoproterozoic clastic-facies microbiota from the Changzhougou Formation, Changcheng Group, Jixian, North China. *Geological Magazine, 134*(2), 145–150.
Zhang, X. -G., & Pratt, B. R. (1994). Middle Cambrian arthropod embryos and blastomeres. *Science, 266*, 637–639.
Zhegallo, E. A., Rozanov, A Yu, Ushatinskaya, G. T., Hoover, R. B., Gerasimenko, L. M., & Ragozina, A. L. (2000). *Atlas of microorganisms from ancient phosphorites of Khubsugul (Mongolia)*. Huntsville: Paleontological Institute of Russian Academy of Sciences, NASA-Marshall Space Flight Centre. 171 pp.
Zhou, C., Brasier, M. D., & Xue, Y. (2001). Three-dimensional phosphatic preservation of giant acritarchs from the terminal Proterozoic Doushantuo Formation in Guizhou and Hubei Provinces, South China. *Palaeontology, 44*(6), 1157–1178.
Zhou, C., Xie, G., McFadden, K., Xiao, S., & Yuan, X. (2006). The diversification and extinction of Doushantuo-Pertatataka acritarchs in South China: causes and biostratigraphic importance. *Geology Journal, 42*(3–4), 229–262.
Zhu, S., Sun, S., Huang, X., He, Y., Zhu, G., Sun, L., et al. (2000). Discovery of carbonaceous compressions and their mutlicellular tissues from the Changzhougou Formation (1800Ma) in the Yanshan Range, North China. *Chinese Science Bulletin, 45*(9), 841–847.

Chapter 16
Mass Extinctions and Changing Taphonomic Processes

Fidelity of the Guadalupian, Lopingian, and Early Triassic Fossil Records

Margaret L. Fraiser, Matthew E. Clapham, and David J. Bottjer

Contents

Abstract The biotic crisis of the Middle Permian through Early Triassic is unmatched in the Phanerozoic in terms of taxonomic diversity losses and paleoecological reorganization. However, the potential taphonomic bias from post mortem diagenesis for this crucial time has not been evaluated. We assessed the quality of the fossil record during this interval by quantifying the number of Lazarus taxa using our own database, data available in the Paleobiology Database and previous compilations.

M.L. Fraiser (✉)
Department of Geosciences, University of Wisconsin-Milwaukee, Milwaukee, WI 53203, USA
e-mail: mfraiser@uwm.edu

M.E. Clapham
Department of Earth and Planetary Sciences, University of California Santa Cruz, 1156 High Street, Santa Cruz, CA 95064, USA
e-mail: mclapham@es.ucsc.edu

D.J. Bottjer
Department of Earth Sciences, University of Southern California, Los Angeles, CA 90089-0740, USA
e-mail: dbottjer@usc.edu

P.A. Allison and D.J. Bottjer (eds.), *Taphonomy: Process and Bias Through Time*, Topics in Geobiology 32, DOI 10.1007/978-90-481-8643-3_16,

We also quantitatively tested for paleoecological differences between silicified versus non-silicified faunas. Herein we report that there is no major taphonomic bias due to skeletal mineralogy or fossil preservation affecting the Middle and Late Permian fossil record, but that aragonite-shelled molluscs may exhibit a significant Lazarus effect during the Induan. We propose that a variety of mechanisms affected the fossil record of the Paleozoic/Mesozoic transition, including ocean chemistry, paleobiology of the examined groups, and human influences on taxonomic and sampling practices.

1 Introduction

Mass extinctions are geologically short intervals of time when biodiversity losses are significantly elevated above background rates of extinction (e.g. Jablonski 1986a; Sepkoski 1986; Flessa 1990). They are a prominent feature of the fossil record and, along with the rise and fall of the three great evolutionary faunas, shaped the Phanerozoic biodiversity curve (Raup and Sepkoski 1982; Sepkoski 1981, 1984; Courtillot and Gaudemer 1996). Mass extinctions are also important agents of macroevolutionary change because they eliminate successful groups of organisms and create new evolutionary opportunities for previously minor groups (Gould and Calloway 1980; Jablonski 1986a, b, 2001, 2005; Raup 1986, 1994; Erwin 2001; Bambach et al. 2002). A complete understanding of the evolutionary role of a mass extinction must include more than just an analysis of the taxonomic crisis because the effects of mass extinctions extend beyond the biodiversity losses: the aftermaths may be as important as the extinctions themselves because of the new ecological patterns arising from survivors that interact in new ways in less crowded ecological niches (Droser et al. 1997, 2000; Erwin 2001; Bambach et al. 2002; Jablonski 2001, 2002).

Proper interpretation of the duration, magnitude, and causes of mass extinctions and the nature of the survival and recovery of organisms during their aftermaths is contingent upon accurate reconstruction of taxonomic and ecological changes. Artifacts of sampling methods or taxonomic practice can obscure the real trends (Sepkoski 1986; Flessa 1990), whereas taphonomic biases inherent in the geologic record, such as mode of organism preservation (Schubert et al. 1997), rock volume (e.g., Crampton et al. 2003), and preferential loss of organisms with aragonitic shell mineralogy (e.g. Cherns and Wright 2000) may also influence observed patterns. Such taphonomic biases may have obscured the true biotic patterns during the Permian–Triassic extinction and its aftermath. These potentially confounding effects have been inferred from the abundance of Lazarus taxa – taxa that temporarily disappear from the fossil record but reappear later unchanged (Flessa and Jablonski 1983) – and from a decrease in preservation by silicification (Erwin and Pan 1996; Schubert et al. 1997; Twitchett 2001). Lazarus taxa may be an indicator of the quality of the fossil record if the phenomenon reflects a failure of certain organisms to be preserved through taphonomic effects such as the Signor–Lipps effect, outcrop area bias, paleolatitudinal sampling bias, or reduced preservation quality (Signor and Lipps 1982; Allison and Briggs 1993; Erwin and Pan 1996; Smith and McGowan 2007). Lazarus taxa abundance may also be due to biological factors such as reduced population size (which may also affect the chance of sampling a taxon) or reduced geographic range and migration to refugia (Jablonski 1986a,b; Kauffman and Harries 1996; Wignall and Benton 1999; Twitchett 2001; Rickards and Wright 2002). Taxonomic uncertainty can cause an apparent Lazarus effect (Wheeley and Twitchett 2005).

Herein we test two aspects of the quality of the fossil record during the Guadalupian, Lopingian, and Early Triassic. The end-Guadalupian and end-Permian extinctions marked the end of the Paleozoic (Fig. 1) and heralded major changes in benthic marine ecology (e.g. Fraiser and Bottjer 2007; Clapham and Bottjer 2007a, b), but several studies have proposed that taphonomic processes make it difficult to extract real ecological patterns during these key intervals in evolutionary history (e.g. Erwin and Pan 1996; Twitchett 2001). First, we quantified the number of Lazarus taxa among several key taxonomic groups, as an increased number of Lazarus taxa may indicate reduced preservation quality. Second, a potential source of bias in the fossil record for this interval, changes in preservation via silicification, was tested by quantifying the proportion of silicified fossil collections, comparing the alpha diversity of silicified and non-silicified (preserved as molds and casts) collections, and assessing the number of taxa exclusive to silicified collections. Silicification is important because it allows fossils to be acid-etched and freed from calcareous matrix, often preserving very fine morphological details and improving ease of identification by taxonomists (e.g. Holdaway and Clayton 1982). It can also preserve a more faithful record of the original diversity and abundance within an assemblage (Cherns and Wright 2000; Wright et al. 2003, Butts and Briggs, this volume). Results of this test will reveal any temporal trends in silicification and the extent to which silicified faunas preserve a higher fidelity record. Together these tests document the

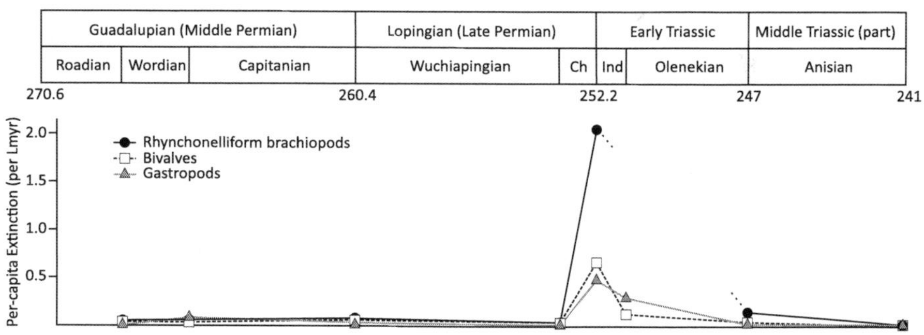

Fig. 1 Geologic timescale of Middle Permian (Guadalupian), Late Permian (Lopingian), and Early Triassic stages. Ch = Changhsingian, Ind = Induan. The lower panel shows the per-capita extinction rates (Foote 2000) for rhynchonelliform brachiopods, bivalves, and gastropods in each stage based on data from Clapham et al. (2009) (Permian invertebrates), Chen et al. (2005) (Early Triassic brachiopods), Gastrobase (Early Triassic gastropods), and the Paleobiology Database (Early Triassic bivalves and gastropods). The per-capita extinction for rhynchonelliform brachiopods is undefined in the Induan because no genera cross both bottom and top boundaries of the stage

taphonomic quality of the Permian–Triassic fossil record in greater detail and elucidate the impact of temporal trends of taphonomic bias on the records of the end-Guadalupian extinction, the end-Permian extinction, and their aftermaths.

2 Previous Understanding of Biases in the Middle Permian to Early Triassic Fossil Record

2.1 End-Guadalupian Extinction and Lopingian Aftermath

The end-Guadalupian extinction, at the end of the Middle Permian (Guadalupian Series), was the first phase of the two-stage taxonomic crisis during the Permian–Triassic interval (Jin et al. 1994; Stanley and Yang 1994). Initial estimates suggested that biodiversity loss during the end-Guadalupian event was severe with as many as 55–60% of all marine invertebrate and fusulinid genera going extinct (Stanley and Yang 1994). However, more recent studies have shown that extinction rates were actually not elevated during the end-Guadalupian interval among most marine invertebrate groups, including brachiopods, bivalves, and gastropods (Shen et al. 2006; Clapham et al. 2009). Nevertheless, the end-Guadalupian extinction remains an especially severe event for fusulinids (Stanley and Yang 1994; Yang et al. 2004). The potential causes of the extinction are unclear, and there may not be a need to invoke serious perturbations given the negligible invertebrate extinctions. Environmental changes during the Guadalupian–Lopingian interval include the Emeishan flood basalts (Wignall 2001), possible climate cooling (Isozaki et al. 2007), and the onset of deep-marine anoxia (Isozaki 1997). Despite the minimal effects on global invertebrate biodiversity, environmental stress during the Guadalupian–Lopingian interval caused profound changes in the habitat distribution of bryozoans during the Lopingian (Powers and Bottjer 2007), shifts in the relative abundance of rhynchonelliform brachiopods and molluscs in offshore habitats (Clapham and Bottjer 2007a, b), and a dramatic reduction in the number and size of reefs in the Wuchiapingian (Weidlich 2002; Weidlich et al. 2003).

The potential influences of taphonomic bias on the apparent severity of the end-Guadalupian extinction were first investigated by Stanley and Yang (1994). They applied three tests and concluded that the end-Guadalupian extinction peak did not result from a poor Lopingian fossil record. The preferential extinction of large, complex fusulinid genera, the inconsistency between observed patterns of extinction and predicted Signor–Lipps effects, and a strong excess of originations relative to extinctions in the Wuchiapingian and early Changhsingian all suggest that taphonomic biases only had minor effects on the end-Guadalupian extinction (Stanley and Yang 1994). However, other taphonomic effects, including changes in the abundance of silicified fossil collections or reductions in preserved rock volume, may have exacerbated the severity of the end-Guadalupian extinction without producing a spurious Signor–Lipps effect or completely masking the Lopingian radiation.

2.2 End-Permian Mass Extinction and Early Triassic Aftermath

The end-Permian mass extinction, approximately 252 million years ago, was the largest biotic crisis of the Phanerozoic (Bambach et al. 2004; Henderson 2005) with 78% of marine genera going extinct (Clapham et al. 2009). For up to 5 million years during the Early Triassic aftermath of the end-Permian mass extinction, benthic marine paleocommunities were characterized by low biodiversity and low ecological complexity compared to pre-extinction Permian and later Triassic paleocommunities (e.g. Fraiser and Bottjer 2005b; Lehrmann et al. 2006). Macroevolutionary changes in benthic marine ecology, such as a shift from primarily non-motile organisms to self-mobile taxa and a switch from rhynchonelliform brachiopod-dominated to bivalve-dominated paleocommunities, were triggered by the end-Permian mass extinction (Bambach et al. 2002; Wagner et al. 2006; Fraiser and Bottjer 2007). Sedimentological and geochemical evidence indicate that much of the latest Permian through the Early Triassic had an atmosphere with elevated CO_2 and low O_2, and an ocean rich in H_2S and depleted in O_2; these conditions were ultimately linked to extensive volcanism and the supercontinent configuration of Pangea (e.g. Wignall and Twitchett 1996; Wignall 2001; Berner 2004; Grice et al. 2005; Huey and Ward 2005; Sephton et al. 2005).

It has been reported that a large portion of Early Triassic taxa are Lazarus taxa (Batten 1973; Erwin and Pan 1996; Twitchett 2001). For example, there are estimates that 69% of gastropod genera are Lazarus genera during the Griesbachian (Erwin 1996), and that 90% of sponge families are Lazarus taxa during all stages of the Early Triassic (Twitchett 2001). Though the Early Triassic Lazarus phenomenon heretofore had been examined for gastropods and sponges only (e.g. Erwin 1996; Erwin and Pan 1996; Twitchett 2001, Wheeley and Twitchett 2005), it has been implied that the Lazarus effect was very large for all groups of skeletoned benthic marine invertebrates during the Early Triassic (e.g. Twitchett 2001; Erwin 2006).

An absence of faunas preserved by silicification has been proposed as a major cause of the Early Triassic Lazarus phenomenon and for the apparent delayed biotic recovery following the end-Permian mass extinction (Erwin 1996, 2006; Erwin and Pan 1996; Kidder and Erwin 2001). This hypothesis is based on studies indicating that silicified faunas have a higher fidelity of fossil preservation than non-silicified faunas preserved as casts and molds (Cherns and Wright 2000; Wright et al. 2003). Furthermore, it has been proposed that the post-Paleozoic fossil record suffers from a taphonomic "megabias" because of low numbers of silicified faunas compared to the Paleozoic (Schubert et al. 1997). Previous studies of the fidelity of the fossil record following the end-Permian mass extinction have focused on only one group of benthic marine organisms (e.g. gastropods, Erwin and Pan 1996; or echinoids, Smith 2007), or have examined data from the Triassic period as a whole (Smith 2007), obscuring any processes that may have been unique to the aftermath of the end-Permian mass extinction.

The extent of silicification during the Early Triassic has not been quantified previously, and the characteristics of silicified faunas have not been statistically compared to those of non-silicified ones.

3 Methods

We compiled a database of Roadian (Middle Permian) through Anisian (Middle Triassic) rhynchonelliform brachiopod, bivalve, gastropod, and demosponge fossil occurrences. This was used to examine two additional taphonomic metrics that test the fidelity of the Permian–Triassic fossil record and its potential influence on the end-Guadalupian and end-Permian extinctions. The dataset includes (1) more than 53,321 Permian marine invertebrate fossil occurrences, including records of all marine invertebrate groups, from 9863 collections (the database used in Clapham et al. 2009); (2) Triassic gastropods modified and updated from Gastrobase, a database of published occurrences of gastropod genera at the stage and substage levels for the Permian and Triassic periods (http://www.earth.cardiff.ac.uk/people/summaries/GASTROBASEdoc.htm); (3) Triassic bivalves and sponges, and Anisian brachiopods, from the Paleobiology Database (www.paleodb.org); and (4) Early Triassic rhynchonelliform brachiopods (Chen et al. 2005). Though the PBDB is not flawless, it is the most complete database available for comparing benthic marine organisms from the Lopingian, Early Triassic, and Middle Triassic. Taxonomic assignments were corrected when necessary and possible.

First, the number of rhynchonelliform brachiopod, bivalve, gastropod, and sponge Lazarus taxa in each stage from the Roadian to Anisian was quantified to test for poor preservation, especially in the Wuchiapingian stage immediately following the traditional end-Guadalupian extinction interval and the Induan and Olenekian stages following the end-Permian extinction (Appendix A). The significance of differences between the proportions of Lazarus taxa between stages was determined using a two-tailed *t*-test.

Second, the number of Permian and Early Triassic silicified collections was tallied using the Clapham et al. (2009) database, 211 Paleobiology Database collections, and 358 additional Induan and Olenekian collections culled from the primary literature to determine whether a reduction in silicification, particularly due to the loss of the rich record from western North America, affected diversity and extinction (Appendix B). Included in the analyses were benthic marine invertebrates from level-bottom marine communities; planktonic, nektonic, and reef collections were excluded in the Triassic but not in the Permian data. Species richness (alpha diversity) of each silicified collection was determined and compared to the richness of non-silicified collections. The number of brachiopod, bivalve, and gastropod genera unique to silicified collections was also quantified to determine the influence of silicification on large-scale compilations of taxonomic diversity.

4 Results

4.1 Guadalupian–Lopingian Lazarus Effect

During the Guadalupian and Lopingian, the prevalence of Lazarus taxa varied significantly among different taxonomic groups. At the genus level, a substantial percentage of gastropod taxa in a given stage, up to 38% of the total genus richness, are actually Lazarus taxa (Fig. 2a). In contrast, only 20–25% of all bivalve genera (Fig. 2b) and 18–20% of all rhynchonelliform brachiopods (Fig. 2c) are Lazarus taxa. Lazarus abundance is calculated by dividing the number of Lazarus taxa by the total diversity (Lazarus taxa plus taxa sampled within the stratigraphic interval) in each stage. Despite the pronounced difference between clades, the proportion of Lazarus taxa within most clades typically exhibits little variation (Fig. 2). There was no statistically significant change in the percentage of gastropod Lazarus taxa from the Roadian through Wuchiapingian stages (varying between 33.3% and 38.6%). Likewise, the number of bivalve Lazarus taxa remained statistically unchanged at 19.6–25.2% from the Roadian to the Wuchiapingian. Both aragonitic bivalves and those with a calcite shell layer (pterioids, pectinoids, and mytiloids) displayed statistically similar patterns and there is no systematic variation in the number of Lazarus genera between the two mineralogies, suggesting that the number of Lazarus taxa is most strongly controlled by the abundance of a group rather than its skeletal mineralogy. Lazarus taxa accounted for 18.4–20.8% of total rhynchonelliform brachiopod diversity in the Roadian–Capitanian interval, with a significant decrease to 10% in the Wuchiapingian ($Z = 3.03$, $p = 0.002$). However, in notable contrast to the other groups, demosponges exhibit dramatic variation in the percentage of Lazarus taxa in a given stage (Fig. 2d). Lazarus genera account for 50.9% of all present or inferred sponges during the Wuchiapingian, but only 10.2% in the Wordian and 17.0% in the Changhsingian.

Although there are few changes in the percentage of Lazarus taxa from the Roadian to Wuchiapingian, all investigated groups have substantially fewer Lazarus genera in the Changhsingian stage (Fig. 2). The percentage of gastropod Lazarus taxa decreased from more than 37% in the Wuchiapingian to only 18.8% in the Changhsingian ($Z = 2.73$, $p = 0.006$), bivalve Lazarus taxa decreased from 19.6% to only 4.9% ($Z = 2.90$, $p = 0.003$), rhynchonelliform brachiopods decreased from 10% to 0% in the Changhsingian ($Z = 4.85$, $p < 0.001$), and demosponges from 50.9% to 17.0% ($Z = 3.69$, $p < 0.001$). However, this dramatic reduction in the percentage of Lazarus taxa does not imply a pronounced increase in the quality of the fossil record or the fidelity of sampling during the Changhsingian. Rather, it reflects edge effects due to the severe taxonomic impact of the end-Permian extinction. Because so many Permian genera became extinct (51% of gastropod genera, 65% of bivalves, and 96% of rhynchonelliform brachiopods, with the remaining brachiopods disappearing in the Griesbachian), the likelihood of Permian taxa occurring in the Triassic was greatly reduced and the latest Permian Changhsingian

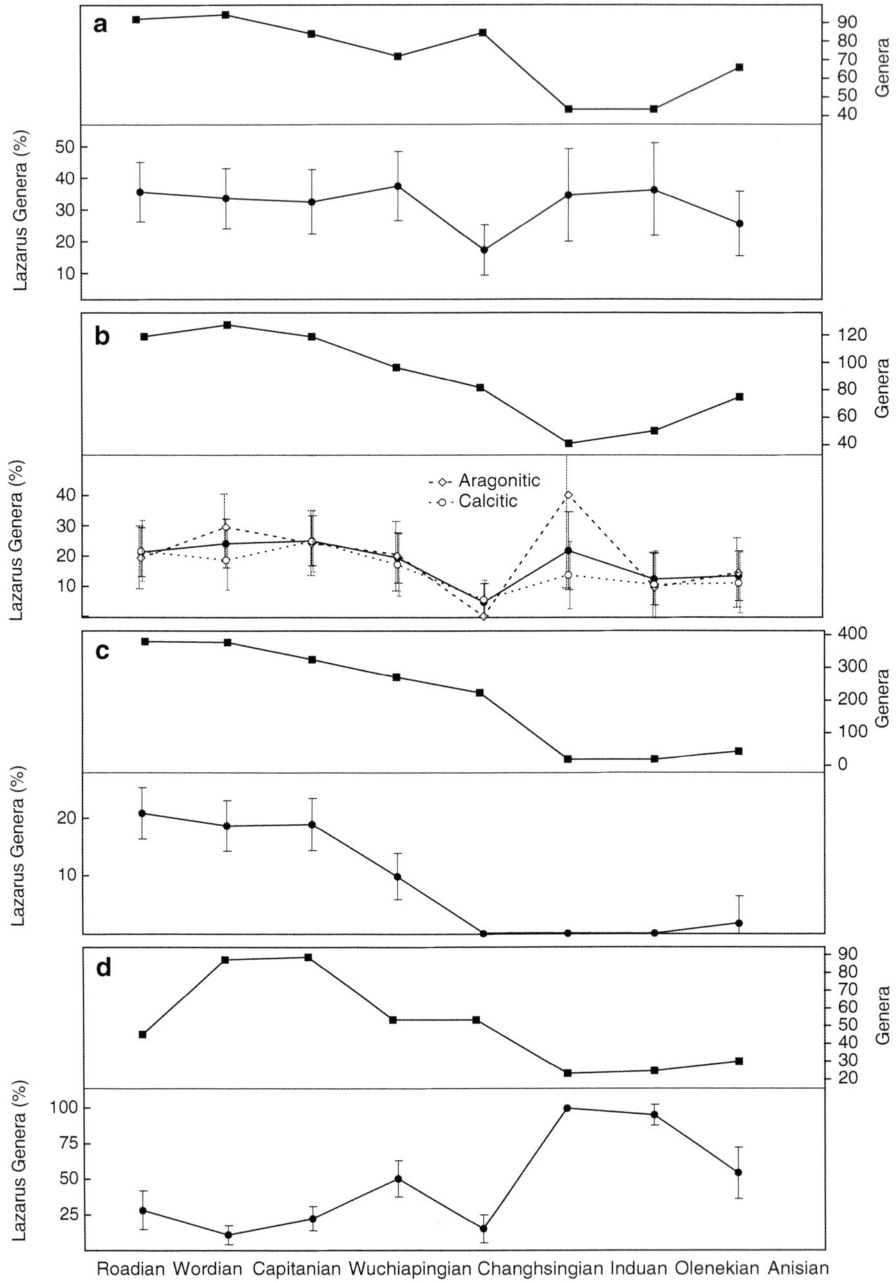

Fig. 2 Total (within-bin and Lazarus) diversity and percentage of Lazarus taxa for gastropods (**a**), bivalves, with aragonitic and calcitic forms plotted separately (**b**), rhynchonelliform brachiopods (**c**), and sponges (**d**) in Middle and Late Permian and Early Triassic stages. Error bars indicate 95% confidence interval for Lazarus percentage

Stage has anomalously low numbers of Lazarus taxa compared to more typical Permian values. For example, there are no rhynchonelliform brachiopod Lazarus genera in the Changhsingian stage due to the extreme severity of the end-Permian extinction event.

The striking stability in the percentage of rhynchonelliform brachiopod, bivalve, and gastropod genera represented by Lazarus taxa during the Roadian–Wuchiapingian interval, and especially across the end-Guadalupian extinction, implies that the quality of the benthic invertebrate fossil record remained consistent across the Guadalupian/Lopingian boundary. Demosponges may be an exception and the significant increase in Lazarus taxa across the end-Guadalupian extinction, from 22.5% in the Capitanian to 50.9% in the Wuchiapingian ($Z = -3.49$, $p < 0.001$), could either reflect poor preservation of sponges or small sponge population sizes in the Wuchiapingian. The Wuchiapingian has few demosponge occurrences compared to the well-sampled surrounding intervals; only 103 generic occurrences of sponges compared to 1,513 in the Capitanian and 223 in the Changhsingian. In addition, the Wuchiapingian is a time of turnover or crisis in the reef ecosystem, and the number of preserved reefs is low compared to the Wordian, Capitanian, or Changhsingian (Weidlich 2002). Thus, the high number of sponge Lazarus taxa is primarily a result of actual decreases in population size rather than taphonomic biases due to poor preservation. The overall lack of substantial variation in the abundance of Lazarus taxa across the end-Guadalupian extinction is consistent with the conclusions of Stanley and Yang (1994) that taphonomic biases did not substantially influence the observed pattern of extinction during the end-Guadalupian crisis.

4.2 Patterns in Permian Silicification

Although variations in Lazarus taxa abundance are not consistent with major taphonomic biases during the Guadalupian–Lopingian interval, shifts in the amount of silicification may have independently affected diversity patterns. Early diagenetic silicification can preserve a higher fidelity record of a fossil assemblage because of enhanced aragonite preservation (Cherns and Wright 2000; Wright et al. 2003, Butts and Briggs, this volume), and temporal variations in the amount of silica-replaced fossils have been argued to influence diversity patterns and extinction estimates (Schubert et al. 1997). To evaluate the potential effects of silicification on the end-Guadalupian extinction, we calculated the percentage of collections with silica-replaced fossils in each Permian stage, quantified the difference in alpha diversity between silicified and non-silicified assemblages, and counted the number of genera that are uniquely found in silicified collections.

Diagenetic silica replacement is thought to be a common phenomenon during much of the Permian, as exemplified by famous silicified localities from Thailand, the Salt Range in Pakistan, and especially from the Glass and Guadalupe Mountains in the United States, among others (e.g. Cooper and Grant 1972; Grant 1968, 1976). However, the number of silicified fossil collections in each stage is actually quite variable (Fig. 3); for example, 7.5% of the 1,280 Wordian collections contain silica-replaced fossils whereas 18.1% of the 1,444 collections in the Capitanian have been silicified. In contrast to the Guadalupian, silicification is much less widespread in the Lopingian. Only 3.3% of the 1,241 Wuchiapingian collections and 1.5% of the 983 Changhsingian collections have been silicified, suggesting that the substantial decline in the proportion of silicified fossils across the end-Guadalupian boundary may contribute to apparent elevated extinction rates.

Collections with silica-replaced fossils also have consistently higher sampled alpha diversity than non-silicified collections (Fig. 4). Note that overall mean alpha

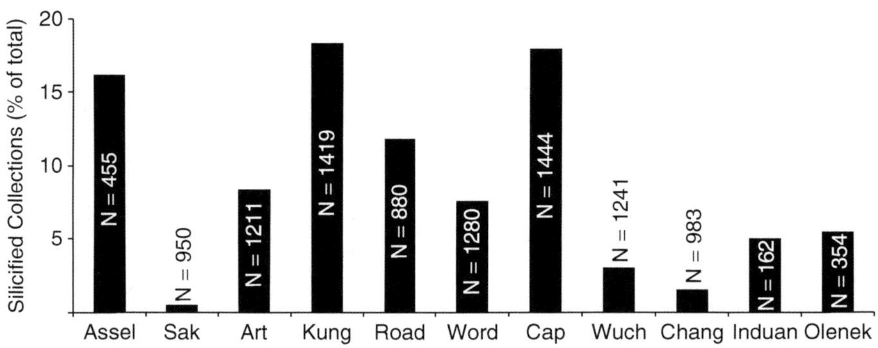

Fig. 3 Percentage of fossil collections containing silicified fossils in each Permian and Early Triassic stage. Assel: Asselian; Sak: Sakmarian; Art: Artinskian; Kung: Kungurian; Road: Roadian; Word: Wordian; Cap: Capitanian; Wuch: Wuchiapingian; Chang: Changhsingian; Ind: Induan; Ole: Olenekian. The n values indicate the total number of collections from each stage

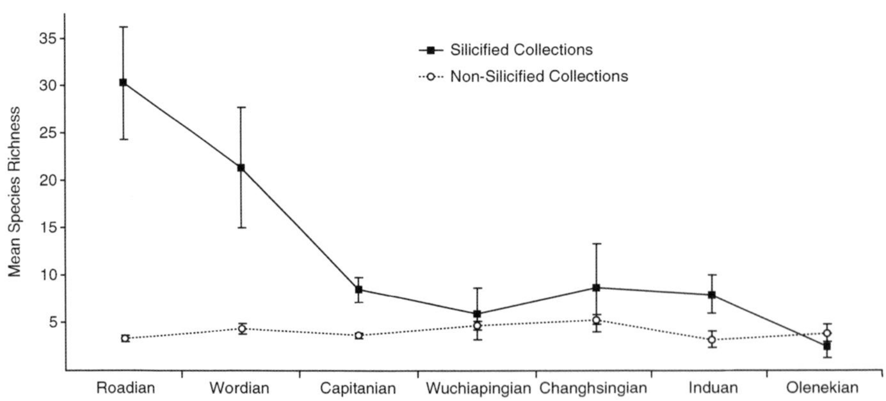

Fig. 4 Mean species richness for collections containing silicified fossils (*solid line and square symbols*) and non-silicified fossils (*dashed line and open circle symbols*) in Middle Permian, Late Permian, and Early Triassic stages. Error bars are 95% confidence intervals

diversity values are a function of the nature of reporting in the published literature and are not representative of actual alpha diversity; many papers are taxonomic descriptions and only consider a single taxonomic group and record one or a few new species of interest from a given locality. In particular, the large discrepancy between silicified and non-silicified alpha diversity in the Middle Permian is primarily a result of the large taxonomic lists reported from the extraordinarily large silicified collections from west Texas. Nevertheless, apparent changes in sampled alpha diversity, whether real biological phenomena or due to changes in the number of taxa actually reported for a collection in published papers, still affect our perception of diversity and extinction in the fossil record. During the Roadian and Wordian, mean silicified alpha diversity is 24.4 species and 14.6 species per collection, compared to only 3.0 and 3.65 species in non-silicified collections from the same stages. The difference between silicified and non-silicified alpha diversity is statistically significant during the Guadalupian, but not in the Lopingian stages (4.5 vs 3.95 species in the Wuchiapingian, $p = 0.51$; 5.95 vs 4.4 species in the Changhsingian, $p = 0.18$). Although the difference in alpha diversity is not always statistically significant, the consistently higher values in silicified collections may have acted in conjunction with the significant decrease in the amount of silicification to exacerbate apparent diversity loss and increase calculated extinction rates. However, the major decrease in alpha diversity in silicified collections occurs from the Roadian to Capitanian stages (Fig. 4), earlier than the traditionally recognized end-Guadalupian extinction. There is a minor but significant decrease in silicified alpha diversity across the Guadalupian/Lopingian boundary (7.4–4.5 species; $p = 0.05$) but non-silicified alpha diversity actually increases significantly (3.45–3.95 species, $p = 0.02$).

Although there were substantial changes in the extent of silicification during the Permian, and silicification may preserve a better record of alpha diversity and relative abundance (Cherns and Wright 2000; Wright et al. 2003), it is not clear to what extent it affects global diversity patterns. If many genera are known exclusively from silicified collections, silica-replacement may exert an important control on global diversity patterns. In contrast, if most genera from silicified assemblages are also found in non-silicified assemblages, the implication is that silicification itself is not important for reconstructing diversity. During the Permian, the percentage of genera uniquely known from silicified assemblages in a given stage is influenced by the percentage of collections that are silicified, and can be as high as 23% of bivalves, 33% of brachiopods, and 60% of gastropods, all during the Roadian Stage. However, overall only a small number of genera are known only from silicified specimens, as many found in silicified collections from one stage are then recorded from non-silicified assemblages at another time. Only 3.2% of Permian bivalves (6 of 190 genera) are exclusive to silicified collections, while 17.4% of gastropods (31 of 178 genera, although several of those may be known from non-silicified collections in the Carboniferous) and 12.9% of brachiopods (94 of 727 genera) are uniquely found in silicified assemblages. Total genus richness is only 5–25% higher when silicified collections are included, compared to the value obtained solely from non-silicified fossils. However, calculated extinction and origination rates vary by no more than 5% if silicified collections are excluded. These results indicate that silicification itself is not necessary for preserving a good record of diversity during the Guadalupian–Lopingian interval and that the severity of the end-Guadalupian extinction is not biased by changes in silicification.

4.3 *Early Triassic Lazarus Effect*

Only 18.8% of gastropod genera during the Changhsingian and 26.9% during the Anisian stage are Lazarus taxa, while 34.9% of Induan and 37.2% of Olenekian gastropod diversity are Lazarus taxa (Fig. 2a). The differences between the proportion of Lazarus gastropod genera from the Changhsingian to Induan is statistically significant ($Z = -2.00$; $p = 0.045$) but the other differences are not significant at $p = 0.05$. The proportions of gastropod Lazarus taxa are also lower than those previously published (Erwin and Pan 1996). Though the proportions of gastropod Lazarus taxa during the Induan and Olenekian are similar to those of the Middle Permian (Fig. 2), the predicted proportion of Lazarus taxa in the Induan would be lower because of extinction edge effects, as in the Changhsingian when less than 20% of taxa were Lazarus genera. The proportion of bivalve Lazarus taxa was 22% in the Induan and 11.3% in the Olenekian. The differences in the proportion of bivalve taxa are not statistically significant between the Induan and Olenekian or from the Olenekian to Anisian, but the change from the Changhsingian to Induan, is ($Z = -2.88$, $p = 0.004$: Fig. 2b). Aragonitic bivalves had a higher proportion of Lazarus taxa compared to calcitic taxa during the Induan (40% versus 13.3%), but the difference is not significant due to the small sample size, especially of aragonitic taxa ($Z = 1.82$, $p = 0.07$). There was also little difference between aragonitic and calcitic mineralogy during the Olenekian. The proportions of Lazarus aragonitic genera were significantly different between the Changhsingian and Induan and the Induan and Olenekian ($Z = -3.65$, $p < 0.001$; $Z = 2.01$, $p = 0.04$). The proportions of calcitic Lazarus taxa did not differ significantly between the Changhsingian through Anisian stages. There are no rhynchonelliform brachiopod Lazarus genera during the Early Triassic stages. One hundred percent and 95.8% of demosponge genera were Lazarus genera during the Induan and the Olenekian, respectively, while many sponges remained known only as Lazarus taxa in the Anisian.

4.3.1 Controls on Early Triassic Lazarus Taxa

Potential controls on the occurrence of Lazarus taxa during the Early Triassic include taphonomic processes, sampling, environmental conditions, paleobiology of the organisms, and taxonomic practices, or a combination thereof.

The Early Triassic Lazarus phenomenon among bivalves and gastropods may have resulted from taphonomic bias related to their aragonitic composition.

Aragonitic shells typically dissolve during meteoric or burial diagenetic processes during early diagenesis because aragonite is less stable than calcite at surface temperatures and pressures, even during "aragonite seas" (e.g. Tucker and Wright 1990). These diagenetic processes may have been compounded by increased acidity and $CaCO_3$ undersaturation of seawater caused by elevated atmospheric CO_2 levels during the Early Triassic (Berner 2004). Thus, the large proportion of aragonitic Lazarus taxa could reflect post mortem taphonomic processes in a high CO_2 world. Similar processes have been proposed to explain Triassic–Jurassic boundary patterns (e.g. Hautmann 2004; Hautmann et al. 2008a), and are observed in the modern ocean and predicted for the future (e.g. Feely et al. 2004). Alternatively, low levels and depth of bioturbation during the Early Triassic (e.g. Twitchett 1999; Pruss and Bottjer 2004; Fraiser and Bottjer 2009, 2010) may have buffered some shell dissolution during the Early Triassic. Shell dissolution is high in areas with well-developed infaunal benthic communities because biogenic reworking of sediments increases oxygen levels in the mixed layer and promotes oxidative decay of organic matter, thereby increasing acidity near the sediment–water interface and causing pore waters to become undersaturated with respect to both aragonite and calcite (Aller 1982; Walter and Burton 1990). Sediment reworking by infaunal organisms also disrupts mold space left after shells dissolve (Cherns and Wright 2000). Even if after death benthic calcareous shells were dissolved on the seafloor in some regions due to a lowered carbonate saturation state of seawater (Berner 2004; Feely et al. 2004), the reduction in bioturbating activity that characterized much of the Early Triassic may have prevented molds from being disturbed.

Sample size, either as a result of actual sampling effort or of true population size, is another potential contributor to the Early Triassic Lazarus phenomenon. A good correlation between a group's abundance, the number of occurrences, and the number of Lazarus taxa can be observed in Middle Permian rhynchonelliform brachiopods, bivalves, and gastropods, confirming the importance of sampling on Lazarus abundance. Reduced Early Triassic sampling between two well-sampled stages tends to increase the number of Lazarus taxa. Very low levels of sampling in the Early Triassic (378 Induan occurrences and 673 Olenekian occurrences of rhynchonelliform brachiopods, gastropods, and bivalves) relative to the Changhsingian (4,755 occurrences) and Anisian (2,439 occurrences) may be influenced by sampling effort but more likely reflect reduced marine invertebrate abundance due to environmental stress following the end-Permian mass extinction (e.g. Fraiser and Bottjer 2007). A specimen of a Griesbachian Lazarus gastropod taxon was found recently in a newly discovered Tethyan section (Wheeley and Twitchett 2005), further highlighting that low numbers of some gastropod taxa contributed to the Lazarus phenomenon (sensu Wignall and Benton 1999) and suggesting that more sampling of Lower Triassic strata, especially in largely ignored regions, could aid in finding missing gastropod taxa that migrated to refugia or that were low in abundance. Future work on physiological and ecological characteristics of Lazarus gastropod genera could determine the extent to which Early Triassic environmental conditions contributed to the low numbers of certain gastropod taxa. The prevalence of sponge Lazarus taxa is also likely tied to sampling; namely, the lack of reef sites during the Early Triassic (a metazoan "reef gap", Flügel and Stanley 1984). Permian–Triassic sponge occurrences are strongly covariant with times of widespread reef-building in the Wordian–Capitanian, Changhsingian, and Late Triassic. Stages with low reef abundance between those reef episodes have many demosponge Lazarus taxa – e.g. the Wuchiapingian (50.9%), Anisian (the beginning of the Triassic reef recovery, but still with 53.3% Lazarus taxa), and to an extreme degree the Induan and Olenekian. As the Early Triassic reef gap may have been due to elevated atmospheric CO_2 and ocean acidification that prevented metazoan reef organisms from forming skeletons (Stanley et al. 2007), extinction-related environmental factors may have contributed to the reduced sampling through reduced population size.

Biological factors may have facilitated the Lazarus effect among some Early Triassic taxa. Most of the aragonitic Induan Lazarus genera had infaunal or semi-faunal lifestyles. During the end-Triassic mass extinction, aragonitic infaunal bivalves suffered greater extinctions than epifaunal bivalves (Hautmann et al. 2008b), and it has been proposed that this pattern indicates a reduction in primary productivity as epifaunal bivalves have physiological characteristics that enabled them to fare better during conditions of reduced food availability (McRoberts and Newton 1995). A decrease in primary productivity has also been proposed for the Early Triassic (Twitchett 2001). It is unclear whether the Lazarus pattern among Early Triassic bivalves resulted more from diagenetic processes or from biological reasons, but both mechanisms were linked to Early Triassic environmental conditions (elevated CO_2). Furthermore, the reason that the Lazarus effect among aragonitic bivalve taxa is more pronounced in Induan versus Olenekian age strata is unknown. However, this temporal pattern supports the argument for environmental conditions contributing to the Lazarus effect because it could reflect an amelioration of some aspect of the global environment later in the aftermath.

Poor taxonomic practice is a plausible hypothesis to explain in part the Early Triassic Lazarus phenomenon among some groups. For example, partial preservation has made it difficult to definitively identify some Early Triassic gastropod taxa (Wheeley and Twitchett 2005). The small size of Early Triassic gastropods could make it difficult to determine gastropod taxonomy; many Early Triassic gastropods are microgastropods <1cm in height (Fraiser and Bottjer 2004; Fraiser et al. 2005), and needles have been required to prepare them to expose areas for proper identification (Batten and Stokes 1986). Some Middle Triassic gastropods have been incorrectly identified as Elvis taxa (Wheeley and Twitchett 2005), taxa that were misidentified as having re-emerged after their presumed extinction, but are not actually descendants of the original taxa (Erwin and Droser 1993). More accurate gastropod taxonomy across the P/T boundary and into the Middle Triassic could determine the extent to which the Lazarus effect among gastropods is real and significant.

4.4 Patterns in Early Triassic Silicification

Contrary to previous reports that shell replacement by silica is absent among Lower Triassic faunas (e.g. Erwin and Pan 1996; Kidder and Erwin 2001; Twitchett 2001), 5% of Early Triassic collections contain fossils preserved via silicification (Fig. 5). Silicified Early Triassic faunas have been reported from Oman, China, and the U.S.A. Although this value likely represents a maximum estimate due to the easily-accessible literature on silicified faunas, the value is broadly similar to the proportion of silicification in Permian stages, in the Lopingian in particular. Of the Early Triassic benthic collections with preservation via silica replacement, 30% are from Induan and 70% are from Olenekian age strata (Fig. 3). Olenekian U.S.A. collections comprise 70% of the silicified collections.

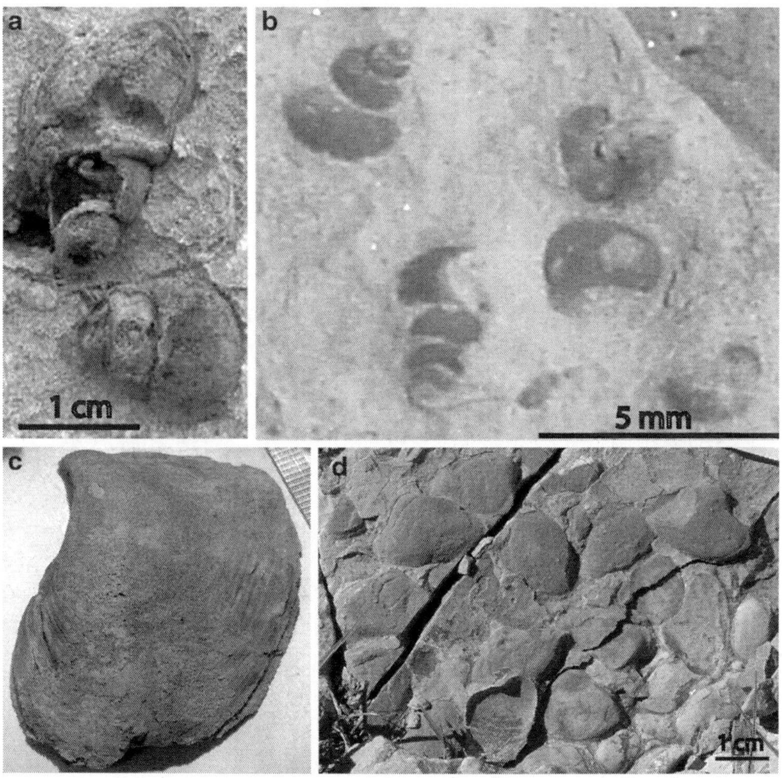

Fig. 5 Preservation of Early Triassic fossils. (**a**) Silicified microgastropods, Virgin Limestone Member, Moenkopi Formation. (**b**) Internal molds of microgastropods, Campil Member, Werfen Formation. (**c**) Silicified *Promyalina*, upper member, Thaynes Formation. Scale in mm. From D. Boyd collection. (**d**) Internal molds of bivalves (*Unionites*), Siusi Member, Werfen Formation (Modified from Fraiser and Bottjer (2005a)

The mean alpha diversity of PBDB Early Triassic collections preserved via silicification is 7.86 for Induan collections, 2.63 for Olenekian collections, and 4.19 for the series. An independent groups t-test of means indicates that the difference between the mean alpha diversities of silicified Induan and Olenekian collections is significant (t = 5.20, $p < 0.0001$). Non-silicified Induan collections have a mean alpha diversity of 3.18, and Olenekian ones have a mean alpha diversity of 4.14 (Fig. 4). The mean alpha diversity for all Early Triassic collections preserved as casts and molds is 3.67. The difference between the means for non-silicified Induan and Olenekian collections is statistically significant (t = 2.50, $p = 0.013$). The difference between means of silicified and non-silicified Induan collections is statistically significant (t = 4.87, $p < 0.0001$), but there is no significant difference between the means for silicified and non-silicified Olenekian collections (t = 1.49, $p = 0.14$), or between the means for silicified and non-silicified Early Triassic collections (t = 0.696, $p = 0.49$).

Fourteen gastropod genera (*Ananias*, *Anomphalus*, *Bellerophon*, *Chartronella*, *Coelostylina*, *Donaldina*, *Jiangxispira*, *Laxella*, *Naticopsis*, *Omphaloptycha*, *Platyzona*, *Streptacis*, *Strobeus*, and *Worthenia*) are found in Induan silicified collections, but all but two of those (*Jiangxispira*, *Laxella*) also occur in non-silicified collections at another time. No Olenekian gastropods are known exclusively from silicified faunas. No Induan bivalve genera are known exclusively from silicified faunas. At least eight bivalve genera (*Elegantinia*, *Entolioides*, *Eumorphotis*, *Leptochondria*, *Neoschizodus*, *Pegmavalvula*, "*Pleuronectites*", *Placunopsis*) are known from Olenekian silicified collections, but *Pegmavalvula* is the only one of 46 Olenekian bivalves exclusively found in silicified localities throughout its entire range. No rhynchonelliform brachiopods are known only from shells preserved via silicification. Therefore, most Early Triassic genera are known from fossil casts and molds.

Documentation of silicified fossil collections from Lower Triassic strata refutes the hypothesis that there is a "complete absence of silicified faunas" during the Early Triassic (Twitchett 2001). Silicified Induan and Olenekian collections are rare in comparison to some Permian stages, but actually occur as frequently as silicified collections in the Wuchiapingian or Changhsingian stages of the Late Permian. There is no change in silicification across the Permian–Triassic boundary, indicating that changes in preservation style are not the primary contributor to the unusual Early Triassic record.

The statistically significant differences between silicified Induan and Olenekian collections and between non-silicified Induan and Olenekian collections supports previous findings that diversity increased through the Early Triassic (e.g. Schubert and Bottjer 1995) and could be an indication of biotic recovery following the end-Permian mass extinction. That there is no statistically significant difference between silicified and non-silicified Olenekian collections does not support the hypothesis that silicified faunas record more information than non-silicified ones. Furthermore, Lazarus taxa have been found in non-silicified collections (Hautmann and Nützel 2004). The statistically significant difference between silicified and non-silicified Induan collections could reflect a real difference in the amount of data preserved in silicified and non-silicified collections. However, 22% of silicified faunas are from one recently discovered Griesbachian-age section in Oman, and these collections have a mean alpha diversity of 8.5 that skews the mean alpha diversity for Induan silicified faunas. When these collections are removed from the analysis, there is no statistically significant difference between the mean alpha diversities of silicified and non-silicified Induan collections. More sampling of Lower Triassic sections around the world would probably lead to the discovery of more silicified faunas.

Only a very small percentage of gastropod, bivalve, or brachiopod genera are known exclusively from silicified collections, whether in the Permian or in the Early Triassic. The impact of silicification, or the lack thereof, on global diversity compilations is therefore minimal.

Differences between silicified and non-silicified collections were likely influenced more by worker-introduced bias and not taphonomic processes. One source of bias could potentially be the focus of many Early Triassic studies. Publications of taxonomic lists from Lower Triassic strata commonly report only one higher taxonomic group, e.g., rhynchonelliform brachiopods (Perry and Chatterton 1979), ostracodes (Crasquin-Soleau and Kershaw 2005), or pectinoid bivalves (e.g. Newell and Boyd 1995), even when other taxa have been collected in the field (D.W. Boyd, pers. comm.). Furthermore, authors commonly report only higher taxa rather than detailed lists of genera or species. Incompletely reported taxonomic lists mean that the alpha diversities of both silicified and non-silicified Induan and Olenekian collections are likely underestimates. Bias also could be introduced by the methods in which data are collected (D.W. Boyd pers. comm.). If only silicified faunas are searched for and collected from the field for analysis, any data from fossils that are not silicified are omitted. It is important to note that the only known Smithian (early Olenekian) collections are from non-silicified collections. Though silicified shells are easily extracted from their encompassing matrix with weak acids (such as hydrochloric or acetic acid), acids also cause the dissolution of any non-silicified material in the sample. Incomplete and incorrectly entered data in the PBDB, i.e., preservation entered as silicification when fossils were not actually silicified, incorrect and outdated taxonomy, and omitted publications, prevent many collections and taxa (e.g., sponges) from being downloaded when certain searches are performed. Outdated taxonomy, incomplete taxonomic lists, and under-sampling likely exert a stronger influence on understanding of Early Triassic ecology than taphonomic processes.

5 Conclusions

A full accounting of the effects of taphonomic biases during mass extinction intervals is critical in any attempt to extract meaningful biological signals from the fossil record during a biotic crisis and its aftermath. Changes in the fidelity or dominant style of fossil preservation can have a substantial impact on the composition and diversity of marine fossil assemblages, and it has been proposed, based on an abundance of Lazarus taxa and a reduction in silicification, that such taphonomic changes irreparably bias the fossil record of the Permian–Triassic mass extinction. A new tabulation of the proportion of Lazarus genera in the major Permian–Triassic taxonomic groups suggests that there was no major bias or change in taphonomic style in the Late Permian but that aragonitic taxa (gastropods and some, primarily infaunal, bivalves) may have suffered from reduced preservation in the Induan. Comparisons between silicified and non-silicified faunas indicate there was little change in the amount of silicification across the Permian–Triassic boundary; regardless, silicification is unlikely to be a major taphonomic bias in global compilations because few taxa are known exclusively from silicified collections.

Documentation of Lazarus taxa does not necessarily indicate that the fossil record is biased. Indeed, taxon outages are a common phenomenon in the fossil record and are caused by a variety of mechanisms. Rickards and Wright (2002) suggest that the concept of a Lazarus taxon is not useful as a taphonomic indicator because it represents nothing more than a taxon's low abundance during a given interval. The Permian–Triassic pattern of Lazarus taxa documented here is partially consistent with this concept; the Early Triassic Lazarus effect is a function of sampling effects, biological and environmental factors, and actual taphonomic degradation.

Though more sampling and refined taxonomy will improve the reconstruction of the end-Permian extinction and its unusual aftermath, our current understanding of the Early Triassic fossil record likely reflects a primary ecological signal (to the extent that any Paleozoic or Mesozoic fossil assemblage reflects a primary biological signal) (Fig. 5). Taphonomy remains an important factor that must be assessed in each fossil assemblage, but analysis of the Permian–Triassic record of Lazarus taxa and silicification demonstrates that the fossil record of the end-Permian extinction and the Early Triassic aftermath is not completely obscured by a taphonomic megabias due to skeletal mineralogy or fossil preservation. Instead of being solely an indication of the poor quality of the fossil record (e.g., Twitchett 2001; Smith 2007), Lazarus taxa could also provide clues about the environmental conditions during deposition.

References

Aller, R. C. (1982). Carbonate dissolution in nearshore terrigenous muds: the role of physical and biological reworking. *Journal of Geology, 90*, 79–95.

Allison, P. A., & Briggs, D. E. G. (1993). Paleolatitudinal sampling bias, Phanerozoic species diversity, and the end-Permian extinction. *Geology, 21*, 65–68.

Bambach, R. K., Knoll, A. H., & Sepkoski, J. J., Jr. (2002). Anatomical and ecological constraints on Phanerozoic animal diversity in the marine realm. *Proceedings of the National Academy of Sciences of the United States of America, 99*, 6954–6959.

Bambach, R. K., Knoll, A. H., & Wang, S. C. (2004). Origination, extinction, and mass depletions of marine diversity. *Paleobiology, 30*, 522–542.

Batten, R. L. (1973). The vicissitudes of the gastropods during the interval of Guadalupian-Ladinian time. In A. Logan & L. V. Hills (Eds.), *The Permian and Triassic systems and their mutual boundary* (Vol. 2, pp. 596–607). Boulder: Canadian Society of Petroleum Geologists Memoir.

Batten, R. L., & Stokes, W. L. (1986). Early Triassic gastropods from the Sinbad Member of the Moenkopi Formation, San Rafael Swell, Utah. *American Museum Novitates, 2864*, 1–33.

Berner, R. A. (2004). *The Phanerozoic carbon cycle: CO_2 and O_2*. New York: Oxford University Press.

Chen, Z. Q., Kaiho, K., & George, A. D. (2005). Early Triassic recovery of the brachiopod faunas from the end-Permian mass extinction: a global review. *Palaeogeography, Palaeoclimatology, Palaeoecology, 224*, 270–290.

Cherns, L., & Wright, V. P. (2000). Missing molluscs as evidence of large-scale, early skeletal aragonite dissolution in a Silurian sea. *Geology, 28*, 791–794.

Clapham, M. E., & Bottjer, D. J. (2007a). Permian marine paleoecology and its implications for large-scale decoupling of brachiopod and bivalve abundance and diversity during the Lopingian (Late Permian). *Palaeogeography, Palaeoclimatology, Palaeoecology, 249*, 283–301.

Clapham, M. E., & Bottjer, D. J. (2007b). Prolonged Permian–Triassic ecological crisis recorded by molluscan dominance in Late Permian offshore assemblages. *Proceedings of the National Academy of Sciences of the United States of America, 104*, 12971–12975.

Clapham, M. E., Shen, S. Z., & Bottjer, D. J. (2009). The double mass extinction revisited: reassessing the severity, selectivity, and causes of the end-Guadalupian biotic crisis (Late Permian). *Paleobiology, 35*, 33–51.

Cooper, G. A., & Grant, R. E. (1972). Permian brachiopods of west Texas, I. *Smithsonian Contributions to Paleobiology, 14*, 1–231.

Courtillot, V., & Gaudemer, Y. (1996). Effects of mass extinctions on diversity. *Nature, 381*, 146–148.

Crampton, J. S., Beu, A. G., Cooper, R. A., Jones, C. M., Marshall, B., & Maxwell, P. A. (2003). Estimating the rock volume bias in paleobiodiversity studies. *Science, 301*, 358–360.

Crasquin-Soleau, S., & Kershaw, S. (2005). Ostracod fauna from the Permian–Triassic boundary interval of South China (Huaying Mountains, eastern Sichuan Province): palaeoenvironmental significance. *Palaeogeography, Palaeoclimatology, Palaeoecology, 217*, 131–141.

Droser, M. L., Bottjer, D. J., & Sheehan, P. M. (1997). Evaluating the ecological architecture of major events in the Phanerozoic history of marine invertebrate life. *Geology, 25*, 167–170.

Droser, M. L., Bottjer, D. J., Sheehan, P. M., & McGhee, G. (2000). Decoupling of taxonomic and ecologic severity of Phanerozoic marine mass extinctions. *Geology, 28*, 675–678.

Erwin, D. H. (1996). Understanding biotic recoveries: extinction, survival, and preservation during the end-Permian mass extinction. In D. Jablonski, D. H. Erwin, & J. Lipps (Eds.), *Evolutionary paleobiology*. Chicago: The University of Chicago Press.

Erwin, D. H. (2001). Lessons from the past: biotic recoveries from mass extinctions. *Proceedings of the National Academy of Sciences of the United States of America, 98*, 5399–5403.

Erwin, D.H. (2006). Extinction: how life on Earth Nearly ended 250 million years ago. Princeton University Press, Princeton. 306 pp.

Erwin, D. H., & Droser, M. L. (1993). Elvis taxa. *Palaios, 8*, 623–624.

Erwin, D. H., & Pan, H. (1996). Recoveries and radiations: gastropods after the Permo-Triassic mass extinction. In M. B. Hart (Ed.), *Biotic recovery from mass extinction events* (Vol. 102, pp. 223–229). London: Geological Society Special Publication.

Feely, R. A., Sabine, C. L., Lee, K., Berelson, W., Kleypas, J., Fabry, V. J., et al. (2004). Impact of anthropogenic CO_2 on the $CaCO_3$ system in the oceans. *Science, 305*, 362–366.

Flessa, K. W. (1990). The "facts" of mass extinctions. In V. L. Sharpton & P. D. Ward (Eds.), *Global catastrophes in Earth history: An interdisciplinary conference on impacts, volcanism, and mass mortality* (Vol. 247, pp. 1–7). London: Geological Society of America Special Paper.

Flessa, K. W., & Jablonski, D. (1983). Extinction is here to stay. *Paleobiology, 9*, 315–321.

Flügel, E., & Stanley, G. D. (1984). Reorganization, development and evolution of post-Permian reefs and reef organisms. *Palaeontographica Americana, 54*, 177–186.

Foote, M. (2000) Origination and extinction components of taxonomic diversity: general problems. In: D.H. Erwin & S.L. Wing (Eds.) *Deep time: Paleobiology's perspective, Paleobiology, 26*, 4:74–102.

Fraiser, M. L., & Bottjer, D. J. (2004). The non-actualistic Early Triassic gastropod fauna: a case study of the Lower Triassic Sinbad Limestone Member. *Palaios, 19*, 259–275.

Fraiser, M. L., & Bottjer, D. J. (2005a). Fossil preservation during the aftermath of the end-Permian mass extinction: taphonomic processes and palaeoecological signals. In J. Morrow, D. J. Over, & P. B. Wignall (Eds.), *Understanding late Devonian and Permian–Triassic biotic and climatic events: towards an integrated approach* (Vol. 20, pp. 299–311). Amsterdam: Developments in Paleontology and Stratigraphy.

Fraiser, M. L., & Bottjer, D. J. (2005b). Restructuring of benthic level-bottom shallow marine communities due to prolonged environmental stress during the aftermath of the end-Permian mass extinction. *Comptes Rendus Palevol, 4*, 515–523.

Fraiser, M. L., & Bottjer, D. J. (2007). When bivalves took over the world. *Paleobiology, 33*, 397–413.

Fraiser, M. L., & Bottjer, D. J. (2009) Opportunistic behavior of the invertebrate benthos following the End-Permian mass extinction. *Australian Journal of Earth Sciences, 56*, 841–857.

Fraiser, M. L., Twitchett, R. J., & Bottjer, D. J. (2005). Unique microgastropod biofacies in the Early Triassic: indicator of long-term biotic stress and the pattern of biotic recovery after the end-Permian mass extinction. *Comptes Rendus Palevol, 4*, 475–484.

Gould, S. J., & Calloway, C. B. (1980). Clams and brachiopods – ships that pass in the night. *Paleobiology, 6*, 383–396.

Grant, R. E. (1968). Structural adaptation in two Permian brachiopod genera, Salt Range, West Pakistan. *Journal of Paleontology, 42*, 1–32.

Grant, R. E. (1976). Permian brachiopods from southern Thailand. *Journal of Paleontology, 50*(3), 1–269.

Grice, K., Cao, C., Love, G. D., Böttcher, M. E., Twitchett, R. J., Grosjean, E., et al. (2005). Photic zone euxinia during the Permian–Triassic superanoxic event. *Science, 307*, 706–709.

Hautmann, M. (2004). Effect of end-Triassic CO_2 maximum on carbonate sedimentation and marine mass extinction. *Facies, 50*, 257–261.

Hautmann, M., Benton, M. J., & Tomasovych, A. (2008). Catastrophic ocean acidification at the Triassic-Jurassic boundary. *Neues Jahrbuch fur Geologie und Palaontologie, Abhandlungen., 249*, 119–127.

Hautmann, M., & Nützel, A. (2004). First record of a heterodont bivalve (Mollusca) from the Early Triassic: palaeoecological significance and implications for the "Lazarus problem". *Palaeontology, 48*, 1131–1138.

Hautmann, M., Stiller, F., Huawei, C., & Jingeng, S. (2008). Extinction-recovery pattern of level-bottom faunas across the Triassic-Jurassic boundary in Tibet: implications for potential killing mechanisms. *Palaios, 23*, 711–718.

Henderson, C. M. (2005). International correlation of the marine Permian time scale. *Permophiles, 46*, 6–9.

Holdaway, H. K., & Clayton, C. J. (1982). Preservation of shell microstructure in silicified brachiopods from the Upper Cretaceous Wilmington Sands of Devon. *Geological Magazine, 119*, 371–382.

Huey, R. B., & Ward, P. D. (2005). Hypoxia, global warming, and terrestrial Late Permian extinctions. *Science, 308*, 398–401.

Isozaki, Y. (1997). Permo-Triassic boundary superanoxia and stratified superocean: records from lost deep sea. *Science, 276*, 235–238.

Isozaki, Y., Kawahata, H., & Minoshima, K. (2007). The Capitanian (Permian) Kamura cooling event: the beginning of the Paleozoic–Mesozoic transition. *Palaeoworld, 16*, 16–30.

Jablonski, D. (1986a). Background and mass extinctions: the alternation of macroevolutionary regimes. *Science, 31*, 129–133.

Jablonski, D. (1986b). Causes and consequences of mass extinctions: a comparative approach. In D. K. Elliot (Ed.), *Dynamics of extinction*. New York: Wiley.

Jablonski, D. (2001). Lessons from the past: evolutionary impacts of mass extinctions. *Proceedings of the National Academy of Sciences of the United States of America, 98*, 5393–5398.

Jablonski, D. (2002). Survival without recovery after mass extinctions. *Proceedings of the National Academy of Sciences of the United States of America, 99*, 8139–8144.

Jablonski, D. (2005). Mass extinctions and macroevolution. *Paleobiology, 31*, 192–210.

Jin, Y. G., Zhang, J., & Shang, Q. H. (1994). Two phases of end-Permian mass extinction. *Canadian Society of Petroleum Geologists Memoir, 17*, 813–822.

Kauffman, E. G., & Harries, P. J. (1996). The importance of crisis progenitors in recovery from mass extinction. In M. B. Hart (Ed.), *Biotic recovery from mass extinction events* (Vol. 102, pp. 15–39). London: Geological Society Special Publication.

Kidder, D. L., & Erwin, D. H. (2001). Secular distribution of biogenic silica through the Phanerozoic: comparison of silica-replaced fossils and bedded cherts at the series level. *The Journal of Geology, 109*, 509–522.

Lehrmann, D. J., Ramezani, J., Bowring, S. A., Martin, M. W., Montgomery, P., Enos, P., et al. (2006). Timing of recovery from the end-Permian mass extinction: Geochronologic and biostratigraphic constraints from south China. *Geology, 34*, 1053–1956.

McRoberts, C. A., & Newton, C. R. (1995). Selective extinction among end-Triassic European bivalves. *Geology, 23*, 102–104.

Newell, N.D., & Boyd, D.W. (1995). Pectinoid bivalves of the Permian–Triassic crisis. Bulletin of the American Museum of Natural History 227, 95 p.

Perry, D. G., & Chatterton, B. D. E. (1979). Late Early Triassic brachiopod and conodont fauna, Thaynes Formation, southeastern Idaho. *Journal of Paleontology, 53*, 307–319.

Powers, C. M., & Bottjer, D. J. (2007). Bryozoan paleoecology indicates mid-Phanerozoic extinctions are the product of long-term environmental stress. *Geology, 35*, 995–998.

Pruss, S. B., & Bottjer, D. J. (2004). Early Triassic trace fossils of the western United States and their implications for prolonged environmental stress from the end-Permian mass extinction. *Palaios, 19*, 551–564.

Raup, D. M. (1986). Biological extinction in earth history. *Science, 231*, 528–1533.

Raup, D. M. (1994). The role of extinction in evolution. *Proceedings of the National Academy of Sciences of the United States of America, 91*, 6758–6763.

Raup, D. M., & Sepkoski, J. J. (1982). Mass extinctions in the marine fossil record. *Science, 215*, 1501–1503.

Rickards, R. B., & Wright, A. J. (2002). Lazarus taxa, refugia and relict faunas: evidence from graptolites. *Journal of the Geological Society, 159*, 1–4.

Schubert, J. K., & Bottjer, D. J. (1995). Aftermath of the Permian–Triassic mass extinction event: paleoecology of Lower Triassic carbonates in the western U.S. *Palaeogeography, Palaeoclimatology, Palaeoecology, 116*, 1–39.

Schubert, J. K., Kidder, D. L., & Erwin, D. H. (1997). Silica-replaced fossils through the Phanerozoic. *Geology, 25*, 1031–1034.

Sephton, M. A., Looy, C. V., Brinkhuis, H., Wignall, P. B., de Leeuw, J. W., & Visscher, H. (2005). Catastrophic soil erosion during the end-Permian biotic crisis. *Geology, 33*, 941–944.

Sepkoski, J. J., Jr. (1981). A factor analytic description of the Phanerozoic marine fossil record. *Paleobiology, 7*, 36–53.

Sepkoski, J. J., Jr. (1984). A kinetic model of Phanerozoic taxonomic diversity. III. Post-Paleozoic families and mass extinctions. *Paleobiology, 10*, 246–267.

Sepkoski, J. J., Jr. (1986). Phanerozoic overview of mass extinction. In D. M. Raup & D. Jablonski (Eds.), *Patterns and processes in the history of life*. Berlin: Springer-Verlag.

Shen, S. Z., Zhang, H., Li, W. Z., Mu, L., & Xie, J. F. (2006). Brachiopod diversity patterns from Carboniferous to Triassic in south China. *Geological Journal, 41*, 345–361.

Signor, P. W., & Lipps, J. H. (1982). Sampling bias, gradual extinction patterns and catastrophe in the fossil record. In L. T. Silver & P. H. Schultz (Eds.), *Geological implications of impacthypothesis of large asteroids and comets on the Earth*. Boulder, CO: Geological Society of America Special Paper.

Smith, A. B. (2007). Intrinsic versus extrinsic biases in the fossil record: contrasting the fossil record of echinoids in the Triassic and early Jurassic using sampling data, phylogenetic analysis, and molecular clocks. *Paleobiology, 33*, 310–323.

Smith, A. B., & McGowan, A. J. (2007). The shape of the Phaerozoic marine palaeodiversity curve: How much can be predicted from the sedimentary rock record of western Europe? *Palaeontology, 50*, 765–774.

Stanley, D. G., Jr., Fine, M., & Tchernov, D. (2007). Ocean acidification and scleractinian corals. *Science, 317*, 1032–1033.

Stanley, S. M., & Yang, X. (1994). A double mass extinction at the end of the Paleozoic Era. *Science, 266*, 1340–1344.

Tucker, M. E., & Wright, V. P. (1990). *Carbonate sedimentology*. London: Blackwell Science.

Twitchett, R. J. (1999). Palaeoenvironments and faunal recovery after the end-Permian mass extinction. *Palaeogeography, Palaeoclimatology, Palaeoecology, 154*, 27–37.

Twitchett, R. J. (2001). Incompleteness of the Permian–Triassic fossil record: a consequence of productivity decline? *Geological Journal, 36*, 341–353.

Wagner, P. J., Kosnik, M. A., & Lidgard, S. (2006). Abundance distributions imply elevated complexity of post-Paleozoic marine ecosystems. *Science, 314*, 1289–1292.

Walter, L. M., & Burton, E. A. (1990). Dissolution of Recent platform carbonate sediments in marine pore fluids. *American Journal of Science, 290*, 601–643.

Weidlich, O. (2002). Permian reefs re-examined: extrinsic control mechanisms of gradual and abrupt changes during 40 my of reef evolution. *Geobios Memoire Special, 24*, 287–294.

Weidlich, O., Kiessling, W., & Flügel, E. (2003). Permian–Triassic boundary interval as a model for forcing marine ecosystem collapse by a long-term atmospheric oxygen drop. *Geology, 31*, 961–964.

Wheeley, J. R., & Twitchett, R. J. (2005). Palaeoecological significance of a new Griesbachian (Early Triassic) gastropod assemblage from Oman. *Lethaia, 38*, 1–9.

Wignall, P. B. (2001). Large igneous provinces and mass extinctions. *Earth-Science Reviews, 53*, 1–33.

Wignall, P. B., & Benton, M. J. (1999). Lazarus taxa and fossil abundance at times of biotic crisis. *Journal of the Geological Society, London, 156*, 453–456.

Wignall, P. B., & Twitchett, R. J. (1996). Oceanic anoxia and the end Permian mass extinction. *Science, 272*, 1155–1158.

Wright, P., Cherns, L., & Hodges, P. (2003). Missing molluscs: field testing taphonomic loss in the Mesozoic through early large-scale aragonite dissolution. *Geology, 31*, 211–214.

Yang, X.-N., Liu, J.-R., & Shi, G.-J. (2004). Extinction process and patterns of Middle Permian fusulinaceans in southwest China. *Lethaia, 37*, 139–147.

Index

D

E

T